Quantitative Functional Brain Imaging with Positron Emission Tomography

Quantitative Functional Brain Imaging with Positron Emission Tomography

Edited by

Richard E. Carson
Margaret E. Daube-Witherspoon
Peter Herscovitch

PET Department
National Institutes of Health
Bethesda, Maryland

Academic Press

San Diego   London   Boston   New York   Sydney   Tokyo   Toronto

Front cover photograph: Midsaggital slice of the brain radioactivity distribution of the opiate antagonist [^{18}-F]cyclofoxy. The original transaxial images were produced by 3D reconstruction of data acquired 40–60 min postinjection. From *NeuroImage*, Vol. 5(4)(1997).

This book is printed on acid-free paper. ∞

Copyright © 1998 by ACADEMIC PRESS

All Rights Reserved.
No part of this publication may be reproduced or transmitted in any form or by any means, electronic or mechanical, including photocopy, recording, or any information storage and retrieval system, without permission in writing from the publisher.

Academic Press
a division of Harcourt Brace & Company
525 B Street, Suite 1900, San Diego, California 92101-4495, USA
http://www.apnet.com

Academic Press Limited
24-28 Oval Road, London NW1 7DX, UK
http://www.hbuk.co.uk/ap/

Library of Congress Catalog Card Number: 98-85439

International Standard Book Number: 0-12-161340-2

PRINTED IN THE UNITED STATES OF AMERICA
98 99 00 01 02 03 QW 9 8 7 6 5 4 3 2 1

Contents

Contributors xi
Preface xix
Acknowledgments xxi

SECTION I

DATA ACQUISITION AND QUANTIFICATION

1. Brain Imaging in Small Animals Using MicroPET 3

   SIMON R. CHERRY, ARION CHATZIIOANNOU, YIPING SHAO, ROBERT W. SILVERMAN, KEN MEADORS, AND MICHAEL E. PHELPS

2. Design of a High-Resolution, High-Sensitivity PET Camera for Human Brains and Small Animals 11

   W. W. MOSES, P. R. G. VIRADOR, S. E. DERENZO, R. H. HUESMAN, AND T. F. BUDINGER

3. The Road to Simultaneous PET / MR Images of the Brain 19

   Y. SHAO, R. SLATES, K. FARAHANI, A. BOWERY, M. DAHLBOM, K. MEADORS, R. W. SILVERMAN, M. SUGITA, AND S. R. CHERRY

4. Brain PET Studies with a High Sensitivity Fully Three-Dimensional Tomograph 25

   D. L. BAILEY, M. P. MILLER, T. J. SPINKS, P. M. BLOOMFIELD, L. LIVIERATOS, R. B. BÁNÁTI, R. MYERS, AND T. JONES

5. Emission–Transmission Realignment Using a Simultaneous Emission–Transmission Postinjection Scan 33

   V. SOSSI, K. S. MORRISON, T. R. OAKES, AND T. J. RUTH

6. Optimization of $H_2^{15}O$ Dose and Data Acquisition in Three-Dimensional Activation Studies Using an ECAT EXACT HR-47 PET Camera and Voxel-by-Voxel t-Statistic 41

   PETER JOHANNSEN, SØREN B. HANSEN, LEIF ØSTERGAARD, AND ALBERT GJEDDE

7. Parametric Image Reconstruction Using Spectral Analysis of (Rebinned) Three-Dimensional Projection Data 45

   S. R. MEIKLE, J. C. MATTHEWS, V. J. CUNNINGHAM, D. L. BAILEY, T. JONES, AND P. PRICE

8. Multislice PET Quantitation Using Three-Dimensional Volumes of Interest 51

   R. H. HUESMAN, G. J. KLEIN, B. W. REUTTER, AND X. TENG

9. Absolute PET Quantification with Correction for Partial Volume Effects within Cerebral Structures 59

   CLAIRE LABBÉ, MATTHIAS KOEPP, JOHN ASHBURNER, TERRY SPINKS, MARK RICHARDSON, JOHN DUNCAN, AND VINCENT CUNNINGHAM

10. Pixel- versus Region-Based Partial Volume Correction in PET 67

    OLIVIER G. ROUSSET, YILONG MA, DEAN F. WONG, AND ALAN C. EVANS

11. Impact of Partial Volume Correction on Kinetic Parameters: Preliminary Experience in Patient Studies   77
F. YOKOI, O. G. ROUSSET, A. S. DOGAN, S. MARENCO, A. C. EVANS, A. H. GJEDDE, AND D. F. WONG

SECTION II

IMAGE PROCESSING

12. Performance Characterization of a Feature-Matching Axial Smoothing Method for Brain PET Images   85
S.-C. HUANG, J. YANG, C. L. YU, AND K. P. LIN

13. Registration of Multitracer PET Data   91
JESPER L. R. ANDERSSON

14. Multimodality Brain Image Registration Using a Three-Dimensional Photogrammetrically Derived Surface   99
OSAMA R. MAWLAWI, BRADLEY J. BEATTIE, STEVE M. LARSON, AND RONALD G. BLASBERG

15. Classification of Dynamic PET Images Using *a Priori* Kinetic Factors   107
JEFFREY T. YAP, VINCENT J. CUNNINGHAM, TERRY JONES, MALCOLM COOPER, CHIN-TU CHEN, AND PAT PRICE

16. Methodology for Statistical Parametric Mapping of [^{18}F]Fluorodopa Uptake Rate Using Three-Dimensional PET   117
J. S. RAKSHI, D. L. BAILEY, K. ITO, T. UEMA, P. K. MORRISH, J. ASHBURNER, K. J. FRISTON, AND D. J. BROOKS

17. Use of Nonlinear Kernel Analysis to Evaluate "Badness-of-Fit" of the Transformation of PET Images into Stereotactic Space: Application to Alzheimer's Disease   125
HAROLD LITT AND BARRY HORWITZ

18. Data Extraction from Brain PET Images Using Three-Dimensional Stereotactic Surface Projections   133
SATOSHI MINOSHIMA, EDWARD P. FICARO, KIRK A. FREY, ROBERT A. KOEPPE, AND DAVID E. KUHL

19. A Method for Surface-Based Quantification of Functional Data from the Human Cortex   139
H.-M. VON STOCKHAUSEN, U. PIETRZYK, K. HERHOLZ, A. THIEL, J. ILMBERGER, H.-J. REULEN, AND W.-D. HEISS

SECTION III

APPLICATIONS

20. Sequential Experimental PET: Voxel-Based Analysis Reveals Spatiotemporal Dynamics of Perfusion in Transient Focal Cerebral Ischemia   145
R. GRAF, A. SCHUSTER, J. LÖTTGEN, U. PIETRZYK, K. OHTA, E. KUMURA, K. WIENHARD, AND W.-D. HEISS

21. Estimation of Ischemic Cerebral Blood Flow Using [^{15}O]Water and PET without Arterial Blood Sampling   151
J. LÖTTGEN, U. PIETRZYK, K. HERHOLZ, K. WIENHARD, AND W.-D. HEISS

22. Suitability of [^{15}O]Water and PET to Detect Activation-Induced Cerebral Blood Flow Changes in Brain Tissue Altered by Brain Tumors   155
A. THIEL, K. HERHOLZ, H.-M. VON STOCKHAUSEN, G. PAWLIK, AND W.-D. HEISS

23. Magnetic Resonance Imaging-Guided Language Activation PET in Patients: Technical Aspects and Clinical Results   159
K. HERHOLZ, A. THIEL, U. PIETRZYK, H.-M. VON STOCKHAUSEN, M. GHAEMI, A. BERZDORF, J. SOBESKY, K. WIENHARD, AND W.-D. HEISS

24. Brain Networks of Motor Behavior Assessed by Principal Component Analysis   165
J. R. MOELLER, C. GHEZ, A. ANTONINI, M. F. GHILARDI, V. DHAWAN, K. KAZUMATA, AND D. EIDELBERG

25. Frequency-Dependent Changes in Cerebral Metabolic Rate of Oxygen during

Activation of Human Visual Cortex Studied by PET   173
M. S. VAFAEE, E. MEYER, S. MARRETT, T. PAUS, A. C. EVANS, AND A. GJEDDE

26. Validation of an [^{18}F]Fluorodeoxyglucose–PET Protocol in Conscious Vervet Monkey   177
A. H. MOORE, M. J. RALEIGH, S. R. CHERRY, S.-C. HUANG, AND M. E. PHELPS

27. Comparison of Ketamine/Midazolam versus Pentobarbital on [^{18}F]Fluorodopa PET Kinetics in Monkeys   183
DAVID B. STOUT, SUNG-CHENG HUANG, MICHAEL J. RALEIGH, MICHAEL E. PHELPS, AND JORGE R. BARRIO

28. Metabolism of [^{18}F]Fluorodopa in Pig Brain Estimated by PET   187
E. H. DANIELSEN, D. F. SMITH, A. D. GEE, T. K. VENKATACHALAM, S. B. HANSEN, AND A. GJEDDE

29. Use of Two- and Three-Dimensional PET and [^{11}C](R)-PK11195 to Image Focal and Regional Brain Pathology   195
RALPH MYERS, RICHARD B. BÁNÁTI, ERALDO PAULESU, JOHN THORPE, DAVID H. MILLER, AND TERRY JONES

30. Noninvasive Imaging of Serotonin Synthesis Rate Using PET and α-Methyltryptophan in Autistic Children   201
OTTO MUZIK, DIANE C. CHUGANI, CHENGGANG SHEN, AND HARRY T. CHUGANI

31. Brain Mapping of the Effects of Aging on Histamine H$_1$ Receptors in Humans: A PET Study with [^{11}C]Doxepin   207
MAKOTO HIGUCHI, MASATOSHI ITOH, KAZUHIKO YANAI, NOBUYUKI OKAMURA, ATSUSHI YAMAKI, TATSUO IDO, HIROYUKI ARAI, AND HIDETADA SASAKI

SECTION IV

STATISTICAL ANALYSIS

32. PET Analysis Using a Variance Stabilizing Transform   217
URS E. RUTTIMANN, DANIEL RIO, ROBERT R. RAWLINGS, PAUL ANDREASON, AND DANIEL W. HOMMER

33. Error and *t* Images Depend on ANOVA Design and Anatomical Standardization in PET Activation Analysis   223
MICHIO SENDA, KENJI ISHII, KEIICHI ODA, NORIHIRO SADATO, RYUTA KAWASHIMA, IWAO KANNO, HINAKO TOYAMA, AND ITARU TATSUMI

34. Calculation of the Probability That an Activation Site Has Occurred by Chance   229
JOHN R. VOTAW, SCOTT T. GRAFTON, AND JOHN M. HOFFMAN

35. Multifiltering Signal Detection and Statistical Power in Brain Activation Studies   237
JOHN DARRELL VAN HORN, TIMOTHY M. ELLMORE, JOHN L. HOLT, GIUSEPPE ESPOSITO, AND KAREN FAITH BERMAN

36. Measuring Activation Pattern Reproducibility Using Resampling Techniques   241
S. C. STROTHER, K. REHM, N. LANGE, J. R. ANDERSON, K. A. SCHAPER, L. K. HANSEN, AND D. A. ROTTENBERG

37. Reproducibility of Regional Metabolic Covariance Patterns: Comparison of Four Populations   247
D. EIDELBERG, J. R. MOELLER, V. DHAWAN, A. ANTONINI, L. MORAN, J. MISSIMER, AND K. L. LEENDERS

38. On the Detection of Activation Patterns Using Principal Components Analysis   253
B. A. ARDEKANI, S. C. STROTHER, J. R. ANDERSON, I. LAW, O. B. PAULSON, I. KANNO, AND D. A. ROTTENBERG

SECTION V

TRACER DEVELOPMENT

39. One for All or One for Each? Matching Radiotracers and Regional Brain Pharmacokinetics   261
MICHAEL R. KILBOURN, THINH B. NGUYEN, SCOTT E. SNYDER, AND ROBERT A. KOEPPE

40. Use of Information Technology in the Search for New PET Tracers 267
A. J. ABRUNHOSA, F. BRADY, S. K. LUTHRA, H. MORRIS, J. J. DE LIMA, AND T. JONES

41. Statistical Power Analysis of *in Vivo* Studies in the Rat Brain Using PET Radiotracers 273
D. HUSSEY, J. N. DASILVA, E. GREENWALD, K. CHEUNG, S. KAPUR, A. A. WILSON, AND S. HOULE

42. A Human Liver Model of Metabolism as a Tool in the Identification of Potential PET Radiotracers 279
A. A. WILSON, T. INABA, N. FISCHER, J. N. DASILVA, AND S. HOULE

43. HPLC Analysis of the Metabolism of 6-[^{18}F]Fluoro-L-DOPA in the Brain of Neonatal Pigs 285
G. VORWIEGER, P. BRUST, R. BERGMANN, R. BAUER, B. WALTER, F. FÜCHTNER, J. STEINBACH, AND B. JOHANNSEN

44. Characterization of the Radiolabeled Metabolites of [^{18}F]Altanserin: Implications for Kinetic Modeling 293
BRIAN LOPRESTI, DANIEL HOLT, N. SCOTT MASON, YIYUN HUANG, JAMES RUSZKIEWICZ, JENNIFER PEREVUZNIK, JULIE PRICE, GWENN SMITH, JAMES DAVIS, AND CHESTER MATHIS

45. Preliminary Evaluation of the Glycine Site Antagonists [^{11}C]L 703,717 and [^{3}H]MDL 105,519 as Putative PET Ligands for Central NMDA Receptors: *In Vivo* Studies in Rats 299
J. OPACKA-JUFFRY, H. MORRIS, S. ASHWORTH, S. OSMAN, E. HIRANI, A. M. MACLEOD, S. K. LUTHRA, AND S. P. HUME

SECTION VI

PARAMETER ESTIMATION

46. Evaluation of the Contribution of Protocol Design to Model Parameter Uncertainty 307
JULIAN C. MATTHEWS, VINCENT J. CUNNINGHAM, AND PAT M. PRICE

47. An Assessment of Optimal Image Sampling Schedule Design in Dynamic PET–FDG Studies 315
D. HO, D. FENG, AND L. C. WU

48. Simultaneous Extraction of Physiological and Input Function Parameters from PET Measurements 321
DAGAN FENG, KOON-PONG WONG, CHI-MING WU, AND WAN-CHI SIU

49. Suppression of Noise Artifacts in Spectral Analysis of Dynamic PET Data 329
VINCENT J. CUNNINGHAM, ROGER N. GUNN, HELEN BYRNE, AND JULIAN C. MATTHEWS

50. Contraints in Spectral Analysis 335
DAVID C. REUTENS AND MARK ANDERMANN

51. Generalized Linear Least-Squares Modeling Algorithm for Optimally Sampled PET Image Data 339
D. FENG, D. HO, K. K. LAU, AND W.-C. SIU

52. Parametric Image Generation with Neural Networks 347
MICHAEL M. GRAHAM, STEVEN B. GILLISPIE, MARK MUZI, AND FINBARR O'SULLIVAN

53. Kinetic Modeling: Achieving Computer Platform Independence with Java 353
SYLVAIN HOULE

SECTION VII

KINETIC MODELING

54. Temporally Overlapping Dual-Tracer PET Studies 359
R. A. KOEPPE, E. P. FICARO, D. M. RAFFEL, S. MINOSHIMA, AND M. R. KILBOURN

55. Simultaneous Estimation of Perfusion (K_1) and Vascular (V_0) Responses in [^{15}O]Water PET Activation Studies 367
P.-J. TOUSSAINT AND E. MEYER

56. Enhancement of the Signal-to-Noise Ratio in [^{15}O]Water Bolus PET Activation Studies Using a Combined Cold-Bolus/Switched Protocol   371

JORGE J. MORENO-CANTÚ, CHRISTOPHER J. THOMPSON, ERNST MEYER, PIERRE FISET, ROBERT J. ZATORRE, DENISE KLEIN, AND DAVID C. REUTENS

57. Neutral Amino Acids Influence [^{18}F]Fluorodopa Quantification   379

P. VONTOBEL, G. KÜNIG, M. BRÜHLMEIER, I. GÜNTHER, A. ANTONINI, M. PSYLLA, AND K. L. LEENDERS

58. Imaging of the Dopamine Presynaptic System by PET: 6-[^{18}F]Fluoro-L-DOPA versus 6-[^{18}F]Fluoro-L-*m*-tyrosine   387

D. J. DOUDET, O. T. DEJESUS, G. L. Y. CHAN, S. JIVAN, J. E. HOLDEN, C. ENGLISH, T. G. AIGNER, AND T. J. RUTH

59. Quantitative Measurement of Acetylcholinesterase Activity in Living Human Brain Using a Radioactive Acetylcholine Analog and Dynamic PET   393

S. NAGATSUKA, H. NAMBA, M. IYO, K. FUKUSHI, H. SHINOTOH, T. SUHARA, Y. SUDO, K. SUZUKI, AND T. IRIE

60. Parametric Imaging of Ligand–Receptor Interactions Using a Reference Tissue Model and Cluster Analysis   401

ROGER N. GUNN, ADRIAAN A. LAMMERTSMA, AND VINCENT J. CUNNINGHAM

61. Limitation of Binding Potential as a Measure of Receptor Function: A Two-Point Correction for the Effects of Mass   407

EVAN D. MORRIS, SVETLANA I. CHEFER, AND EDYTHE D. LONDON

62. Estimation of Nonspecific Binding of [^{18}F]Setoperone, a 5HT$_{2A}$ Receptor PET Radioligand, from Saturation Kinetic Data in Baboon and Human Neocortex   415

M. C. PETIT-TABOUÉ, B. LANDEAU, A. R. YOUNG, P. SCHUMANN, L. BESRET, M. IBAZIZENE, AND J. C. BARON

63. Estimation of Binding Potential for the 5HT$_2$ Receptor Ligand [^{18}F]Setoperone by a Noninvasive Reference Region Graphical Method   421

ALI A. BONAB, ALAN J. FISCHMAN, AND NATHANIEL M. ALPERT

64. [^{18}F]Altanserin PET Studies of Serotonin-2A Binding: Examination of Nonspecific Component   427

J. C. PRICE, B. LOPRESTI, N. S. MASON, Y. HUANG, D. HOLT, G. S. SMITH, AND C. A. MATHIS

65. Characteristics of Neurotransmitter Competition Studies Using Constant Infusion of Tracer   435

CHRISTOPHER J. ENDRES AND RICHARD E. CARSON

66. PET Measurement of Endogenous Neurotransmitter Activity Using High and Low Affinity Radiotracers   441

J. C. PRICE, N. S. MASON, B. LOPRESTI, D. HOLT, N. R. SIMPSON, W. DREVETS, G. S. SMITH, AND C. A. MATHIS

67. Measuring Neurotransmitter Release with PET: Methodological Issues   449

ALAIN DAGHER, ROGER N. GUNN, GEOFF LOCKWOOD, VINCENT J. CUNNINGHAM, PAUL M. GRASBY, AND DAVID J. BROOKS

68. Imaging Receptor Occupancy by Endogenous Transmitters in Humans   455

MARC LARUELLE, ANISSA ABI-DARGHAM, AND ROBERT B. INNIS

69. Quantification of Extracellular Dopamine Release in Schizophrenia and Cocaine Use by Means of TREMBLE   463

DEAN F. WONG, THOMAS SØLLING, FUJI YOKOI, AND ALBERT GJEDDE

SECTION VIII

BRAINPET97 DISCUSSION

Index   497

Contributors

The numbers in parentheses indicate the pages on which the authors' contributions begin.

Anissa Abi-Dargham (455) Brain Imaging Division, New York State Psychiatric Institute, New York, New York 10032

A. J. Abrunhosa (267) MRC Cyclotron Unit, PET Methodology Group, Hammersmith Hospital, London W12 0NN, United Kingdom

T. G. Aigner (387) National Institute of Mental Health, National Institutes of Health, Bethesda, Maryland 20892

Nathaniel M. Alpert (421) PET Imaging Laboratory, Division of Nuclear Medicine, Massachusetts General Hospital, Boston, Massachusetts 02114

Mark Andermann (335) McConell Brain Imaging Centre, Montréal Neurological Institute, Montréal, Québec, H3A 2B4 Canada

J. R. Anderson (241, 253) PET Imaging Center, Veterans Administration Medical Center, Minneapolis, Minnesota 55417

Jesper L. R. Andersson (91) Uppsala University PET Centre, Subfemtomole Biorecognition Project, S-751 85 Uppsala, Sweden

Paul Andreason (217) National Institute of Alcohol Abuse and Alcoholism, National Institutes of Health, Bethesda, Maryland 20892

A. Antonini (165, 247, 379) Department of Neurology, North Shore University Hospital, Manhasset, New York

Hiroyuki Arai (207) Department of Geriatric Medicine, Tohoku University School of Medicine, Sendai 980-77, Japan

B. A. Ardekani (253) Department of Radiology and Nuclear Medicine, Research Institute for Brain and Blood Vessels, Akita City, Akita 010, Japan

John Ashburner (59, 117) Wellcome Department of Cognitive Neurology, London, United Kingdom

S. Ashworth (299) MRC Cyclotron Unit, Clinical Sciences Centre, Hammersmith Hospital, London W12 0NN, United Kingdom

D. L. Bailey (25, 45, 117) MRC Cyclotron Unit, Hammersmith Hospital, London W12 0NN, United Kingdom

Richard B. Bánáti (25, 195) MRC Cyclotron Unit, Hammersmith Hospital, London W12 0NN, United Kingdom

J. C. Baron (415) INSERM U320, CYCERON, 14074 Caen, France

Jorge R. Barrio (183) Department of Molecular and Medical Pharmacology, UCLA School of Medicine, Los Angeles, California 90095

R. Bauer (285) Institut für Pathophysiologie, Friedrich-Schiller-Universität, Jena, Germany

Bradley J. Beattie (99) Department of Neurology, Memorial Sloan-Kettering Cancer Center, New York, New York 10021

R. Bergmann (285) Institut für Bioanorganische und Radiopharmazeutische Chemie, Forschungszentrum Rossendorf, D-013141 Dresden, Germany

Karen Faith Berman (237) Unit on PET, Clinical Brain Disorders Branch, NIMH, Bethesda, Maryland 20892

A. Berzdorf (159) Max-Planck-Institut für neurologische Forschung, D-50931 Cologne, Germany

L. Besret (415) INSERM U320, CEA DSV/DRM, 14074 Caen, France

Ronald G. Blasberg (99) Department of Neurology, Memorial Sloan-Kettering Cancer Center, New York, New York 10021

P. M. Bloomfield (25) MRC Cyclotron Unit, Hammersmith Hospital, London W12 0NN, United Kingdom

Ali A. Bonab (421) Division of Nuclear Medicine, Massachusetts General Hospital, Boston, Massachusetts 02114

A. Bowery (19) Crump Institute for Biological Imaging, Department of Molecular and Medical Pharmacology, UCLA School of Medicine, Los Angeles, California 90095

F. Brady (267) MRC Cyclotron Unit, PET-Methodology Group, Hammersmith Hospital, London W12 0NN, United Kingdom

David J. Brooks (117, 449) MRC Cyclotron Unit, Royal Postgraduate Medical School, Hammersmith Hospital, London W12 0NN, United Kingdom

M. Brühlmeier (379) Kantonsspital Aarau, Switzerland

P. Brust (285) Institut für Bioanorganische und Radiopharmazeutische Chemie, Forschungszentrum Rossendorf, D-013141 Dresden, Germany

T. F. Budinger (11) Lawrence Berkeley National Laboratory, University of California, Berkeley, California 94720

Helen Byrne (329) MRC Cyclotron Unit, Hammersmith Hospital, London W12 0NN, United Kingdom

Richard E. Carson (435) PET Department, National Institutes of Health, Bethesda, Maryland 20892

G. L. Y. Chan (387) Department of Medicine, Division of Neurology and Neurodegenerative Disorders Centre and TRIUMF, University of British Columbia, Vancouver, British Columbia, V6T 2B5 Canada

Arion Chatziioannou (3) Crump Institute for Biological Imaging, Department of Molecular and Medical Pharmacology, UCLA School of Medicine, Los Angeles, California 90095

Svetlana I. Chefer (407) Brain Imaging Center, NIDA Intramural Research Program, National Institutes of Health, Baltimore, Maryland 21224

Chin-Tu Chen (107) Department of Radiology, University of Chicago, Chicago, Illinois 60637

Simon R. Cherry (3, 19, 177) Crump Institute for Biological Imaging, Department of Molecular and Medical Pharmacology, UCLA School of Medicine, Los Angeles, California 90095

K. Cheung (273) PET Centre, Clarke Institute of Psychiatry, Toronto, Ontario, M5T 1R8 Canada

Diane C. Chugani (201) Department of Radiology, Children's Hospital of Michigan PET Center, Detroit, Michigan 48201

Harry T. Chugani (201) Department of Radiology, Children's Hospital of Michigan PET Center, Detroit, Michigan 48201

Malcolm Cooper (107) Department of Radiology, University of Chicago, Chicago, Illinois 60637

Vincent J. Cunningham (45, 59, 107, 307, 329, 401, 449) MRC Cyclotron Unit, Hammersmith Hospital, London W12 0NN, United Kingdom

Alain Dagher (449) McConnell Brain Imaging Centre, Montréal Neurological Institute, Montréal, Québec, H3A 2B4 Canada

M. Dahlbom (19) Crump Institute for Biological Imaging, Department of Molecular and Medical Pharmacology, UCLA School of Medicine, Los Angeles, California 90095

E. H. Danielsen (187) PET Center, Aarhus University Hospital, DK-8000 Aarhus C, Denmark

J. N. DaSilva (273, 279) PET Centre, Clarke Institute of Psychiatry, Toronto, Ontario, M5T 1R8 Canada

James Davis (293) PET Facility, Department of Radiology, University of Pittsburgh Medical Center, Pittsburgh, Pennsylvania 15213

J. J. de Lima (267) IBILI-Faculdade de Medicina, Celas, 3000 Coimbra, Portugal

O. T. DeJesus (387) Department of Medical Physics, University of Wisconsin, Madison, Wisconsin 53706

S. E. Derenzo (11) Lawrence Berkeley National Laboratory, University of California, Berkeley, California 94720

V. Dhawan (165, 247) Department of Neurology, North Shore University Hospital, Manhasset, New York 11030

A. S. Dogan (77) Division of Nuclear Medicine, Department of Radiology, Johns Hopkins Medical Institutions, Baltimore, Maryland 21287

D. J. Doudet (387) Department of Medicine, Division of Neurology and Neurodegenerative Disorders Centre and TRIUMF, and University of British Columbia, Vancouver, British Columbia, V6T 2B5 Canada

W. Drevets (441) Department of Psychiatry, University of Pittsburgh Medical Center, Pittsburgh, Pennsylvania 15213

John Duncan (59) Neuroscience Group of the MRC Cyclotron Unit, Hammersmith Hospital, London W12 0NN, United Kingdom; and Epilepsy Research Group, Institute of Neurology, London, United Kingdom

D. Eidelberg (165, 247) Department of Neurology, North Shore University Hospital, Manhasset, New York 11030

Timothy M. Ellmore (237) Unit on PET, CBDB/NIMH, National Institutes of Health, Bethesda, Maryland 20892

Christopher J. Endres (435) PET Department, National Institutes of Health, Bethesda, Maryland 20892

C. English (387) Department of Medicine, Division of Neurology and Neurodegenerative Disorders Centre and TRIUMF, University of British Columbia, Vancouver, British Columbia, Canada

Giuseppe Esposito (237) Unit on PET, CBDB/NIMH, National Institutes of Health, Bethesda, Maryland 20892

Alan C. Evans (67, 77, 173) McConnell Brain Imaging Centre, Montréal Neurological Institute, McGill University, Montréal, Québec, H3A 2B4 Canada

K. Farahani (19) Department of Radiological Sciences, UCLA School of Medicine, Los Angeles, California 90095

Dagan Feng (315, 321, 339) Biomedical and Multimedia Information Technology (BMIT) Group, Basser Department of Computer Science, University of Sydney, Sydney, New South Wales 2006, Australia

Edward P. Ficaro (133, 359) Division of Nuclear Medicine, Department of Internal Medicine, University of Michigan, Ann Arbor, Michigan 48109

N. Fischer (279) Department of Pharmacology, University of Toronto, Toronto, Ontario, Canada

Alan J. Fischman (421) Division of Nuclear Medicine, Massachusetts General Hospital, Boston, Massachusetts 02114

Pierre Fiset (371) Department of Anaesthesiology, Royal Victoria Hospital, McGill University, Montréal, Québec, Canada

Kirk A. Frey (133) Division of Nuclear Medicine, Department of Internal Medicine, University of Michigan, Ann Arbor, Michigan 48109

K. J. Friston (117) Wellcome Department of Cognitive Neurology, Institute of Neurology, London WC1N 3BG, United Kingdom

F. Füchtner (285) Institut für Bioanorganische und Radiopharmazeutische Chemie, Forschungszentrum Rossendorf, D-013141 Dresden, Germany

K. Fukushi (393) Division of Clinical Research, National Institute of Radiological Sciences, Chiba 263, Japan

A. D. Gee (187) PET Center, Aarhus University Hospital, DK-8000 Aarhus C, Denmark

M. Ghaemi (159) Max-Planck-Institut für neurologische Forschung, D-50931 Cologne, Germany

C. Ghez (165) Department of Neurology, College of Physicians and Surgeons, Columbia University, New York, New York 10032

M. F. Ghilardi (165) Department of Neurology, College of Physicians and Surgeons, Columbia University, New York, New York 10032

Steven B. Gillispie (347) Nuclear Medicine, University of Washington Medical Center, Seattle, Washington 98195

Albert Gjedde (41, 77, 173, 187, 463) PET Center, Aarhus University Hospital, DK-8000 Aarhus C, Denmark

R. Graf (145) Max-Planck-Institut für neurologische Forschung, D-50931 Cologne, Germany

Scott T. Grafton (229) Emory University Center for PET, Departments of Radiology and Neurology, Atlanta, Georgia 30322

Michael M. Graham (347) Nuclear Medicine, University of Washington Medical Center, Seattle, Washington 98195

Paul M. Grasby (449) MRC Cyclotron Unit, Hammersmith Hospital, London W12 0NN, United Kingdom

E. Greenwald (273) PET Centre, Clarke Institute of Psychiatry, Toronto, Ontario, M5T 1R8 Canada

Roger N. Gunn (329, 401, 449) MRC Cyclotron Unit, Hammersmith Hospital, London W12 0NN, United Kingdom

I. Günther (379) PET Department, Paul Scherrer Institute, CH-5232 Villigen PSI, Switzerland

L. K. Hansen (241) Department of Mathematical Modeling, Technical University of Denmark, Lyngby, Denmark DK2800

Søren B. Hansen (41, 187) PET Center, Aarhus University Hospital, DK-8000 Aarhus C, Denmark

W.-D. Heiss (139, 145, 151, 155, 159) Max-Planck-Institut für neurologische Forschung, D-50931 Cologne, Germany

K. Herholz (139, 151, 155, 159) Max-Planck-Institut für neurologische Forschung, D-50931 Cologne, Germany

Makoto Higuchi (207) Department of Geriatric Medicine, Tohoku University School of Medicine, Sendai 980-77, Japan

E. Hirani (299) MRC Cyclotron Unit, Hammersmith Hospital, London W12 0HS, United Kingdom

D. Ho (315, 339) Biomedical and Multimedia Information, Basser Department of Computer Science, University of Sydney, Sydney, New South Wales 2006, Australia

John M. Hoffman (229) Department of Neurology, Emory University, Atlanta, Georgia 30322

J. E. Holden (387) Department of Medical Physics and Radiology, University of Wisconsin-Madison, Madison, Wisconsin 53706

Daniel Holt (293, 427, 441) PET Facility, Department of Radiology, University of Pittsburgh Medical Center, Pittsburgh, Pennsylvania 15213

John L. Holt (237) Unit on PET, CBDB/NIMH, National Institutes of Health, Bethesda, Maryland 20892

Daniel W. Hommer (217) NIAAA, National Institutes of Health, Bethesda, Maryland 20892

Barry Horwitz (125) Laboratory of Neurosciences, National Institute on Aging, National Institutes of Health, Bethesda, Maryland 20892

Sylvain Houle (273, 279, 353) PET Centre, Clark Institute of Psychiatry, Toronto, Ontario, M5T 1R8 Canada

Sung-Cheng Huang (85, 177, 183) Division of Nuclear Medicine and Biophysics, Department of Molecular and Medical Pharmacology, UCLA School of Medicine, Los Angeles, California 90095

Yiyun Huang (293, 427) PET Facility, Department of Radiology, University of Pittsburgh Medical Center, Pittsburgh, Pennsylvania 15213

R. H. Huesman (11, 51) Center for Functional Imaging, Lawrence Berkeley National Laboratory, University of California, Berkeley, California 94720

S. P. Hume (299) MRC Cyclotron Unit, Hammersmith Hospital, London W12 OHS, United Kingdom

D. Hussey (273) PET Centre, Clarke Institute of Psychiatry, Toronto, Ontario, M5T 1R8 Canada

M. Ibazizene (415) CEA DSV/DRM, Caen, France

Tatsuo Ido (207) Cyclotron and Radioisotope Center, Tohoku University, Sendai 980-77, Japan

J. Ilmberger (139) Department of Neurosurgery, Klinikum Großhadern, Ludwig-Maximilians-Universität, Munich, Germany

T. Inaba (279) Department of Pharmacology, University of Toronto, Toronto, Ontario, Canada

Robert B. Innis (455) Department of Psychiatry, Yale University School of Medicine, New Haven, Connecticut 06516

T. Irie (393) Division of Advanced Technology for Medical Imaging, National Institute of Radiological Sciences, Chiba-shi, Chiba 263, Japan

Kenji Ishii (223) PET Center, Tokyo Metropolitan Institute of Gerontology, Tokyo 173, Japan

K. Ito (117) Department of Biofunctional Research, National Institute for Longevity Sciences, Obu, Japan

Masatoshi Itoh (207) Cyclotron and Radioisotope Center, Tohoku University, Aoba-ku, Sendai 980-77, Japan

M. Iyo (393) Division of Clinical Research, National Institute of Radiological Sciences, Chiba 263, Japan

S. Jivan (387) Department of Medicine, Division of Neurology and Neurodegenerative Disorders Centre and TRIUMF, University of British Columbia, Vancouver, British Columbia, Canada

B. Johannsen (285) Institut für Bioanorganische und Radiopharmazeutische Chemie, Forschungszentrum Rossendorf, D-013141 Dresden, Germany

Peter Johannsen (41) PET-Center, Aarhus University Hospital, DK-8000 Aarhus C, Denmark

Terry Jones (25, 45, 107, 195, 267) MRC Cyclotron Unit, Hammersmith Hospital, London W12 0NN, United Kingdom

Iwao Kanno (223, 253) Department of Radiology and Nuclear Medicine, Akita Research Institute of Brain and Blood Vessels, Akita 010, Japan

S. Kapur (273) PET Centre, Clarke Institute of Psychiatry, Toronto, Ontario, M5T 1R8 Canada

Ryuta Kawashima (223) Tohoku University, Japan

K. Kazumata (165) Department of Neurology, North Shore University Hospital, Manhasset, New York 11030

Michael R. Kilbourn (261, 359) Division of Nuclear Medicine, Department of Internal Medicine, University of Michigan, Ann Arbor, Michigan 48109

Denise Klein (371) McConnell Brain Imaging Centre, Montréal Neurological Institute, Montréal, Québec, H3A 2B4 Canada

G. J. Klein (51) Center for Functional Imaging, Lawrence Berkeley National Laboratory, University of California, Berkeley, California 94720

Matthias Koepp (59) MRC Cyclotron Unit, Hammersmith Hospital, London W12 0NN, United Kingdom

Robert A. Koeppe (133, 261, 359) Division of Nuclear Medicine, Department of Internal Medicine, University of Michigan, Ann Arbor, Michigan 48109

David E. Kuhl (133) Division of Nuclear Medicine, Department of Internal Medicine, University of Michigan, Ann Arbor, Michigan 48109

E. Kumura (145) Max-Planck-Institut für neurologische Forschung, D-50931 Cologne, Germany

G. Künig (379) PET Department, Paul Scherrer Institute, CH-5232 Villigen PSI, Switzerland

Claire Labbé (59) Methodology Group of the MRC Cyclotron Unit Hammersmith Hospital, London W12 0NN, United Kingdom

Contributors

Adriaan A. Lammertsma (401) PET Centre, Free University Hospital, 1007 MB Amsterdam, The Netherlands

B. Landeau (415) INSERM U320, 14074 Caen, France

N. Lange (241) Brain Imaging Center, Harvard Medical School and McLean Hospital, Boston, Massachusetts 02178

Steve M. Larson (99) Nuclear Medicine Service, Memorial Sloan Kettering Cancer Center, New York, New York 10021

Marc Laruelle (455) Columbia University College of Physicians and Surgeons, New York State Psychiatric Institute, New York, New York 10032

K. K. Lau (339) Department of Electronic Engineering, The Hong Kong Polytechnic University, Hung Hom, Kowloon, Hong Kong

I. Law (253) Department of Neurology, Rigshospitalet, DK-2100 Copenhagen, Denmark

K. L. Leenders (247, 379) PET Program, Paul Scherrer Institute, CH-5232, Villigen PSI, Switzerland

K. P. Lin (85) Department of Electrical Engineering, Chung-Yuan University, Chung-li, Taiwan, Republic of China

Harold Litt (125) Department of Biophysics, School of Medicine, State University of New York at Buffalo, Buffalo, New York 14214; and Laboratory of Neurosciences, National Institutes of Aging, National Institutes of Health, Bethesda, Maryland 20892

L. Livieratos (25) MRC Cyclotron Unit, Hammersmith Hospital, London W12 0NN, United Kingdom

Geoff Lockwood (449) MRC Cyclotron Unit, Hammersmith Hospital, London W12 0NN, United Kingdom

Edythe D. London (407) Brain Imaging Center, NIDA Intramural Research Program, National Institutes of Health, Baltimore, Maryland 21224

Brian Lopresti (293, 427, 441) PET Facility, Department of Radiology, University of Pittsburgh Medical Center, Pittsburgh, Pennsylvania 15213

J. Löttgen (145, 151) Max-Planck-Institut für neurologische Forschung, D-50931 Cologne, Germany

S. K. Luthra (267, 299) MRC Cyclotron Unit, Hammersmith Hospital, London W12 0NN, United Kingdom

Yilong Ma (67) McConnell Brain Imaging Centre, Montréal Neurological Institute, McGill University, Montréal, Québec, H3A-2B4 Canada

A. M. MacLeod (299) Merck Sharp & Dohme Research Laboratories, Harlow CM20 2QR, United Kingdom

S. Marenco (77) Division of Nuclear Medicine, Department of Radiology, Johns Hopkins Medical Institutions, Baltimore, Maryland 21287

S. Marrett (173) PET Unit, McConnell Brain Imaging Centre, Montréal Neurological Institute, Montréal, Québec, H3A 2B4 Canada

N. Scott Mason (293, 427, 441) PET Facility, Department of Radiology, University of Pittsburgh Medical Center, Pittsburgh, Pennsylvania 15213

Chester A. Mathis (293, 427, 441) PET Facility, Department of Radiology, University of Pittsburgh Medical Center, Pittsburgh, Pennsylvania 15213

Julian C. Matthews (45, 307, 329) Wolfson Brain Imaging Centre, Addenbrookes Hospital, Cambridge CB2 2QQ, United Kingdom; and MRC Cyclotron Unit, Hammersmith Hospital, London W12-0NN, United Kingdom

Osama R. Mawlawi (99) Nuclear Medicine Service, Memorial Sloan Kettering Cancer Center, New York, New York 10021; and Chemical Engineering Department, Columbia University, New York, New York 10027

Ken Meadors (3, 19) Crump Institute for Biological Imaging, Department of Molecular and Medical Pharmacology, UCLA School of Medicine, Los Angeles, California 90095

S. R. Meikle (45) PET and Nuclear Medicine Department, Royal Prince Alfred Hospital, Camperdown, New South Wales 2050, Australia

Ernst Meyer (173, 367, 371) McConnell Brain Imaging Centre, Montréal Neurological Institute, Montréal, Québec, H3A 2B4 Canada

David H. Miller (195) NMR Research Unit, Institute of Neurology, London, WC1N 3BG, United Kingdom

M. P. Miller (25) MRC Cyclotron Unit, Hammersmith Hospital, London W12 0NN, United Kingdom

Satoshi Minoshima (133, 359) Division of Nuclear Medicine, Department of Internal Medicine, University of Michigan, Ann Arbor, Michigan 48109

J. Missimer (247) PET Program, Paul Scherrer Institute, CH 5232 Villigen PSI, Switzerland

J. R. Moeller (165, 247) Department of Psychiatry, College of Physicians and Surgeons, Columbia University, New York State Psychiatric Institute, New York, New York 10032

A. H. Moore (177) Department of Medical and Molecular Pharmacology, UCLA School of Medicine, Los Angeles, California 90095

L. Moran (247) Department of Neurology, North Shore University Hospital, Manhasset, New York 11030

Jorge J. Moreno-Cantú (371) McConnell Brain Imaging Centre, Montréal Neurological Institute, Montréal, Québec, H3A 2B4 Canada

Evan D. Morris (407) Brain Imaging Center, National Institute on Drug Abuse, National Institutes of Health, Baltimore, Maryland 21224

H. Morris (267, 299) MRC Cyclotron Unit, Hammersmith Hospital, London W12 0NN, United Kingdom

P. K. Morrish (117) MRC Cyclotron Unit, Hammersmith Hospital, London W12 0NN, United Kingdom

K. S. Morrison (33) University of British Columbia, TRIUMF/PET Centre, Vancouver, British Columbia, V6T 2A3 Canada

W. W. Moses (11) Lawrence Berkeley National Laboratory, University of California, Berkeley, California 94720

Mark Muzi (347) Nuclear Medicine, University of Washington Medical Center, Seattle, Washington 98195

Otto Muzik (201) Department of Radiology, Children's Hospital of Michigan PET Center, Detroit, Michigan 48201

Ralph Myers (25, 195) MRC Cyclotron Unit, Hammersmith Hospital, London W12 0NN, United Kingdom

S. Nagatsuka (393) Division of Clinical Research, National Institute of Radiological Sciences, Chiba 263, Japan

H. Namba (393) Division of Clinical Research, National Institute of Radiological Sciences, Chiba 263, Japan

Thinh B. Nguyen (261) Division of Nuclear Medicine, Department of Internal Medicine, University of Michigan, Ann Arbor, Michigan 48109

Finbarr O'Sullivan (347) Department of Statistics, University of Washington, Seattle, Washington 98195

T. R. Oakes (33) University of British Columbia, TRIUMF/PET Centre, Vancouver, British Columbia, V6T 2B5 Canada

Keiichi Oda (223) PET Center, Tokyo Metropolitan Institute of Gerontology, Tokyo 173, Japan

K. Ohta (145) Max-Planck-Institut für neurologische Forschung, D-50931 Cologne, Germany

Nobuyuki Okamura (207) Department of Geriatric Medicine, Tohoku University School of Medicine, Sendai 980-77, Japan

J. Opacka-Juffry (299) MRC Cyclotron Unit, Hammersmith Hospital, London W12 0HS, United Kingdom

S. Osman (299) MRC Cyclotron Unit, Hammersmith Hospital, London W12 0HS, United Kingdom

Leif Østergaard (41) PET-Center, Aarhus University Hospital, DK-8000 Aarhus C, Denmark

Eraldo Paulesu (195) Neurologo, Neuropsicologo Centro PET/Ciclotrone, Instituto Scientifico H San Raffaele, 20132 Milan, Italy

O. B. Paulson (253) Neurobiology Research Unit 9201, Rigshospitalet, DK-2100 Copenhagen, Denmark

T. Paus (173) PET Unit, McConnell Brain Imaging Centre, Montréal Neurological Institute, Montréal, Québec, H3A 2B4 Canada

G. Pawlik (155) Max-Planck-Institut für neurologische Forschung, D-50931 Cologne, Germany

Jennifer Perevuznik (293) PET Facility, Department of Radiology, University of Pittsburgh Medical Center, Pittsburgh, Pennsylvania 15213

M. C. Petit-Taboué (415) INSERM U320, CYCERON, 14074 Caen, France

Michael E. Phelps (3, 177, 183) Crump Institute for Biological Imaging, Department of Molecular and Medical Pharmacology, UCLA School of Medicine, Los Angeles, California 90024

U. Pietrzyk (139, 145, 151, 159) Max-Planck-Institut für neurologische Forschung, D-50931 Cologne, Germany

Julie C. Price (293, 427, 441) PET Facility, Department of Radiology, University of Pittsburgh Medical Center, Pittsburgh, Pennsylvania 15213

Pat M. Price (45, 107, 307) MRC Cyclotron Unit, Hammersmith Hospital, London W12 0NN, United Kingdom

M. Psylla (379) PET Department, Paul Scherrer Institute, CH-5232 Villigen PSI, Switzerland

D. M. Raffel (359) Division of Nuclear Medicine, Department of Internal Medicine, University of Michigan, Ann Arbor, Michigan 48109

J. S. Rakshi (117) MRC Cyclotron Unit, Hammersmith Hospital, London W12 0NN, United Kingdom

Michael J. Raleigh (177, 183) Department of Psychiatry and Biobehavioral Sciences, University of California, Los Angeles, Los Angeles, California 90095

Robert R. Rawlings (217) National Institute on Alcohol Abuse and Alcoholism, National Institutes of Health, Bethesda, Maryland 20892

K. Rehm (241) Department of Radiology, University of Minnesota, Minneapolis, Minnesota 55417

H.-J. Reulen (139) Department of Neurosurgery, Klinikum Großhadern, Ludwig-Maximilians-Universität, Munich, Germany

David C. Reutens (335, 371) McConnell Brain Imaging Centre, Montréal Neurological Institute, Montréal, Québec, H3A 2B4 Canada

B. W. Reutter (51) Center for Functional Imaging, Lawrence Berkeley National Laboratory, University of California, Berkeley, California 94720

Mark Richardson (59) MRC Cyclotron Unit, Hammersmith Hospital, London, W12 0NN United Kingdom; and Epilepsy Research Group, Institute of Neurology, London, United Kingdom

Daniel Rio (217) National Institute on Alcohol Abuse and Alcoholism, National Institutes of Health, Bethesda, Maryland 20892

D. A. Rottenberg (241, 253) PET Imaging Center, Veterans Administration Medical Center, Minneapolis, Minnesota 55417

Olivier G. Rousset (67, 77) McConnell Brain Imaging Centre, Montréal Neurological Institute, McGill University, Montréal, Québec, H3A 2B4 Canada

James Ruszkiewicz (293) PET Facility, Department of Radiology, University of Pittsburgh Medical Center, Pittsburgh, Pennsylvania 15213

T. J. Ruth (33, 387) University of British Columbia TRIUMF/PET Centre, Vancouver, British Columbia, V6T 2A3 Canada

Urs E. Ruttimann (217) National Institute on Alcohol Abuse and Alcoholism, National Institutes of Health, Bethesda, Maryland 20892

Norihiro Sadato (223) Fukui Medical School, Biomedical Imaging Research Center, Yoshida, Fukui 910-11, Japan

Hidetada Sasaki (207) Department of Geriatric Medicine, Tohoku University School of Medicine, Sendai 980-77, Japan

K. A. Schaper (241) PET Imaging Center, Veterans Administration Medical Center, Minneapolis, Minnesota 55417

P. Schumann (415) CNRS UMR 6551, University of Caen, France

A. Schuster (145) Max-Planck-Institut für neurologische Forschung, D-50931 Cologne, Germany

Michio Senda (223) PET Center, Tokyo Metropolitan Institute of Gerontology, Tokyo 173, Japan

Yiping Shao (3, 19) Crump Institute for Biological Imaging, Department of Molecular and Medical Pharmacology, UCLA School of Medicine, Los Angeles, California 90095

Chenggang Shen (201) Department of Radiology, Children's Hospital of Michigan PET Center, Detroit, Michigan 48201

H. Shinotoh (393) Division of Clinical Research, National Institute of Radiological Sciences, Chiba 263, Japan

Robert W. Silverman (3, 19) Crump Institute for Biological Imaging, Department of Molecular and Medical Pharmacology, UCLA School of Medicine, Los Angeles, California 90095

N. R. Simpson (441) PET Facility, Department of Radiology, University of Pittsburgh Medical Center, Pittsburgh, Pennsylvania 15213

Wan-Chi Siu (321, 339) Department of Electronic Engineering, the Hong Kong Polytechnic University, Hung Hom, Kowloon, Hong Kong

R. Slates (19) Crump Institute for Biological Imaging, UCLA School of Medicine, Los Angeles, California 90095

D. F. Smith (187) PET Center, Aarhus University Hospital, DK-8000 Aarhus C, Denmark

Gwenn S. Smith (293, 427, 441) Department of Psychiatry, University of Pittsburgh Medical Center, Pittsburgh, Pennsylvania 15213

Scott E. Snyder (261) Division of Nuclear Medicine, Department of Internal Medicine, University of Michigan, Ann Arbor, Michigan 48109

J. Sobesky (159) Max-Planck-Institut für neurologische Forschung, D-50931 Cologne, Germany

Thomas Sølling (463) PET Center, Aarhus University Hospital, DK-8000 Aarhus C, Denmark

V. Sossi (33) University of British Columbia, Vancouver, TRIUMF/PET Centre, British Columbia, V6T 2A3 Canada

Terry J. Spinks (25, 59) MRC Cyclotron Unit, Hammersmith Hospital, London W12 0NN, United Kingdom

J. Steinbach (285) Institut für Bioanorganische und Radiopharmazeutische Chemie, Forschungszentrum Rossendorf, D-01314 Dresden, Germany

David B. Stout (183) Department of Molecular and Medical Pharmacology, UCLA School of Medicine, Los Angeles, California 90095

S. C. Strother (241, 253) PET Imaging Center, Veterans Administration Medical Center, Minneapolis, Minnesota 55417

Y. Sudo (393) Division of Clinical Research, National Institute of Radiological Sciences, Chiba 263, Japan

M. Sugita (19) Department of Surgery, UCLA School of Medicine, Los Angeles, California 90095

T. Suhara (393) Division of Clinical Research, National Institute of Radiological Sciences, Chiba 263, Japan

K. Suzuki (393) Division of Clinical Research, National Institute of Radiological Sciences, Chiba 263, Japan

Itaru Tatsumi (223) PET Center, Tokyo Metropolitan Institute of Gerontology, Tokyo 173, Japan

X. Teng (51) Center for Functional Imaging, Lawrence Berkeley National Laboratory, University of California, Berkeley, California 94720

A. Thiel (139, 155, 159) Max-Planck-Institut für neurologische Forschung, D-50931 Cologne, Germany

Christopher J. Thompson (371) McConnell Brain Imaging Centre, Montréal Neurological Institute, Mon-tréal, Québec, H3A 2B4 Canada

John Thorpe (195) NMR Research Unit, Institute of Neurology, London WC1N 3BG, United Kingdom

P.-J. Toussaint (367) McConnell Brain Imaging Centre, Montréal Neurological Institute, Montréal, Québec, H3A 2B4 Canada

Hinako Toyama (223) Pet Center, Tokyo Metropolitan Institute of Gerontology, Tokyo 173, Japan

T. Uema (117) MRC Cyclotron Unit, Hammersmith Hospital, London W12 0NN, United Kingdom

M. S. Vafaee (173) PET Unit, McConnell Brain Imaging Centre, Montréal Neurological Institute, Montréal, Québec, H3A 2B4 Canada

John Darrell Van Horn (237) Unit on PET, CBDB/NIMH, National Institutes of Health, Bethesda, Maryland 20892

T. K. Venkatachalam (187) PET Center, Aarhus University Hospital, DK-8000 Aarhus C, Denmark

P. R. G. Virador (11) Lawrence Berkeley National Laboratory, University of California, Berkeley, California 94720

H.-M. von Stockhausen (139, 155, 159) Max-Planck-Institut für neurologische Forschung, D-50931 Cologne, Germany

P. Vontobel (379) PET Department, Paul Scherrer Institute, CH-5232 Villigen PSI, Switzerland

G. Vorwieger (285) Institut für Bioanorganische und Radiopharmazeutische Chemie, Forschungszentrum Rossendorf, D-013141 Dresden, Germany

John R. Votaw (229) Division of Nuclear Medicine, Emory University Center for PET, Atlanta, Georgia 30322

B. Walter (285) Institut für Pathophysiologie, Friedrich-Schiller-Universität, 07740 Jena, Germany

K. Wienhard (145, 151, 159) Max-Planck-Institut für neurologische Forschung, D-50931 Cologne, Germany

A. A. Wilson (273, 279) PET Centre, Clarke Institute of Psychiatry, Toronto, Ontario, M5T 1R8 Canada

Dean F. Wong (67, 77, 463) Division of Nuclear Medicine, Department of Radiology, Johns Hopkins Medical Institutions, Baltimore, Maryland 21287

Koon-Pong Wong (321) Biomedical and Multimedia Information Technology (BMIT) Group, Basser Department of Computer Science, University of Sydney, Syndey, New South Wales 2006, Australia

Chi-Ming Wu (321) Department of Electronic Engineering, The Hong Kong Polytechnic University, Hung Hom, Kowloon, Hong Kong

L. C. Wu (315) National PET/Cyclotron Center, Taipei Veterans General Hospital, Taiwan

Atsushi Yamaki (207) Laboratory of Neuroinformation Science, Tohoku Gakuin University, Sendai, Japan

Kazuhiko Yanai (207) Department of Pharmacology, Tohoku University School of Medicine, Sendai 980-77, Japan

J. Yang (85) Division of Nuclear Medicine and Biophysics, Department of Molecular and Medical Pharmacology, University of California, Los Angeles, School of Medicine, Los Angeles, California 90095

Jeffrey T. Yap (107) MRC Cyclotron Unit, Hammersmith Hospital, London W12 0NN, United Kingdom

Fuji Yokoi (77, 463) Division of Nuclear Medicine, Department of Radiology, Johns Hopkins Medical Institutions, Baltimore, Maryland 21287

A. R. Young (415) INSERM U320, 14074 Caen, France

C. L. Yu (85) Department of Electrical Engineering, Chung-Yuan University, Taiwan, Republic of China

Robert J. Zatorre (371) McConnell Brain Imaging Centre, Montréal Neurological Institute, Montréal, Québec, H3A 2B4 Canada

Preface

Over the past 20 years, positron emission tomography (PET) has matured into a powerful tool with which to study the physiology of the brain *in vivo*. The PET signal originates from tracers that are labeled with radionuclides such as ^{15}O, ^{11}C, and ^{18}F, permitting the use of biologically relevant chemical compounds. Innovative design of these tracers permits selective measurement of many aspects of brain function, and their short half-lives permit rapid, repeated scans over a time period that is practical for studies in patients. PET tomographs ("scanners") now offer the highest resolution available for nuclear medicine devices. Furthermore, quantitatively accurate reconstruction algorithms permit measurements of radioactivity concentrations in small regions of the brain with high precision and temporal sampling. To interpret these data, tracer kinetic models are developed for each radiopharmaceutical to convert the radioactivity images into measurements of physiologically relevant variables such as blood flow, metabolism, and receptor number. Powerful statistical and image analysis techniques have been developed to analyze data from multiple subjects rapidly, to extract the subtle biological signals that are produced by activation of the brain, and to identify sites of impaired function.

The research areas that encompass brain PET (i.e., radiochemistry, instrumentation and quantification, modeling, and image analysis) interact in the development and application of PET methods. Productive PET centers are composed of experts in each field leading active research programs. Although each of these categories represents a discipline unto itself, it is essential that PET scientists, usually experts in one field, achieve a high level of understanding of all of these research areas. In this way, all the components of PET can be optimally applied to develop truly sensitive and specific assays.

To help advance brain PET research and to improve scientific "fluency" in the many "languages" of PET, BRAINPET97, the Third International Conference on Quantification of Brain Function with PET, was held in Bethesda, Maryland, on June 20–22, 1997, on the campus of the National Institutes of Health. This meeting followed in the footsteps of two previous conferences, the first held in Akita, Japan, in 1993 and the second in Oxford, England, in 1995. BRAINPET97 was also a satellite meeting of the XVIIIth International Symposium on Cerebral Blood Flow and Metabolism (Brain '97).

This book presents the scientific developments that were discussed at the meeting. The chapters are divided into sections on data acquisition, image processing, applications, statistical analysis, tracer development, parameter estimation, and kinetic modeling. Most of these works, however, include important aspects of more than one of the brain PET disciplines. Many chapters focus on improved methodology, some explore the application of previously developed techniques to new areas, and others raise issues of concern about current PET approaches. The final section of the book is an edited transcript of the scientific discussion that occurred at BRAINPET97 following the oral presentations. These discussion periods were an important component of the conference, as many new ideas result from this type of challenging scientific interchange.

Because of the broad range of issues addressed, we believe that this volume will be a useful reference for the active brain PET scientist, as well as a valuable introduction for students and researchers who wish to take advantage of the capabilities of PET to study the normal and diseased brain.

Acknowledgments

The editors gratefully acknowledge the indispensable work of Brenda Allen, both in the production of this book and in the planning and execution of the BRAINPET97 conference. We are also indebted to Jamie Everett for his development of the BRAINPET97 web page (http://www-pet.cc.nih.gov/brainpet97) and to the staff of Circle Solutions and the NIH Natcher Conference Center for producing a wonderfully successful conference.

We gratefully acknowledge our honorary chairmen, Dr. Seymour Kety and Dr. Louis Sokoloff, for their important contributions. Finally, the members of the Scientific Advisory Committee, listed below, were critically responsible for the high quality of the conference abstracts that resulted in the contents of this book.

BRAINPET97 Scientific Advisory Committee

Nathaniel Alpert, United States
Jean-Claude Baron, France
Bernard Bendriem, France
Ronald G. Blasberg, United States
Simon Cherry, United States
Vincent Cunningham, United Kingdom
Robert Dannals, United States
William C. Eckelman, United States
Lars Eriksson, Sweden
Alan Evans, Canada
Ian Ford, United Kingdom
Joanna Fowler, United States
J. James Frost, United States
Albert Gjedde, Denmark
Christer Halldin, Sweden
Karl Herholz, Germany
Sung-cheng Huang, United States

Hidehiro Iida, Japan
Terry Jones, United Kingdom
Iwao Kanno, Japan
Michael Kilbourn, United States
Robert A. Koeppe, United States
Adriaan A. Lammertsma, Holland
Bengt Langstrom, Sweden
Niels A. Lassen, Denmark
Chester A. Mathis, United States
Bernard Mazoyer, France
Mark Mintun, United States
Stephen Moerlein, United States
Joel S. Perlmutter, United States
Michio Senda, Japan
Stephen Strother, United States
Christopher Thompson, Canada
Klaus Wienhard, Germany

SECTION I

DATA ACQUISITION AND QUANTIFICATION

CHAPTER 1

Brain Imaging in Small Animals Using MicroPET[1]

SIMON R. CHERRY, ARION CHATZIIOANNOU, YIPING SHAO, ROBERT W. SILVERMAN,
KEN MEADORS, and MICHAEL E. PHELPS

*Crump Institute for Biological Imaging
Department of Molecular and Medical Pharmacology
UCLA School of Medicine
Los Angeles, California 90095*

MicroPET is a prototype high-resolution positron emission tomography (PET) scanner designed for imaging small laboratory animals. It is the first PET system to make use of the new scintillator lutetium oxyorthosilicate and also the first system to employ fiber-optic coupling as a means for reading out very small scintillator elements with minimal deadspace. MicroPET consists of a total of thirty, 8×8 element detectors in a ring of diameter 17.1 cm. The animal port is 16 cm in diameter, and the useful field of view measures 11 cm transaxially by 1.8 cm axially. There are no interplane septa and data are acquired exclusively in three-dimensional mode. The system also includes a computer-controlled animal bed with built-in wobble motion. MicroPET produces images with a spatial resolution of < 2 mm in all three axes and has sufficient sensitivity to realize this resolution for many applications. For studying biological systems that are easy to saturate, improved sensitivity could be realized by the addition of a second or third detector ring, which would also serve to increase the axial coverage of the system. The use of microPET to image the brain of several species of animals is presented. Based on the success of this first prototype system, a second, higher resolution, higher sensitivity system is now being developed exclusively for use in mice and rats.

I. INTRODUCTION

The development of higher resolution detector technology has led to substantial interest in the use of positron emission tomography (PET) for applications in animals, particularly in smaller laboratory animals such as mice and rats. The combination of PET with animal models of human disease could provide a new and very powerful tool to the modern biomedical researcher, allowing longitudinal studies to be performed in the same animal. The ability to image biological function repeatedly in a single animal through disease progression or before and after some form of intervention is unique and would be of particular utility in situations where the animal model is highly variable from one animal to another, or when individual animals are extremely valuable.

Recent developments in photon detectors and scintillators have led to a number of designs for high-resolution animal PET systems (Watanabe *et al.*, 1992, 1995; Zhang *et al.*, 1994; Bloomfield *et al.*, 1995; Schmelz *et al.*, 1995; Lecomte *et al.*, 1996; Weber *et al.*, 1997), which are in various stages of development in research groups around the world. A dedicated animal PET system called microPET, which was designed to have 2-mm spatial resolution and be applicable to organ level studies in mice and rats and regional brain imaging in cats and small primates, has been developed at UCLA. This chapter presents an overview of the microPET development program and shows some of the first examples of brain imaging using this new dedicated animal PET scanner.

---

[1] Transcripts of the BRAINPET97 discussion of this chapter can be found in Section VIII.

II. MicroPET DETECTORS

The detectors used in microPET are very different from those found in conventional clinical PET systems. They consist of an 8 × 8 array of 2 × 2 × 10-mm lutetium oxyorthosilicate (LSO) crystals, individually coupled by 2-mm-diameter, 24-cm-long optical fibers (Kuraray Corp., Japan) onto a 64-element multichannel photomultiplier tube (MC-PMT) (Philips XP1722). The detector has been described in detail previously (Cherry *et al.*, 1997). The 64 channels are read out using a specially designed charge division circuit (Siegel *et al.*, 1996), giving an output that mimics that obtained from conventional block PET detectors. This allows conventional PET electronics to be used to read out these detectors, thus reducing the development time for the first prototype system.

The one-to-one coupling results in excellent intrinsic spatial resolution [full-width-at-half-maximum (FWHM) = 1.4 mm], and the optical fiber coupling eliminates dead space due to the MC-PMT packaging when these detectors are combined into a ring. Despite the loss of scintillation light in the optical fiber coupling (both at the fiber/crystal interfaces and transmission losses in the fiber), the timing resolution (2.4 nsec) and the energy resolution (20%) are very acceptable for PET applications (Cherry *et al.*, 1997).

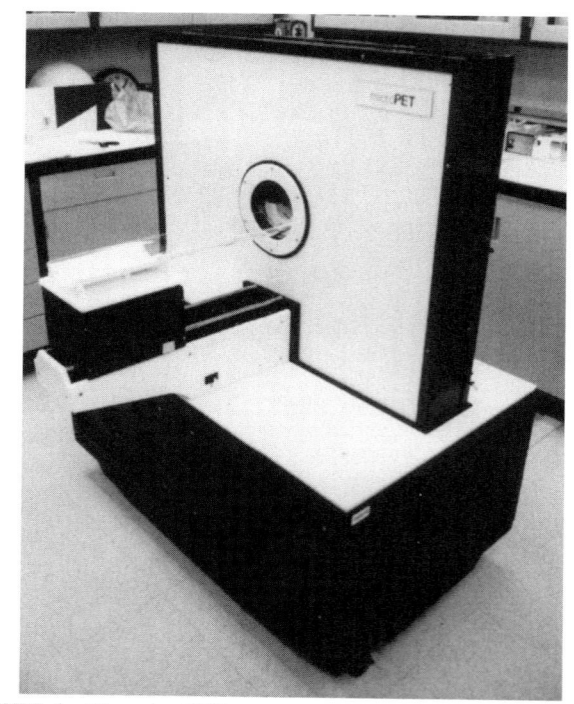

FIGURE 1 The microPET scanner. The detectors reside inside the rectangular housing (90 × 90 cm) and the electronics are housed in the cabinet underneath. The port size is 16 cm, permitting studies of rodents, cats, and small primates.

III. MicroPET SCANNER

The microPET scanner consists of thirty, 64-element LSO detectors in a ring of diameter 17.1 cm. The useful transaxial field of view is 11 cm (although the field of view over which depth of interaction effects are negligible is limited to approximately 6 cm) and the axial field of view is 18 mm. Further rings of detector modules could be added to improve sensitivity and axial coverage, but in this prototype system, one ring of modules was deemed sufficient to test the feasibility of this new approach.

The detectors are mounted in a gantry measuring 90 × 90 cm, which sits on top of a cabinet that houses the electronics and power supplies as shown in Fig. 1. The majority of the electronics and data acquisition system are identical to those used in the most recent generation of ECAT PET systems (CTI PET Systems Inc., Knoxville, TN). The coincidence window is set at 12 nsec, and the integration time on the PMT signals is set to 256 nsec. These are not optimal for the LSO detectors but are determined by limitations in the commercial PET electronics, which were designed to operate with bismuth germanate (BGO) scintillators.

MicroPET acquires data exclusively in three-dimensional (3D) mode as there are no interplane septa. The small size of the data sets and the capacity of the data acquisition system allow rapid dynamic studies to be performed. The animal bed is computer controlled, allowing accurate in and out positioning for whole body studies. The bed also incorporates a 0.7-mm diameter circular wobble motion designed to improve spatial sampling, which will be used to acquire higher resolution images when the counting statistics warrant its use. A laser alignment system is incorporated into the scanner to help in accurate positioning, which is critical given the small size of the axial field of view.

MicroPET data are normalized with the scan from a small centered cylinder, using symmetry arguments to calculate individual detector efficiencies (Hoffman *et al.*, 1989). No attenuation correction was applied to the studies presented in this chapter. Random events are subtracted on line using the delayed coincidence method. Dead-time correction is based on the recorded singles events on each bucket (set of six detectors) and closely follows the correction implemented on the clinical PET systems. Reconstruction uses the 3D reprojection algorithm (Kinahan and Rogers, 1989), although the implementation of a fully 3D statistical algorithm is being investigated (Mumcuoglu *et al.* 1996).

IV. PERFORMANCE MEASUREMENTS

The reconstructed spatial resolution was measured using a 0.5-mm-diameter ^{22}Na point source reconstructed with a ramp filter. Profiles through the images yielded a FWHM of 1.8 mm at the center of the scanner, degrading to no worse than 2 mm at a radial offset of 2.5 cm for both radial and tangential resolution components. The axial component was also less than 2 mm FWHM. The volume resolution of the system is below 8 mm^3 for a 5-cm-diameter field of view, which is sufficient to encompass the brain of small primates such as vervet or squirrel monkeys. This resolution represents almost an order of magnitude improvement over state-of-the-art clinical PET systems.

To illustrate the resolving power more clearly, a baby monkey brain phantom filled with ^{18}F solution was scanned. The phantom is constructed in an analogous fashion to the 3D Hoffman brain phantom (Hoffman *et al.*, 1991) from a stack of lucite sheets. The phantom was imaged on both microPET and the ECAT EXACT HR+ scanner (CTI/Siemens, Knoxville, TN) and sufficient statistics acquired to ensure that neither data set was count limited. The reconstructed images (ramp filter) are shown in Fig. 2 and clearly show the dramatic resolution improvement of microPET over the clinical system.

The absolute sensitivity of the scanner was measured by scanning a point source across the axial field of view and measuring the count rate per μCi of activity as a function of axial position. The sensitivity at the center of the field of view is 145 cps/μCi and the average sensitivity across the whole field of view is 77 cps/μCi. This compares favorably with the sensitivity of UCLA's large animal PET scanner (Cutler *et al.*, 1992), which uses 3-cm-deep BGO detectors and is also similar to that measured in 2D for a range of clinical PET systems (Bailey *et al.*, 1991; S. R. Meikle, unpublished results) using a similar method. In practice, this sensitivity gives sufficient statistics per resolution element to reconstruct with little or no smoothing in the majority of animal studies.

Figure 3 shows the count rate performance for a 6.2-cm-diameter by 6.0-cm-long cylinder (corresponding roughly to the size of a cat head). The scatter fraction for this cylinder was measured by extrapolation of the scatter tails in the sinogram to be 18%. Figure 3 also shows the noise equivalent count (NEC) curve

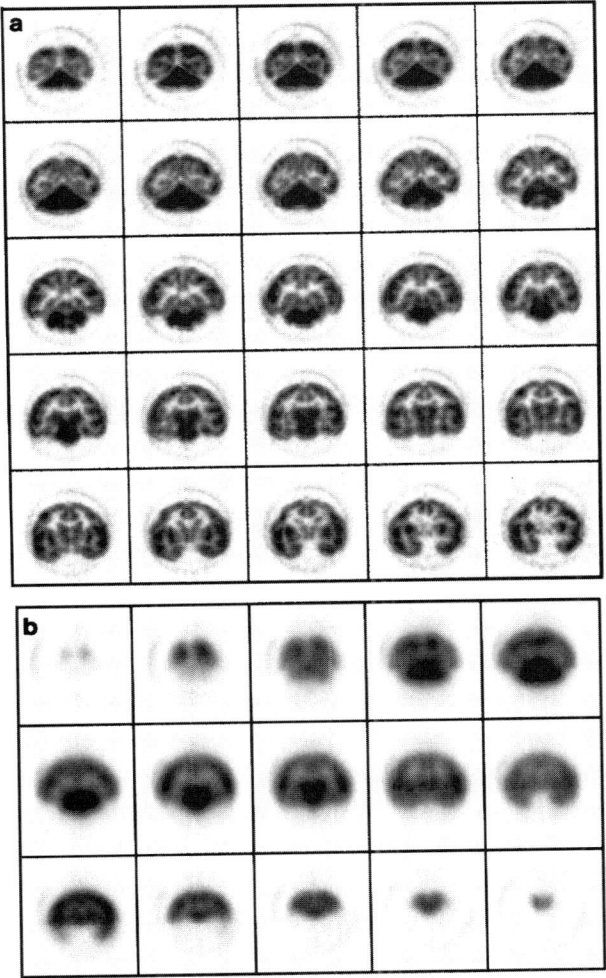

FIGURE 2 Phantom based on the brain of a baby monkey in coronal section. Data sets acquired on microPET (a) and the ECAT EXACT HR+ (b) were reconstructed at the highest resolution possible for each scanner. Note the dramatic increase in detail and the reduction in partial volume effects due to the improved resolution in microPET.

(Strother *et al.*, 1990) for this cylinder. The optimum energy window for most studies on microPET has been determined to be 350–650 keV. Data from this and other experiments reveal that the NEC performance in microPET is largely limited by singles from activity outside the field of view, which in turn are a result of the difficulty of effectively shielding small-diameter PET systems from such activity.

V. BRAIN IMAGING IN ANIMALS

Several examples are presented to show the quality of the images that can be obtained using the microPET

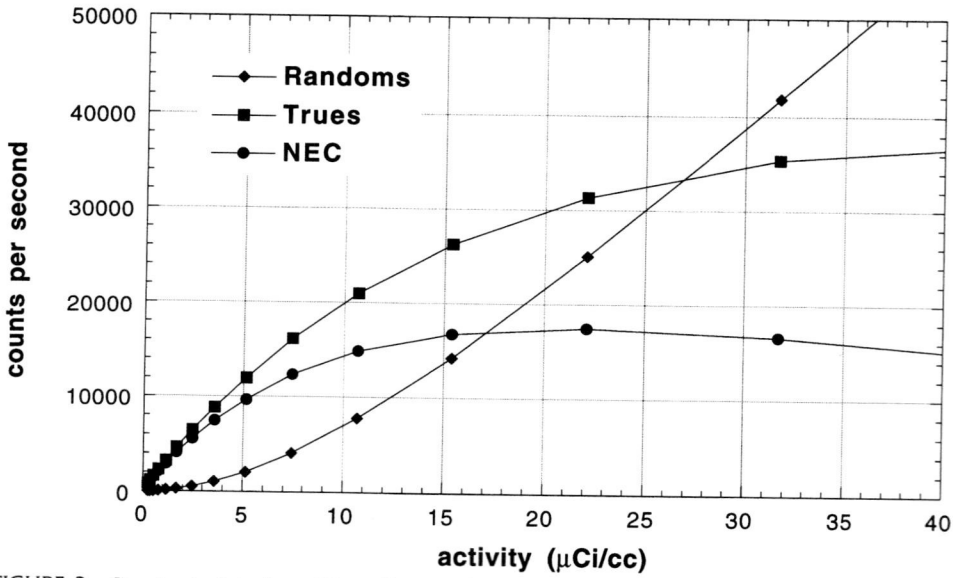

FIGURE 3 Count rate data for a 6.2-cm-diameter by 6.0-cm-long cylinder centered in the field of view. Noise equivalent count rates, which account for the effects of scattered and random coincidences, are also shown.

system when imaging the brain in a number of animal species. Figure 4 shows a [^{18}F]fluorodeoxyglucose (FDG) study in a 3-month-old vervet monkey. The brain is approximately 6 cm across its longest axis. Figure 5 represents an adult vervet monkey after the administration of [^{18}F]fluoroethylspiperone (FESP). Figure 6 shows [^{68}Ga]EDTA images from a rabbit brain tumor model, showing increased blood–brain barrier permeability in the area of the tumor. Finally, Fig. 7 (see color insert) shows a FDG data set from a

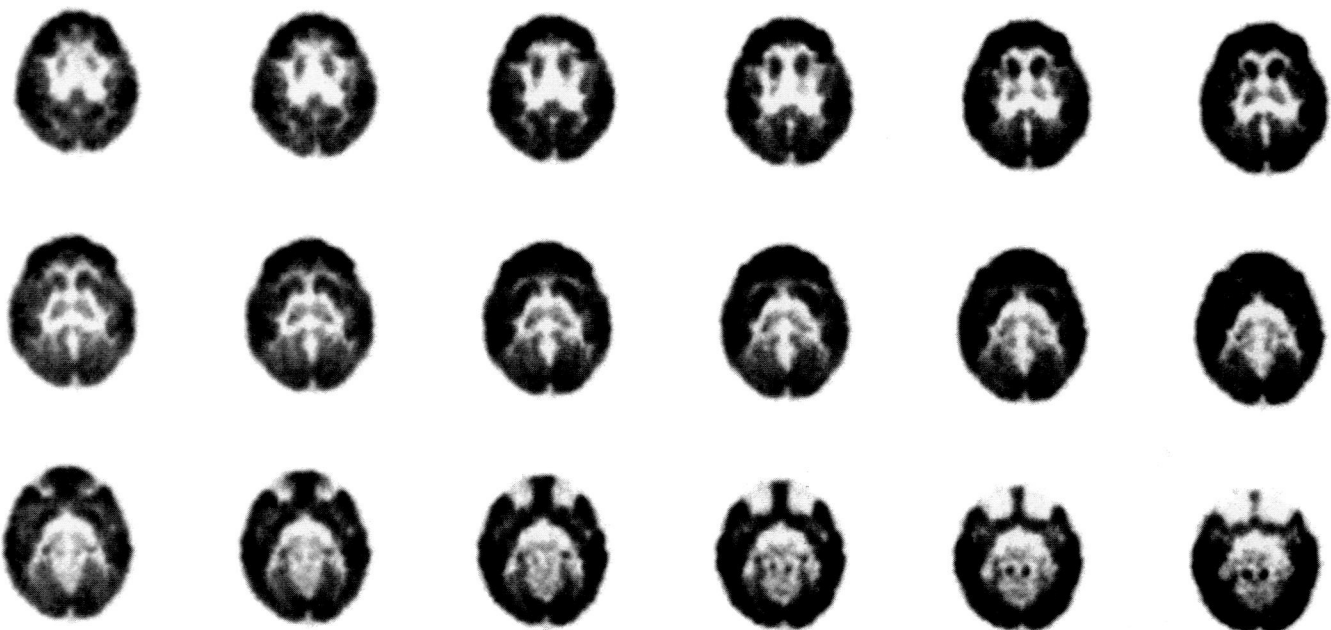

FIGURE 4 Transverse sections showing images of the FDG distribution in the brain of a 3-month-old vervet monkey. The injected dose was 2.1 mCi and the imaging time was 60 min.

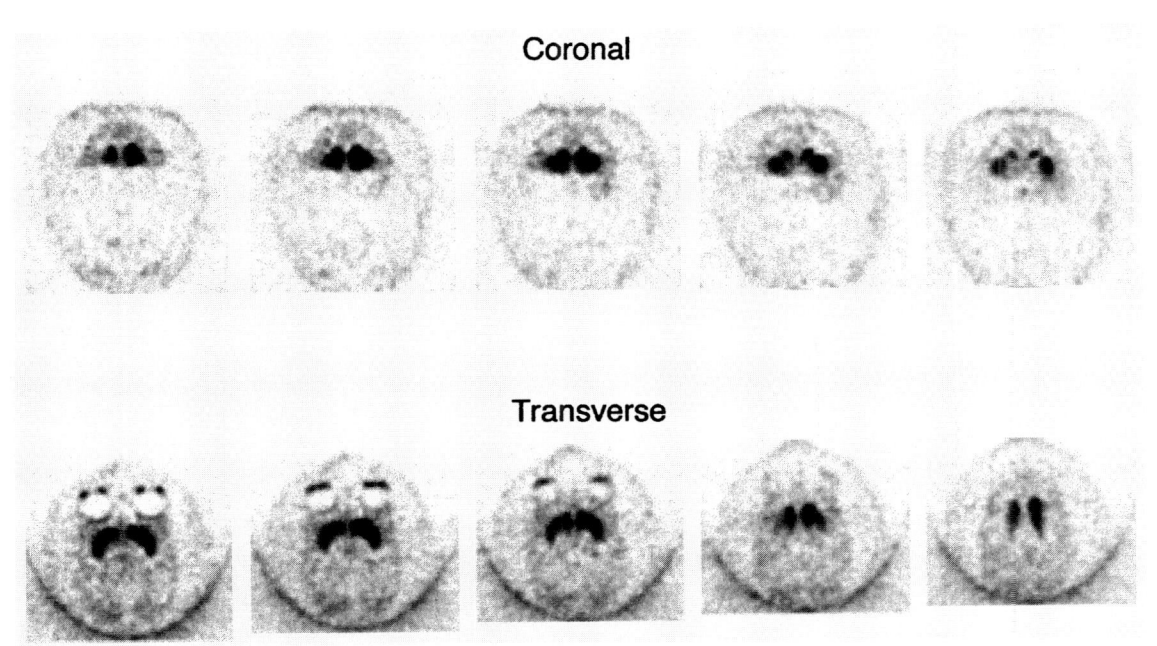

FIGURE 5 Transverse and coronal sections showing distribution of FESP in the brain of an adult vervet monkey. Notice the clear separation of caudate and putamen in the coronal planes. Imaging started 60 min after tracer administration, and six overlapped bed positions were acquired for 20 min each to cover an axial extent of 8.1 cm. The injected dose was 5 mCi.

rat, the corresponding brain atlas picture (Toga *et al.*, 1995), and a composite of the two showing excellent correspondence between the PET signal and structures in the rat head. Full details of the imaging parameters are given in the legend of each figure.

VI. DISCUSSION

The resolution improvement of microPET over state-of-the-art clinical PET systems allows many more details to be discerned in brain studies across a variety of animal species. All the studies shown here used reasonable injected doses and imaging times, and sufficient counts were acquired, so that little or no smoothing had to be applied in the image reconstruction process. However, in some biological systems, more sensitivity may be required to avoid saturating the system under investigation. The sensitivity of the scanner could be improved easily by adding additional detector rings, which would also improve axial coverage. Count rate performance is also adequate for all studies performed so far; the main limitation is high singles and randoms rates, relative to the true coincidence rate, due in large part to activity outside the field of view, which is difficult to shield in a small-diameter PET system.

FIGURE 6 [^{68}Ga]EDTA distribution in a rabbit brain tumor model. Accumulation of tracer in the tumor due to increased blood–brain barrier permeability is clearly seen. The injected dose was 1.8 mCi, and the images were acquired over 30 min.

Future work will concentrate on validating all the data correction methods to permit quantitative dynamic studies. The effect of wobbled acquisitions on spatial resolution and the use of statistical reconstruction algorithms (Mumcuoglu *et al.*, 1996; Qi *et al.*, 1998) to help compensate for blurring due to detector penetration, noncolinearity, and positron range are also being investigated (although the latter effects are still relatively minor on this system when using tracers labeled with ^{18}F or ^{11}C). To provide more robust analysis of animal PET studies, PET and magnetic resonance imaging compatible stereotactic head holders (Myers *et al.*, 1996) are being developed for animal models that will allow PET data to be placed into a reference orientation and permit atlas-based region of interest definition. This will be particularly useful in the rat brain where structures are still small in size compared with the resolution of the imaging system.

Based on the success of this first microPET system, the authors are in the process of designing and developing microPET II where the goal is to push the current detector technology to its limits, resulting in better spatial resolution. The new system will also have higher sensitivity and better axial coverage by incorporating multiple detector rings. This system is expected to be ready for testing in the year 2000.

VII. SUMMARY

MicroPET is a high-resolution PET scanner dedicated to imaging small laboratory animals. It is the first PET system to employ the new scintillator LSO and the first to use fiber-optic readout of small scintillator arrays successfully. The volume resolution of ~ 8 mm^3 is sufficient to permit organ level studies in mice and rats and regional brain imaging in cats and primates. The resolution is also expected to be adequate for performing studies in rat brain using highly specific ligands such as [^{18}F]FESP or [^{11}C]raclopride. MicroPET is seen as a prototype for a new class of compact, low-cost, high-resolution PET scanners, designed for use in biomedical laboratories and in the biotechnology and pharmaceutical industries.

Acknowledgments

We thank the following individuals who have all contributed their expertise to this large project: Abdel Boutefnouchet, Andrew Bowery, David Binkley, Mike Casey, Judy Edwards, Esso Flyckt, Sam Gambhir, Kent Gardner, Edward Hoffman, David Hovda, Andrew Hufford, Bill Jones, Richard Leahy, Waldi Ladno, Stan Majewski, Bill Melega, Amy Moore, Clif Moyers, Erkan Mumcuoglu, Danny Newport, Ron Nutt, Alan Oshiro, Jinyi Qi, Michael Raleigh, N. Satyamurthy, Parag Shah, Ron Sumida, David Truong, and John Young. This project is supported by the Ahmanson Foundation, the UCLA Cancer Center, the Norton Simon Fund, Department of Energy Contract DE-FC03-87-ER60615, and by grants from the Whitaker Foundation (J100193) and the National Institutes of Health (R01 CA69370).

References

Bailey, D. L., Jones, T., and Spinks, T. J. (1991). A method for measuring the absolute sensitivity of positron emission tomographic scanners. *Eur. J. Nucl. Med.* **18**: 374–379.

Bloomfield, P. M., Rajeswaran, S., Spinks, T. J., Hume, S. P., Myers, R., Ashworth, S., Clifford, K. M., Jones, W. F., Byars, L. G., Young, J., Andreaco, M., Williams, C. W., Lammertsma, A. A., and Jones, T. (1995). The design and physical characteristics of a small animal positron emission tomograph. *Phys. Med. Biol.* **40**: 1105–1126.

Cherry, S. R., Shao, Y., Silverman, R. W., Chatziioannou, A., Meadors, K., Siegel, S., Farquhar, T., Young, J., Jones, W. F., Newport, D., Moyers, C., Andreaco, M., Paulus, M., Binkley, D., Nutt, R., and Phelps, M. E. (1997). MicroPET: A high resolution PET scanner for imaging small animals. *IEEE Trans. Nucl. Sci.* **44**: 1161–1166.

Cutler, P. D., Cherry, S. R., Hoffman, E. J., Digby, W. M., and Phelps, M. E. (1992). Design features and performance of a PET system for animal research. *J. Nucl. Med.* **33**: 595–604.

Hoffman, E. J., Guerrero, T. M., Germano, G., Digby, W. M., and Dahlbom, M. (1989). PET system calibration for quantitative and spatially accurate images. *IEEE Trans. Nucl. Sci.* **36**: 1108–1112.

Hoffman, E. J., Cutler, P. D., Guerrero, T. M., Digby, W. M., and Mazziotta, J. C. (1991). Assessment of accuracy of PET utilizing a 3-D phantom to simulate the activity distribution of [^{18}F]fluorodeoxyglucose uptake in the human brain. *J. Cereb. Blood Flow Metab.* **11**: A17–A25.

Kinahan, P. E., and Rogers, J. G. (1989). Analytic 3D image reconstruction using all detected events. *IEEE Trans. Nucl. Sci.* **36**: 964–968.

Lecomte, R., Cadorette, J., Rodrique, S., Lapointe, D., Rouleau, D., Bentourkia, M., Yao, R., and Msaki, P. (1996). Initial results from the Sherbrooke avalanche photodiode positron tomograph. *IEEE Trans. Nucl. Sci.* **43**: 1952–1957.

Mumcuoglu, E. U., Leahy, R. M., and Cherry, S. R. (1996). Bayesian reconstruction of PET images: Quantitative methodology and performance analysis. *Phys. Med. Biol.* **41**: 1777–1807.

Myers, R., Hume, S. P., Ashworth, S., Lammertsma, A. A., Bloomfield, P. M., Rajeswaran, S., and Jones, T. (1996). Quantification of dopamine receptors and transporter in rat striatum using a small animal PET scanner. *In* "Quantification of Brain Function using PET" (R. Myers, V. Cunningham, D. Bailey, and T. Jones, eds.), pp. 11–15. Academic Press, San Diego, CA.

Qi, J., Leahy, R. M., Cherry S. R., Chatziioannou, A., and Farqhar, T. H. (1998). High resolution 3D Bayesian image reconstruction using the small animal microPET scanner. *Phys. Med. Biol.* (submitted for publication).

Schmelz, C., Bradbury, S. M., Holl, I., Lorenz, E., Renker, D., and Ziegler, S. (1995). Feasibility study of an avalanche photodiode readout for a high resolution PET with nsec time resolution. *IEEE Trans. Nucl. Sci.* **42**: 1080–1084.

Siegel, S., Silverman, R. W., Shao, Y., and Cherry, S. R. (1996). Simple charge division readouts for imaging scintillator arrays using a multi-channel PMT. *IEEE Trans. Nucl. Sci.* **43**: 1932–1937.

Strother, S. C., Casey, M. E., and Hoffman, E. J. (1990). Measuring PET scanner sensitivity: Relating count rates to image signal-to-noise ratios using noise equivalent counts. *IEEE Trans. Nucl. Sci.* **37**: 783-788.

Toga, A. W., Santori, E. M., Hazani, R., and Ambach, K. (1995). A 3D digital map of rat brain. *Brain Res. Bull.* **38**: 77-85.

Watanabe, M., Uchida, H., Okada, H., Shimizu, K., Satoh, N., Yoshikawa, E., Ohmura, T., Yamashita, T., and Tanaka, E. (1992). A high resolution PET for animal studies. *IEEE Trans. Med. Imag.* **11**: 577-580.

Watanabe, M., Omura, T., Kyushima, H., Hasegawa, Y., and Yamashita, T. (1995). A compact position-sensitive detector for PET. *IEEE Trans. Nucl. Sci.* **42**: 1090-1094.

Weber, S., Terstegge, A., Engels, R., Herzog, H., Reinartz, R., Reinhart, P., Rongen, F., Müller-Gärtner, H. W., and Halling H. (1997). The KFA TierPET: Performance characteristics and measurements. In "1996 IEEE Nuclear Science Symposium Conference Record" (A. DelGuerra, ed.), pp. 1117-1119. IEEE, Piscataway, NJ.

Zhang, S., Bruyndonckx, P., Goldberg, M. B., and Tavernier, S. (1994). Study of a high resolution 3D PET Scanner. *Nucl. Instrum. Methods* **A348**: 607-612.

CHAPTER 2

Design of a High–Resolution, High–Sensitivity PET Camera for Human Brains and Small Animals[1]

W. W. MOSES, P. R. G. VIRADOR, S. E. DERENZO, R. H. HUESMAN, and T. F. BUDINGER

Lawrence Berkeley National Laboratory
University of California
Berkeley, California 94720

Design parameters for a three-dimensional positron emission tomography camera with high sensitivity (35-cm detector ring diameter, 15-cm axial field of view) and isotropic high resolution provided by detector modules capable of depth of interaction (DOI) measurement are presented. Detector modules are made of lutetium oxyorthosilicate (LSO) crystals (3 mm square by 30 mm deep). The small module size and short decay time of LSO reduce the detector dead time by a factor of 14 compared to conventional bismuth germanate detector modules and narrow the coincidence window width to 4 ns. This yields an expected peak noise equivalent count rate of 800 kcps and a noise equivalent sensitivity of 1370 kcps/μCi/cc with a 20-cm diameter phantom—three to five times higher than conventional scanners. With 5-mm full-width-at-half-maximum (FWHM) DOI resolution, the expected reconstructed spatial resolution is < 3.0 mm FWHM throughout the entire field of view. Depth of interaction measurement information is incorporated into the reconstruction algorithm by rebinning onto a regularly spaced grid. Attenuation correction is performed with an orbiting singles transmission source. A complete treatment of this work appears in Moses et al. (1997) and is reprinted herein with the permission of the journal. Reference to the work should be made to the original publication.

I. INTRODUCTION

Although the factors controlling the spatial resolution and efficiency of positron emission tomography (PET) cameras have long been understood (Derenzo, 1986), performance limitations of 511-keV photon detectors force compromises in the design of practical PET imagers that prevent the fundamental limits of efficiency and resolution from being realized. As most 511-keV photon detectors utilize a common scintillator material, bismuth germanate (BGO), a common set of trade-offs are imposed on PET camera designers, resulting in similar PET camera designs. Incorporating a lutetium oxyorthosilicate (LSO) scintillator (Melcher and Schweitzer, 1992) into PET detector modules, which is becoming an increasingly realistic option, can improve PET detector performance significantly. The fourfold increase in light output and 7.5 times decrease in decay time change the set of compromises imposed on the PET camera designer and alter the resulting camera design significantly. This chapter presents a design study for one such camera that is predicted to obtain significantly higher spatial resolution and efficiency than present BGO-based cameras. These performance improvements should enhance the quantitative accuracy and clinical utility of the reconstructed images.

The relevant scintillation properties of LSO and BGO are listed in Table 1. The main advantage of LSO over BGO is its faster decay time, which leads to a significant reduction in dead time and an improvement

---

[1] Transcripts of the BRAINPET97 discussion of this chapter can be found in Section VIII.

TABLE 1 Comparison of BGO and LSO Scintillator Properties

	BGO	LSO
Chemical formula	$Bi_4Ge_3O_{12}$	$Lu_2SiO_5{:}Ce$
Decay time (nsec)	300	40
Light output (photons/MeV)	8200	30,000
Attenuation length (cm)	1.1	1.2
Photoelectric fraction	43%	34%
Energy resolution (511 keV)	10%	<10%
Emission wavelength (nm)	480	415
Index of refraction	2.15	1.82
Radioactive background?	No	Yes
Hygroscopic?	No	No

in coincidence timing resolution. The higher light output of LSO allows development of depth of interaction measurement ability but, for reasons that are not understood, does not improve the energy resolution greatly.

II. GENERAL TOMOGRAPH DESIGN

The design details for this high-sensitivity, high-resolution PET camera are summarized in Table 2. The patient port dimensions (30 cm diameter, 15 cm axial extent) are suitable for imaging the human brain and limbs or small animals (such as dogs or monkeys), but are too small to accommodate the human thorax. No space for interplane septa is provided so the camera can operate only in "fully 3D" mode. The lack of septa allows a 35-cm detector ring diameter that is only slightly larger than the patient port. This reduces both the number of detector modules (and hence the cost) and the resolution degradation due to annihilation photon acollinearity. Reducing the diameter (while keeping the 15 cm axial extent) also increases the solid angle coverage, yielding a significant increase in sensitivity.

The detector requirements of this design exceed those of conventional cameras. Placing the detector modules closer to the patient port requires detectors with very high count rate performance to minimize

TABLE 2 Comparison between a Conventional PET Camera and the Proposed PET Camera

Quantity	Conventional tomograph (ECAT EXACT HR) (Wienhard et al., 1994)	Proposed tomograph
Crystal size	5.9 × 2.9 × 30 mm	3 × 3 × 30 mm
Crystal material	$Bi_4Ge_3O_{12}$ (BGO)	$Lu_2SiO_5{:}Ce$ (LSO)
Patient port diameter	56 cm	30 cm
Detector ring inner diameter	72 cm	35 cm
Block size	25 × 50 mm	25 × 25 mm
PMT size	25 × 25 mm (dual)	25 × 25 mm (single)
Crystals per block	56	64[a]
Blocks per ring	112	44
Number of rings (15 cm axial FOV)	3	6
Blocks per system	336	264
Crystals per system	18,816	16,896[a]
PMTs per system	672 (dual anode)	264 (single anode)
Volume of scintillator crystal	12,600 ml (90 kg)	4,950 ml (37 kg)
Coincidence window width	12 nsec	4 nsec
Dead time per block	2.5 μsec	0.3 μsec
Maximum coincidence rate	1.4 MHz (ACS I)	10 MHz
Image resolution (^{18}F)	3.6 mm FWHM	2.2 mm FWHM[a]
Maximum noise equivalent event rate (in 20-cm-diameter phantom)	2D 110 k/sec (at 2.0 μCi/cc) 3D 90k/sec (at 0.4 μCi/cc)	2D N/A 3D 730 k/sec (at 1 μCi/cc)

[a] Parameters that will change if the crystal size is other than 3 mm.

dead time losses. The patient port nearly fills the detector ring, so severe radial elongation artifacts will be present unless the detector module can measure the interaction depth of individual 511-keV photons. The system is only capable of septaless operation, so good energy resolution is necessary to reject patient Compton scatter events, and good timing resolution is necessary to reduce random coincidences. Several LSO-based design concepts will soon provide ≤ 3-mm axial and transaxial crystal size, ≤ 6.25-cm² front surface area, ≤ 750-psec timing resolution, ≤ 20% FWHM energy resolution, and ≤ 10-mm FWHM DOI measurement resolution (Moses et al., 1995; Worstell et al., 1996; Moisan et al., 1996; Casey et al., 1997; Huber et al., 1997), so this chapter derives performance estimates based on these detector parameters.

This design shares the small ring diameter and lack of septa with the NaI(Tl)-based PENN-PET camera (Karp et al., 1990), but the thicker scintillator and higher count rate capability allow the proposed design to utilize the increased solid angle coverage more fully, resulting in significantly increased sensitivity. The BGO-based ECAT EXACT3D camera (Spinks et al., 1997) has a septaless design and thick scintillator crystal, but the depth of interaction capability in the proposed design allows the detector diameter to be reduced substantially without degrading spatial resolution, yielding equivalent solid angle coverage and higher spatial resolution with significantly less detector material. Although it could be used to image smaller animals (such as rats or mice), the cost and performance for this task would not be as favorable as for cameras specifically designed for this purpose (Bloomfield et al., 1995; LeComte et al., 1996; Cherry et al., 1997).

III. PERFORMANCE ESTIMATES

A. Spatial Resolution

The contributions to spatial resolution near the center of a tomograph with adequate sampling are the crystal size, crystal decoding process, annihilation photon acollinearity, positron range, and reconstruction algorithm. A reasonably accurate estimate of the reconstructed image resolution (FWHM) Γ of an arbitrary PET camera can be made using the formula

$$\Gamma = 1.25\sqrt{(d/2)^2 + (0.0022D)^2 + R^2 + b^2}, \quad (1)$$

where d is the detector width, D is the detector cylinder diameter, R is the effective positron range, and b is a factor due to the crystal decoding process (0 if the crystal of interaction is individually coupled to a photodetector and 2.2 otherwise), with units of millimeters for all quantities. The b factor, while empirically determined for BGO detector modules, has been found to be necessary to "predict" the performance of several new high-resolution PET cameras (Moses and Derenzo, 1993).

With the design proposed, the crystal width is 3.0 mm, the detector ring diameter is 350 mm, and the crystals would preferably be individually coupled to photodetectors. This yields contributions of 1.5 mm for crystal size, 0.8 mm for acollinearity, and 0 mm for crystal decoding. When imaging ^{18}F ($R = 0.5$ mm), a reconstructed spatial resolution of 2.2 mm is predicted, whereas a design with 2.0-mm² crystals would yield 1.8-mm FWHM resolution.

B. Radial Elongation

The spatial resolution predicted by Eq. (1) applies only near the center of the tomograph (i.e., for distances less than 10% of the ring diameter), where penetration of 511-keV photons into adjacent crystals is minimal, and radial elongation is not present. Measuring the interaction depth on an event by event basis can significantly reduce this elongation; the amount of reduction depends both on the detector ring diameter and on the resolution with which the interaction depth is measured.

To predict the spatial resolution for a single plane PET camera with 3-mm LSO crystals, a Monte Carlo simulation is performed. This simulation assumes the detector module design in Huber et al. (1997) and includes annihilation photon acollinearity, energy-dependent Compton and photoelectric cross sections, and finite (20% FWHM at 511 keV) detector energy resolution, with the measured interaction depth assumed to be the true interaction depth blurred by a Gaussian with fixed width. If Compton interactions cause energy to be deposited in more than one crystal, the crystal with the higher energy is assigned as the crystal of interaction, and if there are multiple energy depositions in a single crystal, the true position is assumed to be the center of gravity of the energy depositions. Data are reconstructed using filtered backprojection with the depth of interaction information incorporated by the rebinning algorithm described later.

The results of this simulation are shown in Fig. 1, which plots the FWHM of the reconstructed line source image (in the radial direction) as a function of distance from the tomograph center. Tomographs with several different DOI measurement resolutions are simulated. Without DOI information, the familiar radial elongation artifact degrades the spatial resolution by a factor of 3 at a distance of 10 cm from the tomograph center.

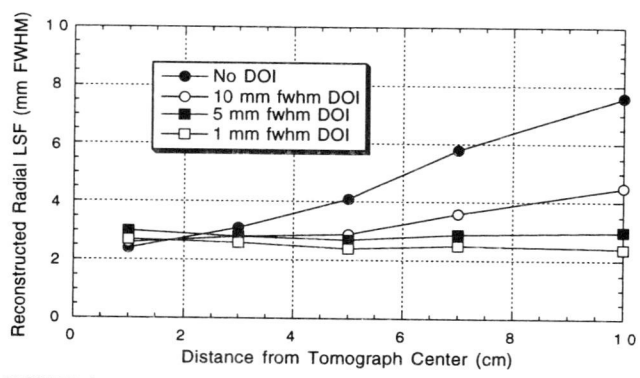

FIGURE 1 The radial component of the reconstructed point spread function as a function of position from the tomograph center for several depth of interaction measurement resolutions. The scintillator crystals modeled are 3 mm square by 30 mm long. © 1997 IEEE. Reprinted, with permission, from *IEEE Transactions on Nuclear Science*, Vol. 44, No. 4, pp. 1487–1491, August 1997.

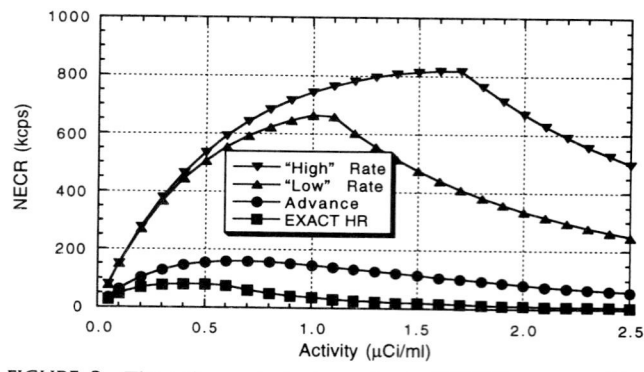

FIGURE 2 The noise equivalent count rate (NECR) versus activity concentration in a 20-cm-diameter phantom for several tomograph designs. © 1997 IEEE. Reprinted, with permission, from *IEEE Transactions on Nuclear Science*, Vol. 44, No. 4, pp. 1487–1491, August 1997.

However, this effect is significantly reduced even with 10-mm FWHM DOI measurement resolution, which is relatively easily obtained. With a DOI measurement resolution of 5 mm FWHM (which is feasible, although with greater difficulty) the spatial resolution becomes essentially uniform at 2.5–3.0 mm FWHM throughout the field of view. Similar simulations were performed with 30- and 40-cm detector ring diameters, and the predicted spatial resolution is insensitive to ring diameter.

C. Noise Equivalent Count Rate

The small ring diameter, septaless operation, and the low dead time of LSO give this design an extremely high efficiency. A measure of the capability of a PET camera to image activity in the field of view accurately is the noise equivalent count rate (NECR) (Strother *et al.*, 1990). The NECR for a 20-cm-diameter water-filled phantom is estimated with an analytic computation that includes dead time in the front end and coincidence processing electronics, scatter and attenuation in the phantom, and the effects of energy discrimination in the detector modules. Scatter subtraction is not explicitly performed; the number of scatter events is assumed to be a fixed fraction of the true coincidence events. This fraction is determined by Monte Carlo simulation to be 75% for an 85-cm-diameter ring and 120% for a 35-cm-diameter ring. To validate this analysis, the computed NECR versus activity concentration was compared to published values for the CTI/Siemens ECAT EXACT HR (Wienhard *et al.*, 1994) and the GE Advance (DeGrado *et al.*, 1994) operated in septaless mode. The predicted curves, plotted in Fig. 2, agree with published data to within 10%.

Data in Fig. 2 do not necessarily represent present capabilities of these manufacturers but are used only to validate the analysis.

Figure 2 also shows the computed NECR versus activity concentration for the proposed detector assuming a 35-cm detector diameter. Two sets of assumptions for dead time are modeled: a low rate design (300 nsec detector dead time and a 10-MHz coincidence processing limit) and a high rate design (200 nsec detector dead time and a 20-MHz coincidence processing limit). A maximum NECR of 700–800 kcps is obtained with activity densities of 1.0–1.5 μCi/ml. Thus, the proposed design has roughly a factor of five higher peak NECR than either of these existing scanners, mostly due to its lower dead time, high coincidence processor throughput, and narrow (4 nsec) coincidence window.

Similar computations predict that increasing the ring diameter to 40 cm increases the NECR approximately 10%, while reducing the diameter to 30 cm decreases the NECR approximately 40%. These differences in NECR are mostly due to significant differences in the randoms rate. With a 30-cm-diameter patient port, 100% of the possible detector module pairs in a 30-cm-diameter detector ring form valid coincidences, whereas only 54% of the possible detector pairs form valid coincidences with a 40-cm-diameter detector ring. The singles rate is nearly independent of ring diameter, so the randoms rate is roughly proportional to the fraction of detector module pairs that form valid coincidences.

D. Noise Equivalent Sensitivity

A measure of the efficiency of a PET camera for detecting activity in the field of view is the noise equivalent sensitivity (NES) (Stearns et al., 1995), which is the slope of the NECR versus activity density curve at zero activity density. The NES of this design is predicted to be 1370 kcps/μCi/cc—the same analytic computation obtains an accurate NES for the CTI/Siemens ECAT EXACT HR (710 kcps/μCi/cc) and the GE Advance (770 kcps/μCi/cc) operated in septaless mode. Thus, the NES of the proposed design is approximately two times higher than existing designs. This factor is attributed to the increased solid angle; the reduced random event contribution enabled by the narrow coincidence window is canceled in part by the increased scatter fraction contribution due to the larger solid angle coverage. The 4-nsec coincidence window in this design comfortably accommodates the expected 2-nsec time of flight difference across the detector ring and the 1-nsec FWHM coincidence time resolution.

IV. ADDITIONAL CONSIDERATIONS

A. Attenuation Correction

The lack of interplane septa (and hence the inability to take *any* data in 2D mode) makes attenuation correction difficult, as attenuation maps for 3D reconstruction are usually derived from transmission scans taken in 2D mode. The authors anticipate following the lead of other groups and using a high activity point source that emits 662-keV photons to perform transmission measurements (DeKemp and Nahmias, 1994; Karp et al., 1995). Although such data have a higher scatter contamination than data collected with windowed coincidence sources, the relatively uniform attenuation coefficient in the human head and low attenuation and scatter fraction in small animals are likely to minimize the effect.

B. Including Depth of Interaction Information

Previous reconstruction algorithms (Moses et al., 1991) have incorporated depth of interaction measurement information in the following manner. A line is constructed joining the measured interaction points, and the r and θ values of the resulting chord are computed. This event is then assigned to the nearest "physical" chord, where a "physical" chord is defined to be a line connecting the center of the front faces of two crystals. Due to the circular shape of the tomograph ring, these "physical" chords become more closely spaced in r as they move away from the center of the tomograph. This change in spacing is more pronounced for our design than for conventional PET cameras, as the patient port diameter fills a significantly larger fraction of the detector ring diameter. As conventional reconstruction algorithms use evenly spaced bins, the "physical" chords are then interpolated onto a grid of evenly spaced "ideal" chords, which are then reconstructed with filtered backprojection.

A variant of this rebinning algorithm where the line joining the measured interaction points is assigned directly to the "ideal" chord is proposed. This obviates the interpolation to "physical" and then to "ideal" chords. The chord spacing in r need not be constrained to half the detector width ($d/2$), and so a smaller spacing (e.g., $d/4$) can be used to improve the radial sampling and so the reconstructed spatial resolution. For chords that pass near the center of the tomograph, some interpolation is necessary to avoid unfilled chords, and subsequent normalization (performed in the same way as with conventional scanners) is necessary to avoid aliasing artifacts caused by nonuniform sampling. It is possible that rotating the detector modules slightly so that their front faces are not normal to the line connecting them to the tomograph center (Derenzo, 1984) could avoid unfilled chords, but the other ramifications of this approach have not been explored.

The left-hand image in Fig. 3 shows a point source located 10 cm from the center of the tomograph, reconstructed using the previously described algorithm. Data were generated using the Monte Carlo simulation with $3 \times 3 \times 30$-mm crystals, a 35-cm-diameter detector ring, and 5-mm FWHM DOI measurement resolution, and data were reconstructed in 2D with a ramp filter. The image is free of artifacts from penetration of 511-keV photons in the detector ring or from the reconstruction algorithm, and the point spread function is 3.0 mm FWHM in the radial direction and 2.9 mm FWHM in the tangential direction. The image at the right in Fig. 3 includes no DOI information and shows

FIGURE 3 The image of a point source displaced 10 cm horizontally from the center of the tomograph, reconstructed with the rebinning algorithm described herein. (Left) The image is with 5 mm FWHM DOI measurement resolution. (Right) The image is without DOI information. The field of view is 1 cm^2. © 1997 IEEE. Reprinted, with permission, from *IEEE Transactions on Nuclear Science*, Vol. 44, No. 4, pp. 1487–1491, August 1997.

severe radial elongation artifacts. Figure 1 shows the expected reconstructed spatial resolution as a function of DOI measurement resolution. This algorithm also produces artifact-free images of extended source phantoms.

Although these images are reconstructed with a 2D algorithm, the concepts for including depth of interaction measurement information and directly rebinning onto a regularly spaced grid can be extended to 3D reconstruction algorithms. The lines of response from a cylindrical tomograph are actually neither evenly spaced nor parallel; the transverse separation of the lines decreases and the elevation angles of the lines increase with distance from the central axis of the cylinder, as shown in Fig. 4. The rebinning onto a regularly spaced grid will be applied to both in-plane (r, θ) and cross-plane (ϕ) coordinates, eliminating the geometrical errors caused by nonuniform sampling in both the r and ϕ directions. This rebinning would be performed before data are histogrammed (or data could be taken in list mode) so the number of chords that are backprojected is similar to that of a conventional 3D scanner. This implies that although the address defining a single crystal must contain three additional bits (for depth of interaction information) and the coincidence processor must be more sophisticated in order to perform this rebinning, no additional memory is necessary.

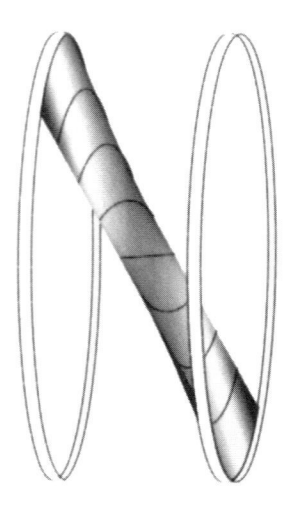

FIGURE 4 Geometric lines of response from oblique (cross-plane) slices. As adjacent detector elements within a plane form arcs, the lines connecting them are neither parallel nor coplanar. Current 3D reconstruction algorithms ignore this effect. © 1997 IEEE. Reprinted, with permission, from *IEEE Transactions on Nuclear Science*, Vol. 44, No. 4, pp. 1487–1491, August 1997.

V. CONCLUSIONS

A conceptual design is presented for a PET camera designed to image the human brain and small animals. Removing the interplane septa and reducing the detector ring diameter so that it is slightly larger than the field of view affords a dramatic increase in solid angle coverage compared to conventional PET cameras and reduces the number of detector modules significantly, potentially reducing the cost. Using LSO scintillator material in the detector modules results in low dead time, which when coupled with a high coincidence processor throughput allows the increased solid angle coverage to be converted into a significantly increased noise equivalent count rate. The detector modules must be able to measure the depth of interaction on an event by event basis in order to eliminate radial elongation artifacts, but such depth information can be incorporated into the reconstruction algorithm in an artifact-free way with a simple rebinning method.

Acknowledgments

This work is supported in part by the Director, Office of Energy Research, Office of Health and Environmental Research, Medical Applications and Biophysical Research Division of the U. S. Department of Energy under Contract DE-AC03-76SF00098 and in part by Public Health Service Grants No. P01-HL25840 and No. R01-NS29655, awarded by the National Heart, Lung, and Blood Institute, and the Neurological Disorder and Stroke Institute, Department of Health and Human Services.

References

Bloomfield, P. M., Rajeswaran, S., Spinks, T. J., and Hume, S. P. (1995). The design and physical characteristics of a small animal positron emission tomograph. *Phys. Med. Biol.* **40**: 1105–1126.

Casey, M. E., Eriksson, L., Schmand, M., Andreaco, M. S., Paulus, M., Dahlbom, M., and Nutt, R. (1997). Investigation of LSO crystals for high spatial resolution positron emission tomography. *IEEE Trans. Nucl. Sci.* **44**: 1109–1113.

Cherry, S. R., Shao, Y., Silverman, R. W., Meadors, K., Siegal, S., Chatziioannou, A., Young, J., Jones, W. F., Moyers, C., Newport, D., Boutefnouchet, A., Farquhar, T., Andreaco, M., Paulus, M., Binkley, D., Nutt, R., and Phelps, M. E. (1997). MicroPET: A high resolution PET scanner for imaging small animals. *IEEE Trans. Nucl. Sci.* **44**: 1161–1166.

DeGrado, T. R., Turkington, T. G., Williams, J. J., Stearns, C. W., Hoffman, J. M., and Coleman, R. E. (1994). Performance characteristics of a whole-body PET scanner. *J. Nucl. Med.* **35**: 1398–1406.

DeKemp, R. A., and Nahmias, C. (1994). Attenuation correction in PET using single photon transmission measurement. *Med. Phys.* **21**: 771–778.

Derenzo, S. E. (1984). Initial characterization of a BGO-silicon photodiode detector for high resolution PET. *IEEE Trans. Nucl. Sci.* **31**: 620–626.

Derenzo, S. E. (1986). Recent developments in positron emission tomography (PET) instrumentation. *Proc. SPIE* **671**: 232–243.

Huber, J. S., Moses, W. W., Derenzo, S. E., Ho, M. H., Andreaco, M. S., Paulus, M. J., and Nutt, R. (1997). Characterization of a 64 channel PET detector using photodiodes for crystal identification. *IEEE Trans. Nucl. Sci.* **44**: 1197–1201.

Karp, J. S., Muehllehner, G., Mankoff, D. A., Ordonez, C. E., Ollinger, J. M., Daube-Witherspoon, M. E., Haigh, A.T., and Beerbohm, D. J. (1990). Continuous-slice PENN-PET: A positron tomograph with volume imaging capability. *J. Nucl. Med.* **31**: 617–627.

Karp, J. S., Muehllehner, G., Qu, H., and Yan, X. H. (1995). Singles transmission in volume imaging with a Cs-137 source. *Phys. Med. Biol.* **40**: 929–944.

LeComte, R., Cadorette, J., Rodrigue, S., and LaPointe, D. (1996). Initial results from the Sherbrooke avalanche photodiode positron tomograph. *IEEE Trans. Nucl. Sci.* **43**: 1952–1957.

Melcher, C. L., and Schweitzer, J. S. (1992). Cerium-doped lutetium orthosilicate: A fast, efficient new scintillator. *IEEE Trans. Nucl. Sci.* **39**: 502–504.

Moisan, C., Tsang, G., Rogers, J. G., and Hoskinson, E. M. (1996). Performance studies of a depth encoding multicrystal detector for PET. *IEEE Trans. Nucl. Sci.* **43**: 1926–1931.

Moses, W. W., Huesman, R. H., and Derenzo, S. E. (1991). A new algorithm for using depth-of-interaction measurement information in PET data acquisition. *J. Nucl. Med.* **32**: 995 (abstr.).

Moses, W. W., and Derenzo, S. E. (1993). Empirical observation of performance degradation in positron emission tomographs utilizing block detectors. *J. Nucl. Med.* **34**: 101P (abstr.).

Moses, W. W., Derenzo, S. E., Melcher, C. L., and Manente, R. A. (1995). A room temperature LSO/PIN photodiode PET detector module that measures depth of interaction. *IEEE Trans. Nucl. Sci.* **42**: 1085–1089.

Moses, W. W., Virador, P. R. G., Derenzo, S. E., Huesman, R. H., Budinger, T. F. (1997). Design of a high-resolution, high-sensitivity PET camera for human brains and small animals. *IEEE Trans. Nucl. Sci.* **44**: 1487–1491.

Spinks, T. J., Bailey, D. L., Bloomfield, P. M., Miller, M , Murayama, H., Jones, T., Jones, W., Reed, J., Newport, D., Casey, M. E., and Nutt, R. (1997). Performance of a new 3D-only PET scanner—the EXACT3D. In "1996 IEEE Nuclear Science Symposium Conference Record" (A. DelGuerra, ed.), pp. 1275–1279. IEEE, Piscataway, NJ.

Stearns, C. W., Cherry, S. R., and Thompson, C. J. (1995). NECR analysis of 3D PET scanner designs. *IEEE Trans. Nucl. Sci.* **42**: 1075–1079.

Strother, S. C., Casey, M. E., and Hoffman, E. J. (1990). Measuring PET scanner sensitivity: Relating count rates to image signal-to-noise ratios using noise equivalent counts. *IEEE Trans. Nucl. Sci.* **37**: 783–788.

Wienhard, K., Dahlbom, M., Eriksson, L., Michel, C., Bruckbauer, T., Pietrzyk, U., and Heiss, W.-D. (1994). The ECAT EXACT HR: Performance of a new high resolution positron scanner. *J. Comput. Assist. Tomogr.* **18**: 110–118.

Worstell, W., Johnson, O., and Zawarzin, V. (1996). Development of a high-resolution PET detector using LSO and wavelength-shifting fibers. In "1995 IEEE Nuclear Science Symposium and Medical Imaging Conference Record" (P. A. Moonier, ed.), pp. 1756–1760. IEEE, Piscataway, NJ.

CHAPTER 3

The Road to Simultaneous PET/MR Images of the Brain[1]

Y. SHAO, R. SLATES, K. FARAHANI,* A. BOWERY, M. DAHLBOM, K. MEADORS,
R. W. SILVERMAN, M. SUGITA,† and S. R. CHERRY

*Crump Institute for Biological Imaging
Department of Molecular and Medical Pharmacology
* Department of Radiological Sciences
† Department of Surgery
UCLA School of Medicine
Los Angeles, California 90095*

There are compelling reasons for merging high-resolution anatomical information from magnetic resonance (MR) imaging with the functional images attainable with positron emission tomography (PET). The ideal solution to this intermodality image registration is to acquire PET and MR scans simultaneously with imaging systems that are fixed relative to each other, which will permit near-perfect image registration. This is particularly important for studies of small brain structures or in animal studies. A prototype PET system has been developed that permits PET and MR images to be acquired simultaneously. It consists of seventy-two 2 × 2 × 5-mm lutetium oxyorthosilicate crystals coupled by 2-mm-diameter, 4-m-long double-clad optical fibers to three multichannel photomultiplier tubes shielded inside an aluminum enclosure. The detector ring diameter is 56 mm. The scanner provides a single-image plane with a transaxial field of view of 28 mm and 1-mm full-width-at-half-maximum slice thickness. The reconstructed resolution is 2 mm. This PET system is shown to be compatible with a clinical 1.5-T MR scanner. Possible image degradation caused by simultaneous imaging has been measured. The PET and MR phantom images show no noticeable image artifacts or distortions resulting from dual modality image acquisition. The first in vivo simultaneous PET and MR images have also been obtained from a rat brain, demonstrating the feasibility of simultaneous PET and MR studies of in vivo systems.

I. INTRODUCTION

The accurate registration of positron emission tomography (PET) and magnetic resonance (MR) images provides important information on structure/function relationships, permits precise anatomically based region of interest definition, and may allow partial volume correction of the PET study. To date, intermodality registration is done mathematically using separate scans with a typical accuracy of ~2 mm (Strother et al., 1994). This may not be sufficient for small brain structures or for animal studies where registration errors are typically larger. The ideal solution is to acquire PET and MR images simultaneously, avoiding any geometric misalignment and providing near-perfect image registration once image distortions caused by each individual modality alone are corrected. In addition, such technology could be used for direct comparison of [^{15}O]water PET with functional MR imaging protocols and for the temporal correlation of PET and MR spectroscopy.

This chapter describes (1) the development of a prototype PET scanner compatible with conventional clinical MR scanners, (2) preliminary studies to assess

---

[1] Transcripts of the BRAINPET97 discussion of this chapter can be found in Section VIII.

possible image artifacts or distortions that may result from dual modality image acquisition using the PET scanner developed, and (3) an initial *in vivo* study of simultaneous PET and MR images of a rat brain.

II. MATERIALS AND METHODS

A. PET Scanner Design

A prototype single-slice PET scanner has been designed and developed to be compatible with typical clinical MR imaging systems for acquiring simultaneous images. The design concept is an extension of previous work (Cherry et al., 1996). Small scintillator crystals placed inside the MR scanner are connected via optical fibers to photodetectors located outside the main magnetic field. With this approach, no conducting or ferromagnetic materials are placed inside the MR magnet so that the main magnetic field homogeneity is maintained. Electromagnetic interference (EMI) is minimized by the separation between the MR scanner magnet and the PET photodetectors and associated electronics. The detailed design and development of the scanner are described elsewhere (Shao et al., 1997a). The PET system consists of seventy-two 2 × 2 × 5-mm lutetium oxyorthosilicate (LSO) crystals (Melcher and Schweitzer, 1992) coupled by 2-mm-diameter, 4-m-long double-clad optical fibers (Kuraray Inc., Japan) to three XP1722 model multichannel photomultiplier tubes (MC-PMT) (Philips Photonics, France). The detector ring diameter is 56 mm. MC-PMTs and associated electronics are enclosed inside an aluminum enclosure that provides shielding for both ambient light and radiofrequency radiation. A NIM and CAMAC-based portable data acquisition system is controlled by Kmax (Sparrow Corp., USA) programs on a Power Macintosh 7100/80. The scanner provides a single image plane with a transaxial field of view of 28 mm, ~1 mm FWHM slice thickness, and ~2 mm in-plane reconstructed spatial resolution.

B. Experimental Setup

A 1.5-T clinical MR whole body scanner (Magnetom Vision, Siemens) was used for testing the MR-compatible PET system. The PET detector ring was placed inside the MR scanner at the center of the magnet and surrounded by a 20-cm-diameter extremity coil. The MC-PMTs and associated electronics were housed more than 3 m away from the center of magnet where the strength of the residual magnetic field is around 0.1 mT. This field strength does not affect the performance of the MC-PMTs (Shao et al., 1997b). Data acquisition and computer systems were outside the MR scanner room to prevent possible EMI.

C. Image Artifact and Distortion Studies

Experiments were conducted to investigate image artifacts and distortions in both PET and MR images that may possibly result from simultaneous data acquisition. To assess possible MR image degradation due to the presence of the PET detector system, a 2-cm-diameter, 5-mm axial extent cylindrical phantom was used. This uniform phantom was chosen in order to have a relatively large object volume for examining any possible artifact or distortion inside the imaging volume. The phantom was filled with a mixture of $NiSO_4$ and NaCl solution to enhance the MR signal. The phantom was scanned first with the MR imaging system only, then with the PET detector ring placed inside the MR imaging volume and surrounding the phantom. The same MR image acquisition parameters were used for both scans and are listed in Table 1.

To assess possible PET image artifacts due to the MR imaging system, a simple phantom was constructed, which consisted of a 21-mm-diameter, 2.2-mm-wide C-shape channel cut into a lucite block, roughly simulating the cortex of a rat brain. In addition, 3- and 2-mm-diameter holes, with a separation of 4 mm, were drilled into the middle of the phantom. This phantom was used in order to test the imaging capability of the PET system. The phantom was filled with a mixture of positron-emitting isotope (^{68}Ga) and $NiSO_4$ solution. PET images of the phantom were acquired with the PET detector ring placed inside and outside of the MR magnet, respectively. Table 2 lists the data acquisition and reconstruction parameters.

D. *In Vivo* Study

Simultaneous PET and MR images were acquired for a 200-g rat with the setup just described. In addi-

TABLE 1 MR Image Data Acquisition Parameters[a]

2D spin echo sequence
TE = 12 msec
TR = 280 msec
FA = 90°
Slice thickness = 3 mm
Image size = 140 × 140 pixels
NEx = 1
Acquisition time = 39 sec

[a] Abbreviations: TE, echo time; TR, repetition time; FA, flip angle.

TABLE 2 PET Data Acquisition and Image Reconstruction Parameters

Phantom source activity	~ 1 mCi
Coincidence count rate	~ 115 per second (at beginning)
Total events collected	~ 58,000
Slice thickness	1 mm (FWHM)
Reconstruction method	Filtered backprojection
Reconstruction filter	Ramp (Nyquist frequency)

tion, a circular plastic tube filled with NiSO$_4$ solution was aligned with the PET imaging plane and attached to the outside of the PET detector ring holder. This allows the PET image plane to be identified in the MR images. The rat was placed on a thin lucite bed and its head was positioned at the center of the PET detector ring. [^{18}F]fluorodeoxyglucose (FDG) was injected (1.3 mCi) into a tail vein and imaging started about 30 min later. Total PET acquisition time was 30 min with ~ 140,000 recorded events. PET data were normalized and reconstructed using filtered backprojection. In the MR imaging acquisition, a total of 75 sec of data with a spin echo sequence of 12 msec TE and 280 msec TR was collected; the slice thickness was 4 mm. The MR image field of view was 100 × 100 mm.

III. RESULTS AND DISCUSSION

Figure 1c shows the subtraction of MR images acquired with the MR scanner only (Fig. 1a) and with the PET detector ring placed inside the MR magnet (Fig. 1b), which results in a random noise distribution in the object area. The noise distribution pattern in Fig. 1c was thought to be solely due to the subtraction of the two MR images, as the signal and noise levels inside and outside the object area are quite different. In order to prove this, the same phantom was scanned twice with the MR scanner only and the two images were subtracted. Figure 1d shows the image after the subtraction, which is essentially the same as in Fig. 1c. The residual noise level after subtraction is less than 5% of the mean signal level in Figs. 1a and 1b. These measurements indicate that there are no noticeable artifacts or distortions observed in the MR images due to the presence of the PET detector system. It is possible that the PET detector ring may introduce a small degradation in magnetic field homogeneity due to its low magnetic susceptibility, particularly in a system containing many crystals and optical fibers. This small nonhomogeneity in the magnetic field can, in principle, be corrected by shimming the magnetic field and should not affect MR imaging studies. Experiments also reveal

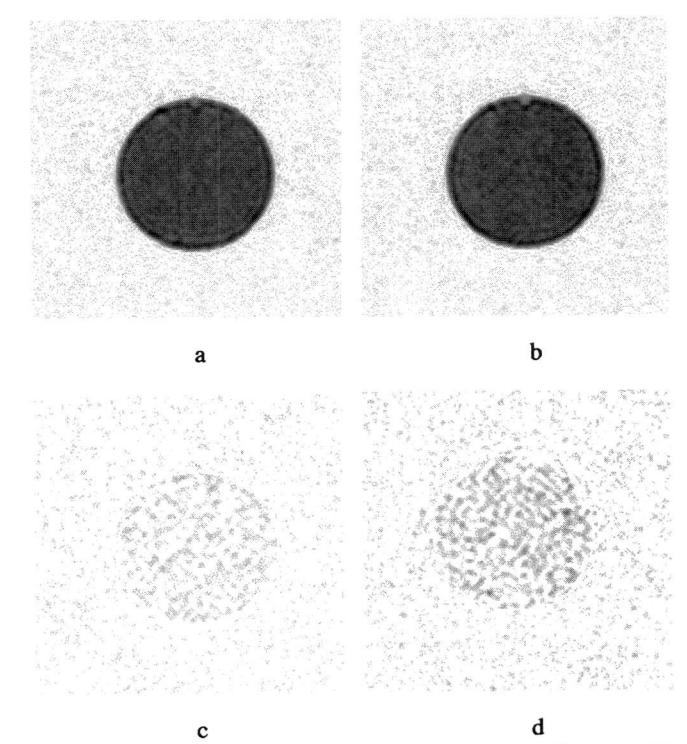

FIGURE 1 MR images of a cylindrical phantom: (a) with MR imaging only and (b) with the PET detector ring placed inside the MR magnet. The subtraction of images (a) and (b) is shown in (c). (d) The subtraction of two MR images of the same phantom acquired with MR imaging only.

that EMI from the PET system is minimal because of the distance between the MR imager coil and the MC-PMTs and the use of proper shielding. Other MR pulse sequences (e.g., gradient echo) have also been tried with the same setup and no noticeable distortion has been found with these sequences either.

Figure 2c shows the subtraction of PET images acquired with PET only (Fig. 2a) and with the PET detector ring and the phantom placed inside the MR magnet (Fig. 2b). The subtracted image shows a random noise distribution. A ramp filter without smoothing was used to allow the noise structure to be seen most clearly. In addition, different MR pulse sequences were tried with the extremity coil and no effects were found on the PET image. These experiments demonstrate the suitability of the PET detector design for compatibility with the MR imaging system.

The simultaneously acquired FDG PET and MR images of a rat brain are shown in Fig. 3 (see color insert). The PET image was reconstructed with a Hanning filter and smoothed due to limited counts. The MR image resolution is 0.5 mm. Both images have similar anatomic structure, but careful examination

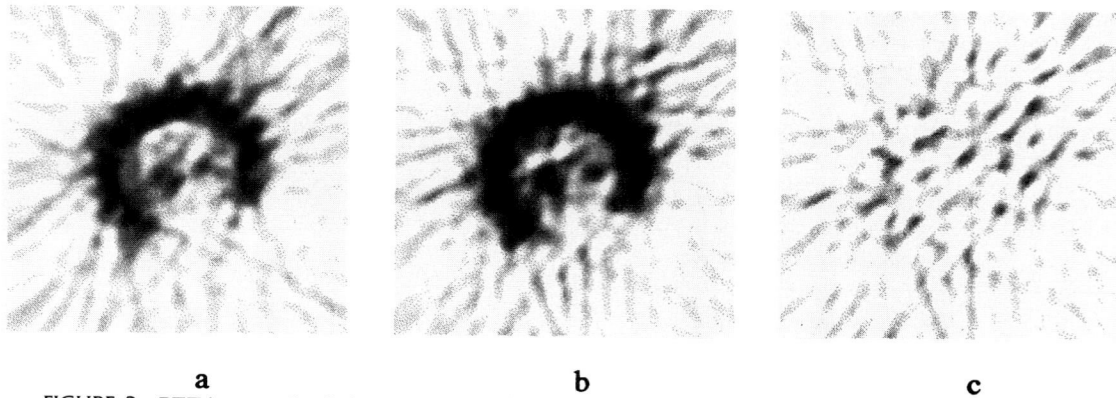

FIGURE 2 PET images of a C-shaped phantom: (a) with PET only and (b) with the PET detector ring placed inside MR magnet. The subtraction of images (a) and (b) is shown in (c).

indicates that they may not be in the exact same imaging plane. This misalignment of the imaging plane was due to a shift of the fiducial plane marker relative to the PET detector ring during the experiment. Besides normalization, no other corrections were applied to PET data in this first *in vivo* study. Nevertheless, the PET image still clearly shows FDG uptake in the area corresponding to the brain in the MR image.

These imaging experiments demonstrate the feasibility of simultaneous PET and MR imaging by using the current approach. Future studies will include:

1. Further optimization of the PET system by increasing the detection efficiency (by increasing the crystal transaxial length) and increasing the ring diameter to allow a wider range of studies.

2. Improvements to the detector system design to ensure alignment of PET and MR image planes. It is critical for a single-slice system that the plane of interest can be aligned easily in the field of view.

3. Improvements to the PET image quality by shielding activity from outside the field of view. This can be achieved by adding high stopping power, low magnetic susceptibility materials, such as BGO, outside the detector plane.

4. Design changes to the MR coil to be compatible with the PET system.

Designs for a single-slice MR-compatible PET system large enough for brain imaging are under investigation. The system will consist of three concentric rings of ~ 3 × 3 × 5-mm LSO crystals, with one 3 × 5-mm side of each crystal connected to an optical fiber. This design results in relatively high efficiency for a single-slice system because of the 15-mm total thickness of LSO and the small ring diameter that is made possible by having depth of interaction information. For example, Fig. 4 shows a simulation of the radial resolution change versus the distance from the center of the detector ring for a 30-cm-diameter system, with three concentric detector rings consisting of 3 × 3-mm cross-section and 5-mm-length LSO crystals. Using the depth of interaction information, the change in radial resolution as a function of distance from the center is quite small compared with the rapid degradation when no depth of interaction information is available. A fiducial ring will be added to the crystal ring to enable

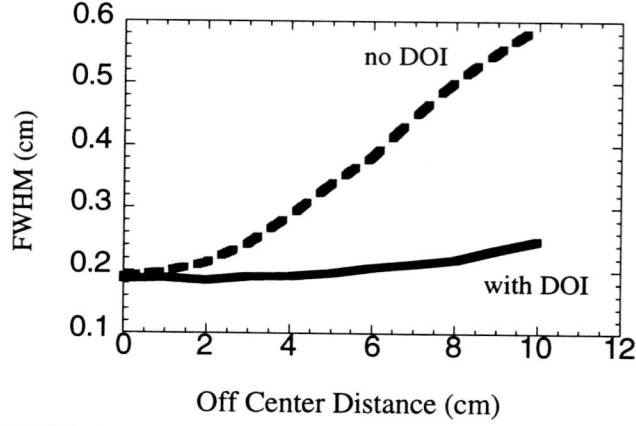

FIGURE 4 Simulated radial resolution as a function of distance from the scanner center. The ring diameter is 30 cm. The dashed curve is calculated for a one-ring detector system consisting of 3 × 3 × 15-mm^3 LSO crystals, which has no depth of interaction (DOI) information available. The solid curve is calculated for a three concentric ring detector system, consisting of 3 × 3 × 5-mm^3 LSO crystals, where DOI information is available and used in the resolution calculation.

the PET image plane to be visualized on MR images, allowing the plane of interest to be positioned in the PET scanner prior to tracer injection. Using this approach, a MR-compatible PET system with high and uniform resolution (~ 3 mm) and adequate sensitivity can be developed quite inexpensively. Many challenges remain, however, in achieving good energy and timing resolution and in finding ways to handle large numbers of optical and electronic channels efficiently in a clinical MR imaging environment.

Acknowledgments

We gratefully acknowledge Ron Sumida and the PET technical staff for supplying PET isotopes. This work was partially supported by a pilot research grant from the Education and Research Foundation of the Society of Nuclear Medicine and by DOE Contract DE-FC03-87-ER60615.

References

Cherry, S. R., Shao, Y., Siegel, S., and Silverman, R. W. (1996). Optical fiber readout of scintillator arrays using a multi-channel PMT: A high resolution PET detector for animal imaging. *IEEE Trans. Nucl. Sci.* **43**: 1932–1937.

Melcher, C. L., and Schweitzer, J. S. (1992). Cerium-doped lutetium oxyorthosilicate: A fast, efficient new scintillator. *IEEE Trans. Nucl Sci.* **39**: 502–505.

Shao, Y., Cherry, S. R., Farahani, K., Slates, R., Silverman, R. W., Meadors, K., Bowery, A., and Siegel, S. (1997a). Development of a PET detector system compatible with MRI/NMR systems. *IEEE Trans. Nucl. Sci.* **44**: 1167–1171.

Shao, Y., Cherry, S. R., Siegel, S., Silverman, R. W., and Majewski, S. (1997b). Evaluation of multi-channel PMT's for readout of scintillator arrays. *Nucl. Instrum. Methods A* **390**: 209–218.

Strother, S. C., Anderson, J. R., Xu, X.-L., Liow, J.-S., Bonar, D. C., and Rottenberg, D. A. (1994). Quantitative comparisons of image registration techniques based on high resolution MRI of the brain. *J. Comput. Assist. Tomogr.* **18**: 954–962.

CHAPTER 4

Brain PET Studies with a High Sensitivity Fully Three-Dimensional Tomograph[1]

D. L. BAILEY, M. P. MILLER, T. J. SPINKS, P. M. BLOOMFIELD, L. LIVIERATOS, R. B. BÁNÁTI, R. MYERS, and T. JONES

MRC Cyclotron Unit
Hammersmith Hospital
London W12 0NN, United Kingdom

This chapter reports on a fully three-dimensional (3D) large axial field of view bismuth germanate positron tomograph. The aim was to develop a high sensitivity tomograph for both brain and whole body studies. The system operates exclusively in 3D mode and incorporates a number of innovative features to optimize and enhance performance for brain studies that were not necessary with previous 2D or combined 2D/3D systems. The absolute sensitivity of this tomograph is almost 6% for coincidences arising from within the field of view. Resolution is 4.5 mm full-width-at-half-maximum (FWHM) isotropic. Transmission scanning is accomplished with a 150-MBq ^{137}Cs point source. The axial field of view for single photons measures approximately 80 cm beyond the ends on the tomograph. To shield against these unwanted single photons in brain studies, detachable lead shields have been employed, which reduce the subject aperture from 65 to 35 cm. When using the extra shields, in the early phase of ^{11}C-labeled neuroreceptor ligand studies, a 70% reduction in randoms rates and a corresponding increase in true event rates due to lower detector dead time have been observed. The sensitivity of this tomograph has reduced the optimal administered dose to < 200 MBq for individual [^{15}O]H$_2$O rCBF "activation" studies. For ^{18}F-labeled tracers, the benefit is improved image quality, whereas for ^{11}C-labeled radiotracers, useful scanning periods have been extended up to 120 min and profit from the higher sensitivity for radiotracers with low cerebral uptake.

I. INTRODUCTION

The move to positron emission tomography (PET) operating exclusively in three-dimensional (3D) mode is accompanied by a number of issues necessary for maintaining quantitative accuracy while exploiting the inherent sensitivity advantage of this approach. The EXACT 3D scanner (CTI, Knoxville, TN) used in the authors' institute is a 23.4-cm axial field of view, wide aperture (whole body) scanner composed of bismuth germanate (BGO) detectors that operates exclusively in 3D mode (Jones et al., 1996; Spinks et al., 1997). This chapter presents issues that are encountered intrinsically in a 3D-only tomograph designed to maximize sensitivity throughout the axial length of the human brain and to report on the solutions that have been implemented to optimize performance and produce quantitative cerebral studies.

II. MATERIALS AND METHODS

A. System Description

The EXACT 3D tomograph was designed with the aim of acquiring PET data with high sensitivity and high spatial and temporal resolution. For brain studies, the goal was to measure whole brain, including cerebellum and brain stem, with high sensitivity. The tomograph uses conventional high-resolution BGO block detectors used in the commercial EXACT HR−. The detectors are sectioned into 8 × 8 elements per block

---

[1] Transcripts of the BRAINPET97 discussion of this chapter can be found in Section VIII.

with each detector element measuring 4.0 × 4.25 × 30 mm (transaxial × axial × depth). The system is composed of 432 blocks in total arranged into 48 rings of detectors. The block detectors are organized into a cassette unit that contains 6 blocks axially by 2 blocks radially. There are 576 detector elements per ring (ring diameter = 82 cm), giving 27,648 individual detectors. The absolute sensitivity of this tomograph is almost 6% for coincidences arising from within the field of view. Resolution is 4.5 mm FWHM isotropic (Spinks et al., 1997). On either end of the detectors there is 2.5-cm-thick lead side shielding that extends 8 cm beyond the ends of the detectors into the field of view. The penumbra for single photons for this device has been measured with point sources and extends approximately 80 cm beyond the end of the coincidence field of view; however, 50% of the single photons contributing to this originate from within the first 10 cm.

The acquisition system allows data to be acquired in frame or list mode. Without any radial or axial rebinning a single frame would occupy ~ 324 Mbytes of storage; however, by employing data averaging, this can be reduced to less than 50 Mbytes per frame. Data are routinely acquired in frame mode with a maximum polar acceptance angle (θ_{max}) of 11°, which corresponds to a maximum ring difference (Δp_{max}) of 40 detectors axially. Polar averaging over a range of up to ± two detectors (8 mm) is employed. In addition, projection data are averaged in the radial dimension over adjacent projections. For brain studies this does not cause any loss of resolution (Spinks et al., 1997). The energy window setting used for all emission studies is 350–650 keV. List mode acquisition was implemented for efficient data storage but has the additional benefit of permitting high temporal sampling with optional frame length rebinning post hoc and multiple physiological gating triggers. There are 32 Mbytes of fast list mode memory partitioned into two 16 Mbyte buffers that fill and write to disk alternately. Sustained disk write rates of ~ 17 Mbytes·sec^{-1} (≤ 4M coincidence events per second) are possible to the 34-Gbyte RAID disk system.

Single photon transmission scanning with a point source has been implemented similar to previously suggested methods (Derenzo et al., 1975; deKemp and Nahmias, 1994; Yu and Nahmias, 1995). The source consists of 150 MBq of ^{137}Cs (E_γ = 0.663 MeV, $t_{1/2}$ = 30.2y). It is contained in a small pellet that is driven at approximately 1 m·sec^{-1} in a fluid-filled steel tube wound into a helix positioned just inside the detector ring. There are 48 turns of the helix within the field of view, corresponding to 1 turn per detector ring. The tubing and liquid in the system attenuate approximately 10% of emission coincidences. A single pass of the transmission source takes approximately 150 sec.

The energy window used for the transmission scanning is 500–800 keV.

All data are reconstructed with the reprojection algorithm (Kinahan and Rogers, 1989; Townsend et al., 1989) using four quad i860 vector processors (CSPI Corp, Billerica, MA) controlled by a Sun Sparc 20 CPU, which reconstructs a volume in ~ 8 min without any corrections or ~ 18 min with normalization, attenuation, and scatter (Watson et al., 1996) corrections. The reconstructed volume contains 95 planes separated by a spacing of ~ 2 mm, giving a total axial coverage of approximately 23.4 cm.

B. Issues Related to 3D Operation

1. Impact of Out of Field of View Radioactivity

Radioactivity from outside the field of view may have an impact on the data acquisition in the following ways:

- increased random coincidence rate from single photons
- increased detector dead time from single photons
- increased true coincidence events where one or both photons has been scattered.

In an effort to limit the acceptance of photons from outside the coincidence field of view for brain studies, three lead annuli of 8 mm thickness were constructed to extend the shielding on the end of the tomograph from an aperture measuring 65 cm wide to one measuring 35 cm. Measurements have been made both with and without the shields in cerebral [^{15}O]H$_2$O and [^{11}C]diprenorphine scans, in the same individuals and with almost identical administered doses, to evaluate their impact on performance in clinical studies.

2. Transmission Scanning

Single photon scanning with the ^{137}Cs point source produces high event rates (1.5–2.5 Mcps). However, as the system contains no septa and the source is uncollimated, there is very high acceptance of scattered photons. The impact of these scattered photons on reconstructed attenuation coefficient (μ) values has been measured. As a pragmatic initial approach to correcting for scatter in transmission data, a local threshold segmentation approach has been implemented (Xu et al., 1994). Data are acquired in 3D mode and reconstructed using single slice rebinning (Daube-Witherspoon and Muehllehner, 1987), segmented with the local threshold technique, and then forward projected to form the required attenuation correction files. The entire process takes approximately 10 min on a combination of Sparc 5 workstation and the vector processors.

3. Scatter Correction

A vector processor implementation of model-based scatter correction (Ollinger, 1996; Watson *et al.*, 1996) supplied with the system by CTI has been evaluated for this tomograph. As this approach uses the transmission distribution as well as the emission distribution to calculate the scatter contribution, it is necessary to have accurate attenuation reconstructions when employing this method. Tests with the Utah phantom (Townsend *et al.*, 1994) with radioactivity outside the coincidence field of view and with the lead shields in place were performed to assess the accuracy of the correction and susceptibility to extraneous radioactivity. The main sections of the phantom inside the field of view contained ^{18}F in various concentrations in the different compartments (often referred to as regions B–E) with one of the small cylinders containing water without any radioactivity, and ^{11}C was placed in the section of the cylinder outside the field of view (region A). Aliquots were taken to determine the radioactivity concentrations in a well counter. Data were acquired as a dynamic study giving a differential contribution from radioactivity outside the field of view as the ^{11}C decayed at a greater rate than the ^{18}F. Data were reconstructed both with and without scatter correction. Regions of interest in the different compartments gave average reconstructed count rates per cm^3, and these were compared to determine the accuracy in the "hot" and "cold" cylinders.

C. Clinical Studies

Examples are shown with the EXACT 3D for cerebral studies using [^{15}O]H$_2$O ("activation" scans), [^{11}C]diprenorphine (an opioid receptor antagonist), [^{11}C]PK11195 (for studying activated microglia), [^{11}C]WAY100635 (a 5HT$_{1A}$ antagonist), [^{11}C]raclopride (a dopamine D$_2$ antagonist), [^{18}F]fluorodeoxyglucose (FDG), and [^{18}F]fluorodopa. All of these studies used dynamic frame acquisition protocols of up to 40 frames of data (~ 1.9 Gbytes total). The doses for these radiotracers that can be used with this highly sensitive tomograph have been determined pragmatically and are reported in Section III.

III. RESULTS AND DISCUSSION

A. Issues Related to 3D Operation

1. Impact of Out of Field of View Radioactivity

In the human studies with 150-MBq doses of [^{15}O]H$_2$O delivered over 20 sec, two features are noted (Fig. 1, upper row). First, without the lead shields the randoms profile is seen to peak some 10 s before the true events, strongly suggesting that the events are coming from outside the field of view, in this case the heart and great vessels of the upper thorax and neck region; after the addition of the shields, the randoms rate is lower and peaks at the same time as the peak for the true events. Second, the maximum true coincidence count rate is increased with the lead shields. The probable explanation for the greater count rate is lower detector dead time due to a large reduction in single photons (including those below the coincidence energy threshold) from outside the field of view. However, other possible causes that cannot be excluded are changes in global cerebral blood flow or slight changes in the dose to the subject, as the radioactivity administered by the [^{15}O]H$_2$O delivery system is accurate to only 20 MBq. In the paired studies with [^{11}C]diprenorphine the ratio of randoms to trues is decreased by 70% by the addition of the shields (Fig. 1, lower row). Again, this is attributed predominantly to reducing the single photon flux registered by the detectors originating from outside the coincidence field of view.

2. Transmission Scanning

Figure 2 shows a representative slice from the single slice rebinned 2D reconstruction of the attenuation reconstruction, the result after segmentation, and comparison with a coincidence measurement with rotating rod sources on the ECAT 953B (10-min acquisition). The attenuation reconstruction exhibits low variance and high bias and thus is easily amenable to segmentation and reassignment of the μ values. The μ values are underestimated by up to 50% in the initial reconstruction. After segmentation and rescaling by a single factor (1.48) the values are now approximately those expected with narrow-beam attenuation and correspond closely to those measured on the ECAT 953B in 2D with ^{68}Ge rotating rod sources.

3. Scatter Correction

The relative accuracy of the scatter correction method with the extra lead end shielding is shown in Fig. 3. Without scatter correction the activity outside the field of view causes inaccurate estimation of the known ratio in both "hot" and "cold" cylinders relative to the surroundings. In addition, a bias increases with higher radioactivity levels outside the field of view. After scatter correction, this is seen to be greatly improved, although further work aimed at absolute quantification, as opposed to relative concentrations, is needed to confirm these results. There is a bias of approximately 5% in the "cold" cylinder due to residual reconstructed counts in this region.

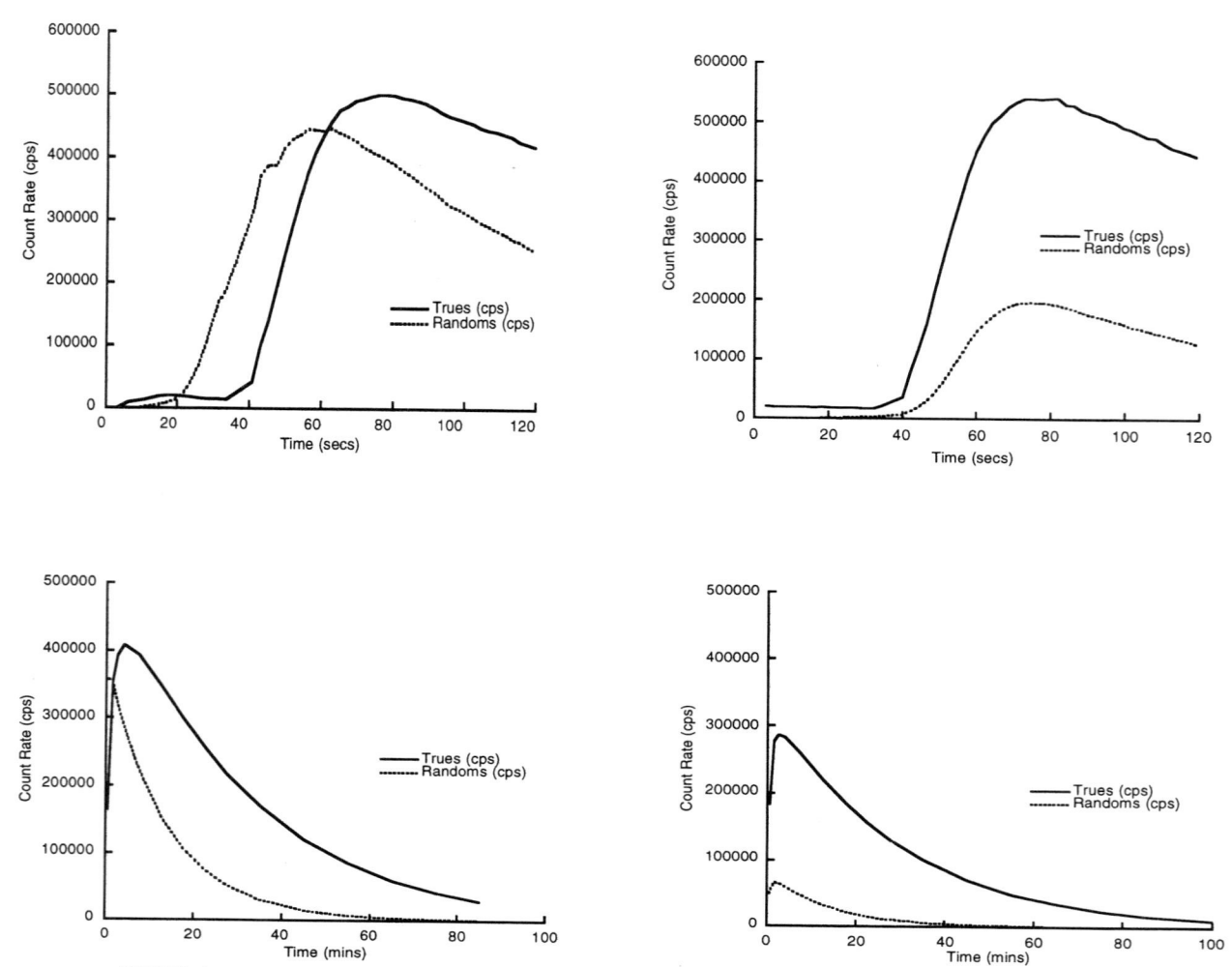

FIGURE 1 The effect of the extended lead end shielding on the coincidence count rates from the EXACT 3D is shown. In a [^{15}O]water "activation" study (top row), a subject was administered the same amount of radiotracer on two occasions separated by approximately 10 min. Without the lead shields (top left), the count rates observed for true events and random events demonstrate that the randoms curve increases approximately 15–20 sec *before* the true coincidences curve, strongly suggesting that the randoms contribution is mostly originating from outside the field of view. After the addition of the lead shields (top right), the randoms peak count rate is much lower and is in phase with the trues count rate peak. In addition, the trues count rate peak is greater than that without the shields, which presumably is due to lower dead time resulting from the lower single photon flux reaching the detectors from outside the field of view. In the [^{11}C]diprenorphine study (bottom row), without the shields (lower left), the randoms and trues rates are approximately identical shortly after the injection of the tracer. After the shields were added (lower right), the randoms as a fraction of the trues count rate are decreased by 70%. The trues rate here is decreased by approximately 25% and may be due to the reduction of scattered true coincidences that originate from outside the field of view or differences in biodistribution of [^{11}C]diprenorphine and [^{15}O]water.

B. Clinical Studies

Figure 4 shows some examples of cerebral radiotracers that have been studied on the EXACT 3D to date. The sensitivity of this tomograph has reduced the optimal administered dose to < 200 MBq for individual [^{15}O]H$_2$O rCBF "activation" studies while maintaining data quality, allowing many more observations in the same or multiple sessions. For ^{18}F-labeled tracers, the benefit is improved image quality compared with previous 2D/3D tomographs, whereas for ^{11}C-labeled radiotracers, useful scanning periods have been extended up to 120 min and profit from the higher sensitivity for radiotracers with low uptake such as [^{11}C]PK11195 studies of activated microglia (see Myers *et al.*, this volume, Chapter 29).

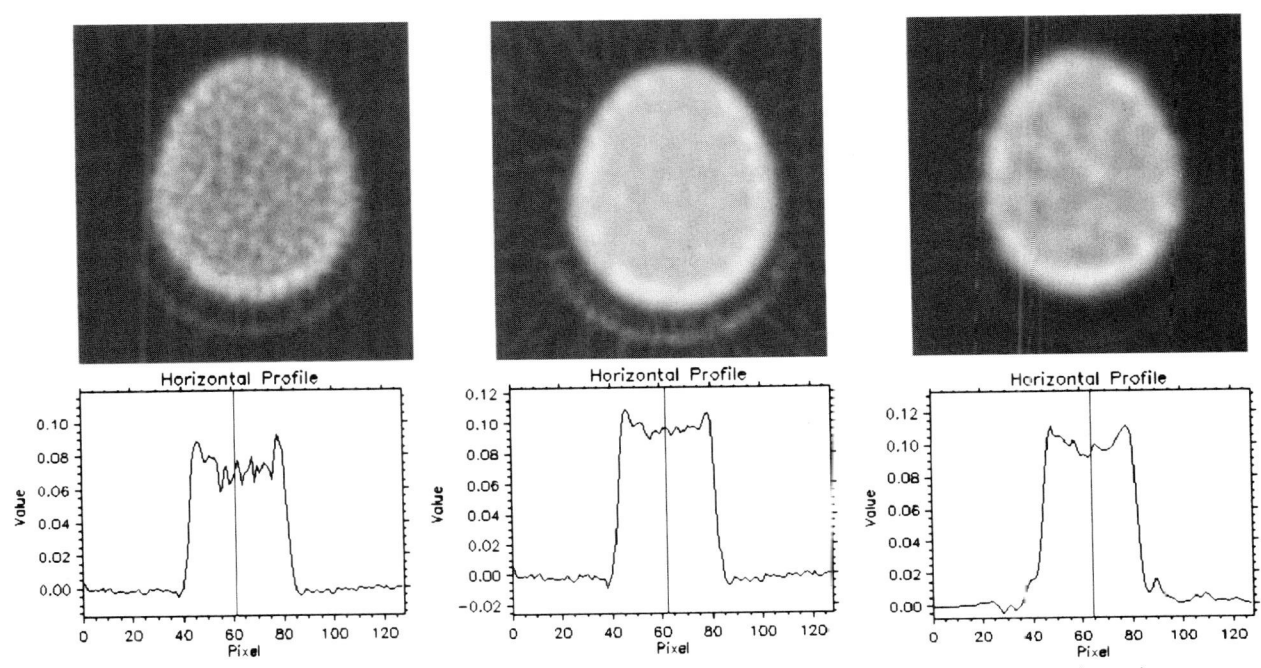

FIGURE 2 Comparison of transmission scans: (left) single photon transmission scan on EXACT 3D; (center) segmented version of the same scan; and (right) approximately the same slice measured with conventional rotating coincidence rods on the ECAT 953B. The y-axis units of the profiles are attenuation coefficients in cm^{-1}.

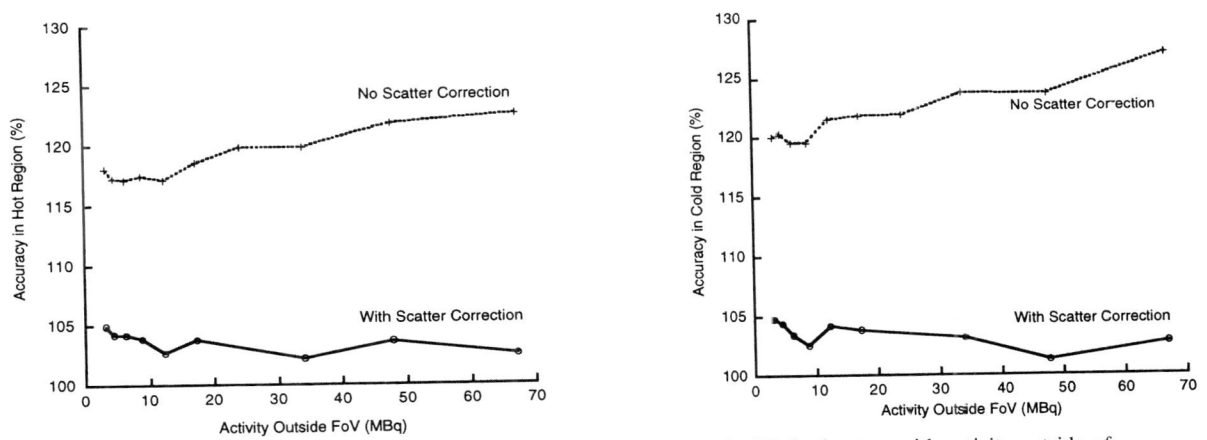

FIGURE 3 The performance of model-based scatter correction in the Utah phantom with activity outside of the coincidence field of view is shown. The plots show the average reconstructed count rate for a region of interest relative to a reference region. The scanner had extra lead shields attached. The graph on the left shows the accuracy in a region of the phantom with a radioactivity concentration higher than the surrounding regions. There is an increase in the value without scatter correction when the activity outside the field of view is greater. With scatter correction, this effect is greatly diminished. In the graph on the right, which shows the accuracy in a region devoid of radioactivity, a similar pattern is seen: without scatter correction there appears to be some spill-in when the activity outside the field of view is greater, whereas after scatter correction almost no trend is discernible. The correction leaves a small (5%) bias in this "cold" region in absolute terms.

I. Data Acquisition and Quantification

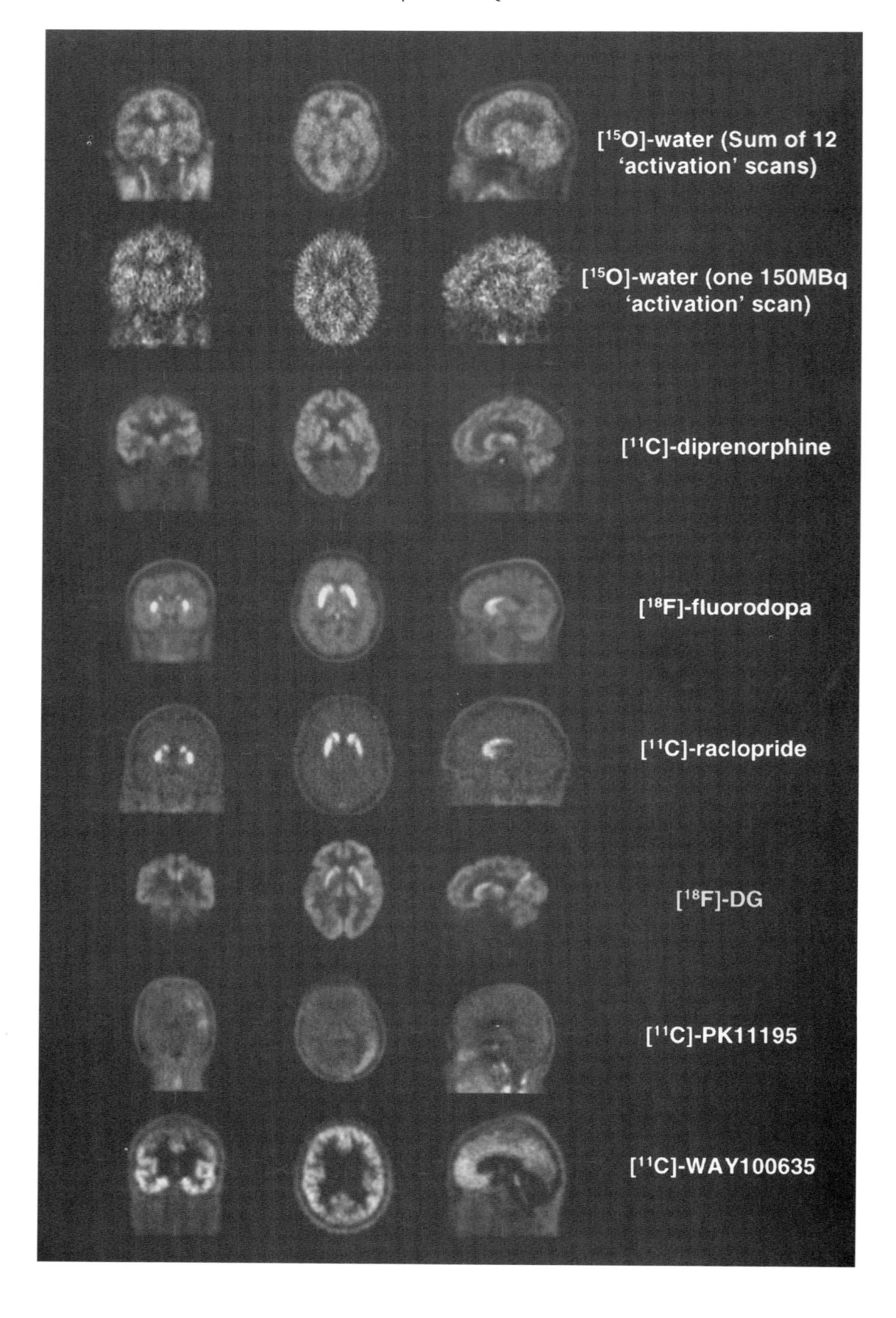

[^{15}O]-water (Sum of 12 'activation' scans)

[^{15}O]-water (one 150MBq 'activation' scan)

[^{11}C]-diprenorphine

[^{18}F]-fluorodopa

[^{11}C]-raclopride

[^{18}F]-DG

[^{11}C]-PK11195

[^{11}C]-WAY100635

IV. CONCLUSIONS

The goal of this chapter was to present issues that are encountered intrinsically in 3D-only tomographs and to report on the solutions that have been implemented to produce quantitative studies. Many of these issues, exclusive to 3D operation, will be of a similar nature in the next generation of PET systems using faster detectors, and thus the EXACT 3D can be seen as a prototype for these new devices. The main issues related to the use of a 3D tomograph with a large axial field of view are dominated by the effects of photons originating from outside the coincidence field of view, especially with a moderately slow scintillator such as BGO or NaI(Tl). Simple 8-mm lead shielding has been extremely effective in reducing these effects. Further, scattered photons in emission and transmission measurements do not appear to present insurmountable problems. The solutions employed in this chapter should be equally applicable when using new, faster detectors such as lutetium oxyorthosilicate.

References

Daube-Witherspoon, M. E., and Muehllehner, G. (1987). Treatment of axial data in three-dimensional PET. *J. Nucl. Med.* **28**(11): 1717–1724.

deKemp, R. A., and Nahmias, C. (1994). Attenuation correction in PET using single photon transmission measurement. *Med. Phys.* **21**(6): 771–778.

Derenzo, S. E., Zaklad, H., and Budinger, T. F. (1975). Analytical study of a high-resolution positron ring detector system for transaxial reconstruction tomography. *J. Nucl. Med.* **16**(12): 1166–1173.

Jones, T., Bailey, D. L., and Bloomfield, P. M. (1996). Performance characteristics and novel design aspects of the most sensitive PET camera built for high temporal and spatial resolution. *J. Nucl. Med.* **37**(5): 85P.

Kinahan, P. E., and Rogers, J. G. (1989). Analytic 3-D image reconstruction using all detected events. *IEEE Trans. Nucl. Sci.* **36**: 964–968.

Ollinger, J. M. (1996). Model-based scatter correction for fully 3D PET. *Phys. Med. Biol.* **41**(1): 153–176.

Spinks, T. J., Bailey, D. L., Bloomfield, P. M., Miller, M., Murayama, H., Jones, T., Jones, W., Reed, J., Newport, D., Casey, M. E., and Nutt, R. (1997). Performance of a new 3D-only PET scanner—the EXACT3D." In "1996 IEEE Nuclear Science Symposium Conference Record," (A. Del Guerra, ed.), pp. 1275–1279. IEEE, Piscataway, NJ.

Townsend, D. W., Spinks, T. J., Jones, T., Geissbühler, A., Defrise, M., Gilardi, M.-C., and Heather, J. D. (1989). Three dimensional reconstruction of PET data from a multi-ring camera. *IEEE Trans. Nucl. Sci.* **36**: 1056–1065.

Townsend, D. W., Choi, Y., Sashin, D., and Mintun, M. (1994). An investigation of practical scatter correction techniques for 3D PET. *J. Nucl. Med.* **35**(5): 50P.

Watson, C. C., Newport, D., and Casey, M. E. (1996). A single scatter simulation technique for scatter correction in 3D PET. *In* "Three-Dimensional Image Reconstruction in Radiology and Nuclear Medicine" (P. Grangeat and J.-L. Amans, eds.), pp. 255–268. Kluwer Academic Publishers, Dordrecht, The Netherlands.

Xu, M., Luk, W. K., Cutler, P. D., and Digby, W. M. (1994). Local threshold for segmented attenuation correction of PET imaging of the thorax. *IEEE Trans. Nucl. Sci.* **41**: 1532–1537.

Yu, S. K., and Nahmias, C. (1995). Single-photon transmission measurements in positron emission tomography using ^{137}Cs. *Phys. Med. Biol.* **40**: 1255–1266.

FIGURE 4 Examples of cerebral studies on the EXACT 3D. From the top in descending order are: (i) sum of 12 [^{15}O]H$_2$O "activation" blood flow studies (total activity ~ 1.8GBq), note the small infarct in the left occipital lobe; (ii) a single activation run in a different individual (injected dose = 150 ± 20 MBq); (iii) sum image from 10 to 90 min of [^{11}C]diprenorphine (injected dose = 220 MBq) in a normal subject; (iv) sum image from 25 to 90 min of [^{18}F]fluorodopa (injected dose = 110 MBq) in a normal subject; (v) sum image from 10 to 90 min of [^{11}C]raclopride (injected dose = 350 MBq) in a normal subject; (vi) sum image from 30 to 120 min of FDG (injected dose = 190 MBq) in a subject with mild Parkinson's disease; (vii) sum image from 10 to 90 min of [^{11}C]PK11195 (injected dose = 305 MBq) in a subject with a stroke following a hypotensive episode during abdominal surgery, note the normal uptake in the anterior pituitary and vertebrae; and (viii) sum image from 10 to 90 min of [^{11}C]WAY100635 (injected dose = 290 MBq) in a normal subject. All sections are approximately 2 mm thick, reconstructed with a ramp filter cutoff at 2.5 cycles/cm, without scatter correction, and using a segmented attenuation correction based on single photon measured transmission data.

CHAPTER 5

Emission–Transmission Realignment Using a Simultaneous Emission–Transmission Postinjection Scan

V. SOSSI, K. S. MORRISON, T. R. OAKES, and T. J. RUTH

University of British Columbia / TRIUMF PET Centre
Vancouver, British Columbia, V6T 2A3 Canada

Misalignment between emission and transmission data has been shown to produce quantitative and qualitative degradation of the attenuation corrected reconstructed images. Methods that align transmission data to emission data have become available; however, they are based on the assumption that there is no movement between the transmission scan and the first emission scan. The hypothesis of no movement between the transmission and emission scan cannot always be satisfied. A method that eliminates this requirement has been developed. A postinjection simultaneous emission and transmission scan (PITsim) is acquired, before or after the emission scan series. The emission part of the PITsim (E), which is intrinsically aligned to the transmission scan (T), is reconstructed and used as a reference for the alignment of the emission scan images. The subsequent registration between the transmission and the emission sequence sinogram data follows the method of Smith et al. (1997). The aligned transmission part of the PITsim is used for attenuation correction. This method was validated on the CTI/Siemens ECAT 953B brain tomograph with phantom scans and human and monkey brain studies. The alignment between E and the emission studies was accurate to within 1 mm and 1° in the case of the phantom studies. For human FDG studies the validation was performed by demonstrating a decrease in the interframe variability after registration. Tests on monkeys and human neuroligand studies indicated that the E provides a reliable reference image even when the ratio between the number of acquired events in T and E is large (> 45:1).

I. INTRODUCTION

Patient motion during a scanning sequence is known to introduce image blurring and degradation of quantitation accuracy. Several image registration algorithms have been developed to align reconstructed attenuation corrected images of a study composed of several time frames. Less attention has been devoted to the alignment of data obtained from the transmission scan to data obtained from the emission scans, even though a mismatch between transmission and emission data is known to produce image artifacts. There are many scanning situations where such a mismatch can occur, e.g., long dynamic scanning sequences, scanning of movement disorder patients for whom lying still for a prolonged time can be difficult, and situations where patients are taken away from the scanning bed for a short period of time before being rescanned. Performing transmission scans that spatially match emission data is often impossible or technically difficult.

Methods that register the transmission scan to the emission data have been developed (Andersson et al., 1995; Smith et al., 1997). It is a common assumption in these methods that there is no patient motion between the transmission scan and the first emission scan. Because this assumption cannot always be satisfied, a method that does not have this constraint has been developed. This method involves a postinjection transmission scan with simultaneous acquisition of transmission and emission data (PITsim) where the emission part of the scan (E) serves as an alignment reference for the emission scans that are part of the study sequence. Factors that could limit the reliability of this method are degradation of the reconstructed E scan image quality due to data contamination from the transmission source and varying tracer distribution. This can be of special concern in neuroligand studies that are characterized by a varying tracer distribution and by emission scans spanning several half-lives of the radiotracer. The reliability of the PITsim E part as a reference image has been tested with phantom studies, human [^{18}F]fluorodeoxyglucose (FDG) scans, and human and monkey neuroligand studies. FDG studies are typically characterized by a relatively short scanning time, so that the decay of the injected radiotracer is not an issue, and by a static tracer distribution, so that the registration of the emission study to the E scan is expected to be fairly reliable. The more serious challenge offered by the neuroligand studies was addressed using [^{18}F]fluoro-L-dopa (FDOPA) and [^{11}C]raclopride (RAC) monkey studies and a RAC human study. The monkey studies were selected to provide a baseline test, as no movement occurred during the course of the studies because the animals were anesthetized. By choosing two different tracers with different half-lives, a wide range of scanning situations was tested.

II. MATERIALS AND METHODS

A. Realignment Method

The emission–transmission realignment follows the method developed by Smith *et al.* (1997), which assumes no motion between the transmission scan and the first emission scan. Emission data are reconstructed without attenuation correction, and the subsequent emission scans are aligned to the first emission scan. The transformation matrices, obtained from aligning the subsequent emission images to the first one, are applied when forward-projecting the attenuation image into sinogram space. In this way, forward-projected transmission data match the spatial position of acquired emission data. The images are then reconstructed with matched emission and transmission data and realigned again in image space.

B. Postinjection Transmission Method

The postinjection transmission method that was optimized and implemented is based on the protocol developed by Meikle *et al.* (1995; Hooper *et al.* 1996). The scanner, the Siemens/CTI ECAT 953B (Spinks *et al.*, 1992), has the capability of acquiring postinjection transmission data simultaneously into two sets of sinograms. The first set contains transmission data that are acquired in a windowed mode, whereas the second set contains data outside the transmission acquisition windows. These data originate primarily from the emission source. Data used for attenuation correction are obtained similarly as in Meikle *et al.* (1995):

$$T_i = W_i^{tr} - \left[(0.95*(1 - r_i)(W_i^{em} - fW_i^{tr}))/(r_i + 0.95*f(r_i - 1))\right], \quad (1)$$

where T_i are events arising from the transmission source in the ith line of response (LOR); W_i^{tr} and W_i^{em} are events recorded in the transmission and the emission windows, respectively; r_i is the relative efficiency of the ith LOR for measuring emission events; and f is the fraction of the transmission events that contaminate the emission window. 0.95 is an empirically determined factor that accounts for different dead time in the two acquisition windows. The emission part of data (E_i) is obtained as

$$E_i = (W_i^{em} - r_i^* W_i^{tr})/(r_i + f(r_i - 1)). \quad (2)$$

These data are not used as part of the study, as the radioactivity of the rod sources used for the transmission scan contaminates emission data excessively and degrades their quantitative accuracy. It is expected that the contamination will increase as the ratio between the number of acquired events in the T and E window increases. If this ratio becomes too high, the image quality of E scan data might degrade enough to make it unusable even for registration purposes.

C. Phantom Study: Acquisitions

A hot 2-cm-diameter sphere was immersed in a warm elliptical cylinder with dimensions representative of an average human head (14.5 × 19 × 14 cm) (radioactivity ratio 5:1, determined with a calibrated well counter) and scanned in five different positions. Between each scan the phantom was moved radially by 5 mm. PITsim attenuation scans were performed in the first positions (pos_0) and in the last position (pos_4). Data were reconstructed in three different ways for positions 0 and 4: (a) with matching attenuation data, (b) with mismatched attenuation data, and (c) with registered attenuation data. For the scans taken in positions 1, 2, and 3, data were reconstructed in two different ways: (a) with mismatched attenuation data and (b) with registered attenuation data.

D. Phantom Study: Data Analysis

A region of interest (ROI) was placed within the sphere image boundaries, and six larger ROIs were placed on the background region around the sphere. The ratio between the image intensity in the small ROI and each of the background ROIs (S/B) was calculated. Mean values and the standard deviation of S/B over all background regions were estimated in the slice where the small ROI image intensity was the highest. The magnitude of the standard deviation was taken as the figure of merit when estimating the image artifact and quantitation degradation introduced by a mismatch between emission and transmission data.

E. Human Scans: Acquisitions

1. FDG Study

The FDG study was performed on a volunteer subject with a 10-min PITsim scan performed 20 min after injection followed by a 4 × 5-min scanning sequence starting 40 min after injection (first scanning sequence). During this time no change in the activity distribution is expected, so all four frames should produce the same image pattern within statistical limitations. The patient was then taken out of the scanner and repositioned several hours later, and the same scanning sequence was performed after a second injection (second scanning sequence).

2. RAC Study

A 16-time frame scanning sequence was performed: 4 × 60 sec, 3 × 120 sec, 8 × 300 sec, 1 × 600 sec. A 10-min PITsim scan was performed immediately following the end of the emission sequence. Only limited patient motion was observed during the scan.

F. Human Scans: Data Analysis

1. FDG Study

Data from all time frames from the first scanning sequence were reconstructed with and without alignment between transmission and emission data. The reliability of the emission part of the PITsim as a registration reference was tested by estimating the standard deviation of a grid of ROIs placed over the images obtained from the four time frames. Had the registration reference frame been inadequate, an increase in standard deviation would be expected. A decrease in standard deviation, however, indicates an improvement in accuracy. The first frame of the second sequence was reconstructed in three different ways: with its own transmission data set (no alignment) and using the transmission data from the first scanning sequence with and without alignment. In this case, concentration values from data reconstructed with the corresponding attenuation scan were considered as reference, to which concentration values obtained from data processed with the first attenuation scan with and without alignment were compared.

2. RAC Study

Data were reconstructed with and without the alignment between the E and T. Four ROIs were placed on the striatum, three on the occipital cortex, two on the frontal cortex, and two on the cerebellum. Time activity curves (TACs) obtained from the two images sets and realignment parameters were compared.

G. Monkey Studies: Acquisitions

These studies were performed to provide a baseline estimate of the effects of a varying tracer distribution and statistical content of the E scan compared to the T scan on the realignment accuracy. Two different scanning sequences with two different tracers were performed on anesthetized rhesus monkeys: a 16-frame [^{11}C]raclopride scan with the following time sequence: 4 × 30 sec, 3 × 1 min, 5 × 2 min, 1 × 5 min, 3 × 10 min (total 1 hr, three radiotracer half-lives), followed by a 10-min PITsim scan; and a 20-time frame FDOPA scan with the time sequence: 6 × 30 sec, 2 × 60 sec, 1 × 5 min, 11 × 10 min (total 2 hr, one radiotracer half-life), followed by a 10-min PITsim scan. The monkeys were immobilized in a stereotaxic frame.

H. Monkey Studies: Data Analysis

Data were reconstructed in two different ways. In the first case, images were reconstructed using the transmission part of the PITsim scan, and the results of the analysis performed on these images were used as the gold standard, as no motion occurred between frames. In the second case, data were processed with the E–T realignment. Because no movement occurred during the scanning sequence, the realignment parameters were expected to be very close to zero and the resulting images qualitatively and quantitatively identical to the reference images. Small ROIs were placed on the frontal cortical region, the striata, the occipital cortex, and the cerebellar region on both sets of images, and TACs were compared for the two cases. Finally, an additional test of the reliability of E scan data as reference image was performed. The E scan was reconstructed, introducing a 5-mm offset in the x direction and a 2-mm offset in the y direction, which is equivalent to scanning the subject shifted by the same amount. The realignment parameters were compared to the actual image shift.

III. RESULTS AND DISCUSSION

A. Phantom Study

Figure 1 shows the phantom image obtained when the phantom was located in position 4, and data were processed with the proper attenuation scan (left), mis-

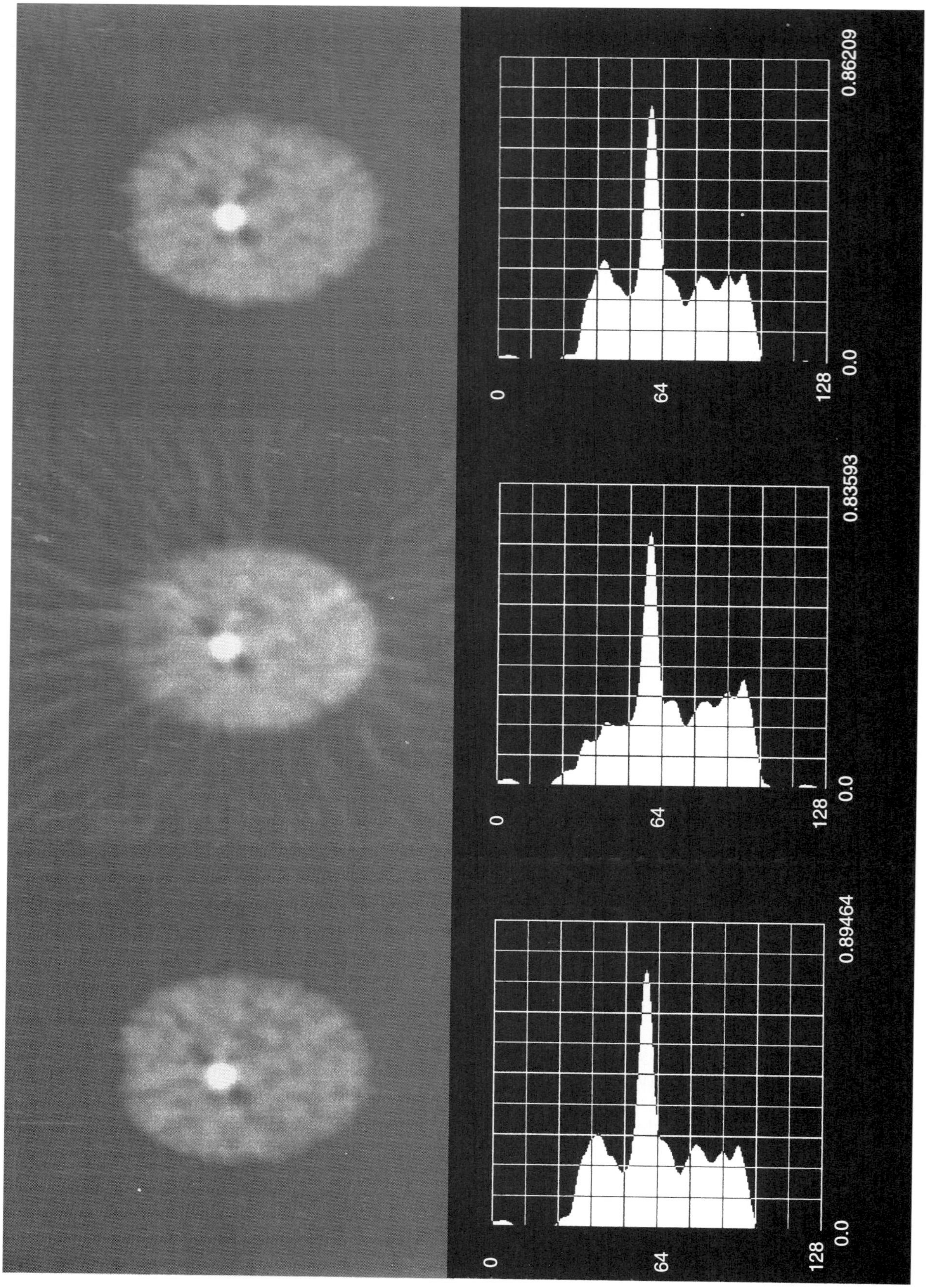
0
64
128
0.0
0.89464
0
64
128
0.0
0.83593
0
64
128
0.0
0.86209

TABLE 1 Comparison of Actual Shift with Estimated Shift in the Phantom Study

	Real shift (mm)	Estimated shift (mm)
pos_0	0	—
pos_1	5	5.06
pos_2	10	9.84
pos_3	15	14.68
pos_4	20	20.27

TABLE 2 Standard Deviation of the Signal / Background Ratio Averaged over All Background Regions[a]

Scan	Ratio STD (%)	Phantom position (MM)	Atten scan position
pos0_att0	4.8	0	0
pos0_att4	16.0	0	+20
pos1_att0	5.0	+5	0
pos1_reg0	7.0	+5	0
pos2_att0	8.0	+10	0
pos2_reg0	4.0	+10	0
pos3_att0	16.0	+15	0
pos3_reg0	4.5	+15	0
pos4_att0	25.0	+20	0
pos4_reg0	8.0	+20	0
pos4_att4	4.5	+20	+20

[a] A high value indicates poor uniformity of the background region (see text and Fig. 1). The notation _reg0 (_reg4) indicates the transmission scan that was taken with the phantom in position 0 (position 4) and aligned to the position where emission data were collected. The notation _att0 (_att4) indicates the transmission scan that was taken with the phantom in the position 0 (position 4) without alignment to emission data.

matched attenuation scan (center), and registered attenuation data (right). A strong artifact is clearly visible when there is a mismatch between attenuation and emission data. There is almost no residual artifact present after registration of transmission data to emission data using the transformation matrix obtained by registering the emission part of the PITsim. Table 1 compares the actual distances by which the phantom was moved to the shift estimated by the registration of each emission scan to the emission part of the PITsim. There is excellent agreement between the two, demonstrating that the emission part of the PITsim scan is an adequate registration reference. Table 2 shows the values of the standard deviation of S/B averaged over all background regions. The standard deviation increases with an increased mismatch between transmission and emission data, indicating increased image nonuniformity (see Fig. 1). After registering the emission and transmission scans, the standard deviation decreases, but never to the baseline level, where the transmission and emission scans were performed with the phantom in the same position. This indicates that it is always preferable to perform the emission and transmission scans in the same position. It was also found that a mismatch between emission and transmission scans of less than ~5 mm leads only to a small image degradation, consistent with results reported by Andersson et al. (1995).

B. Human Scans

1. FDG Study

The standard deviation of the ROI time activity curves averaged over the 31 ROIs constituting the grid decreased by 2% when transmission and emission scan data were aligned using the emission part of the PITsim scan. Although the realignment procedure is based on a variance reduction algorithm and as such is expected to reduce the variance, an inadequate reference frame would introduce image artifacts, thus degrading data reproducibility from frame to frame. A reduction in standard deviation confirms the feasibility of using the emission part of the PETsim as the reference image in the registration protocol. The analysis of the first frame of the second scanning sequence showed similar results; concentration values obtained from ROIs placed on data reconstructed with attenuation correction determined in the first sequence were 3% lower than the concentration values obtained from data reconstructed with attenuation correction determined in the second sequence. When the first attenuation scan was registered to the emission scan, the concentration values were 1.5% lower, indicating a slightly better quantitation recovery. The registration, however,

FIGURE 1 Images of the hot sphere–warm background phantom. (Left) Attenuation and emission scans in position 4. (Center) Attenuation scan in position 0, emission scan in position 4, no alignment; image artifacts are clearly visible. (Right) Attenuation scan in position 0, emission scan in position 4, alignment before reconstruction; visible improvement in image quality compared to the center image. Shown below the images are vertical profiles through the hot sphere.

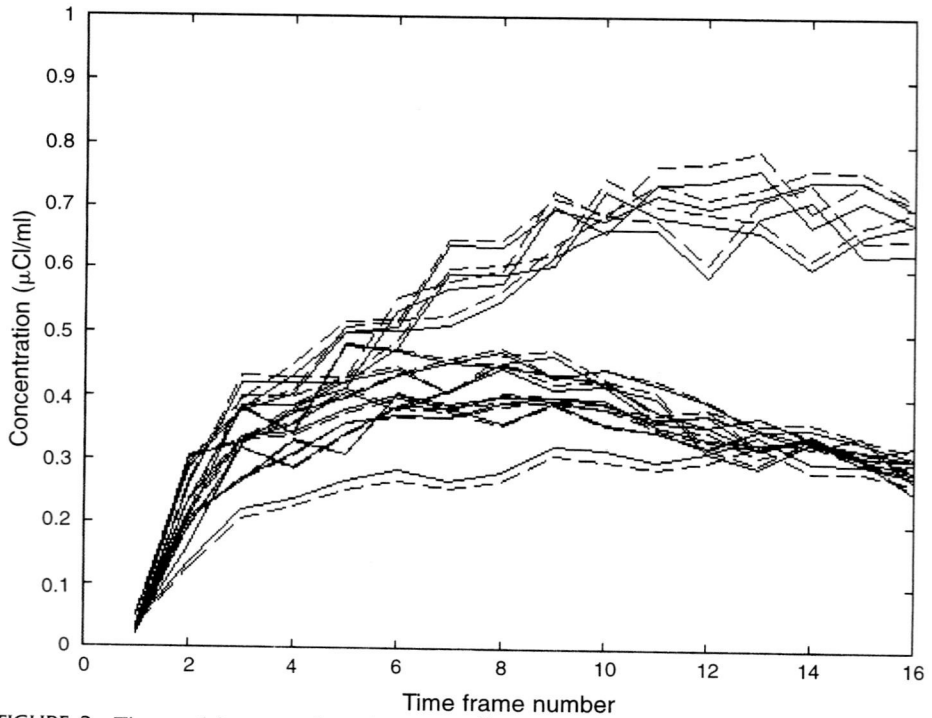

FIGURE 2 Time–activity curves from the human [^{11}C]raclopride study. Results from original data (solid lines) and from realigned data (dashed lines) are shown for ROIs drawn on the frontal cortex, striata, occipital cortex, and cerebellum.

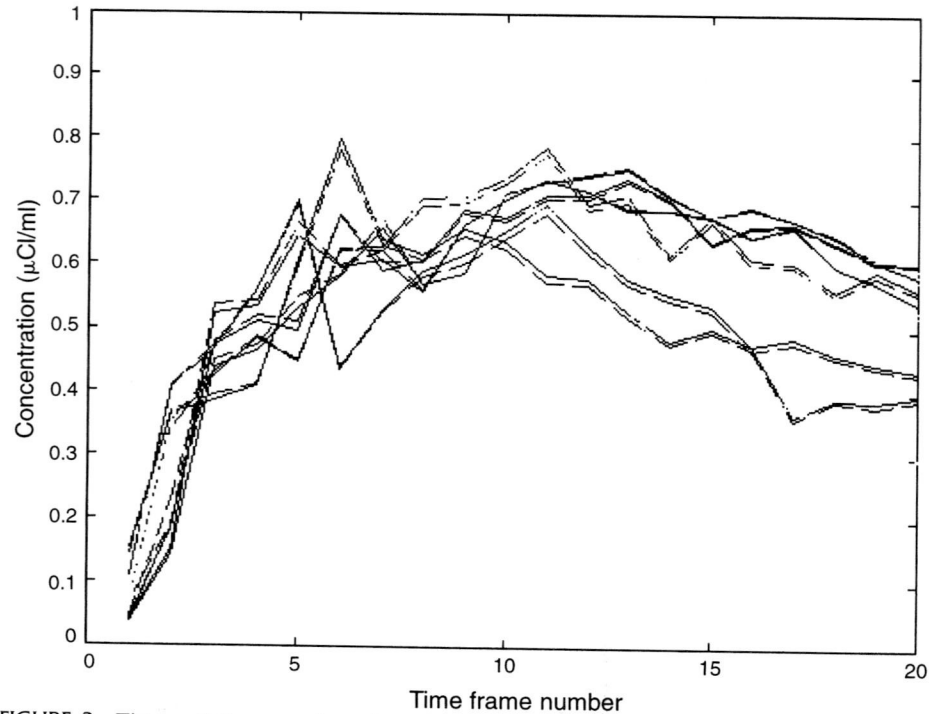

FIGURE 3 Time–activity curves from the monkey FDOPA study. Results from original data (solid lines) and from realigned data (dashed lines) are shown for ROIs drawn on the frontal cortex, striata, occipital cortex, and cerebellum.

indicated that the mismatch between the first emission frame of the second scanning sequence and the first transmission scan was approximately 1.5 mm in the x and y directions and 0.4 mm in the z direction while the rotation was less than 1° around each axis, so no major image degradation was expected even without registration. The ratio between the number of events in the T and E windows was approximately 10:1.

2. RAC Study

Figure 2 shows the TACs obtained from the images of the first study reconstructed with and without alignment. There is very good quantitative agreement, which is consistent with the magnitude of the realignment parameters, that were found to be of the order of 1–2 mm and 1–2°. The ratio between the number of events in the T and E windows was approximately 37:1.

C. Monkey Studies

1. F-DOPA

Figure 3 shows the TACs obtained from the ROIs placed on the gold standard images and on the realigned images. The alignment of the first frame to the reference frame was particularly challenging due to the poor statistical content of the first frame and to the difference in the tracer distribution between earlier and later time frames. Nevertheless, the realignment parameters were inaccurate only for the first two time frames, while for all the other time frames the alignment was accurate within 1 mm and 1°. When the shifted E image (see Section II,H) was used as the reference, the realignment parameters agreed with the actual shift within 1 mm and 1°. Figure 3 also shows that image quantification was very well preserved by the E–T realignment method, within approximately 1%, if the first two time frames are excluded from the comparison. In this case the ratio between the acquired events in the T and E windows was approximately 8:1.

2. [^{11}C]Raclopride

This study presented a more challenging test due to the greater radiotracer decay at the time of the time of the PITsim scan and to the more varying tracer distribution, as indicated by the TACs in Fig. 4. Nevertheless, the registration between the E and the later time frames was accurate within less than 1 mm even when the E data shifted image was used as the reference, confirming the reliability of E scan data even in this situation. The registration was less accurate between the E scan image and the earlier time frames, which resulted in ~5% quantification errors in the TACs (Fig. 4). This inaccuracy is, however, not introduced by

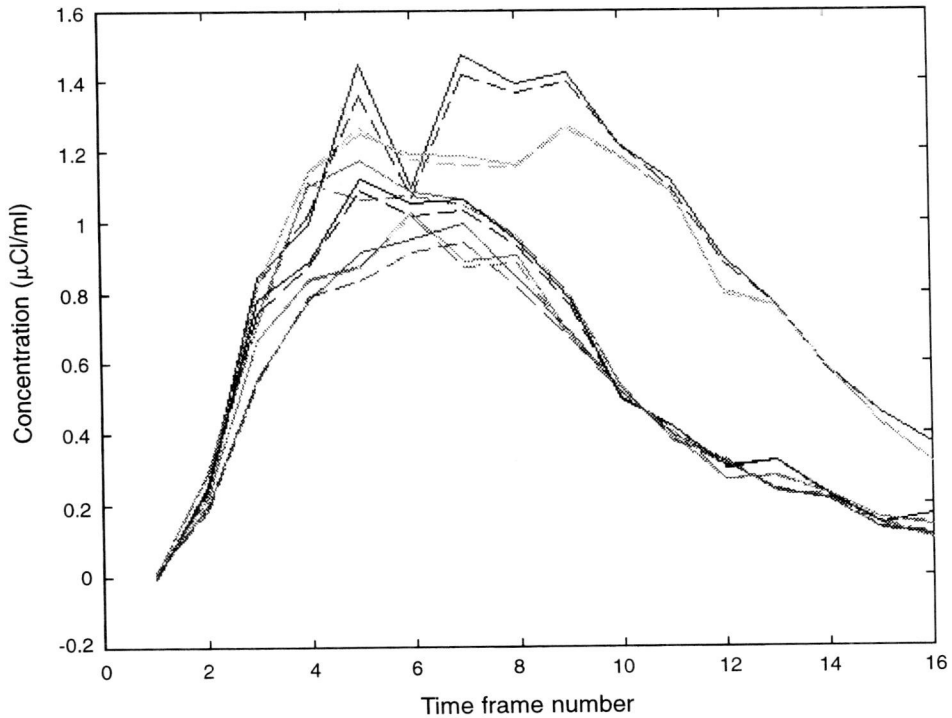

FIGURE 4 Time–activity curves from the monkey [^{11}C]raclopride study. Results from original data (solid lines) and from realigned data (dashed lines) are shown for different ROIs.

using the image of the E scan as the realignment reference, but it is likely due to varying tracer distribution. In this case a recursive alignment method as described by Andersson *et al.* (1995) might prove helpful. In this study the ratio between the acquired events in the T and E windows was 47:1.

IV. CONCLUSION

It has been demonstrated that the emission part of the PITsim scan is a reliable reference image when aligning brain FDG images and for a large range of neuroreceptor studies. The method presented here is based on a previously developed method for alignment of transmission and emission scan data, but it eliminates the requirement of no motion between the transmission scan and the first scan of the emission sequence. This method can thus be used even in those cases where the patient needs to exit the scanner between successive scanning sequences or when the patient moves between transmission and emission scans. Phantom studies have indicated that the most reliable quantitative results are obtained when a transmission scan is acquired in the same position as the emission scan. When this is not possible, a registration between transmission and emission scans noticeably improves the qualitative and quantitative image degradation due to the spatial mismatch between emission and transmission scans.

Acknowledgments

The authors thank Drs. Doudet, Stoessl, and Liddle for sharing monkey and human data and Dr. Smith for providing the E–T realignment code. This work was supported by an MRC group grant and by the NIH NCI 1F32 CA 67486-01 grant.

References

Andersson, J. L. R., Vagnhammar, B. E., and Schneider, H. (1995). Accurate attenuation correction despite movement during PET imaging. *J. Nucl. Med.* **36**: 670–678.

Hooper, P. K., Meikle, S. R., Eberl, S., and Fulham, M. J. (1996). Validation of postinjection transmission measurements for attenuation correction in neurological FDG-PET studies. *J. Nucl. Med.* **37**: 128–136.

Meikle, S. R., Bailey, D. L., Hooper, P. K., Eberl, S., Hutton, B. F., Jones, W. F., Fulton, R. R., and Fulham, M. L. (1995). Simultaneous emission and transmission measurements for attenuation correction in whole-body PET. *J. Nucl. Med.* **36**: 1680–1688.

Smith, A. M., Bruckbauer, T., Wienhard, K., Pietrzyk, U., and Byars, L. G. (1997). Spatial transformation during 3D reconstruction in PET. *Eur. J. Nucl. Med.* **24**: 1413–1417.

Spinks, T. J., Jones, T., Bailey, D. J., Townsend, D. W., Grootoonk, S., Bloomfield, P. M., Gilardi, M.-C., Casey, M. E., Sipe, B., and Reed, J. (1992). Physical performance of a positron tomograph for brain imaging with retractable septa. *Phys. Med. Biol.* **37**(8): 1637–1655.

CHAPTER 6

Optimization of $H_2{}^{15}O$ Dose and Data Acquisition in Three-Dimensional Activation Studies Using an ECAT EXACT HR-47 PET Camera and Voxel-by-Voxel t-Statistic

PETER JOHANNSEN, SØREN B. HANSEN, LEIF ØSTERGAARD, and ALBERT H. GJEDDE

PET Center
Aarhus University Hospital
DK-8000 Aarhus C, Denmark

To improve the sensitivity in positron emission tomography (PET) activation studies, the optimal dose of $H_2{}^{15}O$ and the optimal onset and duration of data acquisition in the three-dimensional (3D) mode of the ECAT EXACT HR-47 PET camera were evaluated. Six healthy right-handed volunteers participated. After an intravenous bolus injection of 25, 50, 100, 250, 500, or 1000 MBq $H_2{}^{15}O$, baseline and motor tasks, apposition of right-hand fingers, were completed six times each. Data were acquired as dynamic scans of 2-min duration from time of injection. For the 500 MBq $H_2{}^{15}O$ injections, sinograms were added for time intervals of 30, 40, 60, and 90 sec, starting at -5 sec before, $+5$ sec after, and just at the time of bolus arrival to the brain. After alignment of PET images to a standard coordinate system and normalization, t-statistic maps were calculated for subtracted PET volumes. Peak t values were evaluated for five cortical sites. Although noise equivalent counts peaked above 1 GBq, the optimal dose for the t-statistic was 500 MBq. Of the 30-, 40-, 60-, and 90-sec, durations, the optimal data acquisition lasted 60 sec, with onset at time of bolus arrival. The peak t value was more influenced by variation of acquisition duration than by onset of data acquisition.

I. INTRODUCTION

Previous studies have identified very different durations of data acquisition as the optimal for positron emission tomography (PET) activation studies; from 40 sec, 60 sec, to more than 90 sec (Fox and Mintun, 1989; Mintun et al., 1989; Kanno et al., 1991: Volkow et al., 1991). The rationale behind brief data acquisitions is the linear relation between the $H_2{}^{15}O$ uptake and the cerebral blood flow (CBF) during first-pass extraction of the tracer. Longer scanning favors the counting statistic and hence increased signal-to-noise ratio.

The evaluation of the optimal data acquisition time was different in each study. Most studies used region of interest (ROI) analysis with evaluation of relative CBF increase or signal-to-noise ratios in a ROI. One previous study by Sadato et al. (1997) used voxel-by-voxel z-statistical analysis to determine the optimal duration of data acquisition for the GE Advance tomograph. Of 60, 90, and 120 sec an acquisition interval of 60 sec was found to be optimal; however, the study did not assess the optimal onset of data acquisition. Some studies (Mintun et al., 1989; Volkow et al., 1991) evaluated acquisitions from the time of tracer injection, whereas others (Kanno et al., 1991) commenced scanning 20–25 sec after the injection.

The aim of the present study was to address the sensitivity of PET activation analysis of the relative accumulation of $H_2{}^{15}O$. The optimal $H_2{}^{15}O$ dose (bolus injection) and the optimal onset and duration of data acquisition in three-dimensional (3D) mode were evaluated, using the ECAT EXACT HR-47 PET camera (Siemens/CTI, Knoxville, TN) (Wienhard et al., 1994).

II. METHODS

Six healthy right-handed volunteers participated after written informed consent. The study protocol was approved by the Aarhus County Research Ethics Committee. After an intravenous bolus injection of 25, 50, 100, 250, 500, or 1000 MBq (0.67, 1.35, 2.7, 6.8, 13.5, or 27 mCi), baseline and motor tasks, the latter requiring apposition of right-hand fingers, were completed six times each. Conditions and doses were counterbalanced across subjects. The task commenced 10 sec prior to injection. Data were acquired in 3D mode as dynamic scans over 2 min (16 frames of 5 sec; 4 frames of 10 sec), starting at injection. For the 500 MBq $H_2^{15}O$ injections, sinograms were added for time intervals of 30, 40, 60, or 90 sec, starting 5 sec before, 5 sec after, or at the moment of bolus arrival to the brain. The moment of bolus arrival to the brain was determined from curves of the true coincidence count rate (head curve) using the following limits: 6000 cps for 25 MBq, 10,000 cps for 50 MBq, 30,000 cps for 100 MBq, 40,000 cps for 250 MBq, 60,000 cps for 500 MBq, and 100,000 cps for 1 GBq. These threshold rates represented approximately 20% of the peak true coincidence count rate.

Images were reconstructed with measured attenuation and scatter corrections and filtered to 12 mm isotropic resolution full-width-at-half-maximum (Hann filter, cutoff = 0.15 cycles/mm). PET volumes were coregistered (Collins *et al.*, 1994) to each subject's own MR brain image and to the Talairach coordinate system (Talairach and Tournoux, 1988). After normalization to the average intracranial voxel count and subtraction of baseline volumes from motor task volumes, a map of the *t*-statistic was calculated on a voxel-by-voxel basis (Worsley *et al.*, 1992) using the pooled intracranial voxel SD and assuming a mean of zero.

Peak *t* values for left and right primary motor cortex (M1), cerebellum, supplementary motor area (SMA), and left premotor cortex were compared for the 12 activation analyses performed on the 500 MBq $H_2^{15}O$ data set. After determining the optimal frame duration and onset from these analyses, these parameters were used in the analysis of the other five different injected doses of $H_2^{15}O$ and for calculation of noise equivalent count rate (NEC) (Strother *et al.*, 1990), given the following formula and assumptions:

$$NEC = (T(1 - SF))^2/(T + 2fR),$$

where T is the count rate for trues and R is the count rate for randoms. In the calculation of NEC the scatter fraction (SF) was set to 0.40, and the fraction of the projection originating from the head (f) was estimated to be 0.32.

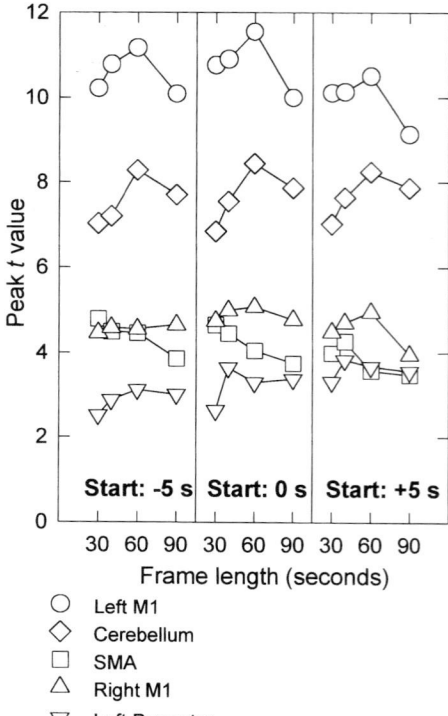

FIGURE 1 Peak *t* value for the five activated regions as a function of frame length (30 to 90 sec) at three different start points of the frame: 5 sec before, and 5 sec after bolus arrival to the brain.

○ Left M1
◇ Cerebellum
□ SMA
△ Right M1
▽ Left Premotor

III. RESULTS

Among all 72 injections the time of bolus arrival to the brain varied from 10 to 21 sec but varied little within each subject, with subject averages for the 12 injections of (mean ± SD) 11.1 ± 0.9 sec, 11.8 ± 1.4 sec; 12.4 ± 1.8 sec; 12.7 ± 18 sec; 15.5 ± 2.3 sec; and 18.2 ± 2.1 sec. The grand mean was 13.6 ± 0.69 sec.

The peak *t* value of the 500 MBq injection of $H_2^{15}O$ for the five cortical areas (listed earlier) at different durations of data acquisition (30, 40, 60, or 90 sec) and with three different starting times (−5, 0, or +5 sec) relative to bolus arrival is shown in Fig. 1. The maximum *t* value was obtained for the 60-sec acquisition with onset at the time of bolus arrival to the brain. Applying these parameters to the different doses resulted in a maximum *t* value for 500 MBq (Fig. 2). Figure 3 shows the number of noise equivalent counts accumulated over 60 sec as a function of the injected dose normalized to 70 kg body weight for all six subjects. Figure 4 shows the mean ± SD percentage ran-

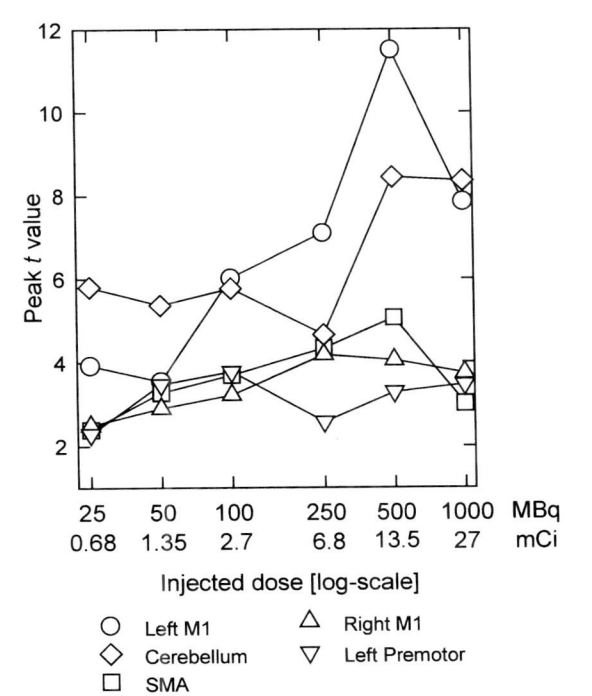

FIGURE 2 Peak t value for the five activated regions as a function of injected dose.

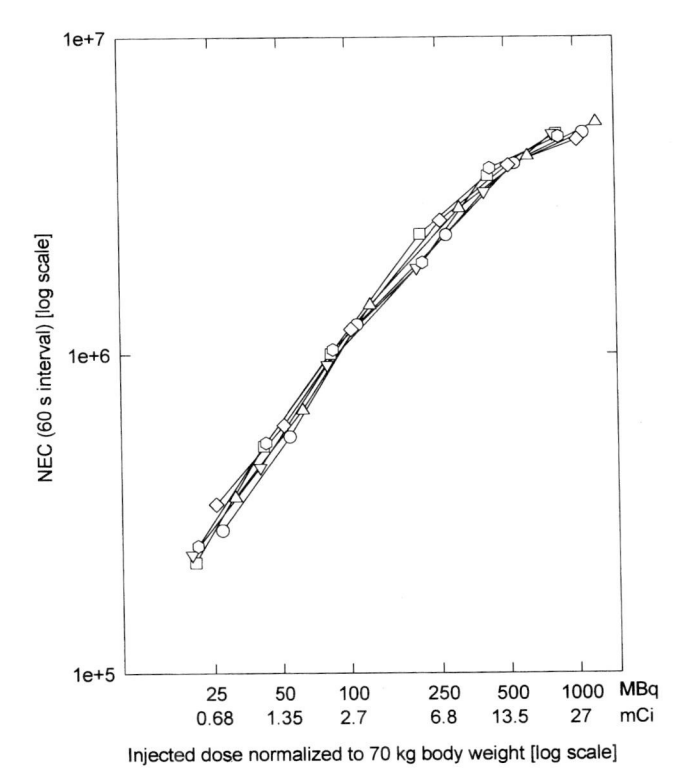

FIGURE 3 NEC as a function of the injected dose (60-sec frame) for the six subjects in a log–log coordinate system.

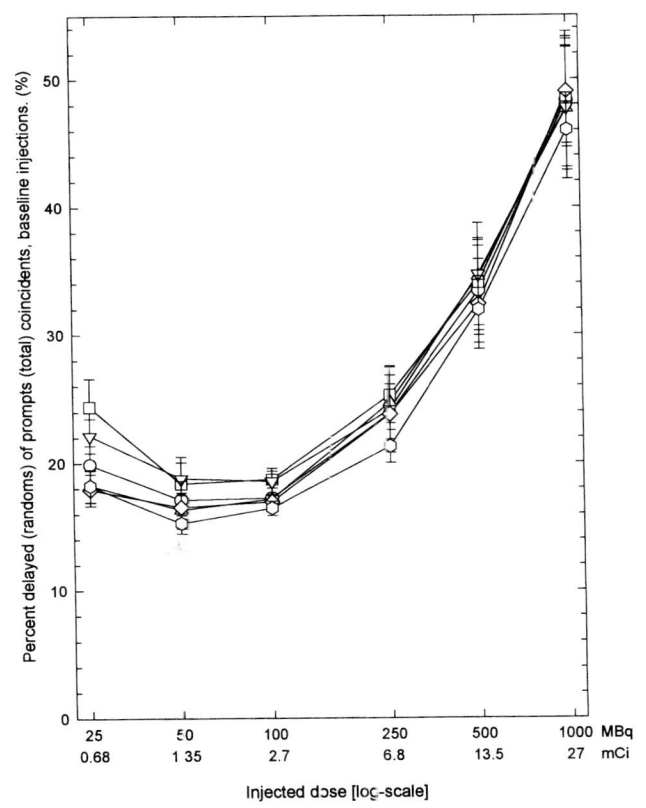

FIGURE 4 Percentage random counts of true coincident counts as a function of the injected dose. The percentage is calculated as a mean ± SD of the ratio from bolus arrival until 120 sec after injection (see Section II).

dom counts (delayed coincidence events) of the total number of coincidence events (prompts) from bolus arrival until the end of scanning for each dose.

IV. DISCUSSION

The initial t-statistic analysis of the optimal time interval was based on the 500 MBq $H_2^{15}O$ injection. This dose was chosen after evaluation of the noise equivalent count curves for the six subjects, indicating that the 500 MBq dose represents a good balance between statistics and radiation dose. Based on the comparison of the t values obtained for the different time intervals and dosages (Figs. 1 and 2), it is believed that the choice of another dose for the time interval analysis would not change the optimal frame length found for the 500 MBq dose. The variation in onset of the frame (−5, 0, or +5 sec) relative to the bolus arrival may seem small compared to the length (30 to 90 sec) of the frame, but this small variation in onset was chosen because the change in counting statistics varies most at the time of bolus arrival.

The optimal frame length determined in different studies seems to depend on the variable chosen for evaluation (i.e., relative rCBF change, S/N ratio, or t value). As activation studies most often are evaluated by voxel-based t- or z-statistical analysis, it is most appropriate to use this variable for optimization of the data acquisition parameters. The statistical outcome of the activation analysis has previously only been used in one study (Sadato et al., 1997) (together with the NEC) to define the optimal dose and duration of data acquisition. The result for the optimal frame length in that study is similar to the results of the present study.

The results replicate the outcome of previous studies in which the optimal frame start coincided with the arrival of the tracer to the brain. The present study shows that a variation of ±5 sec has no effect on the result. The use of the volume of the significantly activated clusters as the outcome variable did not help identify any specific data acquisition parameters as being optimal (data not shown).

As the NEC rate peaks for injected activity levels above 1 GBq whereas the t-statistic peaks at 500 MBq, the NEC may not accurately describe the statistical quality of the image in relation to relative rCBF changes. The discrepancy between the dose optimum selected from the NEC and the dose optimum selected on the basis of the t value may be explained by the unequal counting statistics in the transaxial planes of the scanner. In the calculation of the NEC, all counts from all planes are included, whereas the t-statistic value is based on data from the middle of the axial field of view, which has better counting statistics than the ends.

This study used rapid bolus injections (within 5 sec). Slow bolus injections (over 20–40 sec) may change the ratio between random and true coincident counts. Judged from the standard deviations in Fig. 4, this ratio stayed stable from bolus arrival with a high count rate to scan stop with a low count rate. This indicates that a slow bolus would only yield minor changes in the random/trues ratio, although the effect on the t-statistic remains uncertain.

The choice of activation task may have an impact on the results obtained. In a motor task, as in the present study, the peak percentage increase in normalized rCBF in the primary motor cortex was between 27 and 53%. In studies of cognitive tasks, the peak percentage rCBF increase is often less than 10%. Whether this difference affects optimal data acquisition parameters has not been tested.

To increase the sensitivity, the tomograph was used in 3D mode. The results of this study therefore do not apply to acquisitions taken in 2D mode.

V. CONCLUSIONS

In conclusion, although the NEC peaked for injected activity levels above 1 GBq, the optimal dose from the t-statistic was found to be 500 MBq. The optimal data acquisition length was seen to be 60 sec, starting at the time of bolus arrival in the brain. The peak t value was more influenced by variation of acquisition length than by the starting time of data acquisition.

Acknowledgments

The study was supported by Grants 12-1633 and 12-1634 from the Danish Medical Research Council, Aarhus University Research Foundation, and "Eilif Trier Hansens og Hustrus Legat."

References

Collins, D. L., Neelin, P., Peter, T. M., and Evans, A. C. (1994). Automatic 3D intersubject registration of MR volumetric data in standardized Talairach space. *J. Comput. Assist. Tomogr.* **18**: 192–205.

Fox, P. T., and Mintun, M. A. (1989). Noninvasive functional brain mapping by change-distribution analysis of averaged PET images of $H_2^{15}O$ tissue activity. *J. Nucl. Med.* **30**: 141–149.

Kanno, I., Iida, H., Miura, S., and Murakami, M. (1991). Optimal scan time of oxygen-15-labeled water injection method for measurement of cerebral blood flow. *J. Nucl. Med.* **32**: 1931–1934.

Mintun, M. A., Raichle, M. E., and Quarles, R. P. (1989). Length of PET data acquisition inversely affects ability to detect focal areas of brain activation. *J. Cereb. Blood Flow Metab.* **9**(Suppl. 1): S349 (abstr.).

Sadato, N., Carson, R. E., Daube-Witherspoon, M. E., Campbell, G., Hallet, M., and Herscovitch, P. (1997). Optimization of noninvasive activation studies with ^{15}O-water and three-dimensional positron emission tomography. *J. Cereb. Blood Flow Metab.* **17**: 732–739.

Strother, S. C., Casey, M. E., and Hoffman, E. J. (1990). Measuring PET scanner sensitivity: Relating countrates to image signal-to-noise using noise equivalent counts. *IEEE Trans. Nucl. Sci.* **37**: 783–788.

Talairach, J., and Tournoux, P. (1988). "Co-Planar Stereotaxic Atlas of the Human Brain." Thieme, Stuttgart.

Volkow, N. D., Mullani, N., Gloud, L. K., Adler, S. S., and Gatley, S. J. (1991). Sensitivity of regional brain activation with oxygen-15-water and PET to time of stimulation and period of image reconstruction. *J. Nucl. Med.* **32**: 58–61.

Wienhard, K., Dahlbom, M., Eriksson, L., Michel, C., Bruckbauer, T., Pietrzyk, U., and Heiss, W. D. (1994). The ECAT EXACT HR: Performance of a new high resolution positron scanner. *J. Comput. Assist. Tomogr.* **18**: 110–118.

Worsley, K. J., Evans, A. C., Marrett, S., and Neelin, P. (1992). A three-dimensional statistical analysis for CBF activation studies in human brain. *J. Cereb. Blood Flow Metab.* **12**: 900–918.

CHAPTER 7

Parametric Image Reconstruction Using Spectral Analysis of (Rebinned) Three-Dimensional Projection Data[1]

S. R. MEIKLE,* J. C. MATTHEWS,[†] V. J. CUNNINGHAM,[†] D. L. BAILEY,[†] T. JONES,[†] and P. PRICE[†]

*PET and Nuclear Medicine Department
Royal Prince Alfred Hospital
Camperdown, New South Wales 2050, Australia
[†] PET Oncology and Methodology Groups
MRC Cyclotron Unit
Hammersmith Hospital
London W12 0NN, England

This chapter describes methodology that enables efficient calculation of parametric images with a high signal-to-noise ratio (SNR) from three-dimensional (3D) dynamic positron emission tomography (PET) data. The 3D projections are rebinned into an equivalent 2D data set using single slice or Fourier rebinning. Spectral analysis is then applied directly to rebinned projection data, yielding projections of parameters of interest that are reconstructed using the iterative ordered subsets expectation maximization (OS-EM) algorithm. The processing time is an order of magnitude less than that for the conventional approach of fully 3D filtered backprojection (FBP) and image-based spectral analysis. The methodology was tested using the labeled anticancer drug [^{11}C]temozolomide. Data were acquired in 3D over 90 min on an ECAT 953B tomograph (CTI/Siemens, Knoxville, TN). Parametric images proportional to flow × extraction (K_1) and net rate of influx (K_i) were calculated using the new method, as well as the conventional approach. Using FBP for both image-based and projection-based methods yielded qualitatively similar results, although the new method resulted in slightly noisier images. The relative differences between the two methods, averaged over all brain voxels, were +1.6% for K_1 and −2.4% for K_i. When OS-EM (1 iteration, 16 subsets) was used instead of FBP, the bias was not affected and the SNR was improved appreciably. Thus, spectral analysis of rebinned 3D projection data, combined with OS-EM reconstruction, enables efficient calculation of parametric images with high SNR.

I. INTRODUCTION

Functional parametric images, where each voxel contains a physiological or pharmacokinetic parameter estimate, are often used to summarize the spatial distribution and kinetic behavior of positron-emitting tracers *in vivo*. Conventionally, they are calculated by performing parameter estimation on individual voxel activity versus time data in a reconstructed dynamic image sequence. The statistical reliability of parameter estimates may be enhanced by acquiring the dynamic data on a multiring scanner operated in volumetric (3D) mode (Townsend *et al.*, 1989; Bailey *et al.*, 1991; Cherry *et al.*, 1991). However, this greatly increases the size of the data set, requiring new strategies for efficiently performing reconstruction and modeling tasks. This chapter describes methodology that enables efficient calculation of parametric images with a high signal-to-noise ratio (SNR) from 3D dynamic positron emission tomography (PET) data. The methodology is applicable to a wide range of tracers and is illustrated here with data obtained from a patient study using the labeled anticancer drug [^{11}C]temozolomide.

---

[1] Transcripts of the BRAINPET97 discussion of this chapter can be found in Section VIII.

II. MATERIALS AND METHODS

Several steps are involved in processing 3D dynamic PET data. The method described in this chapter differs from the convention by performing rebinning of the 3D projections into an approximately equivalent 2D set of projections and by applying the tracer kinetic model at an earlier stage in the process, i.e., before image reconstruction (Fig. 1). This approach reduces the reconstruction task to one of reconstructing $M \times P$ 2D sinograms, where M is the number of slices and P is the number of kinetic parameters. As a result, it is practical to use iterative reconstruction, taking advantage of the favorable bias-variance characteristics of such algorithms compared with filtered backprojection (FBP).

A. 3D Rebinning

The 3D projections are first normalized to correct for differences in detection efficiency among the lines of response (LoR) that connect pairs of detectors. Normalized projection data are then corrected for scatter using the convolution-subtraction method (Bailey and Meikle, 1994). For data acquired on the ECAT 953B scanner, a scatter fraction of 0.3 and a radially symmetric exponential convolution kernel with slope = 0.08 cm^{-1} are used. The projections are then rebinned into an approximately equivalent 2D set of projections. This can be done using either the single slice rebinning (SSRB) algorithm or the Fourier rebinning (FORE) algorithm. The SSRB algorithm approximates an oblique LoR by the direct LoR that intersects it at the corresponding z coordinate (Daube-Witherspoon and Muehllehner, 1987):

$$p(s, \phi, z, \delta) \approx p(s, \phi, z, 0), \quad (1)$$

where s is the radial displacement of the LoR in the transaxial plane, ϕ is the angle formed between the LoR (projected onto the transaxial plane) and the y axis of the scanner, z is the axial coordinate of the point midway between the two detectors forming the LoR, and δ is an index of the ring difference (obliqueness) between the two detector rings forming the LoR.

For scanners with a small axial acceptance angle, such as the ECAT 953B, and a source concentrated near the center of the field of view, the SSRB approximation provides a good compromise between speed and accuracy compared with fully 3D reconstruction (Kinahan and Karp, 1994). However, for larger axial acceptance angles or more distributed sources, SSRB can lead to distortion and loss of spatial resolution near the periphery of the field of view. In such cases, the FORE algorithm provides a more accurate rebinning of the 3D projections (Defrise, 1995; Defrise et al., 1997). This algorithm relates the 2D Fourier transform of an oblique sinogram to the 2D Fourier transform of intersecting direct sinograms via a frequency-dependent shift in the axial direction:

$$P(\omega, k, z, \delta) \approx P\left(\omega, k, z - \frac{k\delta}{\omega}, 0\right), \quad (2)$$

where ω and k are the frequency variables corresponding to s and ϕ, respectively. The derivation of these rebinning algorithms and their relationship to an exact rebinning formulation are discussed in Defrise et al. (1997).

B. Spectral Analysis

Using the discretized variable i to indicate the projection bin that records data on the continuous domain (s, ϕ, z, δ), the rebinned dynamic sequence of projections can be related to the time-dependent image function $y(t)$ by the attenuated Radon transform:

$$p_i(t) = \sum_{j \in J} b_{ij} y_j(t), \quad (3)$$

where $p_i(t)$ is the number of counts recorded in projection bin i at time t, J is the subset of pixels that contribute to bin i, and b_{ij} is the probability of a

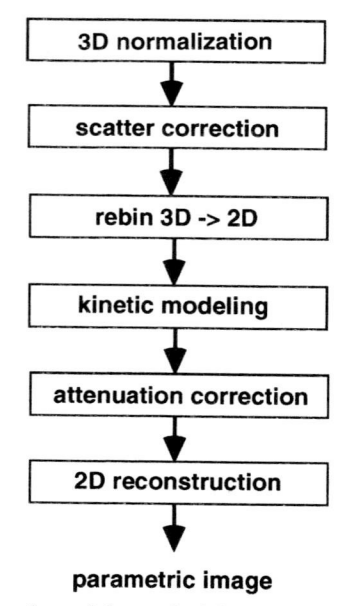

FIGURE 1 Flow chart of the methodology for generating functional parametric images from dynamic 3D PET data.

coincidence event originating from pixel j being recorded in bin i.

The time–activity function can be substituted with any tracer kinetic model that is linear in the unknown parameters (Tsui and Budinger, 1978). Examples that satisfy this requirement include weighted integral models (Huang *et al.*, 1982; Alpert *et al.*, 1984), the Patlak graphical method (Patlak *et al.*, 1983), and spectral analysis (Cunningham and Jones, 1993). In this work, the spectral analysis approach was chosen because of its general applicability. The spectral analysis model describes the tissue time–activity curve y by a linear combination of basis functions, each of which is a single exponential in time convolved with the arterial input function q:

$$y_j(t) = \sum_{k=1}^{N} q(t) \otimes \alpha_{jk} \exp(-\beta_k t), \quad \begin{array}{c} \alpha_k \geq 0 \\ \lambda \leq \beta_k \leq 1 \end{array} \quad (4)$$

where N is the maximum number of basis functions allowed in the model (typically 100) and λ is the decay constant of the radioisotope. Thus, substituting Eq. (4) into Eq. (3), the expression for projection data becomes

$$p_i(t) = \sum_{j \in J} b_{ij} \left\{ \sum_{k=1}^{N} q(t) \otimes \alpha_{jk} \exp(-\beta_k t) \right\}. \quad (5)$$

In solving for the unknown coefficients α, the β values are fixed and are chosen to cover the spectrum of expected kinetic behavior, from the slowest possible clearance (λ) to the fastest measurable dynamic (e.g., 1 sec^{-1}). The values of α are determined by fitting the model to projection data using the nonnegative least-squares algorithm (NNLS) (Lawson and Hanson, 1974), which yields an estimate of the impulse response function (IRF) for the heterogeneous mix of tissues lying along a given projection ray. This is equivalent to summing the IRF over all pixels that contribute to bin i:

$$\hat{h}_i(t) = \sum_{j \in J} b_{ij} \hat{h}_j(t). \quad (6)$$

C. Parametric Image Reconstruction

Because Eq. (6) is in the general form of the Radon transform, it can be inverted using FBP or an iterative reconstruction technique to produce parametric images of the IRF at any specified time point. Various kinetic parameters can be derived from the IRF. This chapter refers to images of the IRF at 1 min, which is proportional to K_1, the product of blood flow and extraction ratio, and the IRF at 60 min, which is proportional to K_i, the irreversible disposal rate constant, assuming steady state is achieved by reversible compartments within 60 min of tracer injection. Strictly, K_1 is defined as the IRF at time 0, but the intercept is very sensitive to measurement noise and transient vascular effects as discussed by Cunningham *et al.* (1993).

D. Patient Study

The methodology described was tested using the labeled anticancer drug [^{11}C]temozolomide. Temozolomide is an oral cytotoxic prodrug that degrades in physiological solutions to form the reactive methylating species MTIC [5-(3-methyltriazen-1-yl)imidazole-4-carboxamide], which is thought to achieve antitumor activity by binding to DNA guanine bases (Stevens *et al.*, 1987). The purpose of the PET study, which was conducted in parallel with a phase II clinical trial by the UK Cancer Research Campaign, was to evaluate the pharmacokinetics of temozolomide *in vivo* and gain insight into its assumed mechanism of action.

Data were acquired for 90 min on an ECAT 953B tomograph (CTI/Siemens, Knoxville, TN) with septa retracted. This scanner has 16 detector rings, a ring diameter of 76 cm, and an axial field of view of 10.8 cm. In this study, coincidence LoRs were formed from detector pairs with a maximum ring difference of 11, resulting in a maximum axial acceptance angle of 5.6°. The arterial concentrations of [^{11}C]temozolomide and its radioactive metabolites were assayed throughout the PET study using an on-line BGO counter (Ranicar *et al.*, 1991) and HPLC.

The 3D sinograms were normalized, corrected for scatter (Bailey and Meikle, 1994) and attenuation, and rebinned into a 2D dataset using the SSRB approximation (Daube-Witherspoon and Muehllehner, 1987). Parametric projections proportional to flow × extraction (K_1) and net rate of influx (K_i) were calculated using the method described and reconstructed using both 2D FBP (Hann filter, $f_{cutoff} = f_{nyquist} = 1.6$ cm^{-1}) and the ordered subsets expectation maximization (OS-EM) algorithm (Hudson and Larkin, 1994) (1 iteration, 16 subsets). For comparison, the same parametric images were also calculated using the conventional approach of fully 3D FBP (Kinahan and Rogers, 1989) and image-based spectral analysis.

III. RESULTS AND DISCUSSION

Parametric images of K_1 calculated using the various methods described are shown in Fig. 2 for four consecutive slices through the brain, including the tu-

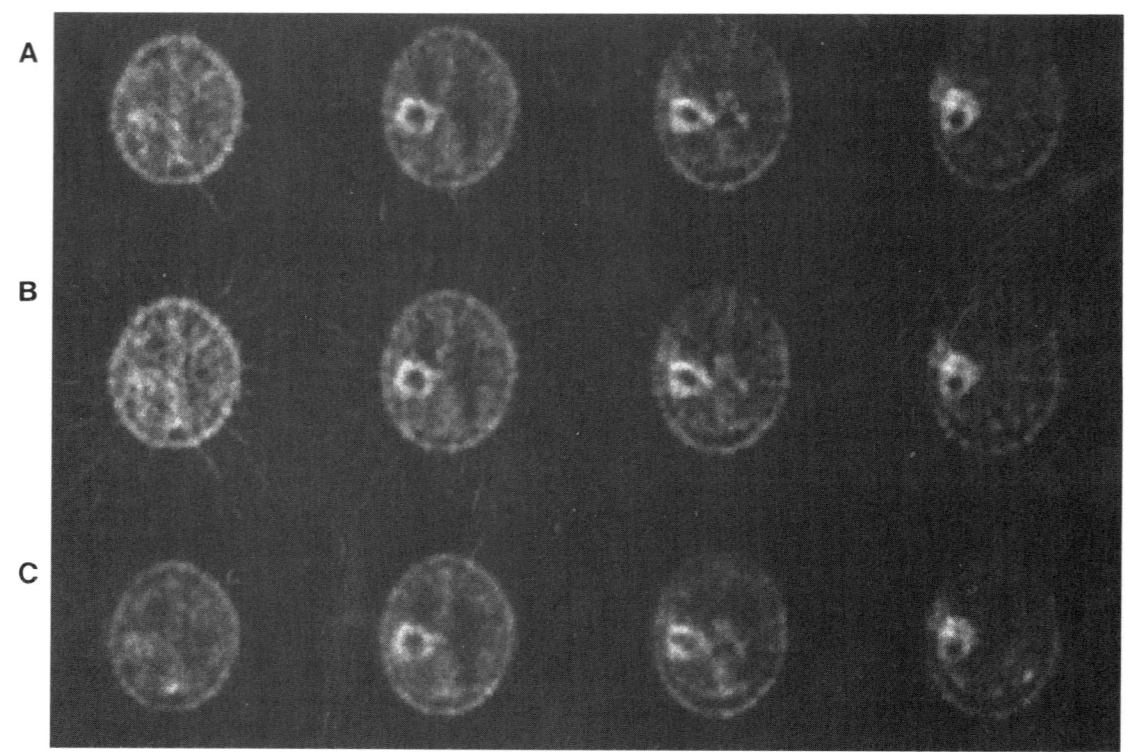

FIGURE 2 Parametric images proportional to K_1 (IRF at 1 min) for four slices through the brain, including tumor. The images were calculated using (A) fully 3D FBP followed by image-based spectral analysis, (B) spectral analysis of rebinned projection data followed by 2D FBP, and (C) spectral analysis of rebinned projection data followed by 2D OS-EM reconstruction.

mor. The K_1 values represented by the images in the two top rows (both using FBP) were plotted against each other for all voxels contained within the brain. The regression line was $y = 1.5 \times 10^{-5} + 0.999x$ and the relative differences, averaged over all brain voxels, were 1.6%. The corresponding images of K_i for the same brain slices are shown in Fig. 3. In this case, the regression line was $y = 1.7 \times 10^{-7} + 0.968x$ and the relative differences, averaged over all brain voxels, were −2.4%. Thus, the agreement between parameter estimates obtained by projection-based and image-based spectral analysis was very good.

There was also good qualitative agreement between the parametric images obtained using the two approaches, although with FBP the new method resulted in slightly noisier images. When OS-EM reconstruction of parametric projections was used (bottom row of Figs. 2 and 3), the SNR was noticeably improved compared with the images obtained using FBP. It should be noted that the noise characteristics of OS-EM images depend on the number of iterations and subsets chosen. One iteration and 16 subsets, which is approximately equivalent to 16 iterations of conventional EM, were used (Meikle *et al.*, 1994). Although this is a relatively small number of iterations and is likely to produce low noise images in favor of accuracy, it has been shown previously that EM reconstructions of kinetic parameter estimates converge rapidly, and after 16 iterations the bias is typically < 5% (Meikle *et al.*, 1997). It is also important to recognize that other iterative reconstruction algorithms may be more appropriate for reconstructing kinetic parameter estimates, as the noise distribution of parametric projections is almost certainly not Poisson. Other algorithms that may be better suited to this application include penalized weighted least squares (Fessler, 1994), provided a suitable noise model can be developed for parametric projection data.

The processing time to generate the parametric images shown in the bottom row of Figs. 2 and 3 was approximately 115 min on a Sun UltraSparc 170 (Sun Microsystems, Mountain View, CA), mainly due to the initial 3D normalization and scatter correction (90 min). This compares with over 7 hr to generate the same images using fully 3D FBP on a quad i860 SuperCard (Byars Consulting, Knoxville, TN) and image-based spectral analysis. Further gains in efficiency could be achieved by performing 3D rebinning prior to scatter correction. This would involve approximating the scat-

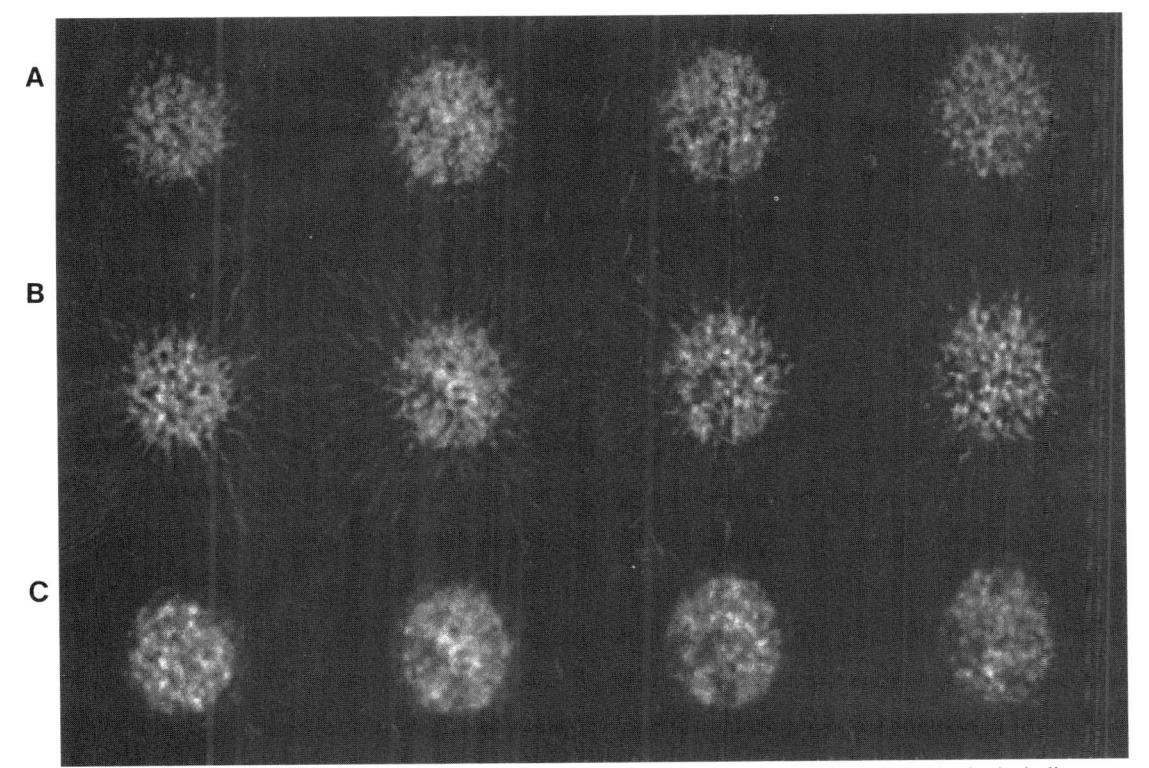

FIGURE 3 Parametric images proportional to K_i (IRF at 60 min) for four slices through the brain, including tumor. The images were calculated using (A) fully 3D FBP followed by image-based spectral analysis, (B) spectral analysis of rebinned projection data followed by 2D FBP, and (C) spectral analysis of rebinned projection data followed by 2D OS-EM reconstruction.

ter contribution to oblique planagrams (group of projections with the same polar angle ϕ and the same ring difference δ) by the scatter estimate obtained for direct planagrams, an approximation that is commonly made when implementing 3D scatter correction in practice (Bailey and Meikle, 1994). It should be noted that the exact order of performing these processing steps may depend on the particular algorithms employed. In particular, the commutability of the FORE algorithm has not yet been fully investigated.

IV. CONCLUSIONS

Spectral analysis of rebinned 3D projection data, combined with OS-EM reconstruction, enables efficient calculation of parametric images with high SNR. Furthermore, the method is applicable to a wide range of tracers due to the generality of the spectral analysis model. Future work will attempt to model the noise distribution of parameter estimates calculated from projections and explore alternative reconstruction algorithms incorporating such noise models.

Acknowledgments

This work was supported by the UK Cancer Research Campaign (CRC) (Grant SP2 193/0101). Temozolomide was developed by the CRC and is now licensed to Shering Plough. The authors thank Dr. Cathryn Brock for use of the patient data and the chemistry staff of the MRC Cyclotron Unit for preparation and metabolite analysis of [^{11}C]temozolomide.

References

Alpert, N. M., Eriksson, L., Chang, J., Bergstrom, M., Litton, J. E., Correia, J. A., Bohm, C., Ackerman, R. H., and Taveras, J. M. (1984). Strategy for the measurement of regional cerebral blood flow using short-lived tracers and emission tomography. *J. Cereb. Blood Flow Metab.* **4**: 28–34.

Bailey, D. L., and Meikle, S. R. (1994). A convolution-subtraction scatter correction method for 3D PET. *Phys. Med. Biol.* **39**: 411–424.

Bailey, D. L., Jones, T., Spinks, T. J., Gilardi, M. C., and Townsend, D. W. (1991). Noise equivalent count measurements in a neuro-PET scanner with retractable septa. *IEEE Trans. Med. Imag.* **10**: 256–260.

Cherry, S. R., Dahlbom, M., and Hoffman, E. J. (1991). 3-D positron emission tomography using a conventional multi-slice tomograph without septa. *J. Comput. Assist. Tomogr.* **15**: 655–668.

Cunningham, V. J., and Jones, T. (1993). Spectral analysis of dynamic PET studies. *J. Cereb. Blood Flow Metab.* **13**: 15–23.

Cunningham, V. J., Ashburner, J., Byrne, H., and Jones, T. (1993). Use of spectral analysis to obtain parametric images from dynamic PET studies. *In* "Quantification of Brain Function. Tracer Kinetics and Image Analysis in Brain PET" (K. Uemura, N. A. Lassen, T. Jones, and I. Kanno, eds.), pp. 101–108. Elsevier, Amsterdam and New York.

Daube-Witherspoon, M. E., and Muehllehner, G. (1987). Treatment of axial data in three dimensional PET. *J. Nucl. Med.* **28**: 1717–1724.

Defrise, M. (1995). A factorization method for the 3D x-ray transform. *Inverse Probl.* **11**: 983–994.

Defrise, M., Kinahan, P. E., Townsend, D. W., Michel, C., Sibomana, M., and Newport, D. F. (1997). Exact and approximate rebinning algorithms for 3-D PET data. *IEEE Trans. Med. Imag.* **16**: 145–158.

Fessler, J. A. (1994). Penalized weighted least-squares image reconstruction for positron emission tomography. *IEEE Trans. Med. Imag.* **13**: 290–300.

Huang, S. C., Carson, R. E., and Phelps, M. E. (1982). Measurement of local blood flow and distribution volume with short-lived isotopes: A general input technique. *J. Cereb. Blood Flow Metab.* **2**: 99–108.

Hudson, H. M., and Larkin, R. S. (1994). Accelerated image reconstruction using ordered subsets of projection data. *IEEE Trans. Med. Imag.* **13**: 601–609.

Kinahan, P. E., and Karp, J. S. (1994). Figures of merit for comparing reconstruction algorithms with a volume-imaging PET scanner. *Phys. Med. Blol.* **39**: 631–642.

Kinahan, P. E., and Rogers, J. G. (1989). Analytic 3-D image reconstruction using all detected events. *IEEE Trans. Nucl. Sci.* **36**: 964–968.

Lawson, C. L., and Hanson, R. J. (1974). "Solving Least Squares Problems." Prentice-Hall, Englewood Cliffs, NJ.

Meikle, S. R., Hutton, B. F., Bailey, D. L., Hooper, P. K., and Fulham, M. J. (1994). Accelerated EM reconstruction in total body PET: Potential for improving tumour detectability. *Phys. Med. Biol.* **39**: 1689–1704.

Meikle, S. R., Matthews, J., Cunningham, V. J., Bailey, D. L., Livieratos, L., Jones, T., and Price, P. (1997). Spectral analysis of PET projection data. *In* "1996 IEEE Nuclear Science Symposium Conference Record" (A. Del Guerra, ed.), pp. 1888–1892. IEEE, Piscataway, NJ.

Patlak, C. S., Blasberg, R. G., and Fenstermacher, J. D. (1983). Graphical evaluation of blood-to brain transfer constants from multiple-time uptake data. *J. Cereb. Blood Flow Metab.* **3**: 1–7.

Ranicar, A. S. O., Williams, C. W., Schnorr, L., Clark, J. C., Rhodes, C. G., Bloomfield, P. M., and Jones, T. (1991). The on-line monitoring of continuously withdrawn arterial blood during PET studies using a single BGO/photomultiplier assembly and non-stick tubing. *Med. Prog. Technol.* **17**: 259–264.

Stevens, M. F. G., Hickman, J. A., and Langdon, S. P. (1987). Antitumour activity and pharmacokinetics in mice of 8-carbamoyl-3-methyl-imidazo [5,1-d]-1,2,3,5-tetrazin-4 (3H)-one (CCRG 91045; M & B 39831), a novel drug with potential as an alternative to dacarbazine. *Cancer Res.* **47**: 5846–5852.

Townsend, D. W., Spinks, T., Jones, T., Geissbuhler, A., Defrise, M., Gilardi, M. C., and Heather, J. (1989). Three dimensional reconstruction of PET data from a multi-ring camera. *IEEE Trans. Nucl. Sci.* **36**: 1056–1065.

Tsui, E., and Budinger, T. F. (1978). Transverse section imaging of mean clearance time. *Phys. Med. Biol.* **23**: 644–653.

CHAPTER 8

Multislice PET Quantitation Using Three-Dimensional Volumes of Interest

R. H. HUESMAN, G. J. KLEIN, B. W. REUTTER, and X. TENG

Center for Functional Imaging
Lawrence Berkeley National Laboratory
University of California
Berkeley, California 94720

Volume of interest extraction for radionuclide and anatomical measurements requires correct identification and delineation of the anatomical feature being studied. A toolset for specifying three-dimensional (3D) volumes of interest (VOIs) on a multislice positron emission tomography (PET) data set has been developed. The software is particularly suited for specifying cerebral cortex VOIs that represent a particular gyrus or deep brain structure. A registered 3D magnetic resonance (MR) image data set is used to provide high-resolution anatomical information, both as oblique 2D sections and as volume renderings of a segmented cortical surface. VOIs are specified indirectly in 2D by drawing a stack of 2D regions on the MR data. The regions are tiled together to form closed triangular mesh surface models that are subsequently transformed into the observation space of the PET scanner. Quantitation by this method allows calculation of radionuclide activity in the VOIs, as well as their statistical uncertainties and correlations. A complete treatment of this work, including validation, appears in Klein et al. (1997).

I. INTRODUCTION

Quantitative analysis of multislice positron emission tomography (PET) data sets using regions of interest is a standard technique for studying brain function. A significant aspect of this technique is the process by which one identifies a desired portion of anatomy and then specifies its boundaries as a region of interest on a single slice, or a volume of interest (VOI) on a stack of slices. In particular, the availability of multislice scanners with true 3D imaging capabilities poses new challenges for 3D data analysis. Numerous issues related to this process exist that could skew quantitative results if not addressed properly. Because PET data represent functional, not necessarily anatomical, information, data from a PET scanner are not always appropriate for specifying regions. Functional boundaries seen on PET may not correlate with anatomical boundaries, hence relying on these boundaries while specifying regions of interest could easily introduce quantitation biases. Even where functional boundaries correspond to anatomical boundaries, the resolution obtained from the highest resolution PET scanners is often inadequate to identify desired anatomical boundaries with confidence. For these reasons, many researchers have relied on other modalities, such as magnetic resonance (MR) imaging, to identify anatomy (Meltzer et al., 1990a; Strother et al., 1994; Pietrzyk et al., 1994). However, even with this multimodality approach, the problem is not solved completely. Typical MR data sets consist of 256 × 256 × 96 voxels and are usually displayed as 256 × 256 pixel images, one slice at a time. Trying to identify a region of anatomy, e.g., a particular gyrus in the cortex, from this slice-based data can be quite difficult even for experienced clinicians. Additional information is required to aid in the 3D navigational task and to convey appropriate cues about the 3D nature of the anatomy.

This chapter describes a methodology for specifying meaningful 3D VOIs on PET data sets. The methodology is particularly suited for the calculation of radiotracer activity and statistical uncertainty in cerebral cortex VOIs representing particular gyri or deep brain structures. The approach addresses two problems that have plagued the specification of such regions in the past: (1) the proper identification of a desired anatomical object from a functional PET image, and (2) the specification of a true 3D boundary around that object once it is identified using conventional X-Windows interfaces. Unique in the approach is the method by which the regions can be used to model accurately the statistical uncertainty of quantified activity in a 3D acquisition environment.

II. METHODS

The procedure for obtaining quantitative PET VOI values can be summarized as follows. Three-dimensional regions of interest are identified and specified using high resolution anatomical data from MR images. VOI boundaries are specified on the MR data set by drawing 2D regions on a set of parallel image planes. The parallel set of planes may be chosen at any oblique slicing orientation through the MR image volume such that cross sections most clearly show features of the desired anatomical object. Two-dimensional regions are specified on these planes by drawing a freehand polygon or by laying out points that are then connected as a continuous cubic spline. Region drawing on this set of planes is facilitated by the display of a "3D cursor," which shows the position not only on the current drawing plane, but also on a number of slices orthogonal to the parallel planes, and on one or more volume renderings of a segmented cortical surface. Further feedback is provided by displaying the intersections of selected regions with these orthogonal slices. Three-dimensional VOIs are formed by tiling each set of parallel 2D regions into a closed triangular mesh surface model.

The MR data are registered to multislice PET data using a sequence of manual and automated techniques. A 4×4 matrix is calculated that describes the transformation in homogeneous coordinates between the registered MR image and PET, as well as between original and obliquely resliced MR images. This matrix is used to transform the surface model from the resliced MR coordinate system into the coordinate system of the PET gantry. Once transformed, the VOI surface models may be used to calculate the activity within a 3D volume inside the PET gantry. In the application described in this chapter, the surface models are currently projected onto the originally acquired PET slices, resulting in a series of labeled 2D regions. Quantitation of activity within these 2D regions and their uncertainty is achieved by projecting the regions into a tomographic sinogram (i.e., Radon, or projection) space and then directly evaluating the counts in this space. Because the full region covariance matrix is available after this calculation, the 2D region values can be added together, giving an activity value and uncertainty for each PET VOI.

Data used for this research were acquired using a CTI/Siemens ECAT EXACT HR PET scanner (Wienhard et al., 1994) and a 1-m-bore 0.5-T Oxford MR scanner magnet with a spectrometer built at the authors' laboratory (Roos et al., 1988). MR image volumes comprised a 3D data set of T1-weighted images (voxel size $1 \times 1 \times 2$ mm, volume size $256 \times 256 \times 96$ voxels) acquired using a 3D gradient-recalled echo sequence (TE = 14.3 msec, TR = 30.0 msec). PET data were obtained using the 47-slice scanner in 2D acquisition mode (slice separation = 3.125 mm) imaging the radiotracer [^{18}F]fluorodeoxyglucose (FDG). Data were reconstructed for use in the segmentation and registration process using standard 2D-filtered backprojection techniques (voxel size $2.4 \times 2.4 \times 3.1$ mm, volume size $128 \times 128 \times 47$ voxels) and a Hanning filter with 0.4 cycle/pixel cutoff, corresponding to a 5.5-mm transaxial resolution in the center. To correct for the effects of attenuation, data were obtained using a ^{68}Ge rod source in a 20-min transmission scan and a 60-min blank scan and combined to produce appropriate correction factors. A normalization file was used to correct each emission, transmission, and blank sinogram on a bin-by-bin basis.

A. Segmentation/Registration

In order to relate MR-based regions to PET measurements, it is necessary to register the two data sets spatially. Automated methods for this process exist, but most require that the brain be segmented from non-brain regions in the MR data (Pelizzari et al., 1989; Woods et al., 1993). Some have claimed success using completely automated methods to perform this segmentation (Ardekani et al., 1994); however, such techniques require special pulse sequences or complex clustering algorithms. In the authors' experience, finely tuning parameters to obtain a successful segmentation via completely automated results can be quite tedious. A simpler approach can be taken by relying on the facts that the outer cortex in FDG PET images can be segmented easily from the background using image thresholding and that the PET and MR data sets can be approximately registered relatively quickly using manual techniques.

Once an approximate registration is found, PET data are resliced at the sampling resolution of the MR image volume and are thresholded to form a binary mask, as suggested by Pietrzyk (1994). To prevent slight misregistration from masking away brain regions in the

MR image, the PET mask is usually dilated using a morphological operator (Peters, 1991). Further, to prevent masking the inner portions of the MR brain, the outer boundaries of the PET mask are filled using a 2D filling operation. The resulting masked MR data set is a nearly complete segmentation; however, because the PET mask generally includes some portions of the outer tissue, a 3D region-growing operation seeded from the interior of the cortex is used to obtain the final result. A final step in obtaining an accurate registration is the use of the segmented MR data set in an automated minimum variance of ratios registration technique (Woods et al., 1993). Note that this technique requires that the brain is entirely within the field of view of the PET scanner.

B. Volume of Interest Construction

Rather than attempting to create a sophisticated virtual reality environment for sculpting 3D VOIs directly, the proposed approach uses conventional 2D X-Windows interfaces that indirectly specify surfaces through a sequence of 2D operations, i.e., VOIs are constructed by drawing stacks of 2D regions. The region drawing environment makes use of two main principles to carry out this task. First, because the cross-sectional 2D geometry of an object boundary can usually be simplified just by reslicing along a different orientation, the user can select a set of parallel slicing planes at an angle different from the original acquisition planes. For cortical regions, the typical reslicing orientation is the coronal view of transaxially acquired MR images. However, in general, these reslice orientations can be at any oblique angle with respect to the original acquisition orientation. Second, to aid in the 3D navigational task, simultaneous views of data are provided in different formats: volume-rendered surfaces, orthogonal slices, or registered PET slices, on which corresponding points can be visually related.

Figure 1 shows an example of the region drawing environment. Structured around the VIDA software package (Hoffman et al., 1992), a main window (Fig. 1a), hereafter called the drawing plane, is provided for the user to draw 2D regions. Regions may be drawn using freehand polygons, laying out points connected via a cubic spline algorithm, or a number of other techniques. Auxiliary viewing planes sliced at orthogonal angles (e.g., sagittal and transverse when the drawing plane is coronal) can be seen as well. A 3D cursor reflecting the position of the drawing cursor is projected on these views using a parallel projection technique. As a stack of 2D regions is drawn, their position with respect to one another may be displayed by showing the intersection with the auxiliary orthogonal planes (Fig. 1b). Additionally, to provide guiding points while drawing regions on the drawing plane, curves may be drawn on the auxiliary views, and their intersection with the drawing plane will be displayed.

For specifying cortical regions, probably the most useful visual cue is a rendering of the cortical surface. Volume renderings are calculated using a parallel projection gradient-shaded technique on segmented MR data. Because a depth map calculated during the rendering is retained along with the corresponding transformation matrix, the 3D position of each point in the rendered surface can be calculated. Therefore, visual cues may be provided in two ways. First, a 3D polyline may be drawn on the rendered brain surface, and its intersections with the 2D drawing plane will be shown (Fig. 2). This technique is useful for following a specific cortical gyrus through subsequent 2D slices. A second method provides real-time feedback between the drawing cursor and the volume rendering. As the cursor is moved in the drawing plane, its position in the rendering may be displayed as the projection of the cursor position along the line of sight used to obtain the rendering. Therefore, as the cursor is moved along the outer boundary of the cortex in the drawing plane, its correct position is seen on the volume rendering of the cortical surface, permitting accurate identification of the cortical gyrus.

A typical set of 2D contours drawn for a brain data set is seen in Fig. 3a. The contours are tiled together using the NUAGES (Boissannat and Geiger, 1992) algorithm to produce a triangular mesh surface model (Fig 3b). The surface mode is integrated into an Inventor Toolkit 3D graphical display environment (Wernecke, 1994).

C. PET Quantitation

Quantitation of PET activity could take place directly without construction of VOI surface models by reslicing calibrated PET data into the voxel space of MR data and summing voxels contained within the boundaries of the 2D regions. Of course, care must be taken to scale calibration factors properly in accord with the new voxel size of the resliced PET data and to treat voxels on the border of the region suitably. This is the approach, for example, taken by Resnick and co-workers (1993). Besides the possible errors induced from summing edge voxels, there is one main disadvantage to this technique: the uncertainty of the activity data can no longer be characterized accurately; only an approximation is possible (Carson et al., 1993). If an accurate estimate of the activity within the region *and* its uncertainty are desired, calculations are easier in the projection, or sinogram, space of the tomograph,

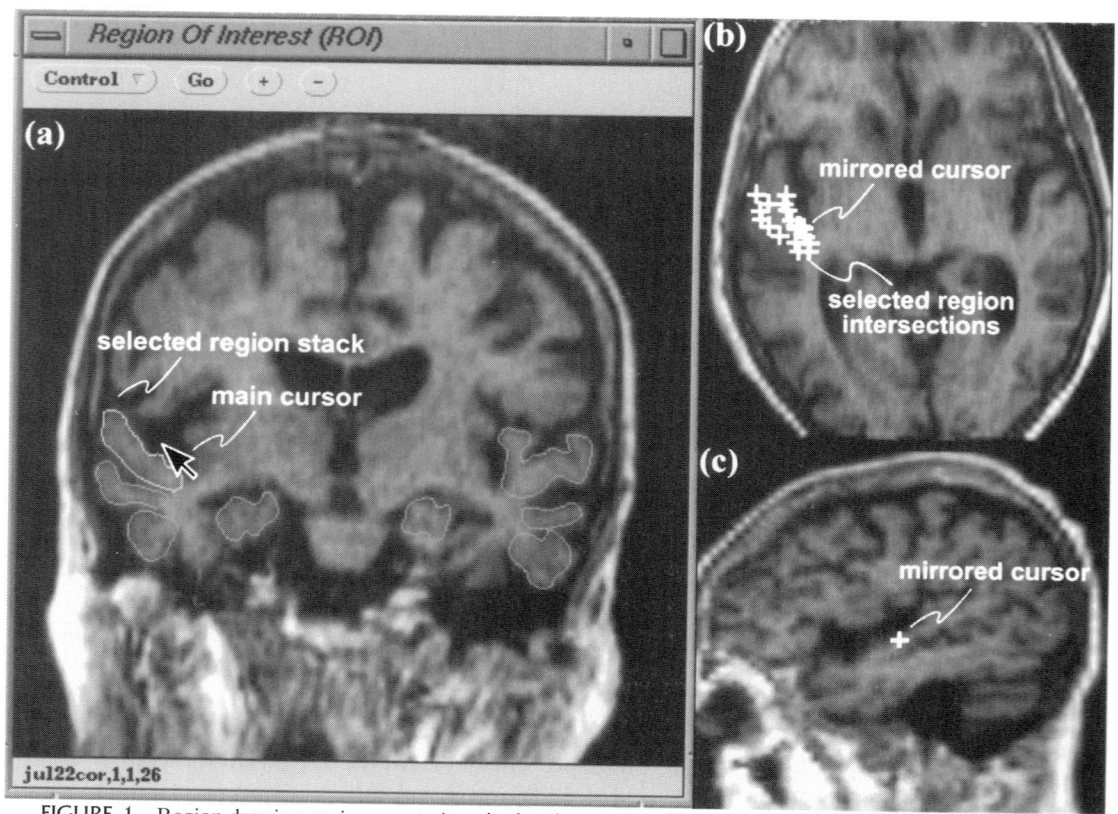

FIGURE 1 Region drawing environment. A main drawing window (a) is used to draw 2D regions on obliquely sliced MR sections. The main window cursor position is mirrored in real time on auxiliary views (b, c) as a projected 3D crosshair cursor to aid visualization of the 3D anatomy (cursor size is enlarged here for emphasis). Intersections of a selected region stack with the auxiliary plane (b) give the user intuition of the 3D shape of the resulting VOI.

where the statistical properties are well established (Huesman, 1984).

The approach described in this chapter resembles the formulation of Votaw (1996), which generalized Huesman's 2D ROI algorithm (1984). Define E_{lmz}, T_{lmz}, B_{lmz}, and N_{lmz} as the projection values at bin l, angle m, and slice z for the emission, transmission, blank, and normalization sinograms, respectively. A normalization factor for each slice, k_z, incorporates corrections for dead time, radiotracer decay, and scan duration and is used to convert reconstructed units into calibrated PET counts/sec. Attenuation factors, A_{lmz}, are calculated from the ratio of smoothed, normalized copies of B_{lmz} and T_{lmz}. The corrected projection bin value is then defined by

$$p_{lmz} = E_{lmz} N_{lmz} A_{lmz} k_z. \quad (1)$$

A voxel in the image space of a reconstructed PET volume is given by

$$I_{xyz} = \sum_{km} F_{xy}^{km} \sum_j C_k^j \sum_l R_j^l p_{lmz}, \quad (2)$$

where F_{xy}^{km} are the 2D backprojection factors, I_{xyz} is the voxel value at location (x, y, z), C_k^j is the convolution kernel, R_j^l are the arc correction rebinning factors, and p_{lmz} are projection data. Volume-of-interest activity in this voxelized space is then given by

$$V_\alpha = \sum_z V_\alpha^z = \sum_z \sum_{x,y \in \alpha_z} I_{xyz}, \quad (3)$$

where α_z denotes the intersection of the VOI indicated by α and the transverse section indicated by z. Changing the order of summation and rearranging as in Huesman (1984),

$$V_\alpha^z = \sum_{lm} \sum_j R_l^{Tj} \sum_k C_j^k \sum_{x,y \in \alpha_z} F_{xy}^{km} p_{lmz} = \sum_{lm} h_{\alpha_z}^{lm} p_{lmz}, \quad (4)$$

where

$$h_{\alpha_z}^{lm} = \sum_j R_l^{Tj} \sum_k C_j^k \sum_{x,y \in \alpha_z} F_{xy}^{km}. \quad (5)$$

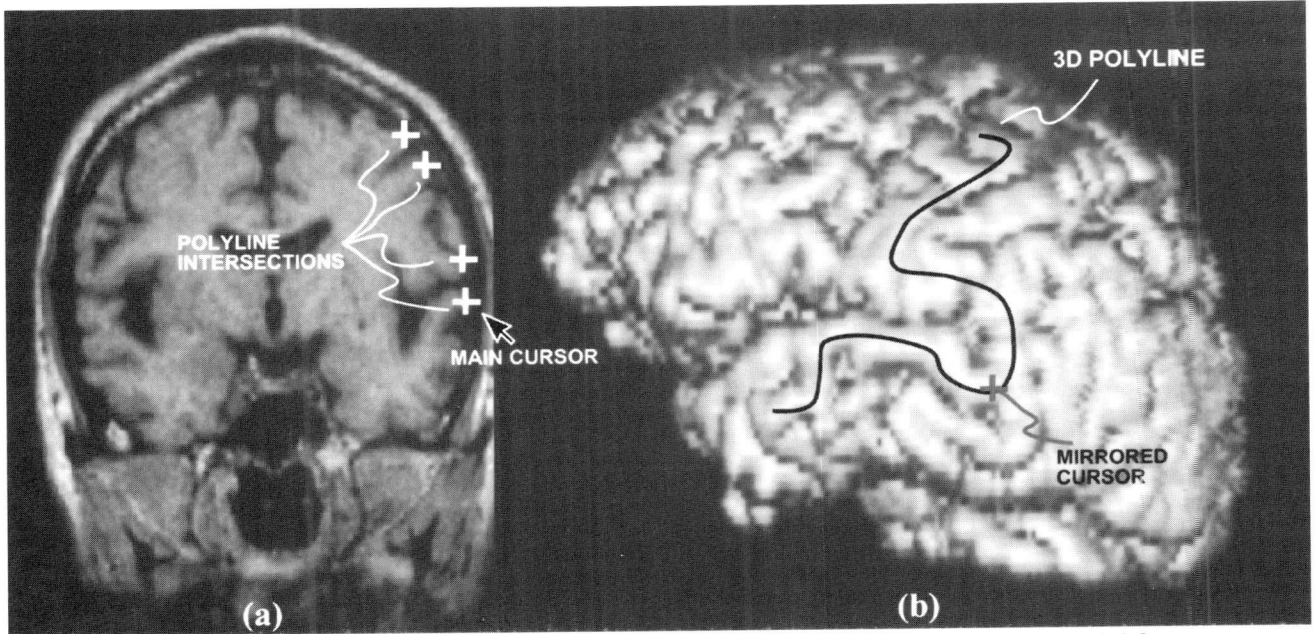

FIGURE 2 Volume rendering-based navigation. Three-dimensional positional information is recorded for each pixel in the volume rendering (b), allowing an interface tying together 2D sectional (a) with rendered information. Intersections of a polyline drawn on the rendering are shown as crosses on the 2D view. Position of the cursor in the 2D view is reflected in the rendering in real time.

To obtain a suitable description of each processed VOI in projection space, i.e., a set of sinograms containing the factors $h^{lm}_{\alpha_z}$, the 2D intersection of the VOI surface model with each acquisition plane is calculated (after PET-MR registration), resulting in another set of 2D regions described as closed polygons. Note at this point that the regions are not linked to the voxel space of a reconstructed PET volume at all but, instead, are real-valued x, y coordinate descriptions in the space of the PET scanner. As was suggested by Huesman (1984),

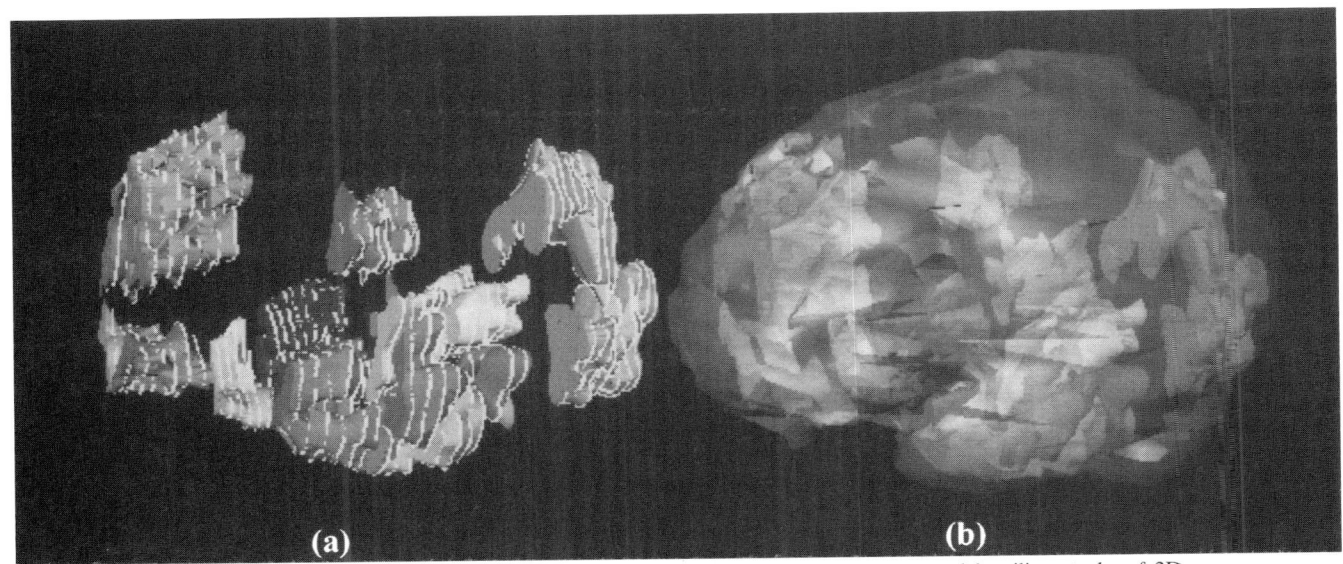

FIGURE 3 Volumes of interest. Three-dimensional volumes of interest are created by tiling stacks of 2D contours. (a) Contours are overlaid on the resulting VOI surfaces. (b) A typical set of VOIs drawn for a brain study in a schematic rendering of the cortical surface.

the forward projection of a given region in this description becomes a continuous integral over the uniformly weighted interior of the polygons,

$$\sum_{x,y \in \alpha_z} F_{xy}^{km} \to g_{\alpha_z}^{km} = \int_{\alpha_z} F^{km}(x,y) dx\, dy, \quad (6)$$

where $F^{km}(x, y)$ denotes an indicator function for the ray in projection space at bin k and angle m. In practice, this integral can be calculated quickly by taking each polygon line segment in turn and summing the signed area of the trapezoidal region defined by the boundaries of line segment and the projection bins, as seen in Fig. 4a. Therefore, Eq. (5) becomes

$$h_{\alpha_z}^{lm} = \sum_j R_l^{Tj} \sum_k C_j^k g_{\alpha_z}^{km}. \quad (7)$$

Figure 4b summarizes the overall calculation.

It is assumed that each individual bin in the sinograms is an independent random variable modeled as a Poisson counting process and the approximation is made that the normalization and smoothed attenuation factors are without statistical variation. The emission sinogram values are collected as $E_{lmz} = E_{lmz}^P - E_{lmz}^R$, where E_{lmz}^P and E_{lmz}^R are the emission prompt and random values, respectively. To estimate the number of random coincidences in each projection bin, E_{lmz}^R, total random events, E_z^{RTOT}, are recorded for each 2D sinogram so that $E_{lmz}^R \approx E_z^{RTOT}/LM$, where L and M are the dimensions of the 2D sinogram. The variance of a single corrected bin is thus

$$\text{var}(p_{lzm}) = N_{lmz}^2 A_{lmz}^2 k_z^2 \left(E_{lmz} + \frac{2E_z^{RTOT}}{LM} \right), \quad (8)$$

so that for two regions on the same slice, the covariance is given by

$$\text{cov}(V_\alpha^z, V_\beta^z) = \sum_{lm} h_{\alpha_z}^{km} h_{\beta_z}^{km} \text{var}(p_{lmz}). \quad (9)$$

Regions on different slices are uncorrelated so that the covariance for two multislice VOIs is

$$\text{cov}(V_\alpha, V_\beta) = \sum_z \text{cov}(V_\alpha^z, V_\beta^z). \quad (10)$$

Calculation of VOI activity by this method has a number of other advantages besides the capability of obtaining statistical properties. Because high resolution anatomical data were used to define VOI boundaries, PET data do not need to be smoothed greatly to obtain suitable visual image quality. Only a ramp filter is used, preserving spatial resolution during quantitation. Also,

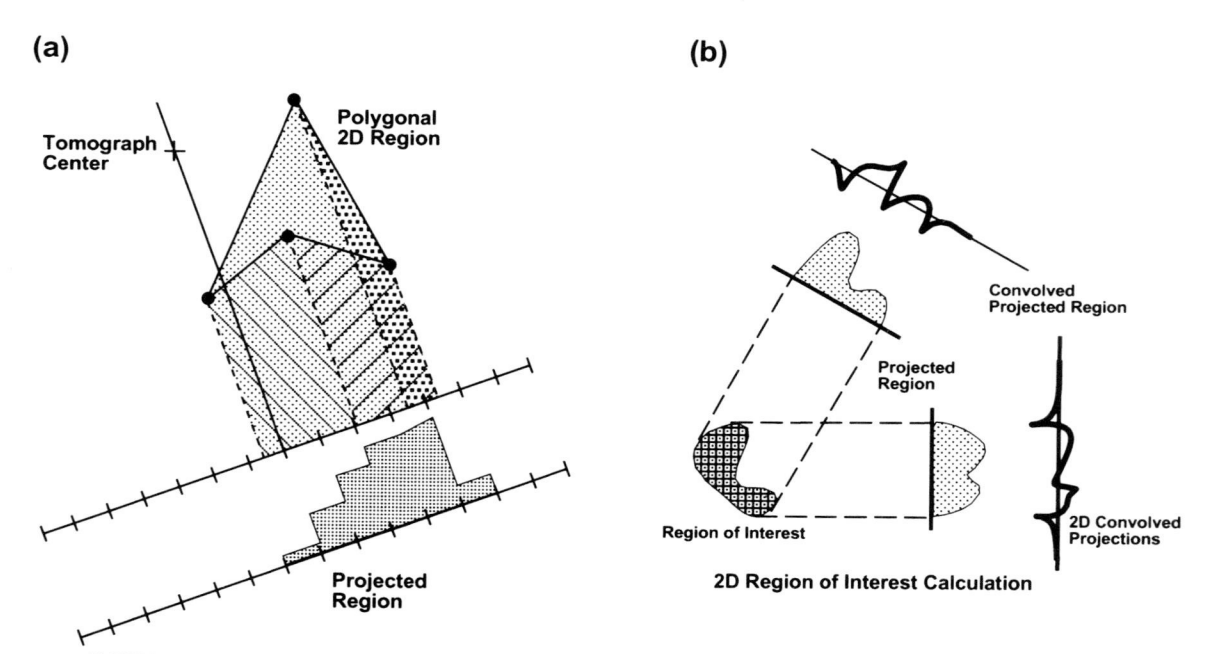

FIGURE 4 Projection of a uniform polygonal region. The uniformly weighted interior of a 2D polygonal region is computed by calculating the signed area of the trapezoids formed by each line segment of the polygon and the projection bins. A simple polygon is shown to convey this idea (a). This procedure is carried out for each projection angle, rebinned, and convolved to obtain a sinogram representing the VOI mask in projection space (b). In practice, the 2D regions are defined by numerous short line segments so that the region boundaries approximate smooth curves.

because the technique effectively performs a fast reconstruction and summing of data, reconstruction of a PET image volume is required only to register data. For dynamic PET acquisitions, it is therefore unnecessary to reconstruct every time point in data acquisition. Finally, a third advantage is that calculation of VOI values for a 3D PET acquisition without septa would proceed in a straightforward manner. Computational load and memory requirements grow considerably in the case of 3D analysis, however, so the overall task of quantifying 3D would not be trivial.

III. DISCUSSION

This research investigates several related aspects of PET VOI quantitation. An approach that allows reasonably fast manual drawing of 3D VOIs suitable for subsequent calculation of PET activity, uncertainties, and VOI correlations has been described. The approach makes the implicit assumption that using anatomical information from MR images is desirable while obtaining VOIs. It is recognized however, that there are other PET analysis applications where use of this anatomical information may not be warranted. Nevertheless, anatomical information is often desirable and even necessary for some forms of data analysis, including dynamic studies, neuroreceptor studies, and testing specific anatomically driven neuroscience hypotheses.

An issue that exists once a structure has been confidently identified is the strategy for sampling PET data from that region. Due to the limited resolution of the PET scanner, activity seen at a given point in a PET image is the spatial convolution of activity in the neighborhood of that point. For example, in typical VOIs of cortical gray matter, there is both spill out of gray matter activity within the borders of the region as well as spill in from the activity of nearby gray and white matter. One approach to dealing with this problem is to define regions well within the anatomical borders to avoid the spillover effects on the boundaries (Hoffman et al., 1979; Mazziotta et al., 1991). However, this strategy may not be adequate because PET activity is seldom uniform throughout an anatomical structure, and thus sampling throughout the entire structure is usually desired to characterize it. An alternate region placement approach that deals with both spillover effects and the uniform sampling requirement is a partial volume correction technique. These techniques define a region along the anatomical borders and model the spatial convolution using prior knowledge of cerebrospinal fluid and brain matter distributions. Partial volume correction approaches have been described by Meltzer and others (1990b, 1996; Müller-Gärtner et al., 1992). In view of these partial volume issues, VOIs specified along true anatomical boundaries should therefore be thought of as a starting point for VOI analysis. For accurate quantitation of tissue within these boundaries, the VOIs can be uniformly "shrunk" or eroded to avoid spillover or corrected using prior information for partial volume effects.

This research is concerned with how to obtain the actual 3D boundaries of anatomical structures. While manually drawing 3D regions can be a time-consuming process (1–2 hr for a set of 48 VOIs on a brain data set) and also has the potential for introducing operator biases, the authors believe it has advantages over the other approach of using a standard region template. Although a template may attempt to describe a given anatomical region set more completely and uniformly, unsuitable elastic transformations to best fit the anatomy could produce unacceptable position errors for some regions. This is a problem in analysis of brain data from patients with significant anatomical variability due to cerebral atrophy, infarction, or tumors. It is for this reason that the manual technique was chosen.

IV. CONCLUSIONS

The technique reported for specifying and analyzing VOIs on PET data sets demonstrates an approach for analyzing complex 3D data sets using common 2D interfaces. Navigation through the data set to find a desired anatomical structure can be greatly simplified using a registered MR anatomical volume showing multiple simultaneous oblique sections and volume renderings of data. Because most clinicians can readily identify specific sulci from high-quality renderings of the cortical surface, a crucial step in quickly identifying sulci in 2D sectional data is providing a feedback mechanism between renderings and sectional data. Once identified, 3D VOIs may be specified on anatomical data sets efficiently by drawing a stack of 2D regions subsequently tiled together to form a VOI surface model. The voxel-independent description of the VOIs allows a quantitative analysis in the observation space of the PET scanner for characterization of both VOI radiotracer activity and statistical properties.

Results obtained from typical brain analyses indicate that the stack of regions defining a VOI may be drawn on slices oriented at the oblique slicing direction that best allows visualization of the cross section for a desired structure. A 3- to 7-mm slicing separation appears sufficient to capture the salient shape features of regions in the cortex. VOIs drawn using a finer slice operation produced similar quantitative results at the cost of increased manual intervention.

Registration and segmentation steps are seen to be crucial preprocessing steps in the VOI analysis. An accurate segmentation of brain from nonbrain structures is needed for high-quality surface renderings and for automated registration routines. Validation results not shown here (Klein *et al.*, 1997) imply that registration between PET and MR data sets must be achieved to an accuracy better than 2 mm. The authors' experiences show that this level of accuracy is difficult to obtain quickly using purely manual techniques. However, by combining manual with automated registration techniques, it is possible to obtain reliable registration with minimal manual burden.

Acknowledgments

The authors thank Uwe Pietrzyk and Roger Woods, who were extremely helpful in providing their image registration software. This work was supported in part by the Director, Office of Energy Research, Office of Biological and Environmental Research, Medical Applications and Biophysical Research Division of the U.S. Department of Energy under contract No. DE-AC03-76SF00098 and in part by the U.S. Department of Health and Human Services under Grants HL25840, AG12435, and AG05890.

References

Ardekani, B. A., Braun, M., Kanno, I., and Hutton, B. F. (1994). Automatic detection of intradural spaces in MR images. *J. Comput. Assist. Tomogr.* **18**(6): 963–969.

Boissannat, J., and Geiger, B. (1992). "Three-dimensional Reconstruction of Complex Shapes Based on the Delaunay Triangulation," Tech. Rep. No. 1697. Institut National De Recherche En Informatique et en Automatique (INRIA), Sophia-Antipolis, France.

Carson, R. E., Yan, Y., Daube-Witherspoon, M. E., Freedman, N., Bacharach, S. L., and Herscovitch, P. (1993). An approximation formula for the variance of PET region-of-interest values. *IEEE Trans. Med. Imag.* **12**: 240–250.

Hoffman, E. A., Gnanaprakasam, D., Gupta, K. B., Hoford, J. D., Kugelmass, S. D., and Kulawiec, R. S. (1992). VIDA: An environment for multidimensional image display and analysis. *Commun. ACM* **20**: 693–702.

Hoffman, E. J., Huang, S., and Phelps, M. E. (1979). Quantiation in positron emission computed tomography. 1. Effect of object size. *J. Comput. Assist. Tomogr.* **3**: 299–308.

Huesman, R. H. (1984). A fast new algorithm for the evaluation of regions of interest and statistical uncertainty in computed tomography. *Phys. Med. Biol.* **25**: 543–552.

Klein, G. J., Teng, X., Jagust, W. J., Eberling, J. L., Acharya, A., Reutter, B. W., and Huesman, R. H. (1997). A methodology for specifying PET VOIs using multimodality techniques. *IEEE Trans. Med. Imag.* **16**: 405–414.

Mazziotta, J. C., Pelizzari, C. C., Chen, G. T., Bookstein, F. L., and Valentino, D. (1991). Region of interest issues: The relationship between structure and function in the brain. *J. Cereb. Blood Flow Metab.* **11**: A51–A56.

Meltzer, C. C., Bryan, R. N., Holcomb, H. H., Kimball, A. W., Mayberg, H. S., Sadzot, B., Leal, J. P., Wagner, H. N., Jr., and Frost J. J. (1990a). Anatomical localization for PET using MR imaging. *J. Comput. Assist. Tomogr.* **14**: 418–426.

Meltzer, C. C., Leal, J. P., Mayberg, H. S., Wagner, H. N., Jr., and Frost, J. J. (1990b). Correction of PET data for partial volume effects in human cerebral cortex by MR imaging. *J. Comput. Assist. Tomogr.* **14**: 561–570.

Meltzer, C. C., Zubieta, J. K., Links, J. M., Brakeman, P., Stumpf, M. J., and Frost, J. J. (1996). MR-based correction of brain PET measurements for heterogeneous gray matter radioactivity distribution. *J. Cereb. Blood Flow Metab.* **16**: 650–658.

Müller-Gärtner, H. W., Links, J. M., Prince, J. L., Bryan, R. N., McVeigh, E., Leal, J. P., Davatzikos, C., and Frost, J. J. (1992). Measurement of radiotracer concentration in brain gray matter using positron emission tomography: MRI-based correction for partial volume effects. *J. Cereb. Blood Flow Metab.* **12**: 571–583.

Pelizzari, C. A., Chen, G. T. Y., Spelbring, D. R., Weichselbaum, R. R., and Chen, C. T. (1989). Accurate three-dimensional registration of CT, PET, and/or MR images of the brain. *J. Comput. Assist. Tomogr.* **13**: 20–26.

Peters, R. A., III, (1991). "Software for 2d and 3d Mathematical Morphology," tech. rep. Air Force Office of Scientific Research. Arnold Engineering and Development Center, Arnold Air Force Base, TN.

Pietrzyk, U. (1994). A rapid three-dimensional visualisation technique to evaluate function in relation with anatomy of the human cortex. *In* "1993 IEEE Nuclear Science Symposium and Medical Imaging Conference Record" (L. Klaisner, ed.), Vol. 3, pp. 1810–1812. IEEE, Piscataway, NJ.

Pietrzyk, U., Herholz, K., Fink, G., Jacubs, A., Mielke, R., Slansky, I., Wurker, M., and Heiss, W. (1994). An interactive technique for three-dimensional image registration: Validation for PET, SPECT, MRI and CT brain studies. *J. Nucl. Med.* **35**: 2011–2018.

Resnick, S. M., Karp, J. S., Turetsky, B., and Gur, R. E. (1993). Comparison of anatomically-defined versus physiologically-based regional localization: Effects on PET-FDG quantitation. *J. Nucl. Med.* **34**: 2201–2207.

Roos, M. S., Mushlin, R. A., Veklerov, E., Port, J. D., Ladd, C., and Harrison, C. G. (1988). An instrument control and data analysis program configured for NMR imaging. *IEEE Trans. Nucl. Sci.* **36**: 988–992.

Strother, S. C., Anderson, J. R., Xu, X., Liow, J., Bonar, D. C., and Rottenberg, D. A. (1994). Quantitative comparisons of image registration techniques based on high-resolution MRI of the brain. *J. Comput. Assist. Tomogr.* **18**: 954–962.

Votaw, J. R. (1996). Signal-to-noise ratio in neuro activation PET studies. *IEEE Trans. Med. Imag.* **15**: 197–205.

Wernecke, J. (1994). "The Inventor Mentor: Programming Object-Oriented 3D Graphics with Open Inventor," Release 2. Addison-Wesley, Reading, MA.

Wienhard, K., Dahlbom, M., Eriksson, L., Michel, C., Bruckbauer, T., Pietrzyk, U., and Heiss, W. (1994). The ECAT EXACT HR: Performance of a new high resolution positron scanner. *J. Comput. Assist. Tomogr.* **18**: 110–118.

Woods, R. P., Mazziota, J. C., and Cherry, S. R. (1993). MR-PET registration with automated algorithm. *J. Comput Assist. Tomogr.* **17**: 536–546.

CHAPTER 9

Absolute PET Quantification with Correction for Partial Volume Effects within Cerebral Structures[1]

CLAIRE LABBÉ,* MATTHIAS KOEPP,[†,‡] JOHN ASHBURNER,[¶] TERRY SPINKS,*
MARK RICHARDSON,[†,‡] JOHN DUNCAN,[†,‡] and VINCENT CUNNINGHAM*

*Methodology and ‡Neuroscience Groups of the MRC Cyclotron Unit
Hammersmith Hospital, London W12 0NN, United Kingdom
†Epilepsy Research Group, Institute of Neurology, London, United Kingdom
¶Wellcome Department of Cognitive Neurology, London, United Kingdom

The accurate quantification of positron emission tomography (PET) data is limited by partial volume effects (PVE), which depend on the size of the structure studied in relation to the spatial resolution and which may lead to over- or underestimation of the true tissue tracer concentration. This study describes a new approach to measure accurately the true tracer activity within multiple volumes of interest (VOI) of any size and shape. This approach to PVE correction is based on (1) segmenting magnetic resonance (MR) images into anatomical probabilistic maps of cerebrospinal fluid and white and gray matter, with further subdivision into multiple VOIs; (2) convolving each segmented probabilistic MR map with a PET point spread function; and (3) solving the resultant linear set of equations, $Ax = b$, where b is the PET image, A contains all the convolved MR VOI images, and x contains the unknown tracer concentrations to be estimated. The solution is obtained using a linear least-squares (LLS) approach giving simultaneous estimates of all parameters corrected for PVE, without restriction on the number of VOIs. The method and the algorithm were validated with computer simulations and an anthropomorphic phantom model. The LLS method was applied to human [^{11}C]flumazenil (FMZ) PET studies of two patients with temporal lobe epilepsy and histologically verified unilateral hippocampal sclerosis. After PVE correction, the in vivo measurement of reduced FMZ volume of distribution within the hippocampus corresponded to the ex vivo, autoradiographically determined reduction of FMZ receptor availability. This PVE correction method allows accurate PET quantification in multiple small cerebral structures and has been directly validated with independent autoradiographic data.

I. INTRODUCTION

Positron emission tomography (PET) provides *in vivo*, quantitative measurements of physiological, biochemical, or pharmacological processes. Because of limited spatial resolution of PET, the accurate interpretation of PET data is obscured by partial volume effects (PVE) (Hoffman *et al.*, 1979; Mazziotta *et al.*, 1981). The cerebral PET signal reflects the average tracer concentration in each region, derived from different cerebral tissue compartments: gray matter (GM), white matter (WM), and cerebrospinal fluid (CSF). PVE are especially important in the case of cerebral atrophy. As a result, it is often impossible to determine the extent to which PET abnormalities represent changes of functional activity in the atrophic structure, structural changes, or a combination of the two.

Assessment of PVE requires a knowledge of the size and shape of the structures of interest, of their relative position in the scanner gantry, and of the presence of surrounding high or low activity (Hoffman *et al.*, 1979; Kessler *et al.*, 1984; Mazziotta *et al.*, 1981; Müller-Gärtner *et al.*, 1992). An improved three-compartment (i.e., GM, WM, and CSF) method, which took into account the inhomogeneity of PET resolution in all

[1] Transcripts of the BRAINPET97 discussion of this chapter can be found in Section VIII.

directions across the field of view (FOV) and which significantly affected the estimation of GM radioactivity, especially in small structures has, already been described (Labbé *et al.*, 1996). Frost *et al.* (1996) and Meltzer *et al.* (1996) extended this three-compartment model to include a fourth compartment, a GM volume of interest (VOI) that could be delineated on magnetic resonance (MR) images and had a local tissue concentration different from the rest of the GM. This four-compartment model still had several constraints, including an *a priori* estimate of tracer concentration in WM, only one VOI processed, assumption of homogeneous activity within the remaining GM, and MR image segmentation by binarizing MR image compartments. Using automated MR image segmentation based on anatomical probability maps, a new method has been developed for PVE correction, based on a linear least-squares (LLS) method that allows the simultaneous measurement of an unlimited number of VOIs. In this method, VOIs may be delineated the on individual MR images or using a standard template. This new LLS method has been investigated using computer simulations, the three-dimensional (3D) Hoffman brain phantom, and human [^{11}C]flumazenil PET studies from 12 healthy controls and 2 patients with unilateral hippocampal sclerosis.

II. METHODS AND MATERIALS

A. MR Image Acquisition and Segmentation

Volumetric MR data of 1 mm^3 was acquired for the brain phantom and human subjects on a Picker 1.0-T MR scanner using a radiofrequency-spoiled, T$_1$-weighted sequence. Automated MR image segmentation based on a clustering algorithm similar to the maximum likelihood "mixture model" algorithm (Hartigan, 1975) was carried out using routines from the SPM96 package (Friston *et al.*, 1995b). The method takes into account knowledge of the spatial distribution of clusters in the form of probability images that have been derived from MR images of a large number of subjects (Evans *et al.*, 1991). The probability images were normalized to the original individual MR images using a nine parameter (three translations, three rotations, and three orthogonal zooms) affine transformation. These images then represent the *a priori* probability of a voxel being GM, WM, or CSF. The image segments were then used for convolution with the local 3D point spread function (PSF) of the PET scanner, as explained later.

B. PET Acquisition

PET images were obtained using a CTI/SIEMENS 953B (CTI, Knoxville, TN) camera with retractable septa in 3D mode (Spinks *et al.*, 1992) and were reconstructed on a 128 × 128 matrix with a ramp filter (cutoff frequency = 0.5 mm^{-1}). Physical characteristics of the PET scanner (i.e., PSF) were determined using a line source of 1 mm diameter placed at three different radical positions from the center of the FOV in one image plane and a point source of 1 mm moving in 2-mm steps in the z-axis direction (Labbé *et al.*, 1996). Functional images were derived from PET data.

C. Methods and Calculation

Segmenting MR images into anatomical probabilistic maps of GM, WM, and CSF with further subdivisions into multiple VOIs results in N distinct images I_{VOI_k} ($k = 1, \ldots, N$), denoting the probability maps of regions VOI$_k$. Assume that each image I_{VOI_k} is associated with a parameter C. C may represent a concentration of tracer or a parameter derived from the analysis of PET data. Now, let PSF denote the PET point spread function; then the PET image (I_{PET}) may be expressed as

$$I_{\text{PET}} = \left(\sum_{k=1}^{N} C_{\text{VOI}_k} I_{\text{VOI}_k} \right) \otimes \text{PSF}, \quad (1)$$

where the symbol $\otimes$ denotes convolution.

Because convolution is a linear operator, Eq. (1) can be expressed for each pixel (i, j) as a set of linear algebraic equations or, more generally, can be written in matrix form as

$$Ax = b \quad (2)$$

with

$$b = \begin{bmatrix} \text{PET}_1 \\ \text{PET}_2 \\ \vdots \\ \text{PET}_j \\ \vdots \\ \text{PET}_M \end{bmatrix};$$

$$A = \begin{bmatrix} A_{\text{VOI}_{1_1}} & A_{\text{VOI}_{2_1}} & \cdots & A_{\text{VOI}_{k_1}} & \cdots & A_{\text{VOI}_{N_1}} \\ A_{\text{VOI}_{1_2}} & A_{\text{VOI}_{2_2}} & \cdots & A_{\text{VOI}_{k_2}} & \cdots & A_{\text{VOI}_{N_2}} \\ \vdots & \vdots & & \vdots & & \vdots \\ A_{\text{VOI}_{1_j}} & A_{\text{VOI}_{2_j}} & \cdots & A_{\text{VOI}_{k_j}} & \cdots & A_{\text{VOI}_{N_j}} \\ \vdots & \vdots & & \vdots & & \vdots \\ A_{\text{VOI}_{1_M}} & A_{\text{VOI}_{2_M}} & \cdots & A_{\text{VOI}_{k_M}} & \cdots & A_{\text{VOI}_{N_M}} \end{bmatrix};$$

$$x = \begin{bmatrix} C_{VOI_1} \\ C_{VOI_2} \\ C_{VOI_k} \\ C_{VOI_N} \end{bmatrix}$$

where b is the PET image and M is the total number of pixels in the whole image volume (i.e., $M = dim_x \cdot dim_y \cdot N_{plane}$, with dim_x and dim_y, respectively, the X and Y image dimension, and N_{plane} the number of PET planes); $A = [B_{VOI_1} B_{VOI_2} \cdots B_{VOI_N}]$ is a matrix of $M \times N$, where N is the total number of VOIs, each column of A representing the convolution of the probabilistic segmented MR image with one VOI (i.e., $A_{VOI} = I_{VOI} \otimes PSF$); and x is the vector of unknown cerebral tissue radioactivity concentrations to be estimated (e.g., $C_{VOI_1}, C_{VOI_2}, \ldots C_{VOI_N}$). The best solution of the matrix equation $Ax = b$ is the one that comes closest to satisfying all equations simultaneously. If this is defined in the least squares sense, i.e., that the sum of the squares of the differences between the left- and right-hand sides of equation $Ax = b$ be minimized, then the linear set problem reduces to a solvable LLS problem. The computation of $x = A^{-1}b$ was done using singular value decomposition (Press *et al.*, 1988).

D. Handling Large Data Sets

The problem of solving equation $Ax = b$ where A corresponds to a huge matrix containing many convolved images has been encountered previously. This can be solved by estimating the parameters plane by plane and weighting as follows:

$$K_{mean} = \left(\sum_{i=1}^{N_{plane}} K_i \right)^{-1} \left(\sum_{i=1}^{N_{plane}} K_i x_i \right)$$

with $K = [A^{-1}.A]$ (3)

where A_i is the A matrix of plane I. In addition, an optional interface allows for VOIs to be merged (e.g., GM + WM), essentially by row addition in the matrix equation.

E. Multidata Analysis Processing

Processing steps for PVE correction are summarized as follows. MR images were segmented into GM and WM probability maps so that they contained values from 0 to 1, representing 0 to 100% probability of belonging to WM or GM, respectively, and these images were subtracted from the total image to derive the rest of the brain. MR images as well PET images were partitioned into different tissue types and coregistered to templates of the same modalities as the images being coregistered (i.e., MR and PET) (Friston *et al.*, 1995a) in order to find the affine transformation matrix (TM). This coregistration is based on the minimization of the sum of squares difference between the image segments. MR images were then reoriented into PET space by applying this TM without reslicing in the z-axis. Each single MR pixel with a 0 to 100% probability of belonging to the same compartment within the VOI was then convolved with the z-axis PSF by weighting 11 contiguous 1-mm-thick probability maps (equal to two times the FWHM of the PET resolution in the z-axis). These summed, weighted images were then convolved with the 3D PSF on a voxel-by-voxel basis, taking into consideration the pixel localization from the FOV center as described previously (Labbé *et al.*, 1996). The LLS PVE correction algorithm was then implemented by solving Eq. (3). The coregistration was used twice: (1) to align the MR image in respect to the center of the PET FOV and (2) after MR image convolution, to ensure that the convolved GM MR and PET images were in the same space. All programs were written in C language, combined with MATLAB (The Mathworks Inc., Natick, MA), ANALYZE image format (Robb and Hanson, 1990).

F. Computer Simulation

A T_1-weighted volumetric MR image of a patient with unilateral hippocampal sclerosis was used to simulate PET images. The MR images were segmented into probabilistic GM and WM images. Then, a template consisting of 27 anatomical cortical and subcortical regions was defined on the patient MR image volume. Different values were assigned to each of these 27 VOIs, resulting in a color-coded image with each color representing an arbitrary fixed value between 1 and 255 (Fig 1). Simulated PET images were then generated by convolving the color-coded image with the 3D PET PSF on a voxel-by-voxel basis (Fig. 1).

G. Hoffman Brain Phantom Imaging

The method was validated further with a 3D Hoffman brain phantom to simulate a human brain image (Hoffman *et al.*, 1979). MR image data for the phantom (Hoffman *et al.*, 1990) were obtained by filling it with copper sulfate, and PET data were obtained by filling it with a [^{11}C]carbon solution of 10.82 kBq/ml (0.29 μCi/ml) at the start of the acquisition, which was obtained in static mode over 20 min. Three samples from the solution were counted in a well counter. Images contained about 1.2 million single events. The counts in each sample were corrected for decay, dead time, and sample volume effects in the well counter.

H. Human Studies

The LLS model was applied to data from 12 healthy volunteers and 2 patients with mesial temporal lobe epilepsy and a volumetric MR imaging diagnosis of unilateral diffuse hippocampal sclerosis that had been histologically verified. [^{11}C]Flumazenil (FMZ) scans were acquired as described previously (Koepp et al., 1996). High specific activity FMZ (370 MBq) was injected intravenously. Arterial blood was sampled in order to determine a metabolite-corrected arterial input function. A dynamic 3D series, consisting of 20 frames over 90 min, was acquired over the full brain volume. Voxel-by-voxel parametric images of FMZ volume of distribution (FMZ-V_d), reflecting binding to central benzodiazepine receptors, were produced from the brain uptake and plasma input functions using spectral analysis (Cunningham and Jones, 1993). In the case of FMZ, which has little nonspecific binding, V_d is closely correlated with receptor availability (B_{max}). The hippocampi, as the VOI, were outlined on the original high-quality MR images by one investigator using accepted criteria (Cook et al., 1992), and head hippocampus was divided longitudinally into three parts of equal length. For surgically resected temporal lobe specimens, B_{max} was assessed autoradiographically in the midbody of the hippocampus and compared with the PVE-corrected FMZ-V_d.

III. RESULTS

A. Computer Simulation

In the simulated PET image containing 27 VOIs, most of the VOIs were affected by PVE, with under- or overestimation of the assigned value depending on the size and shape of the structure and the surrounding activity distribution (Table 1). The effect was most marked in the caudate nucleus, which is completely surrounded by WM. The loss of activity was not marked the smaller the structure, the higher the concentration, and the greater the difference in concentration between neighboring structures. The LLS method fully recovered the assigned values in all VOIs without any error, regardless of the size or shape of the structure, or of the surrounding tracer concentration (Table 1). This analysis of a large, complex set of noise-free data served to validate the algorithm and the code used for subsequent analysis.

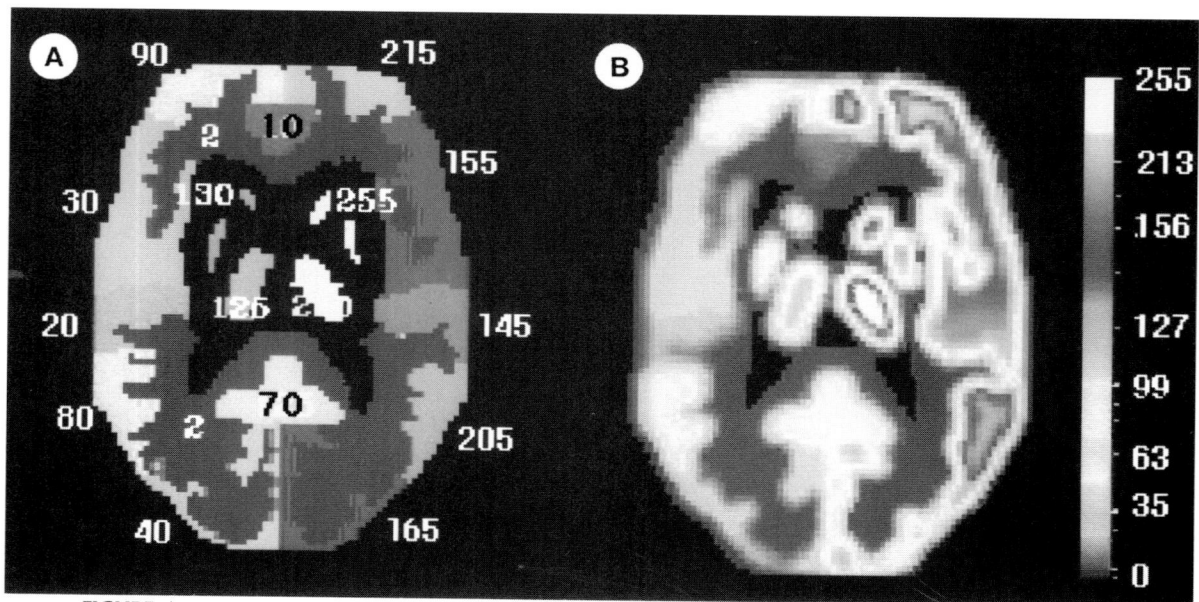

FIGURE 1 Computer simulation for the validation of the LLS algorithm: (A) Template containing 27 cortical and subcortical regions where each VOI was assigned an arbitrary fixed value, resulting in a color-coded image. (B) Simulated PET image generated by convolving (A) with the PET PSF, which is equivalent to the one obtained with the PET scanner. Most of VOIs in (B) are over- or underestimated due to PVE.

TABLE 1 PVE Correction with LLS Method Using Computer Simulations of Multiple VOIs from an Individual Template Containing 27 Cortical and Subcortical Regions[a]

VOI name	Number of pixels	Assigned PET value	Observed PET value	Difference (%)
White matter	1587	2	17.3	+765
Anterior cingulate	88	10	30.7	+207
Orbitofrontal				
L	139	90	65.9	−26.8
R	119	215	146.5	−31.9
Anterotemporal				
L	128	20	19.9	−0.5
R	129	145	131.2	−9.5
Mesiotemporal				
L	304	30	25.7	−14.3
R	276	155	123.8	−20.1
Inferoparietal				
L	127	80	57.4	−28.3
R	126	205	155.7	−24.1
Occipital				
L	137	40	33.4	−16.5
R	100	165	84.4	−48.9
Posterior cingulate	148	70	57.1	−18.4
Caudate nucleus				
L	25	130	91.9	−29.3
R	28	255	113.1	−55.7
Thalamus				
L	64	125	91.8	−26.6
R	88	250	183.7	−26.5

[a] Tabulation of size of VOI and assigned and observed PET values for the 17 regions seen in Fig. 1. PVE correction with LLS method resulted in full recovery of the assigned value (assigned = LLS value).

B. Hoffman Brain Phantom

This phantom was very useful in estimating the limits of accuracy of PVE correction in actual brain imaging. The GM mean value on the PET image was 71.3% of the actual activity, ranging from 58% in the caudate nucleus to 97.6% in central GM. The LLS method recovered the GM concentration to 101% compared with a 8% overcorrection with the three-compartment method (Labbé et al., 1996). This analysis of noisy data further validated our LLS method.

C. Human Studies

Before PVE correction, FMZ-V_d values in the group of 12 controls were similar in the hippocampal VOI and in the surrounding neocortex (Table 2). After PVE correction, the mean FMZ-V_d within GM was significantly higher compared to the uncorrected value ($p < 0.001$), reflecting a mean 37% underestimation of FMZ-V_d in uncorrected GM. There was no difference within the hippocampal VOI before and after PVE correction, between left and right, or between larger and smaller hippocampi. The mean FMZ-V_d asymmetry index (AI = FMZ-V_d of smaller hippocampus divided by FMZ-V_d of larger hippocampus) was similar before and after PVE correction (Table 2) (Koepp et al., 1997a). In two patients with unilateral diffuse hippocampal sclerosis, uncorrected FMZ-V_d of the surrounding GM was slightly lower on the epileptogenic side than contralaterally. PVE correction led to a symmetrically higher FMZ-V_d in the surrounding neocortex compared to uncorrected data. In hippocampal VOIs, PVE correction led to reductions of FMZ-V_d on both sides, being more marked in the epileptogenic hippocampus, resulting in a lower asymmetry index. Additionally, hippocampal VOIs were divided into three

TABLE 2 Comparison of Uncorrected and LLS-PVE Corrected Mean FMZ-V_d Reduction in the Body of the Hippocampus in 2 Patients with mTLE (Left/Right) and Unilateral Diffuse Hippocampal Sclerosis (HS)

| | Controls | | Patients with diffuse HS | | | |
| | | | Uncorrected | | LLS | |
Regions	Uncorrected	LLS	Pat 1 Left/right	Pat 2 Left/right	Pat 1 Left/right	Pat 2 Left/right
White matter	—	0.97 (0.11)	1.87/1.89	1.41/1.42	1.0/1.1	0.59/0.61
Gray matter	4.09 (0.38)	6.48 (0.45)	5.13/5.30	4.15/4.48	7.70/7.75	5.95/6.33
Body hippocampus	—	4.23 (0.50)	3.12/4.18	3.74/4.75	**2.33**/5.15	**2.36**/4.13
Total hippocampus	3.99 (0.33)	3.96 (0.24)	3.86/4.67	3.33/4.27	**2.36**/4.13	**1.58**/3.65
% Red. body	—	—	—	—	44.9	44.2

[a] Patient values in bold = 2 SD below the normal mean. Uncorrected values were not calculated for white matter. Standard deviation values are expressed in parentheses only for the 12 controls.
[b] Reduction percentage within body compared to controls.

subvolumes of equal length along the long axis of the hippocampus (head/body/tail) (Table 2). FMZ-V_d was significantly (e.g., below 2 SD of the control mean) reduced in the head, body, and tail of the epileptogenic hippocampus (Table 2). Autoradiography with [^{3}H]FMZ in these two patients on the midbody of the hippocampus gave a B_{max} of 158.4 and 169.8, corresponding to a 46.8 and 42.9% reduction, respectively, from the mean obtained from 6 postmortem controls (Koepp et al., 1997b).

IV. DISCUSSION

All methods of PVE correction depend critically on the accuracy of MR image segmentation and MR-PET coregistration, on the inhomogeneity of the PSF, and on the assumptions made during PVE correction.

A. Effect of MR Image Segmentation

Automated MR image segmentation based on probability maps was used to reduce any bias introduced by thresholding MR images into single tissue compartments. The convolution processes were performed on each pixel according to its probability of belonging to GM, WM, and CSF compartments. This is probably a more realistic approach than previous methods that used binarized, segmented MR images for this convolution (Meltzer et al., 1996; Müller-Gärtner et al., 1992). Segmentation accuracy is limited mainly by the PVE, which is unavoidable because of the convoluted structure of a 2- to 4-mm-thick cortical ribbon and which leads to blurring of tissue boundaries. Voxels at the structure boundaries exhibit multiple properties due to variability in intensity distribution within a single tissue compartment and magnetic field inhomogeneity. As a result, binarizing pixels into pure tissue classes over- or underestimates the GM volume, leading to an under- or overcorrection of its tracer activity because voxels of mixed tissue composition are not accounted for.

B. MR to PET Coregistration

Misregistration of MR to PET images is an important source of error for PVE correction, depending on the size of the structure (Labbé et al., 1996). For this reason, the coregistration process was run twice in order to obtain a maximal registration between the two image modalities. This resulted in a mean error of less than 1% all directions. The first coregistration minimized the sum of squares difference between two signals of different resolutions: a high (MR) and a low (PET) resolution pixel (Friston et al., 1995a). When MR images were smoothed locally according to the PET resolution, the second coregistration gave more accurate results as it minimized the sum of squares difference between sets of images of similar resolution.

C. Inhomogeneity of the PET Point Spread Function

The impact of PSF distortions across the FOV has been described (Labbé et al., 1996). When using three- or four-compartment correction, not including 3D PET PSF variations in the convolution process results in a error in corrected PET data, increasing locally by up to 244% and being more marked in small structures and in pixels at the edge of cerebral structure (Labbé et al., 1996). Although most authors have neglected the effect of inhomogeneity of the PSF (Frost *et al.*, 1996; Meltzer *et al.*, 1996), it is crucial to recognize that correction for PSF inhomogeneity in LLS PVE allows full recovery of the tracer activity in any pixel within the FOV.

D. LLS Method Advantages

The four-compartmental PVE correction method (Meltzer *et al.*, 1996; Müller-Gärtner *et al.*, 1992; Frost *et al.*, 1996) is limited by the constraints of restricted numbers of compartments (CSF, WM, GM, single VOI) and by the accumulated errors of its assumptions, including (1) subjective segmentation of MR images on a threshold basis into binary anatomical maps; (2) WM concentration measured from a large ROI on the original PET image, which is assumed to be constant; (3) four-compartment correction to determine VOI concentration follows an initial three-compartment correction; (4) the assumption that GM is homogeneous throughout the brain; (5) measurement of only one VOI at a time; and (6) division of the observed PET images by its convolved MR image, which may lead to high corrected pixel values due to misregistration, missegmentation, or both.

A new MR imaging-based approach for correction of PVE, which has considerable advantages over previously described methods has been developed (Frost *et al.*, 1996; Meltzer *et al.*, 1996; Müller-Gärtner *et al.*, 1992). The major advantage of the LLS algorithm is the simultaneous correction of multiple VOIs of any size and of any kind of tissue or structure. These VOIs can be outlined on high-quality MR images. No *a priori* assumptions of radioactivity concentration within any of these VOI are needed. The accurate estimation of each single compartment can be performed at one time. The LLS algorithm was superior to previously described three- and four-compartment methods; it gave full recovery of the activity in computer simulations in an unlimited number of VOIs. The LLS algorithm was validated using the Hoffman brain phantom. Although the activity within the anthropomorphic phantom was fully recovered, such a model can never precisely simulate human brain imaging due to the heterogeneity in actual cerebral structures.

In common with the method of Rousset *et al.* (1993), the correction of the measured activity for each VOI depends on the geometrical relationships between the VOI and the image resolution, but is independent of the tracer activity. A template of several anatomical VOIs of any size and shape has been created from either a PET or an MR image of an individual or from MR images of a large number of subjects. The use of a VOI template is one of the preliminary steps to extend this algorithm to a voxel-by-voxel measurement (i.e., each voxel is considered as one VOI), but this will make calculations unstable and will also render the model more susceptible to noise. In addition, because tracer activity within a VOI is never purely homogeneous, resolving the equation $Ax = b$ by merging or splitting the GM and WM tissue components within one VOI could be done. This could change the estimated parameters due to the different values assigned for each of these tissue VOIs, as each new estimated parameters may have a lower or a higher tracer activity concentration, influencing spillover from all the surrounding VOIs.

E. Clinical Implementation

In the human study, the effect of partial volume averaging was most marked in the GM, with a highly significant underestimation of the FMZ-V_d in uncorrected PET data due to neighboring WM and CSF. In contrast, uncorrected FMZ-V_d within the hippocampus was higher than the PVE corrected value because of spillover from neighboring GM with higher FMZ binding. PVE affected the smaller hippocampus more than the larger. After PVE correction, FMZ-V_d in the body of the atrophic epileptogenic hippocampus was reduced over and above the loss of volume by about 44% compared to controls. This correlated with an average 45% reduction of receptor density in the whole hippocampal cross section determined by *ex vivo* autoradiography in the midbody of the resected hippocampi (Koepp *et al.*, 1997b). In the evaluation of patients with temporal lobe epilepsy, the ability to make measurements from multiple VOIs, including subdivisions of the hippocampus, is necessary. Many patients have sclerosis of the hippocampus confined to the anterior portion (VanPaesschen *et al.*, 1997). The LLS method allows multiple measurements of tracer binding to be made along the whole length of the hippocampus and will help to define those patients who may benefit from tailored resections, leaving the unaffected parts of the hippocampus intact.

V. CONCLUSION

In summary, the LLS method introduced here allows accurate PET quantification in multiple small cerebral structures such as the hippocampus. The method was validated using both computer simulation and an anthropomorphic model. FMZ PET images coregistered with high-quality MR images and corrected for PVE provide a quantitative and objective measure of the true tissue characteristics in hippocampus, neocortex, and other cerebral structures in patients with epilepsy. PVE correction determines whether the differences in receptor density are the result of functional changes, atrophy, or both. Although the opportunities to validate this method in human studies are limited, it has been shown, in two patients with hippocampal sclerosis, that the correlations of *in vivo* PET data with *ex vivo* measurements of autoradiographically determined FMZ receptor density.

Difficulties in interpreting PET images due to anatomical variability of cerebral structures within a subject and in making comparisons between subjects can now be reduced. PET data can be interpreted accurately from small structures and in areas of the brain affected by atrophy and by spillover of activity from surrounding structures. Further work will evaluate this method of PVE correction with other tracers and in subjects with other diseases.

Acknowledgments

We are grateful to the Medical Research Council who funded this work and to our colleagues at the MRC Cyclotron Unit (Andreanna Williams, Leonhard Schnorr, Ralph Myers) for help in the acquisition and analysis of PET data and the Epilepsy Research Group (Drs Samantha Free), Institute of Neurology, for access to their MRI data.

References

Cook, M. J., Free, S. L., Manford, M. R. A., Fish, D. R., Shorvon, S. D., Straughan, K., and Stevens, J. M. (1992). MR imaging in CT negative frontal lobe epilepsy, *Epilepsia* 33(Suppl. 3): 73.

Cunningham, V. J., and Jones, T. (1993). Spectral analysis of dynamic PET data. *J. Cereb. Blood Flow Metab.* 13: 15–23.

Evans, A. C., Marrett, S., Torrescorzo, J., Ku, S., and Collins, L. (1991). MRI-PET correlation in three dimensions using a volume-of-interest (VOI) atlas. *J. Cereb. Blood Flow Metab.* 11: A69–A78.

Friston, K. J., Ashburner, J., Frith, C. F., Poline, J. B., Heather, J. D., and Frackowiak, R. S. J. (1995a). Spatial registration and normalisation of images. *Hum. Brain Mapp.* 2: 165–189.

Friston, K. J., Holmes, A. P., Worsley, K. J., Poline, J. B., Frith, C. D., and Frackowiak, R. S. J. (1995b). Statistical parametric maps in functional imaging: A general linear approach. *Hum. Brain Mapp.* 2: 189–210.

Frost, J. J., Meltzer, C. C., Zubieta, J. K., Links, J. M., Brakeman, P., Stumpf, M. J., and Kruger, M. (1996). MR-based correction of partial volume effects in brain PET imaging. In "Quantification of Brain Function using PET" (R. Myers, V. Cunningham, D. Bailey and T. Jones, eds.), pp. 152–157. Academic Press, San Diego, CA.

Hartigan, J. A. (1975). "Clustering Algorithms." Wiley, New York.

Hoffman, E. J., Huang, S. C., and Phelps, M. E. (1979). Quantitation in positron emission computed tomography: 1. Effect of object size. *J. Comput. Assist. Tomogr.* 3: 299–308.

Hoffman, E. J., Cutler, P. D., Digby, W. M., and Mazziotta, J. C. (1990). 3D phantom to simulate blood flow and metabolic images for PET. *IEEE Trans. Nucl. Sci.* 37(2): 616–620.

Kessler, R. M., Ellis, J. R., and Eden, M. (1984). Analysis of emission tomographic scan data: Limitations imposed by resolution and background. *J. Comput. Assist. Tomogr.* 8: 514–522.

Koepp, M. J., Richardson, M. P., Brooks, D. J., Poline, J. B., Van Paesschen, W., Friston, K. J., and Duncan, J. S. (1996). Cerebral benzodiazepine receptors in hippocampal sclerosis: An objective in vivo analysis. *Brain* 119: 1677–1687.

Koepp, M. J., Labbé, C., Brooks, D. J., Richardson, M. P., Van Paesschen, W., Cunningham, V. J., and Duncan, J. S. (1997a). Regional hippocampal [^{11}C]flumazenil PET in temporal lobe epilepsy with unilateral and bilateral hippocampal sclerosis. *Brain* 120: 1865–1876.

Koepp, M. J., Hand, K. S. P., Labbé, C., Richardson, M. P., Baird, V., Brooks, D. J., Van Paesschen, W., Bowery, N. G., and Duncan, J. S. (1997b). In-vivo [^{11}C]flumazenil-PET correlates with ex-vivo [^{3}H]flumazenil autoradiography in hippocampal sclerosis. *Epilepsia* (in press).

Labbé, C., Froment, J. C., Kennedy, A., Ashburner, J., and Cinotti, L. (1996). Positron emission tomography metabolic data corrected for cortical atrophy using magnetic resonance imaging. *Alzheimer Dis. Assoc. Disord.* 10: 141–170.

Mazziotta, J. C., Phelps, M. E., Plummer, D., and Kuhl, D. E. (1981). Quantitation in positron emission tomography: 5. Physical-anatomical effects. *J. Comput. Assist. Tomogr.* 3: 299–308.

Meltzer, C. C., Zubieta, J. K., Links, J. M., Brakeman, P., Stumpf, M. J., and Frost, J. J. (1996). MR-based correction for brain PET measurements for heterogeneous gray matter radioactivity distribution. *J. Cereb. Blood Flow Metab.* 16: 650–658.

Müller-Gärtner, H. W., Links, J. M., Prince, J. L., Bryan, R. N., McVeigh, E., Leal, J. P., Davatzikos, C., and Frost, J. J. (1992). Measurement of radiotracer concentration in brain gray matter using positron emission tomography: MRI-based correction for partial volume effects. *J. Cereb. Blood Flow Metab.* 12: 571–583.

Press, W. H., Flannery, B. P., Teukolsky, S. A., and Vetterling, W. T. (1988). Modeling of data: Nonlinear models. In "Numerical Recipes: The Art of Scientific Computing," pp. 521–552. Cambridge University Press, New York.

Robb, R. A., and Hanson, D. P. (1990). ANALYZE: A software system for biomedical image analysis. *Conf. Visual. Biomed. Comput.* 1st, Atlanta, GA, pp. 507–518.

Rousset, O., Ma, Y., Leger, G. C., Gjedde, A. H., and Evans, A. C. (1993). Correction for partial volume effects in PET using MRI-based 3D simulations of individual human brain metabolism, In "Quantification of Brain Function. Tracer Kinetics and Image Analysis in Brain PET" (K. Uemura, N. A. Lassen, T. Jones, and I. Kanno, eds.), pp. 113–125. Elsevier, Amsterdam and New York.

Spinks, T. J., Jones, T., Bailey, D. L., Townsend, D. W., Grootoonk, S., Bloomfield, P. M., Gilardi, M.-C., Casey, M. E., Sipe, B., and Reed, J. (1992). Physical performance of a positron tomograph for brain imaging with retractable septa. *Phys. Med. Biol.* 37: 1637–1655.

VanPaesschen, W., Connelly, A., King, M. D., Jackson, G. D., and Duncan, J. S. (1997). The spectrum of hippocampal sclerosis: A quantitative magnetic resonance imaging study. *Ann. Neurol.* 41: 41–51.

CHAPTER 10

Pixel- versus Region-Based Partial Volume Correction in PET[1]

OLIVIER G. ROUSSET, YILONG MA, DEAN F. WONG,* and ALAN C. EVANS

McConnell Brain Imaging Centre
Montréal Neurological Institute
McGill University
Montréal, Quebec, H3A 2B4 Canada
* Department of Radiology
Johns Hopkins School of Medicine
Baltimore, Maryland 21287-0807

Partial volume correction (PVC) methods require precise determination of structural information from anatomy-oriented devices such as magnetic resonance imaging and the incorporation of positron emission tomography physical characteristics. The primary goal of this work was to study the error introduced by a heterogeneous tracer distribution within a particular tissue component assumed to be homogeneous. Two different approaches were compared: (i) a pixel-based method (PIX-PVC), which is a successive elimination–substitution method that yields corrected tissue maps (Meltzer et al., 1996), and (ii) a region-of-interest-based technique (ROI-PVC), which is a direct method of solving a system of simultaneous linear algebraic equations (Rousset et al., 1993b). To simulate a heterogeneous tracer distribution, the left insula region (lINS) was identified as a distinct volume of interest, and PVC was performed considering three pure tissue components: white matter, cortical gray matter, and basal ganglia. The relative errors made on regional measurements after PVC were found to be very similar between PIX–PVC and ROI–PVC for simulated data of the tracer [^{18}F]fluoro-L-dopa uptake. In addition, both methods were applied to a real data set for this tracer and provided comparable results. However, this study cannot predict the amount of error propagation in the case of a more heterogeneous tracer distribution. In particular, the PIX–PVC method will encounter increasing difficulties to fulfill the assumption of tracer distribution homogeneity. However, the ROI–PVC method appears more flexible, does not require a priori knowledge of some tracer concentration, and is easier and faster to implement.

I. INTRODUCTION

The possibility of incorporating structural information into positron emission tomography (PET) image analysis provides researchers with the possibility to account for the (partial volume) effect of limited PET spatial resolution. There is a great area of research on iterative methods of image reconstruction from projections that are capable of incorporating *a priori* structural information and the physical characteristics of a scanner to account for tissue blurring and finite sampling. These methods include, for example, Bayesian methods derived from the ML-EM[2] algorithm (Ouyang et al., 1994). Such methods are to be distinguished from postreconstruction partial volume correction (PVC) methods based on simulations that mimic the limited resolution inherent to PET. These methods intend to reproduce the interactions between the distributed activity within the brain and the PET detection system during PET data acquisition. Furthermore, because these methods "work" on images obtained using the popular filtered backprojection (FBP) algorithm, they

---

[1] Transcripts of the BRAINPET97 discussion of this chapter can be found in Section VIII.

[2] Maximum-likelihood expectation–maximization.

must also incorporate the smoothing effect of low-pass filtering during image reconstruction.

II. METHODS

A. Theory

For an activity distribution, $f(r)$, and a PET system point spread function (PSF), $h(r)$, the PET image, $g(r)$, can be expressed as

$$g(r) = \int f(r')h(r,r')\,dr' + v, \qquad (1)$$

where r and r' are three-dimensional (3D) vectors in respectively image and object space, and v is the noise in the image.

If it is assumed that $f(r)$ is distributed over N homogeneous functional components of true activity concentration, I_i, each defined over a spatial domain, D_i, Eq. (1) becomes (in the absence of noise)

$$g(r) = \sum_{i=1}^{N} I_i \int_{D_i} h(r,r')\,dr'. \qquad (2)$$

Equation 2 represents the imaging equation when considering N independent tissue components, each of which takes up the tracer uniformly. The integration of the system's PSF over D_i represents the PET response function for this tissue component, or regional spread function (RSF$_i$) image. Assuming a unit activity distribution within D_i (i.e., $I_i = 1$), the RSF represents the fraction of true activity of tissue component i that will be contained in each voxel of the final image

$$\text{RSF}_i(r) = \int_{D_i} h(r,r')\,dr'. \qquad (3)$$

B. Generation of RSF Images

Both approaches studied, PIX–PVC (Meltzer et al., 1996) and ROI–PVC (Rousset et al., 1993b), require the generation of the RSF image of each tissue component, i.e., the individual response function of the tomograph to a unit distribution of activity within each region of the brain. For this purpose, magnetic resonance (MR) imaging and PET data are registered using a linear landmark matching algorithm (Evans et al., 1991b) and segmented using single- or multispectral cluster analysis classification techniques, in combination with the definition of volumes of interest (VOIs) (Rousset et al., 1993a, b).

The generation of the RSF image of a given tissue component is performed by assigning the corresponding tissue code in the segmented MR image with a value of one (1) and then processing through the PET simulator developed in the authors' institute (Ma et al., 1993). Among other physical effects, the simulator incorporates the limitations introduced by the finite resolution of the PET scanner, as well as smoothing performed during image reconstruction (Rousset et al., 1993a). The resolution of the Scanditronix/GE PC2048-15B PET scanner used in this study was approximated by a spatially invariant Gaussian function of $6 \times 6 \times 6$ mm full-width-at-half-maximum (FWHM) (Evans et al., 1991a).

C. Simulation of Tracer Uptake ($N = 3$)

Consider for a given tracer that there are three ($N = 3$) tissue components with different kinetic properties, i.e., white matter (W), cortical gray matter (G_1), and the basal ganglia (G_2), with the rest of the brain not participating in the uptake of the conceptual tracer. Time–activity curves (TACs) derived for each tissue from [^{18}F]fluoro-L-dopa (FDOPA) modeling of normal control data after application of the ROI-PVC algorithm (Rousset et al., 1993b) were used to represent the true tracer course within each tissue component. Those curves were chosen in order to provide a typical range of contrast expected in this type of receptor studies.

To simulate tracer heterogeneity within cortical gray matter (G_1), the gray matter constituting the left insula (lINS) was manually "painted" from MR images using a 3D brush and a threshold procedure (Fig. 1). This component was assigned a TAC derived from the weighted average of cortical gray matter and basal ganglia (G_2) TACs:

$$\text{lINS} = G_1 + \beta(G_2 - G_1). \qquad (4)$$

The activity of lINS was varied from 100% G_1 ($\beta = 0$) to 100% G_2 ($\beta = 1$) in 10% increments (Fig. 2). PVC was then performed according to the original three-tissue component model. The recovery of activity was then assessed within the various regions of the brain for varying β values and as a function of the true contrast χ between G_1 and G_2 regions:

$$\chi = \frac{G_2 - G_1}{G_2 + G_1}. \qquad (5)$$

D. Implementation of Pixel-Based Method (PIX–PVC)

This method consists of solving the imaging equation [Eq. (2)] as it is for the true *activity distributions*, I_i.

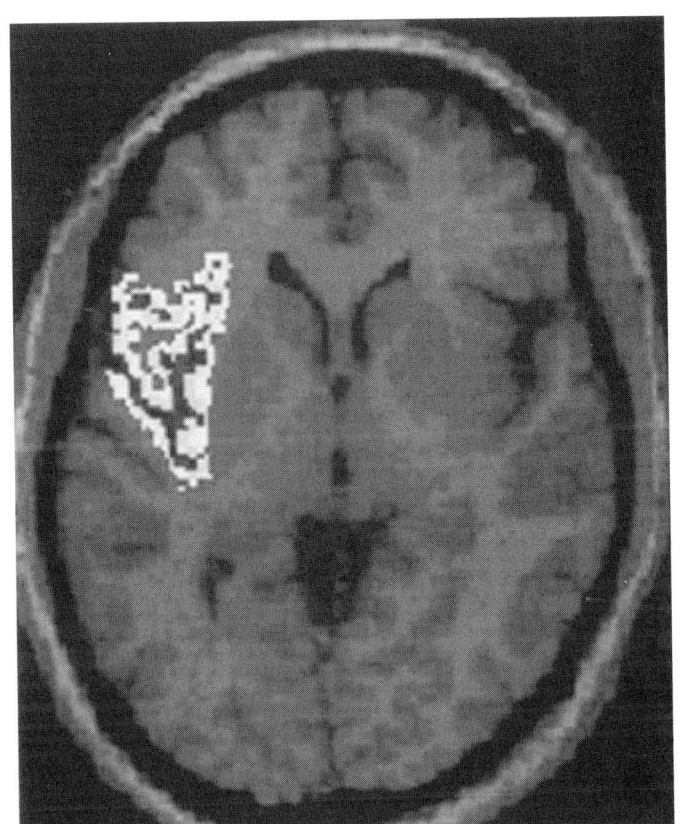

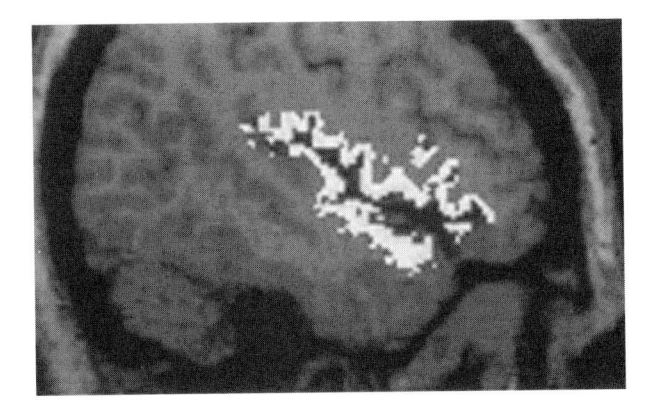

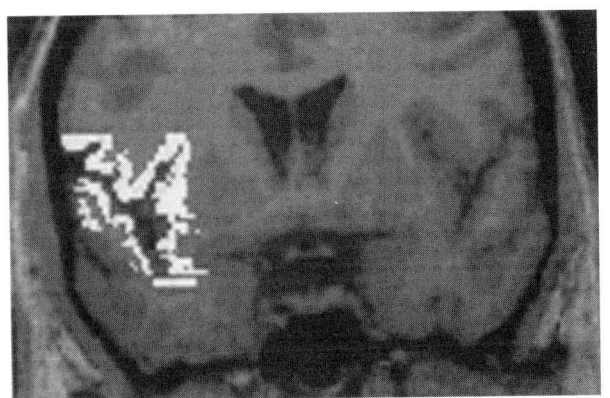

FIGURE 1 Left insula region manually segmented from MR image. This region was used to simulate tracer heterogeneity within gray matter and was assigned a weighted average of cortical gray (G_1) and basal ganglia (G_2) activity concentrations (see Fig. 2).

The obvious difficulty is that there are N unknowns for only one equation. To overcome this problem, some hypotheses regarding the tracer distribution within the brain are made.

First, the brain consists of two homogeneous components: whole gray matter and white matter. To solve Eq. (2), it is necessary to determine one of the two unknowns, e.g., white matter, as follows:

- Estimate the true white matter activity $\tilde{I}_W$
- Assign $\tilde{I}_W$ to RSF_W to estimate the contribution of white matter to the PET image $g(r)$:

$$\tilde{g}_W(r) = \tilde{I}_W \times \text{RSF}_W(r), \quad (6)$$

- Subtract $\tilde{g}_W$ from the original PET image (PET1):

$$g(r) - \tilde{g}_W(r), \quad (7)$$

- PET1 divided by the gray matter regional spread function image $\text{RSF}_{\tilde{G}}(r)$ yields a whole gray matter image, corrected for the partial volume effect:

$$\tilde{f}_G = \frac{g(r) - \tilde{g}_W(r)}{\text{RSF}_{\tilde{G}}}, \quad (8)$$

At this point, the corrected (whole) gray matter image $\tilde{f}_G$ does not differentiate cortical gray matter (G_1) from the basal ganglia (G_2) and, therefore, does not account for cross-contamination between these two tissue components. To correct for this effect, the whole gray matter image is further subdivided into two distinct components (G_1 and G_2). To solve this new two-tissue system, the true activity of one of the components needs to be determined. As it is assumed in the present case that G_1 activity is homogeneous throughout the brain, three separate ROIs are selected to extract the "true" cortical gray matter (G_1) activity, i.e., the left/right (L/R) frontal cortex, the L/R insula, and the L/R occipital cortex.

- The estimate of the "true" activity of cortical gray matter ($\tilde{I}_{G_1}$) is assigned to the RSF of G_1 to yield

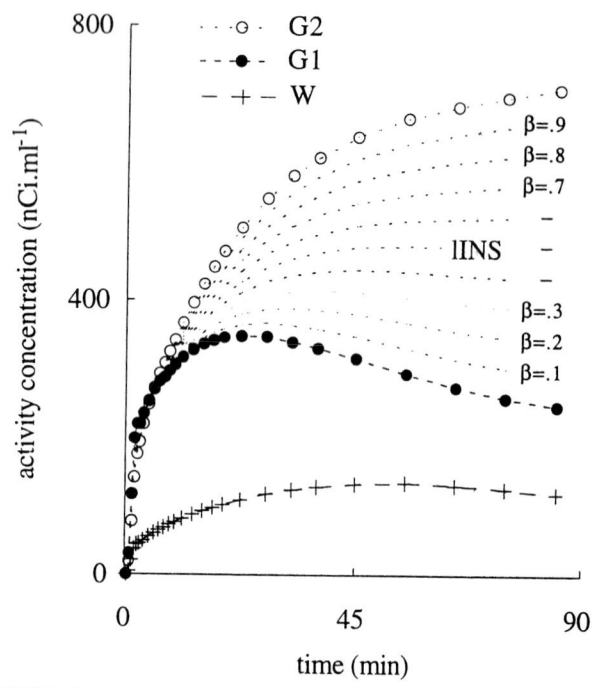

FIGURE 2 Set of time–activity curves used to simulate true tracer course in the brain within white matter (W), cortical gray matter (G_1), and the basal ganglia (G_2). The activity of the left insula (lINS) was set as a mixture of G_1 and G_2 activities, with the heterogeneity factor β indicating the relative proportion of G_2 activity [Eq. (4)].

the contribution of G_1 to the PET image:

$$\tilde{g}_{G_1}(r) = \tilde{I}_{G_1} \times \mathrm{RSF}_{G_1}(r), \qquad (9)$$

- This image is then subtracted from PET1:

$$g(r) - \tilde{g}_W(r) - \tilde{g}_{G_1}(r). \qquad (10)$$

The image resulting from Eq. (10) has been corrected for the contributions of white matter ($\tilde{g}_W$) and cortical gray matter ($\tilde{g}_{G_1}$) and is then divided by the RSF of the basal ganglia ($\mathrm{RSF}_{\tilde{G}_2}$) to yield the corrected basal ganglia image:

$$\tilde{f}_{G_2} = \frac{g(r) - \tilde{g}_W(r) - \tilde{g}_{G_1}(r)}{\mathrm{RSF}_{\tilde{G}_2}}. \qquad (11)$$

At this point the correction for the basal ganglia region (G_2) is completed. If the complete corrected cortical gray matter (G_1) image is needed as well, the same operations, still between G_1 and G_2, need to be performed, but this time by taking true G_2 activity as the known variable. ROIs are placed on the BG area on $\tilde{f}_{G_2}(r)$, and corrected estimates are extracted. The same set of operations are then performed. The "true"

BG value, $\tilde{I}_{G_2}$, is assigned to the basal ganglia regional spread function image (RSF_{G_2}) to yield the image of G_2 contribution to the PET image $\tilde{g}_{G_2}(r)$. The resulting image is then subtracted from PET1 to yield a PET image excluding white matter and basal ganglia contributions. This image is then divided by the RSF mask of G_1 to yield the corrected cortical gray matter image:

$$\tilde{f}_{G_1} = \frac{g(r) - \tilde{g}_W(r) - \tilde{g}_{G_2}(r)}{\mathrm{RSF}_{\tilde{G}_1}}. \qquad (12)$$

At this level, a composite corrected PET image is created by masking the various corrected images with nonoverlapping anatomical masks:

$$\tilde{f} = \tilde{f}_W \times \mathrm{mask}_W + \tilde{f}_{G_1} \times \mathrm{mask}_{G_1} + \tilde{f}_{G_2} \times \mathrm{mask}_{G_2}. \qquad (13)$$

Note that "true" white matter activity necessary to estimate the contribution of white matter to the PET image [Eq. (6)] is usually extracted from a large remote white matter region such as the pons for which partial volume effects are considered negligible (Müller-Gärtner et al., 1992). In the case of simulated data, this value was taken from the true white matter TAC, and in the case of real data, these values were obtained from the ROI–PVC method.

E. Implementation of the ROI–PVC Method

It was suggested previously that instead of considering the whole PET image to recover, the calculation of the PSF convolution with the true objects can be restricted to a limited area that constitutes a ROI and then the *regional activity concentrations* determined independent of tracer levels (Rousset et al., 1993b). As an example, consider the three-tissue system as:

$$\begin{cases} f_1 = \text{white matter (W) of true activity} \\ \quad\quad \text{concentration } \tilde{I}_1 \\ f_2 = \text{cortical gray matter } (G_1) \text{ of true} \\ \quad\quad \text{activity concentration } \tilde{I}_2 \\ f_2 = \text{basal ganglia } (G_2) \text{ of true activity} \\ \quad\quad \text{concentration } \tilde{I}_3 \end{cases} \qquad (14)$$

Note that f_1, f_2, and f_3 could have been defined as any of the three components identified, and in any order, as the correction is performed simultaneously on all the tissue components considered in the system.

The regional activity concentration extracted from a particular ROI_j can be expressed as the sum of fractions of true activity from each tissue component integrated in that ROI. Therefore, for the three-tissue system chosen here,

$$\begin{cases} g_1 = I_1 \times \int_{\text{ROI}_1} \text{RSF}_1(r)\,dr \\ \qquad + I_2 \times \int_{\text{ROI}_1} \text{RSF}_2(r)\,dr \\ \qquad + I_3 \times \int_{\text{ROI}_1} \text{RSF}_3(r)\,dr \\ g_2 = I_1 \times \int_{\text{ROI}_2} \text{RSF}_1(r)\,dr \\ \qquad + I_2 \times \int_{\text{ROI}_2} \text{RSF}_2(r)\,dr \\ \qquad + I_3 \times \int_{\text{ROI}_2} \text{RSF}_3(r)\,dr \\ g_3 = I_1 \times \int_{\text{ROI}_3} \text{RSF}_1(r)\,dr \\ \qquad + I_2 \times \int_{\text{ROI}_3} \text{RSF}_2(r)\,dr \\ \qquad + I_3 \times \int_{\text{ROI}_3} \text{RSF}_3(r)\,dr \end{cases} \quad (15)$$

Equation (15) can be reexpressed by the general formulation

$$g_j = \sum_{i=1}^{N} \omega_{ij} I_i, \qquad (16)$$

where

$$\omega_{ij} = \frac{1}{n_{\text{pix}}} \int_{\text{ROI}_j} \text{RSF}_i(r)\,dr \qquad (17)$$

and n_{pix} is the number of pixels in ROI_j.

The geometric transfer factors ω_{ij} express the contribution of each tissue component i to any region of interest j (ROI_j) in the PET image. They constitute a matrix that is hereafter referred to as the geometric transfer matrix (GTM). The diagonal terms of this matrix represent tissue self-interaction or the regional recovery coefficient (RC) (Hoffman et al., 1979); the other terms $\omega_{ij}(j \neq i)$ refer to the spillover fraction, i.e., fraction of true activity (I_i) of tissue i integrated in ROI_j. The regional values actually observed with PET and the (known) regional GTM then represent a system of linear equations that can be solved for the true regional values, I_i. The GTM for a given PET image includes all the geometrical relationships among the structures of interest, the ROIs used, and the scanner's resolution, and is independent of tracer activity distribution.

In practice, the ROI-based method consists of sampling the various RSF images with a user-defined ROI template normally used for PET data extraction, e.g., MR-based anatomical ROIs. As at least one ROI per active tissue component is needed to fulfill the requirements of Eq. (15), the extraction of the transfer matrix is simply performed by masking the three RSF images with three selected ROIs, yielding a total of nine transfer coefficients.

III. RESULTS AND DISCUSSION

A. Recovery of Simulated FDOPA Time-activity Curves

Corrected TACs from the ROI-based correction were compared with TACs generated by sampling the dynamic images obtained after PIX-based correction with the same ROI template.

When no noise was added to the projections and in the absence of tissue heterogeneity (i.e., $\beta = 0$), both methods gave an accurate correction (errors < 1% of true value, data not shown). However, in the case of some degree of heterogeneity in tracer distribution within cortical gray matter (G_1), regional activity concentrations within the brain were not fully recovered after PVC. This inaccuracy is due to the fact that PVC is performed on the assumption of a three-tissue system instead of the actual four-tissue system [i.e., white matter (W), cortical gray matter minus insula region (G_1 − INS), insula (INS), and basal ganglia (G_2)]. For a given value of the heterogeneity factor β, these errors were found to be linearly proportional to the true contrast χ existing between cortical gray (G_1) and basal ganglia (G_2) regions [Eq. (5)]:

$$\%\varepsilon = \alpha \chi + \varepsilon_0, \qquad (18)$$

where α is the proportionality factor and ε_0 is the intercept or residual error due to the difference in segmentation of the region corresponding to the left insula. This is an experimental error resulting from the use of the automatically classified lINS in the case of homogeneous concentration ($\beta = 0$) and of the manually painted lINS in the case where $\beta \geq 0.1$. The intercepts ε_0 express the susceptibility of both PVC methods toward tissue segmentation and were not found to be significantly different between the PIX–PVC and ROI–PVC methods.

Furthermore, the proportionality factor α was computed for each regional measurement, using various gray matter reference regions. In all cases except the left insula, the value of α was directly proportional to the heterogeneity factor β. For these cases,

$$\alpha = K\beta + \alpha_0, \quad (19)$$

where α_0, like ε_0 in Eq. (18) is the intercept due to errors in segmentation of the lINS. Hence, the absolute percentage error can be estimated directly by computing

$$\%\varepsilon = K \times \beta \times \chi. \quad (20)$$

An example of the error in terms of the susceptibility factor α is plotted as a function of the heterogeneity factor β when the left front cortex (lfrCx) or the left insula was taken as cortical gray matter reference regions (Fig. 3). When using the left frontal cortex as the reference region for cortical gray matter (G_1), the left putamen is the region suffering the most from the tissue heterogeneity present. However, the error was not significant when the left insula was taken as the reference region for G_1. This is not surprising, as the putamen lies closely against the insular region, so the insula is, therefore, a good representation of cortical gray matter activity actually "contaminating" the putaminal activity measurement. Concerning the left insula, the underestimation encountered is due to the difference in self-recovery (RC) of the lINS compared to that of the whole cortical gray matter region (G_1) actually used in the calculation, leading to bigger recovery coefficient values, and, therefore, to an undercorrection for the partial volume effect. These errors are comparable for PIX-PVC and ROI-PVC (Fig. 3). Regions located in the right hemisphere were not significantly affected by tracer heterogeneity in lINS due to their remote spatial location. However, the errors encountered ($< 0.8\%$) represent rounding errors consequent to computation and appeared to fluctuate more in the case of PIX-PVC due to a larger number of arithmetic operations involved (data not shown).

In the presence of statistical noise added to the simulation projections and in the absence of tracer heterogeneity, both methods provided corrected estimates that fluctuated with the same magnitude around the true tissue TACs. An example is given in Fig. 4.

B. Application to Real Data

Both methods were applied to a real data set from a normal subject who received a single bolus injection of 5 mCi of the tracer FDOPA and was then scanned with the Scanditronix/GE PC2048-15B system (Evans *et al.*, 1991a).

Because the true tracer course is not known in the case of real PET data, it is not possible to assess the absolute accuracy of either of the two PVC methods

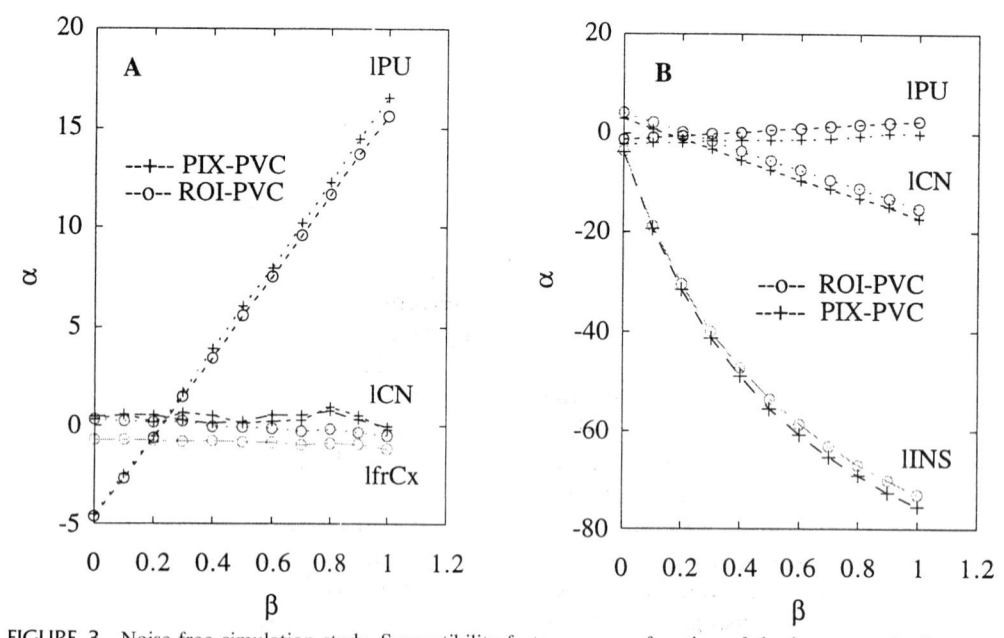

FIGURE 3 Noise-free simulation study. Susceptibility factors α as a function of the heterogeneity factor β for regions located in the left hemisphere: (A) when the left front cortex (lfrCx) was chosen as the reference cortical gray matter (G_1) region and (B) when the left insula (lINS) was used. The other regions plotted are left putamen (lPU) and left caudate nucleus (lCN).

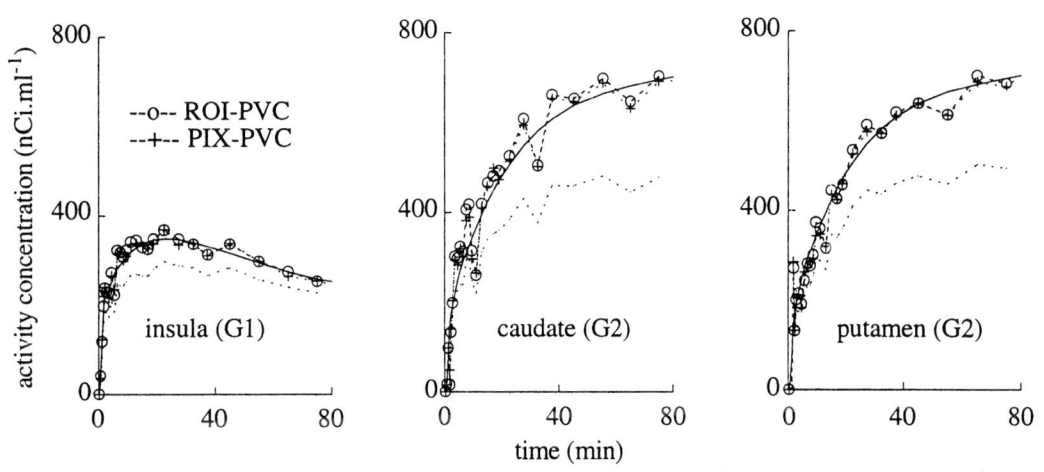

FIGURE 4 Time–activity curves extracted from various regions of the brain (insula, caudate, and putamen) in the case of noisy simulated images obtained after the activity distribution given in Fig. 2. Note the underestimation of observed estimates (dotted curves) and the recovery of true activity concentration (solid and dashed curves) when using the PIX–PVC method (+) and the ROI–PVC method (○).

involved. However, in the case of BG structures (G_2), corrected estimates were found to be very similar, regardless of the cortical gray matter region used as reference region for G_1 (Fig. 6). An example is given in Fig. 5 with the insula taken as the reference gray matter (G_1) region. It can be concluded that cortical gray matter has a fairly homogeneous uptake of the FDOPA tracer.

The largest differences were observed for the corrected TAC of the right occipital cortical region that was corrected less by the PIX–PVC method than when the ROI–PVC technique was used (data not shown). The corrected TACs obtained from other gray matter (G_1) regions suggest that PIX–PVC underestimates the corrected activity concentration of the occipital cortex region.

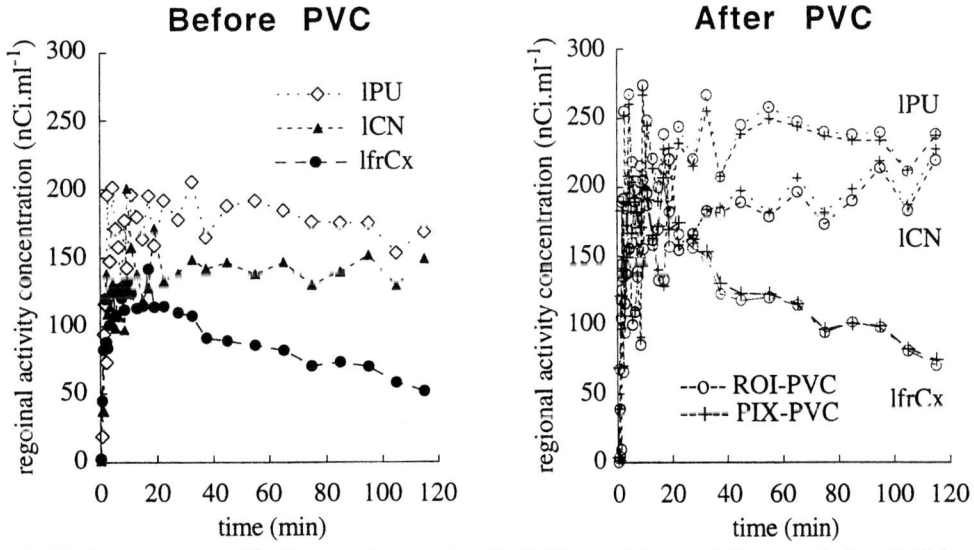

FIGURE 5 Real data of [^{18}F]fluoro-L-dopa uptake. (Left) Observed time–activity curves before PVC for the basal ganglia [left putamen (lPU) and caudate (lCN)] and for cortical gray matter [left frontal cortex (lfrCx)]. (Right) PIX–PVC (+) and ROI–PVC (○) estimates of regional activity concentrations after PVC, with the left front cortex taken (i) in PIX–PVC as a reference G_1 region for the computation of the BG-corrected image [Eq. (11)] and (ii) in ROI–PVC as a reference region for extracting observed cortical gray matter (G_1) activity and corresponding elements in the geometric transfer matrix (GTM) [Eq. (17)].

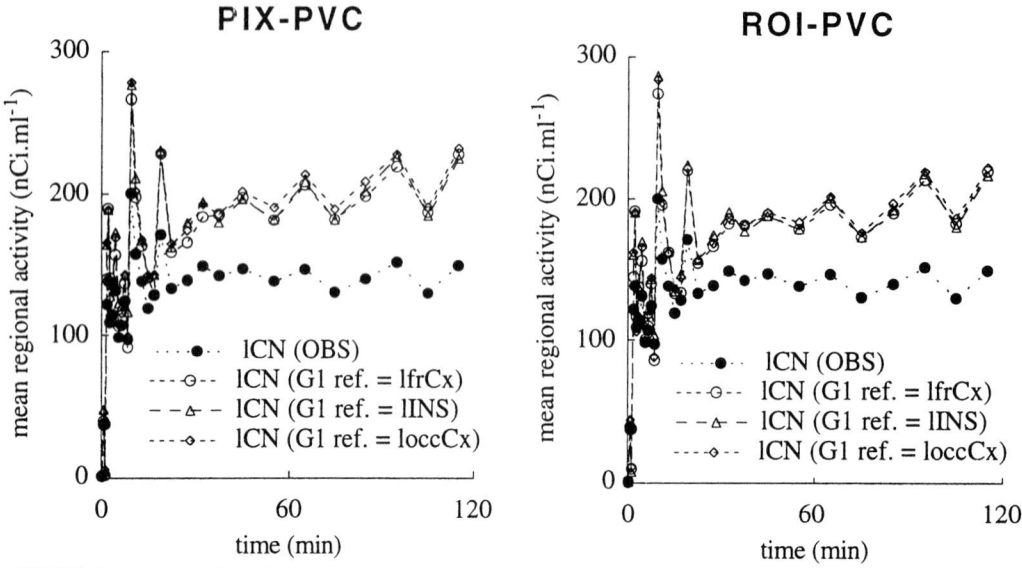

FIGURE 6 Observed (OBS) and PVC-corrected time–activity curves for the left caudate nucleus for various cortical gray matter reference regions [left frontal cortex (lfrCx), left insula (lINS), and left occipital cortex (loccCx)] for PIX–PVC and ROI–PVC techniques.

An interesting finding concerns corrected estimates of cortical gray matter (G_1) regions. Although observed TACs extracted from the insula would suggest that this region has a higher uptake than other cortical regions (Fig. 7), the corrected curves appeared to be of a lower magnitude than that of frontal cortex after PVC. This would result from correction of activity spillover from putaminal regions. Furthermore, although frontal and occipital cortices had a similar (observed) TAC shape, the occipital region reached the lowest level of tracer uptake after PVC (not shown). This corroborates the finding that the occipital cortex region has less dopamine terminals than other cortical gray matter regions of the brain (Berger and Gaspar, 1994).

IV. CONCLUSION

This comparison study indicates that both PIX–PVC and ROI–PVC methods perform corrections with equivalent accuracy in the case of simulated data. The

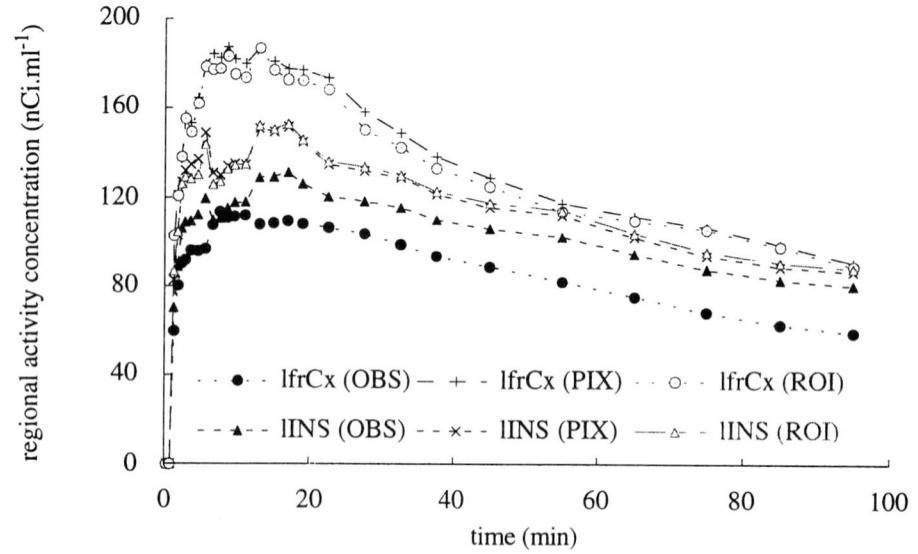

FIGURE 7 Average time–activity curves obtained from four consecutive PET slices in the case of real data of FDOPA uptake for the left frontal cortex (lfrCx) and the left insula (lINS), before PVC (OBS), after PIX–PVC (PIX), and after ROI–PVC (ROI).

application of the PVC methods to real data does not allow to favor one method over the other, although ROI–PVC seems to provide more consistent estimates. The authors also believe that ROI–PVC is easier to implement in the case of more heterogeneous tracer distributions, since it does not require any *a priori* knowledge of tracer levels, as is the case for PIX–PVC. It must also be noted that because the authors were only interested in the propagation of error due to an heterogeneous tracer distribution within cortical gray matter, this study does not include the effect of potential white matter heterogeneity and/or inaccuracy on PIX–PVC estimates.

References

Berger, B., and Gaspar, P. (1994). Comparative anatomy of the catecholaminergic innervation of rat and primate cerebral cortex. In "Phylogeny and Development of Catecholamine Systems in the CNS of Vertebrates" (J. A. Wilhelmus, J. Smeets, and A. Reiner, eds.), pp. 293–324. Cambridge University Press, Cambridge, U.K.

Evans, A. C., Thompson, C. J., Marrett, S., Meyer, E., and Mazza, M. (1991a). Performance evaluation of the PC-2048: A new 15-slice encoded-crystal PET scanner for neurological studies. *IEEE Trans. Med. Imag.* **10**: 90–98.

Evans, A. C., Marrett, S., and Collins, L. (1991b). MRI-PET correlation in three dimensions using a volume-of-interest (VOI) atlas. *J. Cereb. Blood Flow Metab.* **11**(2): A69–A78.

Hoffman, E. J., Huang, S. C., and Phelps, E. (1979). Quantitation in positron emission tomography. 1. Effect of object size. *J. Comput. Assist. Tomogr.* **3**: 299–308.

Ma, Y., Kamber, M., and Evans, A. C. (1993). 3-D simulation of PET brain images using segmented MRI and positron tomograph characteristics. *Comput. Med. Imag. Graphics* **11**(4/5): 365–371.

Meltzer, C. D., Zubieta, J. K., Links, J. M., Brakeman, P., Stumpf, M. J., and Frost, J. J. (1996). MR-based correction of brain PET measurements for heterogeneous gray matter radioactivity distribution. *J. Cereb. Blood Flow Metab.* **16**: 650–658.

Müller-Gärtner, H., Links, J., Leprince, J., Bryan, R., McVeigh, E., Leal, J., Devatzikos, C., and Frost, J. J. (1992). Measurement of radiotracer concentration in brain gray matter using positron emission tomography: MRI-based correction for partial volume effects. *J. Cereb. Blood Flow Metab.* **12**: 571–583.

Ouyang, X., Wong, W. H., Johnson, V. E., Hu, X., and Chen, C. T. (1994). Incorporation of correlated structural images in PET image reconstruction. *IEEE Trans. Med. Imag.* **13**: 627–640.

Rousset, O., Ma, Y., Kamber, M., and Evans, A. C. (1993a). 3-D simulations of radio tracer uptake in deep nuclei of human brain. *Comput. Med. Imag. Graphics* **11**: 373–379.

Rousset, O. G., Ma, Y., Léger, G. C., Gjedde, A. H., and Evans, A. C. (1993b). Correction for partial volume effects in PET using MRI-based 3D simulations of individual human brain metabolism. In "Quantification of Brain Function: Tracer Kinetics and Image Analysis in Brain PET" (K. Uemura, N. A. Lassen, T. Jones, and I. Kanno, eds.), pp. 113–125. Elsevier, Amsterdam and New York.

CHAPTER 11

Impact of Partial Volume Correction on Kinetic Parameters: Preliminary Experience in Patient Studies

F. YOKOI, O. G. ROUSSET,* A. S. DOGAN, S. MARENCO, A. C. EVANS,* A. H. GJEDDE,[†] and D. F. WONG

Department of Radiology
Division of Nuclear Medicine
Johns Hopkins Medical Institutions
Baltimore, Maryland 21287
*Positron Imaging Laboratories
McConnell Brain Imaging Centre
Montréal Neurological Institute
McGill University
Montréal, Québec, H3A 2B4 Canada
[†] PET Center
Aarhus University Hospital
DK8000C Aarhus C, Denmark

Two radioligands specific for the dopamine system were used to study the influence of correction for partial volume effects on quantitative binding data obtained from positron emission tomography from various brain regions. [^{11}C]Raclopride binds to the dopamine D_2 and D_3 receptors, and [^{11}C]WIN 35,428 binds to the dopamine transporter. Both are highly concentrated in the basal ganglia but not in the cerebellum. Normal subjects and patients with Lesch-Nyhan syndrome, a disease in which brain atrophy occurs, were studied before and after correction for the partial volume effect. A dramatic elevation in rate constants, especially the k_3/k_4 ratio, was observed after correction in both normal subjects and patients. The patients, who initially had a decrease in [^{11}C]WIN 35,428 binding at the dopamine transporter site compared to normal subjects, retained this separation after the correction algorithm was applied.

I. INTRODUCTION

The accuracy of positron emission tomography (PET) for measuring *in vivo* radioactivity is affected by the limited spatial resolution of PET scanners. Two components, referred to as partial volume effects (PVE), influence the observed radioactivity measurement in small tissue structures. The first component is a diminished recovery of the signal that originates from radioactivity within the tissue structure (Hoffman et al., 1979). The second is contamination from radioactivity that originates from surrounding tissue structures (spillover effect) (Herrero et al., 1988). The radioactivity in each tissue structure is a mixture of radioactivity originating from the structure itself and from adjacent tissue structures that have different tracer kinetics.

The goal of this chapter is to examine the impact of partial volume correction (PVC) on both the entire time–activity curve (TAC) and the subsequent analysis of kinetic parameters. Kinetic analysis is applied to TACs obtained before and after PVC for two PET radioligands: [^{11}C]raclopride ([^{11}C]RAC), which binds to the dopamine D_2 and D_3 receptors, and [^{11}C]WIN 35,428 ([^{11}C]WIN), which binds to the dopamine transporter. The basal ganglia are rich in both dopamine D_2 and D_3 receptors and the dopamine transporter. Dopamine receptors and the transporter are known to be negligible in the cerebellum.

II. METHODS

A. Subjects

Six normal subjects with no history of psychiatric or neurologic disease (three females and three males; age range 22–42 years) were studied with [^{11}C]RAC. Seven normal subjects (age range 21–32 years) and four patients with Lesch-Nyhan syndrome (LNS; age range 23–32 years) were studied with [^{11}C]WIN. Informed consent was obtained from all subjects prior to the PET studies.

B. Synthesis of [^{11}C]WIN and [^{11}C]RAC

[^{11}C]Carbon dioxide was produced by the ^{14}N(p, α)^{11}C nuclear reaction at Johns Hopkins Medical Institutions, and [^{11}C]methyl iodide was prepared from [^{11}C]carbon dioxide with LiAlH$_4$, tetrahydrofuran and hydriodic acid.

[^{11}C]RAC was synthesized by O-alkylation of a phenolic (S) precursor from raclopride with [^{11}C]methyl iodide. The radiotracer was purified using a reversed-phase semipreparative column (specific activity: 1520–2990 mCi/μmol). Synthesis and preparation procedures are slightly modified from the original method of Ehrin et al. (1987).

[^{11}C]WIN [2β-carbomethoxy-3β-(4-fluorophenyl)-[N-^{11}C-methyl]tropane] was prepared by N-methylation of nor-methyl precursor in dimethylformamide with [^{11}C]methyliodide. The radiotracer was purified using reversed-phase semipreparative column. The specific activity is 1890–3650 mCi/μmol. The time for synthesis, including formulation, was approximately 20 min (Dannals et al., 1993).

C. PET Procedure

Twenty milliCuries of [^{11}C]RAC or [^{11}C]WIN were injected intravenously over 30–40 sec. A 50 time-point dynamic PET scan (obtained with the GE 4096 Plus PET system) was performed over 90 min after tracer injection. The regions of interest (ROIs) on the PET images were drawn on the right and left caudate nucleus (4 × 4 pixels, 8 mm × 8 mm × 6 mm), putamen (4 × 4 pixels, 8 × 8 × 6 mm), and cerebellar cortex (7 × 5 pixels, 14 × 10 × 6 mm). TACs for brain ROIs were obtained with decay correction according to the following dynamic scan mode: eight 15-sec scans, sixteen 30-sec scans, six 60-sec scans, six 120-sec scans, eleven 240-sec scans, and three 360-sec scans. These brain TACs were used for kinetic modeling with metabolite-corrected plasma TACs.

A radial arterial catheter was inserted in order to collect 2 ml whole blood samples according to the following protocol: 3- or 4-sec intervals for the first 2 min after the tracer injection, 1-min intervals from 2 to 10 min, 2-min intervals from 10 to 20 min, 5-min intervals from 20 to 30 min, and 15-min intervals from 30 to 90 min. The heparinized whole blood samples were centrifuged, and the radioactivity in 500 μl of arterial plasma was measured and decay corrected to the time of tracer injection.

The plasma metabolites of [^{11}C]WIN and [^{11}C]RAC were assayed using high-performance liquid chromatography. The plasma metabolite fractions were used to correct the plasma input functions.

D. Correction for Partial Volume Effect

The procedure of Rousset et al. (1993) and Ma et al. (1993), which is described later, was used to correct for the partial volume effect on PET images.

1. Creation of a Customized Three-Dimensional Brain Phantom

High-resolution magnetic resonance (MR) images (SPGR: TE 5, TR 35: 0.96 × 0.96 × 1.5 mm pixel) of the brain were registered with corresponding PET data using a three-dimensional (3D) landmark matching algorithm. The MR image volume was then resliced to provide 50 images (2 mm thick) covering the volume of interest in the same plane as the PET scan. These images were segmented into tissue maps of gray matter, white matter, and cerebrospinal fluid using a 3D supervised decision tree classifier (Kamber et al., 1992). Precise anatomical outlines corresponding to specific structures of interest, i.e., caudate nucleus and putamen, were drawn manually wherever they appeared on the MR images, and all the voxels within each structure were identified. The coded brain structure can be assigned with tracer concentrations to create a 3D brain phantom representing true radioactivity distributions. The tracer was assumed to be distributed uniformly over the volume of each simulated brain tissue. [^{11}C]RAC and [^{11}C]WIN are specifically taken up in the striatum; the gray matter and white matter regions can be used as homogeneous regions of background radioactivity.

2. Acquisition of PET Data

A 90-min [^{11}C]RAC PET scan provided a data set of 15 slices (6 mm thick, 6.5 mm axial and 6 mm transaxial resolution) × 50 frames. Data were reconstructed by filtered backprojection with a 6-mm Hanning filter after correction for attenuation using a transmission scan.

3. Generation of Simulated PET Data

The brain phantom was sampled according to the geometry of the GE 4096 Plus PET system. The 3D point spread function (PSF) of the tomograph was approximated by a spatially invariant Gaussian function of 6 × 6 × 6 mm full-width-at-half-maximum (FWHM). The phantom was smoothed in the axial direction to generate a set of 6-mm-thick images spaced at 6.5 mm. Projections were then computed from each axially weighted slice using a simulation program (Herman et al., 1989). Poisson noise was included at the projection stage for each time point based on count densities equivalent to those found in actual dynamic PET scans. Simulated projections were reconstructed with the filtered backprojection algorithm provided on the GE 4096 Plus PET system.

4. Image Data Analysis

Quantitative analysis of both simulated and real images was performed using anatomical ROIs provided by the correlated MR images. The polygonal ROIs were placed on the striatum to extract the mean radioactivity of [^{11}C]RAC and [^{11}C]WIN within the caudate nucleus and putamen. For the estimation of activity in the background surrounding these structures, multiple large irregular ROIs (two gray matter regions and two white matter regions) surrounding the striatum were chosen. The same ROI templates were also used for the analysis of real data. Uncorrected TACs of real images and corrected TACs from simulated images for caudate, putamen, gray matter (including cerebral and cerebellar cortex), and white matter (including cerebral and cerebellar cortex) were generated using the MR image-based ROI templates.

5. Principle of the Correction Method

The correction method for PVE is derived directly from the properties of the image formation process. The PET image is the result of a convolution of the 3D PSF of the system with the activity distribution. Because convolution is a linear operation, the measured activity within a particular tissue is the weighted sum of the true activity from all the tissues contributing to the measurement. Considering a tissue structure of interest surrounded by $(N - 1)$ functional tissue components that have true activity concentration T_j, the apparent activity t_i of the tissue structure is given by

$$t_i = \sum_{j=1}^{N} \omega_{ij} T_j, \quad (1)$$

where ω_{ij} represent the fractions of activity from each component contributing to the observed measurement for any ROI. ω_{ij} depend on the geometrical relationships between the structure of interest and the ROIs, and on image resolution, but are independent of the tracer activity distributions.

E. Kinetic Modeling

Two models were used in data analysis.

1. Model I

This model was applied to analysis of data obtained with [^{11}C]RAC and [^{11}C]WIN. The model applied to the basal ganglia consisted of three compartments (a plasma compartment, a free compartment, and a specifically bound compartment) and four parameters (K_1, k_2, k_3, and k_4). The nonspecifically bound compartment was included in the free compartment. Parameters K_1 and k_2 are rate constants for the transport of tracer from the plasma compartment to the free compartment and from the free compartment to the plasma compartment, respectively; k_3 is a rate constant for the transport of tracer from the free compartment to the specifically bound compartment; and k_4 is a rate constant from the specifically bound compartment to the free compartment.

Parameters K_1' and k_2' are rate constants for the transport of tracer from the plasma compartment to the free compartment for the cerebellum and from the cerebellar free compartment to the plasma compartment, respectively. Rate constants K_1' and k_2' were estimated using a two-compartment model that consists of a plasma compartment and a free compartment in the cerebellum (C_{cereb}):

$$\frac{dC_{\text{cereb}}(t)}{dt} = K_1' C_p(t) - k_2' C_{\text{cereb}}(t). \quad (2)$$

The K_1/k_2 ratio of basal ganglia was constrained to the cerebellar K_1'/k_2'. This assumption reduces the number of parameters needed to analyze data from basal ganglia and makes it possible to fit TACs to the dynamic parameters using a least-square minimization procedure. In basal ganglia,

$$\frac{dC_f(t)}{dt} = K_1 C_p(t) - (k_2 + k_3) C_f(t) + k_4 C_b(t) \quad (3)$$

and

$$\frac{dC_b(t)}{dt} = k_3 C_f(t) - k_4 C_b(t), \quad (4)$$

where C_p, C_f, and C_b represent the quantity of [^{11}C]RAC in the plasma, free compartment, and specific-bound compartment of basal ganglia, respectively.

The optimal values of the rate constants were estimated using a least-square minimization procedure (a variant of the Levenberg–Marquardt algorithm). The k_3/k_4 ratio was used to evaluate the specific binding of [^{11}C]RAC (Farde et al., 1989; Lammertsma et al., 1996).

2. Model II

A nonlinear regression model was applied to [^{11}C]WIN data that were acquired over 44 min. The model applied to the basal ganglia consists of three compartments and three parameters (K_1, k_2, and k_3) with constraint of the cerebellar K_1'/k_2' ratio. The rate constant k_4 can be ignored. Rate constants K_1' and k_2' were estimated using a two-compartment model that consists of a plasma compartment and a free compartment of the cerebellum (C_{cereb}):

$$\frac{dC_{cereb}(t)}{dt} = K_1' C_p(t) - k_2' C_{cereb}(t). \quad (5)$$

The cerebellar partition coefficient K_1'/k_2', was assumed to be identical to K_1/k_2 of the basal ganglia. For the basal ganglia, the following equations can be written:

$$\frac{dC_f(t)}{dt} = K_1 C_p(t) - k_2 C_f(t) - k_3 C_b(t) \quad (6)$$

$$\frac{dC_b(t)}{dt} = k_2 C_f(t), \quad (7)$$

where C_p, C_f, and C_b represent the quantity of [^{11}C]WIN in the plasma, free compartment, and specific-bound compartment, respectively. The estimated value of rate constant k_3 was used to evaluate the specific binding of [^{11}C]WIN.

III. RESULTS

Typical time–activity curves of [^{11}C]RAC in caudate nucleus and cerebellum before and after PVC are shown in Fig. 1. After correction for the PVE, TACs from the caudate nucleus and putamen are higher than the original, uncorrected time–activity curves. The post-PVC time–activity curve from the cerebellum is almost indistinguishable from its original TAC. Typical TACs for [^{11}C]WIN in caudate nucleus and cerebellum before and after PVC are shown in Fig. 2. The post-PVC time–activity curve from the basal ganglia is also much higher than the original pre-PVC time–activity curve. The cerebellar post-PVC time–activity curve is almost identical to its original TAC.

The percentage changes in the cerebellar K_1/k_2 ratio and the k_3/k_4 ratio for [^{11}C]RAC in the caudate nucleus and putamen are shown in Table 1. The change in the cerebellar K_1/k_2 ratio is approximately 12%, the k_3/k_4 ratio of the caudate nucleus is 71%, and the putamen is 44%.

The change in the k_3/k_4 ratio of [^{11}C]WIN in caudate nucleus and putamen, which was estimated with the three-compartment, four-parameter model, is shown in Table 2. The change in the k_3 value in the caudate nucleus and putamen estimated with a three-compartment, three-parameter model is also shown in Table 2. After correction for partial volume, the k_3/k_4 ratio in the caudate nucleus increased 95% and the k_3 value increased 145%. The k_3/k_4 ratio for the putamen increased 225% and the k_3 value increased 71%.

The k_3/k_4 ratios for the caudate nucleus of patients with LNS before and after PVC are shown in Table 3. The original k_3/k_4 ratio for normal subjects was 6.40 and for LNS patients was 2.51. After correction, k_3/k_4 ratios for normals increased to 12.5 and for LNS patients increased to 7.53. The percentage changes in the k_3/k_4 ratio for normal subjects and LNS patients were 95.3 and 200%, respectively. Both pre- and post-PVC k_3/k_4 ratios of LNS patients are significantly lower than those of the age-matched normal subjects (two-sample t test: $p < 0.05$).

IV. DISCUSSION

This chapter showed the impact and clinical relevance of applying PVC to PET scans of dopamine receptors and dopamine transporter in living human brain.

The activity concentration of [^{11}C]RAC and [^{11}C]WIN and the k_3/k_4 ratio of the basal ganglia showed a dramatic elevation after the correction, both in normal subjects and in patients with Lesch-Nyhan syndrome, a genetic disorder that is clinically characterized by hyperuricemia, choreoathetosis, dystonia, and compulsive self-injury. It was previously reported that dopamin transporters are reduced in LNS relative to normals (Wong et al., 1996). However, there was also a tendency for reduced caudate nucleus and putamen volumes on volumetric MR images compared to control subjects. Therefore, it is possible that the observed reduction of radioactivity and the estimated k_3/k_4 ratio are due to PVE rather than an actual reduction of dopamine transporter. The post-PVC ratio of k_3/k_4 for [^{11}C]WIN appears to be more reliable than the

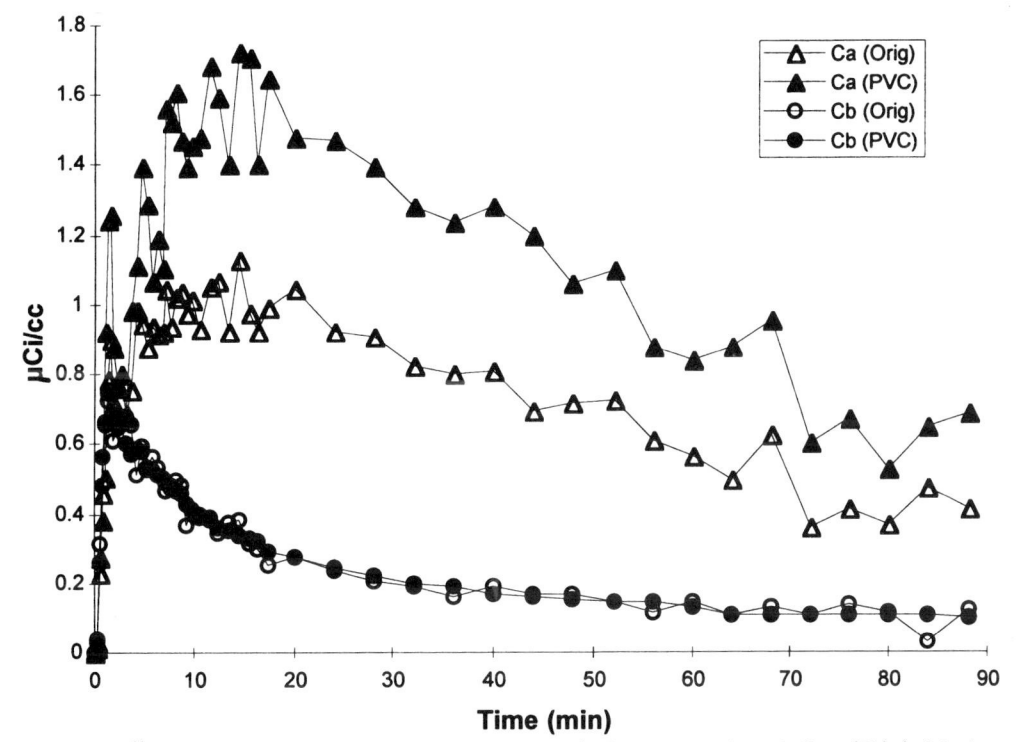

FIGURE 1 [^{11}C]Raclopride: Time-activity curve of caudate (Ca) nucleus and cerebellum (Cb) (original vs PVC).

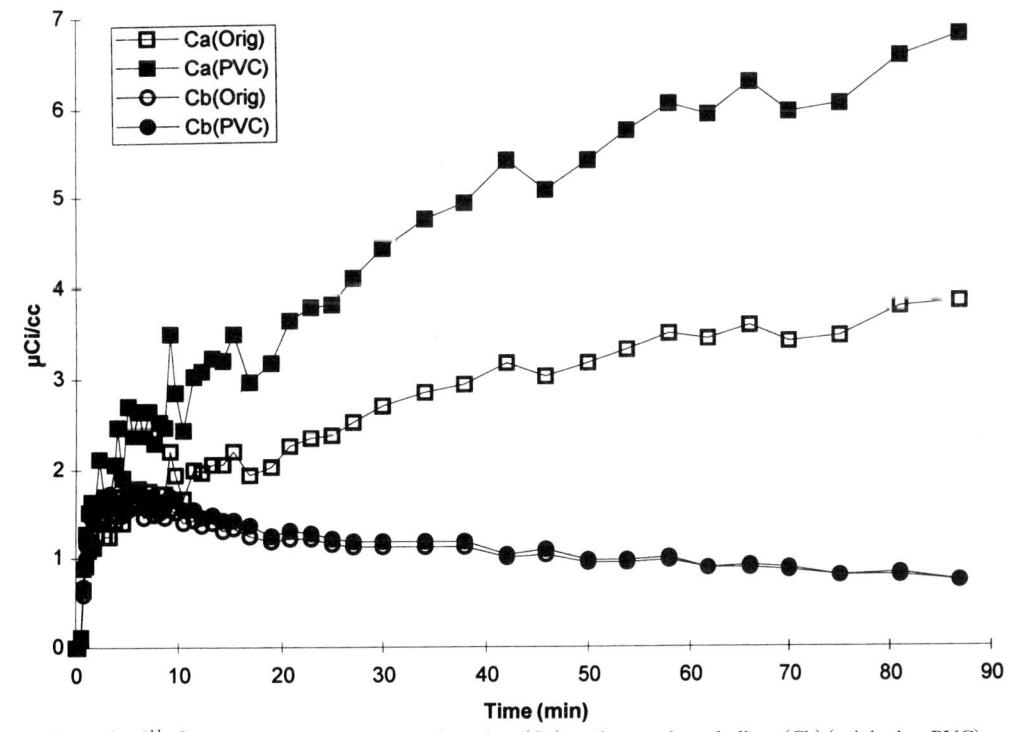

FIGURE 2 [^{11}C]WIN: Time-activity curve of caudate (Ca) nucleus and cerebellum (Cb) (original vs PVC).

TABLE 1 K_1'/k_2' and k_2/k_4 Ratio of Normal Volunteers: [^{11}C]RAC (Mean ± SEM)

	Original	Post-PVC	% change
Cerebellum K_1'/k_2'	0.37 ± 0.02	0.41 ± 0.03	11.54 ± 6.80
Caudate k_3/k_4	2.69 ± 0.22	4.60 ± 0.45	70.55 ± 5.18
Putamen k_3/k_4	3.43 ± 0.26	4.91 ± 0.44	43.68 ± 8.47

TABLE 2 k_3/k_4 Ratio of Normal Volunteers with [^{11}C]WIN (Mean ± SEM)

	Original	Post-PVC	% change
Caudate			
k_3/k_4	6.40 ± 1.0	12.5 ± 1.1	95.3
k_3 (1/min)	0.20 ± 0.04	0.49 ± 0.11	145
Putamen			
k_3/k_4	5.80 ± 0.3	18.9 ± 2.3	225
k_3 (1/min)	0.14 ± 0.01	0.24 ± 0.02	71.4

TABLE 3 Caudate Nucleus k_3/k_4 Values in Lesch-Nyhan Syndrome (LNS) and Normal Volunteers with [^{11}C]WIN (Mean ± SEM)

Caudate/Nucleus	Original	Post-PVC	% change
Normal	6.40 ± 1.0	12.5 ± 1.1	95.3
LNS	2.51 ± 0.3	7.53 ± 1.4	200

original k_3/k_4 ratio to evaluate the specific binding of [^{11}C]WIN. Both pre- and post-PVC k_3 values of LNS are lower than those of normal subjects. A significant loss of dopaminergic nerve terminals and neurons in LNS might reduce the number of dopamine transporter sites. The [^{11}C]WIN binding at the dopamine site of atrophic basal ganglia of LNS is still decreased after PVC compared to that of normal subjects.

Acknowledgments

Special thanks to Dr. James Harris, Dr. Charles Hong, Dr. Gerhard Gründer, and Dr. Sakkubai Naidu who collaborated on patient referral, patient care, and PET scan acquisition. This work was supported in part by Public Health Service Grants PO1 HD24448, PO1 HD24061, RO1 MH42821, and DA 09482.

References

Dannals, R. F., Neumeyer, J. L., Milius, R. A., Ravert, H. T., Wilson, A. A., and Wagner, H. N., Jr. (1993). Synthesis of radiotracer for studying dopamine uptake sites *in vivo* using PET: 2β-carbomethoxy-3β-94-fluorophenyl)-[N-^{11}C-methyl]tropane ([^{11}C]CFT or [^{11}C]WIN-35,428). *J. Labelled Compd. Radiopharm.* **33**: 147–1572.

Ehrin, E., Gawell, L., Hogberg, T., de Paulis, T., and Strom, P. (1987). Synthesis of [methoxy-^{2}H]- and [methoxy-^{11}C]-labelled raclopride. Specific dopamine-D2 receptor ligands. *J. Labelled Compd. Radiopharm.* **26**: 931–940.

Farde, L., Erickson, L., Bloomquist, G., and Hallidin, C. (1989). Kinetic analysis of central [^{11}C]raclopride binding to D$_2$-dopamine receptors studied by PET: A comparison to the equilibrium analysis. *J. Cereb. Blood Flow Metab.* **9**: 696–708.

Herman, G. T., Lewitt, R. M., Odhner, D., and Rowland, S. W. (1989). "Snark 89, a Programming System for Image Reconstruction from Projections." Tech. Rep. No. MIPG160. University of Pennsylvania, Philadelphia.

Herrero, P., Markham, J., Meyars, D. W., Weinheimer, C. J., and Bergman, S. R. (1988). Measurement of myocardial blood flow with positron emission tomography: Correction for count spillover and partial volume effects. *Math. Comput. Model.* **11**: 807–812.

Hoffman, E. J., Huang, S. C., and Phelps, M. E. (1979). Quantification in positron emission computed tomography: 1. Effect of object size. *J. Comput. Assist. Tomogr.* **3**: 299–308.

Kamber, M., Collins, D. L., Shinghal, R., Francis, G. S., and Evans, A. C. (1992). Model-based 3D segmentation of multiple sclerosis lesions in dual-echo MRI data. *Proc. SPIE* **1808**: 590–600.

Lammertsma, A. A., Bench, C. J., Hume, S. P., Osman, S., Gunn, K., Brooks, D. J., and Frackowiak, R. S. J. (1996). Comparison of methods for analysis of clinical [^{11}C]raclopride studies. *J. Cereb. Blood Flow Metab.* **16**: 42–52.

Ma, Y., Kamber, M., and Evans, A. C. (1993). 3D simulation of PET brain images using segmented MRI data and position tomograph characteristics. *Comput. Med. Imag. Graphics* **17**: 365–371.

Rousset, O., Ma, Y., Kamber, M., and Evans, A. C. (1993). 3D simulations of radiotracer uptake in deep nuclei of human brain. *Comput. Med. Imag. Graphics* **17**: 373–379.

Wong, D. F., Harris, J. C., Naidu, S., Yokoi, F., Marenco, S., Dannals, R. F., Ravert, H. T., Yaster, M., Evans, A., Rousset, O., Bryan, R. N., Gjedde, A., Kuhar, M. J., and Breese, G. R. (1996). Dopamine transporters are markedly reduced in Lesch-Nyhan disease *in vivo*. *Proc. Natl. Acad. Sci. U.S.A.* **93**: 5539–5543.

SECTION II

IMAGE PROCESSING

CHAPTER 12

Performance Characterization of a Feature-Matching Axial Smoothing Method for Brain PET Images

S.-C. HUANG, J. YANG, C. L. YU,* and K. P. LIN*

Division of Nuclear Medicine and Biophysics
Department of Molecular and Medical Pharmacology
UCLA School of Medicine
Los Angeles, California 90095
* Department of Electrical Engineering
Chung-Yuan University
Chung-li, Taiwan
Republic of China

A feature-matching axial smoothing method for images obtained with new generation positron emission tomography (PET) scanners has been introduced before and has been shown to enhance visually the signal-to-noise ratio without generating adverse artifacts. This chapter studies the performance of the method quantitatively by applying the method to multiple realizations of computer-simulated brain PET images and to PET images of the Hoffman brain phantom. The pixel-by-pixel noise level was calculated as the sample standard deviation of multiple noise realizations. Spatial resolution degradation was evaluated by the frequency response function of the smoothing procedure. For brain PET images, the results show that the feature-matching method causes a smaller spatial resolution degradation than the conventional axial smoothing method that uses the same smoothing filter. For the conventional method to give a comparable resolution loss, a shorter smoothing filter must be used, which would result in less noise reduction (equivalent to ~ 70% fewer photon counts than with the new method). Moreover, the amount of resolution degradation and/or noise advantage of the feature-matching method does not appear to be dependent on the in-plane resolution or noise level of the original images.

I. INTRODUCTION

A feature-matching axial smoothing method for images obtained with new generation positron emission tomography (PET) scanners has been introduced previously (Huang *et al.*, 1996; Yang *et al.*, 1996). The method smooths each image value across planes along a path that is determined by elastic mapping (Lin *et al.*, 1996) to match closest to the features in the neighborhood of the specific image pixel (Huang *et al.*, 1996; Yang *et al.*, 1996). This new method has been shown to enhance visually the signal-to-noise ratio without generating adverse artifacts (Huang *et al.*, 1996; Yang *et al.*, 1996). However, the characteristics of the method have not been examined extensively. This study investigates the performance of the method quantitatively by assessing its impact on noise and spatial resolution.

II. METHODS

A. Computer Simulations

A set of magnetic resonance (MR) brain images (124 planes of 256 × 256 with a pixel size of 0.109 cm and a slice spacing of 0.14 cm) from a normal subject was segmented into gray matter, white matter, cerebrospinal fluid, and muscle (and others) regions (Zubal *et al.*, 1994). Concentration levels matching the normal brain [^{18}F]fluorodeoxyglucose uptake levels were assigned to the four different regions. These images were then projected at various angular directions to generate a set of sinograms (size 128 × 192 for each of 40 cross-sectional planes) with intrinsic resolution parameterized by the full-width-at-half-maximum (FWHM$_X$ = 0.42 cm and FWHM$_Z$ = 0.35 cm) and sampling distances ($dt = 0.1604$ cm, $dz = 0.28$ cm) that approximate the configuration of a Siemens/CTI EXACT HR+ scanner.

Various noise levels (from ~95,000 to 38 million counts/plane and with variance proportional to the mean) were added to the sinograms, which were then reconstructed using a filtered backprojection algorithm (Hann filter with a cutoff frequency of either 1.0 or 0.25 of the Nyquist frequency). The resultant image resolutions of the two reconstruction filters are about 0.5 and 1.6 cm FWHM and bracket the resolution range of most realistic brain PET images. For each noise level and reconstruction filter, 50 realizations were generated. A noise-free simulation was also generated for each resolution condition to serve as the reference for evaluation of the spatial resolution change of the smoothed images.

For each noise and resolution realization, the images were smoothed across planes by the new method and separately by the conventional method. The same axial smoothing filter ([0.1 0.2 0.4 0.2 0.1]) was used for both methods. The new method was implemented using the neural network-based elastic mapping method reported previously by Kosugi *et al.* (1993) and Lin and Huang (1993).

To analyze image data, images of the pixel-by-pixel mean and standard deviation (SD) of the smoothed images of the 50 realizations of each image noise/resolution condition were first generated. A two-dimensional (2D) fast Fourier transform was applied to the mean images of a plane to give the in-plane frequency spectrum of the smoothed images. The magnitude of the frequency components along the y-frequency axis (or the x-frequency axis) was selected to represent one-dimensional (1D) in-plane frequency spectrum. This representative spectrum was then divided by the spectrum of the corresponding original noise-free image to give the in-plane response function (FRF) of the smoothing method.

A similar procedure was used for calculating the frequency response function along the axial direction, except that the transaxial images were first converted to coronal/sagittal images, and the frequency spectrum along the axial direction was divided by the corresponding one from the noise-free reference image to give the axial frequency response function.

The SD image from the conventional method of axial smoothing represents the pixel-by-pixel noise level of those images. For the new method, the standard deviation images contain, in addition to the image noise level, a component that is due to the variation among different realizations in the small edge shift of the elastic mapping. The amount of this edge shift variation was quantitated by using the following equation that converted the SD of an image pixel value (SD) near an edge to the SD of the edge location shift (SD_{shift}). The equation assumes that the image value crossing an edge changes linearly over the region of largest slope:

$$SD_{shift} = SD_{image}/s_{max}. \quad (1)$$

The largest slope s_{max} was estimated from the image resolution (Gaussian with a FWHM = 2.36σ) and the magnitude of the image transition (dL) across an edge as

$$s_{max} = dL/\sigma\sqrt{2\pi}. \quad (2)$$

Therefore, the SD_{shift} can be estimated as

$$SD_{shift} = SD_{image}\sqrt{2\pi}\,FWHM/(2.36\,dL). \quad (3)$$

To separate out this component, a single elastic mapping was applied to all 50 realizations of each condition. The standard deviation image so obtained for each set gave the noise level of the processed images of the new method.

B. Physical Phantom Studies

The 3D Hoffman brain phantom was filled with ^{18}F solution and scanned by a Siemens/CTI EXACT HR+ scanner using a scanning sequence of 20×2-min, 10×5-min, 10×2-min frames. PET images of each frame were reconstructed using the filtered backprojection algorithm (ramp filter of cutoff frequency equal to the Nyquist frequency) to give transaxial images ($dx = 0.16875$ cm, $dz = 0.2425$ cm) of the ^{18}F radioactivity distribution in the brain phantom. The sinograms of all the frames were also summed and reconstructed to create the reference image (at a low noise level).

The images of a single 2-min frame (the first frame; ~250,000 counts/plane) were smoothed along the axial direction by the new method and the conventional method separately, but with the same axial smoothing filter ([0.1 0.2 0.4 0.2 0.1]). The magnitude of the frequency components of the processed images relative to those of the unprocessed image was calculated as the frequency response function of the processing method, similar to what was described earlier for the computer simulation studies.

III. RESULTS

Figure 1 shows the transaxial, coronal, and sagittal slices of a set of computer-simulated PET images before and after axial smoothing with the conventional

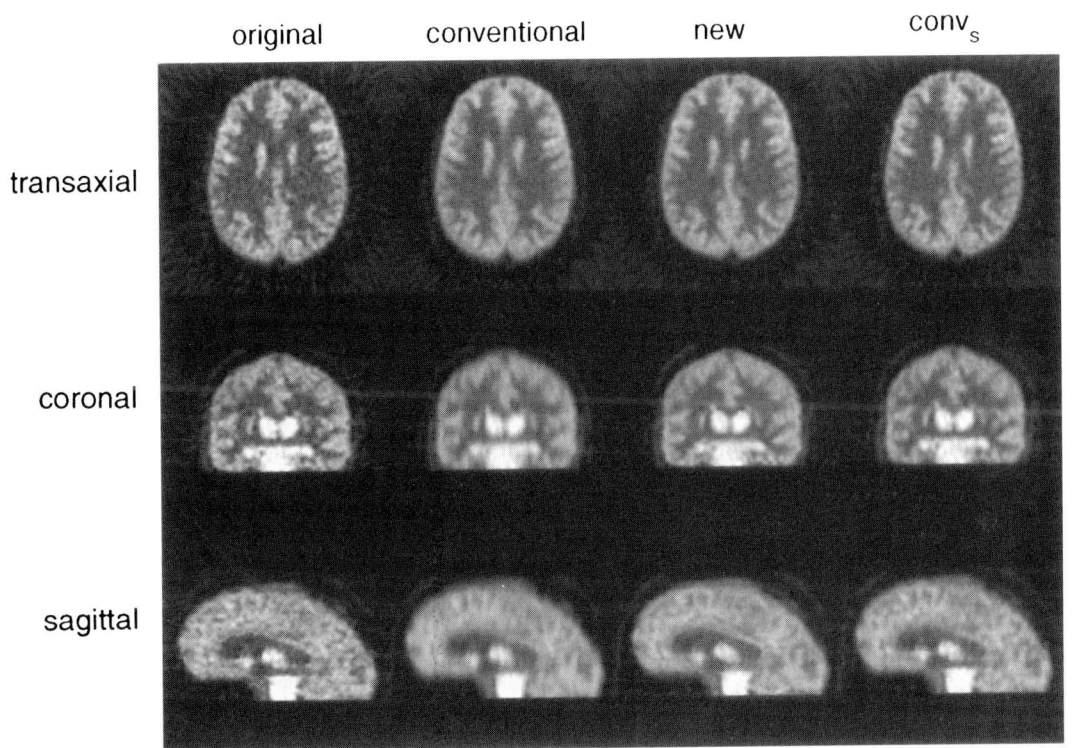

FIGURE 1 Transaxial, coronal, and sagittal slices of a set of computer-simulated PET images before (left column) and after axial smoothing with the conventional method (second column) and with the new method (third column). In the fourth column, conv$_S$ are images smoothed by the conventional method with a shorter filter to match the axial image resolution of the new method. The original transaxial images correspond to ~ 1.5 million counts/plane and were reconstructed using a filtered backprojection algorithm with a Hann filter of cutoff frequency equal to the Nyquist frequency.

method and the new method. Vertical smearing is seen on the coronal and sagittal images that were processed by the conventional method because the spatial resolution of these images is much worse along the axial direction than in the transaxial directions. Even though the same smoothing filter was used, the images processed using the new method do not have the vertical smearing, indicating less spatial resolution degradation with the new method than with the conventional method.

Figure 2 shows the mean and SD images calculated from 50 separate realizations of a set of computer-simulated images. The SD images of the original and the conventional method represent the noise levels before and after the smoothing. As described in the previous section, the SD images of the new method contain, in addition to the image noise level, a component that is due to the variation among different realizations in the edge shift of the elastic mapping. After the edge-shift component is separated out (by using the same elastic mappings for all 50 realizations), the SD image due to noise alone (SD′) is shown in the last row of Fig. 2. In the column "conv$_S$" are images for the conventional method with a shorter filter to match the axial image resolution of the new method (see Fig. 3). Similar magnitudes of SD images of the conventional method and the new method are as expected because the same smoothing filter was used. For the conventional method with a shorter filter (conv$_S$), the corresponding SD image is seen to be higher in magnitude.

Table 1 shows the noise level and the amount of edge shift due to different amounts of axial smoothing on different sets of simulated images. The amount of noise reduction is seen to be independent of image noise level and resolution and is consistent with the theoretical values of 0.51 and 0.67 for long and short filters, respectively. The amount of edge shift due to the elastic mapping is small (< 1 mm) for all the conditions and appears to be more related to image noise level than to image resolution.

Figure 3 shows the frequency response functions of the different axial smoothing methods that allow one to compare quantitatively the spatial resolution degradation of the different methods. For the same smoothing filter, the axial resolution of the resultant images is seen to be much worse for the conventional method

	original	conventional	new	conv$_s$
mean				
s.d.				
s.d.'				

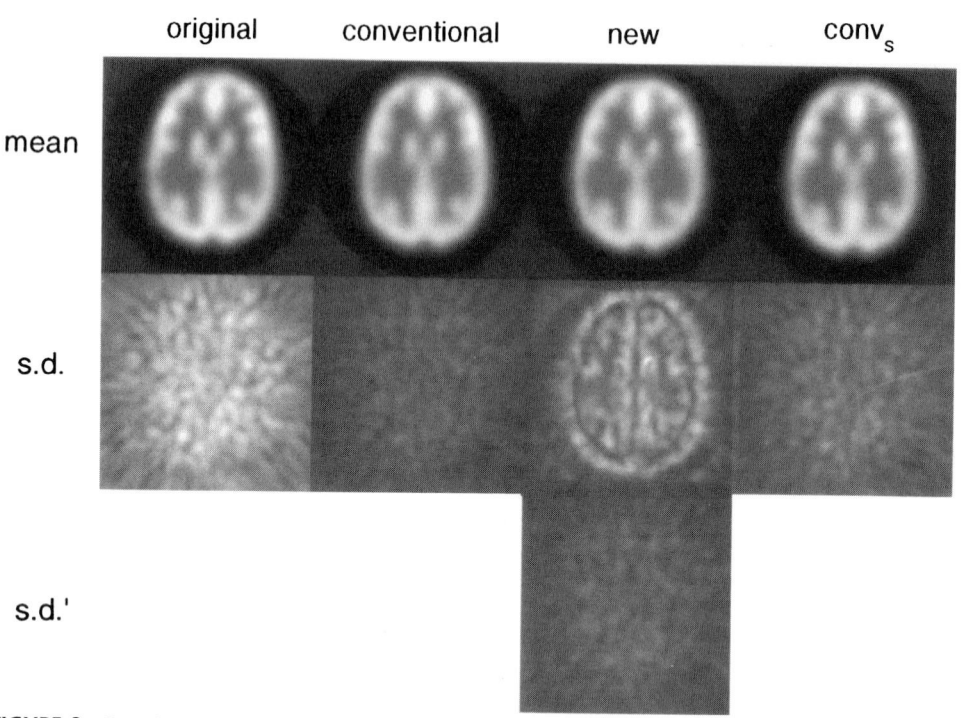

FIGURE 2 Sample mean and standard deviation (s.d.) images in a set of computer simulation studies. The simulated original images correspond to ~ 95,000 counts/plane and were reconstructed with a Hann filter of cutoff frequency equal to one-fourth of the Nyquist frequency. The images were smoothed axially with the conventional method and with the new method separately. The s.d. images of the original and the conventional method represent the noise levels before and after smoothing by the conventional method. The s.d. images of the new method contain, in addition to the image noise level, a component that is due to the variation among different realizations in the edge shift of the elastic mapping in the new method. After the edge-shift component is separated, the s.d. image due to noise alone (s.d.') is shown in the last row. The column conv$_S$ are images for the conventional method with a shorter filter to match the axial image resolution of the new method. All s.d. images were scaled with the same gray level scale so that the magnitudes across different s.d. images (including the s.d.' image) can be compared directly.

than for the new method. To achieve a comparable resolution, the conventional method needs to use a shorter filter ([0.2 0.6 0.2]) that yields less noise reduction (see Fig. 2 and Table 1). The noise reduction advantage (square of noise reduction values in Table 1) of the new method compared to the conventional method with the shorter filter (26% vs 44%) is equivalent to having about 70% more counts in the original image. Furthermore, the similarity of the axial FRFs between high and low in-plane resolutions indicates the insensitivity of the performance of the new method to the in-plane resolution of the original images. FRF results for other noise levels were also found to be similar and indicate that the performance is also independent of the noise level of the original images.

The axial smoothing results for real PET images of the 3D Hoffman phantom are shown in Figs. 4 and 5. The transaxial slices of the smoothed images with the conventional method and the new method are similar, but the coronal images clearly show the higher axial resolution of the images smoothed by the new method. This is shown quantitatively by the FRF analysis in Fig. 5. Furthermore, the similarity of the FRFs from the computer simulation studies (Fig. 3) and from actual PET images (Fig. 5) supports the validity of the computer simulations and further strengthens the conclusions drawn from these studies.

IV. SUMMARY

For brain PET images, the feature-matching method caused a much smaller spatial resolution degradation along the axial direction and a slightly smaller degradation in the transaxial direction as compared to the conventional axial smoothing method with the same smoothing filter. For the conventional method to give a comparable resolution loss, a shorter smoothing filter must be used, which would result in less noise reduction (equivalent to ~ 70% fewer photon counts than

12. Performance of A Feature-Matching Axial Smoothing Method

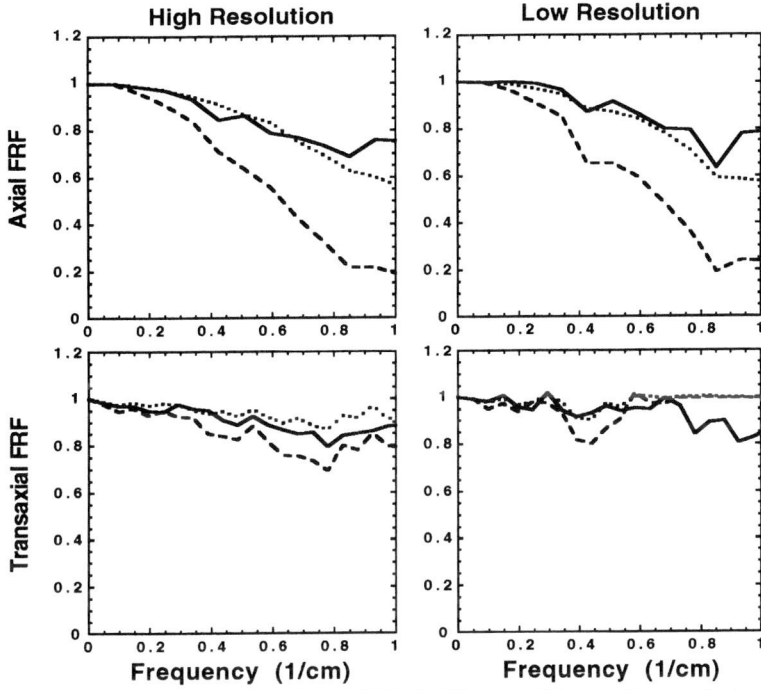

FIGURE 3 Frequency response functions (FRF) of different axial smoothing methods for two image resolution conditions. (Left) Images reconstructed with a Hann filter of 1.0 Nyquist cutoff; (right) images reconstructed with a Hann filter of 0.25 Nyquist cutoff. (Top) Frequencies along the axial direction. (Bottom) Transaxial directions (i.e., in-plane). Only FRF over the meaningful frequency range (i.e., the range with a significant signal component in the original images) are shown. Magnitudes of the FRF for the new method (solid curves) are shown to be higher than those of the conventional method (dashed curves), especially for FRF along the axial direction. Dotted curves represent the conventional method with a shorter filter that gives FRF comparable to those by the new method.

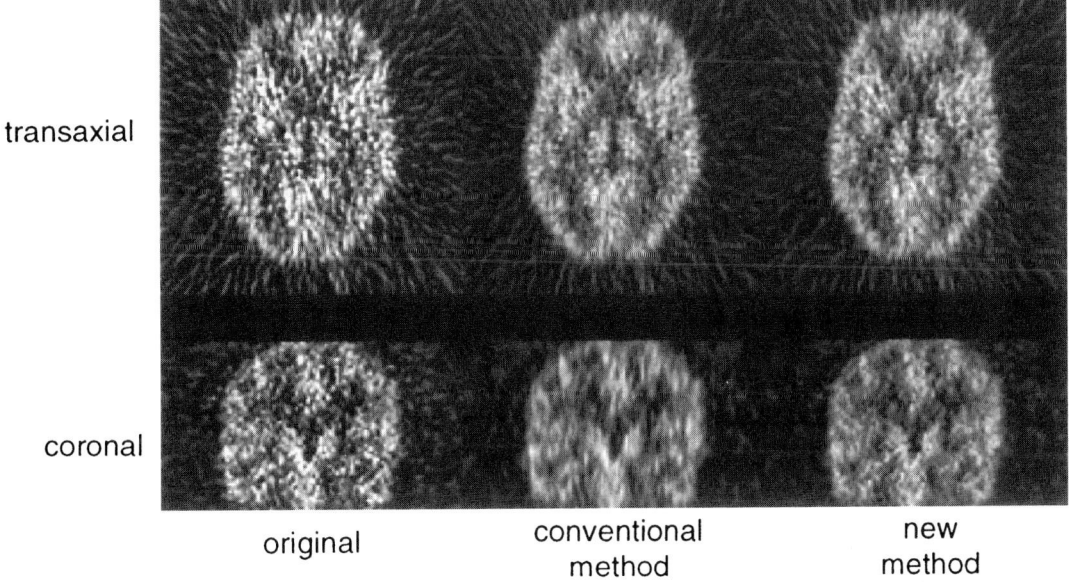

FIGURE 4 Transaxial and coronal images obtained from a 2-min scan ($\sim$ 250,000 counts/plane) of the 3D Hoffman brain phantom in a Siemens/CTI EXACT HR+ PET scanner. Images on the left are original images. Images on the right have been processed by the new method. Images in the middle column have been processed by the conventional method.

TABLE 1 Pixel Noise Levels and Edge Shifts of Different Axial Smoothing for Images of Various Resolution and Noise Level Conditions

	Conventional (long filter)	Feature-matching (long filter)	Conventional (short filter)
Noise	0.51[a]	0.51	0.67
reduction[b]	0.51	0.51	0.66
	0.51	0.54	0.65
	0.50	0.50	0.66
Edge shift (cm)		0.03	
		0.04	
		0.04	
		0.09	

[a] The four numbers in each group are, respectively, for four different image resolution and noise level conditions. The first condition corresponds to an in-plane resolution of 0.5 cm FWHM and about 38 million counts per image plane; the second condition corresponds to 0.5 cm FWHM resolution and 1.5 million counts per plane; the third condition corresponds to 1.6 cm FWHM in-plane resolution and 1.5 million counts per plane; and the fourth condition corresponds to 1.6 cm FWHM resolution and 95,000 counts per plane.

[b] The noise reduction was calculated as the pixel-wise sample standard deviation of the smoothed images divided by the pixel-wise standard deviation of the original unsmoothed images.

with the new method). Furthermore, the amount of axial resolution loss and noise level advantage of the feature-matching method appears to be independent of the in-plane resolution and noise level of the original images. The amount of edge shift caused by the feature-matching method is minimal compared to the in-plane resolution of the images.

Acknowledgment

The authors thank Mr. David Truong for computer system support. This work was partially supported by DOE Contract DE-FC03087ER60615.

References

Huang, S. C., Yang, J., Dahlbom, M., Hoh, C., Czernin, J., Zhou, Y., and Yu, D. C. (1996). Feature-matching axial averaging method for enhancing signal-to-noise ratio of images generated by new generation of PET scanners. In "Quantification of Brain Function Using PET" (R. Myers, V. Cunningham, D. Bailey, and T. Jones, eds.), pp. 147–151. Academic Press, San Diego, CA.

Kosugi, Y., Sase, M., Kuwatani, H., Kinoshita, N., Momose, T., Nishikawa, J., and Watanabe, T. (1993). Neural network mapping for nonlinear stereotactic normalization of brain MR images. J. Comput. Assist. Tomogr. 17: 455–460.

Lin, K. P., and Huang, S. C. (1993). An elastic mapping algorithm for tomographic images. Proc. Annu. Conf. Biomed. Eng., pp. 40–41.

Lin, K. P., Iida, H., Kanno, I., and Huang, S. C. (1996). An elastic image transformation method for 3D intersubject brain image mapping, In "Quantification of Brain Function Using PET" (R. Myers, V. Cunningham, D. Bailey, and T. Jones, eds.), pp. 404–409. Academic Press, San Diego, CA.

Yang, J., Huang, S. C., Lin, K. P., Hoh, C. K., Wolfenden, P., Dahlbom, M., Czernin, J., and Phelps, M. E. (1996). A new axial smoothing method based on elastic mapping. IEEE Trans. Nucl. Sci. 43: 3355–3360.

Zubal, I. G., Harrell, C. R., Smith, E. O., Rattner, Z., Gindi, G. R., and Hoffer, P. B. (1994). Computerized three-dimensional segmented human anatomy. Med. Phys. 21: 299–302.

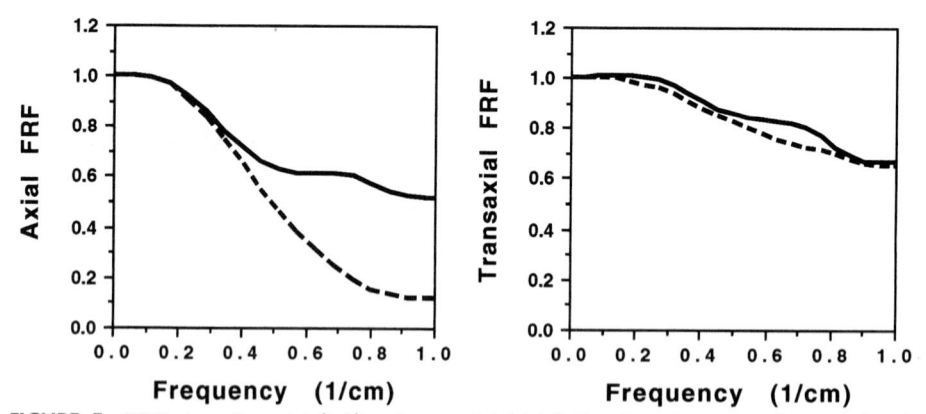

FIGURE 5 FRF along the axial (left) and transaxial (right) directions for new and conventional smoothing methods as estimated from the images shown in Fig. 4.

CHAPTER 13

Registration of Multitracer PET Data[1]

JESPER L. R. ANDERSSON
Uppsala University PET Centre
Subfemtomole Biorecognition Project
Uppsala, Sweden

For many questions the combined use of different tracers in positron emission tomography (PET) may yield important additional information. The evaluation of such studies is facilitated if the examinations are performed with exactly the same positioning of the subject. When this is not the case, a method for retrospective alignment is needed. Previously published methods for PET–PET registration are not generally applicable, as they assume a high degree of inherent similarity between the images. The method presented here uses the first image set to classify the other, attempting to explain the variance encountered in that image set. To maximize the available information, multivariate approaches are used both for the classification and when explaining the variance in the second image set. A principle component analysis is performed on both time series of images to yield a manageable number of images (dimensions) for each tracer. A multivariate classification is performed on the stationary image set and the other image set is translated and rotated relative to that. For each transformation the voxels in the second data set are divided into classes defined by the stationary set and Wilks' lambda (Λ^*) is evaluated, relating the variance remaining after classification to the total variance. The transformation that minimizes Λ^* is postulated to be that which aligns the images correctly. A variety of different tracer combinations was tried, with seemingly good results in a large majority of the cases. An unimpeachable quantification of the accuracy was difficult due to lack of a correct answer, but it was estimated to be approximately 2 mm.

---

[1] Transcripts of the BRAINPET97 discussion of this chapter can be found in Section VIII.

I. INTRODUCTION

When registering an image to another image a cost function is typically defined, and its value is subsequently maximized (or minimized) by means of an iterative method. The choice of cost function is based on the type of assumptions that may be made regarding the images at hand. When matching images with a similar information content, e.g., multiple positron emission tomography (PET) or magnetic resonance (MR) studies, relatively strong assumptions may be made, and a simple similarity measure (Woods *et al.*, 1992; Andersson, 1995) may suffice, whereas when matching dissimilar images that approach is normally not applicable.

Several approaches have been suggested to overcome the problem of registering dissimilar images. One family of algorithms for intermodality registration uses one data set to classify the other data set and assumes that data are aligned correctly when the classification is most successful at explaining the observed variance of the other data set (Woods *et al.*, 1993; Ardekani *et al.*, 1995).

Another situation where registration of dissimilar images is of interest is when performing studies with the same modality, but with different tracers. Although the correlation between, for example, a [^{11}C]raclopride image and a [^{15}O]H$_2$O image is poor, it is still possible even for the untrained eye to distinguish certain structures in both data sets. Hence there is some amount of mutual information that may be used to drive a registration. The information is utilized in this chapter by performing a multidimensional classification of the voxels in an image volume consisting of the first principal components of multitemporal data obtained with one of the tracers. That classification is used subsequently on the other image set, and the optimal orientation is considered to be that which minimizes Wilks' lambda (Λ^*) when performing a multivariate analysis of variance (MANOVA) (Johnson and Wichern, 1988) on the first principal components of the second data set based on the classification from the first set.

II. MATERIALS AND METHODS

Essentially all methods for image registration proceed along the same general path outlined in Fig. 1. One of the image volumes is held fixed, hereafter denoted the "reference" image volume, and the other, hereafter denoted the "reslice" image volume, is translated and rotated by multiplying the position vector of each voxel with a four-by-four transformation matrix. After each transformation the relative positioning is checked by evaluating a "cost function" that should have its minimum (or maximum) when the volumes are aligned correctly. After each evaluation of the cost function, prior to convergence, a new transformation matrix should be decided upon.

The properties of the cost function will determine whether its minimization (or maximization) will yield correctly aligned images, and most previously published papers differ mainly in their choice of cost function. Another potentially influential factor is the choice of minimization algorithm, which may determine whether the global minimum is found.

A. Cost Function

The evaluation of the cost function consists of the following steps:

1. Extraction of tracer kinetics
2. Delineation of the object
3. Classification of the reference image volume
4. Evaluation of Wilks' Λ

Steps one to three are performed once per registration, and step four once per iteration.

1. Extraction of Tracer Dynamics

Data are typically sampled for irregular intervals (frames) following injection of tracer, with interval lengths gradually increasing with time after injection. However, when looking at consecutive frames, a high degree of correlation is evident, which implies that the information might be represented in a more efficient way to facilitate further processing.

A previously published method was used, in which data were represented as the first few principal components of the full dynamics, with little or no loss of information (Pedersen *et al.*, 1994). Consider an $m \times k$ multitemporal image matrix **I** containing mean-adjusted values, where m is the number of voxels belonging to the object and k is the number of frames in the dynamic sequence.

$$\mathbf{I} = \begin{bmatrix} x_{11} - \bar{x}_1 & x_{12} - \bar{x}_2 & \cdots & x_{1k} - \bar{x}_k \\ x_{21} - \bar{x}_1 & x_{22} - \bar{x}_2 & \cdots & x_{2k} - \bar{x}_k \\ \vdots & \vdots & \ddots & \vdots \\ x_{m1} - \bar{x}_1 & x_{m2} - \bar{x}_2 & \cdots & x_{mk} - \bar{x}_k \end{bmatrix}. \quad (1)$$

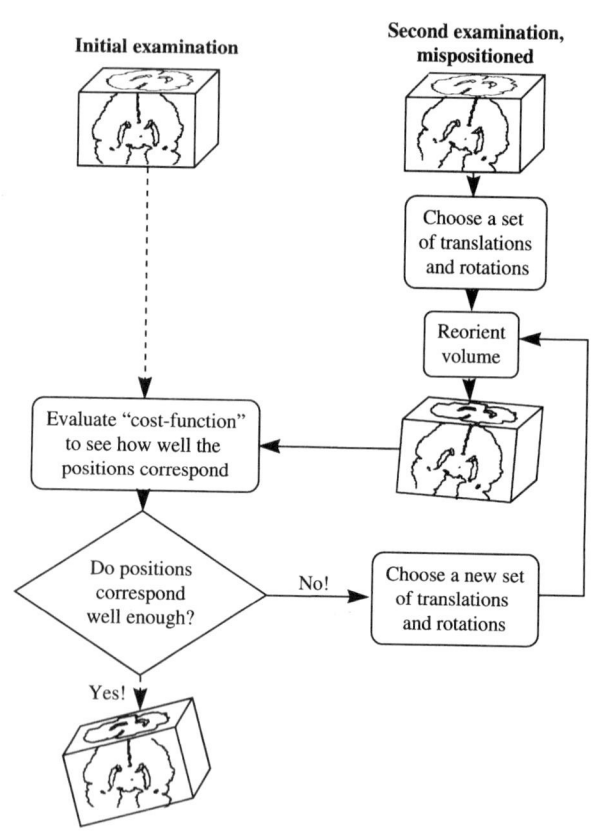

FIGURE 1 Schematic description of a general algorithm for image registration. Published methods differ primarily in how the cost function is calculated, which will determine whether the global minima of the transformation space will be located at the spot that coincides with correctly aligned images. In addition, the strategy for trying a new transformation may determine whether or not the global minimum is found.

A frame–frame covariance matrix **S** was created from

$$\mathbf{S} = \mathbf{I}'\mathbf{I}. \quad (2)$$

The eigenvectors of **S** were calculated by the *QL* algorithm as implemented by Press *et al.* (1992). The eigenvectors were stored as the columns of the matrix **e**, sorted in descending order of their appurtenant eigenvalues. A principal component image matrix **P** was calculated according to

$$\mathbf{P} = \mathbf{I}\mathbf{e}, \quad (3)$$

where the columns of the $m \times k$ matrix **P** are the principal component images (eigenimages) in descending importance (accounting for less variance). In this chapter the three first principal components have been used throughout.

2. Delineation of Object

Images reconstructed from measured projections will contain negative components in parts where the expected value would be zero. By assuming that all negative values originate from areas where the "true" value is zero (outside the object), an estimate of the variance (s^2) of the background outside the object could be achieved from the mean sum of squares of all voxels with negative intensity values. A threshold was defined as $1.28s$ ($p = 0.10$, one-tailed), above which there was a 10% risk of belonging to the background.

A combination of erode and grow operations and a connected component labeling eliminated all reconstruction streaks and holes from the threshold image and produced a binary image of the object.

3. Classification of Reference Volume

The reference volume was classified using an algorithm originally proposed by Milvang (1988). A histogram representation of all voxel values was created along each of the three principal component directions. A measure called *split effect* was defined and described by

$$SE(s,k) = \sum_{i=0}^{n} \left(x_{ik} - \overline{X}_k\right)^2 - \sum_{i=0}^{s} \left(x_{ik} - \overline{X}_{bk}\right)^2$$
$$- \sum_{i=s+1}^{n} \left(x_{ik} - \overline{X}_{ak}\right)^2, \quad (4)$$

where s is the split index along the principal direction k, x_{ik} is the histogram value for index i along k, and $\overline{X}_k$, $\overline{X}_{bk}$, and $\overline{X}_{ak}$ are the total histogram mean, the mean below the split, and the mean above the split, respectively. The split effect was maximized by simply testing all possible combinations of s and k. When the first split line had been obtained, the principal component volume was split into two boxes, each of which was subsequently treated separately, such that in the pursuit of the second split both boxes were searched, and the split was induced in the box where the maximal SE was obtained. A total of seven split lines was calculated, partitioning the principal component volume into eight distinct classes.

As a second step a spatial proximity criterion was introduced by a connected component labeling that was performed such that all connected voxels of the same initial class were lumped into a new class. This typically increased the total number of classes from 8 to somewhere between 20 and 60, depending on the noise level and tracer properties.

4. The "Outside the Object" Class

After the object had been classified, an additional three grow operations were performed, and the voxels affected by these were collected into one class, the "outside the object" class. This is an important step, as without it the information conveyed by the borders of the object would have been poorly used.

5. Evaluation of Wilks' Λ

The merit of a given set of transformation parameters was evaluated by performing a MANOVA on the reslice set based on the classification obtained from the reference set. Let us assume that the reference set has been divided into k classes, each containing m_i voxels, where i is the class number. Let $\mathbf{x}_{ij}$ be an n-dimensional vector (where n is the number of principal components used, three in this case) from voxel j of class i, let $\overline{\mathbf{x}}_i$ be the mean vector for class i, and let $\overline{\mathbf{x}}$ be the mean vector for all voxels. Then the $n \times n$ total variance–covariance matrix may be estimated from

$$\mathbf{T} = \sum_{i=1}^{k} \sum_{j=1}^{m_i} (\mathbf{x}_{ij} - \overline{\mathbf{x}})(\mathbf{x}_{ij} - \overline{\mathbf{x}})' \quad (5)$$

and the residual variance–covariance matrix from

$$\mathbf{E} = \sum_{i=1}^{k} \sum_{j=1}^{m_i} (\mathbf{x}_{ij} - \overline{\mathbf{x}}_i)(\mathbf{x}_{ij} - \overline{\mathbf{x}}_i)'. \quad (6)$$

The generalized variance is given by the determinant of the variance–covariance matrix, and an entity known as Wilks' Λ (Johnson and Wichern, 1988) may be evaluated from

$$\Lambda^* = \frac{|\mathbf{E}|}{|\mathbf{T}|}. \tag{7}$$

Hence, Λ^* relates the residual variance to the sum of the residual variance and the variance explained by the classification. A small value indicates that a large part of the total variance has been explained by the classification and would thus in our case indicate a good registration.

B. Search Algorithm

In addition to Powell's algorithm (Powell, 1964), an algorithm based on "simulated annealing" (Kirkpatrick *et al.*, 1983) was used. The basic difference from other algorithms, which accept or discard a new point in parameter space based on whether it is better or worse than previously tested points, is the concept of "better or not too much worse." The algorithm will always embrace a new point with a lower Λ^*, but may also accept one that gives a higher Λ^*, provided it fulfills the following criterion

$$e^{\frac{\Lambda^*_{\text{old}} - \Lambda^*_{\text{new}}}{T}} > \text{rand}(0, 1), \tag{8}$$

where Λ^*_{old} and Λ^*_{new} are the old and new values for Λ^*, respectively, T is a "temperature" constant, and $\text{rand}(0, 1)$ is a random value with equal probability for all values between zero and one. On initialization, T is chosen such that the algorithm may escape local minima, and the "melt" is subsequently "cooled" after each 100 trial points by

$$T_{n+1} = (1 - \alpha) T_n, \tag{9}$$

where α is used to control the cooling rate.

Throughout this chapter, $T_0 = 0.1$ and $\alpha = 0.05$ has been used, yielding slow but robust convergence to the global minimum. The initial value of 0.1 will allow the algorithm to explore the entire parameter space with essentially no restrictions whatsoever. Therefore, in order to prevent the algorithm from wandering off to where there is no longer any overlap of the examinations, the cost function was multiplied with an additional function to create a bounding box. Hence the resulting cost function was

$$CF = \Lambda^* \prod_{i=1}^{6} e^{0.1 f},$$
$$f = \begin{cases} 0 & (|\text{TR}_i| < \text{limit}_i) \\ |\text{TR}_i| - \text{limit}_i & (|\text{TR}_i| > \text{limit}_i) \end{cases}, \tag{10}$$

where TR_i is the translation or rotation along or around the axis i, and where limit_i is the border of the bounding box.

The simulated annealing scheme has been incorporated into the "downhill slope simplex method" (Nelder and Mead, 1965) and follows the implementation suggested by Press *et al.* (1992) rather faithfully.

C. Experiments

Experiments were performed on a Scanditronix/GEMS PC2048-15B (Holte *et al.*, 1989) or on a Scanditronix/GEMS PC4096-15WB (Kops *et al.*, 1990) scanner. Data collected both in humans and in rhesus monkeys following injection of a variety of tracers were used. Validation data were selected such that they would represent a spectrum of dissimilarities ranging from relatively comparable uptake patterns (e.g., [^{11}C]DOPA and [^{11}C]nomifensine) to highly dissimilar ones (e.g., [^{11}C]salicylate and [^{15}O]CO$_2$).

D. Validation

The validation of an intertracer registration algorithm is complicated by the lack of a correct answer, as is also the case for the validation of intermodality registration. The validation undertaken for the present method may be summarized as follows:

1. When applying the method to data obtained with the same tracer (i.e., regular PET–PET registration) it should yield the same result as previously validated methods.

2. A classification method is inherently nonsymmetric, as one of the data sets is the classifee and the other the classified. When shifting place of the data sets, the results should be the same.

3. By mapping one of the data sets into a red scale and the other into a blue scale and by displaying them on top of each other, using a color-wash technique, the accuracy of the registration may be assessed visually.

4. By using presumably aligned data and imposing known misalignments on one data set, the ability to recover these may be assessed.

III. RESULTS AND DISCUSSION

Experiments performed on intratracer data suggested that the method was able to align these data with an accuracy comparable to a previously validated method for PET–PET registration (Andersson, 1995), albeit with a considerably longer execution time.

When shifting places of the "reference" and "reslice" data sets the differences were within 2 mm and 2° for almost all the tested tracer combinations. However, there were some cases [e.g., when registering [^{15}O]CO (one-dimensional) to [^{15}O]O$_2$ (two-dimensional) data] where registration was achieved only in one direction. Generally, it seems that the data set offering the most structural information should be used for the classification.

The registration accuracy as assessed by the red–blue color-wash technique was quite satisfactory for all the tested cases. An example of [^{11}C]DOPA and [^{11}C]nomifensine data before and after registration is shown in Fig. 2, albeit in a gray-scale representation. The "symmetry" of this example was such that the difference in results when using DOPA versus when using nomifensine for the classification was less than 0.6 mm along and 0.3° around the worst axis.

A rather striking example of the ability of the method to also correct for very large errors in positioning is shown in Fig. 3. The classification was based on data from [^{15}O]CO$_2$, [^{15}O]O$_2$, and [^{15}O]CO scans (i.e., three distinct tracer administrations) that had been displaced 25 mm and rotated 25° along and around all axes in relation to a [^{11}C]salicylate scan from which the three first principal component images were used. The salicylate has a very low intracerebral compared to extracerebral uptake and a high uptake in mucus membranes whereas both CO$_2$ and O$_2$ have a very high intracerebral uptake. The algorithm was able to recover the known transformation with an accuracy of 2.5 mm in the worst direction.

However, when inspecting these data using the red–blue color-wash technique, it appeared that data were better aligned after than before the registration. If this means that there was an initial misalignment that was corrected by the registration, in which case the 2.5-mm accuracy would be on the conservative side, or if it means that the algorithm judges data in a manner similar to a human observer is not clear.

When imposing known translations and rotations along and around all axes the registration frequently failed when using Powell's search algorithm whereas

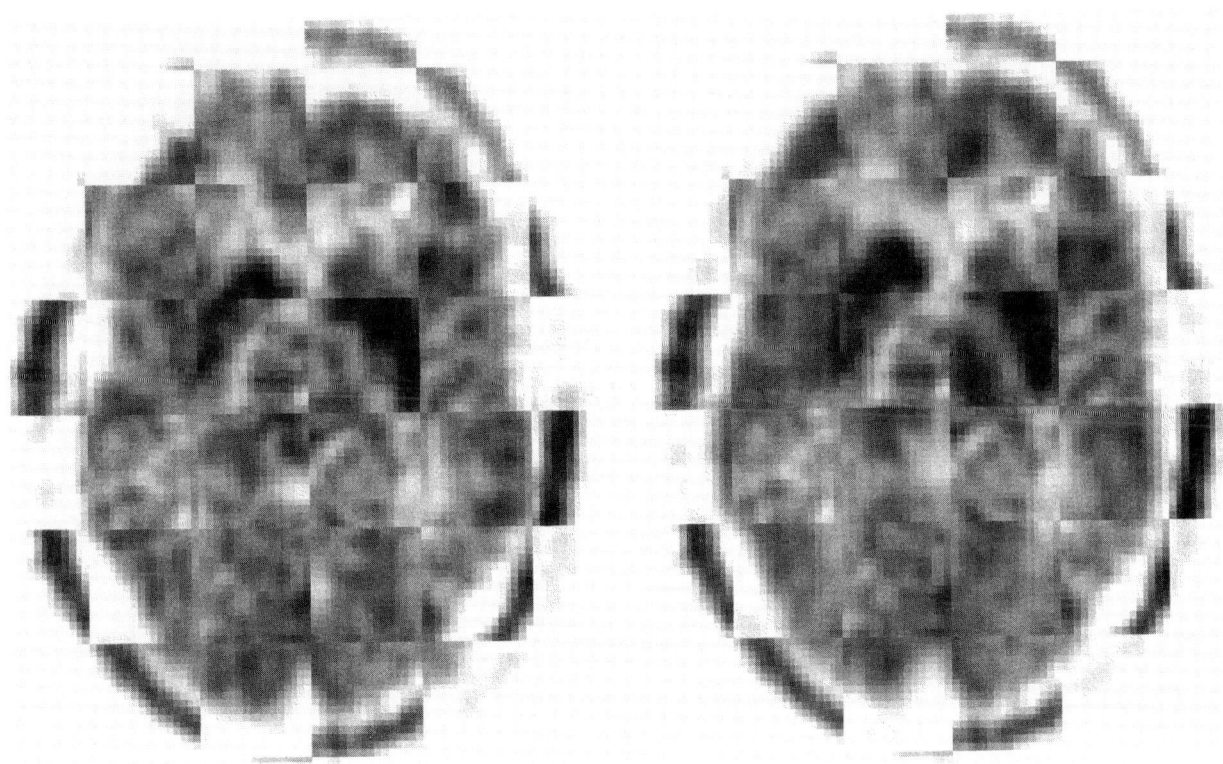

FIGURE 2 A [^{11}C]DOPA and a [^{11}C]nomifensine scan before (left) and after (right) registration. Data have been mapped into a common gray scale with darker gray for higher uptake. The images are alternatively displayed in squares of 16 × 16 voxels each. Note especially the cortical surface before and after registration.

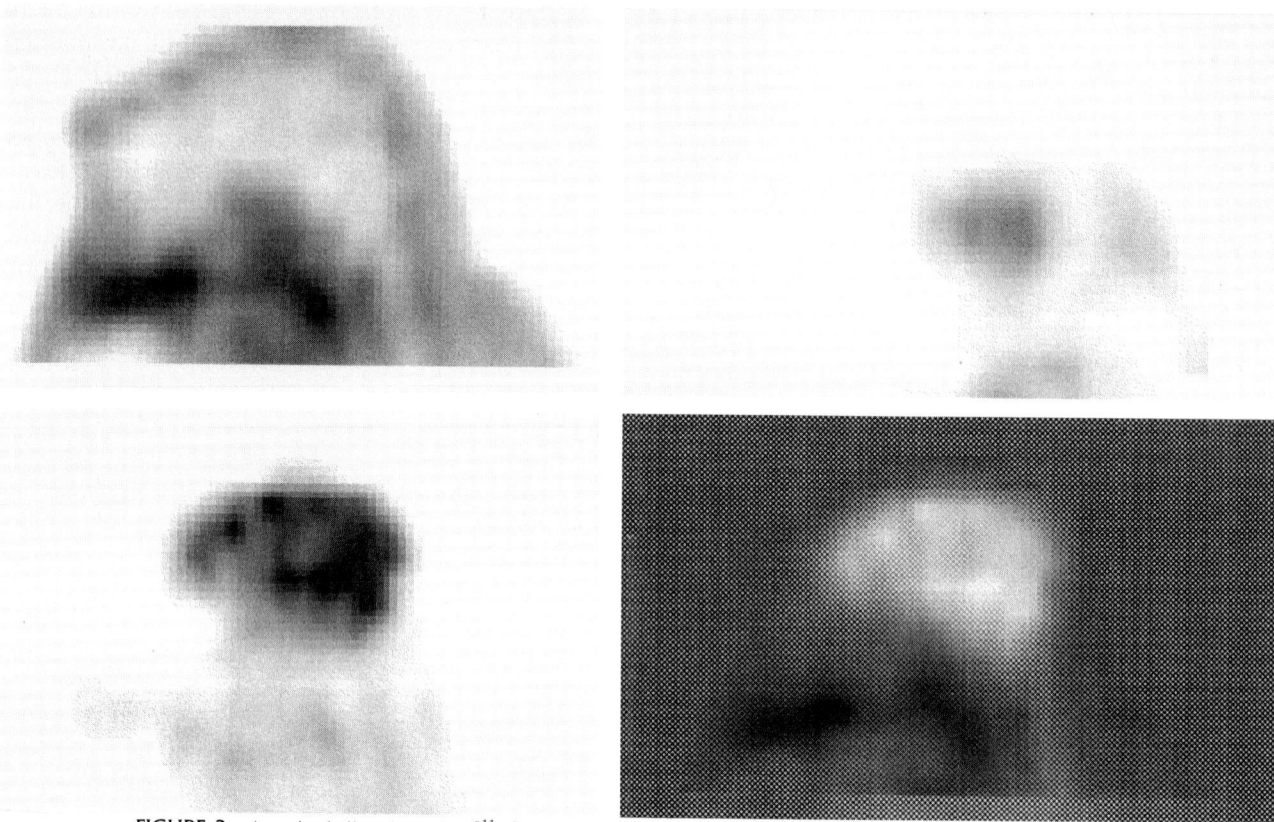

FIGURE 3 A sagittal slice through a [^{11}C]salicylate examination in a rhesus monkey (top left) and a [^{15}O]CO$_2$ examination in the same monkey following translation and rotation 25 mm and 25° along and around all axes (top right). (Lower left) CO$_2$ examination following registration. The gray scale was inverted on the CO$_2$ image and is shown overlaid on the salicylate image (lower right).

the simulated annealing scheme converged reliably to the true transformation. However, when applying the method to actual collected data, where one has attempted to reposition the subject accurately, convergence was also often achieved with Powell's algorithm. It seems that if the initial misalignments are reasonably small, particularly if they are very small in some of the directions, Powell's algorithm will suffice. As the execution time differs quite dramatically, 5 min compared to 1–2 hr, it may be a good choice to try Powell's algorithm first, and if some concern remains following visual inspection of the registration, to use simulated annealing.

It should be noted that while this chapter has concentrated on the application to multitracer registration, the suggested method is really a general method to align sets of image volumes, where images may be assumed to be aligned correctly within each set. The method has, for example, been successfully applied to MR–PET registration where both emission and transmission PET data have been used, thereby using all available information to drive the registration.

References

Andersson, J. L. R. (1995). A rapid and accurate method to realign PET scans utilizing image edge information. *J. Nucl. Med.* **36**: 657–669.

Ardekani, B. A., Braun, M., Hutton, B. F., Kanno, I., and Iida, H. (1995). A fully automatic multimodality image registration algorithm. *J. Comput. Assist. Tomogr.* **19**: 615–623.

Holte, S., Eriksson, L., and Dahlbom, M. (1989). A preliminary evaluation of the Scanditronix PC2048-15B brain scanner. *Eur. J. Nucl. Med.* **15**: 719–721.

Johnson, R. A., and Wichern, D. W. (1988). "Applied Multivariate Statistical Analysis," 2nd ed. Prentice-Hall, Englewood Cliffs, NJ.

Kirkpatrick, S., Gelatt, C. D., Jr., and Vecchi, M. P. (1983). Optimization by simulated annealing. *Science* **220**: 671–680.

Kops, R., Herzog, H., Schmid, A., Holte, S., and Feinendegen, L. (1990). Performance characteristics of an eight-ring whole body PET scanner. *J. Comput. Assist. Tomogr.* **14**: 437–445.

Milvang, O. (1988). An adaptive algorithm for color image quantization. *Proc. Scand. Conf. Image Anal. 5th*, Stockholm, Sweden, 1987, Vol. I, pp. 43–47.

Nelder, J. A., and Mead, R. (1965). A simplex method for function minimization. *Comput. J.* **7**: 308–313.

Pedersen, F., Bergström, M., Bengtsson, E., and Långström, B. (1994). Principal component analysis of dynamic positron emission tomography images. *Eur. J. Nucl. Med.* **21**: 1285–1292.

Powell, M. J. D. (1964). An efficient method for finding the minimum of a function of several variables without calculating derivatives. *Comput. J.* **7**: 155–162.

Press, W. H., Teukolsky, S. A., Wetterling, W. T., and Flannery, B. P. (1992). "Numerical Recipes in C. The Art of Scientific Computing," 2nd ed. Cambridge University Press, New York.

Woods, R. P., Cherry, S. R., and Mazziotta, J. C. (1992). Rapid automated algorithm for aligning and reslicing PET images. *J. Comput. Assist. Tomogr.* **16**: 620–633.

Woods, R. P., Mazziotta, J. C., and Cherry, S. R. (1993). MRI-PET registration with automated algorithm. *J. Comput. Assist. Tomogr.* **17**: 536–546.

CHAPTER 14

Multimodality Brain Image Registration Using a Three-Dimensional Photogrammetrically Derived Surface

OSAMA R. MAWLAWI,*,† BRADLEY J. BEATTIE,‡ STEVE M. LARSON,* and RONALD G. BLASBERG‡

*Nuclear Medicine Service and
‡Department of Neurology
Memorial Sloan Kettering Cancer Center
New York, New York 10021
†Chemical Engineering Department
Columbia University
New York, New York 10027

Registration of functional images (e.g., PET, SPECT) to anatomical images (e.g., MR, CT) of the brain is of importance, particularly when the functional images lack structural information. This chapter proposes a new method of registering such images, which involves adding structural information to functional image data. This additional structural information consists of a three-dimensional (3D) surface of the subject's face (S_{obj}) obtained using photogrammetric techniques. A similar 3D face surface (S_{ref}) is extracted directly from anatomical image data and a constrained nonlinear search algorithm is then utilized to minimize the root mean square distance between the two face surfaces. The method was tested by registering MR and PET image sets of the head for six patients with brain tumors. These registrations were then compared to registrations determined by the Pelizzari-Chen surface-matching method. The two methods produced very similar results (mean difference 3.6 mm); the proposed method was judged by a blinded observer to be more accurate in four out of the six cases and worse in only one case. This technique is also useful in the registration of different functional images.

I. INTRODUCTION

Using current medical imaging techniques it is possible to obtain images of the interior of the brain that are primarily either structural (e.g., standard MR and CT) or functional (e.g., PET glucose metabolism or receptor-binding affinity) in nature. Spatial registration of functional to structural images and of different functional images to one another has been shown to be of great clinical and research value (Pietrzyk et al., 1996). To address this need, several image registration methods have been proposed, but each suffers from drawbacks that have limited the scope of their application.

For example, methods that match internal landmarks, whether the landmarks are points (Ende et al., 1991; Evans et al., 1991), principal axes of the brain (Alpert et al., 1990), or the surface of the brain (Pelizzari et al., 1989; Mangin et al., 1994), all require that the structure be clearly and accurately discernible within both structural and functional image sets. Although these methods may work well with certain functional image types, they do not work when applied to functional images that contain little structural information (e.g., PET images of iododeoxyuridine). Furthermore, these methods often require operator intervention and/or knowledge of brain anatomy.

Voxel correlation methods (Apicella et al., 1989; Woods et al., 1993; Hill et al., 1994; Collignon et al., 1994), however, assume a particular relationship between the voxel intensities within the two image sets being registered. If this assumption is violated, as is often the case (e.g., when there is a brain tumor or other nonbrain tissues present), the effect on the accuracy of the registration is unpredictable.

Methods relying on externally placed fiducial markers (Pohjonen *et al.*, 1996) require an elaborate setup procedure and are unsuitable for retrospective investigations.

Finally, methods that rely on the registration of a proxy transmission scan suffer from potential errors due to subject movement between transmission and emission scans. Furthermore, transmission scans are not always available because of the overhead (i.e., increased scanner time and patient radiation exposure) involved in acquiring transmission data and are not typically acquired in modalities other than PET.

This chapter introduces a new method for image registration that involves adding structural information to the functional images. This additional information is in the form of a three-dimensional (3D) surface of the subject's face, which is obtained using photogrammetric techniques during the acquisition of functional data. The surface is then fit to a similar surface patch of the face determined during another scanning session or, in the case of MR or CT, directly from the image data (i.e., MR/CT scans do not require photogrammetric equipment).

This method avoids the limitations of other registration methodologies. It is useful for functional to structural as well as functional to functional image registrations and will work independent of the tracer(s) used. Moreover, this method does not require knowledge of brain anatomy or significant operator intervention.

II. MATERIALS AND METHODS

To demonstrate this method, [^{18}F]fluorodeoxyglucose PET images of the brain were registered to corresponding MR images (T1 with contrast) in six patients with brain tumors. Photogrammetric techniques were utilized to generate the face surface of each patient while they were being scanned inside the PET scanner. The face surface for each MR image set was generated from the MR images themselves.

A. Photogrammetry

Photogrammetry allows 3D coordinates, in this case defining a surface, to be derived from points within two 2D pictures that have been taken from slightly different positions using charge-coupled device (CCD) cameras. In this case, the points within each of the 2D pictures (i.e., the points known to be the same in the two pictures) were determined by the intersection of horizontal and vertical laser lines that were scanned across the patient's face (Fig. 1). Instead of individual

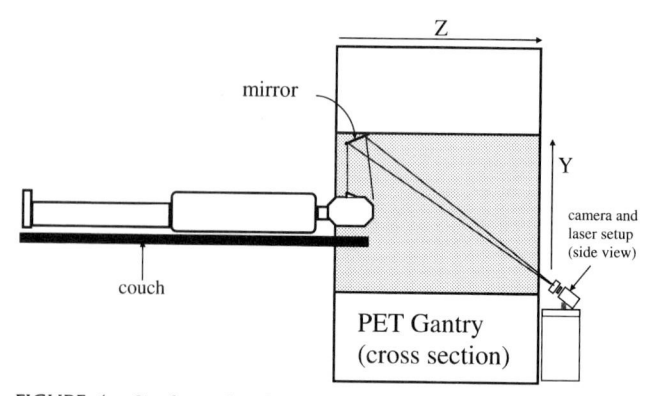

FIGURE 1 Configuration for measurement of photogrammetric information in the PET scanner. Two CCD cameras are positioned behind the PET gantry, toward a mirror which in turn is directed such that it will view the patient's face while lying within the PET field of view. Vertical and horizontal laser lines are scanned across the face to generate the surface of the face.

points or a grid, lines were used as a compromise between speed of acquisition and ease of accuracy of automated identification of the points.

B. Calibration

It is necessary to perform a calibration operation that determines the positions of the two cameras with respect to a general frame of reference. This frame of reference was forced to be the same as that of the PET scanner by performing the calibration with a "calibration frame" whose position was correlated to the reference frame of the PET scanner. The calibration frame consisted of four light-emitting diodes (LEDs) fitted into a wooden strip that could be attached to the couch of the PET scanner. The LEDs were positioned in a straight line, parallel to the x axis, with a 4-cm separation between them. Two CCD cameras were positioned behind the PET gantry, pointing toward a mirror which in turn was directed such that it would view the patient's face as they lay within the PET field of view. The mirror similarly allowed the calibration frame to be seen by the CCD cameras. The cameras and mirror were kept in this position throughout the calibration and surface generation procedures.

The calibration frame was imaged by both of the CCD cameras in several positions along the axial and vertical direction covering the entire field of view of the PET scanner. These positions were quantified using the bed positioning hardware of the scanner (an optical encoder) and thus to the scanner's frame of reference. A personal computer equipped with a frame grabber (Meteor, Matrox, Canada) was used to digitize the images from each camera at each bed position.

Calibration was performed using the direct linear transformation (DLT) method applied to the LED locations in the digitized images at each bed position. These locations were automatically identified by thresholding each of the digitized images. The DLT calibration results in 11 camera calibration parameters.

C. Laser Scanning

Following the calibration, each patient was positioned inside the scanner in preparation for his or her PET scan. The patients were asked to lie still with their eyes closed. Two laser line generators, mounted on stepper motors, were used to scan the patient's face. One laser line generator was oriented to produce horizontal lines across the patient's face whereas the other generator produced vertical lines. First, a pair of images of the patient's face with the laser line generators turned OFF was acquired as a reference image. Then multiple pairs of pictures were taken with the lasers turned ON at each motor location. A total of 120 pairs of pictures (60 pairs with horizontal laser lines and 60 with vertical lines) were acquired while the laser line generators swept across the face from the bridge of the nose to the top of the forehead and from the left ear to the right ear (Fig. 2). Each pair of pictures was then subtracted from the reference images, leaving images each containing a single laser line. The locations of the pixels that were above a specific threshold were taken as a set, defining a laser line. For each horizontal line's set of pixels the intersections with every vertical line's set of pixels (i.e., where the lines crossed) were determined. The pixel locations within each intersection were then averaged to determine the line cross-point locations. Pairs of cross-points, one point from each camera, when combined with the parameters of the spatial calibration procedure produced 3D coordinates within the coordinate system of the PET scanner. Taken together, these points (P_{xi}, P_{yi}, P_{zi}) approximate the face surface, which will henceforth be referred to as the "object" surface (S_{obj}). The total time needed for generating S_{obj} is about 10 sec during which time the patients were asked not to move.

D. Surface Fitting

The "reference" surface (S_{ref}) was determined directly from the axial MR images of the patients. In contrast to S_{obj}, S_{ref} is represented by a two-dimensional matrix $M(x_i, z_i)$ whose elements are the distances of the face surface above a roughly midbrain coronal plane (Fig 3). The patients were positioned in the standard supine (face up) position such that z_i are the slice locations (neck to top of head) and x_i are the pixel locations along the width (ear to ear) of the image. The skin surface was determined using automatic edge detection. The elevations $M(x_i, z_i)$ are therefore simply the y values of the edge in each column x_i and slice z_i minus a constant Y_0 corresponding to the height of the midbrain coronal plane.

The fitting procedure minimizes the RMS distance between S_{ref} and S_{obj} through rotations and translations of the object surface. Because both the elevations of S_{ref} and the points of S_{obj} are calibrated in millimeters, no scaling adjustment is needed. Prior to the application of the fitting procedure, the two surfaces are brought into rough correspondence by translating S_{obj} so that the centroid of its points (P_{xi}, P_{yi}, P_{zi}) is coincident with the centroid of S_{ref}. The distance between S_{ref} and each of the points (P_{xi}, P_{yi}, P_{zi}) in S_{obj} is taken to be the y value of the point P_{yi}, minus the bilinear interpolant of $M(x, z)$ at (P_{xi}, P_{zi}). In effect, the distance between S_{ref} and S_{obj} is measured along lines perpendicular to the midbrain coronal plane. The RMS distance between the two surfaces is, therefore, the square root of the mean of the squares of these distances. Points whose x or z coordinates are outside of the bounds of $M(x, z)$ are omitted from the calculation of the mean.

This distance measure is a variant of the one used by Pelizzari et al. (1989) in their head and hat surface fitting procedure. Other distance measuring methods (e.g., distance map used by Mangin et al. 1994) can also be used with similar results.

The translations and rotations producing a minimal RMS distance were found using a constrained nonlinear optimization procedure. The resultant spatial transformation along with the prefit positioning transformation described earlier were then applied to PET image data, resulting in a registered image set. PET data were then realized to match the MR slice locations.

E. Analysis

To validate the face surface registration method, these results were compared to registrations obtained using the Pelizzari–Chen surface matching method (Pelizzari et al., 1989) applied to the same subjects. The comparison was accomplished by first identifying the locations of voxels within the object brain prior to registration. The transformations determined by the two registration procedures were applied to each of these locations. The mean, standard deviation, minimum, and maximum distances between the resulting pairs of locations were then determined for each patient. Also, a nuclear medicine physician, blinded to the

images from camera 1   images from camera 2

Reference images

Images with lasers ON

Difference images

Concatenation of all
laser lines

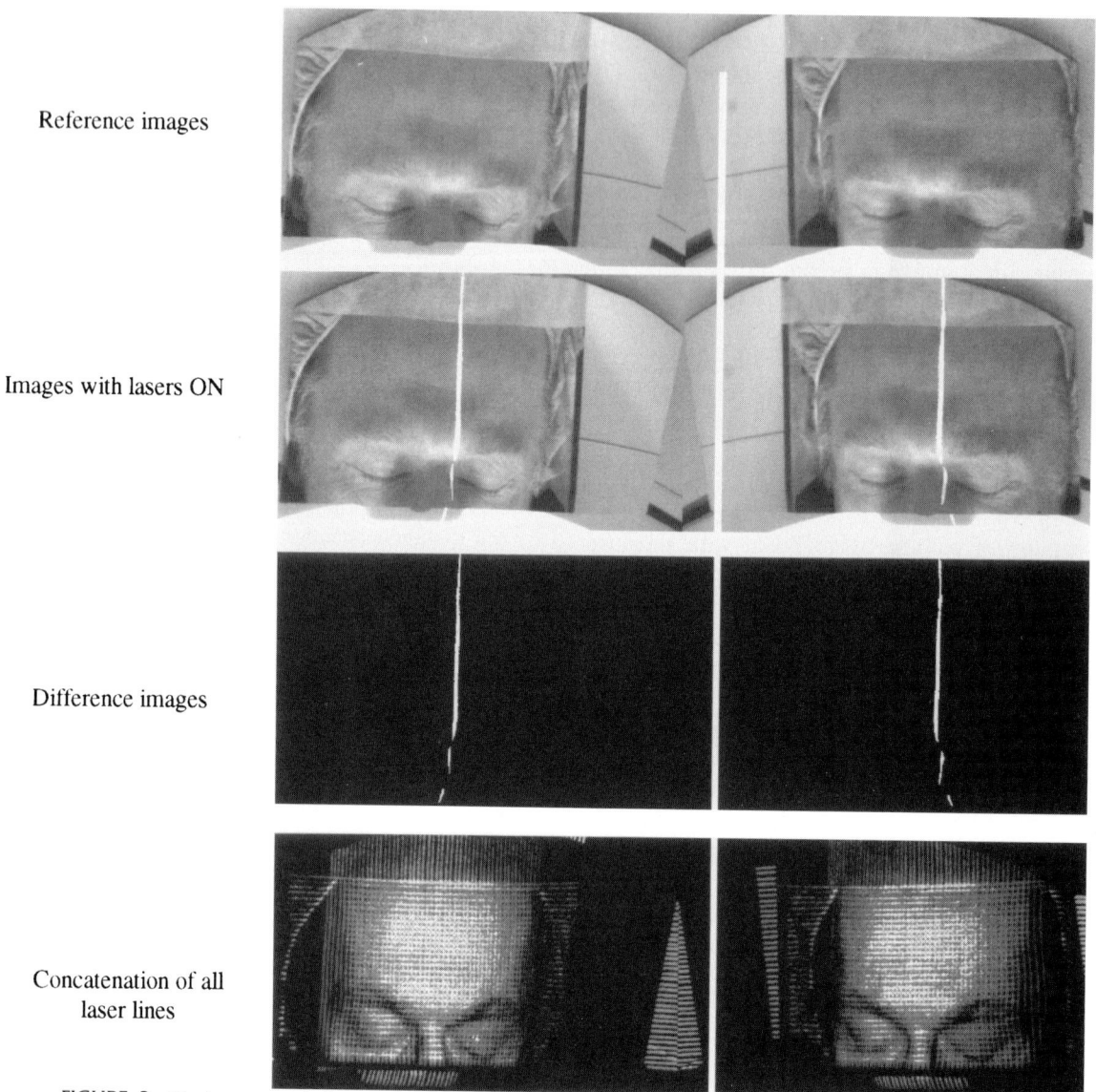

FIGURE 2 Photogrammetric images from the two CCD cameras, showing the laser line generation. (Top) Reference images are obtained with the laser turned off. A total of 120 pairs of pictures are then acquired with the laser lines on. Subtraction of the reference image from each of the images with the laser on results in images of the laser lines. When all images of the laser lines are concatenated (bottom row), the points of intersection of the lines can be determined and a map of the face surface generated.

registration method used, was asked to compare the two registration techniques.

III. RESULTS

The results of the registration procedure for one of the studied patients (patient F) are shown in Fig. 4. The PET 3D surface S_{obj} is shown as points in relation to the MR 3D surface S_{ref}. Figure 4a shows the positions prior to the distance minimization, whereas Fig. 4b shows the two surfaces after registration. Figure 5a shows sample slices from the MR data set, whereas Figs. 5b and 5c show postregistration resliced PET images determined by the photogrammetric and Pelizzari–Chen registration methods, respectively. The reference contour appears at the same location in all images.

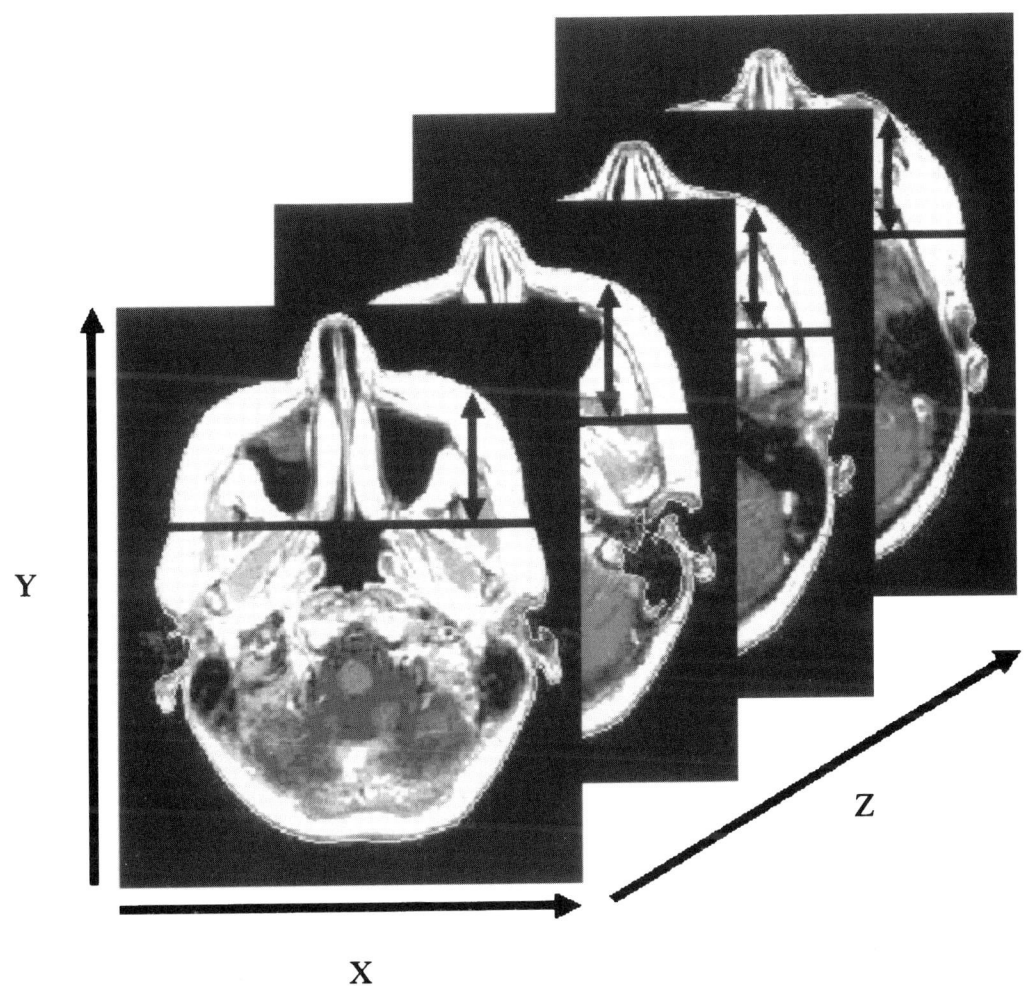

FIGURE 3 Determination of the reference surface from axial MR images. Arrows show the distance from a midbrain coronal plane for a given x and z location. These values are stored in the elevation matrix $M(x_i, z_i)$.

Table 1 summarizes the differences between the two registrations for each subject. Overall, the average distance between the brain positions determined by the two registration methods was 3.6 mm. In all cases but one, the photogrammetry registration was judged to be as good as or better than the registration determined by the surface-matching method. In the one case in which the reverse was true, the registration was to a scanned film of the MR image set.

IV. DISCUSSION

A new method for intrasubject, intermodality three-dimensional brain image registration has been developed. The method has several distinct advantages over other image registration procedures and is of particular benefit when utilized with functional images, which are severely lacking in structural information. This method requires the use of photogrammetric techniques during the functional imaging session and, therefore, cannot be used in a completely retrospective manner; however, the acquisition of the photogrammetric data involves no additional patient procedures and could easily become a routine part of functional image scanning sessions.

There are several other advantages to having photogrammetric equipment (i.e., two video cameras) focused on the subject while being imaged by a functional imaging scanner. In addition to patient monitoring and assisting in patient positioning, the digital images can be stored and utilized for precise subject repositioning in subsequent scanning sessions via real-time image overlay. Furthermore, the cameras can be used to monitor subject motion both within a static frame and between frames of a dynamic study. This movement record can then be utilized to correct directly for subject motion between frames.

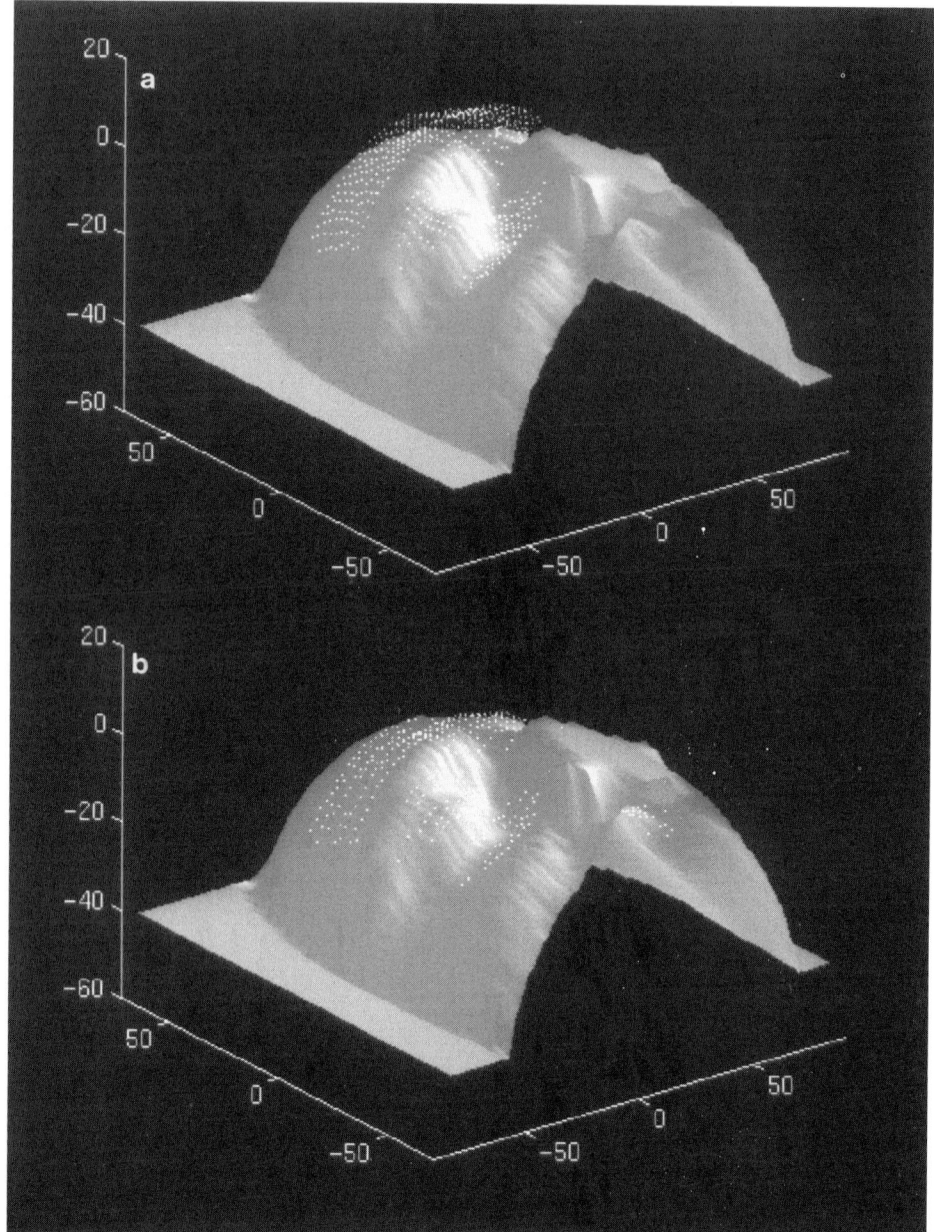

FIGURE 4 Face surfaces before (a) and after (b) registration with the photogrammetric method. The PET 3D surface S_{obj} is shown as points in relation to the MR 3D surface S_{ref}.

There are some circumstances, however, where this method will not be applicable. Because it is dependent on pictures of the patient's face, the shape of the face must be the same during the two scanning sessions. This might not be the case if the patient undergoes surgery or steroid therapy, which can cause edema. This method will not work with head-holding masks or straps (or the presence of bandages) that obscure the view of the subject's face. For some patients, involuntary movement of the face muscles during the 10 sec of laser scanning is problematic.

The fitting technique works best when the surfaces being registered have large topographical variations, such as the bridge of the nose and the eye sockets. Without these features the fitting procedure will have difficulty converging, especially on the translation parameters that control the fit within the plane of the face (i.e., x and z translations) and on the rotation parameter controlling orientation perpendicular to the plane of the face (i.e., rotation about the y-axis).

The placement of the mirror in the field of view of the PET scanner causes a small increase in the mea-

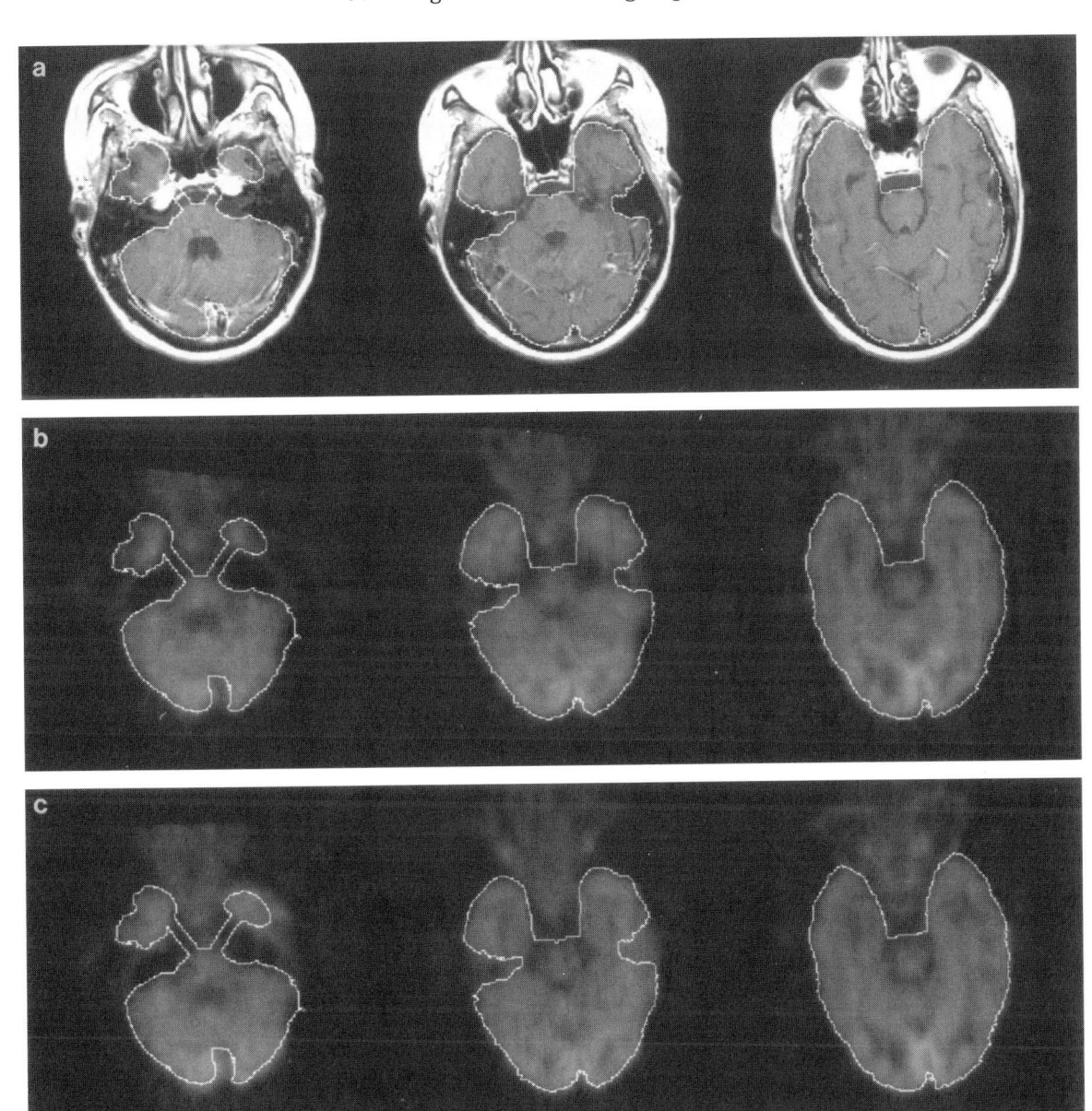

FIGURE 5 (a) Sample slices from the MR data set. (b) PET images after registration with the photogrammetric method. (c) PET images after registration with the Pelizzari Chen method. The reference contour appears at the same location in all images.

TABLE 1 Registration Comparison

Patient	Mean (mm)	SD (mm)	(min, max) (mm)	Judged better
A	1.22	0.212	(0.91, 1.81)	Neither
B	4.12	0.938	(2.02, 6.48)	Photogrammetry
C	2.45	1.049	(0.07, 4.88)	Photogrammetry
D	4.12	1.494	(1.91, 7.97)	Photogrammetry
E	4.59	0.994	(2.96, 6.89)	Surface matching
F	5.22	2.205	(0.96, 10.10)	Photogrammetry

sured attenuation. To quantify this effect, blank scans with and without the mirror in place were compared. The results showed an increase in attenuation of less than 2% for the lines of response passing through the mirror. This suggests that the mirror can be kept permanently in position for all of the PET study procedures.

Acknowledgments

The authors thank Dr. J. Zhang (Nuclear Medicine Service, MSKCC) for performing the Pelizzi–Chen registrations and Dr. T.

Akhurst (Nuclear Medicine Service, MSKCC) for judging the registration results.

References

Alpert, N. M., Bradashaw, J. F., Kennedy, D., and Correia, J. A. (1990). The principal axes transformation—a method for image registration. *J. Nucl. Med.* **31**: 1717–1722.

Apicella, A., Kippenham, J. S., and Nagel, J. H. (1989). Fast multimodality image matching. *Proc. SPIE* **1092**: 252–261.

Collignon, A., Maes, F., Delaere, D., Vandermeulen, D., Suetens, P., and Marchal, G. (1994). Automated multimodality image registration using information theory. *Proc. SPIE* **2359**: 263–274.

Ende, G., Treuer, H., and Boesecke, R. (1991). Optimization and evaluation of landmark-based image correlation. *Phys. Med. Biol.* **37**: 261–271.

Evans, A. C., Marrett, S., Torrescorzo, J., Ku, S., and Collins, L. (1991). MRI-PET correlation in three dimensions using a volume-of-interest (VOI) atlas. *J. Cereb. Blood Flow Metab.* **11**: A69–A78.

Hill, D. L. G., Studholme, C., and Hawkes, D. J. (1994). Voxel similarity measures for automated image registration. *Proc. SPIE* **2359**: 205–216.

Mangin, J. F., Frouin, V., Bloch, I., Bendriem, B., and Lopez-Krahe, J. (1994). Fast nonsupervised 3D registration of PET and MR images of the brain. *J. Cereb. Blood Flow Metab.* **14**: 749–762.

Pelizzari, C. A., Chen, G. T. Y., Spelbring, D. R., Weichselbaum, R. R., and Chen, C. T. (1989). Accurate three-dimensional registration of CT, PET and/or MR images of the brain. *J. Comput. Assist. Tomogr.* **13**: 20–26.

Pietrzyk, U., Herholz, K., and Heiss, W. D. (1996). Clinical applications of registration and fusion of multimodality brain images from PET, SPECT, CT and MRI. *Eur. J. Radiol.* **21**: 174–186.

Pohjonen, H., Nikkinen, P., and Liewendahl, K. (1996). Registration and display of brain SPECT and MRI using external markers. *Neuroradiology* **38**: 108–116.

Woods, R. P., Mazziotta, J. C., and Cherry, S. R. (1993). MRI-PET registration with automated algorithm. *J. Comput. Assist. Tomogr.* **17**: 536–546.

CHAPTER 15

Classification of Dynamic PET Images Using *a Priori* Kinetic Factors

JEFFREY T. YAP,* VINCENT J. CUNNINGHAM,* TERRY JONES,* MALCOLM COOPER,[†]
CHIN-TU CHEN,[†] and PAT PRICE*

*MRC Cyclotron Unit
Hammersmith Hospital
London W12 0NN, United Kingdom
[†] Department of Radiology
The University of Chicago
Chicago, Illinois 60637

A method has been developed for classifying pixel time–activity curves (TACs) using predefined time factors based on a priori knowledge. For studies that have a measured input function and are well described by a kinetic model, the measured input function and the known or assumed parameter values can be used to predefine the time factors. However, for many novel compounds and diseased tissues there is no existing model or an established range of parameter values. If there is no available model but previous data have been collected, a population-based approach has been developed that uses previously measured region of interest TACs. Single-subject methods have also been developed to define factors when no other information is available. Once the time factors have been defined by any of these methods, the corresponding factor images of the weighting coefficients can be computed. A constraint is imposed to restrict each pixel to only be described by the single factor that minimizes the residual sum of squares. An additional iteration allows the time factors to be modified based on real data to account for errors in the predefined factors. The model-based constrained factor method showed marked improvement in the ability to segment gray matter from white matter as the quality of simulated image sequences based on the Hoffman phantom are degraded by the effects of noise, spatial resolution, and incorrect model parameters. The population-based approach was applied to adequately segment cell proliferation studies. Similarly, variations of the single-subject approach were used to classify and segment glucose metabolism and neuroreceptor studies.

I. INTRODUCTION

One of the major limitations in conventional kinetic analysis of dynamic positron emission tomography (PET) images is the determination of an appropriate tissue time–activity curve (TAC) for the desired structure of interest. Many analyses begin with manual definition of regions of interest (ROIs) to generate tissue TACs. Even if the kinetic analysis is performed at the pixel level to produce parametric images, regional parameters are often still desired, requiring ROI definition on the parametric images. For example, quantitative comparison of different subjects cannot be performed at the pixel level for the majority of studies that cannot be accurately coregistered and spatially normalized.

This chapter presents variations of a simple method for classifying pixel TACs in order to segment the desired structures and extract their corresponding kinetics. Principal component analysis (PCA) and factor analysis (FA) have shown the ability to utilize the correlations between pixel TACs to characterize dynamic PET images in terms of the linear combination of a few basis functions, i.e., principal components (PCs), which can then be rotated to final factors that are physiologically meaningful (Yap *et al.*, 1996). However, by definition, the initial PCs are orthogonal and can therefore be limited in cases where the kinetics of different structures have very subtle differences (i.e., highly oblique), especially in the presence of high noise levels. For example, differences in normal [^{18}F]fluorodeoxyglucose (FDG) PET brain studies were found to be largely magnitude effects and thus the PCA typically generates a single component that can be scaled to approximate all of the pixel time–activity curves in the study. However, in reality there may be multiple oblique factors that account for the subtle differences in the kinetics of different structures, e.g., gray matter versus white matter metabolism. For such cases, the number of PCs produced may be less than the desired number of final factors, i.e., the number of underlying physiologic functions. However, there are several other sources of *a priori* knowledge that can be used to constrain the analysis and retrieve the additional factors. To develop this strategy, a general constrained factor method for tissue classification has been developed that uses *a priori* kinetic factors.

II. MATERIALS AND METHODS

A. Definition of Model Factors

Several approaches for defining kinetic factors have been developed to cover a spectrum of available *a priori* knowledge. In each case, a group of kinetic factors is formed to classify each pixel as belonging to one of the desired structures/functions. The factors can be represented as a matrix containing m p-element row vectors, where m is the number of factors and p is the number of time frames:

$$\mathbf{F} = \begin{bmatrix} \mathbf{f}_1 \\ \mathbf{f}_2 \\ \vdots \\ \mathbf{f}_m \end{bmatrix} = \begin{bmatrix} f_{11} & f_{12} & \cdots & f_{1p} \\ f_{21} & f_{22} & \cdots & f_{2p} \\ \vdots & \vdots & \vdots & \vdots \\ f_{m1} & f_{m2} & \cdots & f_{mp} \end{bmatrix}.$$

In order to remove the effect of magnitude and assess differences in shape, each factor is normalized by its integral counts over time:

$$\mathbf{f}_i = \frac{\mathbf{f}_i}{\sum_{j=1}^{p} f_{ij}}.$$

1. Model-Based Approach

For studies that have a measured input function and are well described by a kinetic model, a set of assumed parameter values can be used to predefine the time factor for each tissue type. For example, in FDG studies, a kinetic factor can be defined from the compartmental model using the plasma input function and assumed or previously measured values of the rate constants. A family of factors can be defined using a set of rate constants for each tissue type, such as gray matter and white matter.

2. Population-Based Approach

If there is not an appropriate model but previous data have been collected that demonstrate reproducible kinetics, the model factors can be defined as the mean normalized TAC of the desired tissue structures obtained from conventional ROI analysis:

$$\mathbf{f} = \frac{1}{n} \sum_{i=1}^{n} \frac{\mathbf{x}_i}{\sum_{j=1}^{p} x_{ij}},$$

where n is the number of subjects in the population, p is the number of time frames, and $\mathbf{x}_i = (x_{i1}, x_{i2}, \ldots, x_{ip})$ is the p-element row vector of the tissue TAC for the ith subject. Using raw TACs assumes that the shape of the input functions is similar across the population, which is often the case.

3. Single Subject Approach

For the case where there is either no previously collected data or the kinetics within the population are not reproducible, factors can be defined for a single subject using ROI TACs. These can be defined manually on a single image plane containing the structure(s) and then applied to the entire volume. The advantage of this is that only a small sampling of the involved pixels is needed to obtain the initial estimate of the kinetic factor. In the next step, the noise in the factor is reduced by single averaging over all the TACs from those pixels that are best described by the initial factor. A hybrid approach has also been developed for cases when PCA/FA can be used to define some or all of the factors. For these studies the factor images are first segmented to extract the ROI and then the original pixel TACs within the ROI are summed to create new time factors for each structure. If the desired number of structures is greater than the number of factors extracted by the PCA/FA, additional factors can be obtained by either manually defining ROIs or using the residual factor method (see Section C).

B. Tissue Classification

Once the factors are defined, a fit is performed to determine the weighting coefficient of every factor on each pixel TAC. This results in an image of weighting coefficients similar to factor analysis. Neglecting the error term of the classic factor analysis model, the TAC can be described by the linear combination of the factors:

$$x = \alpha_1 f_1 + \alpha_2 f_2 + \cdots + \alpha_m f_m,$$

where $\mathbf{x}$ is a pixel TAC, $\mathbf{f}_k$ is the ith model factor, and α_k is the weighting coefficient of the factor. However, the goal of this study is to classify each pixel as being a single tissue type and hence be described by a single factor. For this reason a constraint is imposed on the factor model to fit a given pixel TAC to each factor independently:

$$\mathbf{x} = \alpha_k \mathbf{f}_k.$$

For each factor, a least-squares fit to every pixel TAC is performed by projecting the factor vector onto the original data:

$$\alpha_k = \mathbf{x} \mathbf{f}_k^T.$$

The residual sum-of-squares (RSS) error of each fit is also computed for each factor:

$$\text{RSS}_k = \frac{\sum_{j=1}^{p} (x_j - \alpha_k \cdot f_{kj})^2}{p-1}.$$

This process is repeated for each factor to produce a set of factor images (weighting coefficients) and their resulting error maps (Fig. 1). The RSS images are then compared to determine which of the factors provides the best fit for each pixel, i.e., the lowest RSS error. For each pixel, the factor with the minimum RSS is determined and the weights for all other factors are set to zero. Unlike PCA/FA, which allows a pixel to be described by more than one factor (i.e., have loadings on multiple factors), the advantage of this method is to provide a strict cutoff for distinguishing structures. Each factor image is then optionally thresholded, and the mean TAC of the resulting pixels is computed and normalized to generate a new model factor curve. An additional iteration is performed to generate new factor images based on the modified factors.

Although the methods of computation differ from traditional factor analysis by restricting the loading of multiple factors, the end result still conforms to the matrix form of the factor analysis model:

$$[\mathbf{X}]_{n \times p} = [\mathbf{W}]_{n \times m} [\mathbf{F}]_{m \times p}$$

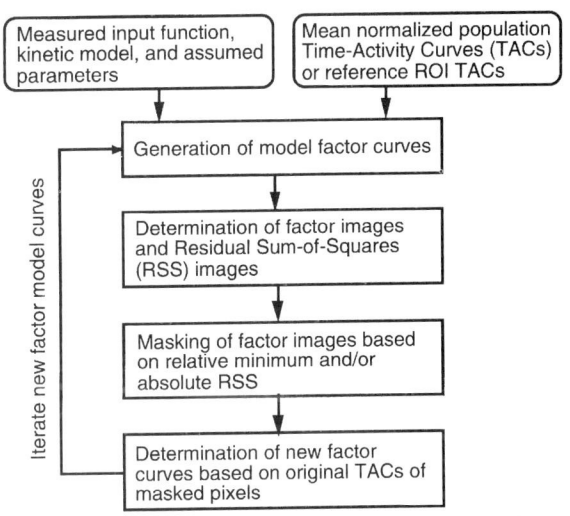

FIGURE 1 The constrained factor method. Depending on the available *a priori* knowledge, model factors can be defined by a kinetic model and assumed parameter, population TACs, or single-subject reference ROIs. Weighting coefficients and the resulting RSS are computed for each factor. Weights with nonminimal RSS and/or above a maximum threshold are set to zero and new model factors are derived from the TACs of all pixels with positive weights.

However, with the constrained method, each row of $\mathbf{W}$ can only have a single positive element with all other elements being zero.

C. Masking and Residual Factors

After the weights with nonminimal RSS have been set to zero, a new RSS image can be formed that accounts for the error of the fit using all of the factors. This composite RSS image can be used to determine how well each pixel is described by the factors, both in the absolute least-squares sense as well as relative to the other pixels in the RSS image. A maximum tolerance can be placed on the error to discard pixels with poor fits by setting their weights to zero. For most interests, this procedure can be used to mask out background pixels and/or undesired structures that are not accounted for by the model. In cases where there are additional functions not accounted for by the model but which are of interest, such as pathology or activation, the RSS can be segmented to generate an additional residual factor.

III. RESULTS

A. Simulations

A series of FDG brain image sequences were simulated based on the Hoffman brain phantom (Hoffman *et al.*, 1990) and the compartmental model for FDG using previously reported parameter values (Phelps *et al.*, 1979). The effects of Poisson noise and finite spatial resolution were modeled by adding various degrees of noise and Gaussian blurring. The noise-free gray and white matter curves show distinct separation; however, normalization by the total counts reveals that the shapes are quite similar (Fig. 2). This similarity in kinetics accounts for the difficulty in distinguishing between gray matter and white PCA/FA, particularly in real data containing noise and partial volume effect.

Model factors were defined using the same input function, kinetic model, and assumed rate constant as were used to simulate the images. In addition, the rate constants were varied to assess the method with known errors in the model factors. Despite the similarity of normalized gray and white matter factors, the constrained factor method showed marked improvement over PCA in the ability to segment gray matter from white matter as the quality of simulated FDG image sequences is degraded by the effects of noise, spatial resolution, and incorrect model parameters (Figs. 3 and 4).

B. FDG Brain

The single-subject reference ROI approach was applied to a normal FDG brain study in Fig. 5. PCA was only able to identify a single factor that accounted for all brain structures, including gray matter and white matter. However, the constrained factor method was able to segment gray and white matter based on subtle differences in the kinetics of the reference ROIs.

C. Radioligand Studies

PCA/FA has shown the ability to distinguish specific from nonspecific binding in dopamine D_2 monkey PET studies using [^{18}F]fallypride (Yap *et al.*, 1994). The hybrid single-subject approach using PCA/FA was used to segment specific binding sites in the striata from nonspecific sites in the rest of the brain (Fig. 6).

[^{11}C]R-PK11195 has been used as a marker to study peripheral benzodiazepine receptors in the brain (Myers *et al.*, this volume, Chapter 29). The hybrid single-subject approach using PCA/FA was applied to segment the scalp and skull from gray matter. The lesion could then be segmented by generating a third factor from either a reference ROI or using the residual factor method (Fig. 7).

D. Cell Proliferation Studies

The population-based method was applied to 2-[^{11}C]thymidine PET studies used to study cell proliferation in cancer patients (Yap *et al.*, 1997). The lung, spleen, liver, and kidney TACs from 18 subjects showed good reproducibility after normalization and were used to define four *a priori* factors (Fig. 8). In contrast, tumor TACs showed a wide range of variability suggestive of differences in tumor proliferation. Four time factors were generated from mean normalized lung, spleen, liver, and kidney TACs. Four factor images corresponding to the known anatomy and physiology were generated for each tomographic slice of a 15 slice study not included in the population database. The spleen, liver, and kidneys were segmented correctly in the slices where they occurred (Fig. 9). In addition, the injection site, aorta, and cardiac blood pool were seg-

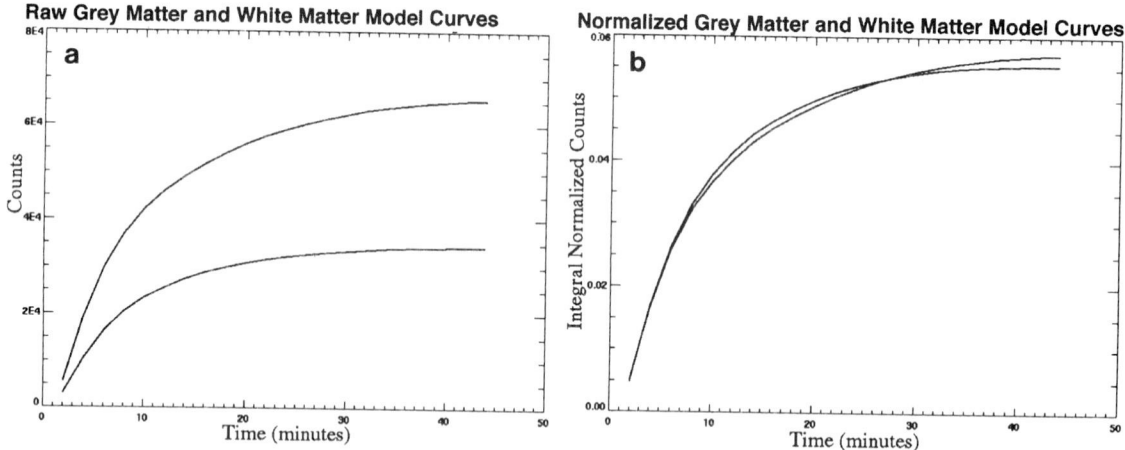

FIGURE 2 Gray and white matter model curves before (a) and after (b) normalization.

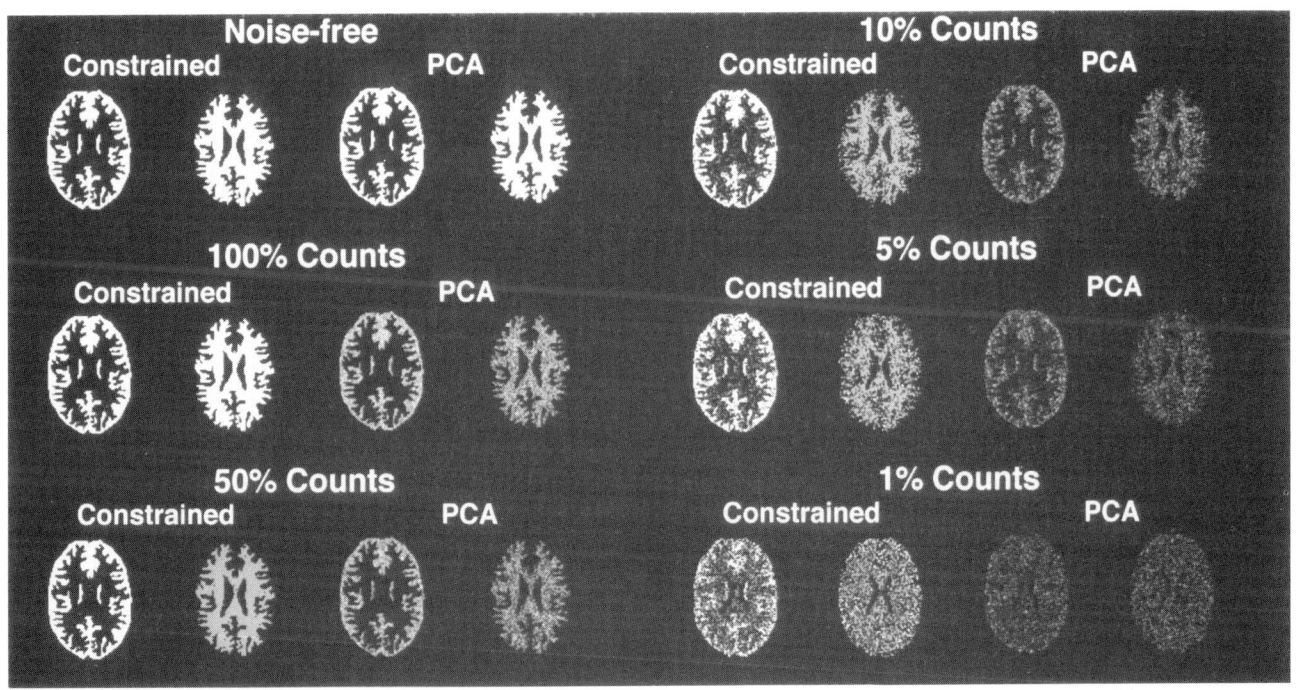

FIGURE 3 Segmentation of gray and white matter in simulated FDG images at different count rates with Poisson noise using the model-based method.

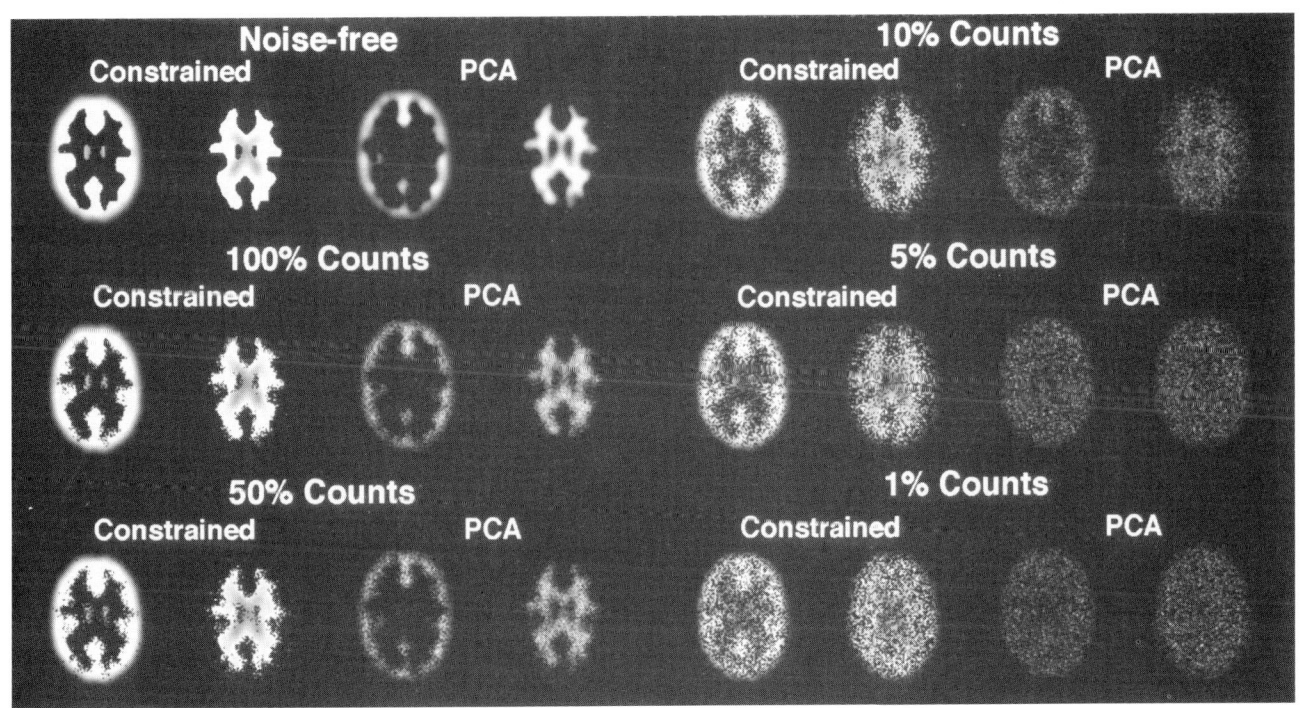

FIGURE 4 Segmentation of gray and white matter in simulated FDG images at different count rates with Gaussian blurring and Poisson noise using the model-based method.

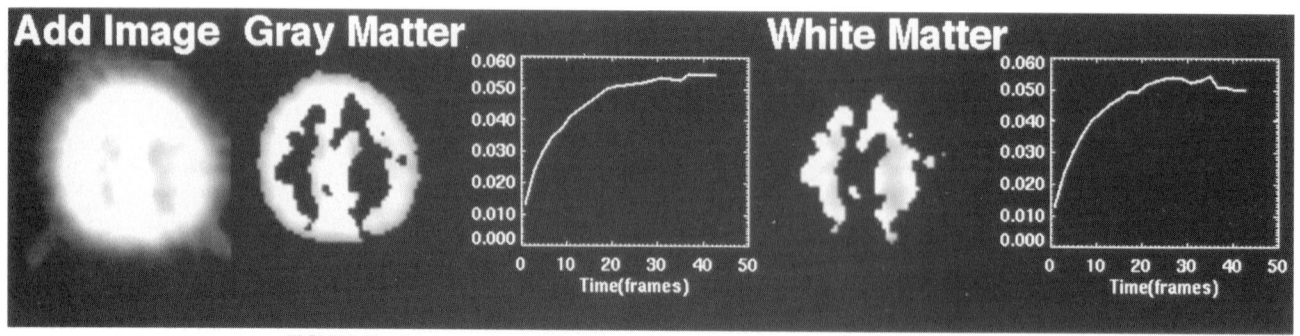

FIGURE 5 Segmentation of FDG images using the single-subject ROI method.

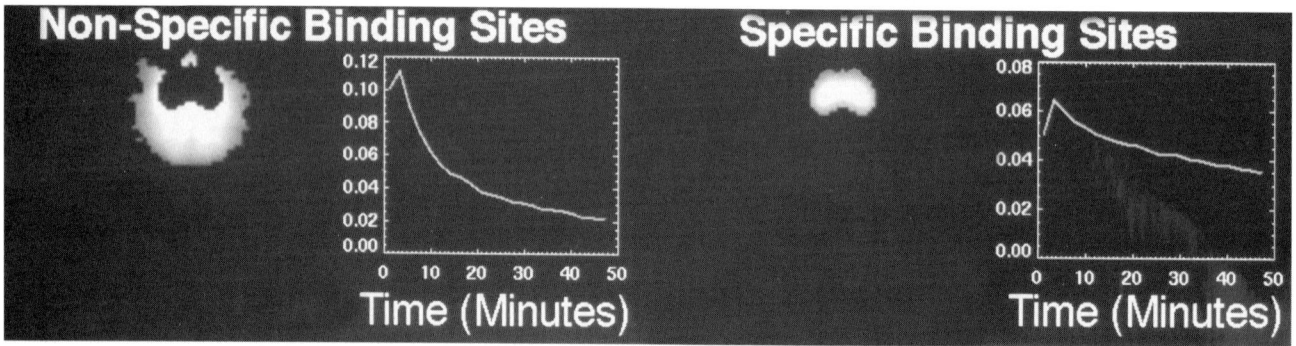

FIGURE 6 Segmentation of [^{18}F]fallypride images using the single-subject hybrid method.

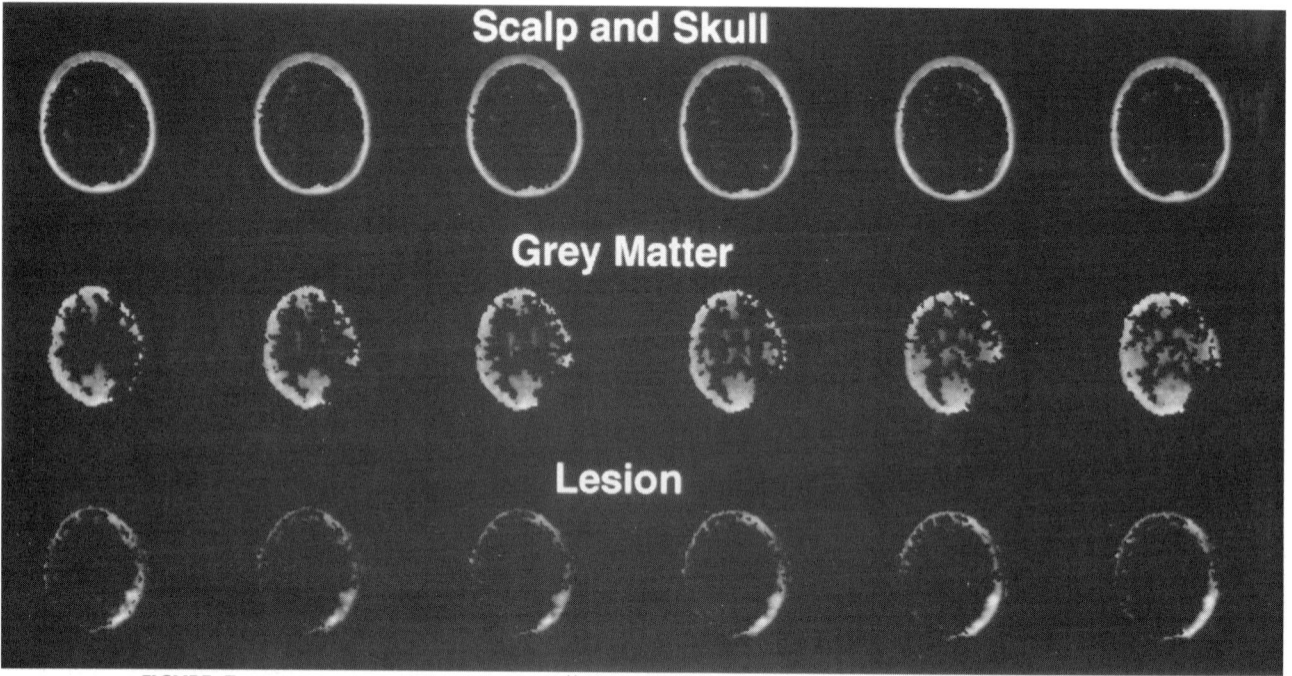

FIGURE 7 Segmentation of slices 40–45 of [^{11}C]R-PK11195 PET image sequence using the single-subject hybrid and residual factor method.

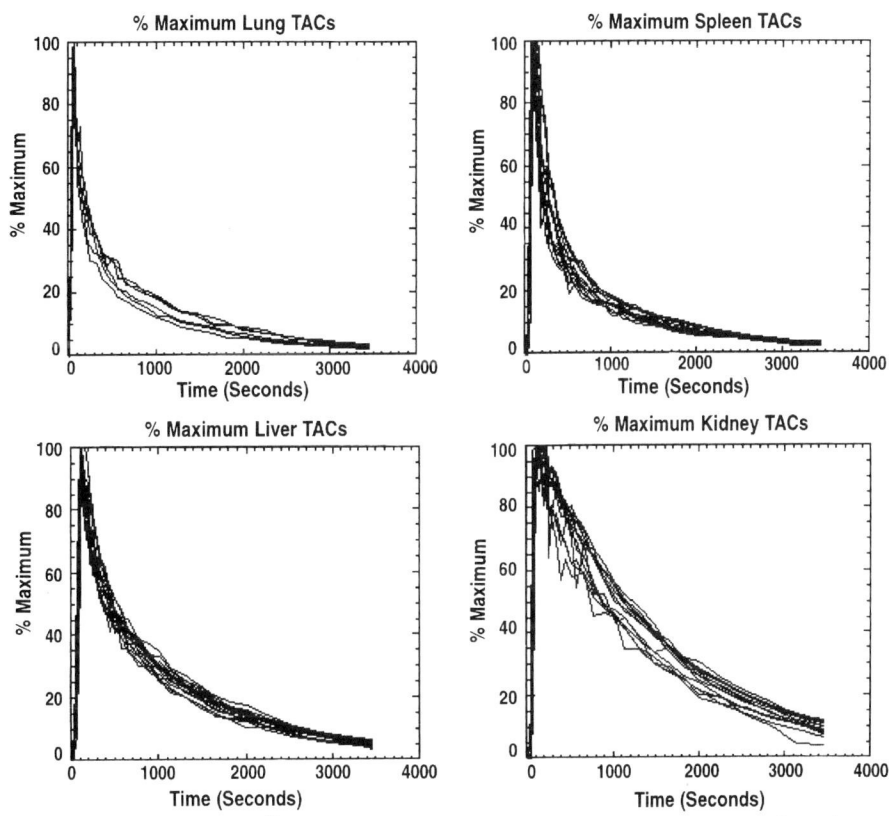

FIGURE 8 Normalized 2-[^{11}C]thymidine organ TACs displayed as percentage of maximum counts.

mented from the first factor and the proliferative tumor regions were segmented from the fourth factor. The iterative procedure also modified the corresponding factors to reflect the kinetics of the different structures.

IV. DISCUSSION

The proposed methods have shown the ability to distinguish subtle kinetic differences in the presence of noise and classify pixels into predefined tissue types. The end result is to provide segmented tissue structures corresponding to the desired anatomic structures and/or physiologic functions that can be used for further kinetic analysis. Each predefined factor does not need to be exact but must be closer to the observed kinetics than any other factor. The iterative procedure corrects for errors in the factor model curves as well as allowing structures to be classified that were not accounted for by the original model. The accuracy of the factors required to achieve proper classification needs to be investigated for each specific application.

Although there are advantages and disadvantages of the various methods presented, the rationale for developing them was not to compare which was best, but rather to provide an overall approach for classifying tissue function that could incorporate the varying degrees of *a priori* information available for different study types. The common concept is to determine the kinetic functions that account for the relevant functional information and to use these to classify each pixel TAC based on errors of the fit to each factor. In general, the kinetic functions should be determined using as much information as possible. Ideally, the use of previously collected studies and a well-defined kinetic model will optimize the ability to classify, characterize, and interpret a particular data set. However, this information is often not available or invalid for a particular study. In this case, the single-scan approaches allow the kinetic factors to be tailored to a specific data set to retrieve information that would be otherwise unavailable. Manual ROIs can be used to define the kinetic factors and, in many cases, PCA/FA can be used to define some or all of the factors with the benefit of being less user dependent and hence more

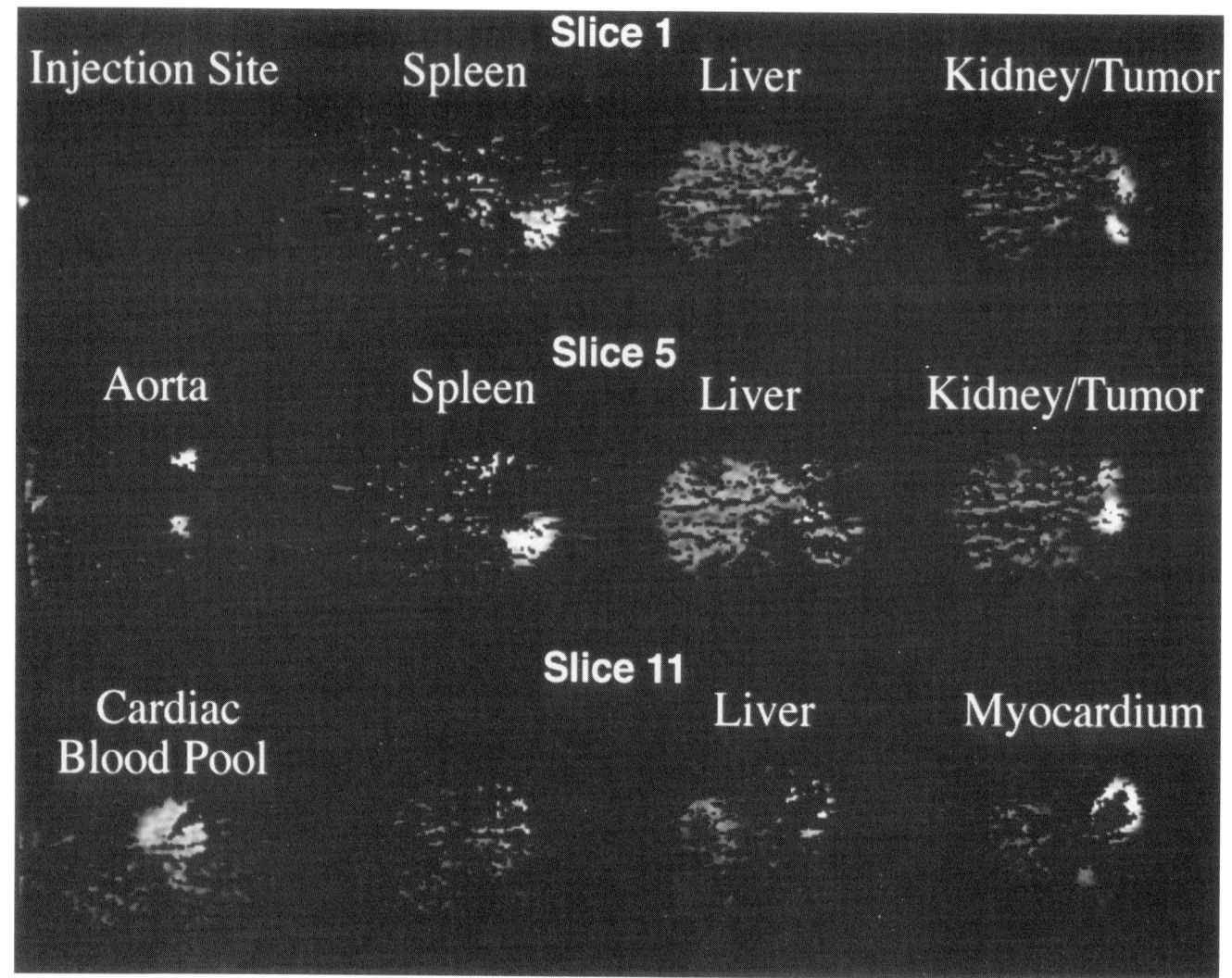

FIGURE 9 Segmentation of three representative slices of a dynamic 2-[^{11}C]thymidine PET scan using the population-based method.

objective. Although the ROI approach can be used to define additional factors not accounted for by the model, the residual factor approach suggests the potential for objectively identifying when additional factors are needed, e.g., for diseased tissues, as well as being able to define the kinetic factor and the pixels involved. Although the cutoff for determining pixels that are not well described by the existing factors is based on the RSS, the particular cutoff has been chosen empirically. However, it may be possible to set absolute cutoffs based on statistical criteria that take into account the goodness of fit and noise properties.

The main limitation of the population ROI approach is that it is only applicable when the normalized raw TACs shown little variability across subjects, e.g., to variable input functions. However, if this is not the case and the measured input functions are available, spectral analysis (Cunningham and Jones, 1993) can be used to first deconvolve the input functions to create a population of impulse response functions (IRFs). From these a mean IRF can be calculated and convolved with the input function of a particular subject to produce a case-specific kinetic factor that accounts for differences in the delivery of the radiotracer. This may improve the results of the existing method as well as enable it to be applied in cases where there are significant differences in ROI TACs due to differences in the input function.

Acknowledgments

The authors thank the following for providing data and related discussions: Drs. Roger Gunn and Paula Wells (2-[^{11}C]thymidine),

Dr. Jogesh Mukherjee ([^{18}F]fallypride), and Drs. Ralph Myers and Richard Banati ([^{11}C]R-PK11195).

References

Cunningham, V., and Jones, T. (1993). Spectral analysis of dynamic PET studies. *J. Cereb. Blood Flow Metab.* **13**: 15–23.

Hoffman, E. H., Cutler, P. D., Digby, W. M., and Mazziotta, J. C. (1990). 3-d Phantom to simulate cerebral blood flow and metabolic images for PET. *IEEE Trans. Nucl. Sci.* **NS-37**: 616–620.

Phelps, M. E., Huang, S. C., Hoffman, E. J., Selin, C. J., Sokoloff, L., and Kuhl, D. E. (1979). Tomographic measurement of local cerebral glucose metabolic rate in humans with [F-18]-fluoro-2-deoxy-d-glucose: Validation of method. *Ann. Neurol.* **6**: 371–382.

Yap, J. T., Chen, C.-T., Cooper, M., and Treffert, J. D. (1994). Knowledge-based factor analysis of multidimensional nuclear medicine image sequences. *Proc. SPIE* **2168**:289–297.

Yap, J. T., Cooper, M., Chen, C.-T., and Cunningham, V. J. (1996). Generation of parametric images using factor analysis of dynamic PET. *In* "Quantification of Brain Function Using PET" (R. Myers, V. Cunningham, D. Bailey, and T. Jones, eds.), pp. 292–296. Academic Press, San Diego, CA.

Yap, J. T., Cunningham, V. J., Jones, T., and Price, P. (1997). Tissue classification and segmentation of dynamic C-11 thymidine PET images using population-based kinetic functions. *J. Nucl. Med.* **38**(5): 205P.

CHAPTER 16

Methodology for Statistical Parametric Mapping of [^{18}F]Fluorodopa Uptake Rate Using Three-Dimensional PET[1]

J. S. RAKSHI,* D. L. BAILEY,* K. ITO,*,† T. UEMA,* P. K. MORRISH,* J. ASHBURNER,‡ K. J. FRISTON,‡ and D. J. BROOKS*,¶

*MRC Cyclotron Unit
Hammersmith Hospital
London W12 0NN, United Kingdom
†Department of Biofunctional Research
National Institute for Longevity Sciences
Obu, Japan
‡The Welcome Department of Cognitive Neurology
Institute of Neurology
London, United Kingdom
¶Institute of Neurology
Queen Square
London, United Kingdom

The conventional method of [^{18}F]fluorodopa analysis uses regions of interest to assess presynaptic dopaminergic dysfunction in Parkinson's disease (PD). This approach, although useful, has certain drawbacks. The selection of region size and shape is arbitrary and placement of regions over the striatum is subjective, as the [^{18}F]fluorodopa signal per se is used to predict the anatomical location of the putamen and caudate. We have developed, therefore, a novel approach to [^{18}F]fluorodopa analysis, by applying statistical parametric mapping (SPM) to parametric images of dopa influx (K_i). This enables one to make a direct statistical comparison of dopa influx in PD patients compared to normal controls across identical brain regions and provides a means of objectively identifying subregional striatal and extrastriatal dopaminergic changes. This chapter presents an overview and discusses methodological issues involved in applying SPM analysis to [^{18}F]fluorodopa data sets. It also discusses the potential applications of this approach to studying dopaminergic dysfunction and progression in PD and to establishing differences between presynaptic dopaminergic dysfunction in secondary parkinsonism and idiopathic PD.

I. INTRODUCTION

The application of statistical parametric mapping (SPM) analysis to positron emission tomography (PET) activation studies to identify significant changes in regional cerebral blood flow (rCBF) is well established (Friston et al., 1991a). Attempts have now been made to apply this approach to ligand studies, which potentially offers the possibility of localizing differences in receptor function in spatially normalized brain images across identical brain regions in different subjects, on a voxel-by-voxel basis. SPM has been applied not only to [^{18}F]fluorodopa PET data (Rakshi et al., 1996a), but also to other radioligands, such as [^{11}C]diprenorphine in Huntington's disease and Tourette's syndrome (Weeks et al., 1995), [^{11}C]flumazenil to study GABA$_A$ receptor binding in focal epilepsy (Richardson et al., 1996), and [^{11}C]WIN 35,428, a cocaine analog, to assess reductions in dopamine transporter binding in Parkin-

---
[1] Transcripts of the BRAINPET97 discussion of this chapter can be found in Section VIII.

reductions in dopamine transporter binding in Parkinson's disease (PD) (J. Frost, personal communication).

With the introduction of three-dimensional (3D) acquisition and reconstruction of dynamic [^{18}F]fluorodopa PET data (Rakshi et al., 1996b; Trebossen et al., 1996) and the development of more advanced PET scanners, which have led to increased sensitivity and higher resolution of dynamic [^{18}F]fluorodopa images, the authors have attempted to study subregional striatal and extrastriatal presynaptic dopaminergic dysfunction employing the SPM method. Extrastriatal midbrain and mesocortical dopaminergic regions have previously been difficult to study in vivo because of the lower specific [^{18}F]fluorodopa uptake in these regions and the limitations of PET resolution and sensitivity.

II. METHODS

A. Data Acquisition and Reconstruction

The 3D [^{18}F]fluorodopa data sets were acquired on the ECAT 953B neuroscanner (CTI/Siemens, Knoxville, TN) using protocols and reconstruction methods described previously (Rakshi et al., 1996b). An additional normal subject was scanned on the latest generation, EXACT 3D PET scanner (CTI/Siemens), (Jones et al., 1996; Bailey et al., this volume, Chapter 4) which has an axial field of view of 23.4 cm and provides 95 transverse planes with an isotropic spatial resolution of approximately 4.5 mm full-width-at-half-maximum (FWHM).

B. Subjects

Dynamic 3D [^{18}F]fluorodopa scans were performed on 7 subjects with early left hemi-PD, 7 subjects with advanced bilateral PD, and a group of 12 normal age-matched controls. The two PD groups were selected to represent the extreme ends of the clinical disease spectrum. The United Kingdom Brain Bank criteria for prospective diagnosis of PD were fulfilled by all patients (Gibb and Lees, 1988). All control subjects had a normal neurological examination with no evidence of rest tremor, rigidity, or bradykinesia.

All patients and normal volunteers gave written informed consent before their PET scan, after a full explanation of the procedure. Permission to perform these studies was granted by the Ethical Committee of the Royal Postgraduate Medical School, London, and by the Administration of Radioactive Substances Advisory Committee, United Kingdom.

C. Scanning Protocol

All patients had their PD medication omitted on the morning of the scan and all subjects were fasted. Each subject received an oral dose of 150 mg carbidopa and 400 mg entacapone, a peripheral catachol-O-methyltransferase inhibitor (Sawle et al., 1994; Ishikawa et al., 1996). Tissue attenuation of 511 keV γ radiation was measured with a 10-min 2D transmission scan performed prior to tracer injection and acquired using retractable rod ^{68}Ga/^{68}Ge sources.

Tracer injections of 80–175 MBq of [^{18}F]fluorodopa in 10 ml of normal saline were infused intravenously over 30 sec. Scanning began at the start of tracer injection with a protocol of 25 time frames over 94 min (4 × 1 min, 3 × 2 min, 3 × 3 min, 15 × 5 min). The subjects were positioned such that the orbitomeatal line was parallel to the transaxial plane of the tomograph. Head position was carefully monitored throughout the scan.

D. Data Analysis

The method of analysis employed was stereotactic spatial normalization of parametric images of [^{18}F]fluorodopa influx K_i with comparison of mean voxel K_i values between PD and control groups using SPM.

The following between-group comparisons were performed:

a. Two groups of 6 randomly selected normal controls.
b. Seven early left hemi-PD patients with 12 normal controls.
c. Seven advanced bilateral PD patients with 12 normal controls.

Parametric images of [^{18}F]fluorodopa influx K_i were produced from each subject's dynamic 3D [^{18}F]fluorodopa study by applying multiple time graphical analysis (MTGA) (Patlak and Blasberg, 1985) on a voxel-by-voxel basis and generating influx rate constants K_i for each individual voxel for all 31 planes. Time frames 14 to 25 (i.e., the last 60 min) postinjection were used for tissue count sampling, with occipital tissue counts (0 to 94 min) as the input function, to calculate K_i. The occipital counts were obtained by sampling right and left occipital lobes with two 32-mm-diameter circular regions placed on the three contiguous striatal planes that would be selected for the conventional regions of interest (ROI) approach.

As there is insufficient anatomical detail in parametric images of K_i to transform these images directly into Talairach stereotactic space (Talairach and Tournoux, 1988; Friston et al., 1991b), an indirect approach was therefore used. An integrated "add image" for each subject was produced by combining all time frames from 0 to 94 min of the dynamic 3D [^{18}F]fluorodopa data set. This "add image" contains rCBF information with sufficient anatomical detail to allow stereotactic normalization (Fig. 1, see color insert). The stereotactic spatial transformation parameters derived from the "add image" were then applied to the corresponding K_i images for each subject so that all parametric images were in the same standard stereotactic space, thereby allowing comparisons of scan data sets in identical brain regions across subjects, on a voxel-by-voxel basis. A diagrammatic overview of the indirect spatial normalization method and a comparison of a normalized [^{18}F]fluorodopa "add image" with the SPM standard cerebral blood flow (CBF) template are shown in Figs. 2 and 3, respectively. Following stereotactic normalization of the K_i images, a Gaussian filter of $8 \times 8 \times 6$ mm (full width half-maximum in the x, y, and z planes, respectively) was applied to remove high-frequency noise in the images.

The study design employed was "single subject:replication of conditions" with two contrasts ($-1, 1; 1, -1$) so that significant decreases and increases in K_i on a voxel-by-voxel basis could be observed in the between-group comparisons. No global normalization or proportional scaling was applied. Significant differences in mean voxel K_i values for the between-group comparisons were localized with SPM. The contrasts were used to derive the between-group Z score (unpaired t-statistic normalized to the unit Gaussian vector) on a voxel-by-voxel basis using the general linear model (Friston et al., 1991a, 1995). The p values associated with these regional effects were corrected for multiple-dependent comparisons implicit in SPM.

Analysis of data was performed on a SUN Sparc 10 workstation (Sun Microsystems, Silicon Valley, CA) using ANALYZE 7.0 (Mayo Foundation, Baltimore,

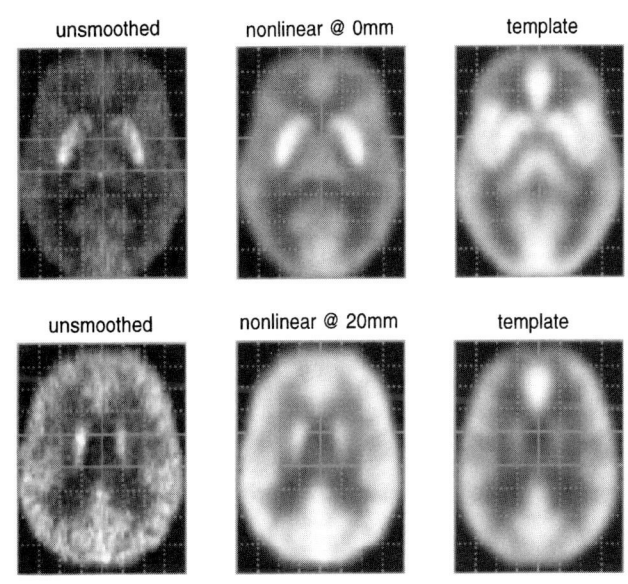

FIGURE 3 Direct comparison of normal subject's spatially normalized [^{18}F]fluorodopa "add image" to the standard CBF template displayed in SPM.

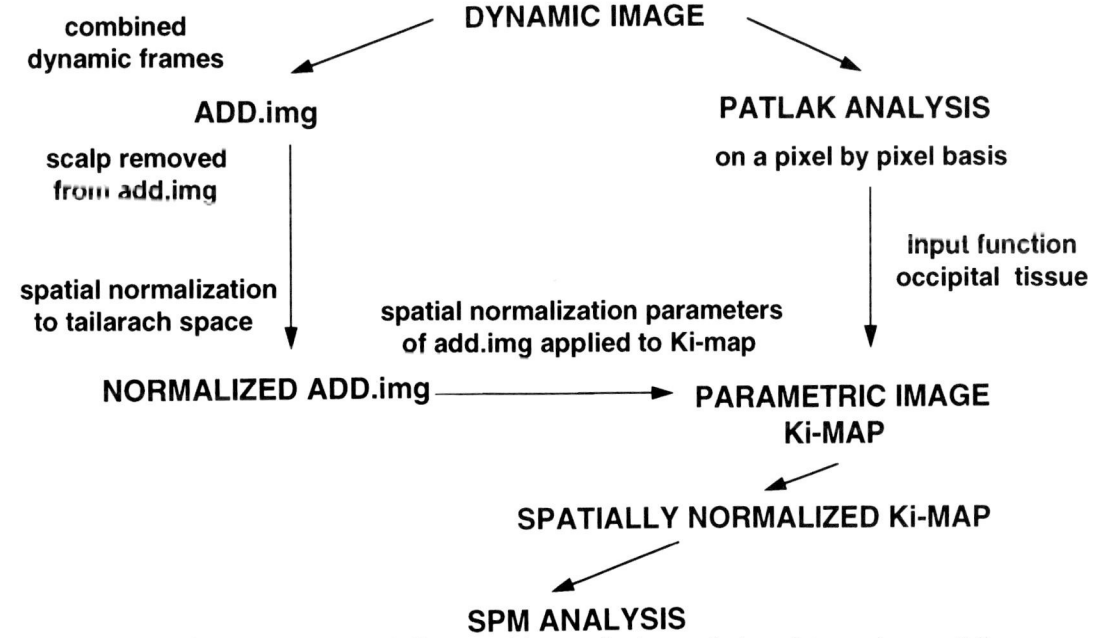

FIGURE 2 Overview of the indirect spatial normalization method used to produce spatially normalized parametric images of [^{18}F]fluorodopa K_i for SPM analysis.

120   II. Image Processing

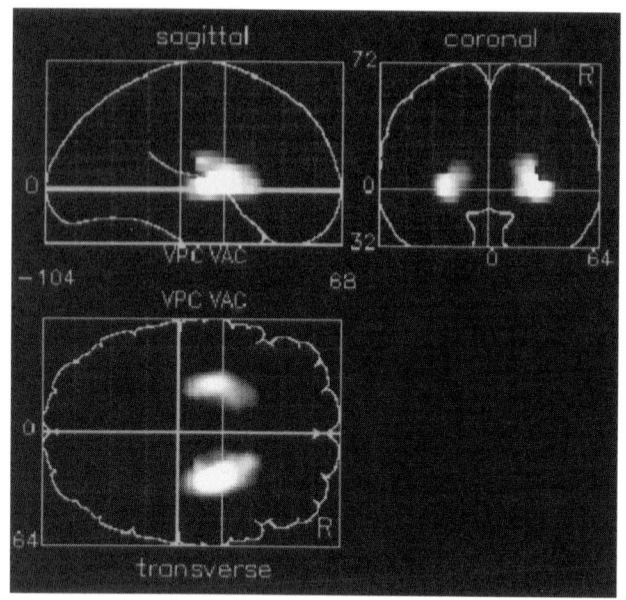

FIGURE 4 Striatal reductions in [^{18}F]fluorodopa K_i in the between-group comparison of 7 early left hemi-PD patients with 12 normal controls. SPM projections of the statistical maps (threshold Z score > 2.3, $p < 0.01$).

MD) (Robb and Hanson, 1991) and in-house software written in IDL image analysis software (Research Systems, Inc., Boulder, CO). Parametric images of influx K_i were generated with IDL image analysis software.

III. RESULTS

A. Comparison of Two Groups of Six Randomly Selected Normal Controls

When compared using SPM, no significant decreases or increases in K_i were found between the two subgroups, which each contained six normal controls.

B. Comparison of 7 Early Left Hemi-PD Patients with 12 Normal Controls

1. Striatal Changes

The comparison of the early left hemi-PD group with the normal control group demonstrated significant reductions in K_i throughout the right and left putamen (Fig. 4) The maximal Z scores were located in the caudal halves of the dorsal putamen ($Z = 5.03$, $p < 0.001$ and $Z = 6.07$, $p < 0.0001$ corrected, respectively), which suggests significant bilateral reductions in putamen K_i in early hemi-PD. Individual K_i values in the PD and normal control groups at the location of the maximal Z scores for right and left putamen are plot-

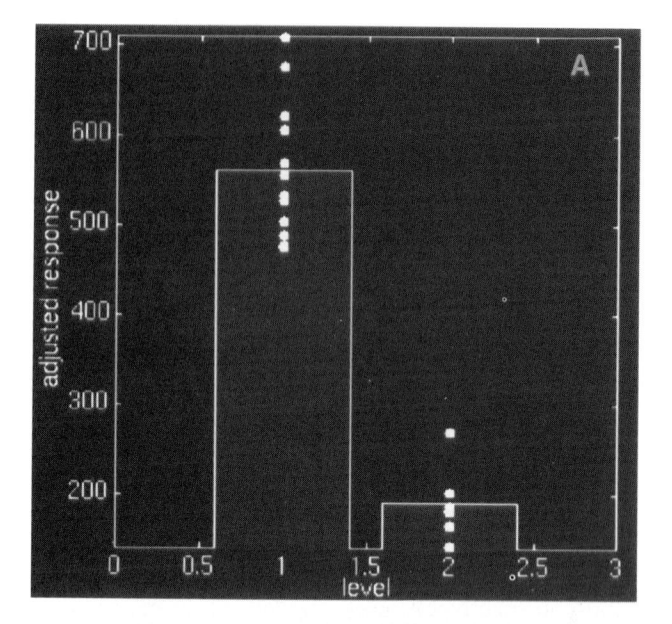

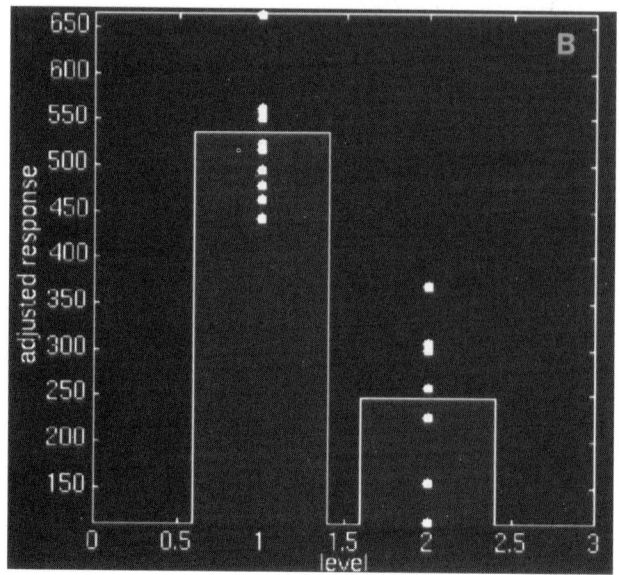

FIGURE 5 Plots of individual right putamen (A) and left putamen (B) K_i values in 7 early left hemi-PD and 12 normal subjects demonstrating complete bilateral separation of putamen ranges.

ted in Fig. 5. There is clear separation of all putamen K_i values between groups (Fig. 5). There was also a second focus located in the right anterior putamen (contralateral to the symptomatic side), extending into the right head of caudate, and another separate focus in the dorsal body of the right caudate ($Z = 4.12$, $p < 0.030$ and $Z = 4.00$, $p < 0.044$ corrected, respectively).

2. Anterior Cingulate and Midbrain Changes

The comparison of the early left hemi-PD group with the normal control group also demonstrated a

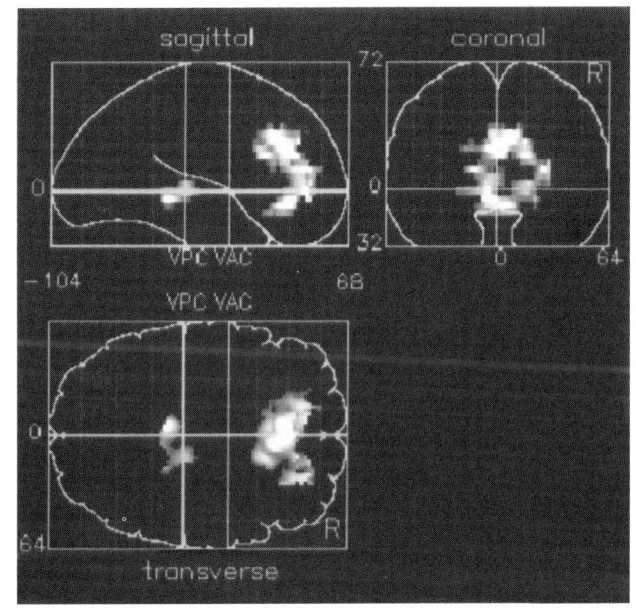

FIGURE 6 Significant prefrontal and midbrain increases in K_i in the early left hemi-PD study (threshold Z score > 2.3, $p < 0.01$). Most significant changes were on the left asymptomatic side.

significant increase in anterior cingulate K_i ($Z = 4.98$, $p < 0.001$ corrected) and dorsal midbrain K_i ($Z = 4.05$, $p < 0.038$ corrected) (Fig. 6). Individual K_i values in each group at the location of the maximal Z scores are plotted in Fig. 7.

C. Comparison of 7 Advanced Bilateral PD Patients with 12 Normal Controls

1. Striatal and Midbrain Changes

In the advanced PD group, more widespread and significant reductions in K_i were found throughout the right and left striatum in both caudate and putamen compared with the normal control group (Fig. 8). Maximal Z scores for right and left putamen were 6.81 and 6.80, $p < 0.0001$ corrected, respectively. There was also a significant reduction in K_i involving both ventral and dorsal midbrain regions $Z = 4.56$, $p < 0.008$ corrected), as shown in Fig. 8. There were no significant increases in K_i.

IV. DISCUSSION

The authors have produced spatially normalized parametric images of [^{18}F]fluorodopa influx K_i from dynamic 3D [^{18}F]fluorodopa PET data sets and have applied statistical parametric analysis to objectively identify subregional striatal and extrastriatal presynap-

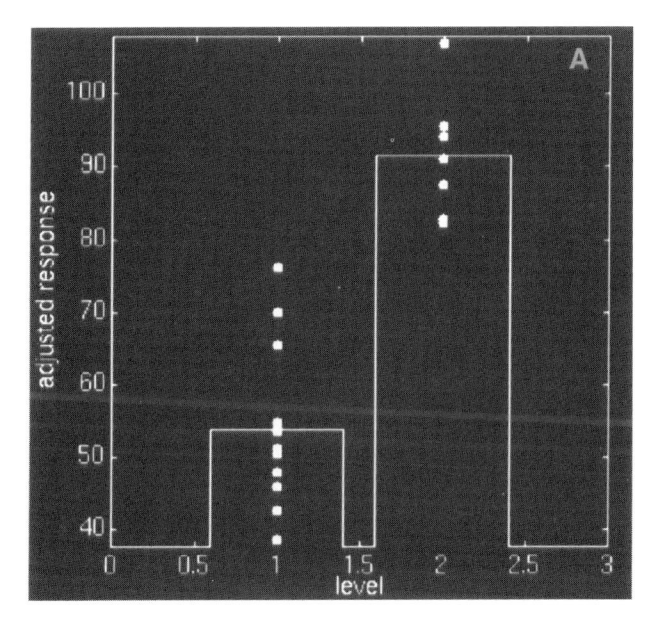

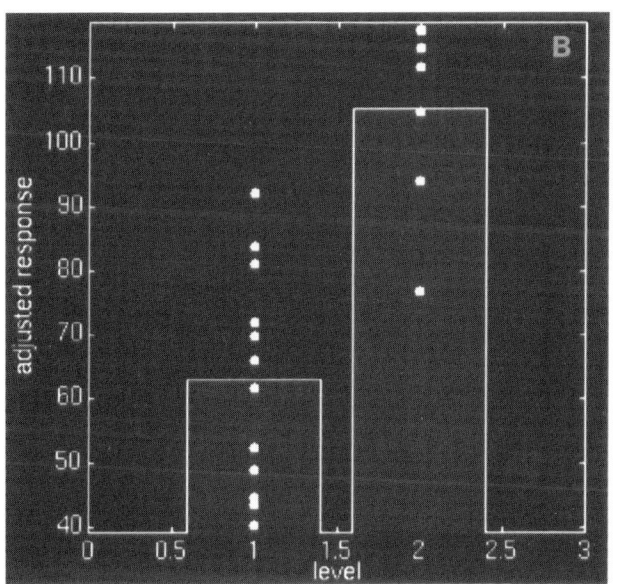

FIGURE 7 Plots of individual left prefrontal (A) and midbrain (B) K_i values in 7 early left hemi-PD and 12 normal subjects.

tic dopaminergic dysfunction in PD. The SPM results are very similar to those obtained using conventional ROI analysis, except for the increases in anterior cingulate K_i in early hemi-PD. These increases have been confirmed retrospectively by the ROI approach on untransformed [^{18}F]fluorodopa data sets (Rakshi et al., 1998).

Although both methods of analysis produced comparable results, there are some important differences between the two approaches. SPM allows objective 3D anatomical identification of subregional dopaminergic dysfunction within the putamen and caudate and also within extrastriatal dopaminergic regions on a voxel-

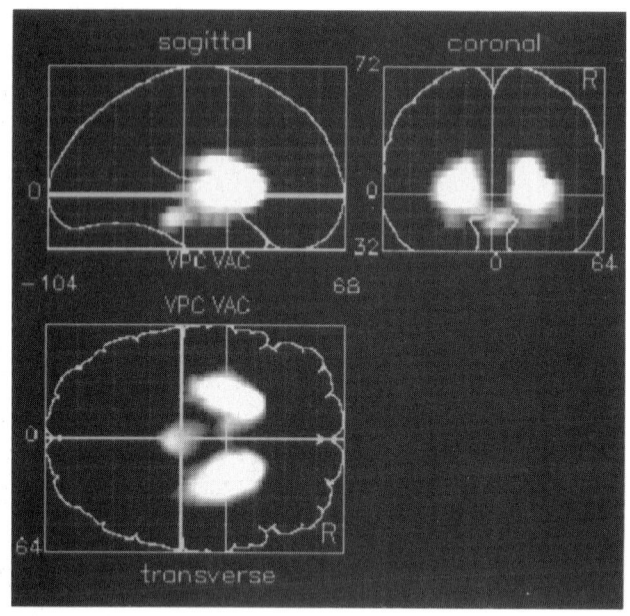

FIGURE 8 Widespread striatal and midbrain reductions in [^{18}F]fluorodopa influx (K_i) in the between-group comparison of 7 advanced bilateral PD patients with 12 normal controls (threshold Z score > 2.3, $p < 0.01$).

by-voxel basis. These regions can then be visualized by rendering them onto a spatially normalized magnetic resonance (MR) image. The ROI method, however, calculates an average value of K_i for each putamen and caudate structure, determined by sampling activity from regions placed on transverse striatal planes. The latter method, however, tells us little about focal or subregional changes in PD. Knowledge of these changes is important in understanding how striatal and extrastriatal dopaminergic dysfunction starts and subsequently develops and in determining whether there is a common pattern of progression in idiopathic PD. It may also help in establishing particular patterns of presynaptic dopaminergic dysfunction associated with somatotopical representation, clinical variants of PD (e.g., tremor dominant or akinetic rigid PD), or secondary parkinsonism.

Another important difference between the two methods is that SPM identifies statistically significant differences in dopaminergic function between subject groups. Dopaminegic dysfunction per se is not being measured, although it may be possible to quantify dopaminergic function in individual subjects on spatially normalized images. Alternatively, SPM could be used as an initial screening tool to identify and locate dopaminergic dysfunction that could then be quantified using the conventional ROI approach.

Two issues concerning the [^{18}F]fluorodopa SPM method require further discussion: (1) the validity of spatially normalizing [^{18}F]fluorodopa "add images" to the SPM standard CBF template and (2) the validity of applying an unpaired "t" test that assumes a normal distribution to K_i voxel values across subjects.

With regard to the first issue, the authors tried spatially normalizing a PD subject's [^{18}F]fluorodopa "add image" to a dopa template that was produced from a normal subject's spatially normalized [^{18}F]fluorodopa "add image." The spatially normalized image produced from the dopa template was indistinguishable from the one produced by direct normalization to the CBF template. On the basis of this empirical observation, the authors therefore decided to spatially normalize directly to the CBF template. It is possible, however, that the relative differences in contrast and intensity between [^{18}F]fluorodopa "add images" and the CBF template might affect spatial normalization. The normalization process is partly dependent on these factors, but the degree to which they might affect spatial normalization will also depend on the particular constraints of the spatial normalization program. Nevertheless, reasonable spatial normalization was achieved in that the SPM maps accurately rendered onto the putamen and caudate regions of the standard normalized MR image, and the maximal Z scores were identified in Talairach space as corresponding to putamen and caudate structures (Rakshi et al., 1998). This is further supported by the similar findings obtained using the ROI approach. In the future, however, there is a strong case for using a dopa template or for coregistering a subject's [^{18}F]fluorodopa PET image to their MR image and then applying the transformation parameters of the spatially normalized MR image to the [^{18}F]fluorodopa PET image.

Addressing the second issue, it is reasonable to assume that the voxel K_i values across the parametric images for both normal and PD groups will be distributed normally. This assumption is supported by no outliers being observed at the maximal Z score plots in the two PD studies (Figs. 5 and 7) and can be inferred from a normal distribution of ROI K_i values in normal and PD subjects. The unpaired "t" test is generally considered to be a robust statistical test, capable of handling suboptimal normally distributed data. At these Z scores, the permutation test is thought to be unnecessary (Holmes et al., 1996).

In conclusion, by applying statistical parametric analysis to [^{18}F]fluorodopa PET data, focal presynaptic dopaminergic dysfunction in PD has been identified objectively. With advances in PET technology and improvements in the methodology (e.g., employing a true dopa template for spatial normalization), this new approach may prove helpful in unraveling striatal and extrastriatal dopaminergic dysfunction in PD.

Acknowledgments

The authors are grateful to their data manager, Leonard Schnorr, for preparation of data for this study. We also thank chemists Nigel Steel and Maria Constantinou for producing [^{18}F]fluorodopa and radiographers David Griffiths, Hope McDevitt, and Andrew Blyth for their assistance in scanning. JSR is supported by SmithKline Beecham.

References

Friston, K. J., Frith, C. D., Liddle, P. F., and Frackowiak, R. S. J. (1991a). Comparing functional PET images: The assessment of significant change. *J. Cereb. Blood Flow Metab.* **11**: 690–699.

Friston, K. J., Frith, C. D., Liddle, P. F., and Frackowiak, R. S. J. (1991b). Plastic transformation of PET images. *J. Comput. Asst. Tomogr.* **15**: 634–639.

Friston, K. J., Holmes, A. P., Worsley, K. J., Poline, J. B., Frith, C. D., and Frackowiak, R. S. J. (1995). Statistical parametric maps in functional imaging: A general linear approach. *Hum. Brain Mapp.* **2**: 189–210.

Gibb, W. R. G., and Lees, A. J. (1988). The relevance of the lewy body to the pathogenesis of idiopathic Parkinson's disease (review). *J. Neurol. Neurosurg. Psychiatry* **51**: 745–752.

Holmes, A. P., Blair, R. C., Watson, J. D. G., and Ford, I. (1996). Non-parametric analysis of statistic images from functional mapping experiments. *J. Cereb. Blood Flow Metab.* **16**: 7–22.

Ishikawa, T., Dhawan, V., Chaly, T., Robeson, W., Belakhlef, A., Mandel, F., Dahl, R., Marouleff, C., and Eidelberg, D. (1996). Fluorodopa positron emission tomography with an inhibitor of catechol-O-methyltransferase: Effect of the plasma 3-O-methyldopa fraction on data analysis. *J. Cereb. Blood Flow Metab.* **16**: 854–863.

Jones, T., Bailey, D. L., Bloomfield, P. M., Spinks, T. J., Jones, W., Vaigneur, K., Reed, J., Young, J., Newport, D., Moyers, C., Casey, M. E., and Nutt, R. (1996). Performance characteristic and novel design aspects of the most sensitive PET camera built for high temporal and spatial resolution. *J. Nucl. Med.* **37**: 85P.

Patlak, C. S., and Blasberg, R. G. (1985). Graphical evaluation of blood-to-brain transfer constants from multiple-time uptake data. Generalizations. *J. Cereb. Blood Flow Metab.* **5**: 584–590.

Rakshi, J. S., Uema, T., Ito, K., Ashburner, J., Morrish, P. K., Bailey, D., Jenkins, I. H., Friston, K. J., and Brooks, D. J. (1996a). Statistical parametric mapping of three dimensional ^{18}F-dopa PET in early and advanced Parkinson's disease. *Movement Disord.* **11**, Suppl. 1: 147.

Rakshi, J. S., Bailey, D. L., Morrish, P. K., and Brooks, D. J. (1996b). Implementation of 3D acquisition, reconstruction and analysis of dynamic fluorodopa studies. In "Quantification of Brain Function Using PET" (R. Myers, V. Cunningham, D. Bailey, and T. Jones, eds.), pp. 82–87. Academic Press, San Diego, CA.

Rakshi, J. S. *et al.* (1998). Submitted for publication.

Richardson, M. P., Koepp, M. J., Brooks, D. J., Fish, D. R., and Duncan, J. S. (1996). Benzodiazepine receptors in focal epilepsy with cortical dysgenesis: An ^{11}C-Flumazanil PET study. *Ann. Neurol.* **40**: 188–198.

Robb, R. A., and Hanson, D. P. (1991). A software system for interactive and quantitative visualization of multidimensional biomedical images. *Australas. Phys. Eng. Sci. Med.* **14**: 9–30.

Sawle, G. V., Burn, D. J., Morrish, P. K., Lammertsma, A. A., Snow, B. J., Luthra, S., Osman, S., and Brooks, D. J. (1994). The effect of entacopone (OR-611) on brain [^{18}F]-6-L-fluorodopa metabolism: Implications for levodopa therapy of Parkinson's disease. *Neurology* **44**: 1292–1297.

Talairach, J., and Tournoux, P. (1988). "Co-Planar Stereotaxic Atlas of the Human Brain." Thieme, Stuttgart.

Trebossen, R., Bendriem, B., Fontaine, A., Frouin, V., and Remy, P. (1996). Implementation of 3D acquisition, reconstruction and analysis of dynamic fluorodopa studies. In "Quantification of Brain Function Using PET" (R. Myers, V. Cunningham, D. Bailey, and T. Jones, eds.), pp. 88–92. Academic Press, San Diego, CA.

Weeks, R. A., Cunningham, V. J., Waters, S., Harding, A. E., and Brooks, D. J. (1995). A comparison of region of interest and statistical parametric mapping analysis in PET ligand work: ^{11}C-diprenorphine in Huntington's disease and Tourette's syndrome. *J. Cereb. Blood Flow Metab.* **15**(Supp. 1): S41.

CHAPTER

17

Use of Nonlinear Kernel Analysis to Evaluate "Badness-of-Fit" of the Transformation of PET Images into Stereotactic Space: Application to Alzheimer's Disease

HAROLD LITT[*,†] and BARRY HORWITZ[*]

[*] Laboratory of Neurosciences
National Institute on Aging
National Institutes of Health
Bethesda, Maryland 20892
[†] Department of Biophysics
School of Medicine
State University of New York at Buffalo
Buffalo, New York 14214

Transformation of images into a standard stereotactic space is often used in intersubject positron emission tomography (PET) studies. The normalization process may be less successful if the image being normalized is significantly different from the stereotactic template, e.g., as a result of atrophy or functional change in Alzheimer's disease (AD). A method that allows quantitative evaluation of a stereotactic normalization using the Volterra model of nonlinear systems has been developed. This model considers the output at a point to be dependent on interactions among points in a region of the input known as the system memory. Volterra kernels for the transformation between a PET template in stereotactic space and FDG-PET images from six AD patients [Mini-Mental Status (MMS) score 0–26] and nine matched controls were calculated. A 7*7 pixel system memory modeled the transformation for controls better than AD patients, as assessed by the difference between the model's predicted output and the template image. There was also a significant correlation between model error and disease severity as measured by the MMS score. Areas of greater error in the model fit included parietal, temporal, and dorsal occipital cortex, as well as parts of the cingulate gyrus, which are regions significantly affected in AD. These results demonstrate that spatial normalization algorithms may not perform as well in patients with AD, especially for those with severe disease.

I. INTRODUCTION

Transformation of images into a standard stereotactic space is often used for intersubject analysis in positron emission tomography (PET) studies. The aim is to match the overall spatial organization of a given brain with a standard brain template while preserving local anatomical and functional differences (Fox et al., 1985). This is accomplished in the statistical parametric-mapping package SPM95 (Friston et al., 1995) by separating the transformation into spatial and intensity components and displaying only the application of the spatial transformation. Although these normalized images look more like the template brain [which is based on the stereotactic atlas of Talairach and Tournoux (1988)] and can be used for intersubject comparisons, they cannot be used to evaluate quantitatively how well the normalized image fits the template, e.g., by calculating a difference image, because the intensity component of the transformation is not available. In addition, the intensity component of the transformation may contain useful information about the patient's pathology, especially for comparing global and local intensity differences in conditions with extensive atrophy, such as Alzheimer disease (AD) and cerebrovascular disease.

There is a high failure rate for the spatial transformation of brains that are very different from the template. Thus, it is important to have a quantitative measure of the success of the spatial normalization. We have implemented a normalization algorithm that allows direct comparison of the normalized and template images, making it possible to evaluate the success of the stereotactic transformation. This chapter presents the details of this method and uses it to compare the efficacy of stereotactic transformation in normal volunteers and in patients with AD.

II. MATERIALS AND METHODS

A. PET Data

Data used in this analysis were initially acquired as part of a study of brain glucose metabolism in normal aging and Alzheimer's disease conducted by the Laboratory of Neurosciences, at the National Institute on Aging (Protocols 80AG26 and 81AG10). They consisted of [^{18}F]fluorodeoxyglucose (FDG) PET images obtained using a Scanditronix PC 1024-7B scanner from six AD patients (48–74 years, mean ± SD of 63 ± 10 years) with a wide range of disease severity [Mini Mental Scores (MMS) (Folstein et al., 1975): 0–26] and nine healthy control subjects (42–72 years, 58 ± 10 years). The PET data set from each subject consisted of 14 slices, each a 128 × 128 × 8-bit matrix. Several preprocessing steps were employed, including roll-yaw correction and alignment to a PET template in stereotactic space using a rigid body transformation (Woods et al., 1992). Images were then resliced to match the 62 slice template, and three slices (20, 29, and 41) were chosen for analysis (Fig. 1; note that all images are presented with the front of the brain at the bottom of the image). A linear two-dimensional scaling transform was then applied to each image by scaling the major and minor axes of an ellipse fit around the acquired brain slices to match those of the template.

B. The Volterra Theory of Nonlinear Systems

A nonlinear system is one in which the response to an input at one point is affected by inputs at a range of other points; this range is called the "memory" of the system. The Volterra theory describes the input–output relationship of a nonlinear system with memory. The Volterra equation for a two-dimensional system with memory (Korenberg, 1988), such as the transformation of one image to another, is

$$y(n_1, n_2) = h_0 + \sum_{i=-R_1}^{R_1} \sum_{j=-R_2}^{R_2} h_1(i,j) x(n_1 - i, n_2 - j)$$

$$+ \sum_{i_1=-R_1}^{R_1} \sum_{j_1=-R_2}^{R_2} \sum_{i_2=-R_1}^{R_1} \sum_{j_2=-R_2}^{R_2}$$

$$\times h_2(i_1, j_1, i_2, j_2) x(n_1 - i_1, n_2 - j_1)$$

$$\times x(n_1 - i_2, n_2 - j_2),$$

where x and y represent the input and output functions. The h terms are known as "Volterra kernels" and describe the contributions of inputs at other points to the output at a given point. Specifically, h_0 (zero order kernel) is related to the mean difference (input–output); $h_1(i, j)$, the first-order kernel, describes the contribution of the input at distance (i, j) to output at point (n_1, n_2); and $h_2(i_1, j_1, i_2, j_2)$, the second-order kernel, describes the contribution of interaction be-

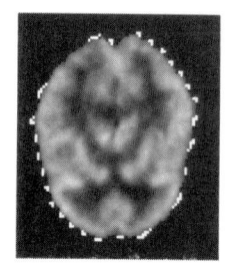

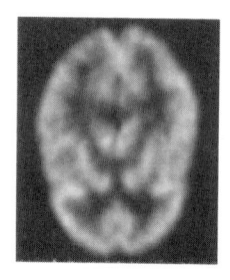

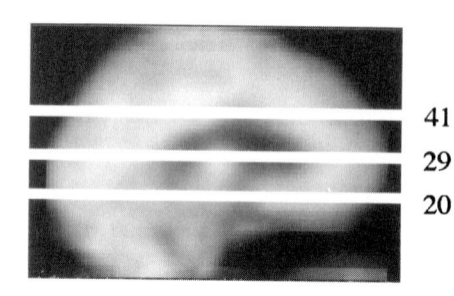

original   **template**   **stretched**   **chosen slices**

41
29
20

FIGURE 1 PET slices chosen for stereotactic transformation. The z level of the three slices on which the analysis was performed is shown on the right. Each slice was scaled individually for each subject using a linear two-dimensional transformation to match the overall size of the template image. Note that all images are presented with the front of the brain at the bottom of the image.

tween inputs at distances (i_1, j_1) and (i_2, j_2) to output at point (n_1, n_2). If the input and output functions are known, a unique set of kernels, which completely characterize the transformation, can be calculated using one of several techniques. The "fast orthogonal" technique (Korenberg, 1988) was used, which allows the rapid calculation of kernels for varying sizes of system memory, i.e., models of the system that take into account interactions among increasing numbers of points in the input image.

C. Kernel Analysis Methods

The number and amplitude of first-order kernel peaks provide a measure of the spatial heterogeneity of the transformation. Each peak can be thought of as the superposition of an offset copy of the input, similar to elements in a linear convolution kernel or filter. A large number of peaks means that pixels in the input had to be moved and stretched in many different ways to match the output. Comparing the relative contribution of the first- and second-order kernels to the total output provides a measure of the overall nonlinearity of the transformation.

Kernels can also be used to construct a predicted output image based on the application of either just the first-order kernels or the entire kernel set to the input. The predicted first order images provide a visual representation of the spatial heterogeneity described earlier. Comparison of the predicted images based on both first- and second- order kernels with the actual output, by taking a difference image, gives a quantitative measure of the fit of the calculated model. Difference images also present a spatial map of the error of the model and can be used to localize areas of poor fit.

III. RESULTS

Figure 2 shows the predicted output for slice 29 based on kernels calculated for the transformation of a single PET image into the template for both a normal control and a patient with severe AD (MMS = 0),

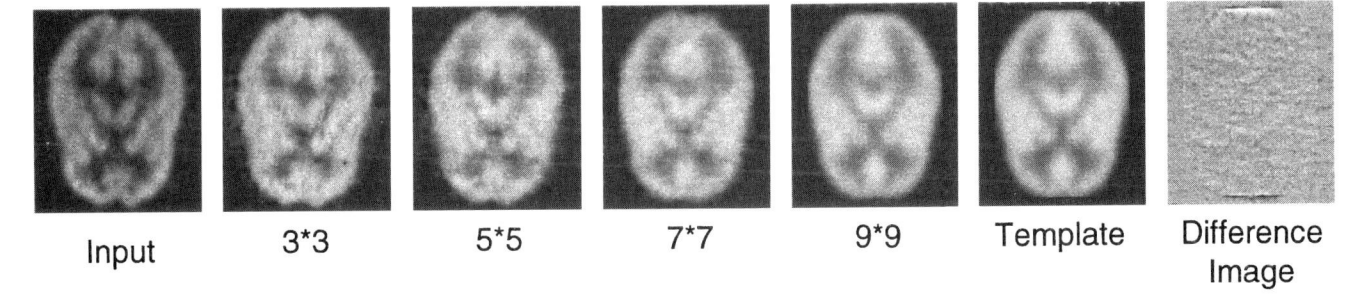

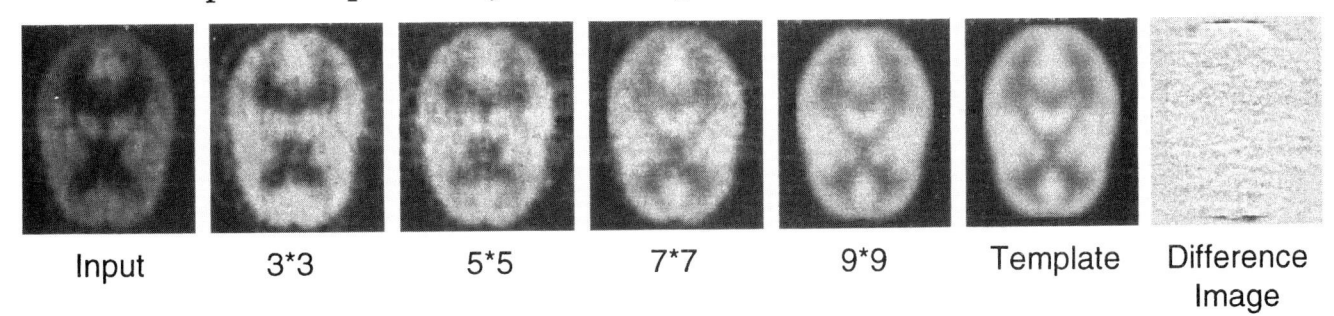

FIGURE 2 Kernel calculation performance. Nonlinear kernel analysis was applied to each PET slice to transform it into the template image in stereotactic space. System memory sizes from 3*3 to 9*9 pixels were used; larger sizes provided a better fit to the template but required longer calculation times. The difference images were obtained by subtracting the 9*9 output from the template image.

using system memory sizes from 3*3 to 9*9 pixels. A 9*9 system memory size modeled the transformation between both control and AD brains and the template quite well, as can be seen in the difference images at the far right of Fig. 2. However, for a given system memory, the fit was always better for controls than AD brains. The mean and variance of the difference images are (0.2, 29) and (0.29, 38) for the control and AD brain, respectively, suggesting that they are composed principally of noise. Although the 9*9 model predicted both transformations accurately, the complexity of this model required extremely long calculation times.

Figure 3 shows the predicted output based on only the first-order kernels for the same control subject. The lack of the image contrast seen in the output image suggests that the intensity component of the transformation occurs primarily in the second-order kernels, whereas the spatial component seems to require both first- and second-order kernels. The ratio of the root mean squared contributions of the first- and second-order kernels to the overall predicted output is 4.25:1, i.e., the first-order kernels are more important, suggesting (1) that the transformation for this brain is not highly nonlinear and (2) that the spatial transformation is the more important of the two components. As the system memory size increases, it appears that more copies of the original image are superimposed slightly offset from the original. This represents the presence of multiple peaks in first-order kernels and reflects the spatial heterogeneity of the transformation, i.e., different areas of the brain need to be moved and stretched by different amounts to make them match the template.

Figure 4 contains density plots of the first-order kernels for the 7*7 transformation of all subjects using the upper slice; the control subjects' density plots are shown on the left and the corresponding plot for each AD patient is on the right. If the input and output had similar spatial characteristics, the density plot for the first-order kernel would have a large positive peak in the center, with small peaks extending in each direction. The control plot shows this roughly: there is a bright central band surrounded by smaller peaks, corresponding to a single large positive kernel. The density plot for AD patients shows more heterogeneity—the central positive band is less distinct and there are more large peaks away from the center, suggesting that more spatial manipulation was required to fit the AD images to the template than for controls.

Given the hypothesis that AD brains would be more difficult to fit to the template, the 7*7 stereotactic transformation for the three slices of the AD patients was compared to those of the controls by calculating difference images between the predicted output of the Volterra transformation and the actual output, i.e., the template image. Figure 5 plots the mean pixel value of the difference images for slice 41 in two ways: against patient status and against the MMS score (an indicator of dementia severity: 30 is normal and 0 is extremely demented). There is no overlap in values between the control and AD populations for this slice, although the plots for slices 20 and 29 showed some overlap. Moreover, a significant correlation ($r = 0.77$) exists between model error and dementia severity in the AD patients, confirming that the transformation into stereotactic space is less reliable in the presence of severe disease. The etiology of the relatively large scatter among the controls is not entirely clear and merits further investigation.

The two central columns of Fig. 6 show an average of all subjects' difference images for each of the three

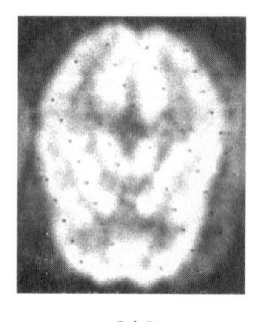

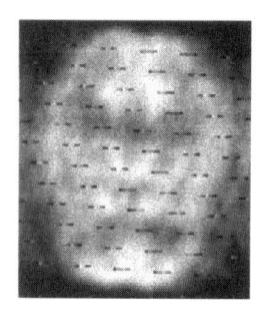

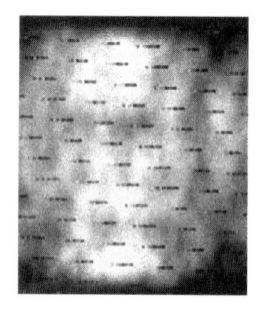

3*3 **5*5** **7*7** **9*9**

FIGURE 3 Predicted outputs based only on the calculated first-order kernels for system memory sizes from 3*3 to 9*9 pixels. The increasing number of offset copies in the larger memory sizes reflects the larger number of peaks in the kernels and thus greater heterogeneity of the spatial transformation. Note the loss of image contrast, suggesting that the intensity component of the transformation occurs principally in the second-order kernels.

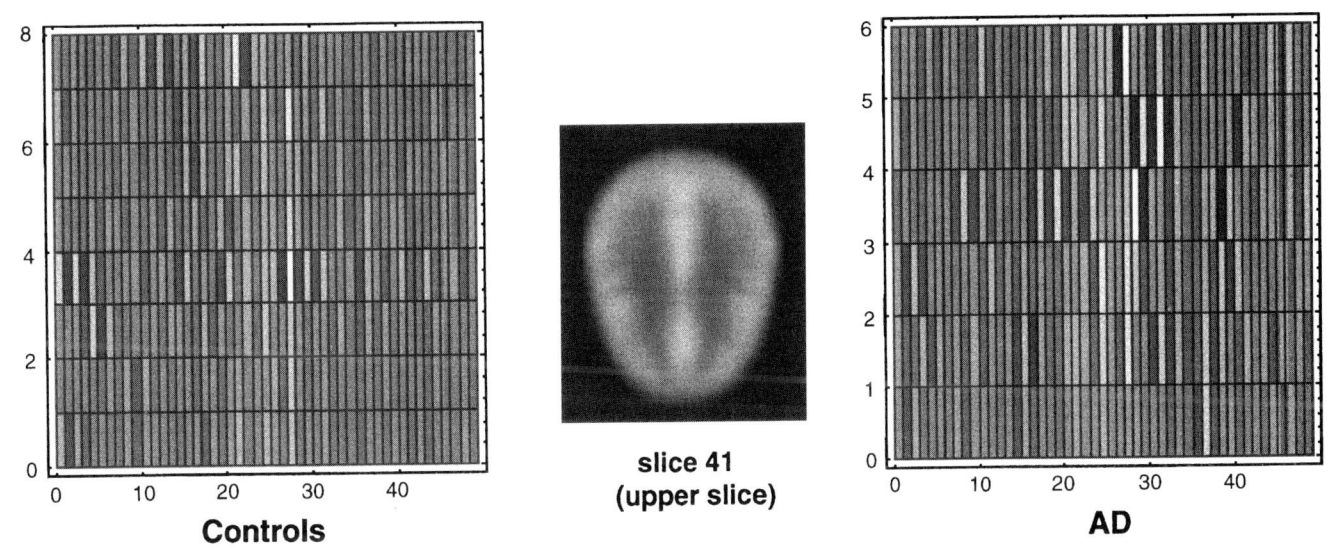

FIGURE 4 First-order kernel density plots for all the subjects for the upper PET slice (slice 41). Each row contains the 7*7 kernel matrix for a single subject (eight controls, six AD patients) transformed into a vector with 49 elements. The color of an individual vertical bar is proportional to the value of the kernel at that point: black represents the most negative values, white the most positive, and gray values near zero. Data for different subjects are on the same scale without normalization.

slices, whereas the right-most column contains the "difference of average difference images" so that darker areas correspond to pixels in which there was a greater error in the AD group than in controls. Areas showing the largest difference in error include parietal and dorsal occipital cortex (slice 41), temporal and parietal cortex (slices 20 and 29), and cingulate cortex (slice 29). Generally, these brain areas are thought to be most affected in AD and whose FDG values are the most hypometabolic in patients with AD (Duara, 1990; Grady et al., 1990; Haxby et al., 1985).

IV. DISCUSSION

It has long been suspected that stereotactic normalization of PET images may be problematic if the im-

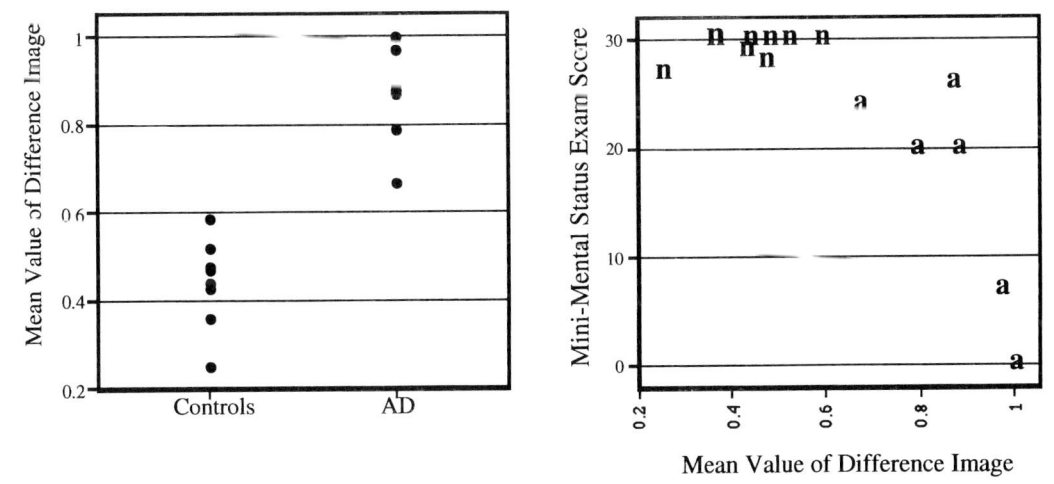

FIGURE 5 Plots of the average model error versus subject status and MMS score. The error is calculated by taking the mean value of the difference image between the template image and the predicted output of the 7*7 kernel model.

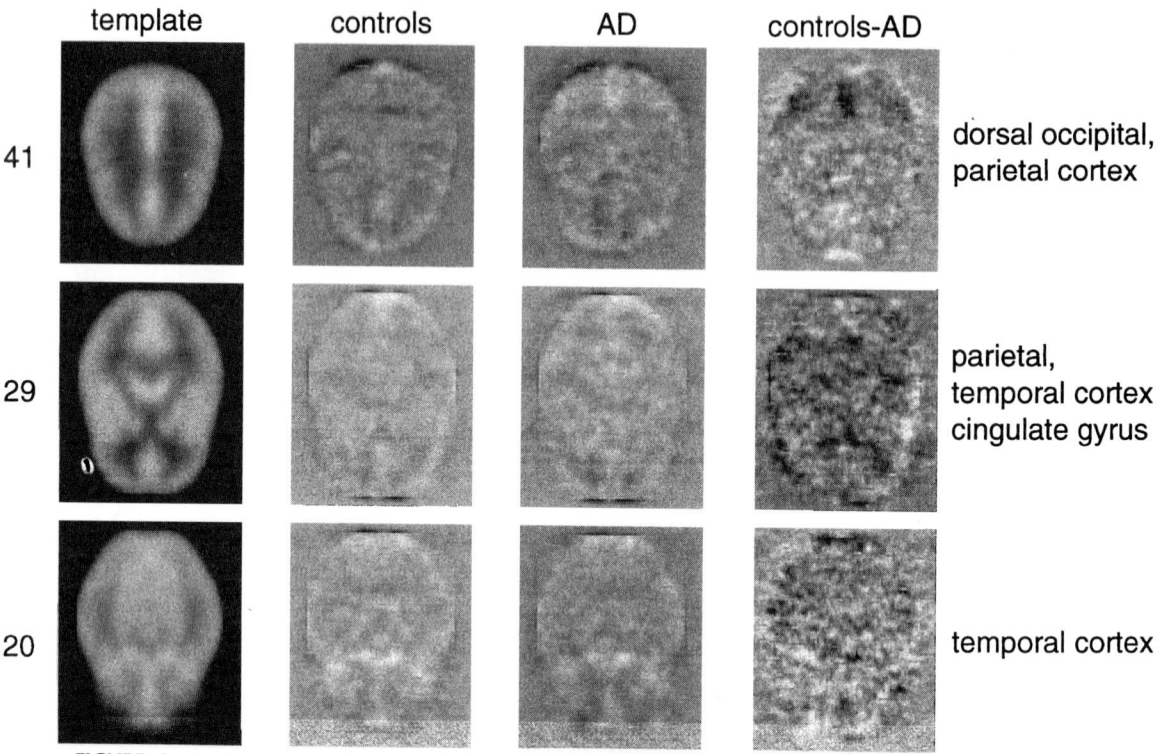

FIGURE 6 Spatial analysis of model error. The two middle columns contain the subject average of all of the calculated difference images for each slice. The column on the right presents difference images for each slice obtained by subtracting the AD average difference image from the control average difference image; dark pixels represent areas of greater error in the AD patients. The labels roughly indicate the anatomic location of these areas, which correspond to those areas known to be affected in AD.

ages come from subjects whose brains are dissimilar to those of young normal subjects on which the stereotactic methods were based. This study confirms that this is indeed the case; in a group of subjects with Alzheimer's disease, we found that the more severely demented the subject, the less reliable is the transformation into stereotactic space. This was demonstrated by employing a nonlinear kernel method that was able to model the transformation accurately to a stereotactic template. The kernel method permitted us to quantitatively assess the success of the transformation (by calculating the difference between the output of the transformation and the template), but also to determine the spatial locations at which the transformation was least successful. The largest errors were found to occur in the very regions in which Alzheimer's brains show the largest hypometabolic foci—parietotemporal cortex and other association areas, and the amount of error correlated with disease severity. These results suggest that transformation into stereotactic space of PET images from AD patients may not be reliable, especially in severe disease. It also may be possible to use the amount of error in this type of transformation, a "badness-of-fit" index, to provide a measure of diffuse cerebral disease that may be useful for discriminating AD and normal PET images.

Acknowledgments

We thank Drs. Stanley Rapoport and Mark Schapiro and their staff in the Laboratory of Neurosciences, NIA, for their support and contribution in acquiring the data used in this study.

References

Duara, R. (1990). "Positron Emission Tomography in Dementia." Wiley-Liss, New York.

Folstein, M. F., Folstein, S. E., and McHugh, P. R. (1975). "Mini Mental State"—a practical method for grading the cognitive state of patients for the clinician. *J. Psychiatr. Res.* **12**: 189–198.

Fox, P. T., Perlmutter, J. S., and Raichle, M. E. (1985). A stereotactic method for anatomical localization for positron emission tomography. *J. Comput. Assist. Tomogr.* **9**: 141–153.

Friston, K. J., Ashburner, J., Frith, C. D., Poline, J.-B., Heather, J. D., and Frackowiak, R. S. J. (1995). Spatial registration and normalization of images. *Hum. Brain Mapp.* **2**: 165–189.

Grady, C. L., Horwitz, B., Schapiro, M. B., and Rapoport, S. I. (1990). Changes in the integrated activity of the brain with healthy aging and dementia of the Alzheimer type. *In* "Aging Brain and Dementia: New Trends, Diagnosis and Therapy" (L. Battistin and F. Gerstenbrand, eds.), pp. 355-369. Liss, New York.

Haxby, J. V., Duara, R., Grady, C. L., Rapoport, S. I., and Cutler, N. R. (1985). Relations between neuropsychological and cerebral metabolic asymmetries in early Alzheimer's disease. *J. Cereb. Blood Flow Metab.* **5**: 193-200.

Korenberg, M. J. (1988). Identifying nonlinear difference equations and functional expansion representations: The fast orthogonal method. *Ann. Biomed. Eng.* **16**: 123-142.

Talairach, J., and Tournoux, P. (1988). "Co-Planar Stereotaxic Atlas of the Human Brain." Thieme, Stuttgart.

Woods, R. P., Cherry, S. R., and Mazziotta, J. C. (1992). Rapid automated algorithm for aligning and reslicing PET images. *J. Comput. Assist. Tomogr.* **16**: 620-633.

CHAPTER

18

Data Extraction from Brain PET Images Using Three-Dimensional Stereotactic Surface Projections[1]

SATOSHI MINOSHIMA, EDWARD P. FICARO, KIRK A. FREY, ROBERT A. KOEPPE, and DAVID E. KUHL

Division of Nuclear Medicine, Department of Internal Medicine
The University of Michigan, Ann Arbor, Michigan 48109

Data extraction is an essential procedure in the analysis of positron emission tomography (PET) images of the human brain. A novel data extraction technique, three-dimensional stereotactic surface projections (3D-SSP), specifically targeting cortical functional analysis, has been developed. In this technique, a reconstructed brain PET image set is first standardized to a stereotactic coordinate system. Cortical activity is searched on a predefined search vector from a predefined stereotactic surface pixel in three dimensions. The peak cortical activity along the vector is then projected back and assigned to the surface pixel. This procedure is repeated on a pixel-by-pixel basis covering the whole cortex, including medial aspects of the hemispheres. The quantitative accuracy of this technique was evaluated using 3D-simulated brain PET images as well as PET images obtained in humans. Results demonstrated that the quantitative accuracy of 3D-SSP data extraction was less biased by the size of regions or by cortical atrophy when compared to standard volume of interest (VOI) analysis. Influence of image noise on 3D-SSP data extraction was no worse than that on VOI-extracted data. Tracer kinetic estimation of rate constants in a neuroreceptor study was improved with 3D-SSP data extraction because of less contamination from white matter and cerebrospinal fluid space. In pixel-by-pixel group subtraction analysis, 3D-SSP minimized subtraction artifacts due to small cortical mismatches between two groups. Three-dimensional SSP-extracted data are readily viewed in three dimensions, which improves region identification and visual interpreta-tion. Three-dimensional SSP will be a useful tool in quantitative PET analysis of various cortical functions.

I. INTRODUCTION

Data extraction is an essential procedure when analyzing brain positron emission tomography (PET) images. Region or volume of interest (ROI or VOI) analysis has long been used for data extraction in statistical and kinetic analyses of PET images. This approach is suitable for discrete structures, such as the caudate nucleus or thalamus, but generally does not completely sample cortical functional information, as the number of regions definable in the cortex is quite limited in many settings. The use of a limited number of regions or volumes may also bias the interpretation of pathophysiology of the diseases due to potential biases in region selection. An alternative approach is to analyze data on a pixel-by-pixel basis, most extensively developed for activation studies (Friston *et al.*, 1990) and further extended to neuroreceptor studies (Frey *et al.*, 1996). The approach, however, is prone to residual regional anatomic variances in the stereotactic system (e.g., varying depth of gyri) and partial volume effects due to image smoothing. Since developments in PET tomographs permit measurement of the entire brain with high spatial resolution, consistent and reproducible data extraction techniques for cortical function are strongly desired. In general, optimal data extraction requires the schemes to be quantitatively accurate, efficient in data reduction, reproducible, unbiased by observers, suitable for visual inspection, suitable for

[1] Transcripts of the BRAINPET97 discussion of this chapter can be found in Section VIII.

further statistical analyses, and applicable to different tracer studies; they must also preserve anatomic information. To satisfy and balance these factors, a novel data extraction technique, three-dimensional stereotactic surface projections (3D-SSP), was developed and its diagnostic use in brain PET imaging has been reported (Minoshima et al., 1995). This study examined the quantitative accuracy of 3D-SSP data extraction in various brain PET image applications, comparing results obtained with conventional VOI analysis.

II. METHODS AND MATERIALS

A. Three-Dimensional Stereotactic Surface Projections

In 3D-SSP, a set of surface pixels covering the entire lateral and medial brain surface is predefined in the stereotactic system with vectors perpendicular to the brain surface. Peak cortical activity is searched into the brain (typically 13.5 mm) on a predefined vector for each stereotactic surface pixel in three dimensions, and the peak value is projected back and assigned to the originating surface pixel. This procedure is repeated on a pixel-by-pixel basis covering the entire cortex of the brain (Minoshima et al., 1995) (Fig. 1, see color insert). The extracted data set can be viewed readily from any projection, including lateral, medial, superior, inferior, anterior, and posterior, and any oblique views if necessary. This greatly improves regional identification in three dimensions, and extracted data are suitable for further statistical assessment either using region of interest or pixel-by-pixel analysis.

B. Three-Dimensional Digital Brain Phantom and Human Brain PET Image Sets

To evaluate the quantitative accuracy of 3D-SSP, both a three-dimensional brain phantom and brain PET images obtained in humans were used. Emission and transmission sinograms were simulated, accounting for activity distributions, acquisition parameters, detector resolution and efficiencies, random coincidence noise, and emission scatter using a cortically delineated, stereotactically aligned, 3D brain phantom with a voxel size of 2.25 mm. The scatter fraction was set to 20% of detected emission events and convolved with a Gaussian function with full-width-at-half-maximum (FWHM) of 36 mm. Images were reconstructed from the simulated sinograms using filtered backprojection with a Hanning filter of cutoff 0.30 cycles/projection element. The emission images were not corrected for photon scatter. [^{18}F]Fluorodeoxyglucose (FDG) image sets with a gray-to-white matter ratio of 4:1 were stimulated. Simulated FDG image sets without noise or scatter were also generated to assess bias due to partial volume effects. Different cortical thicknesses were simulated to assess the effects of atrophy on data extraction. For the dynamic data set, time–activity curves were generated for multiple brain regions simulating [^{11}C]flumazenil (FMZ) kinetics in a normal subject (Koeppe et al., 1991). From the time–activity curves, 17 phantom brain distributions were defined, projected, and reconstructed to produce a realistic estimate of a measured dynamic 3D FMZ image set. Additionally, 3D-SSP data extraction in pixel-by-pixel group comparison analysis using FDG scans from 20 normal subjects and 20 patients with Alzheimer's disease acquired on a Siemens ECAT 931/8-12 scanner was tested.

C. Comparisons of 3D-SSP versus VOI Data Extraction

To compare 3D-SSP and conventional VOI analyses, multiple ROIs in the cortex were defined. Spherical VOIs were placed at the points of peak cortical activity in a PET volume image set. At the equivalent location in 3D-SSP-extracted data, ROIs were defined to include the exact same amount of cortical area in both analyses. Regions were defined bilaterally in the frontal, parietal, and temporal cortices. Different region sizes ranging from one pixel radius (2.25 mm) to six pixel radius were examined.

III. RESULTS AND DISCUSSION

A. Region Size and Partial Volume Effects on 3D-SSP and VOI Analysis

The relationship between the size of region and the magnitude of partial volume effect was examined first using noiseless images. When very small regions were used (e.g., one pixel radius, seven total pixels in VOI, and five total pixels in 3D-SSP), VOI and 3D-SSP resulted in similar fractions of gray matter (GW), white matter (WM), and cerebrospinal fluid (CSF) contributing to measured signals (Fig. 2). However, when region size was increased, contamination from WM and CSF became greater in the VOI analysis. An averaged gray matter fraction in VOIs decreased from 95% (one pixel radius) to 65% (six pixel radius). In contrast, once peak cortical activity was extracted onto the 3D-SSP format, contamination from white matter and CSF space remained small, even when using large ROIs.

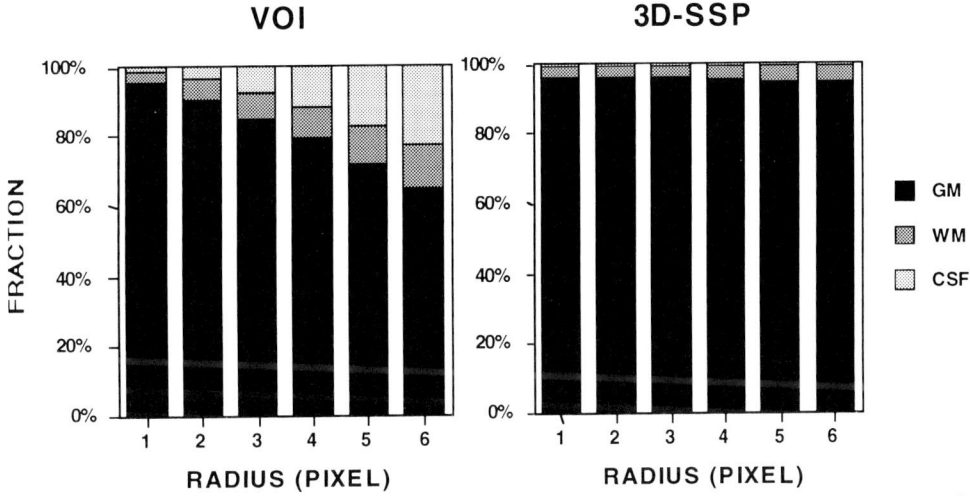

FIGURE 2 Region size and partial volume effects in 3D-SSP and VOI analyses. Values are averaged across six cortical regions. The vertical axis represents fractions of gray matter (GM), white matter (WM), and cerebrospinal fluid space (CSF) contributing to measured signals. Typically, region sizes of three to four pixel radius are used in VOI analysis.

B. Cortical Atrophy and Partial Volume Effects on 3D-SSP and VOI Analyses

The effects of cortical atrophy on VOI and 3D-SSP data extraction were examined using simulated FDG PET image sets with three levels of cortical atrophy (0, 23, and 44% reduction of a whole gray matter volume) generated by geometric dilation and erosion algorithms. When cortical activity was extracted using a four pixel radius region, gray matter fraction decreased from 80 to 57% with 44% cortical atrophy (Fig. 3). In contrast, using 3D-SSP data extraction, gray matter fraction decreased to only 77%, comparable to VOI analysis with no atrophy (80%). This result indicates that 3D-SSP-extracted data are less affected by cortical atrophy, although its effects are not eliminated completely. A combination of 3D-SSP with modeling for partial volumes (Yu *et al.*, 1993) or with the use of magnetic resonance images to provide additional information (Meltzer *et al.*, 1990) should improve this result

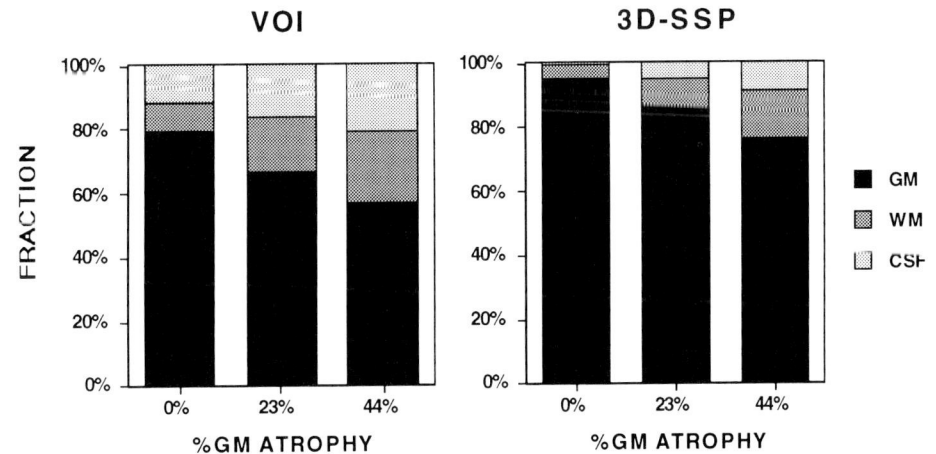

FIGURE 3 Effects of cortical atrophy in 3D-SSP and VOI analyses. A region size of four pixel radius was used. In both analyses, contamination from WM and CSF increases with cortical atrophy. Note that the GM fraction at 44% atrophy in 3D-SSP analysis is approximately equal to that at 0% atrophy in VOI analysis.

further, although such correction methods generally require strict assumptions for tracer distribution and disease-specific alteration in cerebral layers.

C. Effects of Image Noise on 3D-SSP and VOI Data Extraction

Because 3D-SSP extracts only peak cortical activity, the effect of image noise on the accuracy of quantification is of concern. Influence of image noise on VOI and 3D-SSP analyses was examined using simulated PET images with different noise levels. Without image noise, coefficients of variation (COV) increased with larger VOIs, primarily due to varying degrees of contamination from white matter and CSF fractions for different regions of the brain (Fig. 4). In contrast, the increase in COV using 3D-SSP was small at the largest region size. This result indicates that 3D-SSP data extraction is robust in the face of regional variation in cortical anatomy and allows more consistent estimates of cortical activity.

When increasing image noise was simulated by decreasing scan durations, COVs increased, particularly for small VOIs. Large VOI measurements were less affected by image noise because the noise averaged out within the volume. Partial volume effects become much greater with large VOIs (Fig. 2), however, creating the need to choose between increased bias and loss of precision. In contrast, for 3D-SSP, although COVs increased on average with increasing noise level, the increase in COVs was consistently smaller than or comparable to that with VOI analysis, depending on the size of regions. This result indicates that the effects of image noise on 3D-SSP are no worse than that with conventional VOI analysis.

D. Tracer Kinetic Analysis Using 3D-SSP versus VOI

Three-dimensional SSP data extraction used in conjunction with tracer kinetic analysis of a neuroreceptor study by simulating a dynamic FMZ image set with known rate constants was evaluated. Cortical activity was extracted using conventional VOI and 3D-SSP analyses. Cortical peak locations were determined on an averaged image set of the first 10 min of dynamic data. VOI-extracted activity, measured with a region size of four pixel radius, was generally lower than that with 3D-SSP due to contamination from WM and CSF. Because of the time dependence of the contamination, the shapes of the two curves were not identical. When estimating kinetic parameters using a three-compartment model (three parameters with fixed k_4), VOI-extracted data showed a 10% underestimate in K_1 and a 15% underestimate in DV'' as compared to a +3% overestimate in K_1 and only −5% underestimate in DV'' using 3D-SSP. When estimating K_1 and DV using a simplified two-compartment model for FMZ (Koeppe et al., 1991), results were similar, with VOI and 3D-SSP again yielding underestimates in DV of 15 and 5%, respectively.

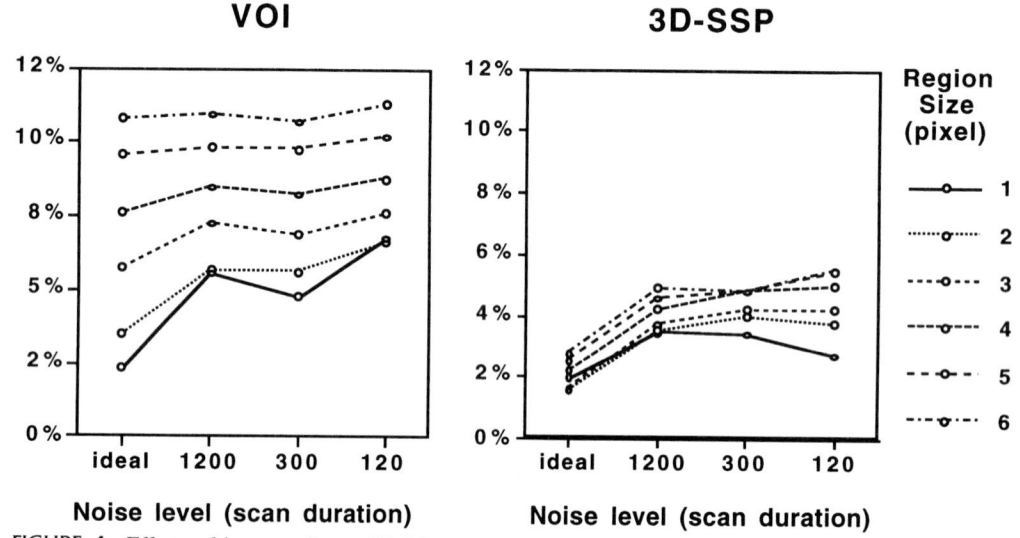

FIGURE 4 Effects of image noise on 3D-SSP and VOI analyses. The horizontal axis represents increasing noise level with shorter scan durations (scan duration in seconds; "ideal" represents noise-free data). The vertical axis represents coefficients of variation calculated across multiple brain regions. Region sizes of one to six pixel radius are examined.

E. Pixel-by-Pixel Group Comparison of Brain PET Image Sets with 3D-SSP

In a conventional approach for pixel-by-pixel group comparison of brain PET images, individual image sets are first standardized anatomically, then summed within groups to produce mean and variance for each pixel, and finally tested for group differences on a pixel-by-pixel basis using a two-sample t test or ANOVA if three or more groups are examined. In this approach, results can then be displayed in three dimensions for presentation purposes. Alternatively, cortical activity can be extracted first using 3D-SSP before summation analysis. This should improve the quantitative accuracy of group comparison analyses in two ways: (1) by minimizing contamination from white matter and CSF fractions as mentioned previously and (2) by compensating for small mismatches of gray matter locations across subjects. For example, when comparing averaged FDG PET image sets from normal subjects and patients with Alzheimer's disease in the standard stereotactic coordinate system, the interhemispheric space in the Alzheimer's group was slightly widened due to the presence of cortical atrophy (11 mm peak-to-peak in the normal group versus 13 mm in the Alzheimer's disease group). Subtraction analysis between these two image sets produced "pseudo" metabolic reduction in the medial frontal cortex (Fig. 5, see color insert). In contrast, when extracting cortical activity using 3D-SSP prior to summation analysis, this artifact was effectively eliminated as 3D-SSP compares peak cortical activity in one group to peak cortical activity in the other group.

IV. SUMMARY

In comparison to a conventional VOI analysis, 3D-SSP data extraction improves quantitative accuracy and consistency across different brain regions by decreasing partial volume and cortical thickness biases. This results in better estimates of regional activity and tracer kinetic parameters. 3D-SSP-extracted data sets are readily viewed in three dimensions for visual interpretation and are suitable for further pixel-by-pixel analyses. One obvious drawback with 3D-SSP is that the method is not suitable for sampling deep structures such as the putamen or insular cortex. In addition, 3D-SSP does not distinguish functional activity at different depths of sulci, although other methods currently available do not accomplish this accurately either. The development of more accurate stereotactic registration at gyral levels may improve this. Nevertheless, 3D-SSP provides a useful alternative to the conventional VOI approach in the quantitative analysis of cortical function measured with PET.

Acknowledgment

This study is supported in part by DE-FG02-87-ER60561 from the Department of Energy and NS24896 from the National Institutes of Health.

References

Frey, K. A., Minoshima, S., Koeppe, R. A., Kilbourn, M. R., Berger, K. L., and Kuhl, D. E. (1996). Stereotaxic summation analysis of human cerebral benzodiazepine binding maps. *J. Cereb. Blood Flow Metab.* **16**: 409–417.

Friston, K. J., Frith, C. D., Liddle, P. F., Dolan, R. J., Lammertsma, A. A., and Frackowiak, R. S. (1990). The relationship between global and local changes in PET scans. *J. Cereb. Blood Flow Metab.* **10**: 458–466.

Koeppe, R. A., Holthoff, V. A., Frey, K. A., Kilbourn, M. R., and Kuhl, D. E. (1991). Compartmental analysis of [^{11}C]flumazenil kinetics for the estimation of ligand transport rate and receptor distribution using positron emission tomography. *J. Cereb. Blood Flow Metab.* **11**: 735–744.

Meltzer, C. C., Leal, J. P., Mayberg, H. S., Wagner, H. N., Jr., and Frost, J. J. (1990). Correction of PET data for partial volume effects in human cerebral cortex by MR imaging. *J. Comput. Assist. Tomogr.* **14**: 561–570.

Minoshima, S., Frey, K. A., Koeppe, R. A., Foster, N. L., and Kuhl, D. E. (1995). A diagnostic approach in Alzheimer's disease using three-dimensional stereotactic surface projections of fluorine-18-FDG PET. *J. Nucl. Med.* **36**: 1238–1248.

Yu, D. C., Huang, S. C., Gratton, S. T., Meluga, W. P., Barrio, J. R., Mazziotta, J. C., and Phelps, M. E. (1993). Methods for improving quantitation of putamen uptake constant of FDOPA in PET studies. *J. Nucl. Med.* **34**: 679–688.

CHAPTER 19

A Method for Surface-Based Quantification of Functional Data from the Human Cortex

H.-M. VON STOCKHAUSEN,* U. PIETRZYK,* K. HERHOLZ,* A. THIEL,* J. ILMBERGER,[†] H.-J. REULEN,[†] and W.-D. HEISS*

Max Planck Institute for Neurological Research and University Clinic
Cologne, Germany
[†] *Department of Neurosurgery*
Klinikum Großhadern
Ludwig-Maximilians-Universität
Munich, Germany

Projection of functional data onto the surface of the brain helps in relating centers of neuronal activation to anatomical structures during the planning of a neurosurgical intervention. Because mental assignment of colors to actual values of underlying data is subjective, those renderings of the brain surface show only qualitative aspects. The presented method allows a surface-based quantitative assessment. Within a rendering of the brain surface, locations are defined by the user that serve as starting points for automatic definition of volumes of interest (VOIs). Definition is done by deforming a geometrical object to match the local shape of the surface. Statistical data about activation can be derived from these VOIs.

I. INTRODUCTION

The development of tomographic methods, especially magnetic resonance (MR) imaging, is the basis for definition and presentation of three-dimensional (3D) anatomic structures. Furthermore, measurement of brain function with positron emission tomography (PET) allows quantification of physiological parameters of those structures, e.g., change in regional cerebral blood flow (rCBF) during a defined task relative to a control condition.

Combined visualization (fusion) of functional data, e.g., activation data from PET imaging and morphological data from MR imaging, allows one to create surface renderings of the human brain that show not only topographical aspects, but also the locations of activated areas by color coding of the brain surface (Fig. 1). These pictures can be used by neurosurgeons in the process of planning an intervention, e.g., resection of a tumor. Unfortunately, those pictures give only qualitative information, as the intensity of coloring depends not only on the magnitude of activation, but also on the inclination of the surface and scaling of the color table. Furthermore, perception of colors is subjective. For quantitative analysis of activation of anatomically defined parts of the cortex, delineation of volumes of interest (VOIs) is necessary.

This chapter describes a novel method for the easy definition of such VOIs using a surface rendering of the human brain created from T1-weighted MR head scans.

II. MATERIALS AND METHODS

A. Data Acquisition

T1-weighted MR images were obtained on a 1.5-T Magnetom Vision (Siemens Medical Systems, Erlangen, Germany) as 128 sagittal slices of 0.98-mm thickness, acquired in MPRAGE (magnetization prepared rapid acquisition gradient echo) sequence (TR/TE = 12/4.4 msec; TA = 13.38 min; AC = 1; FOV = 250 × 250; matrix = 256 × 256).

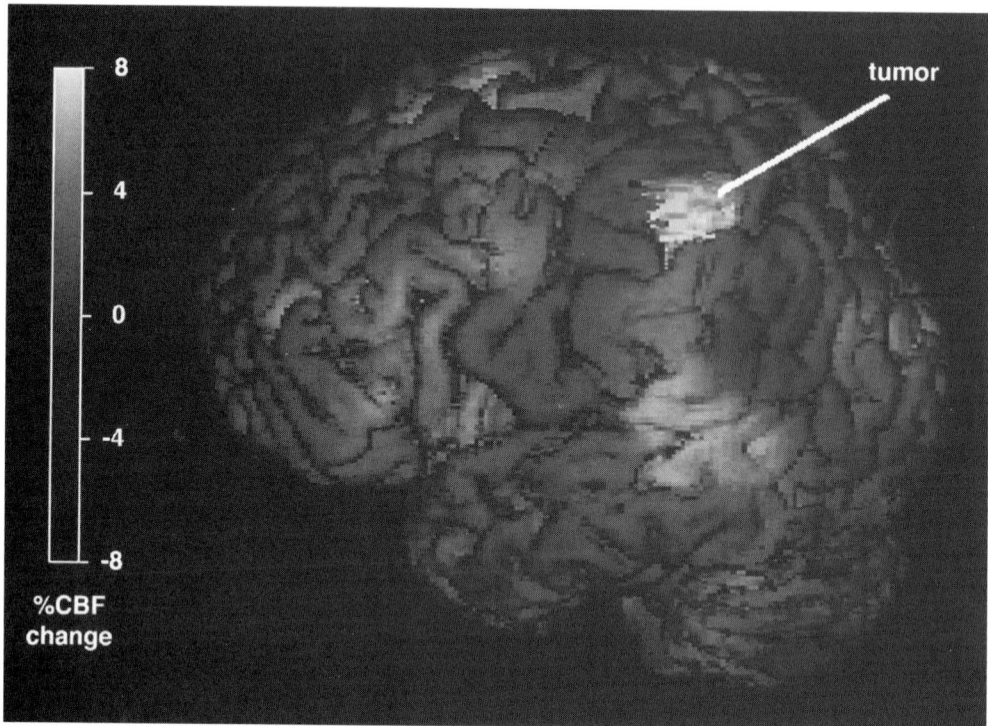

FIGURE 1 Maxima from activation data (percentage change in rCBF) projected from the cortex onto the brain surface.

PET was performed on a high-resolution scanner (ECAT EXACT HR, Siemens-CTI, Knoxville, TN) in 3D acquisition mode. Acquisition over 90 sec was started simultaneously with an intravenous bolus injection of 370 MBq (10 mCi) [^{15}O]water. After reconstruction by filtered backprojection with a Hanning filter (0.4 cycles/pixel) and correction for random coincidences, attenuation, and scatter, 47 transaxial slices of 3.125 mm thickness and 2.166 mm pixel size were obtained.

Images were normalized to constant whole brain counts. The local change of signal during each activation condition, which is closely related to CBF change, was calculated for each voxel as a percentage of the control value. Resulting data were smoothed within the brain by a spherical median filter with 8 mm radius.

B. Preprocessing

Prior to actual surface-based quantification of functional data, three preprocessing steps are performed: (1) alignment of the data sets, (2) segmentation of morphological objects, and (3) rendering of a surface projection.

1. Alignment of Data Sets

The prerequisite for multimodal presentation and analysis of functional data using digital morphological atlases is spatial correspondence between points in the data volumes. A retrospective method (Pietrzyk et al., 1994) was used to register MR data with PET data. Because of its interactive nature, it is applicable even in the presence of large structural brain lesions. Following alignment, both data sets are resliced into the same voxel grid of 1 mm^3.

2. Segmentation

Crucial to the presented method is the volumetric description of the brain (brain mask), which is obtained from the MR head scan. Additionally, brain lesions are segmented in order to visualize the topographic relation of activation data to the lesions. Segmentation is based on thresholding, connected component analysis, and morphological operations (Höhne and Hanson, 1992). Processing is performed slicewise in two steps: after an automatic run, corrections are made interactively (von Stockhausen et al., 1996).

3. Visualization

Rendering of brain surface and lesions is done using ray-casting (Tuy and Tuy, 1984) and gray-level gradient shading (Höhne and Bernstein, 1986). The rendering serves both for anatomical navigation and for an overview of the distribution of activation data. The viewing angle and zoom factor are adjusted to project the surface area under investigation in an optimal way.

Integration of activation data is done by reverse gradient fusion (Stokking *et al.*, 1994). For each visible surface location the color is defined by the absolute maximum in functional data within a distance of 10 mm perpendicular to the brain surface.

Together with the surface rendering, a matrix with the same dimensions as the picture is created that holds the volume coordinate for each visible surface location. This matrix serves as a base for the surface-based navigation in the volume data set (von Stockhausen, 1995).

C. VOI Definition

The initial VOI consists of a cylinder with a defined height h_z and a defined diameter d_z, which is placed at a location P_s on the brain surface by the user. Afterward, the shape of the cylinder will be automatically deformed in two steps to match the shape of the brain surface at P_s by means of ray casting. Prior to the actual deformation, the cylinder is centered on P_s with its bottom tangent to the surface in P_s. In the first step, search rays of length h_F are cast out from the bottom of the cylinder in order to find the surface points of the brain from which the VOI will be defined (Fig. 2a). If the starting point of a search ray lies within the brain (case I), the search direction is parallel to the surface normal n_s, otherwise (case II), the search direction is opposite to n_s. If the brain surface is not reached within the search length h_F, the ray and its associated surface location will be ignored in the second step for case I. In case II, the starting point of the search ray on the bottom of the cylinder will be used.

In the second step, rays will be cast from the defined surface locations around P_s into the brain. All voxels that are connected to the brain surface along a straight line lying parallel to the direction of n_s and whose distance from the brain surface is less or equal to h_z are included in the VOI (Fig. 2b).

Finally, the VOI is masked with the brain volume, and a connected component analysis is performed. The component of the VOI that includes the selected surface point P_s is chosen to make sure that the VOI will not overlap into a neighboring area, e.g., separated by a sulcus (Fig. 2c). These steps of VOI definition are repeated for all user-selected locations on the brain surface. After definition of all VOIs, statistical parameters are calculated from all functional data sets to assess the state of activation within the VOIs.

III. RESULTS

The method was applied to [15O]water PET activation studies in eight glioma patients. VOIs were created from cylinders with 10 mm diameter and with an initial height of 10 mm. The comparison of local cortical CBF increases within those VOIs during preoperative verb generation with results from intraoperative data permitted the determination of an optimum threshold for preoperative identification of areas with intraoperative language disturbance during electrical cortical stimulation (Fig. 3).

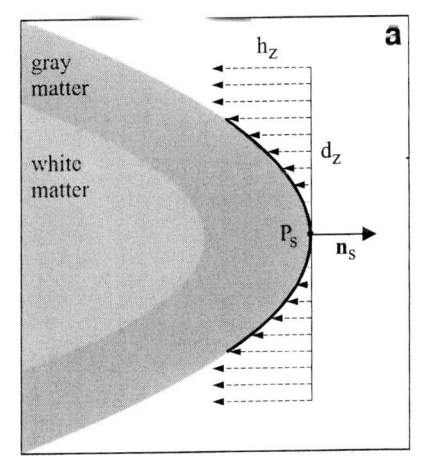

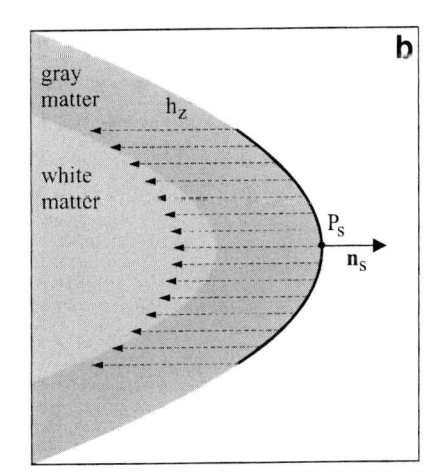

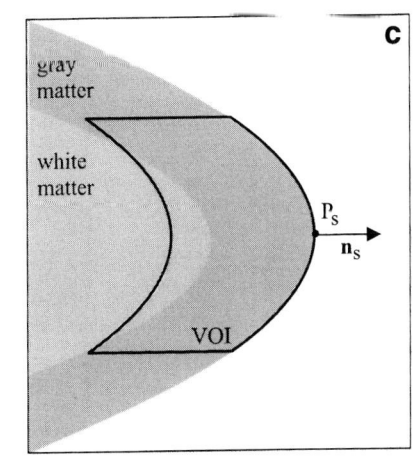

FIGURE 2 Steps for defining a surface based VOI: (a) Search for initial points on the brain surface starting from the bottom of the cylinder (left), (b) definition of VOI voxels (center), and (c) final VOI (right).

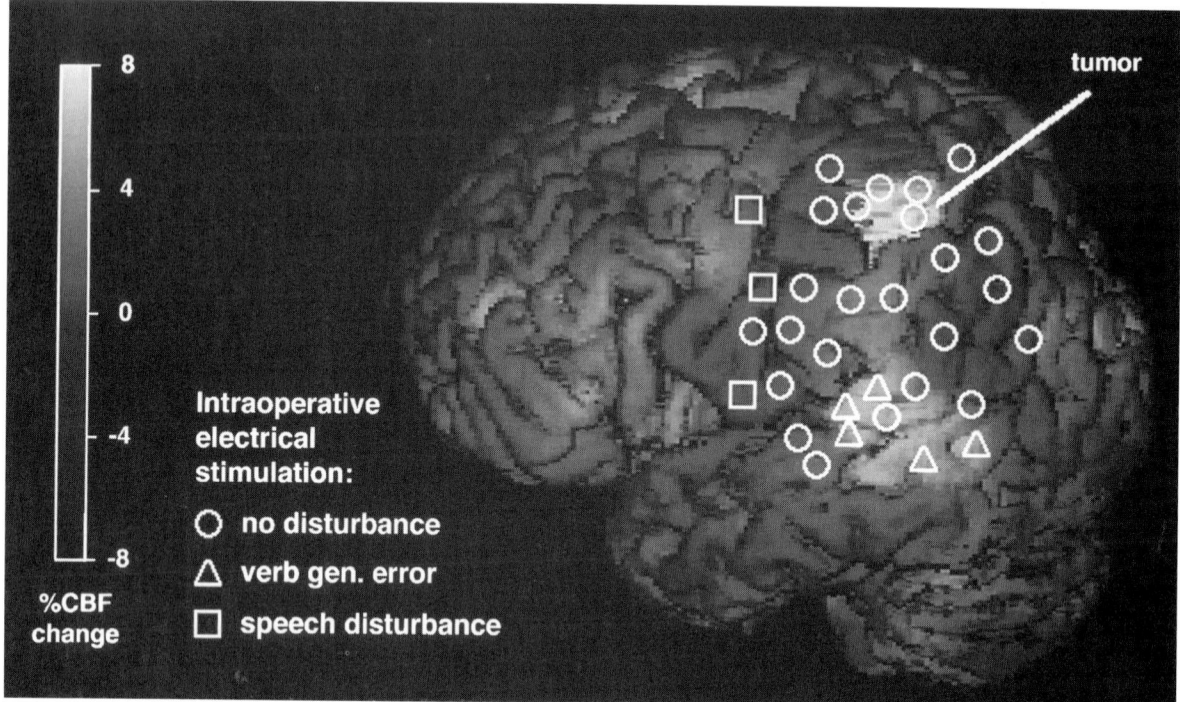

FIGURE 3 Brain surface with projected activation data combined with results from intraoperative electrical cortical stimulation.

IV. CONCLUSION

The described method allows the definition of anatomical VOIs for quantitative assessment of cortical structures guided by a surface projection of the brain. In contrast to stereotactic normalization, this method takes into account individual morphology, activation patterns, and position of a lesion.

References

Höhne, K. H., and Bernstein, R. (1986). Shading 3D-images from CT using gray level gradients. *IEEE Trans. Med. Imag.* **5**: 45–47.

Höhne, K. H., and Hanson, W. A. (1992). Interactive 3D segmentation of MRI and CT volumes using morphological operations. *J. Comput. Assist. Tomogr.* **16**: 285–294.

Pietrzyk, U., Herholz, K., Fink, G., Jacobs, A., Mielke, R., Slansky, I., Würker, M., and Heiss, W.-D. (1994). An interactive technique for three-dimensional image registation: Validation for PET, SPECT, MRI and CT brain studies. *J. Nucl. Med.* **35**: 2011–2018.

Stokking, R., Zuiderveld, K. J., Hulshoff Pol, H. E., and Viergever, M. A. (1994). Integrated visualization of SPECT and MR images for frontal lobe damaged regions. *Proc. SPIE* **2359**, 282–290.

Tuy, H. K., and Tuy, L. T. (1984). Direct 2-D display of 3-D objects. *IEEE Comput. Graphics Appl.* **4**: 29–33.

von Stockhausen, H.-M. (1995). Ein System zur interaktiven dreidimensionalen Volume-of-Interest Segmentierung und Visualisierung medizinischer Bilddaten. Masters Thesis, Universität Heidelberg, Medizinische Informatik.

von Stockhausen, H.-M., Pietrzyk, U., and Herholz, K. (1996). Techniken zur Visualisierung funktioneller tomographischer Daten in der klinischen Forschung. *In* "Digitale Bildverarbeitung in der Medizin" (B. Arnolds, H. Müller, D. Saupe, and T. Tolxdorff, eds.), pp. 46–52. Universitätsklinikum Benjamin Franklin, Freie Universität, Berlin.

SECTION III

APPLICATIONS

CHAPTER 20

Sequential Experimental PET: Voxel-Based Analysis Reveals Spatiotemporal Dynamics of Perfusion in Transient Focal Cerebral Ischemia[1]

R. GRAF, A. SCHUSTER, J. LÖTTGEN, U. PIETRZYK, K. OHTA, E. KUMURA, K. WIENHARD, and W.-D. HEISS

Max Planck Institute for Neurological Research
Cologne, Germany

A voxel-based approach was developed as an exploratory tool to compare ischemic and postischemic positron emission tomography (PET) measurements of cerebral blood flow (CBF) after experimental focal ischemia to preischemic control CBF. It was used to demonstrate the spatiotemporal dynamics of brain perfusion with special emphasis on patterns of postischemic hyperperfusion. CBF was measured repeatedly in 16 halothane-anesthetized cats before and up to 24 hr after 60 min of middle cerebral artery occlusion (MCAO) on a CTI/Siemens EXACT HR PET scanner with a bolus intravenous injection of [^{15}O]H$_2$O. Display of data from single coronal planes in scatter plots allowed separation of clusters of voxels by thresholding or by visual analysis and subsequent reprojection of these clusters onto the corresponding images. The spatiotemporal evolution of CBF after recirculation in regions defined in this manner was analyzed. In the ischemic territory, CBF decreased in all animals to below 45% of preischemic control values. After recirculation, reperfusion was only partially successful in 3 cats, and hyperperfusion was not observed. Continued hyperperfusion covering large parts of the ischemic territory was observed in 5 cats. Transient hyperperfusion, often followed by hypoperfusion, developed in 6 cats. In 2 cats, hyperperfused regions grew over time. Regions with continued hyperperfusion had a bad outcome as shown by histology, as had those with continued incomplete reperfusion. Voxel-based analysis revealed variable spatiotemporal evolution of postischemic perfusion patterns following transient MCAO in cats. This approach can be applied for exploratory purposes to a variety of studies that use sequential PET.

I. INTRODUCTION

From earlier work it is evident that the occlusion of a major cerebral artery initiates a sequence of pathophysiological changes that may include large parts of brain inside and even outside the territory of the occluded artery (Siesjö, 1992; Pulsinelli, 1992; Kogure et al., 1993; Heiss and Graf, 1994). These alterations propagate from the core to regions in the border of the ischemic focus (Hakim and Hogan, 1991; Heiss et al., 1994), suggesting that the so-called ischemic penumbra (Astrup et al., 1981) should be regarded as a highly dynamic process rather than as a discrete condition in a distinct brain region. In transient focal ischemia, similar dynamics have been described for the reperfusion period (Heiss et al., 1997).

Recent developments in functional imaging technology make it possible to follow such changes noninvasively even in small animals. With high-resolution positron emission tomography (PET), regional cerebral blood flow and the regional metabolic rate for oxygen and glucose can be measured repeatedly before and after occlusion of the middle cerebral artery in baboons (Tenjin et al., 1992; Pappata et al., 1993) and in cats (Heiss et al., 1994). The development of the penumbra and the eventual progression to infarction can be followed. To better understand the interrela-

---

[1] Transcripts of the BRAINPET97 discussion of this chapter can be found in Section VIII.

tionship between multiple measurements in the course of transient focal ischemia, a voxel-based approach has been developed as an exploratory tool to compare ischemic and postischemic PET determinations with preischemic control values. This was used to demonstrate the spatiotemporal dynamics of brain perfusion during and after ischemia, with special emphasis on patterns of postischemic hyperperfusion.

II. MATERIALS AND METHODS

Sixteen adult cats of either sex weighing 3.6–5.1 kg were used. Anesthesia was induced with 25 mg/kg of intramuscular ketamine hydrochloride. After catheterization of the left femoral artery and vein, the cats were tracheostomized, immobilized with 0.2 mg/kg of intravenous pancuronium bromide, and artificially ventilated. Anesthesia was continued with 0.8–1.5% halothane in a 70%/30% N_2O/O_2 gas mixture. An intravenous infusion of 2 ml/kg/hr Ringer solution containing 5 mg/kg/hr of gallamine triethiodide for muscle relaxation was maintained throughout the experiment. Physiological variables were kept in the normal range for awake cats (Herbert and Mitchell, 1971). Deep body temperature was controlled at 37°C. An occluding device to achieve reversible middle cerebral artery occlusion (MCAO) for 60 min was implanted (Graf *et al.*, 1986). The study was approved by the local Animal Care Committee and the Regierungspräsident of Cologne and is in compliance with the German laws for animal protection.

Multiple consecutive PET studies were performed in each cat before and up to 24 hr after 60 min of MCAO as described elsewhere (Heiss *et al.*, 1997). In brief, the animals were kept fixed in the scanner gantry throughout the experiments such that coronal brain sections corresponding to a stereotaxic cat brain atlas (Reinoso-Suárez, 1961) were obtained. For the assessment of cerebral blood flow (CBF), bolus intravenous injections of 20 mCi of [^{15}O]H$_2$O were used. Serial PET scanning was performed on a Siemens/CTI EXACT HR (resolution in-plane 3.6 mm, axial 4.0 mm; Wienhard *et al.*, 1994). During scans, radioactivity in arterial blood was measured continuously using an arteriovenous shunt and an LV automatic, calibrated blood sampling system (Eriksson *et al.*, 1988). At the end of the experiments, animals were perfusion-fixed with formalin (4%), and serial hematoxylin and eosin or Luxol Fast blue-stained 7-μm sections were obtained parallel to the PET planes.

To study the spatiotemporal dynamics in sequential data sets, an exploratory tool was used allowing voxel-by-voxel comparison of different studies. The analysis used coronal planes to match the histological sections. In scatter plots derived by comparing two data sets, clusters of voxels were separated from the normal distribution of voxels by defining physiological thresholds or by visual examination (see Fig. 1). In scatter plots derived by comparing the two data sets, clusters of voxels were separated from the normal distribution of voxels (see Fig. 1) by defining physiological thresholds (threshold for morphological integrity, functional integrity, and for postischemic hyperfusion at 10, 20, and 40 ml/100 g/min, respectively) or by visual examination (centering separated clusters of voxels in scatter plots). Temporal and spatial evolutions after recircula-

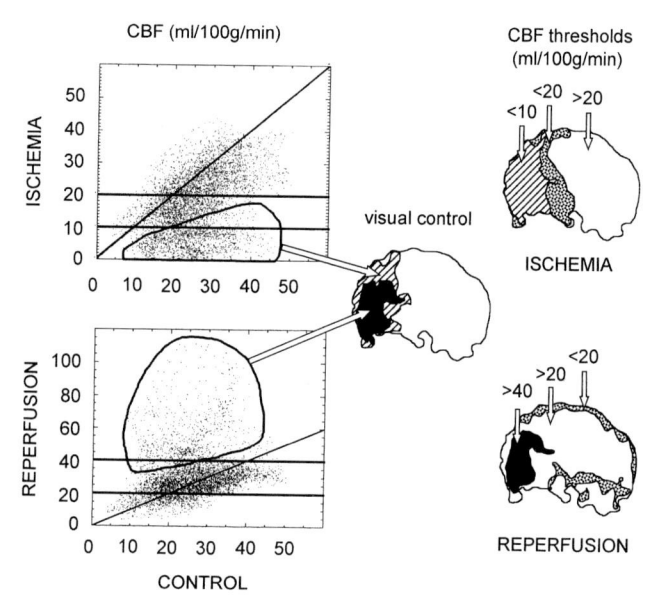

FIGURE 1 Threshold-based (horizontal lines) vs visually oriented definition (see circled clusters of voxels) of hypo- or hyperperfusion during ischemia and early reperfusion in scatter plots and subsequent reconstruction of respective regions. Voxel values of ischemic and postischemic images (*y* axis) were plotted against voxel values of control images (*x* axis). For threshold-based analysis, 10 and 20 ml/100 g/min (see horizontal lines in scatter plots) were chosen as thresholds for ischemic hypoperfusion (thresholds for morphological and functional integrity) and 40 ml/100 g/min for postischemic hyperperfusion. Subsequently, reconstructed areas were marked in the schematic drawings of brain cross sections on the right (see dashed, dotted, and dark areas, respectively). For visually oriented analysis, clusters of voxels separated from the normal distribution around the line of identity were circled in respective scatter plots, and subsequently reconstructed areas of hypo- and hyperperfusion were marked (see dashed and dark areas, respectively, in the schematic cross section in the middle). Note the correspondence between regions of hypo- and hyperperfusion defined by the two methods.

tion of regions defined in this manner were compared to regions of reduced CBF during ischemia.

III. RESULTS AND DISCUSSION

In all animals, CBF decreased in the ischemic territory below 45% of preischemic control values. The investigation revealed interindividual variability of postischemic hyper- and hypoperfusion. In Fig. 2, a case of transient hyperperfusion lasting about 2 hr is shown, comparing postischemic CBF determinations with control measurements in scatter plots and in corresponding schematic coronal planes. With this form of exploratory analysis, an easy qualitative assessment of various types of postischemic hyperperfusion was possible. Figure 3 demonstrates examples of the three main types: persisting hyperperfusion (type 1), transient hyperperfusion (type 2), and progressive hyperperfusion (type 3). It was particularly helpful to perform the analysis using multiple coronal planes as patterns of postischemic perfusion were not homogeneously distributed in individual brains. Variable spatiotemporal dynamics that would have been missed by conventional region of interest analysis were found to be a characteristic of the recirculation period. Additionally, by studying coronal planes, direct comparison with histology was possible. In most cases, simple visual determination of voxel clusters permitted rapid determination of the various types of perfusion and was almost as reliable as threshold-based analysis. The method has also been applied successfully with multiparametric PET studies (data not shown). It may also be helpful in the analysis of sequential data sets derived from other imaging techniques to study experimental ischemia such as functional magnetic resonance imaging (for review, see Hossmann and Hoehn-Berlage, 1995).

Results are summarized in Table 1. Recirculation was only partially successful in three animals, and hypoperfused regions persisted throughout the observation period, resulting in gross infarction at the final stage of the experiment. Type 1 ($n = 5$) showed immediate hyperperfusion lasting for the whole observation period and covering a large part of the ischemic territory. During ischemia, CBF reduction was severe. At the final stage, large infarcts were observed. In type 2 ($n = 6$), transient hyperperfusion was often followed by hypoperfusion in parts of previously hyperperfused regions. During MCAO, CBF reduction in the ischemic territory was more gradual. Infarcts were normally restricted to the center of hyperperfused regions or did not develop. In type 3 ($n = 2$), hyperperfused regions grew progressively over time, CBF reduction was gradual, and infarcts were small.

In conclusion, sequential PET scans revealed variable spatiotemporal patterns of postischemic perfusion following transient MCAO in cats. Because of the small number of animals and the interindividual heterogeneity, it is difficult to assign distinct pathophysiologies to the various types of postischemic hyperperfusion. Comparison with metabolism and histology allows a preliminary interpretation. Regions with continued hyperperfusion seemed to have a particularly bad outcome, as had those with continued incomplete recirculation. Even regions with transient early hyperperfusion often deteriorated if the ischemic CBF reduction was

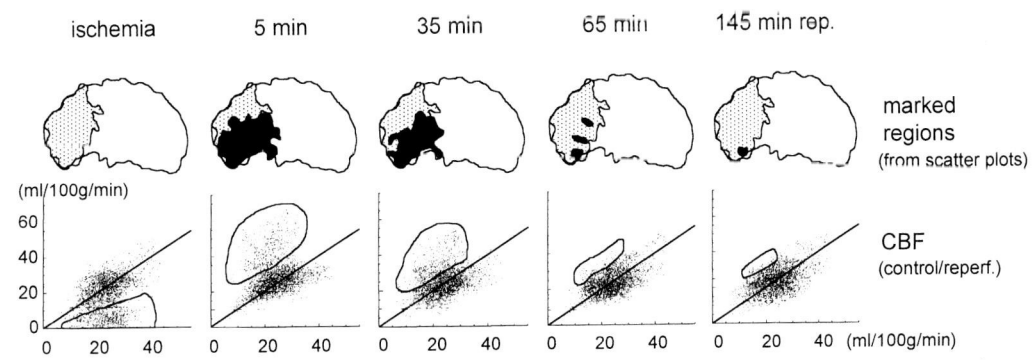

FIGURE 2 Scatter plots derived from sequential PET determinations of postischemic perfusion against control CBF (bottom) and regions reconstructed from the marked clusters of voxels. In this individual animal, hyperperfusion lasted about 2 hr.

TABLE 1 Summary of Postischemic Hyperperfusion Patterns in Individual Planes[a]

Hyperperfusion type	Animal number	No hyperfusion	Type 1	Type 2	Type 3
None	1	16	—	—	—
None	2	16	—	—	—
None	3	16	—	—	—
1	4	10	6	—	—
1	5	10	5	1	—
1	6	9	6	1	—
1	7	8	6	2	—
1	8	7	7	2	—
2	9	12	—	4	—
2	10	12	1	3	—
2	11	11	—	5	—
2	12	10	—	5	1
2	13	10	1	5	—
2	14	9	1	6	—
3	15	11	—	2	3
3	16	9	—	2	4

[a] See Fig. 3 and text.
[b] Numbers of planes per animal showing distinct patterns are given.

severe. Postischemic hyperperfusion seemed to be beneficial, however, in zones with subcritical CBF reduction that is characteristic of border zone areas of ischemic foci.

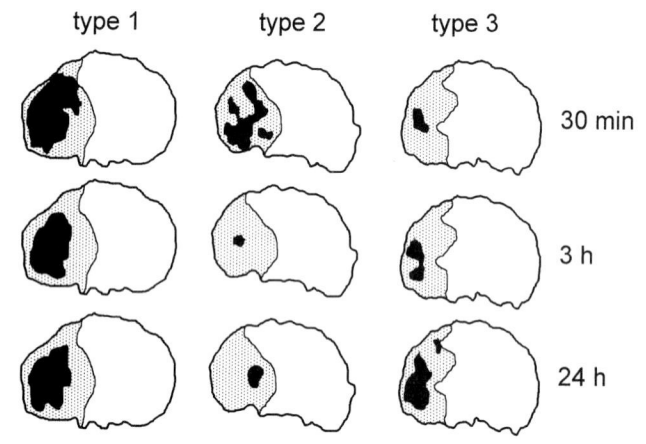

FIGURE 3 Major types of hyperfusion patterns (type 1, persisting hyperperfusion; type 2, transient hyperperfusion; type 3, progressive hyperperfusion) after 60 min occlusion of the middle cerebral artery. Examples of determinations in individual cats derived from voxel-based analysis are shown.

References

Astrup, J., Siesjö, B. K., and Symon, L. (1981). Thresholds in cerebral ischemia—the ischemic penumbra. *Stroke* **12**: 723–725.

Eriksson, L., Holte, S., Bohm, C., Kesselberg, M., and Hovander, B. (1988). Automated blood sampling systems for positron emission tomography. *IEEE Trans. Nucl. Sci.* **35**: 703–707.

Graf, R., Kataoka, K., Rosner, G., and Heiss, W.-D. (1986). Cortical deafferentation in cat focal ischemia: Disturbance and recovery of sensory functions in cortical areas with different degrees of CBF reduction. *J. Cereb. Blood Flow Metab.* **6**: 566–573.

Hakim, A. M., and Hogan, M. J. (1991). In vivo binding of nimodipine in the brain: I. The effect of focal cerebral ischemia. *J. Cereb. Blood Flow Metab.* **11**: 762–770.

Heiss, W.-D., and Graf, R. (1994). The ischemic penumbra. *Curr. Opin. Neurol* **7**: 11–19.

Heiss, W.-D., Graf, R., Wienhard, K., Löttgen, J., Saito, R., Fujita, T., Rosner, G., and Wagner, R. (1994). Dynamic penumbra demonstrated by sequential multitracer PET after middle cerebral artery occlusion in cats. *J. Cereb. Blood Flow Metab.* **14**: 892–902.

Heiss, W.-D., Graf, R., Löttgen, J., Ohta, K., Fujita, T., Wagner, R., Grond, M., and Wienhard, K. (1997). Repeat positron emission tomographic studies in transient middle cerebral artery occlusion in cats: residual perfusion and efficacy of postischemic reperfusion. *J. Cereb. Blood Flow Metab.* **17**: 388–400.

Herbert, D. A., and Mitchell, R.A. (1971). Blood gas tensions and acidic balance in awake cats. *J. Appl. Physiol.* **30**: 434–436.

Hossmann, K.-A., and Hoehn-Berlage, M. (1995). Diffusion and perfusion MR imaging of cerebral ischemia. *Cerebrovasc. Brain Metab. Rev.* **7**: 187–217.

Kogure, K., Hossmann, K.-A., and Siesjö, B. K. (1993). "Neurology of Ischemic Brain Damage." Elsevier, Amsterdam.

Pappata, S., Fiorelli, M., Rommel, T., Hartmann, A., Dettmers, C., Yamaguchi, T., Chabriat, H., Poline, J. B., Crouzel, C., Di Giamberardino, L., and Baron, J. C. (1993). PET study of changes in local brain hemodynamics and oxygen metabolism after unilateral middle cerebral artery occlusion in baboons. *J. Cereb. Blood Flow Metab.* **13**: 416–424.

Pulsinelli, W. (1992). Pathophysiology of acute ischaemic stroke. *Lancet* **339**: 533–536.

Reinoso-Suárez, F. (1961). "Topographischer Hirnatlas der Katze." E. Merck AG, Darmstadt.

Siesjö, B. K. (1992). Pathophysiology and treatment of focal cerebral ischemia. Part 1: Pathophysiology. *J. Neurosurg.* **77**: 169–184.

Tenjin, H., Ueda, S., Mizukawa, N., Imahori, Y., Hino, A., Ohmori, Y., and Yasukochi, K. (1992). Positron emission tomographic measurement of acute hemodynamic changes in primate middle cerebral artery occlusion. *Neurol. Med. Chir.* (*Tokyo*) **32**: 805–810.

Wienhard, K., Dahlbom, M., Eriksson, L., Michel, C. H., Bruckbauer, T., Pietrzyk, U., and Heiss, W.-D. (1994). The ECAT EXACT HR: Performance of a new high resolution positron scanner. *J. Comput. Assist. Tomogr* **18**: 110–118.

CHAPTER 21

Estimation of Ischemic Cerebral Blood Flow Using [^{15}O]Water and PET without Arterial Blood Sampling

J. LÖTTGEN, U. PIETRZYK, K. HERHOLZ, K. WIENHARD, and W.-D. HEISS

MPI for Neurological Research
D-50931 Cologne, Germany

Measurements of regional cerebral blood flow (CBF) using [^{15}O]H$_2$O and positron emission tomography (PET) require the determination of the input function by arterial blood sampling. In some cases this is impossible, however, for example in studies during thrombolytic therapy in stroke patients. Fifteen [^{15}O]H$_2$O PET studies of nine patients with unilateral acute or subacute ischemic infarcts in the middle cerebral artery territory were evaluated. It was shown that the tissue activity normalized to the mean of a region of interest covering the hemisphere contralateral to the ischemic area is strongly related to the absolute CBF value, measured using arterial blood sampling. By a pixel-by-pixel comparison between normalized tissue activities and measured CBF values, this relation could be quantified; for a given value of the normalized tissue activity the corresponding CBF can be estimated and the uncertainty of this value can be determined. The uncertainty of the CBF estimate increases substantially with increasing CBF. However, this study shows that low CBF values can be estimated with reasonable accuracy. This result is useful for the determination and interpretation of CBF thresholds in studies of cerebral ischemia.

I. INTRODUCTION

It is widely accepted that for a reliable and accurate measurement of cerebral blood flow (CBF) with [^{15}O]H$_2$O and positron emission tomography (PET), it is crucial to measure the input function from arterial blood samples. Nevertheless, in some cases this is impossible, e.g., studies in stroke patients before or during thrombolytic therapy. For studies of ischemic brain injury, measurements of CBF are of great interest, especially in regions with reduced perfusion.

Methods are available to correct for the nonlinearity of the relation between tissue activity and CBF for both PET (Meija *et al.*, 1994) and SPECT (Tsuchida *et al.*, 1996). Also, some methods to estimate CBF without arterial sampling have been described (Watabe *et al.*, 1995; Lammertsma, 1994). These algorithms require assumptions about the mean CBF in the whole brain or in selected regions, most often the cortical gray matter. Therefore, it is questionable whether they are suitable for measurements in patients.

The aim of this study was to demonstrate that it is possible to obtain information about absolute CBF values using [^{15}O]H$_2$O and PET without arterial sampling, at least in regions with low CBF. In order to do this, CBF measurements were evaluated in patients with subacute or acute unilateral stroke.

II. MATERIALS AND METHODS

Fifteen [^{15}O]H$_2$O PET scans in nine patients (age 68 ± 8 years) with acute or subacute unilateral ischemic middle cerebral artery infarct were evaluated. The scanner used was an ECAT EXACT HR (Wienhard *et al.*, 1994), which provides 47 contiguous planes of 3.125 mm thickness in a 128 × 128 pixel matrix with a pixel size of 2.1 mm. Forty milliCuries of [^{15}O]H$_2$O were injected intraveneously as a short bolus (about 10 sec). Scan data were collected from 0 to 120 sec postinjection.

Arterial blood was withdrawn continuously from the radial artery by an automated blood sampling device (Eriksson *et al.*, 1988). The blood activity was measured every second by two coincidence counters. The blood

TABLE 1 Definitions of Regions of Interest

ROI 1	Centrum semiovale	Contralateral to ischemic area
ROI 2	Hemisphere (level of basal ganglia)	Contralateral to ischemic area
ROI 3	Cingulum (unaffected gray matter)	Ipsi- and contralateral to ischemic area
ROI 4	Cerebellum	Ipsilateral to ischemic area

activity curve was corrected for delay and dispersion in both the blood sampling device and the body in order to obtain an estimate of the input function. This method has been described and validated previously (Pietrzyk et al., 1993). CBF was calculated using the autoradiographic method (Raichle et al., 1983) with a fixed value for the distribution volume of water (95 ml/100 g).

For each study, four regions of interest (ROIs) were defined on transaxial planes. In the hemisphere contralateral to the ischemic area, two ROIs were defined, the first corresponding to the centrum semiovale the second covering the whole hemisphere at the level of the basal ganglia. The third ROI was defined on both hemispheres, corresponding to the cingulum. Visual examination in each patient showed that the cingulum was not affected by ischemia. The fourth ROI was placed on the cerebellar hemisphere ipsilateral to the ischemic area. The definition of the ROIs is summarized in Table 1. Regional means of the CBF were determined for each study. The coefficient of variation (COV) of these values across the studies was calculated for the four ROIs. The tissue activity was normalized by the mean tissue activity in ROI 2 (contralateral hemisphere), which turned out to have the lowest COV.

Normalized tissue activity c_T and measured tissue activity C_T were compared with the calculated CBF pixel by pixel for all pixels in the field of view with CBF > 3 ml/min/100 g. The pairs of CBF and c_T for each pixel and all studies were grouped over intervals of c_T with an interval width of 1%. In these groups, median and percentiles of the CBF values were calculated.

III. RESULTS AND DISCUSSION

For all pixels in the field of view with CBF > 3 ml/min/100 g, C_T and CBF are plotted in Fig. 1. This plot demonstrates the nonlinear relation between CBF and activity. It also shows that the variance of CBF values for a measured tissue activity increases with increasing CBF. The COV of mean regional CBF was minimal for ROI 2 (contralateral hemisphere), followed by ROI 3 (cingulum, unaffected gray matter), ROI 4 (ipsilateral cerebellum), and ROI 1 (contralateral white matter). For this reason the tissue activity in the hemisphere with the ischemic area was normalized by the regional mean of ROI 2 in order to minimize the variability between the studies. When local tissue activity was then normalized by the mean value of C_T in the contralateral hemisphere (ROI 2), this reduced the variability in the relation between CBF and tissue activity. The data pairs (c_T, CBF) were grouped over c_T as described earlier. Figure 2 shows c_T

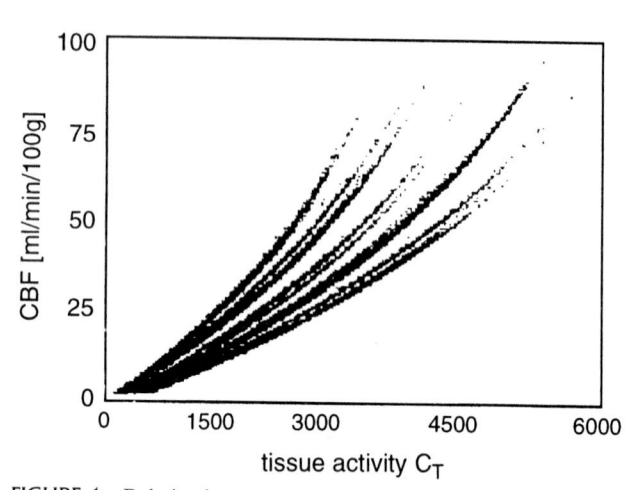

FIGURE 1 Relation between measured tissue activity C_T and calculated CBF pixel by pixel.

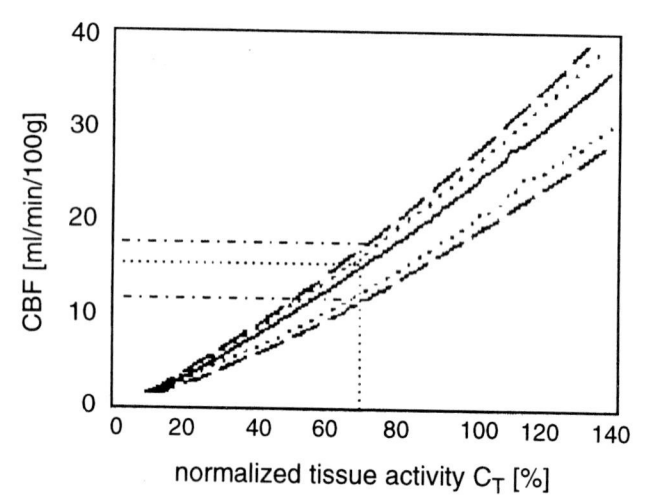

FIGURE 2 Normalized tissue activity c_T vs CBF [median (solid); 10 and 90% percentiles (dashes); and 25 and 75% percentiles (dots)].

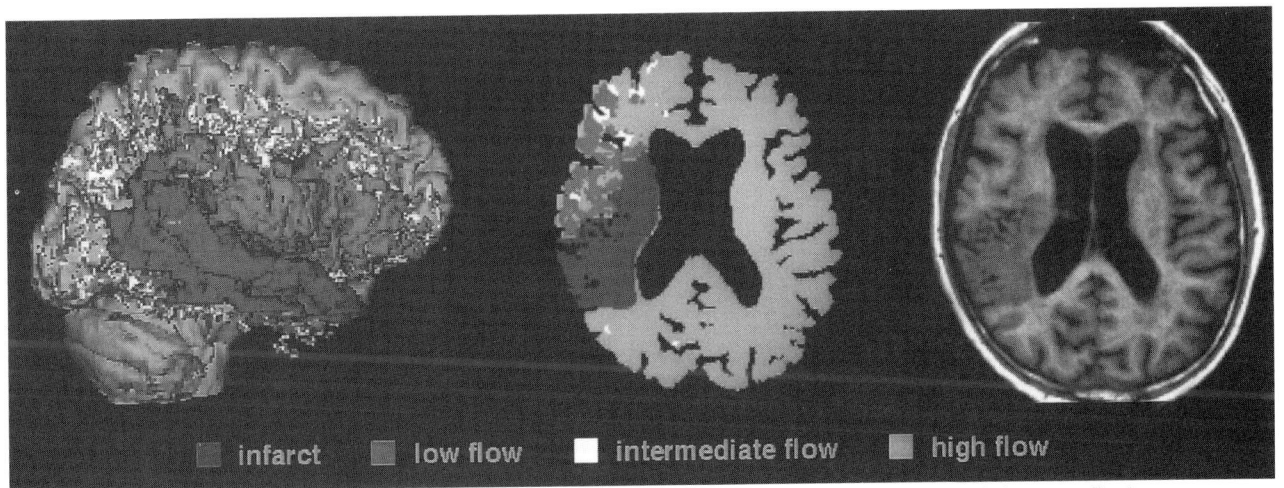

FIGURE 3 Images from a patient with severe stroke. VOIs were defined by thresholds of the normalized tissue activity, corresponding to different ranges of CBF. Medium gray corresponds to low flow with CBF < 12 ml/min/100 g; white corresponds to intermediate flow with 12 ml/min/100 g ≤ CBF ≤ 20 ml/min/100 g; light gray corresponds to high flow with CBF > 20 ml/min/100 g; and dark gray corresponds to the final infarct, defined on a T1-weighted magnetic resonance image.

and the median of CBF(c_T) in these groups; 10, 25, 75, and 90% percentiles are indicated.

Figure 2 (or a look-up table of the corresponding values) can be used to assign an approximate CBF value to a given value of c_T and to quantify the uncertainty of this estimate of CBF. For example, a value of the normalized tissue activity of 70% corresponds to a CBF of 16 ml/min/100 g with a variability of 13 to 19 ml/min/100 g.

An application of this method is illustrated in Fig. 3. The image shows the brain of a patient with severe ischemia who underwent thrombolytic therapy with recombinant tissue plasminogen activator. The volume was segmented using different thresholds of the normalized tissue activity of a pretherapeutic [^{15}O]H$_2$O PET scan. VOIs represent volumes with different CBF values during ischemia, which can be used for studies of tissue outcome. It was compared with the final volume of the infarct, as defined on a coregistered magnetic resonance image.

In conclusion, this study has demonstrated that CBF can be estimated without arterial sampling, at least for studies of cerebral ischemia. The steps involved are

1. selection of an ROI in which the COV of the measured CBF across studies is small;
2. normalization of the tissue activity by the regional mean of that ROI;
3. generating a look-up table of the relation between measured CBF and normalized tissue activity using data from a limited number of [^{15}O]H$_2$O PET studies with measured input functions; and
4. conversion of normalized tissue activities to CBF values using this look-up table in subsequent [^{15}O]H$_2$O PET studies performed without arterial blood sampling.

To apply this method to other groups of patients, it is necessary to choose an appropriate reference ROI for normalization and to determine whether the variability of the relationship between normalized tissue activity and CBF is small enough to obtain meaningful results.

References

Eriksson, L., Holte, S., Bohm, C., Kesselberg, M., and Hovander, B. (1988). Automated blood sampling system for positron emission tomography. *IEEE Trans. Nucl. Sci.* **35**: 703–708.

Lammertsma, A. A. (1994). Noninvasive estimation of cerebral blood flow. *J. Nucl. Med.* **35**: 1878–1879.

Mejia, M. A., Itoh, M., Watabe, H., Fujiwara, T., and Nakamura, T. (1994). Simplified nonlinearity correction of oxygen-15-water regional cerebral blood flow images without blood sampling. *J. Nucl. Med.* **35**: 1870–1877.

Pietrzyk, U., Löttgen, J., Fink, G. R., Wilms, R., Herholz, K., Wienhard, K., and Heiss, W.-D. (1993). Correction for dispersion and delay for PET rCBF measurements in cerebrovascular disease. In "Quantification of Brain Function. Tracer Kinetics and Image Analysis in Brain PET" (K. Uemura, N. A. Lassen, T. Jones, and I. Kanno, eds.), pp. 71–76. Elsevier, Amsterdam and New York.

Raichle, M. E., Martin, W. R., Herscovitch, P., Mintun, M. A., and Markham, J. (1983). Brain blood flow measured with intravenous H$_2^{15}$O. *J. Nucl. Med.* **24**: 790–798.

Tsuchida, T., Yonekura, Y., Nishizawa, S., Sadato, N., Tamaki, N., Fujita, T., Magata, Y., and Konishi, J. (1996). Nonlinearity correction of brain perfusion SPECT based on permeability-surface area product model. *J. Nucl. Med.* **37**: 1237–1241.

Watabe, H., Itoh, M., Fujiwara, T., Jones, T., and Nakamura, T. (1995). Validation of noninvasive quantification of rCBF compared with dynamic/integral method using positron emission tomography and oxygen-15 labeled water. *Ann. Nucl. Med.* **9**: 191–198.

Wienhard, K., Dahlbom, M., Eriksson, L., Michel, C., Bruckbauer, T., Pietrzyk, U., and Heiss, W.-D. (1994). The ECAT EXACT HR: Performance of a new high resolution positron scanner. *J. Comput. Assist. Tomogr.* **18**: 110–118.

CHAPTER 22

Suitability of [¹⁵O]Water and PET to Detect Activation-Induced Cerebral Blood Flow Changes in Brain Tissue Altered by Brain Tumors

A. THIEL, K. HERHOLZ, H.-M. VON STOCKHAUSEN, G. PAWLIK, and W.-D. HEISS

Department of Neurology
Max Planck Institute for Neurological Research and University Clinic
D-50931 Cologne, Germany

[¹⁵O]Water and positron emission tomography have been widely used to map language function in normal volunteers. Based on these studies and the sensitivity increase available with three-dimensional data acquisition, the question arises whether activation-induced cerebral blood flow changes can be detected in brain tissue adjacent to gliomas in individual patients. Image processing methodology may be critical with respect to the localization of activation. Twelve brain tumor patients performing a verb generation task were examined and filtered and unfiltered images with respect to the location and extent of activations were compared.

I. INTRODUCTION

Sensitivity increases due to advances in positron emission tomography (PET) scanner technology such as three-dimensional data acquisition (Townsend *et al.*, 1991) now permit [¹⁵O]H$_2$O PET measurements of cerebral blood flow (CBF) in single subjects performing neurobehavioral tasks. These advances and new methods of PET–magnetic resonance (MR) image coregistration (Pietrzyk *et al.*, 1994) not only had a great impact on cognitive neuroscience, but also offer new possibilities with respect to the application of PET as a clinical diagnostic tool. Single-subject speech activation studies are of special clinical interest in patients with brain tumors to facilitate planning of therapy, including tumor resection and implantation of radioactive seeds. In addition to determining hemispheric language dominance, information on the exact topographical location of activated brain tissue in relation to the brain tumor is needed. Here the border zone between tumor and normal brain tissue is of special interest because functional brain tissue in this area may limit the extent of resection. Studies of intraoperative mapping procedures gave evidence of preserved language function in brain tissue altered by infiltration, edema, or compression (Ojemann *et al.*, 1996; Skirboll *et al.*, 1996). Current methods to analyze [¹⁵O]H$_2$O PET activation studies in normal volunteers involve some steps that may prove critical when applying the method to patients with brain tumors. First, the assumption of a global uniform variance of the activations may be violated in brain tissue altered by brain tumors. Therefore, thresholds for detection of activated pixels may need adjustment to ensure good sensitivity. Second, data from single subject studies require secondary filtering in order to improve the signal-to-noise ratio. This process of filtering results in a loss of localizing power and may lead to a shifting or growing of activations in normal unaffected brain into the critical borderzone. This study examined whether activations in altered brain tissue can be detected despite these possible limitations in the analysis of [¹⁵O]H$_2$O PET data.

II. PATIENTS AND METHODS

Twelve right-handed patients (23 to 45 years) with gliomas (10 astrocytomas and 2 oligodendrogliomas) in the left hemisphere and mild to moderate aphasic symptoms were studied. All patients had four measurements of relative CBF under resting conditions (darkened room, eyes closed, ears unplugged) and four scans while performing a verb generation task (finding verbs

TABLE 1   Volumes of Affected Tissue and Volumes of Activated Tissue

Patient	Volume of the affected tissue (cm³)	Activated volume of the affected tissue (cm³) Filter	Activated volume of the affected tissue (cm³) No filter	Activated volume of normal brain tissue (cm³) Filter	Activated volume of normal brain tissue (cm³) No filter
1	57.63	5.71	2.12	5.46	25.47
2	27.99	2.57	1.11	34.11	29.46
3	122.75	0.14	1.90	10.05	14.04
4	131.67	7.11	4.78	43.26	36.59
5	166.76	0.91	2.23	49.02	35.44
6	351.41	4.21	6.30	40.20	28.81
7	191.32	0.71	3.08	37.09	24.97
8	101.21	2.48	2.14	29.25	25.37
9	301.89	34.10	15.20	26.41	11.21
10	180.57	1.32	1.95	23.29	16.96
11	210.13	3.83	5.18	28.45	26.71
12	256.52	6.36	8.15	16.98	15.30

to match acoustically presented nouns). PET measurements were performed on an ECAT EXACT HR camera (CTI/Siemens Medical Systems) (Wienhard *et al.*, 1994) following a bolus injection of 370 MBq of [^{15}O]H$_2$O. Data were acquired in 3D mode for 90 sec starting with the intravenous injection of the tracer. After correction for scatter and measured attenuation, images were reconstructed to a matrix of 128 × 128 × 47 voxels with 2.17 × 2.17 × 3.125 mm voxel size applying a Hanning filter of 0.4 cycles/pixel. For morphological localization of PET-measured changes in relative CBF, MR imaging was performed with a 1-T Magnetom (Siemens Medical Systems) producing 64 transaxial T1-weighted slices. After initial alignment (Woods *et al.*, 1992) of all PET scans of each subject, an average PET image was calculated and coregistered to the subject's MR images using the MPI-Tool (multipurpose imaging tool, Pietrzyk *et al.*, 1994) and all scans were resliced to this average image. All coregistered scans of each condition were averaged and ratio normalized to global mean activity. Difference images from normalized average images were calculated. For data analysis, these difference images were directly Z transformed and were also Z transformed after filtering with a spherical median filter of 8 mm radius within a brain mask defined on the MR images. All voxels with differences in relative CBF above 2 standard deviations were considered as activated and will be further referred to as the "activated volume." Using interactive software written in IDL (Interactive Data Language Research Systems Inc.), the MR images were segmented into an atlas of three volumes of interest (VOIs): (1) tumor and adjacent brain tissue altered by infiltration or compression ("affected tissue"), (2) ipsilateral hemisphere (excluding 1), and (3) contralateral hemisphere.

This atlas was combined with the "activated volume" in both filtered and unfiltered PET difference images in order to determine the volume of voxels activated within the atlas VOIs.

III. RESULTS

The volume of the affected tissue ranged from 28 to 351 cm³ (Table 1). Analysis of CBF changes within the VOIs revealed consistently lower standard deviations within the affected region than in the ipsi- or contralateral hemisphere for the filtered images (Table 2). In all

TABLE 2   Average Standard Deviations of Changes within Regions across Subjects

Region	Average standard deviation Filtered	Average standard deviation Unfiltered
Contralateral hemisphere	0.89 ± 0.085[a]	0.94 ± 0.100
Ipsilateral hemisphere	0.91 ± 0.110	0.95 ± 0.115
Affected tissue	0.84 ± 0.103	0.94 ± 0.094
Whole brain	0.93 ± 0.171	0.95 ± 0.106

[a] Standard deviation of the average standard deviation values.

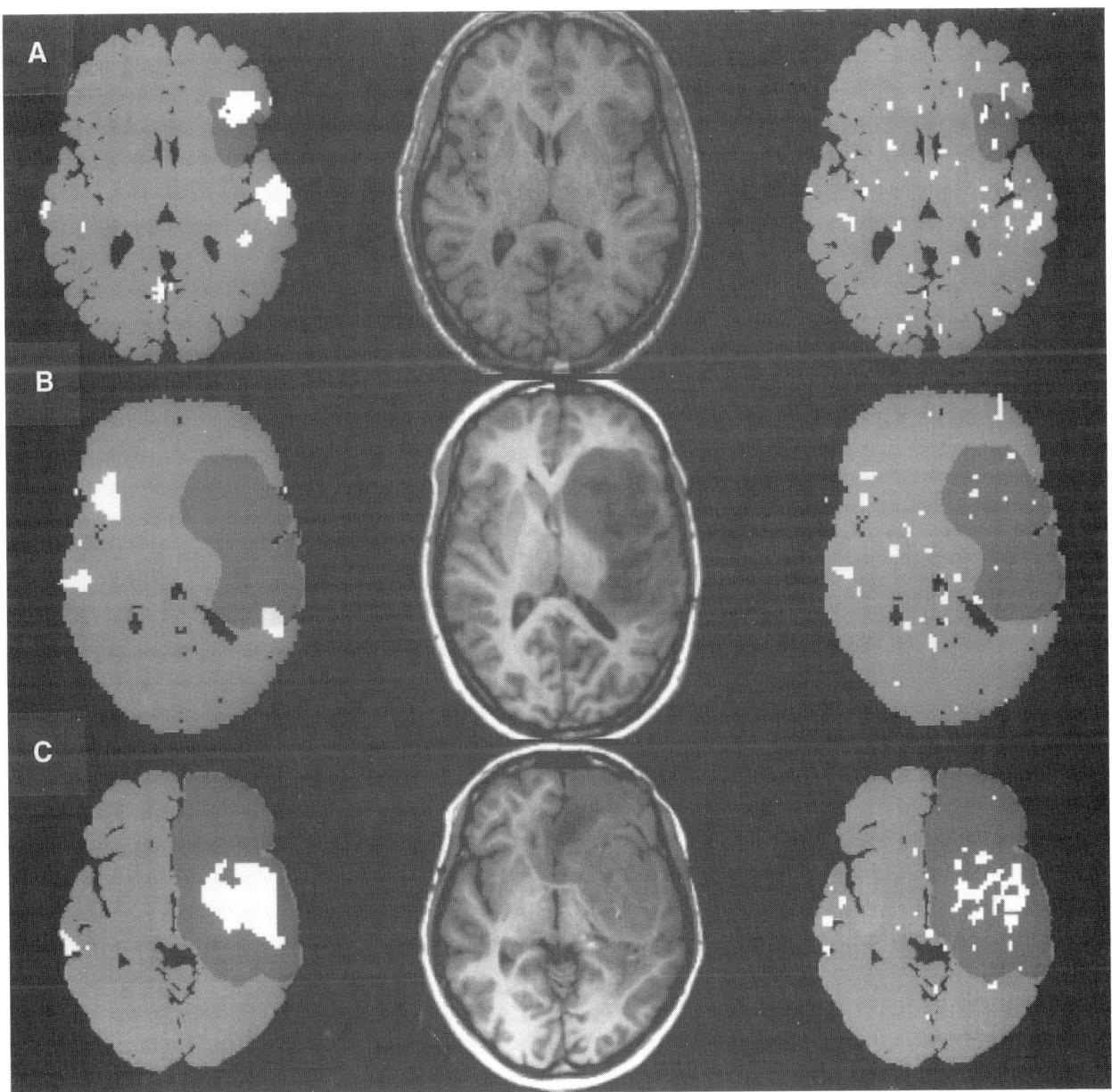

FIGURE 1 The left column shows filtered images, the right column shows unfiltered images, and the middle column shows MR images. Activations (white) in altered brain tissue (dark gray) can be found in both filtered and unfiltered images in A. (B) Activations (white) in normal tissue (light gray) appear in altered tissue only in the filtered image. (C) Effect of filtering when disseminated activations are present.

patients, activated voxels within the affected tissue were found (Table 1). The average activated volume within the affected tissue was 4.5 ± 4.00 cm^3 in unfiltered images and only slightly differed from the average volume of 5.8 ± 9.20 cm^3 in filtered images. Activations in filtered images were also found in unfiltered images (Fig 1A). Thus these activations must have been generated in the altered tissue and do not merely reflect spillover from activations in adjacent tissue. In other cases (Fig. 1B), an activated area in normal tissue was expanded by filtering in such a way as to also reach the affected tissue, where no activated voxels were found in the unfiltered image. The largest activated volume within affected tissue was found in patient 9 (Fig. 1C); this case also exhibited the most striking difference between filtered and unfiltered images. This patient's tumor was an infiltrating oligodendroglioma (WHO II) that had developed over many years. The presence of disseminated activations within the tumor was compatible with the clinical symptoms; the patient had a hemiparesis but only mild aphasic symptoms that did not severely impair normal communication.

IV. DISCUSSION

These results suggest that activation-induced changes in relative CBF do exist in brain tissue altered by tumor infiltration or compression. Furthermore, it was documented that some of these activations cannot be simply explained as artifacts of image processing steps such as secondary filtering and must originate from altered brain tissue. Nevertheless, filtering can have a significant influence on the shape and extent of activations, depending on the distribution of activated voxels in the original image (e.g., see Fig. 1C). Thus it is necessary to verify the spatial extent of activations by inspection of unfiltered images. A more appropriate solution may be a multifiltering approach as proposed by Poline and Mazoyer (1994) or even more sophisticated approaches such as scale space transformations (Poline et al., 1995).

A second result is that the variances within the affected tissue differ from those in ipsi- and contralateral hemispheres. This finding suggests that global estimates of variance, as often used in studies with normal volunteers, may not be appropriate in the case of altered brain tissue. The use of lower thresholds obtained from regional estimates of variance may be indicated to increase sensitivity.

References

Ojemann, J. G., Miller, J. W., and Silbergeld, D. L. (1996). Preserved function in brain invaded by tumor. *Neurosurgery* **39**: 253–258.

Pietrzyk, U., Herholz, K., Fink, G., Jacobs, A., Mielke, R., Slansky, I., Wuerker, M., and Heiss, W.-D. (1994). An interactive technique for three-dimensional image registration: Validation for PET, SPECT, MRI and CT brain studies. *J. Nucl. Med.* **35**: 2011–2018.

Poline, J. B., and Mazoyer, B. M. (1994). Enhanced detection in brain activation maps using a multifiltering approach. *J. Cereb. Blood Flow Metab.* **14**: 639–642.

Poline, J. B., Worsley, K. J., Holmes, A. P., Frackowiak, R. S., and Friston, K. J. (1995). Estimating smoothness in statistical parametric maps: Variability of p values. *J. Comput. Assist. Tomogr.* **19**: 788–796.

Skirboll, S. S., Ojemann, G. A., Berger, M. S., Lettich, E., and Winn, H. R. (1996). Functional cortex and subcortical white matter located within gliomas. *Neurosurgery* **38**: 678–684.

Townsend, D. W., Geissbuhler, A., Defrise, M., Hoffman, E. J., Spinks, T. J., Bailey, D. L., Gilardi, M. C., and Jones, T. (1991). Fully 3-dimensional reconstruction for a PET camera with retractable septa. *IEEE Trans. Med. Imag.* **10**: 505–512.

Wienhard, K., Dahlbom, M., Eriksson, L., Michel, C., Bruckbauer, T., Pietrzyk, U., and Heiss, W. D. (1994). The ECAT EXACT HR: Performance of a new high resolution positron scanner. *J. Comput. Assist. Tomogr.* **18**: 110–118.

Woods, R. P., Cherry, S. R., and Mazziotta, J. C. (1992). Rapid automated algorithm for aligning and reslicing PET images. *J. Comput. Assist. Tomogr.* **16**: 620–633.

CHAPTER 23

Magnetic Resonance Imaging-Guided Language Activation PET in Patients: Technical Aspects and Clinical Results

K. HERHOLZ, A. THIEL, U. PIETRZYK, H.-M. VON STOCKHAUSEN, M. GHAEMI,
A. BERZDORF, J. SOBESKY, K. WIENHARD, and W.-D. HEISS

Max Planck Institute for Neurological Research
Cologne, Germany

One hundred and ten stimulation studies with three-dimensional positron emission tomography (PET) and coregistered magnetic resonance images in 71 subjects using six language stimulation paradigms were performed. At a threshold of 10% signal increase or more in an area comprising several slices, left Broca's area was activated most consistently in 88 to 100% of right handers with verb generation tasks, with auditory as well as visual input, and with spoken as well as silent verb production. Significant asymmetry was also seen in inferior temporal gyrus. Thus, this type of task is particularly suited for investigation of language lateralization and for localization of the anterior language center. Object naming and semantic judgement tasks produced somewhat less intense and more variable activations. Simple word repetition mainly activated the superior temporal cortex, including and adjacent to Heschl's gyrus. Aphasia tended to increase rather than decrease activation responses, predominantly in the right hemisphere. Activations were also observed in the immediate vicinity of brain tumors. Thus, such PET activation studies cannot only be performed in normal subjects, but are also clinically feasible and yield consistent results in patients.

I. INTRODUCTION

Stereotactic averaging of brain anatomy, as used commonly in normals, is not possible in patients with infarcts or tumors because of the distortion of normal anatomy by disease. Data acquisition in three-dimensional mode with multiple replication of conditions reduces the technical variability of [^{15}O]water positron emission tomography (PET) studies to a degree that permits mapping of cognitive functions with typical regional cerebral blood flow (CBF) increases of 5–15% in individual subjects. Anatomical localization is possible by coregistration with magnetic resonance image (Mazoyer et al., 1993; Seitz et al., 1995; Herholz et al., 1996; Schmahmann, 1996). Thus, preoperative localization of language function in patients with brain tumors and studying recovery from brain infarcts has become possible.

Clinical demands include high reproducibility and ease of use. Therefore, this study used quantitative visual identification of activation foci that can be performed in about half an hour, rather than laborious delineation of anatomical structures by volumes of interest. The frequency of activation foci in relevant anatomical structures was examined in order to identify paradigms that activate distinct anatomical structures in a highly reproducible manner and that therefore can be used in clinical studies.

II. METHODS

A. Subjects

The present analysis comprises 110 stimulation studies in 71 sequential subjects (age 43 ± 15 years) with the following characteristics. There were 39 male and 32 female subjects. Fourteen were normal, 20 had ischemic brain infarct, 36 had a brain tumor, 1 had an angioma; 65 were right handed and 6 left; 32 subjects were aphasic.

B. Activation Studies

Language activation studies were performed on a high-resolution PET scanner (Siemens/CTI EXACT HR), operating in 3D mode. Images were corrected for randoms, scatter, and measured attenuation. At each activation, 370 MBq [^{15}O]water was administered as a rapid bolus, with 90 sec data acquisition and integration time (in the most recent studies reduced to 45 sec, as indicated) and three or four replications per condition in a balanced order. Activations were usually self-paced (rate approximately 10–20/min) because aphasics often required more time to complete a task.

The following stimulation paradigms were used (number of right and left handers indicated in parentheses):

A (R40/L4): repeating nouns aloud after auditory presentation; reference condition, resting

B (R18/L4): generating verbs aloud after auditory presentation of nouns; reference condition, resting, usually in combination with A

C (R16/L2): generating verbs aloud after visual presentation of nouns; reference condition, nonsense answer "pataka" after presentation of kana letters

D (R12/L1): naming common objects aloud after visual presentation of line drawings, in combination with C using the same reference condition

E (R4/L1): identification of the unrelated word out of a visual presentation of three words, in combination with C using the same reference condition, 45 sec acquisition time

F (R8/L0): generating verbs silently after auditory presentation of nouns at fixed rate (6/min); reference condition, resting (used in normals only)

C. Data Analysis

PET scans of each subject were coregistered with each other and with the individual's MR images (T1-weighted volume data, 128 or 256 slices, 2.5 or 1 mm thickness), oriented in parallel to the anterior commissure–posterior commissure line. After normalization of each scan to brain average activity, the signal change due to activation was calculated voxel by voxel as a percentage of the reference control value. The result was smoothed by an 8-mm-radius median filter applied only within brain. Activation images were displayed using a 10-level color scale ranging from 3 to 23% signal increase with overlay of individual MR images (Fig. 1).

Only activation foci with peak values of 10% or higher that extended over several slices were reported. They were localized anatomically for each individual on the PET–MR fusion images using a display in three orthogonal planes. Locations were coded using 81 different anatomical sites.

III. RESULTS

Results are listed for right handers. Activations in left handers were less clearly lateralized. The number of activation foci in each subject was first studied as a parameter that provides some information on the extent and lateralization of the network that is intensely involved in a particular task. The number (mean ± SD) of supratentorial activation foci per subject is shown in Table 1.

Verb generation aloud (paradigms B and C) produced the largest number of activation foci with strong lateralization to the left hemisphere. Activation foci tended to cluster in four major areas of interest: the anterior language cortex, comprising Brodman areas 44 to 47 and the anterior part of the insula ("Broca"); the posterior language cortex, comprising Heschl's gyrus and adjacent superior and middle temporal gyrus ("Wernicke"); motor cortex, comprising the inferior and middle part of the precentral gyrus ("motor"); and visual association cortex, comprising cuneus, fusiform and lingual gyrus ("visual"). The frequency of activations (% of subjects) within these clusters is shown in Table 2.

Thus, the most reliable activation of left Broca's area was obtained with verb generation aloud and auditory input (paradigm B). The other verb generation paradigms (C, F) came close to that result and may have a slight advantage with respect to detecting L/R asymmetry. Repeating words (paradigm A) showed strong motor and temporal activations with little asymmetry. The other paradigms did not produce highly reproducible activation foci, with the exception of bilateral inferior visual association cortex with object naming (paradigm D).

Other frequent activation foci were:

• left supplementary motor cortex (66% of all studies), most frequently with spoken (paradigm B, 83%) and silent (paradigm F, 88%) verb generation
• left and right cerebellum (R, 55%; L, 43% of all studies), with a strong laterization effect (R, 100%; L, 25%) during silent verb generation (paradigm F)
• left inferior temporal gyrus was activated in 61% of subjects with auditory (paradigm B) and in 50% with

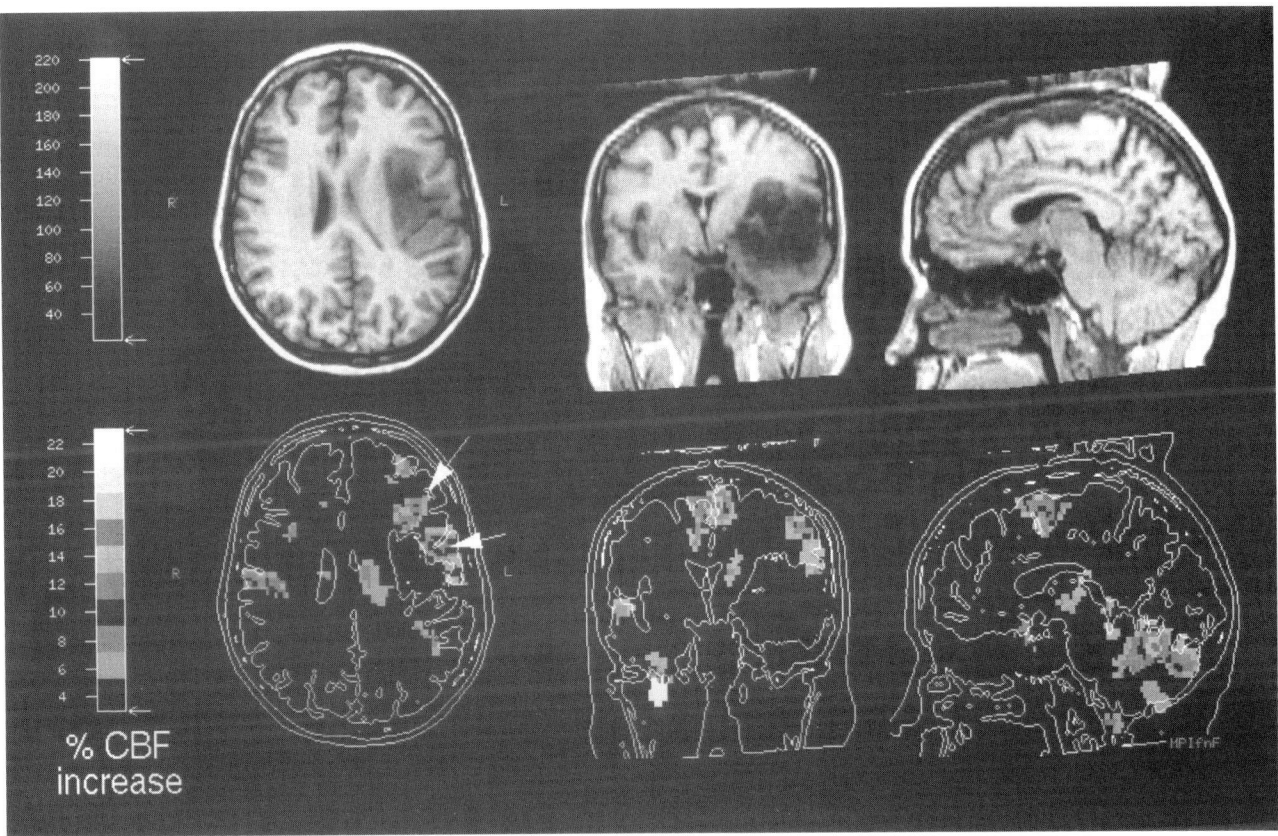

FIGURE 1 Verb generation (paradigm B) in a patient with a large left insular glioma. Selected orthogonal cuts (left, transaxial; middle, coronal; right, sagittal) of T1-weighted MR images (top) and PET activations with superimposed MR image contours (bottom) are shown. Activations include Broca's area and Wernicke's area immediately adjacent to tumor (arrows).

visual verb generation (paradigm C); the respective frequencies for the right side were both 6%; the differences between left and right hemisphere were highly significant (B, $p = 0.0004$; C, $p = 0.006$, χ^2 test).

- the precentral part of left middle frontal gyrus (75%) and right anterior cingulate cortex (75%) were often activated with silent verb generation (paradigm F)

Activated brain was often seen closely adjacent to tumors, and there was apparently no major decrease of activation intensity caused by these tumors (Fig. 1).

The effect of aphasia could be studied for paradigms A and B. During auditory verb generation (paradigm B), the right precentral cortex was activated in 80% of aphasics ($n = 10$), but only in 25% of nonaphasics ($n = 8$) ($p = 0.02$, χ^2 test). A related effect was seen

TABLE 1 Number of Supratentorial Activation Foci per Subject

Task	A	B	C	D	E	F
Response	Repeat nouns aloud	Generate verbs aloud	Generate verbs aloud	Naming digits aloud	Select unrelated words aloud	Generate verbs silently
Stimulus	Auditory presentation	Auditory nouns	Visual nouns	Visual objects	Visual list	Auditory nouns
Number of subjects	40	18	16	12	4	8
Left	6.7 ± 3.1	8.5 ± 2.1	7.1 ± 2.5	4.9 ± 2.8	5.0 ± 1.4	6.3 ± 1.8
Right	5.5 ± 1.9	4.3 ± 1.6	3.0 ± 1.7	3.5 ± 2.0	3.5 ± 1.7	2.9 ± 1.6

TABLE 2 Frequency of Activations within Different Brain Regions (Expressed as a Percentage of Subjects) for Different Activation Paradigms

Task	A	B	C	D	E	F
Response	Repeat nouns aloud	Generate verbs aloud	Generate verbs aloud	Name objects aloud	Select unrelated words aloud	Generate verbs silent
Stimulus	Auditory	Auditory	Visual	Visual	Visual	Auditory
Number of subjects	40	18	16	12	4	8
Left						
Broca	55	100	88	42	25	88
Wernicke	88	94	25	42	50	63
motor	73	56	31	17	25	0
Visual	18	33	50	75	50	38
Right						
Broca	40	33	13	25	0	13
Wernicke	98	83	13	0	50	75
Motor	75	56	19	0	25	0
Visual	13	11	38	83	0	25

during word repetition (paradigm A): aphasic patients tended to show a higher frequency of right Broca's activation (47%, n = 30) than nonphasics (20%, n = 10). Thus, aphasia tended to cause increased right precentral activation.

IV. DISCUSSION

Determination of the hemisphere that is dominant for language is a potentially important clinical application of PET activation studies in patients with brain tumors or focal epilepsy. Its validity has been demonstrated by correspondence with results of the Wada test (Pardo and Fox, 1993). In the present study, the verb generation tasks were generally superior to the other tasks with respect to lateralization.

Functional MR (fMR) imaging studies have also been used successfully to determine language dominance (Desmond et al., 1995; Binder et al., 1996). Despite more widespread availability, fMR imaging studies are subject to some limitations: a high level of background noise that makes auditory stimulation difficult, and high sensitivity to movement artifacts that may be induced by active speech. Another potential advantage of PET is its higher sensitivity to detect CBF changes (Kraut et al., 1995). Data from this study demonstrate the robustness of the PET technique in a large series of sequential patients, whereas in most fMR imaging studies some patients could not be evaluated because of technical difficulties. The ability of the subject to perform active speech in the PET scanner facilitates monitoring task performance, which is particularly important in aphasic patients.

Localization of left Broca's area was successful in 88 to 100% of subjects with the verb generation tasks, which therefore can be used clinically for that purpose. It remains to be determined whether the interindividual variability of localization in the subdivisions of the inferior frontal gyrus and adjacent structures, which has already been demonstrated earlier in normals with paradigm (Herholz et al., 1996) and which was seen also with the other verb generation tasks, is of clinical significance. Obviously, activation of Broca's area did not depend on actually speaking aloud, as it was seen with similar frequency during silent verb generation.

Activation of superior temporal cortex, particularly primary auditory cortex and surrounding association cortex, was most frequently seen with the auditory stimulation tasks, especially in connection with active speech, i.e., also hearing one's own voice, as expected from previous studies (Price et al., 1996). Yet it is difficult with these paradigms to separate general bilateral auditory processing from language-specific functions in these areas, although significant asymmetries have been detected with metabolic imaging of word repetition (Karbe et al., 1995). Lateralized effects were seen more clearly in the inferior temporal gyrus, which was activated by visual and auditory verb generation tasks on the left side in at least half of the subjects, but rarely on the right side. This strong lateralization was somewhat unexpected, as the inferior temporal gyrus is usually not included in the classical depictions of Wernicke's area, although language function has been demonstrated there for lexical retrieval with visual presentation of objects (Damasio *et al.*, 1996).

Increased activation in the right cerebral hemisphere in aphasics has been described previously (Weiller *et al.*, 1995). It is thought to represent a compensatory mechanism within the language network, although its contribution to recovery seems to be small (Heiss *et al.*, 1997). Further studies are required to establish its clinical significance.

V. CONCLUSION

MR imaging-guided 3D PET using verb generation tasks yields consistent focal activations of 10% or more that extend over several slices in most subjects. Thus, determination of language dominance and localization of language-related cortex can be performed in clinical investigations with this technique.

References

Binder, J. R., Swanson, S. J., Hammeke, T. A., Morris, G. L., Mueller, W. M., Fischer, M., Benbadis, S., Frost, J. A., Rao, S. M., and Haughton, V. M. (1996). Determination of language dominance using functional MRI: A comparison with the Wada test. *Neurology* **46**: 978–984.

Damasio, H., Grabowski, T. J., Tranel, D., Hichwa, R. D., and Damasio, A. R. (1996). A neural basis for lexical retrieval. *Nature (London)* **380**: 499–505.

Desmond, J. E., Sum, J. M., Wagner, A. D., Demb, J. B., Shear, P. K., Glover, G. H., Gabrieli, J. D. E., and Morrell, M. J. (1995). Functional MRI measurement of language lateralization in Wada-tested patients. *Brain* **118**: 1411–1419.

Heiss, W.-D., Karbe, H., Weber-Luxenburger, G., Herholz, K., Kessler, J., Pietrzyk, U., and Pawlik, G. (1997). Speech-induced cerebral metabolic activation reflects recovery from aphasia. *J. Neurol. Sci.* **145**: 213–217.

Herholz, K., Thiel, A., Wienhard, K., Pietrzyk, U., von Stockhausen, H. M., Karbe, H., Kessler, J., Bruckbauer, T., Halber, M., and Heiss, W. D. (1996). Individual functional anatomy of verb generation. *NeuroImage* **3**: 185–194.

Karbe, H., Wurker, M., Herholz, K., Ghaemi, M., Pietrzyk, U., Kessler, J., and Heiss, W. D. (1995). Planum temporale and Brodmann's area 22. Magnetic resonance imaging and high-resolution positron emission tomography demonstrate functional left-right asymmetry. *Arch. Neurol. (Chicago)* **52**: 869–874.

Kraut, M. A., Marenco, S., Soher, B. J., Wong, D. F., and Bryan, R. N. (1995). Comparison of functional MR and H2-15O positron emission tomography in stimulation of the primary visual cortex. *Am. J. Neuroradiol.* **16**: 2101–2107.

Mazoyer, B. M., Tzourio, N., Frak, V., Syrota, A., Murayama, N., Levrier, O., Salamon, G., Dehaene, S., Cohen, L., and Mehler, J. (1993). The cortical representation of speech. *J. Cognit. Neurosci.* **5**: 467–479.

Pardo, J. V., and Fox, P. T. (1993). Preoperative assessment of the cerebral hemisphere dominance for language with CBF PET. *Hum. Brain Mapp.* **1**: 57–68.

Price, C. J., Wise, R. J., Warburton, E. A., Moore, C. J., Howard, D., Patterson, K., Frackowiak, R. S., and Friston, K. J. (1996). Hearing and saying. The functional neuro-anatomy of auditory word-processing. *Brain* **119**: 919–931.

Schmahmann, J. D. (1996). From movement to thought—anatomic substrates of the cerebellar contribution to cognitive processing. *Hum. Brain Mapp.* **4**: 174–198.

Seitz, R. J., Huang, Y., Knorr, U., Tellmann, L., Herzog, H., and Freund, H. J. (1995). Large-scale plasticity of the human motor cortex. *NeuroReport* **6**: 742–744.

Weiller, C., Isensee, C., Rijntjes, M., Huber, W., Muller, S., Bier, D., Dutschka, K., Woods, R. P., Noth, J., and Diener, H. C. (1995). Recovery from Wernicke's aphasia: a positron emission tomographic study. *Ann. Neurol.* **37**: 723–732.

CHAPTER 24

Brain Networks of Motor Behavior Assessed by Principal Component Analysis

J. R. MOELLER,* C. GHEZ,[†] A. ANTONINI,[‡] M. F. GHILARDI,[†] V. DHAWAN,[‡] K. KAZUMATA,[‡] and D. EIDELBERG[‡]

*Departments of * Psychiatry and [†] Neurology*
College of Physicians and Surgeons
and the [†] Center for Neurobiology and Behavior, Columbia University
New York, New York 10032
[‡] Department of Neurology
North Shore University Hospital
Manhasset, New York 11030
and New York University School of Medicine, New York, New York 10003

Network-related changes in regional cerebral blood flow during execution and learning of reaching movements to visual targets using principal component analysis (PCA) of $H_2^{15}O$/PET subtraction images were examined. Ten healthy control subjects were studied while moving their right hand on a digitizing tablet from a central location to equidistant targets in synchrony with a 1/sec tone over a 90-sec scan. Success was signaled when the correct target was reached in a time window centered on the tone. A sensory control without movement (S, scrambled visual and auditory display) and three movement conditions were examined: M1, movements to predictable targets; M2, adaptation to a rotated hand-path display; and M3, discovered target sequence by trial and error using visuospatial working memory. Improvement in spatial and timing errors and a global performance score were computed for each test. Subtraction images were analyzed by the subprofile scaling model (SSM) with principal component analysis to identify regional covariance patterns. Individual scores for pattern expression in SSM were correlated with changes in errors and performance scores. M1-S: Improvements in spatial accuracy were predicted by pattern-related increases on the contralateral striatum and caudate covarying with reductions in contralateral premotor, lateral prefrontal cortex, and SMA. M2-M1: Adaptive changes in mean direction errors were predicted by network-related increases on the contralateral SMA and pre-SMA with covariant decreases in contralateral putamen. M3-M1: Rate of sequence learning was correlated with increased activation in ipsilateral hippocampal and perihippocampal areas and reductions in an ipsilateral striatum and medial prefrontal regions. Specific aspects of motor learning are expressed by distinct brain networks that are evident using SSM/PCA of subtraction [^{15}O]water/PET images.

I. INTRODUCTION

The execution of skilled limb movements is understood to require the cooperative contributions of multiple brain regions and visuomotor learning, to be associated with both reorganization and changes in the interplay among these regions (Georgopoulos, 1994). Using positron emission tomography (PET) and [^{15}O]water, the relationship between the expression of specific movement-related brain networks and quantitative measurements of the accuracy of the execution and learning of controlled motor tasks was examined.

Each of 10 subjects was scanned during the performance of three motor tasks and a sensory reference task. In all four experimental conditions the subjects moved their dominant right hand on a digitizing tablet out and back from a central location to eight equidistant targets in synchrony with a brief tone presented once per second. In the baseline motor task, targets appeared predictably in a counterclockwise sequence and were displayed, along with a cursor, on a computer screen. Correct targets were grayed when hit within a narrow time window around the tone. Improvements in the accuracy of motor performance were determined by quantifying spatial errors. New motor learning was also studied. First, subjects were scanned while learning a visuomotor transformation where the same targets were presented but cursor motion was rotated by 30–60°. In addition, spatial sequence learning was studied by presenting subjects with a novel order of the targets. In this learning task, the subject had to discover and remember the correct order of targets by trial and error using visuospatial working memory. Finally, in a sensory control condition, subjects observed the same screen elements and tones, without moving.

Although all subjects achieved comparable levels on a global measure of motor performance (final hit rate), more detailed analyses of performance changes revealed substantial intersubject differences in each task (Ghez *et al.*, 1995, 1996). The subprofile scaling model (SSM; Moeller *et al.*, 1987; Moeller and Strother, 1991; Eidelberg *et al.*, 1994, 1996) was used to identify the functional interactions that occurred among brain regions during the three motor tasks. By this means, it was possible to acquire information concerning the neural basis for individual differences in motor performance in normal subjects.

II. MATERIALS AND METHODS

A. Task Description

While lying in a supine position, subjects were instructed to move a hand-held cursor on a horizontally positioned digitizing tablet (200-Hz sampling rate) while viewing a video display that showed their cursor position overlayed on a circular array of computer-generated targets (uniformly arrayed at 45° intervals). Individuals performed a sequence of cursor movements, moving to successive targets with an out and back motion from a central circle. With each target, individuals were instructed to make a single uncorrected motion of the hand, reversing direction without stopping at the target. The extent of the hand movement was 9.6 cm. Movements to successive targets were made in a continuous fashion and were paced by a 60-msec tone, occurring at a rate of one per second. Task duration was 90 sec.

In the first of three motor tasks (M1), eight targets appeared sequentially in a prelearned order. Each successive target temporarily turned black at the tone, then gray when the cursor entered the circular target area within the 600-msec window centered at the tone. The second motor task (M2) was similar to M1 but with one difference: correspondence between the direction of cursor movement on the screen relative to that of the hand movement was rotated by 30–60°. In the third motor task (M3), as in the previous two tasks, the target array displayed on the monitor was eight white unfilled circles. However, unlike M1 and M2, at each sound of the pacing tone, the correct target remained white, unless the cursor entered the target area within the 600-msec window, at which time the target temporarily turned gray. Subjects had not only to learn, but first discover the order of five to seven successive targets by making correct movements. In the M2 and M3 tasks, respectively, rotation angle and target set size were titrated for each subject so that he/she achieved a final hit rate of better than 90% correct. A sensory control task (S) was also included in which subjects remained immobile while being exposed to the same visual stimuli as those experienced in the motor activation tasks. The same number of targets and cursor images were displayed on the screen and the same tones were sounded. However, these sensory events were divorced from their prior task implications by appearing asynchronously and irregularly.

The trajectories of the digitized hand movements were analyzed off-line. Time and position at movement onset, movement reversal at the target, and trajectory peak velocities were quantified. These performance data were used to calculate learning curves for M1–M3 where the main parameter of interest was rate of improvement. In M1, spatial error—the linear distance between the point at which the hand motion was reversed and the center of the correct target—was computed for successive blocks of 20 target excursions. In M2, directional error—the arc subtended (in degrees) between the point at which the hand motion was reversed and the center of the correct target—was computed for blocks of 20 target excursions. Finally, in M3, the hit rate—the average percentage of correct responses—was computed for blocks of 20 target excursions. In M1, rate of improvement was calculated as one minus the ratio of the spatial error rates for the first and final blocks; in M2, the ratio of the directional errors for the first and final blocks; and in M3, the difference in hit rates between first and final blocks.

B. PET Data

Ten right-handed normal volunteer subjects (six men, four women; 35 ± 10.1 mean age and SD, range 25–57 years) were studied with [^{15}O]water and PET using the GE Advance tomograph. The performance characteristics of this instrument have been described previously (De Grado *et al.*, 1994). All subjects fasted overnight prior to PET scanning.

Subjects were studied while executing the motor tasks described earlier. Subjects were positioned in the Laitinen stereoadapter to minimize repositioning errors. Sequential [^{15}O]water boluses were given for each of three repetitions of the four experimental conditions: M1–M3 and S. All motor tasks were performed with the dominant hand with an intravenous and catheter placed in the nondominant hand. Subjects were asked to perform each of these tasks in an arbitrary order for a duration of 90 sec. Relative cerebral blood flow was estimated using a modification of the slow bolus method of Silbersweig et al. (1993) in which a dose of up to 12 mCi of [^{15}O]water in 4 ml saline is injected by automatic pump in 16 sec (15 ml/min) followed by a manual 3-ml saline flush. Using this injection protocol, a time delay in the arrival of radioactivity in brain of approximately 17 sec was observed, and the time from rise to peak count rate was 35–40 sec. The timing of task initiation was adjusted individually so that the arrival of radioactivity occurred approximately 10 sec after the start of the task. Dynamic PET data acquisition began at the time of radioactivity arrival in the brain and continued for 80 sec. (The end of task thus coincided with the end of data acquisition.) The interval between successive [^{15}O]water administrations was 12 min. A maximum of 16 injections per subject was performed.

Region of interest (ROI) analysis was performed on PET reconstructions using Scan/VP software (Spetsieris et al., 1993, 1995) on a SUN microcomputer (SUN Microsystems, Mountain View, CA). Twenty-six (13 per hemisphere) standardized cortical and subcortical gray matter ROIs and two cerebellar and two brain stem ROIs (Eidelberg et al., 1994; Moeller et al., 1996) were selected. The coordinates of these ROIs are provided elsewhere (Eidelberg et al., 1997). Because of intersubject differences in brain morphology, ROI boundaries were adjusted manually to conform to individual gyral anatomy utilizing 3D MR–PET coregistration (Woods et al., 1993). ROI analysis was performed on composite PET images such that the identical regional coordinates were selected for all conditions within individual subjects. Total counts in each global image were estimated and images were rescaled to a mean of 50 ml/min/100 g.

C. SSM Statistical Analysis

In separate between-task comparisons of M1 and S, M2 and M1, and M3 and M1, the first step of the SSM network analysis was to identify candidate patterns of covarying regional activity that manifested large subject-by-task interactions in pattern expression. Principal component analysis (PCA) was applied to subject profiles of between-task differences in ROI activity (the average of three task repetitions). A logarithmic transformation of the raw ROI profiles of the control and experimental tasks was made before profile subtraction. Prior to the application of PCA, both subject variation in region-independent global activity and regional variation contributed by the group mean subtraction profile were subtracted from regional subtraction data of individual subjects. PCA was applied to the matrix of covariances calculated between subject residual subtraction profiles.

The second step of the SSM network analysis was to calculate a map of regional covariance that was associated with the learning rate. The SSM map of motor behavior was restricted to being either the first or the second principal component pattern or a linear combination thereof. The best linear combination was determined by using the first two PCA subject scores in a multiple linear regression prediction of the subject differences in learning rate. (These two PCA subject scores each represented the shift in the expression of the associated regional covariance pattern that occurred between experimental conditions.) The statistical significance reported for each regional covariance map is based on the squared multiple regression coefficient.

III. RESULTS AND DISCUSSION

The results of statistical parameter mapping analysis (SPM; Friston et al., 1995a, b), involving task comparisons of M1 and S, M2 and M1, M3 and M1, and M3 and M2, have been reported elsewhere (Ghez et al., 1996). By way of review, in the SPM comparison of the motor execution condition M1 with the sensory control condition S, significant differences in activation (omnibus $p < 0.001$) were found contralaterally in striatum, primary motor cortex (4), and superior parietal lobe (7); in ipsilateral cerebellum (and bilaterally in cerebellar vermis); and bilaterally in SMA and an occipital area (17). In the SPM comparison of the M2 learning condition with M1, significant differences in activation (omnibus $p = 0.05$) were found bilaterally in cerebellum (with greater ipsilateral activation) and the visual association area (19) and in the contralateral premotor region (6). In the SPM comparison of the M3 learning condition with M1, significant differences in activation (omnibus $p = 0.01$) were found ipsilaterally in cerebellum, prefrontal (46 and 9: DLPFC) and superior parietal areas; and bilaterally in the premotor region (with a greater ipsilateral activation) and the visual association area. Finally, in the SPM comparison of M3 with M2, there was a statistical trend (omnibus $p = 0.1$) toward greater activity in the ipsilateral hemisphere in areas of M3 activation.

SSM analysis revealed substantial differences in the patterns of activation associated with the two forms of visuomotor learning. In M2, the rate at which directional errors decreased with the number of targets attempted was correlated significantly ($R^2 = 0.86$, $p < 0.001$: Fig. 1A) with the expression of a regional covariance pattern (Fig. 1B) characterized by relative increases in the contralateral SMA and medial prefrontal regions and a relative decrease in contralateral striatum, i.e., SSM revealed a significant shift toward cortical areas—away from subcortical sites—in subjects with the highest rates of adaptation to imposed rotations. In M3, the rate at which accuracy increased with the number of targets attempted was correlated significantly ($R^2 = 0.64$, $p < 0.005$: Fig. 2A) with the expression of the covariance pattern (Fig. 2B) characterized by increased activation in ipsilateral hippocampus and relative deactivations in ipsilateral striatum and medial prefrontal regions, i.e., SSM revealed a significant shift toward hippocampus and perihippocampal regions—away from basal ganglia and prefrontal areas—in subjects with the highest rates at which a new target sequence was learned.

Even in the execution of the prelearned target sequence in M1, linear errors decreased with the number of targets attempted. This decrease in error rate was significantly correlated ($R^2 = 0.53$, $p < 0.02$: Fig. 3A) with the expression of a regional covariance pattern (Fig. 3B) characterized by a relative increase in the contralateral striatum and caudate and relative decreases in contralateral premotor, lateral prefrontal cortex, and SMA, i.e., SSM revealed a significant shift from cortical areas to basal ganglia in subjects exhibiting the largest improvements in spatial accuracy.

In each of the three visuomotor tasks, a major source of covariation in regional blood flow was found to be correlated significantly with intersubject differences in a specific dimension of motor learning. In M1–S, the SSM regional covariance effect accounted for 33% of the region × subject variance. In M2–M1, the SSM regional covariance effect accounted for 24%, and in M3–M1 it accounted for 33%. In all three tasks, these SSM covariance effects were nearly as large, or larger than any other regional covariance effect in the blood flow data. In addition, in the M1 and M2 tasks, the reduction in spatial errors with the number of targets attempted was the predominant dimension of subject differences in motor performance. In M3, the rate of increase in target hits as a function of the number of targets attempted was the major dimension of subject performance differences.

A. Interpretations of Task-Related Covariance Patterns

Results from the SSM analyses of visuomotor tasks have been interpreted as indicating significant changes in the use of brain circuitry that occur during 90 sec of visuomotor learning. First, response production in M1 appears to be mediated by a left hemisphere corticobasal ganglionic motor loop, where the enhancement of a subject's rate of learning is associated with decreased activation in the cortical components of the loop and increased activation in the striatum. This enhancement is interpreted to mean that less control of response production is taken by the motor planning circuits of the SMA, the lateral premotor areas and lateral prefrontal regions, whereas more control is exercised by the circuits of the striatum. One specific conjecture is that this implicit reduction of spatial errors is achieved by reducing the contributions of the stimulus-controlled, feedback processes of cortical circuits, while reciprocally increasing the contributions of an open loop, feedforward system in the striatum (operating by means of a learned internal model of the dynamics of limb movements). Second, response learning in M3 appears to be mediated both by working memory circuits of the right hemisphere, dorsal lateral prefrontal cortex and by the right hippocampal (and perihippocampal) circuits that also are associated with spatial memory processing. The enhancement of a subject's rate of learning is associated with decreased activation in the prefrontal working memory components and increased activation in the memory processing components of the hippocampus. One particular conjecture is that enhanced rates of learning are achieved by reducing the influence of the semantic control processes of a prefrontal working memory while increasing the influence of hippocampal areas in the creation of nonsemantic, pneumonic representations of sequential limb movements. Finally, response learning in M2 appears to be mediated by a left hemisphere corticobasal ganglionic motor loop, where the enhancement of a subject's rate of learning is associated with decreased activation in the striatum and increased activation in the SMA and pre-SMA components of the loop. Although it is thought that the internal model of the M2 motor adaptation is learned implicitly, SSM results suggest that the circuits which mediate this kind of learning reside within frontal cortex, i.e., the SMA and pre-SMA, and not within the circuits of the striatum (or the sensory-motor cortex). Moreover, in contrast to the efficient reduction of spatial errors of the M1 kind in the striatum, apparently the striatum cannot achieve a similar level of efficiency in the reduction of the M2 type of spatial error.

24. Brain Networks of Motor Behavior

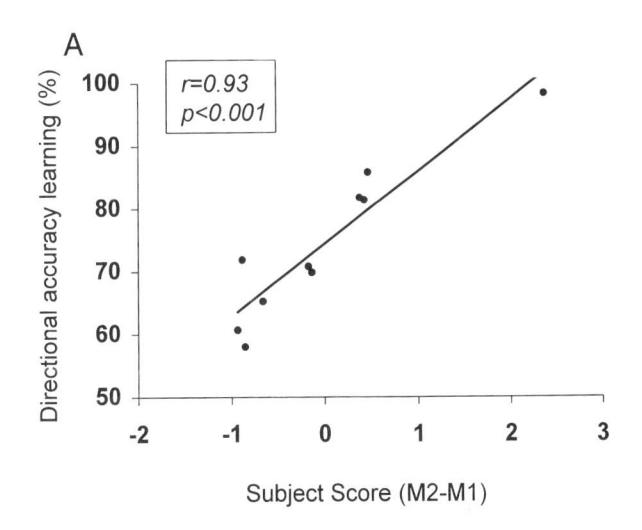

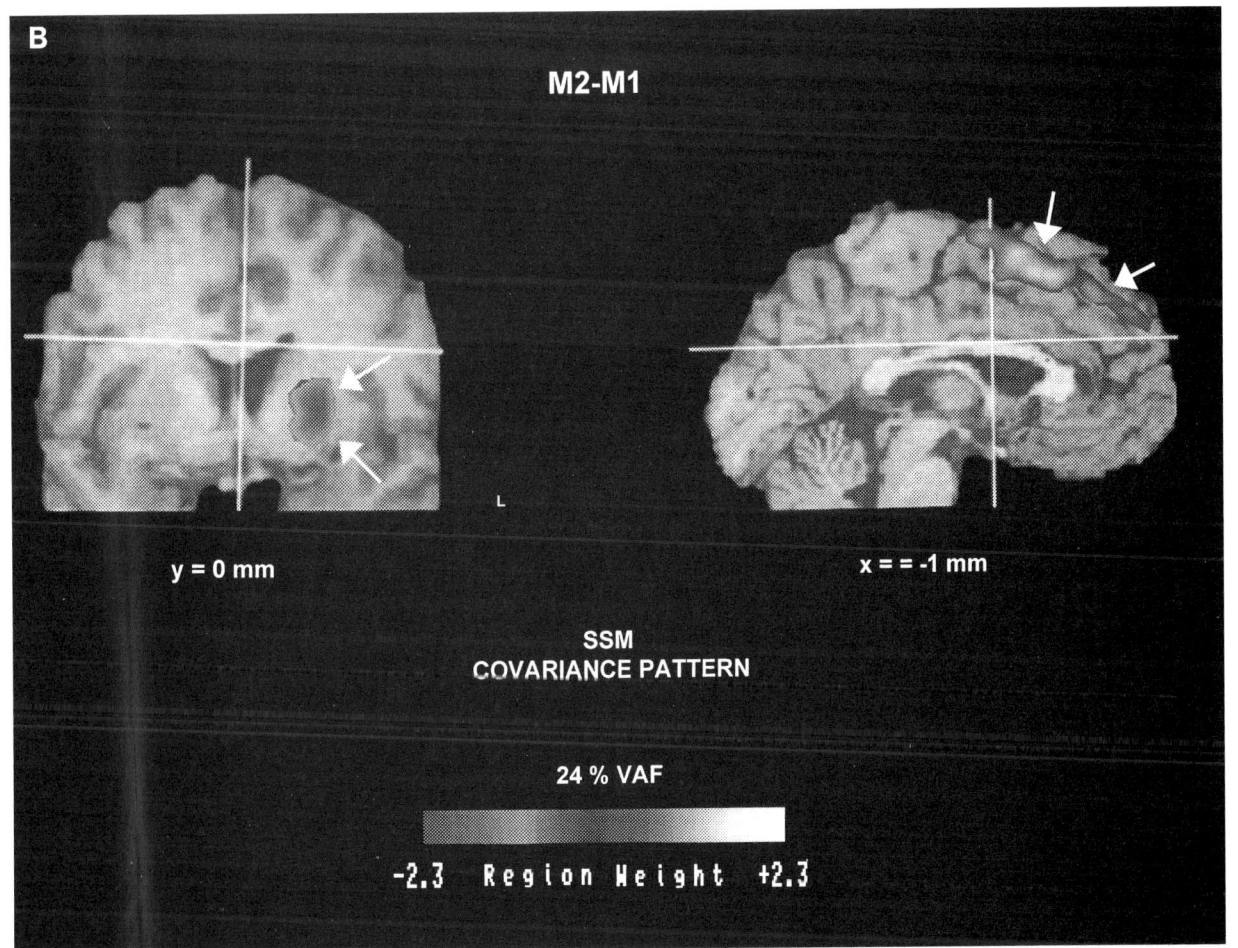

FIGURE 1 (A) Scatter plot of the M2 rate of directional accuracy learning (i.e., the rate of adaptation to imposed rotations) and its SSM subject score predictor obtained from the between-task comparison of M2 and M1. This predictor of motor performance was obtained from the SSM between-task comparison calculated for the hemisphere contralateral to hand movement. (B) Display of thresholded regional weights of the covariance pattern whose expression was the SSM subject score predictor. Regional weights have been overlaid on standard Talairach-transformed MRI sections. The pattern is characterized by relative increases in the SMA and medial prefrontal regions and a relative decrease in striatum.

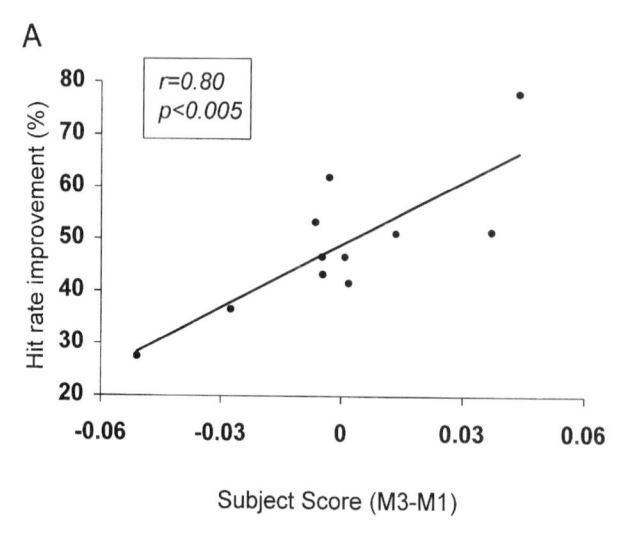

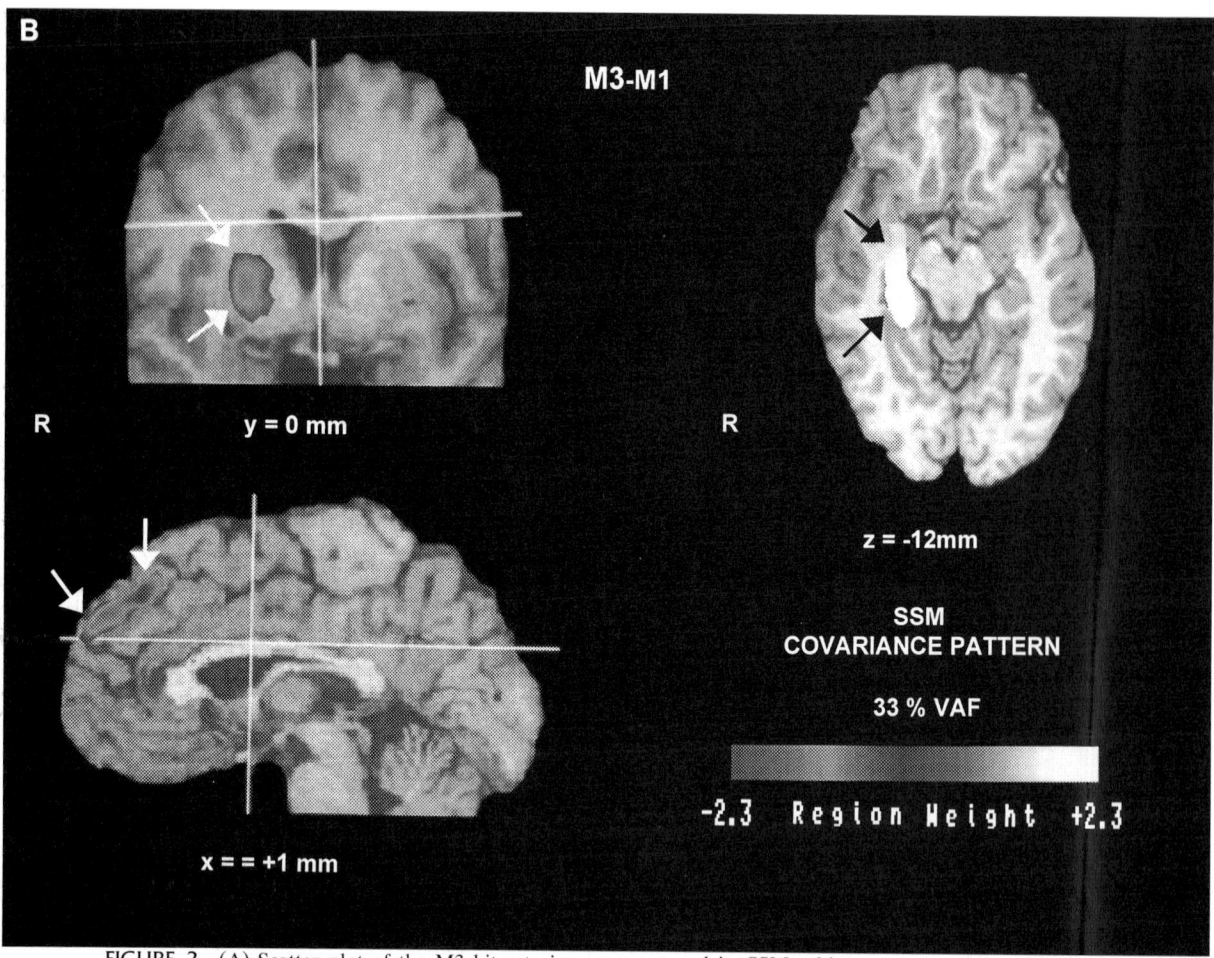

FIGURE 2 (A) Scatter plot of the M3 hit rate improvement and its SSM subject score predictor obtained from the between-task comparison of M3 and M1. This predictor of motor performance was obtained from the SSM between-task comparison calculated for the hemisphere ipsilateral to hand movement. (B) Display of thresholded regional weights of the covariance pattern whose expression was the SSM subject score predictor. The pattern is characterized by relatively increased activation in hippocampus and relative deactivations in striatum and medial prefrontal regions.

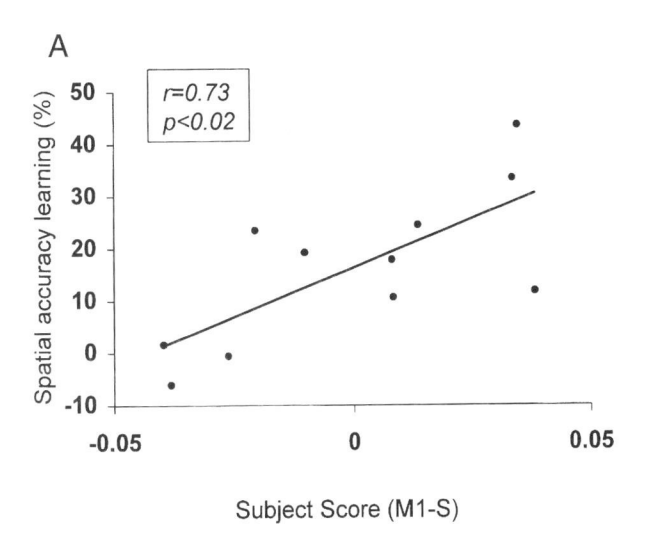

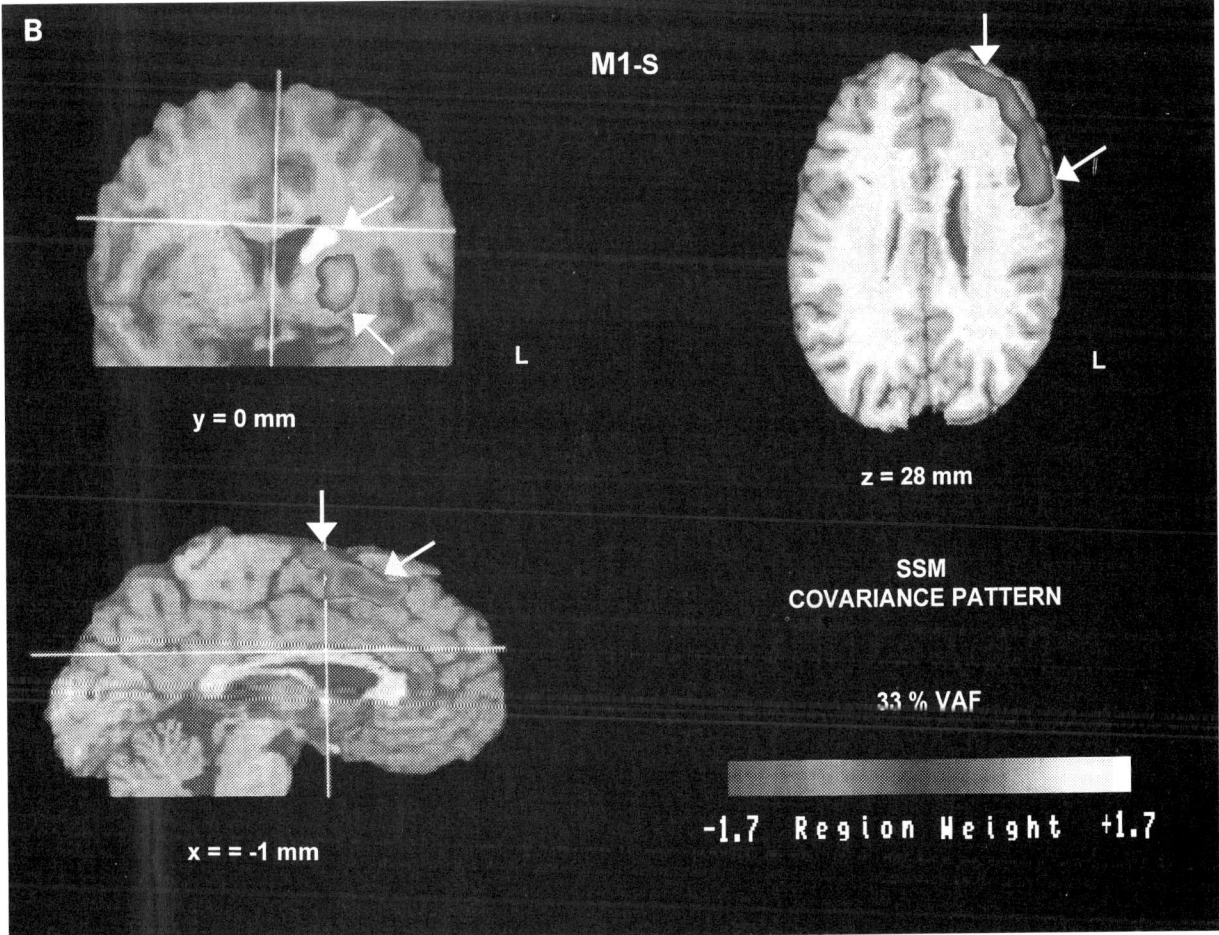

FIGURE 3 (A) Scatter plot of M1 spatial accuracy learning and its SSM subject score predictor obtained from the between-task comparison of M1 and S. This predictor of motor performance was obtained from the SSM between-task comparison calculated for the hemisphere contralateral to hand movement. (B) Display of thresholded regional weights of the covariance pattern whose expression was the SSM subject score predictor. The pattern is characterized by a relative increase in the striatum and caudate and by relative decreases in contralateral premotor, lateral prefrontal cortex, and SMA.

Because blood flow images were acquired using a standard [^{15}O]water bolus technique (Silbersweig *et al.*, 1993), the network effects analyzed must necessarily represent subject differences in the efficient use of motor-related networks that are present at the initial stages of learning, i.e., as early as 30 to 40 sec after initiation of an experimental task. However, because of the high correlations between these subject differences and individual learning rates, the authors suggest that this continuum of differential network activations obtained across subjects is likely to be representative of the changes in any one subject, i.e., the changes in the use of brain circuitry that occur during the course of visuomotor learning.

Acknowledgments

This work was supported by NIH NS RO1-35069. We acknowledge the important contributions of Dr. Phoebe Spetsieris and Dr. Thomas Chaly to this work. We thank Claude Margouleff for help with the PET studies and Dr. Robert Dahl and Mr. Ralph Matacchieri for cyclotron support.

References

De Grado, T. R., Turkington, T. G., Williams, J. J., Stearns, C. W., Hoffman, J. M., and Coleman, R. E. (1994). Performance characteristics of a whole-body PET scanner. *J. Nucl. Med.* **35**: 1398–1406.

Eidelberg, D., Moeller, J. R., Dhawan, V., Spetsieris, P., Takikawa, S., Ishihawa, T., Chaly, T., Robeson, W., Margouleff, D., Przedborski, S., and Fahn, S. (1994). The metabolic topography of parkinsonism. *J. Cereb. Blood Flow Metab.* **14**: 783–801.

Eidelberg, D., Moeller, J. R., Ishihawa, T., Dhawan, V., Spetsieris, P., Silbersweig, D., Stern, D., Woods, R., Fazzini, E., Dogali, M., and Beric, A. (1996). Regional metabolic correlates of surgical outcome following unilateral pallidotomy for Parkinson's disease. *Ann. Neurol.* **39**: 450–459.

Eidelberg, D., Moeller, J. R., Kazumata, K., Antonini, A., Sterio, D., Dhawan, V., Spetsieris, P., Alterman, R., Kelly, P. J., Dogali, M., Fazzini, M., Fazzini, E., and Beric, A. (1997). Metabolic correlates of pallidal neuronal activity in Parkinson's disease. *Brain* **120**: 1315–1324.

Friston, K. J., Ashburner, J., Poline, J.-P., Frith, C. D., Heather, J. D., and Frackowiak, R. S. J. (1995a). Spatial realignment and normalization of images. *Hum. Brain Mapp.* **2**: 165–189.

Friston, K. J., Holmes, A. P., Worsley, K. J., Poline, J.-P., Frith, C. D., and Frackowiak, R. S. J. (1995b). Statistical parametric maps in functional imaging: A general linear approach. *Hum. Brain Mapp.* **2**: 189–210.

Georgopoulos, G. (1995). Motor cortex and cognitive processing. In "The Cognitive Neurosciences" (M. S. Gazzaniga, ed.), pp. 507–517. MIT Press, Cambridge, MA.

Ghez, C., Gordon, J., Ghilardi, M. F., Sainburg, R. (1995). Contribution of vision and proprioception to accuracy in limb movements. In "The Cognitive Neurosciences" (M. S. Gazzaniga, ed.), pp. 534–541. MIT Press, Cambridge, MA.

Ghez, C., Ghilardi, M. F., Moeller, J. R., Dhawan, V., and Eidelberg, D. (1996). Patterns of regional brain activation associated with different aspects of motor learning. *Soc. Neurosci. Abstr.* **22**(Part 2): 899.

Moeller, J. R., and Strother, S. C. (1991). A regional covariance approach to the analysis of functional in positron emission tomographic data. *J. Cereb. Blood Flow Metab.* **11**: A121–A135.

Moeller, J. R., Strother, S. C., Sidtis, J. J., and Rottenberg, D. A. (1987). The scaled subprofile model: A statistical approach to the analysis of functional patterns in positron emission tomographic data. *J. Cereb. Blood Flow Metab.* **7**: 649–658.

Moeller, J. R., Ishihawa, T., Dhawan, V., Spetsieris, P., Mandel, P., Alexander, G. E., Grady, C. L., Pietrini, P., and Eidelberg, D. (1996). The metabolic topography of normal aging. *J. Cereb. Blood Flow Metab.* **16**: 385–398.

Silbersweig, D. A., Stern, E., Frith, C. D., Cahill, C. Schnorr, L., Grootoonk, S., Spinks, T., Clark, J., Frackowiak, R. S. J., and Jones, T. (1993). Detection of thirty second cognitive activations in single subjects with positron emission tomography: A new low dose H$_2^{15}$O regional cerebral blood flow three dimensional imaging technique. *J. Cereb. Blood Flow Metab.* **13**: 617–629.

Spetsieris, P. G., Dhawan, V., Takikawa, S., Margouleff, D., and Eidelberg, D. (1993). Imaging cerebral function. *IEEE Comput. Graphics Appl.* **13**: 15–26.

Spetsieris, P. G., Moeller, J. R., Dhawan, V., Takikawa, S., and Eidelberg, D. (1995). Visualizing the evolution of abnormal metabolic networks in the brain using PET. *Comput. Med. Imag. Graphics* **19**: 295–306.

Woods, R. P., Mazziotta, J. C., and Cherry, S. R. (1993). MRI-PET registration with automated algorithm. *J. Comput. Assist. Tomogr.* **17**: 436–546.

CHAPTER

25

Frequency-Dependent Changes in Cerebral Metabolic Rate of Oxygen during Activation of Human Visual Cortex Studied by PET[1]

M. S. VAFAEE, E. MEYER, S. MARRETT, T. PAUS, A. C. EVANS, and A. GJEDDE

McConnell Brain Imaging Centre
Positron Emission Tomography Laboratories
Montreal Neurological Institute, McGill University
Montreal, Quebec, H3A 2B4 Canada

This study investigated the effect of stimulus presentation frequency on the regional cerebral metabolic rate of oxygen ($rCMR_{O_2}$) in human visual cortex. $rCMR_{O_2}$ was measured in 12 healthy normal volunteers with the ECAT EXACT HR$^+$ (CTI/Siemens) three-dimensional whole body tomograph. In seven successive activation conditions, the subjects were shown a blue and yellow annular checkerboard, reversing its contrast at frequencies of 0, 1, 4, 8, 16, 32, and 50 Hz. Stimulation began 4 min before the onset of the dynamic positron emission tomography (PET) scan and continued throughout the 3-min scan. In the baseline condition, the subjects were asked to fixate on a cross-hair 30 sec before and during the scan. At the start of each scan, the subjects inhaled 20 mCi of $[^{15}O]O_2$ in a single breath. CMR_{O_2} was calculated using a two-compartment, weighted integration method. Normalized PET images were averaged across subjects and coregistered with the subjects' magnetic resonance images in stereotaxic space. Mean subtracted image volumes (activation minus baseline) of CMR_{O_2} were then obtained and converted to z-statistic volumes by division by the pooled standard deviation. This allowed a demonstration of a statistically significant focal change of CMR_{O_2} in the striate cortex, which reached a maximum at 4 Hz and dropped off sharply at higher stimulus frequencies.

I. INTRODUCTION

The human brain consumes glucose and oxygen to sustain its function. These substrates are supplied continuously via the cerebral circulation, which accounts for 20% of the cardiac output. The mechanism linking the circulation to neuronal activity is generally believed to constitute a flow metabolism couple (Gjedde, 1997). The purpose of this couple is thought to satisfy the principle of Roy and Sherrington (1890). This has been interpreted to mean that cerebral blood flow (CBF) changes must subserve a tight coupling between cellular energy requirements and supplies of glucose and oxygen to the brain. It is generally believed that such a homeostatic mechanism maintains a constant concentration of adenosine triphospate (ATP), the compound that ties together the processes that deplete the energy potential of brain tissue to the processes that restore it.

Reports have claimed a mismatch between changes of CBF and oxygen utilization during functional activation of the human brain, despite a match between the observed changes in regional glucose consumption and CBF (Fox and Raichle, 1986). Significant increases of the cerebral metabolic rate of oxygen (CMR_{O_2}) in the striate cortex in response to a specific visual stimulus were observed (Marrett *et al.*, 1993; Vafaee *et al.*, 1996), along with a commensurate elevation of CBF in agreement with the principle of Roy and Sherrington.

A direct relationship between the stimulus rate and the cerebral metabolic rate of glucose (CMR_{glc}) has

---

[1] Transcripts of the BRAINPET97 discussion of this chapter can be found in Section VIII.

been reported in both the central nervous system (Toga and Collins, 1981) and the peripheral nervous system of rats (Yarowsky et al., 1983). The effect of stimulus rate on CBF in humans was also studied by several investigators. Thus, Fox and Raichle (1984) investigated the relationship between stimulus rate and CBF change in human visual cortex, using positron emission tomography (PET).

The goal of the present experiment was to determine the frequency dependence of CMR_{O_2} changes using a previously tested visual stimulus, the yellow-blue reversing contrast checkerboard, which changes CMR_{O_2} in the visual cortex under specific conditions (Vafaee et al., 1996).

II. MATERIALS AND METHODS

Twelve healthy normal volunteers (six males and six females), aged 22 to 32 years (mean = 25 ± 3.5), were studied for this protocol, which was approved by the Research Ethics Committee of the Montreal Neurological Institute and Hospital. Written consent was obtained from each volunteer.

PET studies were performed on the ECAT EXACT HR$^+$ (CTI/Siemens) three-dimensional (3D) whole body tomograph with a transverse resolution of 4.5–5.8 mm and an axial resolution of 4.9–8.8 mm (Adam et al., 1997). The images were reconstructed as 128 × 128 matrices of 2 × 2-mm pixels using filtered backprojection and blurred with a 22-mm FWHM filter. Reconstruction software included corrections for random and scattered events, detector efficiency variations, and dead time.

The subjects were positioned in the tomograph with their heads immobilized by means of a customized headholder (bean bag). A short indwelling catheter was placed into the left radial artery for blood sampling and blood gas examination. Arterial blood radioactivity was sampled automatically, corrected for delay and dispersion, and calibrated with respect to the tomograph using samples obtained manually during the last 60 sec of each 3-min scan. At the start of each scan, the subjects inhaled 20 mCi of [^{15}O]O$_2$ in a single breath. CMR_{O_2} was calculated using the two-compartment, weighted integration method (Ohta et al., 1992). Each subject also underwent magnetic resonance (MR) imaging for structural–functional (MR imaging–PET) correlation.

In the baseline condition, the subjects were asked to fixate on a cross-hair in the center of the screen 30 sec before the scan and throughout the subsequent 3-min scan. In seven successive activation conditions, the subjects were shown a blue and yellow annular checkerboard with the contrast reversed at frequencies of 0, 1, 4, 8, 16, 32, and 50 Hz. Stimulation began 4 min before the start of the dynamic PET scan and continued throughout the following 3-min scan for a total of 7 min. Black drapes were used to create a dark environment around the screen.

The MR images were transferred into stereotaxic coordinates (Talairach and Tournoux, 1988) by means of an automatic registration program developed at the Montreal Neurological Institute (Collins et al., 1994). The reconstructed PET images were coregistered with the subjects' MR images using an automatic registration program based on the automatic image registration algorithm (Woods et al., 1992). To correct for between-scan subject movement, PET-to-PET automatic registration was also performed (Woods et al., 1993). The PET images were then normalized for global CMR_{O_2} and averaged across subjects. Mean subtracted image volumes (stimulation minus baseline) were then obtained and converted into z-statistic volumes by dividing each voxel by the mean standard deviation (SD) of the normalized CMR_{O_2} subtraction image obtained by pooling the SD across all intracerebral voxels. Significant focal changes were defined by a method based on 3D Gaussian random field theory (Worsley et al., 1992). Values equal to or exceeding a criterion of $z = 3.5$ were deemed statistically significant ($p < 0.00046$, two-tailed, uncorrected). Correcting for multiple comparisons, a z value of 3.5 yields a false positive rate of 0.85 in 250 resolution elements, which approximates the total volume of cortex scanned.

III. RESULTS AND DISCUSSION

The mean global CMR_{O_2} was calculated for all measurements made in the 12 subjects at the different stimulus frequencies. Fluctuations in global CMR_{O_2} occurred from scan to scan within a single individual session as well as between subjects for different frequencies. However, there was no significant effect of frequency on global CMR_{O_2} (Fig. 1).

In contrast, regional CMR_{O_2} was altered in primary visual cortex (V1) for different frequencies. As shown in Fig. 2, CMR_{O_2} in primary visual cortex increased as the stimulus frequency was increased up to 4 Hz at which frequency it peaked and then dropped off at higher frequencies.

It has been shown previously that stimulus frequency is a significant determinant of rCBF in the visual as well as in auditory and motor cortex during

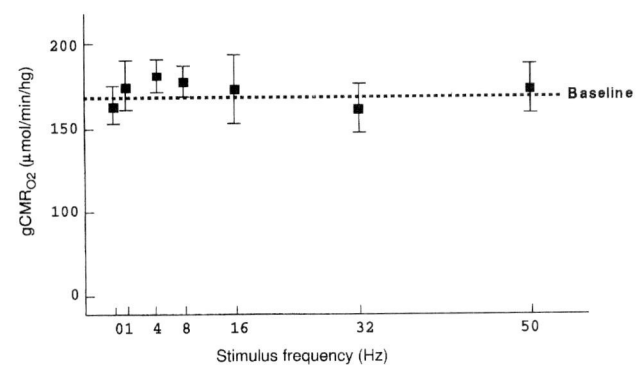

FIGURE 1 Global CMR_{O_2} ($\pm SD$) as a function of stimulus frequency.

physiological activation (Price *et al.*, 1992; Sadato *et al.*, 1996). The purpose of this study was to determine what relation, if any, exists between stimulus frequency and $rCMR_{O_2}$. Significant CMR_{O_2} changes in the primary visual cortex during the presentation of a chromatic stimulus (blue and yellow) have been shown previously (Vafaee *et al.*, 1996).

Data show that CMR_{O_2} varies in the visual cortex as a function of stimulus frequency. Unlike CBF, which peaks at 8 Hz (Fox and Raichle, 1984), CMR_{O_2} reaches its peak at 4 Hz and drops off at the higher frequencies. Based on these findings, the authors speculate that a primary block, the oxygen diffusibility, is overcome whenever blood flow increases sufficiently. Subsequently, an enzymatic block prevents the brain from enjoying the increase of CMR_{O_2} allowed by the blood flow increase, but this block is overcome with an increase in stimulus load, the variable that integrates length, strength, and kind of stimulation applied. Thus, two blocks operate, a functional enzymatic block that is lifted in proportion to the stimulus load and a basic oxygen diffusibility block that is lifted when blood flow is raised.

In conclusion, the ability of brain neurons to increase their oxygen use may vary, depending on the type of task a neuron performs. Moreover, the stimulus load imposed on brain tissue must exceed a certain threshold before glycolysis is augmented by an increase in oxidative metabolism.

Acknowledgments

We thank our cyclotron staff for providing the radioisotopes. We also thank our technicians for their assistance with the PET experiments. The helpful comments of Mr. Peter Neelin with regard to image analysis are particularly appreciated. This work was supported by Medical Research Council (Canada) Grant SP-30, the Isaac Walton Killam Fellowship Fund of the Montreal Neurological Institute, and the McDonnell-Pew Program in Cognitive Neuroscience.

References

Adam, L. E., Zaers, J., Ostertag, H., Troban, H., Bellemann, M. E., Brix, G., and Lorenz, W. J. (1997). Performance evaluation of the whole-body PET scanner ECAT EXACT HR+. In "1996 IEEE Nuclear Science Symposium Conference Record" (A. Del Guerra, ed.), pp. 1270–1274. IEEE, Piscataway, NJ.

Collins, D. L., Neelin, P., Peters, T. M., and Evans, A. C. (1994). Automatic 3D intersubject registration of MR volumetric data in standardized Talairach space. *J. Comput. Assist. Tomgr.* **18**: 192–205.

Fox, T. P., and Raichle, M. (1984). Stimulus rate dependence of regional blood flow in human striate cortex, demonstrated by positron emission tomography. *J. Neurophysiol.* **51**: 1109–1120.

Fox, T. P., and Raichle, M. E. (1986). Focal physiological uncoupling of cerebral blood flow and oxidative metabolism during somatosensory stimulation in human subjects. *Proc. Natl. Acad. Sci. U.S.A.* **83**: 1140–1144.

Gjedde, A. (1997). The relation between brain function and cerebral blood flow and metabolism. In "Cerebrovascular Disease" (H. Hunt Batjer, ed.), pp. 23–40. Lippincott-Raven Press, Philadelphia.

Marrett, S., Fujita, H., Ribeiro, L., Kuwabara, H., Meyer, E., Evans, A. C., and Gjedde, A. (1993). Evidence for stimulus-specific changes in oxidative metabolism. *Ann. Nucl. Med.* **7**: S120.

Ohta, S., Meyer, E., Thompson, C. J., and Gjedde, A. (1992). Oxygen consumption of the living human brain measured after a single inhalation of positron emitting oxygen. *J. Cereb. Blood Flow Metab.* **12**: 179–192.

Price, C., Wise, R., Friston, K., Howard, D., Patterson, K., and Frackowiak, R. (1992). Regional response differences within the human auditory cortex when listening to words. *Neurosci. Lett.* **146**: 179–182.

Roy, C. S., and Sherrington, C. S. (1890). On the regulation of the blood supply of the brain. *J. Physiol. (London)* **11**: 85–108.

Sadato, N., Ibanez, V., Deiber, M. P., Campbell, G., Leonardo, M., and Hallett, M. (1996). Frequency-dependent changes of regional cerebral blood flow during finger movements. *J. Cereb. Blood Flow Metab.* **16**: 23–33.

Talairach, J., and Tournoux, P. (1988). "Co-Planar Stereotactic Atlas of the Human Brain." Thieme, Stuttgart.

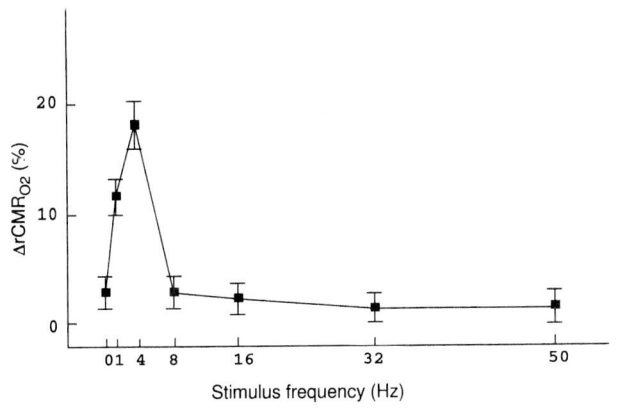

FIGURE 2 Change in $rCMR_{O_2}$ ($\pm SD$) in primary visual cortex as a function of stimulus frequency.

Toga, A. W., and Collins, R. W. (1981). Metabolic response of optic centers to visual stimuli in the albino rat: Anatomical and physiological considerations. *J. Comp. Neurol.* **199**: 443–464.

Vafaee, M., Paus, T., Gjedde, A., Evans, A. C., Ptito, A., and Meyer, E. (1996). Oxidative metabolism in human visual cortex during physiological activation studied by PET. *Soc. Neurosci. Abstr.* **22**: 1060.

Woods, R. P., Cherry, S. R., and Mazziotta, J. C. (1992). Rapid automated algorithm for aligning and reslicing PET images. *J. Comput. Assist. Tomogr.* **16**: 620–633.

Woods, R. P., Mazziotta, J. C., and Cherry, S. R. (1993). MRI-PET registration with automated algorithm. *J. Comput. Assist. Tomogr.* **17**: 536–546.

Worsley, K. J., Evans, A. C., Marrett, S., and Neelin, P. (1992). A three-dimensional statistical analysis for CBF activation studies in human brain. *J. Cereb. Blood Flow Metab.* **12**: 900–918.

Yarowsky, P., Kadekaro, M., and Sokoloff, L. (1983). Frequency-dependent activation of glucose utilization in the superior cervical ganglion by electrical stimulation of cervical sympathetic trunk. *Proc. Natl. Acad. Sci. USA* **80**: 4179–4183.

CHAPTER

26

Validation of an [¹⁸F]Fluorodeoxyglucose–PET Protocol in Conscious Vervet Monkey

A. H. MOORE,* M. J. RALEIGH,† S. R. CHERRY,*,‡ S.-C. HUANG,* and M. E. PHELPS*,‡

Departments of *Medical and Molecular Pharmacology and †Psychiatry and Biobehavioral Sciences
‡Crump Institute for Biological Imaging
University of California, Los Angeles
Los Angeles, California 90095

In order to document the temporal changes in local cerebral metabolic rates of glucose (lCMRGlc) during development, a method that permits reliable and repeatable measurement and analysis of lCMRGlc in conscious monkeys using position emission tomography (PET) and [¹⁸F]fluorodeoxyglucose (FDG) has been established. This protocol avoids the confounding effects of anesthesia on lCMRGlc. Socially living male vervet monkeys ranging from 2 to 8 months of age were used. All studies were quantitative and conducted with the subjects' mothers anesthetized and within close proximity. After receiving an intravenous bolus injection of FDG (1.5 mCi/kg), subjects remained in a resting, conscious state during the 42-min uptake. Subjects were then anesthetized, and data acquisition began at 60 min post-FDG injection. Regions of interest (ROIs) were defined on the FDG images with reference to age-matched magnetic resonance images. Intrarater reliability of the ROI analysis method was determined and showed high correlation (r = 0.962). Reproducibility of lCMRGlc patterns was assessed by obtaining multiple scans on an adult subject over a short period of time and comparing ROI values. No significant difference was found in the ROI values of normalized FDG uptake across time ($p > 0.05$), and the average coefficient of variation of ROI values between scans was $5.3 \pm 1.6\%$. Comparison of CMRGlc of neonatal (≤ 4 months) and infant (6–8 months) monkeys show a significant increase in glucose utilization in the infant group ($p < 0.05$). These results indicate that this method is reliable, reproducible, and sensitive to developmental changes in cerebral glucose metabolism. It can be used in future studies to determine region-specific patterns of lCMRGlc during nonhuman primate maturation, without the confounding effects of anesthesia.

I. INTRODUCTION

The correlation between glucose metabolic changes in the mammalian brain and behavioral maturation has been well documented in several animal species using autoradiography (Kennedy et al., 1982; Abrams et al., 1984; Nehlig et al., 1988; Chugani et al., 1991). [¹⁸F]Fluorodeoxyglucose (FDG) and positron emission tomography (PET) have been used to study the cerebral metabolic rate of glucose (CMRGlc) during human development (Chugani et al., 1987; Kinnala et al., 1996). Although FDG–PET is a noninvasive method that allows multiple and longitudinal studies of CMRGlc maturation, its application in animal models is limited due to resolution of the PET scanner and to the complexity of the experimental method required to obtain quantitative scans.

A main concern of using FDG/PET in animal studies is the effect of anesthesia on resting state cerebral glucose utilization during the tracer uptake period. Previous reports show that a variety of sedatives have region-specific effects on lCMRGlc (Crosby et al., 1982; Saija et al., 1989) that may cause misinterpretation of results. This concern was addressed in a study that attempted to show a developmental pattern of lCMRGlc in sedated rhesus macaques and vervet monkeys (Jacobs et al., 1995). Similar changes in the author's PET studies were observed when ketamine anesthesia was used to sedate an adult monkey during the tracer uptake period (Fig. 1, see color insert).

Instead of sedating subjects during FDG uptake, other groups have trained animals to sit in a restraining chair (Kennedy et al., 1978; Eberling et al., 1995). Even though this alternative method avoids the confounding effects of anesthesia, it introduces the factor of stress. Physical restraint has been shown to affect physiological measures in nonhuman primates (pH, pO_2, pCO_2) (Bush *et al.*, 1977) and may alter the subjects' anxiety, which has been reported to decrease CMRGlc in humans (Gur *et al.*, 1987).

The present study established a method that permits reliable and repeatable measurement and analysis of absolute lCMRGlc in conscious infant monkeys using FDG–PET. All subjects were awake and resting during the tracer uptake period with minimal restraint, thereby avoiding the confounding factors of stress and anesthesia. Even though this method is limited to monkeys less than 12 months of age (due to the increase of activity and aggression of older animals), this time span contains the period of greatest neuroplasticity and brain development as reported in histological and autoradiographic studies in nonhuman primates (Kennedy *et al.*, 1982; Rakic *et al.*, 1986; Zecevic *et al.*, 1989; Zecevic and Rakic, 1991) and as documented in the emergence of social and motor skills (Fairbanks and McGuire, 1988; Struhsaker, 1967, 1971).

II. MATERIALS AND METHODS

All procedures were conducted in accordance with the institutional Animal Research Committee guidelines. A time line of this method is presented in Fig. 2.

A. Subjects

Thirteen male vervet monkeys (*Cercopithecus aethiops sabeus*) from the Sepulveda VAMC/UCLA Nonhuman Primate Laboratory were used as subjects in these studies. Twelve monkeys, ranging in age from 2 to 8 months, were scanned once to obtain cross-sectional data. One adult monkey was scanned six times between 103 and 109 months of age to test reproducibility. Monkeys were born and raised in species-typical social groups that consist of three or more adult males, five or more adult females, and their immature offspring. Groups are housed in outside enclosures that contain play structures, visual barriers, protection from inclement weather, and unlimited access to commercial monkey chow and water.

B. Animal Preparation

On the day prior to the scan, the subject and its mother are fasted and transported to the UCLA vivarium and caged overnight (maximum 16 hr). The authors chose to transfer and house the mother and the infant together in order to minimize any stress associated with maternal separation.

On the day of the study, the mother is anesthetized with ketamine (15 mg/kg) and her sedation is continued throughout the FDG uptake period. The subject and mother were moved to a moderately lit, sound-attenuated room. The subject is sedated briefly (1% isoflurane) for a maximum of 15 min. An arterial or venous line is established rapidly and is maintained with a heparinized saline solution (0.9% NaCl and 0.05% heparin). During sedation recovery (> 30 min), the subject is placed on his mother and remains there with little to no restraint.

C. PET Methodology

When the subject is alert and has acclimated to the position on its mother, the subject receives a bolus FDG (1.5 mCi/kg) injection over 30 sec. Ten blood samples (~ 2 cc) are collected during the initial 2 min and then subsequently at 3, 4, 5, 7, 10, 20, 30, 40, 60, 90, and 120 min following FDG administration.

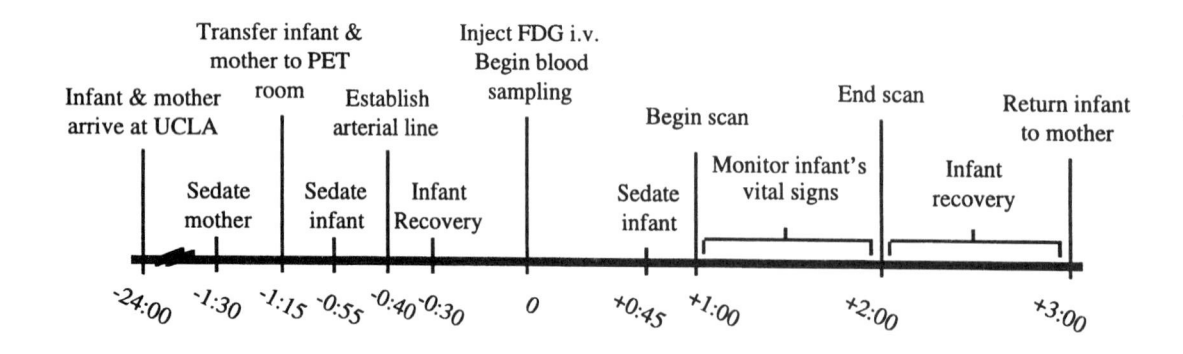

FIGURE 2 Time line of conscious monkey FDG–PET protocol. Scale in hour:minutes.

The subject is in a conscious resting state during the FDG uptake period and remains in close contact with its anesthetized mother. Behavior is monitored and recorded as high, moderate, or low agitation level by a primatologist familiar with species-typical behavior. Forty-two minutes after FDG injection, the subject is sedated with ketamine (15 mg/kg). The head of the subject is positioned in the scanner with the orbitocanthomeatal line parallel to the imaging planes and the top of the head at the front of the field of view. Positioning is achieved with reference to laser axes. A 2-min scout scan is conducted to validate positioning and symmetry of the head. The Anesthesia is maintained throughout the scan using a perfusion pump delivering ketamine (12 mg/kg/hr) in a heparinized saline solution. Midazolam hydrochloride is administered as needed with dosage ranging from 0.05 to 0.10 mg/kg/hr. The heart rate, respiration, and O_2 saturation of the subject is monitored and recorded continuously using a pulse oximeter (Nellcor Instruments).

Scanning starts at 60 min post FDG injection. These PET studies were conducted on the ECAT-713 (CTI PET Systems Inc., Knoxville, TN), a high-resolution PET tomograph dedicated to animal imaging with resolutions of 4 mm in plane and 4.5 mm axially (Cutler et al., 1992). The scanner has a transaxial field of view of 40 cm and an axial field of view of 5.4 cm, dimensions that are sufficient to image the entire monkey brain. In standard mode, the ECAT-713 produces 15 transaxial images with plane separation of 3.38 mm. An interleaved data acquisition mode using two bed positions 1.69 mm apart was used. Twelve 5-min frames were acquired alternating between bed positions, producing 30 min of data in each position. A calculated geometric attenuation method (Siegel and Dahlbom, 1992) was used to correct for photon attenuation. Data were reconstructed using filtered backprojection with a Shepp filter (cutoff = 0.6x Nyquist frequency) for interleaved dynamic data sets and a Shepp filter (cutoff = Nyquist frequency) for summed data sets.

D. Calculation of CMRGlc

Calculations of CMRGlc for monkey FDG studies are based on the operational equation of Phelps et al. (1979). Blood samples are centrifuged immediately, and plasma supernatant is used to measure radiotracer concentration using an automated gamma well counter (Searle Analytic, Inc.). The typical rate constants used in the equation are those obtained from dynamic human studies at UCLA (gray matter [min^{-1}]: $k_1 = 0.102$, $k_2 = 0.130$, $k_3 = 0.062$, $k_4 = 0.0068$; white matter [min^{-1}]: $k_1 = 0.054$, $k_2 = 0.109$, $k_3 = 0.045$, $k_4 = 0.0058$) (Phelps et al., 1979). The lumped constant is 0.340 (Kennedy et al., 1978).

E. Region of Interest Selection

Regions of interest (ROIs) were determined by an investigator trained in neuroanatomy. Template ROIs were created based on age-matched magnetic resonance (MR) images and primate neuroanatomical atlases of serial transaxial Nissl-stained slices. Taking into account the possible partial volume (PV) effects introduced by the resolution limitations of the scanner, ROIs were drawn on structures that were large enough to be clearly identified and relatively insensitive to PV effects. ROIs include frontal, parietal, temporal, and occipital cortices, body of caudate, striatum, thalamus, cerebellar hemispheres, and whole brain. Template ROIs are viewed and realigned onto the summed PET images and then copied onto parametric images. If the ROI template does not outline the structure appropriately, the template may be rotated to achieve the best visual fit. The average pixel value of all pixels contained within the ROI was taken as the ROI value. If a region appears in more than one plane, a weighted average (based on area) is calculated from the ROI values of planes where the region was clearly visible. Whole brain CMRGlc is calculated as the weighted average of all planes found in the supratentorial region.

F. Plasma Glucose Measurement

Plasma glucose concentrations were determined using blood samples obtained at 0, 5, 10, 20, and 40 min post FDG injection by a glucose diagnostic assay (Sigma, St. Louis, MO) as measured on a spectrophotometer (Spectonic 100, Bausch and Lomb, Rochester, NY). An average of the plasma glucose measurements at 0 and 5 min post injection was used in the calculation of CMRGlc.

G. Data Analysis

Intrarater reliability was assessed by having one rater (A.M.) redraw the 20 ROIs on eight summed FDG-PET image sets and correlating the lCMRGlc values to those initially obtained, using a Pearson product-moment correlation.

To determine if the method provided reproducible data, multiple qualitative scans of an adult male subject over a 6-month period were obtained. In order to compare scans, each ROI was normalized to the whole brain value. Normalized ratios were evaluated using a two, way ANOVA with repeated measures.

Subjects were separated into two age groups, neonatal (≤ 4 months of age) and infant (6-8 months of age), based on the development of social and motor skills (Fairbanks and McGuire, 1988). Age-related changes of CMRGlc were determined using a two-tailed, two-sample Student's t test with equal variance.

All statistical analyses were evaluated with $\alpha = 0.05$. Values are presented as mean ± standard deviation.

III. RESULTS

All subjects completed the study without any complications and were returned safely to the social colony. Agitation levels of the subjects were recorded as low or moderate.

A. ROI Analysis

Intrarater analysis showed a high correlation between ROI values from the same images analyzed on separate days ($r = 0.962$, $p < 0.05$).

Current research using PET to determine changes in function of specific brain structures usually define these areas by registration of the PET images with the MR images of the subject. Even though this method provides accurate results for human studies, its application in nonhuman primate studies is questionable due to the small brain size. Studies (Black et al., 1996) show that coregistration of [^{15}O]H$_2$O PET and MR images of baboon brain using automated image registration (AIR) (Woods et al., 1993) introduce a mean registration error of 2.9 mm, an error that is unacceptable in these studies. Also, it is impractical to conduct MR imaging on each subject at each age in longitudinal studies due to the high cost and the possible compromise of the conscious FDG-PET protocol.

B. Reproducibility of CMRGlc Pattern in Adult Vervet Monkey

In order to claim that this method is sensitive to age-related changes in CMRGlc, it is necessary to demonstrate reproducible data in normal adult animals that should have stable patterns of lCMRGlc. ROI values from multiple FDG-PET scans of one adult subject were evaluated by a two-way ANOVA with repeated measures and showed no significant main or interactive effects between ROI values across time ($F_{5,45} = 1.626$, $p > 0.05$). Coefficient of variation of each ROI over time ranged from 1.6 (frontal cortex) to 9.3% (cerebellar hemispheres) with an average of 5.3 ± 1.6% (Fig. 3).

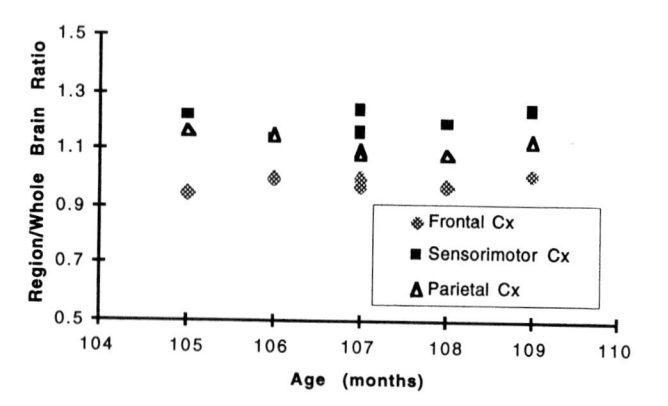

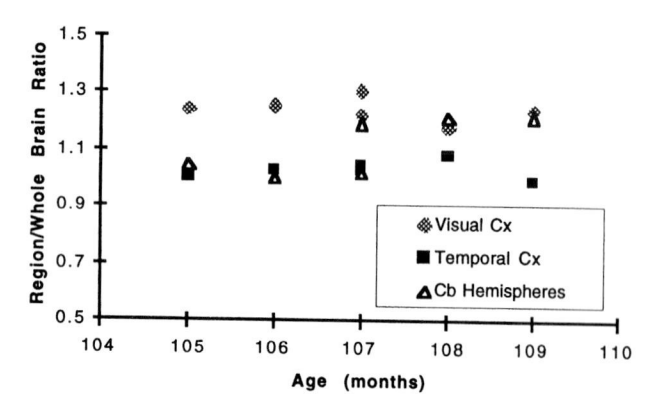

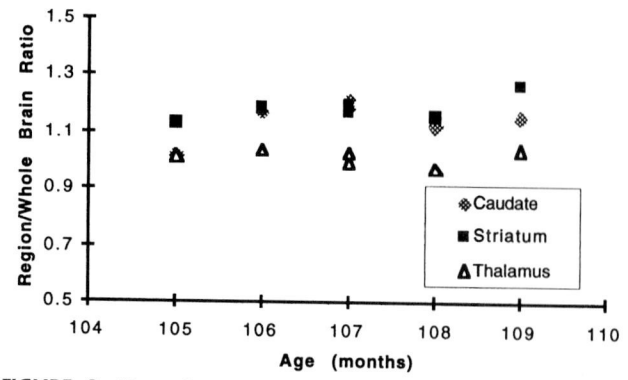

FIGURE 3 Normalized ROI values from an adult male monkey scanned multiple times over a 6-month period. Due to the age of the animal, quantitative scans were not attained. All ROI values were normalized to the whole brain value. (Top) Frontal cortex, sensorimotor cortex, and parietal cortex. (Center) Visual cortex, temporal cortex, and cerebellar hemisphere. (Bottom) Body of caudate, striatum, and thalamus. Two-way ANOVA with repeated measures showed no significant main effect or interactions ($p > 0.05$). Coefficient of variation of each ROI value across scans ranged from 1.6 to 9.3% with an average of 5.3% ± 1.6.

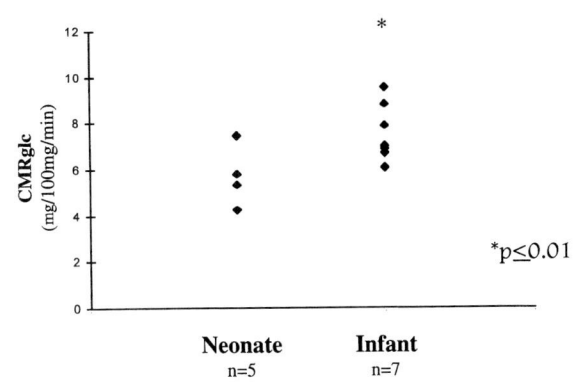

FIGURE 4 Absolute whole brain cerebral metabolic rate of glucose in conscious neonatal (≤ 4 months) and infant (6–8 months) monkeys. Analysis with a two-tailed, two-sample Student's *t* test indicated that infant whole brain glucose metabolism is significantly higher (*$p \leq 0.01$*).

C. CMRGlc in Neonates and Infants

Analysis of CMRGlc of neonates (≤ 4 months) compared to infants (6–8 months) demonstrates a significant increase in the infant group (Figs. 4 and 5, see color insert). These results agree with previous studies in humans (Chugani and Phelps, 1986, Kinnala *et al.*, 1996) and other animal models (Abrams *et al.*, 1984; Chugani *et al.*, 1991; Kennedy *et al.*, 1982; Nehlig *et al.*, 1988) that show a developmental increase in CMRGlc. This increase in glucose utilization corresponds to a period of neuroplasticity characterized by synaptogenesis (Rakic *et al.*, 1986; Zecevic *et al.*, 1989; Zecevic and Rakic, 1991), dendritic growth, and development of complex behaviors (Fairbanks and McGuire, 1988; Lancaster, 1971; McGuire, 1974; Struhsaker, 1971).

IV. CONCLUSION

The present study is the first to validate an FDG–PET protocol in conscious infant monkeys, to demonstrate reproducibility of adult monkey lCMRGlc data, and to present absolute CMRGlc values in conscious neonatal and infant vervet monkeys. The protocol used in this study was designed to avoid factors (sedation and stress) that are known to affect CMRGlc. Results indicate that instrumentation and analysis provide reliable and reproducible data and that this method is sensitive to changes in CMRGlc. Therefore, this method may be used to determine the normal metabolic development of the brain in nonhuman primates.

Acknowledgments

We thank D. Pollack, J. Edwards, W. Ladno, A. Oshiro, R. Sumida, N. Doshi, M. Prins, and D. Kozlowski for their time and assistance during these studies.

References

Abrams, R. M., Ito, M., Frisinger, J. E., Patlak, C. S., Pettigrew, K. D., and Kennedy, C. (1984). Local cerebral glucose utilization in fetal and neonatal sheep. *Am. J. Physiol.* **246**: R608–R618.

Black, K. J., Videen, T. O., and Perlmutter, J. S. (1996). A metric for testing the accuracy of cross-modality image registration: Validation and application. *J. Comput. Assist. Tomogr.* **20**: 855–861.

Bush, M., Custer R., Smeller, J., and Bush, L. M. (1977). Physiological measures of non-human primates during physical restraint and chemical immobilization. *J. Am. Vet. Med. Assoc.* **171**: 866–870.

Chugani, H. T., and Phelps, M. E. (1986). Maturational changes in cerebral function in infants determined by 18FDG positron emission tomography. *Science* **231**: 840–843.

Chugani, H. T., Phelps, M. E., and Mazziotta, J. C. (1987). Positron emission tomography study of human brain functional development. *Ann. Neurol.* **22**: 487–497.

Chugani, H. T., Hovda, D. A., Villablanca, J. R., Phelps, M. E., and Xu, W. F. (1991). Metabolic maturation of the brain: A study of local cerebral glucose utilization in the developing cat. *J. Cereb. Blood Flow Metab.* **11**: 35–47.

Crosby, G., Crane, A. M., and Sokoloff, L. (1982). Local changes in cerebral glucose utilization during ketamine anesthesia. *Anesthesiology* **56**: 437–443.

Cutler, P. D., Cherry, S. R., Hoffman, E. J., Digby, W. M., and Phelps, M. E. (1992). Design features and performance of a PET system for animal research. *J. Nucl. Med.* **33**: 595–604.

Eberling, J. L., Roberts, J. A., De Manincor, D. J., Brennan, K. M., Haurahan, S. M., Vanbrocklin, H. F., Roos, M. S., and Jagust, W. J. (1995). PET studies of Cerebral glucose metabolism in conscious rhesus macaques. *Neurobiol. Aging* **16**: 825–832.

Fairbanks, L. A., and McGuire, M. T. (1988). Long-term effects of early mothering behavior on responsiveness to the environment in vervet monkeys. *Dev. Psychobiol.* **21**: 711–724.

Gur, R. C., Gur, R. E., Resnick, S. M., Skolnick, B. E., Alavi, A., and Reivich, M. (1987). The effect of anxiety on cortical cerebral blood flow and metabolism. *J. Cereb. Blood Flow Metab.* **7**: 173–177.

Jacobs, B., Chugani, H. T., Allada, V., Chen, S., Phelps, M. E., Pollack, D. B., and Raleigh, M. J. (1995). Developmental changes in brain metabolism in sedated rhesus macaques and vervet monkeys revealed by positron emission tomography. *Cereb. Cortex* **5**: 222–233.

Kennedy, C., Sakurada, O., Shinohara, M., Jehle, J., and Sokoloff, L. (1978). Local cerebral glucose utilization in the normal conscious macaque monkey. *Ann. Neurol.* **4**: 293–301.

Kennedy, C., Sakurada, O., Shinohara, M., and Miyaoka, M. (1982). Local cerebral glucose utilization in the newborn macaque monkey. *Ann. Neurol.* **12**: 333–340.

Kinnala, A., Suhonen-Polvi, H., Aarimaa, T., Kero, P., Korvenranta, H., Ruotsalainen, U., Bergman, J., Haaparanta, M., Solin, O., Nuutila, P., and Wegelius, U. (1996). Cerebral metabolic rate for glucose during the first six months of life: An FDG positron emission tomography study. *Arch. Dis. Child. Fetal Neonatal. Ed.* **74**: F153–F157.

Lancaster, J. B. (1971). Play-mothering: The relations between juvenile females and young infants among free-ranging vervet monkeys (*Cercopithecus aethiops*). *Folia Primatol.* **15**: 163–182.

McGuire, M. T. (1974). The St. Kitts vervet (*Cercopithecus aethiops*). *J. Med. Primatol.* **3**: 285–297.

Nehlig, A., Pereira de Vasconcelos, A., and Boyet, S. (1988). Quantitative autoradiographic measurement of local cerebral glucose utilization in freely moving rats during postnatal development. *J. Neurosci.* **8**(7): 2321–2333.

Phelps, M. E., Huang, S. C., Hoffman, E. J., Selin, C., Sokoloff, L., and Kuhl, D. E. (1979). Tomographic measurement of local cerebral glucose metabolic rate in humans with (F-18)2-fluoro-2-deoxy-D-glucose: Validation of method. *Ann. Neurol.* **6**: 371–388.

Rakic, P., Bourgeois, J. P., Eckenhoff, M. F., Zecevic, N., and Goldman-Rakic, P. S. (1986). Concurrent overproduction of synapses in diverse regions of the primate cerebral cortex. *Science* **232**: 232–235

Saija, A., Princi, P., De Pasquale, R., and Costa, G. (1989). Modifications of the permeability of the blood-brain barrier and local cerebral metabolism in pentobarbital- and ketamine-anaesthetized rats. *Neuropharmacology* **28**: 997–1002.

Siegal, S., and Dahlbom, M. (1992). Implementation and evaluation of a calculated attenuation correction for PET. *IEEE Trans. Nucl. Sci.* **39**: 1117–1121.

Struhsaker, T. T. (1967). Social structure among vervet monkeys (*Cercopithecus aethiops*). *Behaviour* **29**: 6–121

Struhsaker, T. (1971). Social behavior of mother and infant vervet monkey (*Cercopithecus aethiops*). *Anim. Behav.* **19**: 223–250.

Woods, R. P., Mazziotta, J. C., and Cherry, S. R. (1993). MRI-PET registration with automated algorithm. *J. Comput. Assist. Tomogr.* **17**: 536–546.

Zecevic, N., and Rakic, P. (1991). Synaptogenesis in monkey somatosensory cortex. *Cereb. Cortex* **1**: 510–523.

Zecevic, N., Bourgeois, J. P., and Rakic, P. (1989). Changes in synaptic density in motor cortex of rhesus monkey during fetal and postnatal life. *Brain Res. Dev. Brain Res.* **50**: 11–32.

CHAPTER 27

Comparison of Ketamine/Midazolam versus Pentobarbital on [¹⁸F]Fluorodopa PET Kinetics in Monkeys

DAVID B. STOUT,*,† **SUNG-CHENG HUANG,***,† **MICHAEL J. RALEIGH,**‡,¶,§
MICHAEL E. PHELPS,*,† **and JORGE R. BARRIO***,†

Departments of **Molecular and Medical Pharmacology*
†*Laboratory of Structural Biology and Molecular Medicine*
‡*Brain Research Institute,* ¶*Psychiatry*
UCLA School of Medicine, Los Angeles, California 90095
§*Sepulveda Veterans Administration Medical Center*
Sepulveda, California 91343

A combination of ketamine and midazolam (K/M) produced a safer, shorter-acting anesthesia than pentobarbital (PB) in both vervet and squirrel monkeys. [¹⁸F]Fluorodopa positron emission tomography kinetics were examined in 23 vervet and 18 squirrel monkeys using PB and K/M. Relative to PB, K/M reduced recovery times by 32% in squirrel monkeys and 48% in vervets. Systemic metabolism of [¹⁸F]fluorodopa was unchanged in both species. K_1, a measure of the blood–brain barrier transport rate of [¹⁸F]fluorodopa, was slightly lower with K/M compared to PB in vervet (10%, $p = 0.05$) and significantly lower in squirrel monkeys (32%, $p < 0.01$). K_i, the [¹⁸F]fluorodopa uptake constant, was unchanged in vervet and slightly lower with K/M in squirrel monkeys (19%, $p = 0.10$). K/M produced effective immobilization, was suitable for measuring [¹⁸F]fluorodopa kinetics, and significantly reduced the time needed for animal recovery.

I. INTRODUCTION

Pentobarbital (PB) is widely used to provide anesthesia and inhibit movement for animal positron emission tomography (PET) research. Unfortunately, PB has several disadvantages, including respiratory depression and a relatively long recovery time, approximately 3 hr in vervet monkeys and 4 hr in squirrel monkeys. A shorter-acting, safer anesthetic is the commonly used drug ketamine. Used alone, ketamine does not completely inhibit movement; however, coadministration with midazolam effectively produces complete sedation (Jacobs *et al.*, 1993). Nonetheless, there are concerns that different anesthetics may differentially affect tracer kinetics in the brain. Therefore, [¹⁸F]fluorodopa metabolism, tracer kinetics, and recovery times were compared following K/M and PB anesthesia in two commonly used monkey species.

II. MATERIAL AND METHODS

Twenty-three vervet monkeys (*Cercopithecus aethiops sabaeus*) and 18 squirrel monkeys (*Saimiri sciureus*) received an initial dose of ketamine (10 mg/kg im) in order to establish venous and arterial catheters. Eighteen vervets and 8 squirrel monkeys were scanned using intravenous PB (10–15 mg/kg/hr). Five vervets and 10 squirrel monkeys received intravenous K/M (20–35 and 0.2–0.3 mg/kg/hr of anesthetic, respectively). For vervets, approximately every 15 min the anesthetics were diluted into 3 cc of heparinized saline and injected over 10 sec. Squirrel monkeys received an infusion of ketamine (35 and 25 mg/kg/hr for the first and second hours, respectively) with intravenous midazolam injections every 15 min. [¹⁸F]Fluorodopa PET scans were acquired using a standard protocol for all animals (Barrio *et al.*, 1990). Plasma samples were obtained to measure plasma ¹⁸F radioactivity as a function of time and to determine large neutral amino acid concentrations. *In vivo* tracer metabolism was determined by HPLC separation of several plasma samples and was used to separate total ^{18}F plasma curves into [^{18}F]fluorodopa and peripherally formed 3-O-methyl-[^{18}F]fluorodopa components (Huang *et al.*, 1991; Melega *et al.*, 1991). K_1 and K_i kinetic values were determined by Patlak analysis (Patlak *et al.*, 1983) using the plasma curve and PET region of interest data from 0 to 7.5 and 30 to 120 min, respectively. The relationship between [^{18}F]fluorodopa uptake kinetics and large neutral amino acid (LNAA) concentrations has been characterized (Stout *et al.*, 1998) as a linear relationship between kinetic estimates (K_1 and K_i) and LNAA concentrations. This linear relationship can be used to adjust the kinetic estimates of K_1 and K_i to a common reference LNAA value, thus removing the effects of LNAAs on these estimates. Correlations of LNAAs to K_1 and K_i and the adjustment factors were determined separately for each species and anesthetic agent.

III. RESULTS

Recovery times and K_1 and K_i values for both species are shown in Tables 1 and 2. Animal recovery times, as defined by the end of the scan to the time when the animal could walk and retrieve food, decreased by 32% ($p = 0.01$) for squirrel monkeys and 48% ($p < 0.01$) for vervet monkeys using K/M compared to pentobarbital. K_1 values using K/M were slightly lower (10%, $p = 0.05$) in vervets and significantly lower in squirrel monkeys (36%, $p < 0.01$) compared to values using PB. K_i, the [^{18}F]fluorodopa uptake constant [$K_i = (K_1^* k_3)/(k_2 + k_3)$], showed no significant difference between the two anesthetic agents ($p > 0.05$ for both squirrel and vervet monkeys). In both species, systemic metabolism of [^{18}F]fluorodopa showed no difference in the metabolic profile between anesthetic agents (Fig. 1). All tracer metabolite fractions using K/M were within one standard deviation of the values obtained using PB. Large neutral amino acid levels were also unaffected by the anesthetic agent [vervet: PB, 424 μM (127 SD), K/M; 527 μM (137), $p > 0.05$; squirrel: PB, 338 μM (127), K/M, 373 μM (128), $p > 0.05$].

IV. DISCUSSION

Ketamine/midazolam anesthesia is effective and safer than PB for maintaining anesthesia in vervet and squirrel monkeys. Although the effect of ketamine on cerebral blood flow (CBF) is variable (Bjorkman *et al.*, 1992; Strebel *et al.*, 1995; Werner, 1995), the effects of CBF on [^{18}F]fluorodopa kinetics are small due to the small first-pass extraction fraction of this tracer (Huang and Phelps, 1986).

Kinetic analysis showed a difference in K_1 estimates between species. The vervet monkeys had a slight K_1 reduction with K/M compared to PB (10%), whereas the squirrel monkey had a much larger K_1 reduction (36%). The estimates of K_i were not significantly different between the two anesthetic agents for either species; however, the larger reduction (19%) in the

TABLE 1 Squirrel Monkey Recovery Times and K_1 and K_i Values[a]

Pentobarbital				Ketamine/midazolam			
Animal no.	Recovery time (hr)	K_1 (ml/min/g)	K_i (ml/min/g)	Animal no.	Recovery time (hr)	K_1 (ml/min/g)	K_i (ml/min/g)
182	2.8	0.059	0.020	316	2.8	0.020	0.008
191	2.7	0.063	0.016	317	3.7	0.047	0.010
219	4.6	0.059	0.014	360	2.2	0.039	0.023
221	3.4	0.061	0.016	361	1.9	0.060	0.017
224	5.2	0.059	0.018	363	2.4	0.043	0.014
225	2.3	0.053	0.015	366	1.5	0.046	0.014
233	4.6	0.052	0.012	426	3.0	0.041	0.011
660	4.7	0.099	0.023	427	2.1	0.034	0.012
				428	3.6	0.037	0.014
Average	3.8	0.063	0.017	429	2.7	0.033	0.012
SD	1.1	0.015	0.003				
T test				Average	2.6	0.040	0.014
$p =$	0.01	0.004	0.10	SD	0.7	0.010	0.004

[a] T test p values are shown for comparisons made between the two anesthetic agents.

TABLE 2 Vervet Monkey Recovery Times and K_1 and K_i Values[a]

| Pentobarbital |||| Ketamine/midazolam ||||
Animal No.	Recovery time (hr)	K_1 (ml/min/g)	K_i (ml/min/g)	Animal No.	Recovery time (hr)	K_1 (ml/min/g)	K_i (ml/min/g)
7909	2.5	0.033	0.0066	8933	1.2	0.031	0.0086
9032	1.6	0.030	0.0101	8614	0.8	0.030	0.0079
8822	1.5	0.032	0.0091	9199	1.4	0.026	0.0092
8718	3.0	0.031	0.0078	9140	1.5	0.028	0.0086
9024	2.7	0.032	0.0085	8614	1.9	0.027	0.0081
9006	2.9	0.027	0.0057				
8616	4.1	0.034	0.0099	Average	1.4	0.028	0.0085
8719	2.4	0.026	0.0067	SD	0.4	0.002	0.0005
8930	2.0	0.032	0.0083	T test			
8718	2.4	0.030	0.0074	$p =$	0.004	0.05	0.42
8616	3.2	0.036	0.0095				
8939	2.6	0.030	0.0067				
8820	2.0	0.030	0.0072				
8823	3.0	0.034	0.0084				
8618	4.7	0.025	0.0065				
8820	2.5	0.034	0.0087				
7905	1.5	0.036	0.0086				
9037	2.6	0.034	0.0084				
Average	2.6	0.031	0.0080				
SD	0.8	0.003	0.0013				

[a] T test p values are shown for comparisons made between the two anesthetic agents.

squirrel monkey K_i estimate was probably related to the significant decrease in K_1. Mechanisms that cause the difference in K_1 between the two anesthetics are still unclear and further study is needed to address this issue. Changing anesthetic agents during a series of experiments might be feasible with vervet monkeys; however, it would not be recommended with squirrel monkeys. Although there are anesthesia-induced differences in K_1 values, K_i values are still expected to be valid for assessing the functional activity of the dopaminergic system as long as a single anesthetic agent is used for all experiments.

Systemic [^{18}F]fluorodopa metabolism showed no change using K/M compared with PB. Other investigators have also shown that ketamine does not alter striatal dopamine metabolism (Koshikawa et al., 1988). Recovery times were substantially shortened with K/M (squirrel monkey, 32%; vervet monkey, 48%), reducing the time the animal is sedated and the time needed to complete the research protocol. In conclusion, K/M is a shorter-acting anesthetic agent than PB, does not alter the peripheral or central metabolism of [^{18}F]fluorodopa, and can be used in experiments to measure cerebral dopaminergic function in squirrel and vervet monkeys.

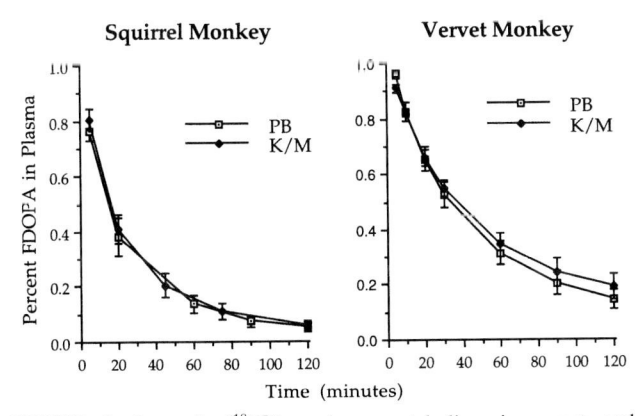

FIGURE 1 Systemic [^{18}F]fluorodopa metabolism in vervet and squirrel monkeys anesthetized with pentobarbital (PB) and ketamine/midazolam (K/M).

Acknowledgments

The authors thank Dr. Satyamurthy, Joe Cook, and the cyclotron staff for the production of [^{18}F]fluorodopa. We also thank Ron Sumida, Judy Edwards, Waldemar Ladno, and Alan Oshiro for their long hours and expertise with the PET procedures. This work was made possible by the financial support from the National Institutes of Health (RO1 NS 33356), the Department of Energy (DE FC0387-ER60615), and the Dana Foundation.

References

Barrio, J. R., Huang, S. C., Melega, W. P., Yu, D. C., Hoffman, J. M., Schneider, J. S., Satyamurthy, N., Mazziotta, J. C., and Phelps, M. E. (1990). 6-[F-18]Fluoro-L-DOPA probes dopamine turn-over rates in central dopaminergic systems. *J. Neurosci. Res.* **27**: 487–493.

Bjorkman, S., Akeson, J., Nilsson, F., Messeter, K., and Roth, B. (1992). Ketamine and midazolam decrease cerebral blood flow and consequently their own rate of transport to the brain: An application of mass balance pharmacokinetics with a changing regional blood flow. *J. Pharmacokinet. Biopharm.* **20**: 637–652.

Huang, S. C., and Phelps, M. E. (1986). Principles of tracer kinetic modeling in positron emission tomography and autoradiography. *In* "Positron Emission Tomography and Autoradiography" (M. E. Phelps, J. C. Mazziotta, and H. R. Schelbert, eds.), pp. 316–317. Raven Press, New York.

Huang, S. C., Yu, D. C., Barrio, J. R., Grafton, S., Melega, W. P., Hoffman, J. M., Satyamurthy, N., Mazziotta, J. C., and Phelps, M. E. (1991). Kinetics and modeling of L-6-[F-18]fluoro-dopa in human positron emission tomographic studies. *J. Cereb. Blood Flow Metab.* **11**: 898–913.

Jacobs, B., Chugani, H. T., Allada, V., Chen, S., Phelps, M. E., Pollack, D. B., and Raleigh, M. J. (1993). Developmental changes in brain metabolism in sedated rhesus macaques and vervet monkeys revealed by positron emission tomography. *Am. J. Primatol.* **29**: 291–298.

Koshikawa, N., Tomiyama, K., Omiya, K., and Kobayashi, M. (1988). Ketamine anesthesia has no effect on striatal dopamine metabolism in rats. *Brain Res.* **444**: 394–396.

Melega, W. P., Grafton, S. T., Huang, S. C., Satyamurthy, N., Phelps, M. E., and Barrio, J. R. (1991). L-6-[F-18]Fluoro-DOPA metabolism in monkeys and humans: Biochemical parameters for the formulation of tracer kinetic models with positron emission tomography. *J. Cereb. Blood Flow Metab.* **11**: 890–897.

Patlak, C. S., Blasberg, R. G., and Fenstermacher, J. D. (1983). Graphical evaluation of blood-to-brain transfer constants from multiple-time uptake data. *J. Cereb. Blood Flow Metab.* **3**: 1–7.

Stout, D. B., Huang, S.-C., Melega, W. P., Raleigh, M. J., Phelps, M. E., and Barrio, J. B. (1998). Effects of large neutral amino acid concentrations on 6[F-18]fluoro-L-DOPA kinetics. *J. Cereb. Blood Flow Metab.* **18**: 43–51.

Strebel, S., Kaufmann, M., Maitre, M., and Schaefer, H. G. (1995). Effects of ketamine on cerebral blood flow velocity in humans. Influence of pretreatment with midazolam or esmolol. *Anaesthesia* **50**: 223–229.

Werner, C. (1995). Effects of analgesia and sedation on cerebral circulation, cerebral blood volume, cerebral metabolism and intracranial pressure. *Anaesthesist* **44** (Suppl.): S566–S572.

CHAPTER

28

Metabolism of [^{18}F]Fluorodopa in Pig Brain Estimated by PET

E. H. DANIELSEN,* D. F. SMITH,*,† A. D. GEE,* T. K. VENKATACHALAM,* S. B. HANSEN,*
and A. GJEDDE*

*PET Center, Aarhus University Hospitals
DK-8000 Aarhus C, Denmark
†Department of Biological Psychiatry
Aarhus University Hospitals
DK-8240 Risskov, Denmark

The use of animals with large brains has technical, economic, and ethical ramifications in neuroscience. The aims of this study were (1) to determine the suitability of the pig as an experimental model in brain research and (2) to establish normal values of [^{18}F]fluoroDOPA (FDOPA) uptake and metabolism in the brain of the pig. Female Danish country-bred pigs were scanned dynamically using FDOPA and the Siemens/CTI ECAT EXACT HR47 positron emission tomography (PET) scanner. Volumes of interest were identified by region growing from the maximum value of striatal activity. They contained all connected voxels with greater than 90% of the maximum voxel activity. The FDOPA net influx rate constant (k_3^s) was calculated for the striatum and frontal cortex, using the cerebellum as reference and multiple-time graphical analysis, and was found to be 0.0079 ± 0.001 (min^{-1}). In striatum the net plasma clearance rate $K[= K_1^D k_3^D / (k_2^D + k_3^D)]$ at steady state equaled 0.0077 ± 0.0009 (ml g^{-1} min^{-1}). In striatum the unidirectional tracer transfer coefficient (K_1^D) was 0.029 ± 0.003 (ml g^{-1} min^{-1}), and the metabolic rate constant (k_3^D) was 0.018 ± 0.002 (min^{-1}). These parameters were estimated by multicompartmental analysis using the individual estimates of the cerebellum partition (V_e) and striatal plasma (V_p) volumes, corrected for the presence of [^{18}F]3-O-methyl-fluoroDOPA. Results show that anesthetized pigs are appropriate experimental animals for PET imaging of decarboxylation in the living brain. The striatal DOPA decarboxylation rate was found to range between those of rats and humans and to be equal to the rate constants found in primates. Results suggest that the partial volume effect in the striatum is directly related to the size of the striatum.

I. INTRODUCTION

The use of animals with large brains in neuroscience has technical, economic, and ethical ramifications (Carey, 1996). The use of pigs as neuronal xenotransplant donors (Deacon *et al.*, 1997) and as models of normal and pathological cerebral metabolism (Andrews *et al.*, 1993) gives this species a special role in studies of central dopamine metabolism. Specifically, the authors wished to test the claim that the size and physiology of the animal makes it a suitable model of DOPA metabolism in humans.

The radiolabeled compound [^{18}F]6-fluoro L-3,4-dihydroxyphenylalanine (FDOPA) is a tracer of choice for the study of normal and pathological dopamine (DA) synthesis and metabolism with positron emission tomography (PET) in humans (Leenders, 1990; Gjedde *et al.*, 1991; Kuwabara *et al.*, 1993; Hoshi *et al.*, 1993), in primates (Firnau *et al.*, 1987; Doudet *et al.*, 1991, 1992; Melega *et al.*, 1991; Pate *et al.*, 1993), and in rodents (Cumming *et al.*, 1987, 1988, 1992, 1994; Reith *et al.*, 1990). Following intravenous administration of FDOPA, radioactivity accumulates in the striatum. This accumulation is influenced by diseases in the basal ganglia (Brooks, 1993; Snow and Calne, 1995).

The aims of this study were (1) to determine the suitability of the pig as an experimental model in brain research and (2) to establish normal values of FDOPA uptake and metabolism in the brain of the pig.

II. METHODS

A. Experimental Animals

Female Danish country-bred pigs (immature cross of Yorkshire/Landrace/Hampshire/Dyroc), weighing 39–44 kg and aged 2–4 months, were housed singly in stalls in a thermostatically controlled (20°C) animal colony with natural lighting. The pigs were deprived of food 24 hr prior to the experiments but had access to water.

B. Operative Procedures

On arrival at the scanning suite, the eight pigs were sedated by a subcutaneous (sc) injection of a mixture of 4 ml (0.5 mg/kg) midazolam (5 mg/ml) and 8 ml (10 mg/kg) ketamine HCl (50 mg/ml). The pigs were then intubated and continuously ventilated with air (Engstrøm Erica) mixed with 1–2.5% isoflurane. Catheters (Avanti) size 4F-7F) were surgically installed in a femoral artery for blood sampling and blood pressure monitoring and in a femoral vein for tracer injection and continuous infusion of isotonic saline (100 ml/hr) and 5% glucose (20 ml/hr). Animals were placed supine in the tomograph using a custom-made head-holding device. Physiological functions (i.e., blood pressure, heart rate, expired air CO_2, ECG) were monitored (Kiwex instruments). Body temperature was thermostatically maintained in the normal range (38.5–39.5°C) using a thermostatically controlled heating blanket. At the end of the experiments, the anesthetized pigs were sacrificed by an intravenous (iv) injection of 20 ml saturated potassium chloride in accordance with the regulations of the Danish National Committee for Ethical Animal Experiments. Inhibition of extracerebral aromatic amino acid decarboxylase (AAAD) activity was performed by two methods: (1) 150 mg carbidopa in tablets (kindly provided by Merck, Sharp & Dohme) dissolved in water and administered through a gastric tube 1–1.5 hr before FDOPA injection or (2) 150 mg carbidopa in powder (Merck, Sharp & Dohme) dissolved by slow heating in water and administered iv 30 min before FDOPA injection.

C. Tomography

The animals were studied using a Siemens/CTI ECAT EXACT HR47 tomograph. This tomograph allows simultaneous acquisition of two-dimensional (2D) or 3D data from 47 sections each 3.1 mm thick. Each PET frame was reconstructed to a 128 × 128 matrix of 2 × 2-mm pixels with a hanning filter (cutoff of 0.5 pixels^{-1}), for an image resolution of 4.6 mm full-width-at-half-maximum (FWHM) in 3D mode. A transmission scan was performed prior to injection. Scanning commenced upon injection of 200–346 MBq FDOPA into the femoral vein. A total of 28 frames (6 × 30 sec, 7 × 60 sec, 5 × 2 min, 4 × 5 min, 5 × 10 min, 1 × 30 min) were acquired during 120 min.

Arterial blood samples were drawn for determination of total plasma radioactivity concentration of ^{18}F-labeled compounds (40 samples). Selected samples were drawn for measurement of pH, pCO_2, PO_2, HCO_3 concentration, and O_2 saturation during the 120 min following tracer injection to ensure uniform conditions.

D. HPLC

Arterial plasma samples were collected at 2.5, 5, 15, 25, 35, 60, 90, and 120 min after injection for HPLC separation of tracer and metabolites. Cold 0.5 M perchloric acid [0.5 ml containing 0.05% (w/v) sodium thiosulfate] was added to 0.5 ml of each plasma sample, vortexed, placed on ice for at least 15 min, and centrifuged for 10 min at 14,000 g. Supernatants were analyzed by isocratic reversed-phase HPLC (4.6 × 250 mm, 4 μm, Spherisorb).

E. Data Analysis

Image analysis of brain regions was carried out using ECAT software version 7.0. Brain regions of interest (ROIs) were selected from the neuroanatomical atlas of the pig brain by Yosikawa (1968) and with the aid of MR images of the pig brain. Volumes of interest (VOIs) were created by region growing from the maximal value of striatal activity and contained all connected voxels with values greater than 90% of the maximum value. Mean VOI values were used in the analysis.

The FDOPA net influx rate constant (k_3^s) was calculated for the striatum and frontal cortex, using the cerebellum as the reference region for multiple-time graphical analysis, as described by Hoshi et al. (1993). The steady-state plasma clearance rate K was calculated as $K_1^D k_3^D / (k_2^D + k_3^D)$. The unidirectional tracer transfer coefficient (K_1^D) and metabolic rate constant (k_3^D) were estimated by multicompartmental analysis, using the individual estimates of the cerebellum partition volume (V_e) and striatal plasma volume (V_p). The latter were corrected for the presence of [^{18}F]3-O-methyl-fluoroDOPA (3OMFD) by measuring the rate constants of plasma metabolite synthesis and clearance, k_0^D and k_{-1}^M. The use of these biological constraints has been discussed in detail elsewhere (Gjedde *et al.*, 1991; Kuwabara *et al.*, 1993).

III. RESULTS AND DISCUSSION

Only two radiolabeled compounds, FDOPA and 3OMFD, were found and identified in plasma samples collected up to 120 min after injection of FDOPA. At 35 min after injection, the two compounds were equal in concentration. The time–activity curves of FDOPA and 3OMFD in plasma for the young pig are shown in Fig. 1. The relative amounts of plasma FDOPA and 3OMFD as a function of time are shown in Fig. 2.

The radioactivity recorded in three regions in a pig brain after injection of FDOPA is shown in Fig. 3. Maximum FDOPA uptake was noted in striatum. The mean estimates of K_1^D were 0.029, 0.032, and 0.026 (ml g^{-1} min^{-1}) in striatum, frontal cortex, and cerebellum, respectively. The mean estimates of k_3^D were 0.018, 0.002, and 0.00 (min^{-1}), respectively. Using the cerebellum as reference for the multiple-time graphical analysis, the FDOPA net influx rate constant (k_3^s) was

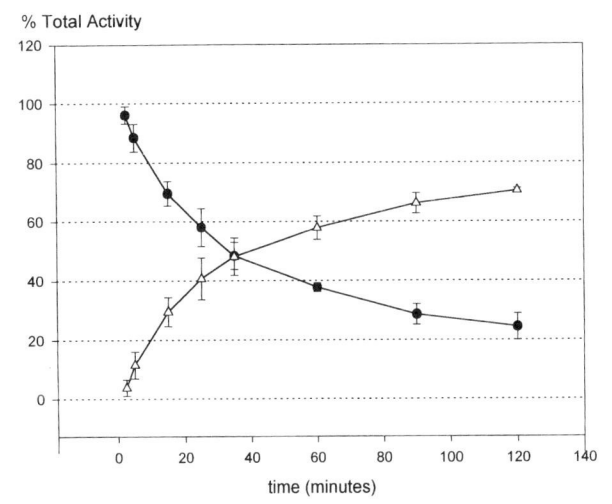

FIGURE 2 Relative amounts (mean values ± SD) of plasma FDOPA (●) and 3OMFD (△) as a function of time. FDOPA and 3OMFD were the only two radiolabeled compounds detected in plasma up to 120 min after injection of FDOPA. At 35 min after injection the two compounds were equal in concentration.

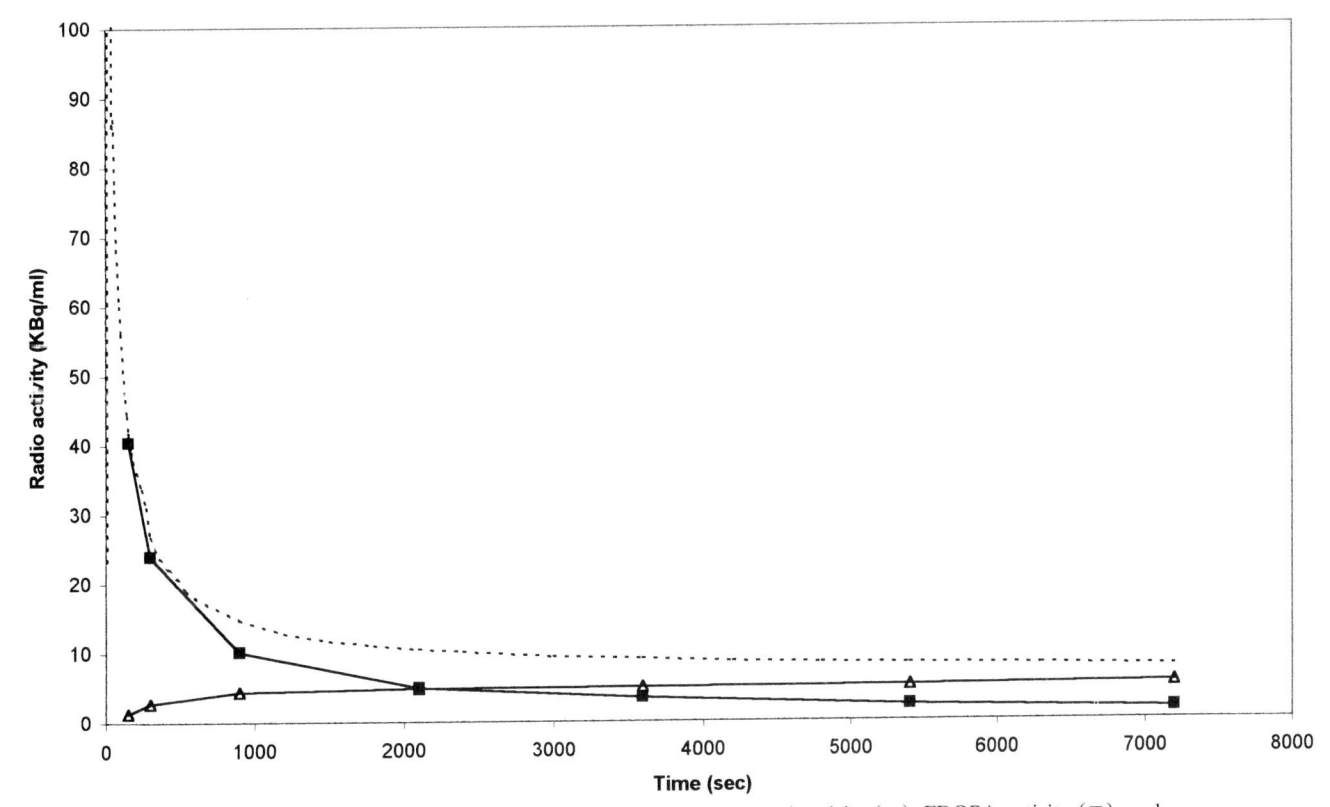

FIGURE 1 Time–activity curves in plasma for a young pig. Total activity (---), FDOPA activity (■), and 3OMFD (△) are shown as a function of time after injection. No other metabolites were detected and identified in plasma up to 120 min after injection of FDOPA.

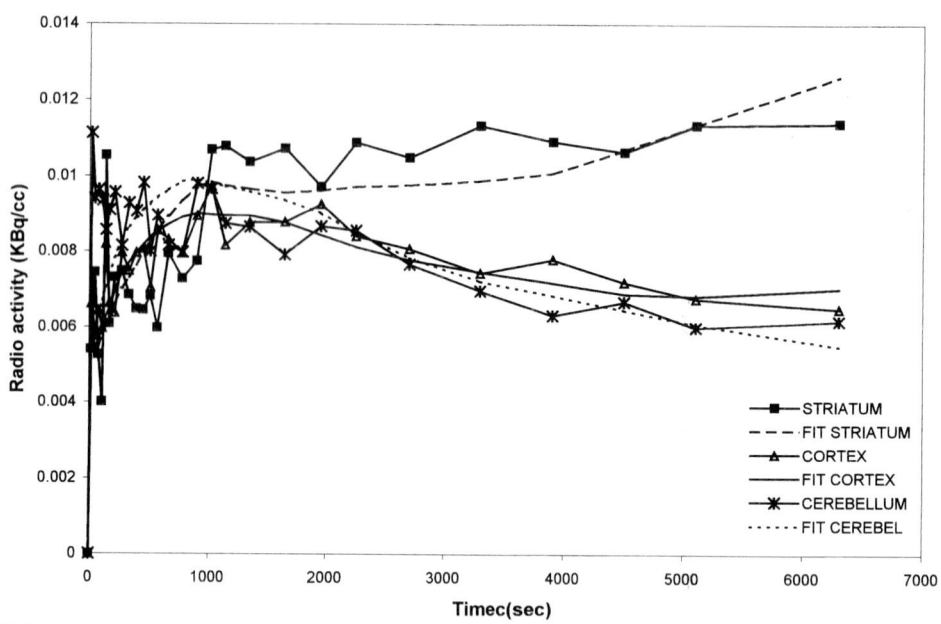

FIGURE 3 Radioactivity recorded in three regions in a pig brain as a function of time after injection of FDOPA. Using the values $q = 2.3$, $V_e = 0.593$, and $V_p = 0.03$, the tissue radioactivity in the three VOIs was subjected to nonlinear least-squares regression analysis. Lines were drawn on the basis of the parameter estimates. In this pig the estimates in striatum, frontal cortex, and cerebellum of K_1^D were 0.025, 0.026, and 0.032 (ml g^{-1} min^{-1}) and k_3^D were 0.027, 0.004, and 0.001 (min^{-1}), respectively.

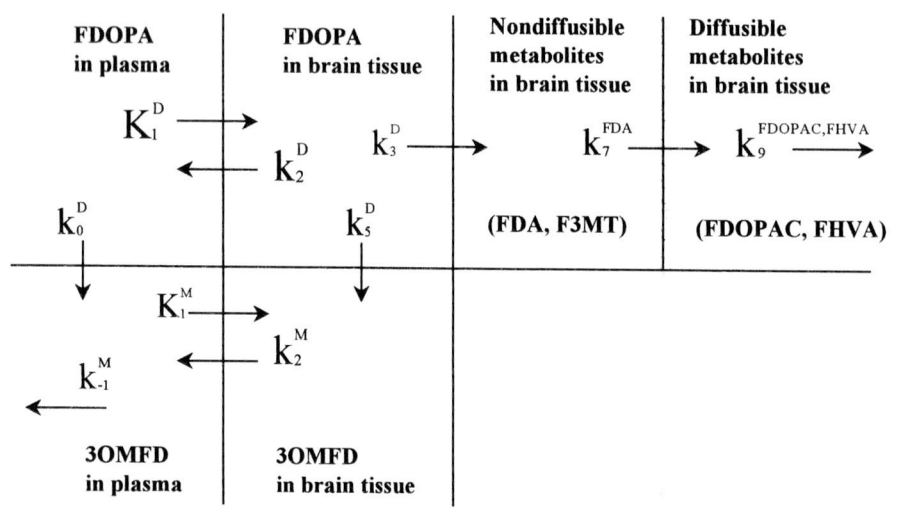

FIGURE 4 Compartments, transfer coefficients, and steps in the metabolic pathway of FDOPA in the brain *in vivo*. FDOPA in circulation is *O*-methylated by COMT at the rate of k_0^D (min^{-1}), and the product 3OMFD is eliminated from plasma at the rate of k_{-1}^M (min^{-1}). 3OMFD and FDOPA are reversibly transferred across the blood–brain barrier by saturable, sodium-independent facilitated transporters for the large neutral amino acids. The transport of 3OMFD over the blood–brain barrier is described by the rate constants K_1^M (ml g^{-1} min^{-1}) and k_2^M and of FDOPA by K_1^D (ml g^{-1} min^{-1}) and k_2^D (min^{-1}). Within the brain tissue, FDOPA is a substrate for COMT described by the rate constant k_5^D; this step is considered to be negligible in this compartment analysis. FDOPA in the brain is decarboxylated to [^{18}F]dopamine (FDA) by L-amino acid decarboxylase at the rate k_3^D (min^{-1}), corresponding to the activity of the enzyme. FDA is stored in intracellular vesicles. FDA is converted by MAO or COMT to form the fluorinated acidic metabolites [^{18}F]DOPAC and [^{18}F]HVA, respectively, at the rate of k_7^{FDA} (min^{-1}). These two acidic metabolites are eliminated to the circulation at the rate $k_9^{FDOPAC,FHVA}$.

TABLE 1   Mean FDOPA Uptake and Metabolic Rates from the Pig Brain Striatum, Cortex, and Cerebellum[a]

	Species and brain weight	Pig brain (100 g) Mean	± SEM	Rat brain (2 g) Mean	Nonhuman primate (150–200 g) mean	Human brain (1500 g) Mean
Plasma	k_0^D (ml g^{-1} min^{-1})	0.013	0.004	0.01		0.01
Plasma	k_{-1}^M (min^{-1})	0.053	0.048	0.005		0.01
Striatum	V_p (ml g^{-1})	0.032	0.007			
Striatum	k_3^s (min^{-1})	0.0079	0.0014		0.008	0.013[b]
Striatum	K (ml g^{-1} min^{-1})	0.0077	0.0009		0.0067	
Striatum	K_1^D (ml g^{-1} min^{-1})	0.029	0.003	0.04	0.0331	0.03
Striatum	k_3^D (min^{-1})	0.018	0.002	0.01	0.015	0.057
Cortex	K_1^D (ml g^{-1} min^{-1})	0.032	0.008			0.024
Cortex	k_3^D (min^{-1})	0.002	0.001			
Cerebellum	K_1^D (ml g^{-1} min^{-1})	0.026	0.003		0.038	0.03
Cerebellum	k_3^D (min^{-1})	0.0	0.001			
Cerebellum	V_e (ml g^{-1})	0.63	0.06	0.57		0.64

[a] Plasma results show the O-methylation rate of FDOPA to 3OMFD (k_0^D) and 3OMFD elimination from plasma (k_{-1}^M). For comparison, previous results from studies on rats, nonhuman primates, and humans are listed.
[b] From Danielson et al. (1997).

calculated to be 0.0079 ± 0.0014 (min^{-1}) for striatum. The steady-state plasma clearance rate in striatum was 0.0077 ± 0.0009 (ml g^{-1} min^{-1}).

Figure 4 shows the compartmental model for FDOPA in the brain *in vivo*. The mean uptake and metabolic rates for the three brain regions and plasma are shown in Table 1. Brain weights of the pig brain were approximately 100 g. The brains were approximately 8 cm long, 5 cm wide, and 4 cm high.

In rats the major metabolites from FDOPA *in vitro* have been shown to be [^{18}F]dopamine (FDA), [^{18}F]3-4-dihydroxyphenylacetic acid (FDOPAC), and [^{18}F]homovanillic acid (FHVA). In plasma, 3OMFD was also the main metabolite, whereas FDOPA disappeared rapidly. In striatum, 3OMFD was the major metabolite but was less than 50% of total activity. FDA was the next greatest metabolite followed by FDOPAC. FHVA and [^{18}F]3-methoxytyramine remained under 10% of the total activity. Using HPLC the metabolic rate of L-amino acid decarboxylase k_3^D was found to be 0.17 min^{-1} in extracts from rat brain. Data obtained by compartmental autoradiographic analysis showed k_3^D values that were 10 times lower than data obtained by HPLC (Cumming et al., 1987, 1988, 1992, 1994).

In primate brain extracts, the main metabolites of FDOPA in striatum have also been shown to be FDA (up to 90%), FDOPAC (up to 20%), and FHVA (35%) during a 3-hr study period. 3OMFD accounted for < 5% of activity in striatum and frontal cortex but was the main metabolite (60%) in occipital cortex. In temporal muscle, 79% of the activity remained from 3OMFD and inactive sulfated FDOPA. In plasma, 3OMFD was the major metabolite and reached 80% of plasma activity 120 min after injection of FDOPA. Ten percent of the activity was due to the sulfated metabolite (Firnau et al., 1986, 1987; Melega et al., 1990, 1991). PET data from vervet monkeys showed the FDOPA net influx rate constant (k_3^s) to be 0.008 (min^{-1}) (Hartvig et al., 1992) and the steady-state plasma clearance rate K to be 0.0067 ± 0.002 (ml g^{-1} min^{-1}). K_1 has been found to be 0.033 ± 0.003 (ml g^{-1} min^{-1}), and k_3^D to equal 0.015 ± 0.003 (min^{-1}) using a three-compartment model (Barrio et al., 1996).

The condition that activity in a brain region is altered by the spillover from the structure into adjacent tissue or contamination of activity from surrounding tissue is known as the partial volume effect (PVE). Activity measured in structures smaller than the resolution of the tomograph is sensitive to the PVE. Measurements of metabolic rates and tracer transfer coefficients in the striatum have been found to be lower in nonhuman primates and pigs compared to normal humans. The small size of the nonhuman striatum makes the measurement of the time–activity curves sensitive to the intrinsic resolution of the tomograph. Registration of PET images to magnetic resonance images makes it possible to correct the time–activity curves for the PVE (Rousset et al., 1993, 1996; Frost et al., 1996). Systematic investigations of the influence of the PVE on the measured metabolic rates and tracer transfer coefficients are still needed, as well as implementation of a PVE correction for the PVE into routine kinetic analysis.

IV. CONCLUSIONS

Results show that anesthetized pigs are appropriate experimental animals for PET imaging of dopamine synthesis in the living brain. Striatal DOPA decarboxylation rates were found to be between those reported in rats and humans (Reith et al., 1990; Gjedde et al., 1991; Cumming et al., 1987, 1988, 1992, 1994). Results indicate that the partial volume effect in the striatum is directly related to the size of the region.

Acknowledgments

We thank the technical staff (Bente, 2x Helle, Vikie, Gloria, John, Niels, and Torben) at the PET Center for their skillful assistance. A special thanks to Professor P. Cumming in Montreal for advice and help. This work was kindly supported by grants from the Danish Parkinson Society, Aarhus University Research Foundation, the Danish Medical Society's Research Foundation, Novo Nordic Foundation, and Institute of Experimental Clinical Research at Aarhus University.

References

Andrews, R. J., Bringas, J. R, Alonzo, G., Salamat, M. S., Khoshyomn, S., and Gluck, D. S. (1993). Corpus callosotomy effects on cerebral blood flow and evoked potentials (transcallosal diaschisis). *Neurosci. Lett.* **154**: 9–12.

Barrio, J. R., Huang, S. C., Yu, D. C., Melega, W. P., Quintana, J., Cherry, S. R., Jacobson, A., Namavari, M., Satyamurthy, N., and Phelps, M. E. (1996). Radiofluorinated L-m-tyrosines: New *in vivo* probes for central dopamine biochemistry. *J. Cereb. Blood Flow Metab.* **16**: 667–678.

Brooks, D. J. (1993). PET studies on the early and differential diagnosis of Parkinson's disease. *Neurology* **43**: S6–S16.

Carey, M. E. (1996). Lessons learned from a scientist's personal confrontation with animal zealots. *Physiologist* **39**: 1–11.

Cumming, P., Boyes, B. E., Martin, W. R., Adam, M., Grierson, J., and Ruth, T. (1987). The metabolism of [18F]6-fluoro-L-3,4-dihydroxyphenylalanine in the hooded rat. *J. Neurochem.* **48**: 601–608.

Cumming, P., Hausser, M., Martin, W. R., Grierson, J., Adam, M. J., and Ruth, T. J. (1988). Kinetics of *in vitro* decarboxylation and the *in vivo* metabolism of 2-18F- and 6-18F-fluorodopa in the hooded rat. *Biochem. Pharmacol.* **37**: 247–250.

Cumming, P., Brown, E., Damsma, G., and Fibiger, H. C. (1992). Formation and clearance of interstitial metabolites of dopamine and serotonin in rat striatum: An *in vivo* microdialysis study. *J. Neurochem.* **59**: 1905–1914.

Cumming, P., Kuwabara, H., and Gjedde, A. (1994). A kinetic analysis of 6-[18F]fluoro-L-dihydroxyphenylalanine metabolism in the rat. *J. Neurochem.* **63**: 1675–1682.

Danielsen, E. H., Smith, D. F., Bender, D., Gee, A. D., Hansen, S. B., and Gjedde, A. (1997). No up or down regulation of dope decarboxylase act. In short term dope treatment of normal volunteers and dope holiday in parkinson's disease. *J. Cereb. Blood Flow Metab.* **17**: S684.

Deacon, T. W., Schumacher, J., Dinsmore, J., Thomas, C., Palmer, P., Kott, S., Edge, A., Penny, D., Kassissieh, S., Dempsey, P., and Isacson, O. (1997). Histological evidence of fetal pig neural cell survival after transplantation into a patient with Parkinson's disease. *Nat. Med.* **39**(3): 350–353.

Doudet, D. J., Mclellan, C. A., Aigner, T. G., Wyatt, R., Adams, H. R., and Miyake, H. (1991). Postinjection L-phenylalanine increases basal ganglia contrast in PET scans of 6-18F-DOPA. *J. Nucl. Med.* **32**: 1408–1413.

Doudet, D. J., Aigner, T. G., McLellan, C. A., and Cohen, R. M. (1992). Positron emission tomography with 18F-dopa: Interpretation and biological correlates in nonhuman primates. *Psychiatry Res.* **44**: 153–168.

Firnau, G., Garnett, E. S., Chirakal, R., Sood, S., Nahmias, C., and Schrobilgen, G. (1986). [18F]fluoro-L-dopa for the *in vivo* study of intracerebral dopamine. *Int. J. Radiat. Appl. Instrum. A* **37**: 669–675.

Firnau, G., Sood, S., Chirakal, R., Nahmias, C., and Garnett, E. S. (1987). Cerebral metabolism of 6-[18F]fluoro-L-3,4-dihydroxyphenylalanine in the primate. *J. Neurochem.* **48**: 1077–1082.

Frost, J. J., Meltzer, C. C., Zubieta, J. K., Links, J. M., Brakeman, P., Stumpf, M. J., and Kruger, M. (1996). MR based correction of partial volume effects in brain PET imaging. In "Quantification of Brain Function Using PET" (R. Myers, V. Cunningham, D. Bailey, and T. Jones, eds.), pp. 152–157. Academic Press, San Diego, CA.

Gjedde, A., Reith, J., Dyve, S., Leger, G., Guttman, M., and Diksic, M. (1991). Dopa decarboxylase activity of the living human brain. *Proc. Natl. Acad. Sci U.S.A.* **88**: 2721–2725.

Hartvig, P., Lindner, K. J., Tedroff, J., Bjurling, P., Hörnefelt, K., and Laengstrom, B. (1992). Regional brain kinetics of 6-fluoro-(b-11C)-L-dopa and (b-11C)-L-dopa following COMT inhibition. A study *in vivo* using positron emission tomography. *J. Neural Transm., Gen. Sect.* **87**: 15–22.

Hoshi, H., Kuwabara, H., Leger, G., Cumming, P., Guttman, M., and Gjedde, A. (1993). 6-[^{18}F]fluoro-L-dopa metabolism in living human brain: A comparison of six analytical methods. *J. Cereb. Blood Flow Metab.* **13**: 57–69.

Kuwabara, H., Cumming, P., Reith, J., Leger, G., Diksic, M., and Evans, A. C. (1993). Human striatal L-dopa decarboxylase activity estimated *in vivo* using 6-[^{18}F]fluoro-dopa and positron emission tomography: Error analysis and application to normal subjects. *J. Cereb. Blood Flow Metab.* **13**: 43–56.

Leenders, K. L. (1990). Parkinsonism and PET scanning. In "The Assessment and Therapy of Parkinsonism" (C. D. Marsden and S. Falner, eds.), pp. 31–52. Parthenon, Carnforth, UK.

Melega, W. P., Hoffman, J. M., Luxen, A., Nissenson, C. H., Phelps, M. E., and Barrio, J. R. (1990). The effects of carbidopa on the metabolism of 6-[18F]fluoro-L-dopa in rats, monkeys and humans. *Life Sci.* **47**: 149–157.

Melega, W. P., Grafton, S. T., Huang, S. C., Satyamurthy, N., Phelps, M. E., and Barrio, J. R. (1991). L-6-[^{18}F]fluoro-dopa metabolism in monkeys and humans: biochemical parameters for the formulation of tracer kinetic models with positron emission tomography. *J. Cereb. Blood Flow Metab.* **11**: 890–897.

Pate, B. D., Kawamata, T., Yamada, T., McGeer, E. G., Hewitt, K. A., Snow, B. J., Ruth, T. J., and Calne, D. B. (1993). Correlation of striatal fluorodopa uptake in the MPTP monkey with dopaminergic indices. *Ann. Neurol.* **34**: 331–338.

Reith, J., Dyve, S., Kuwabara, H., Guttman, M., Diksic, M., and Gjedde, A. (1990). Blood-brain transfer and metabolism of 6-[18F]fluoro-L-dopa in rat. *J. Cereb. Blood Flow Metab.* **10**: 707–719.

Rousset, O. G., Ma, Y., Léger, G. C., Gjedde, A., and Evans, A. (1993). Correction for partial volume effects in PET using MRI-based 3D simulations of individual human brain metabolism. *In* "Quantification of Brain Function. Tracer Kinetics and Image Analysis in Brain PET." (K. Uemura, N. A. Lassen, T. Jones, and I. Kanno, eds.), pp. 113–125. Elsevier, Amsterdam and New York.

Rousset, O. G., Ma, Y., Marenco, S., Wong, D. F., and Evans, A. (1996). *In vivo* correction method for partial volume effects in positron emission tomography, accuracy, and precision. *In* "Quantification of Brain Function Using PET" (R. Myers, V. Cunningham, D. Bailey, and T. Jones, eds.), pp. 158–165. Academic Press, San Diego, CA.

Snow, B. J., and Calne, D. B. (1995). Movement disorders. *In* "Principles of Nuclear Medicine" (H. N. Wagner, Z. Szabo, and J. W. Buchanan, eds.), pp. 557–564. Saunders, Philadelphia.

Yosikawa, T. (1968). "Atlas of the Brains of Domestic Animals. The Brain of the Pig." Pennsylvania State University Press, University Park.

CHAPTER

29

Use of Two- and Three-Dimensional PET and [¹¹C](R)-PK11195 to Image Focal and Regional Brain Pathology[1]

RALPH MYERS, RICHARD B. BÁNATI, ERALDO PAULESU,* JOHN THORPE,[†]
DAVID H. MILLER,[†] and TERRY JONES

MRC Cyclotron Unit
Hammersmith Hospital
London W12 0NN, United Kingdom
*Neurologo, Neuropsicologo Centro PET / Ciclotrone
Instituto Scientifico H San Raffaele
INB-CNR, Universita degli Studi di Milano
Milano, Italy
† NMR Research Unit
Institute of Neurology
London WC1N 3BG, United Kingdom

The ligand PK11195 is selective for the peripheral benzodiazepine binding site and exhibits minimal binding in normal brain. In brain lesions, however, there is a massive increase in binding, which in vitro studies have demonstrated is associated with infiltrating macrophages and activated microglia. Here, the binding of [¹¹C]PK11195 was studied in the brains of multiple sclerosis patients using a CTI/Siemens 953B positron emission tomograph with retractable septa, acquiring data in either two-dimensional (2D) or 3D mode. Images acquired in 2D following injection of ~ 370 MBq were generally featureless. In some cases, poorly circumscribed signals were observed in brain stem or thalamus but with a minimal signal-to-noise ratio. However, regions showing gadolinium enhancement in magnetic resonance (MR) imaging, indicating a blood–brain barrier disturbance and known from histology to contain macrophages, never showed an elevated [¹¹C]PK11195 signal. Three dimensional positron emission tomography (PET), using similar doses of radioligand, gave a fivefold increase in counts and produced a marked improvement in image quality. Gadolinium-enhancing lesions were clearly visualized, together with hot spots of binding in regions with apparently normal MR imaging appearance. The PET signals were often well delineated and, following coregistration with MR images, could be localized to well-defined anatomical structures such as individual thalamic nuclei and white matter tracts. In several cases, these structures identified pathways correlating with the clinical condition of the patient. It was concluded that while 2D PET and [¹¹C]PK11195 can be used to image gross macrophage infiltration in the central nervous system, more subtle pathology requires 3D scanning in order to realize the full potential of the methodology.

I. INTRODUCTION

The ligand PK11195 [1-(2-chlorophenyl)-N-methyl-N-(1-methylpropyl)-3-isoquinolinecarboxamide] is selective for the peripheral benzodiazepine binding site (PBBS) (Le Fur et al., 1983) and has been labeled with ¹¹C for use in positron emission tomography (PET) (Camsonne et al., 1984). The PBBS exhibits minimal numbers in normal central nervous system (CNS), being present only in periventricular regions such as the ependymal cell layer and in organs devoid of a blood–brain barrier such as the area postrema (Bénavidès et al., 1983). In CNS that has been damaged, however, PBBS numbers are greatly increased (Dubois

---

[1] Transcripts of the BRAINPET97 discussion of this chapter can be found in Section VIII.

et al., 1988). Immunohistochemical studies have demonstrated that this increase is associated both with macrophages (Myers *et al.*, 1991a), which are known to possess a high density of PBBS (Zavala *et al.*, 1984), and with activated microglia (Myers, 1993; Stephenson *et al.*, 1995; Banati *et al.*, 1997).

Microglia are the resident macrophages of the CNS (for a review, see Kreutzberg, 1996). In normal CNS, they roughly equal neurons in number and have a highly ramified form; they are quiescent, with no known function. Following even a mild pathological stimulus, microglia are rapidly activated, exhibiting hypertrophy and an upregulation of, among other molecules, a number of macrophage-related antigens. More severe stimulation or neuronal cell death results in these cells becoming rounded, amoeboid and phagocytic, and thus indistinguishable from circulating macrophages. PBBS expression by microglia is increased above baseline at an early stage in this activation process.

The first successful PET study using [^{11}C]PK11195 in the brain (Junck *et al.*, 1989) reported binding in various brain tumors. Subsequently, [^{11}C]PK11195 binding associated with primary stroke lesions was reported by both Junck *et al.* (1990) and Ramsay *et al.* (1992). Pappata *et al.* (1993) reported additional binding in distant secondary thalamic lesions, consistent with earlier experimental results in an animal model (Myers *et al.*, 1991b). Groom and colleagues (1995), however, reported that the PBBS in mild-to-moderate Alzheimer's disease was undetectable using PET and [^{11}C]PK11195.

The aim of the present study was to explore the potential of PET and [^{11}C]PK11195 in multiple sclerosis, an immune system-related disease characterized by microglial activation (Sriram and Rodriguez, 1997) and focal macrophage infiltration (Nesbit *et al.*, 1991).

II. MATERIALS AND METHODS

A. Patients

Patients (male: $n = 6$, average age 39 years, range 27–50 years; female: $n = 6$, average age 38 years, range 27–48 years) were diagnosed as suffering from either chronic progressive or relapsing remitting multiple sclerosis (MS). Two normal male volunteers were scanned in 3D mode for comparison.

B. Ligand Synthesis and Administration

Shah *et al.* (1994) demonstrated that the *R*-enantiomer of PK11195 was preferable to either the *S*-enantiomer or the racemate for *in vivo* studies. This compound, labeled with ^{11}C, was synthesized according to the method they described. For each subject, 370 MBq ($\pm 8\%$) with a specific activity of 36 GBq/μmol ($\pm 47\%$) was injected intravenously over 20 sec. Full arterial sampling and metabolite analyses were carried out, but these results are not reported here.

C. PET Image Data Acquisition

PET data were acquired using a CTI/Siemens 953B PET scanner operating with either septa extended (2D mode) or septa retracted (3D mode). The latter is known to produce a sevenfold increase in counts at the center of the field of view compared with the 2D mode, falling to a fivefold increase following correction for scatter (Spinks *et al.*, 1992). In the 2D mode, emission data were acquired in 39 time frames (1×30 sec, 12×5 sec, 6×10 sec, 6×30 sec, 5×60 sec, 2×150 sec, 5×30 sec, 2×600 sec) following ligand injection. In the 3D mode, 8 frames of data were acquired (1×30 sec, 2×120 sec, 3×300 sec, 1×600 sec, 1×1200 sec) in the initial studies reported here, although full dynamic acquisition has been introduced subsequently. An [^{15}O]H$_2$O scan was acquired shortly before each [^{11}C](*R*)-PK11195 scan in order to facilitate image coregistration with MR imaging. Images were reconstructed using either filtered backprojection (2D mode) into a matrix with voxel dimensions of $2.05 \times 2.05 \times 3.43$ mm using a Hanning filter at Nyquist cutoff or a reprojection algorithm (3D mode) (Townsend *et al.*, 1991) into a matrix with voxel dimensions of $2.09 \times 2.09 \times 3.43$ mm using a ramp filter at Nyquist cutoff. For the purpose of visualization of PK11195 signals, summed images from 10 to 50 min postinjection were created. In the 3D mode, correction for scatter was applied using the dual window method of Grootoonk *et al.* (1996).

D. MRI Image Acquisition and Coregistration

Close to the time of the PET scan, each patient underwent three MR scans as part of their routine clinical assessment, proton-density, T2-weighted, and T1-weighted gadolinium-enhanced. These image volumes were each coregistered to the patient's own [^{15}O]H$_2$O scan using software developed by Woods *et al.* (1993), and the resulting transformation parameters were used to reslice the [^{11}C](*R*)-PK11195 summed image. Accuracy of coregistration was carefully checked by visual inspection. This and all other image manipulations were carried out using ANALYZE software (Robb and Hanson, 1991).

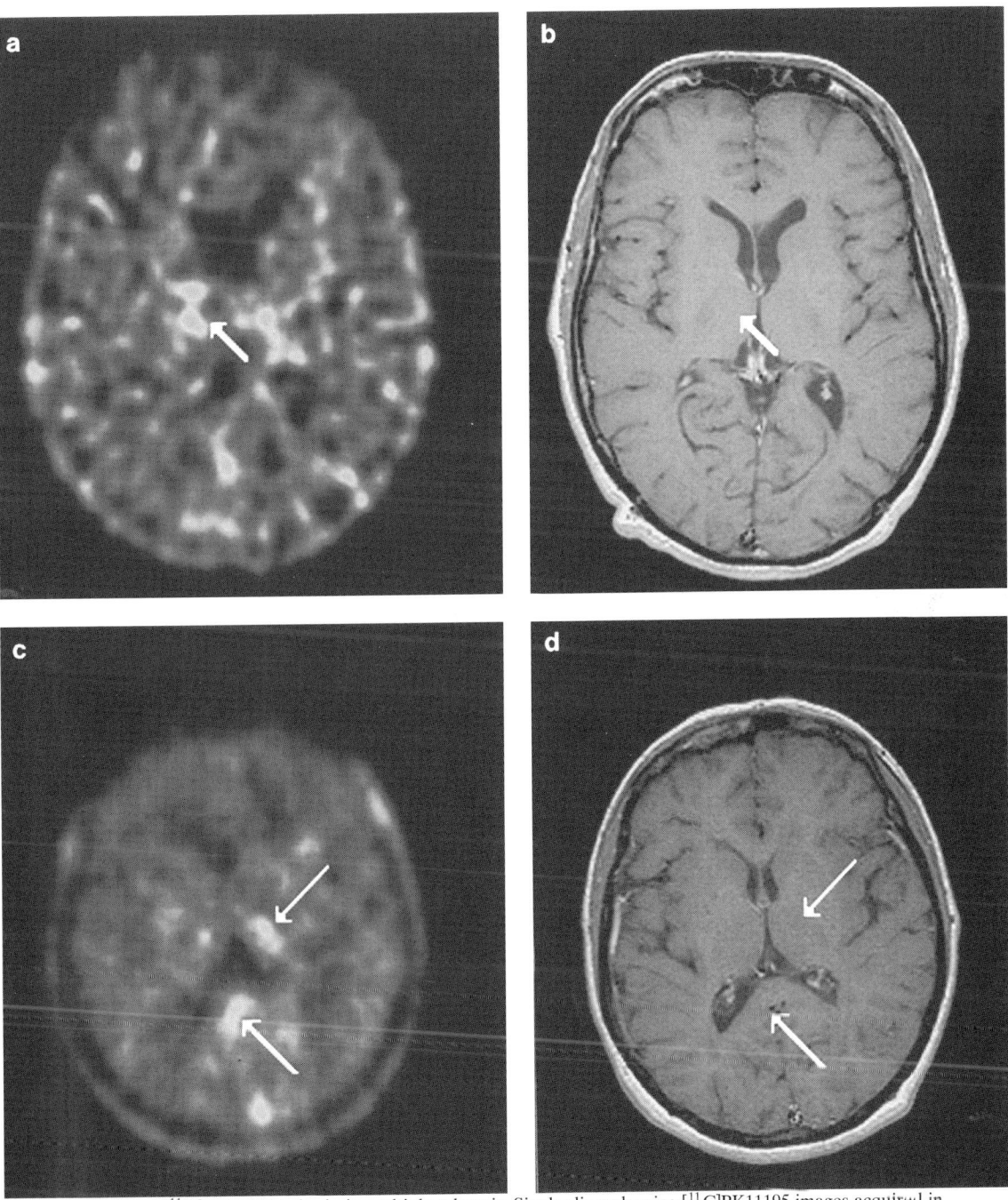

FIGURE 1 [¹¹C]PK11195 in brain in multiple sclerosis. Single slices showing [¹¹C]PK11195 images acquired in either 2D mode (a) or 3D mode (c) are shown. The images are not quantitatively normalized to each other. The image volumes from which the slices were taken were coregistered to the subject's own MR images (Woods *et al.*, 1993), and the corresponding MR images are shown in b and d. Note that the 2D mode image is of relatively poor quality; the arrow indicates a poorly delineated thalamic signal. Foci of [¹¹C]PK11195 binding in the 3D mode image are more clearly described; the upper arrow indicates a thalamic focus and the lower arrow a hot spot in the corpus callosum.

III. RESULTS AND DISCUSSION

A. Results in 2D Mode

Following acquisition in 2D mode, images were frequently featureless with generally poor signal-to-noise ratios. Elevated [^{11}C]PK11195 signals were apparent in brain stem and in thalamus in some MS patients (Fig. 1), but few other convincing signals could be identified. Most notably, regions identified on gadolinium-enhanced MR images as having a compromised blood–brain barrier, and expected from earlier experimental studies to contain macrophages (Nesbit *et al.*, 1991) and to give a high PK11195 signal (Bénavidès *et al.*, 1988), were never associated with specific binding of [^{11}C]PK11195.

B. Results in 3D Mode

[^{11}C]PK11195 images acquired in 3D had an improved signal-to-noise ratio compared with 2D acquisition, although no formal estimate of smoothness was made. High binding associated with gadolinium-enhancing lesions observed on MR images was frequently seen (Fig. 2). However, many other [^{11}C]PK11195 signals, such as those in thalamus and brain stem, were observed in regions identified as normal in the three MR imaging sequences used. These regions of high signal were frequently well defined and, following coregistration with MR images, could be accurately identified in terms of their anatomical location. In the case of the thalamus, distribution of signal could be unilateral or bilateral and did not necessarily involve the whole thalamus, in some cases permitting the thalamic nuclei involved to be identified.

It is important to note that PET signals in regions of normal MR images did not necessarily identify regions of primary damage but more likely indicated a secondary response tracking along connected neuronal pathways. This tracking, representing activated microglia rather than macrophages, could be observed in regions such as the spinothalamic tracts and the internal capsule. Most significantly, foci of binding were observed along neuronal pathways that correlated with the clinical condition of the patient. For example, two patients suffering from oscillopsia, a symptom producing oscillating vision, had a high signal in the oculomotor nucleus in the brain stem, whereas those patients suffering from optic neuritis had signals in the lateral geniculate bodies (two patients), the visual cortex (two patients), and the optic nerve (one patient). One patient showed a clear focal signal in the medial longitudinal fasciculus (Fig. 3), a unilateral lesion of which causes a denervation of the medial rectus muscle, a condition known as internuclear opthalmoplegia, from which the patient was suffering close to the time of the scan. This strong association between [^{11}C]PK11195 signals and clinical symptoms contrasts with the limited

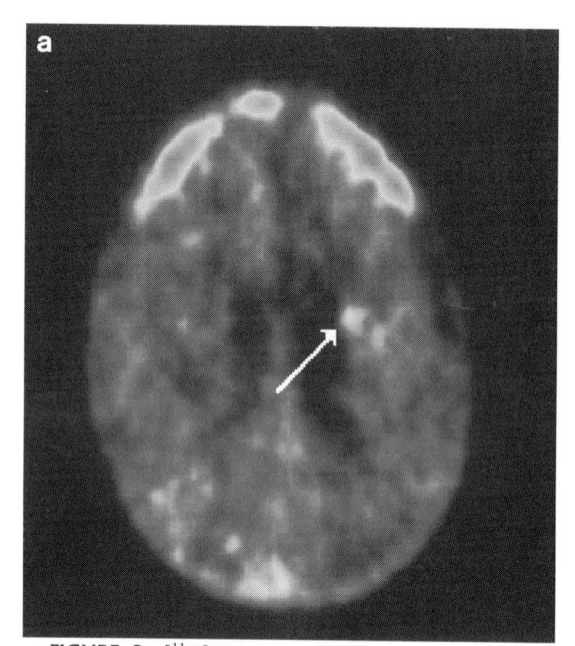

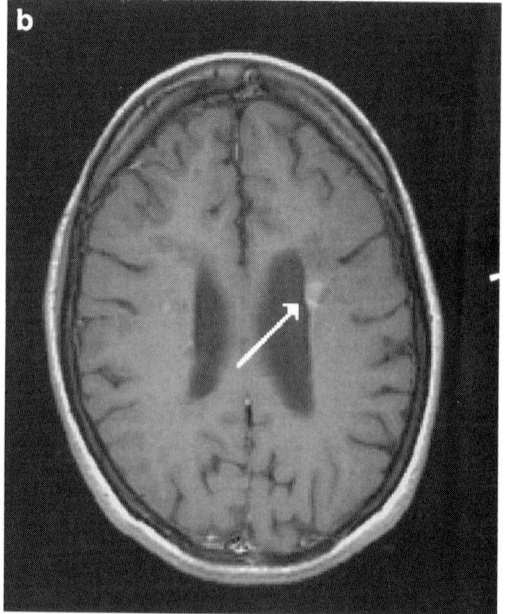

FIGURE 2 [^{11}C]PK11195 colocalizing with an MR imaging gadolinium-enhancing lesion in the brain of a multiple sclerosis patient. (a) [^{11}C]PK11195 acquired in 3D mode; (b) coregistered, T1-weighted, gadolinium-enhanced MR image.

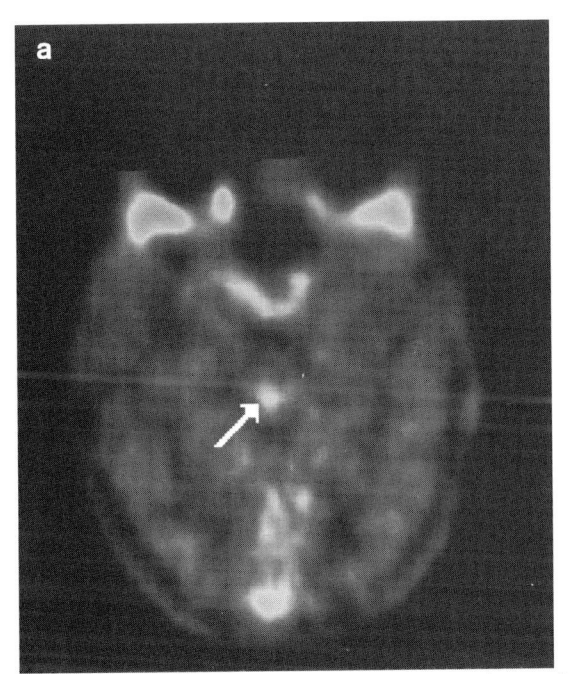

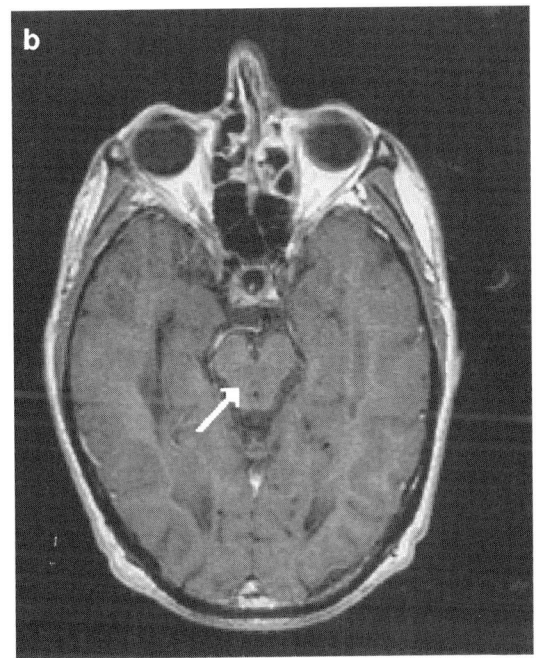

FIGURE 3 [^{11}C]PK11195 image and internuclear opthalmoplegia in multiple sclerosis. The arrow indicates focal binding in the medial longitudinal fasciculus a lesion of which causes internuclear opthalmoplegia from which this patient suffered. (a) [^{11}C]PK11195 acquired in 3D mode; (b) coregistered, T1-weighted, gadolinium-enhanced MR image.

correlation with disability in established MS seen using MR imaging (Miller et al., 1996).

No notable foci of [^{11}C]PK11195 signal were observed in the brains of control subjects. However, high signals were consistently observed in several extracerebral structures, including sagittal sinus, scalp and marrow in the skull, pituitary gland, and tissue behind the orbits of the eyes.

No model for the quantification of specific binding of [^{11}C]PK11195 in brain has yet been validated. "Hot spots," such as those identified in the brain stem nuclei or thalamus, however, had a radioactivity concentration two to three times that in the white matter. The physical size of the structures involved, often smaller than 1 ml in volume, demands the high sensitivity of 3D PET to maximize image quality. In addition to focal binding, it is likely that more global increases in [^{11}C]PK11195 binding are present in such diseases as MS and Alzheimer's disease, due to a widespread microglial response. The possible presence of such a global change emphasizes the need for full quantification of [^{11}C]PK11195 binding in the brain.

Using 2D PET, Groom and colleagues (1995) attempted to image "microgliosis" in the brains of patients suffering from Alzheimer's disease. These authors reported that PBBS known to be associated with microgliosis in this disease were "undetectable by PET using [^{11}C]PK11195." Although Alzheimer's disease patients were not scanned in the present study in 2D mode, results in MS would appear to support their conclusion in that subtle signals associated with activated microglia cannot be detected in 2D mode. The greater sensitivity of 3D PET, however, appears to be an essential requirement for imaging activated microglia using [^{11}C]PK11195 and PET.

IV. CONCLUSIONS

In conclusion, 2D PET and [^{11}C]PK11195 can be used to visualize gross macrophage infiltration into damaged CNS; the high sensitivity of 3D PET is essential to visualize more subtle pathologies, including activated microglia; [^{11}C]PK11195 signals may occur along connected neuronal pathways and correlate well with clinical symptoms; and accurate coregistration with structural imaging is essential for anatomical identification.

Acknowledgments

RBB received support from the Multiple Sclerosis Society of Great Britain and Northern Ireland, the Max-Planck Institute of Psychiatry (Martinsreid, Germany), and the Deutsche Forschungsgemeinschaft grant "The mitochondrial benzodiazepine receptor as indicator of early CNS pathology, clinical application in PET."

References

Banati, R. B., Myers, R., and Kreutzberg, G. W. (1997). PK ('peripheral benzodiazepine')-binding sites in the CNS indicate early and discrete brain lesions: Microautoradiographic detection of [^{3}H]PK11195 binding to activated microglia. *J. Neurocytol.* **26**: 77–82.

Bénavidès, J., Quarteronet, D., Imbault, F., Malgouris, C., Uzan, A., Renault, C., Dubroeucq, M. C., Guérémy, C., and Le Fur, G. (1983). Labelling of 'peripheral-type' benzodiazepine binding sites in the rat brain by using [^{3}H]PK11195, an isoquinoline carboxamide derivative: Kinetic studies and autoradiographic localization. *J. Neurochem.* **41**: 1744–1750.

Bénavidès, J., Cornu, P., Dennis, T., Dubois, A., Hauw, J.-J., MacKenzie, E. T., Sazdovitch, V., and Scatton, B. (1988). Imaging of human brain lesions with an ω_3 site radioligand. *Ann. Neurol.* **24**: 708–712.

Camsonne, R., Cronzel, C., Comar, D., Mazière, M., Prenant, C., Sastre, J., Moulin, M. A., and Syrota, A. (1984). Synthesis of N-[^{11}C]methyl, N-(methyl-1 propyl), (chloro-2 phenyl)-1 isoquinoline carboxamide-3 (PK11195): A new ligand for peripheral benzodiazepine receptors. *J. Labelled Compd. Radiopharm.* **21**: 985–991.

Dubois, A., Bénavidès, J., Peny, B., Duverger, D., Fage, D., Gotti, B., MacKenzie, E. T., and Scatton, B. (1988). Imaging of primary and remote ischaemic and excitotoxic brain lesions. An autoradiographic study of peripheral type benzodiazepine binding sites in the rat and cat. *Brain Res.* **445**: 77–90.

Groom, G. N., Junck, L., Foster, N. L., Frey, K. A., and Kuhl, D. E. (1995). PET of peripheral benzodiazepine binding sites in the microgliosis of Alzheimer's disease. *J. Nucl. Med.* **36**: 2207–2210.

Grootoonk, S., Spinks, T. J., Sashin, D., Spyrou, N. M., and Jones, T. (1996). Correction for scatter in 3D brain PET using a dual energy window method. *Phys. Med. Biol.* **41**: 2757–2774.

Junck, L., Olsen, J. M. M., Ciliax, B. S., Koeppe, R. A., Watkins, G. L., Jewett, D. M., McKeever, P. E., Wieland, D. M., Kilbourn, M. R., Starosta-Rubinstein, S., Mancini, W. R., Kuhl, D. E., Greenberg, H. S., and Young, A. B. (1989). PET imaging of human gliomas with ligands for the peripheral benzodiazepine binding site. *Ann. Neurol.* **26**: 752–758.

Junck, L., Jewett, D. M., Kilbourn, M. R., Young, A. B., and Kuhl, D. E. (1990). PET imaging of cerebral infarcts using a ligand for the peripheral benzodiazepine binding site. *Neurology* **40**(Suppl. 1): 265.

Kreutzberg, G. W. (1996). Microglia: A sensor for pathological events in the CNS. *Trends Neurosci.* **19**: 312–318.

Le Fur, G., Perrier, M. L., Vaucher, N., Imbault, F., Flamier, A., Bénavidès, J., Uzan, A., Renault, C., Dubroeucq, M. C., and Guérémy, C. (1983). Peripheral benzodiazepine binding sites: Effect of PK11195, 1-(2-chlorophenyl)-N-methyl-(1-methylpropyl)-3-isoquinolinecarboxamide. I. *in vitro* studies. *Life Sci.* **32**: 1839–1847.

Miller, D. H., Albert, P. S., Barkof, F., Francis, G., Frank, J. A., Hodgkinson, S., Lublin, F. D., Paty, D. W., Reingold, S. C., and Simon, J. (1996). Guidelines for the use of magnetic resonance techniques in monitoring the treatment of multiple sclerosis. *Ann. Neurol.* **39**: 6–16.

Myers, R. (1993). Mitochondrial benzodiazepine receptor ligands as indicators of damage in the CNS: Their application in positron emission tomography. In "Peripheral Benzodiazepine Receptors" (E. Giesen-Crouse, ed.), pp. 235–273. Academic Press, London.

Myers, R., Manjil, L. G., Cullen, B. M., Price, G. W., Frackowiak, R. S. J., and Cremer, J. E. (1991a). Macrophage and astrocyte populations in relation to [^{3}H]PK11195 binding in rat cerebral cortex following a local ischaemic lesion. *J. Cereb. Blood Flow Metab.* **11**: 314–322.

Myers, R., Manjil, L. G., Frackowiak, R. S. J., and Cremer, J. E. (1991b). [^{3}H]PK11195 and the localization of secondary thalamic lesions following focal ischaemia in rat motor cortex. *Neurosci. Lett.* **133**: 20–24.

Nesbit, G. M., Forbes, G. S., Scheithauer, B. W., Okazaki, H., and Rodriguez, M. (1991). Multiple sclerosis: Histopathologic and MR and/or CT correlation in 37 cases at biopsy and three cases at autopsy. *Radiology* **180**: 467–474.

Pappata, S., Levasseur, M., Samson, Y., Cornu, P., Boullais, N., Crouzel, C., and Syrota, A. (1993). Thalamic and peri-infarct ^{11}C-PK11195 uptake in patients with chronic middle cerebral artery infarcts: A PET study. *J. Cereb. Blood Flow Metab.* **13**(Suppl. 1): S346.

Ramsay, S. C., Weiller, C., Myers, R., Cremer, J. E., Luthra, S. K., Lammertsma, A. A., and Frackowiak, R. S. J. (1992). Monitoring by PET of macrophage accumulation in brain after ischaemic stroke. *Lancet* **339**: 1054–1055

Robb, R. A., and Hanson, D. P. (1991). A software system for interactive and quantitative visualization of multidimensional images. *Australas. Phys. Eng. Med.* **14**: 9–30.

Shah, F., Hume, S. P., Pike, V. W., Ashworth, S., and McDermott, J. (1994). Synthesis of the enantiomers of [N-methyl-^{11}C]PK11195 and comparison of their behaviors as radioligands for PK binding sites in rats. *Nucl. Med. Biol.* **21**: 573–581.

Spinks, T. J., Jones, T., Bailey, D. L., Townsend, D. W., Grootoonk, S., Bloomfield, P. M., Gilardi, M.-C., Casey, M. E., Sipe, B., and Reed, J. (1992). Physical performance of a positron tomograph for brain imaging with retractable septa. *Phys. Med. Biol.* **37**: 1637–1655.

Sriram, S., and Rodriguez, M. (1997). Indictment of the microglia as the villain in multiple sclerosis. *Neurology* **48**: 464–470.

Stephenson, D. T., Schober, D. A, Smalstig, E. B., Mincy, R. E., Gehlert, D. R., and Clemens, J. A. (1995). Peripheral benzodiazepine receptors are co-localized with activated microglia following transient global forebrain ischaemia in the rat. *J. Neurosci* **15**: 5263–5274.

Townsend, D. W., Geissbuhler, A., Defrise, M., Hoffman, E. J., Spinks, T. J., Bailey, D. L., Gilardi, M.-C., and Jones, T. (1991). Fully three-dimensional reconstruction for a PET camera with retractable septa. *IEEE Trans. Med. Imag.* **MI-10**: 505–512.

Woods, R. P., Mazziotta, J. C., and Cherry, S. R. (1993). MRI-PET registration with automated algorithm. *J. Comput. Assist. Tomogr.* **17**: 536–546.

Zavala, F., Haumont, J., and Lenfant, M. (1984). Interaction of benzodiazepines with mouse macrophages. *Eur. J. Pharmacol.* **106**: 561–566.

CHAPTER 30

Noninvasive Imaging of Serotonin Synthesis Rate Using PET and α-Methyltryptophan in Autistic Children[1]

OTTO MUZIK,* DIANE C. CHUGANI,*,† CHENGGANG SHEN,* and HARRY T. CHUGANI,*,†,‡

*Departments of * Radiology, † Pediatrics, and ‡ Neurology*
Wayne State University
Children's Hospital of Michigan
Detroit, Michigan 48201

α-Methyl-[^{11}C]tryptophan (AMT) was previously developed as a tracer for serotonin synthesis with positron emission tomography (PET). Because of the low uptake of AMT in the brain (about 2%) and the radiation dose restriction in children (0.5 rem/year), a combined two-dimensional (2D)/3D PET protocol was designed that allows noninvasive acquisition of the arterial input function and uses the Patlak-plot together with a nonstationary spatial filter to create parametric images of the serotonin synthesis rate. Seven autistic boys (ages 4.1–11.1 years) and four of their male siblings (ages 8.2–14.4 years) were studied with the EXACT HR PET scanner. A dynamic heart scan in 2D mode was acquired for the first 20 min following injection of 0.1 mCi/kg of AMT. This was followed by a dynamic brain scan in 3D mode 25–60 min after tracer injection. The complete arterial input function was obtained by combining the time–activity curve of a small region in the center of the left ventricle with four venous samples drawn between 30 and 60 min postinjection. The Patlak graphical analysis included a nonstationary spatial filter that averaged locally only those pixels that had a high probability to differ only by noise. Simulation studies were performed to optimize the filter parameter. This method resulted in noise reduction and contrast preservation in comparison to conventionally smoothed images, allowing better detection of focal abnormalities in the frontal cortex of autistic children.

I. INTRODUCTION

α-Methyl[^{11}C]tryptophan (AMT) is an analog of tryptophan that was originally developed by Diksic and colleagues as a tracer for serotonin synthesis with positron emission tomography (PET) in animals (Diksic et al., 1990) and humans (Nishizawa et al., 1997). Animal studies have demonstrated that AMT is converted to α-methyl-[^{11}C]serotonin in serotonergic neurons (Cohen et al., 1995) and does not significantly enter into other metabolic pathways (Missala and Sourkes, 1988). Although this tracer has many attractive properties for imaging of serotonin synthesis due to its accumulation in serotonergic nerve terminals, problems remain, the most important being the low extraction in brain tissue.

AMT has been studied in healthy adults (Muzik et al., 1997a; Chugani et al., 1998), and the rank order of the unidirectional rate constant for AMT (also termed the K complex) in various brain regions showed highly significant agreement with the rank order of regional serotonin content in human brain. The calculation of the absolute serotonin synthesis rate requires multiplication of the K complex with values for the concentration of tryptophan available for transport across the blood–brain barrier and the inverse of the "lumped constant" (which combines values for the brain/plasma partition coefficient for tryptophan and AMT, as well as the K_m and V_{max} of these substrates for tryptophan hydroxylase). However, within an individual subject's brain, regional differences in the serotonin synthesis rate are directly related to the K com-

---

[1] Transcripts of the BRAINPET97 discussion of this chapter can be found in Section VIII.

plex, as other factors are merely scaling factors for all regions (Shoaf and Schmall, 1996). Hence, the K complex value for AMT has been termed as the "serotonin synthesis capacity." In addition, regional values for the K complex and standard uptake values (SUV) were highly correlated in different brain structures within a subject (Chugani et al., 1998), suggesting that tissue accumulation is similarly a measure of regional serotonin synthesis capacity.

The application of this tracer to children required some methodological changes to the previously applied protocol used in adults. Because of the low uptake of AMT in the brain (about 2%) and the radiation dose restriction in normal children younger than 18 years of age (0.5 rem to the critical organ per year), it was necessary to increase the sensitivity of the scanning procedure by performing the studies in three-dimensional (3D) mode. In addition, pediatric protocols demand a noninvasive approach with minimal blood drawing. The authors, therefore, designed a combined 2D/3D mode PET protocol that allows noninvasive acquisition of the left ventricular input function from a dynamic 2D mode scan sequence of the heart (eliminating the need for an arterial catheter) and the determination of parametric images of the K complex derived from a 3D image sequence of the brain (allowing a decrease in injected dose to comply with pediatric radiation dose limits). The dynamic heart scan was performed in 2D mode in order to use acquisition disk space efficiently, as the heart scan does not require the same quality of image statistics as the brain scan. The dynamic brain scan was processed using the Patlak graphical approach in order to obtain parametric images representing the K complex. The quality of parametric images was further optimized using a nonstationary spatial filter (Herholz, 1988).

II. MATERIALS AND METHODS

A. PET Scanning Procedure

AMT was produced using a synthesis module designed and built in-house (Chakraborty et al., 1996). Dynamic PET measurements were performed using the SIEMENS/CTI EXACT HR PET scanner, which allows simultaneous acquisition of 47 contiguous transaxial images with a slice thickness of 3.125 mm. The reconstructed image resolution obtained in the study was 7.5 ± 0.38 mm at full width at half-maximum (FWHM) in-plane and 7.0 ± 0.49 mm FWHM in axial direction (reconstruction parameters: Parzen filter with 2.53 cycles/cm cutoff frequency).

Seven autistic boys (ages 4.1–11.1, mean age 6.6 years) and four of their male siblings (ages 8.2–14.4, mean age 9.9 years) were studied. Studies were performed in compliance with the regulations of the Human Investigation and Radioactive Drug Research Committees of Wayne State University and informed consent of parent or guardian was obtained. In addition, the assent of the siblings was obtained. Subjects were fasted for 6 hr prior to the PET procedure in order to obtain stable plasma tryptophan levels during the course of the study. All autistic children and three of the four siblings required sedation with nembutal (5 mg/kg iv) or midazolam (0.2–0.4 mg/kg iv). Prior studies performed in the authors' laboratory on five adults scanned twice with and without sedation have found global differences within the accepted test/retest range for PET tracers ($< 10\%$).

Following a 15-min transmission scan of the chest to correct for photon attenuation within the child's body, a 20-min dynamic scan sequence of the myocardium (12×10 sec/3×60 sec/3×300 sec) was initiated in 2D mode starting with the administration of activity. In order to stay within the dose limits of 0.5 rem to the critical organ per year, only 0.1 mCi/kg of AMT was administered (total dose 1.0–3.5 mCi). After repositioning the patient with the brain being in the field of view, a 35-min dynamic brain scan (7×300 sec) was started 25 min following tracer injection in 3D mode with septa retracted from the field of view. Four timed venous blood samples were drawn during the brain scan (20, 40, 50, 60 min after injection) to obtain a complete arterial input function. At these late times after tracer injection, venous activity is approximately in equilibrium with arterial activity. All brain images were corrected for photon attenuation using calculated attenuation correction.

B. Dosimetry

Radiation dosimetry for AMT was calculated based on the activity distribution of AMT in the organs of four rats at 60 min after injection, normalized to injected activity and body weight. Human organ activity and residence times were estimated from rat data using organ weights from ICRP Publication 53 (1988). The doses were calculated for the adult and for three different pediatric age groups (5, 10, and 15 years old) using the MIRDOSE program (Oak Ridge Institute for Science and Education, Oak Ridge, TN, IBM PC version 3.0).

C. Extraction of the Arterial Input Function

The arterial input function was derived from a small region in the center of the left ventricle (LV) of the heart. It was previously validated that such a region is free from partial volume distortions in children older than 2 years of age due to the size of the end-diastolic and end-systolic heart diameter (Muzik et al., 1994). Anatomical information derived from the transmission scan was used to define such an LV ROI, as myocardial retention of AMT is very low and the heart muscle itself cannot be visualized (Muzik et al., 1997b). In short, a center plane through the LV is estimated based on the contour of the heart against lung tissue in the transmission scan. The coordinates of this plane are then transformed to the dynamic heart emission scan and a small LV region is defined guided by the appearance of the tracer bolus first in the right ventricle and then in the LV.

D. Nonstationary Spatial Filter

Patlak graphical analysis was performed for brain data for times 25 to 60 min to obtain parametric images of the K complex. Previous compartmental analysis has shown that dynamic equilibrium has been approached at this time for most brain regions (Chugani et al., 1998). Prior to the Patlak analysis, a nonstationary spatial filter was implemented to improve the quality of these parametric images. The method uses spatial averaging of pixel time–activity curves in each plane with a Gaussian filter to increase the number of counts contributing to each pixel (Herholz, 1988). The range of this filter is controlled locally by an algorithm that is designed to discriminate between different tissue types based on dynamic image information. To discriminate if two pixel curves ($P1_i, P2_i$) represent the same tissue type and thus can be averaged, a difference curve D_i ($= P1_i - P2_i$) is calculated. If both pixel curves represent the same tissue type, the difference curve D_i represents only random fluctuations around a mean value of zero and the maximum run length (MRL), i.e., the largest number of subsequent elements D_i with the same sign, will be small. In contrast, if the two pixel curves represent different tissue types, the MRL will be large as pixel curves cross only once or may not cross at all. Starting from a central pixel, all pixel time–activity curves with a MRL less than a predefined value are averaged in a 9×9 pixel neighborhood. Thus, the larger the MRL, the more pixel time–activity curves will be incorporated into the average, resulting in a smoother image.

In order to optimize the MRL for the given study design, simulation studies were performed. The objective was to create low-noise parametric images but to preserve contrast between different tissue types. Four different curves representing AMT kinetics in frontal cortex, thalamus, caudate, and white matter were arranged in four quadrants. A curve representing the

TABLE 1   Dosimetry of α-Methyl-[^{11}C]tryptophan in Children and Adults[a]

Organ	5 year old (19.8 kg)	10 year old (33.2 kg)	15 year old (56.8 kg)	Adult (70 kg)
Brain	12.4	10.2	9.0	8.8
Heart	44.0	28.4	18.6	14.4
Kidney	**229.0**	**150.0**	**110.0**	**91.4**
Liver	52.2	34.7	23.2	17.6
Adrenals	35.2	23.2	15.4	13.0
Muscle	77.8	27.4	14.3	9.6
Lungs	41.7	27.3	19.2	13.3
Red marrow	48.7	27.4	16.7	14.5
Urinary bladder	78.5	50.0	33.1	27.0
Testes	133	113	18.6	9.5
Ovaries	72.0	42.0	17.6	17.5
Effective dose equiv.	79.6	57.0	23.9	20.0
Total body	33.3	20.9	13.0	10.4

[a] Based on the MIRDOSE program provided by ORISE. Organ residence time calculation was based on the distribution of α-methyl-[^{11}C]tryptophan in four rats. All values are in units of mrem/mCi. The dose to the critical organ, kidney, is in bold.

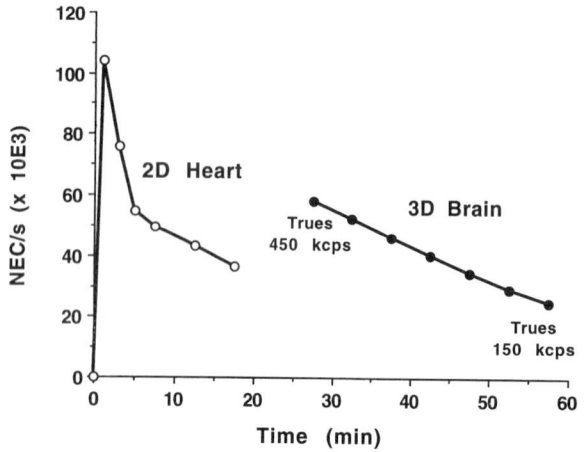

FIGURE 1 Noise equivalent count rate (NEC/s) obtained in the combined 2D/3D protocol. Acquisition of PET data in 3D mode results in a significant increase of NEC/s, despite the fact that a much smaller organ (brain) is in the field of view instead of the patient's body. The true count rate at the beginning of the 3D protocol was about 450 kcps, which represents the limit for 3D data acquisition (about 25% dead time).

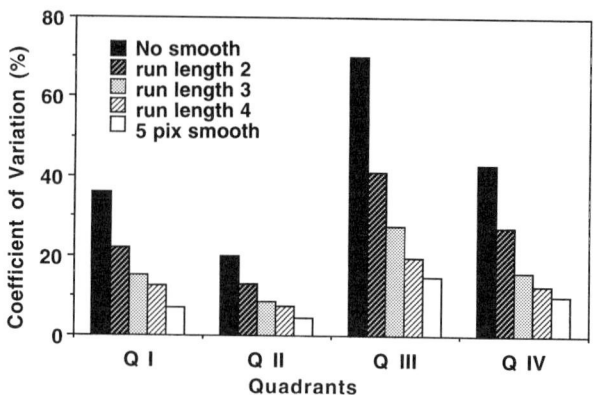

FIGURE 2 Coefficient of variation (COV) within quadrants I–IV in parametric K complex images obtained from simulated data. The COV in each quadrant depended on the kinetics of the underlying curve; however, the COV decreased with increasing run length in all cases. The graph shows the gradual smoothing effect of the nonstationary spatial filter. The higher the run length, the more pixel time–activity curves are incorporated into the average.

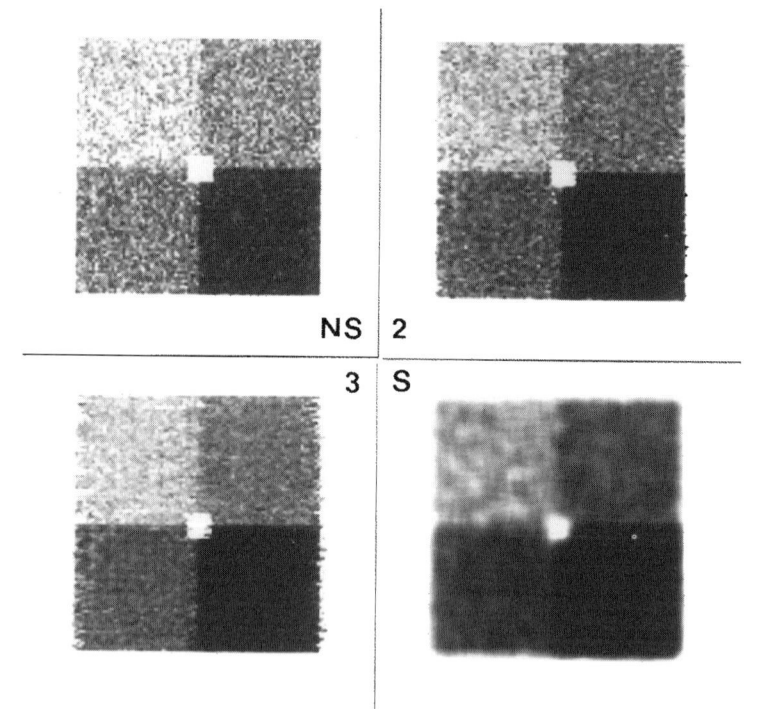

FIGURE 3 The four quadrants represent K complex images obtained from simulated data processed with different values of the maximum run length (MRL) (NS, no smoothing; MRL, 2; MRL, 3; S, 5 pixel smooth). It can be seen that a compromise has to be accepted between homogeneity within individual quadrants and the contrast at the borders between quadrants. An MRL of 3 was chosen for application in patient studies.

input function was written into a 5 × 5 pixel ROI in the image center simulating a major vessel. Within each quadrant all pixels carried curves created by the same parameters but with independently superimposed random Gaussian noise of 15%. The simulated image sequence was written into a new dynamic file using the same scan times as the emission study. Parametric images were then computed with different values for MRL and the contrast between the sectors as well as the coefficients of variation within the quadrants examined.

III. RESULTS AND DISCUSSION

Table 1 shows the calculated doses for selected organs in the three pediatric groups and in the adult group. The critical organ is the kidney. The injected activity of 0.1 mCi/kg resulted in a total dose to the kidneys of less than 0.5 rem in all three pediatric groups.

The image quality of the cardiac image sequence in 2D mode was adequate and allowed the accurate determination of the LV input function. The average noise equivalent count (NEC) rate (2D, scatter fraction = 0.12) was about 100 kcps during the blood peak phase and declined to about 35 kcps by the end of the 20-min scan (Fig. 1). For the 3D brain sequence, the NEC rate (scatter fraction = 0.35) started at 60 kcps at 25 min following tracer injection and declined to approximately 25 kcps at the end of the 60-min scan. This compares favorably to clinical pediatric [^{18}F]fluorodeoxyglucose studies in 2D mode with a NEC rate of 30–50 kcps.

The simulated image sequence was processed with an MRL of 2, 3, and 4, and a 5 pixel boxcar smooth. The coefficient of variation (COV) within quadrants declined in the same order, with the COV of the unsmoothed parametric images being the highest in all quadrants and being lowest for the boxcar smooth (Fig. 2). However, the visual contrast between quadrants declined in the same order, as can be seen in Fig. 3, indicating that a compromise between image contrast and image homogeneity has to be accepted. Based on these results, a value for MRL of 3 was chosen for all further studies.

The quality of parametric images of the K complex remains suboptimal despite the improvement achieved with the nonstationary spatial filter algorithm. However, anatomical information is retained in the parametric images, e.g., caudate, thalamus, and cortical regions can be identified. Furthermore, the decrease in

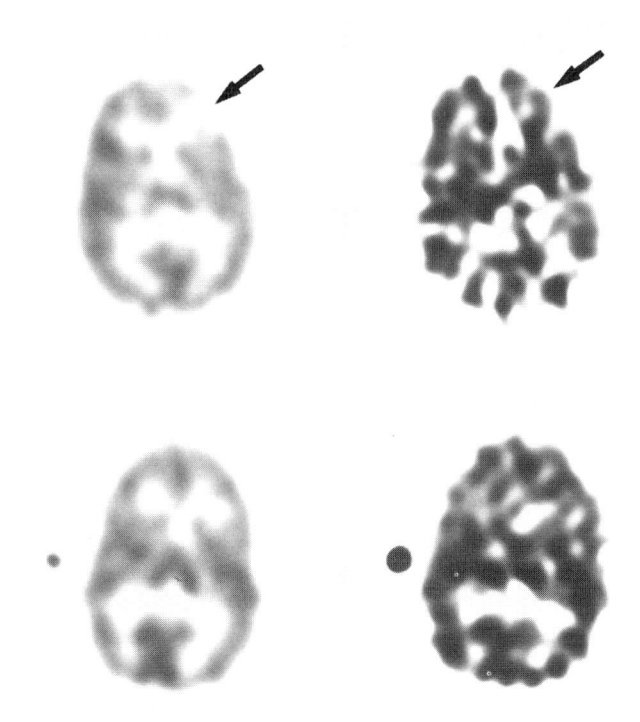

FIGURE 4 Summed AMT dynamic images and parametric images in a boy with autism (upper row) and in his normal male sibling (lower row). (Left) Summed images obtained 25–60 min after injection of AMT; (right) corresponding parametric images representing the K complex. A decrease in the left frontal cortex can be seen in images of the autistic boy in both summed and K complex images.

left frontal cortical regions in autistic boys, which has been reported previously using SUV analysis (Chugani et al., 1998), can be appreciated on the parametric images of autistic boys, but not in their male siblings (Fig. 4). The observed focal alterations in AMT synthesis rate may represent either aberrant innervation by serotonergic terminals or altered function in anatomically normal pathways. Further studies in a larger number of patients are needed to assess the clinical significance of this finding.

The results of this study demonstrate that regional differences in serotonin synthesis rate can be assessed noninvasively in brains of children and that the method is sensitive to differences in regional serotonin synthesis between autistic boys and their male siblings. The protocol can be performed easily and is well tolerated by children. The high sensitivity of PET in 3D mode combined with a nonstationary spatial filter algorithm allows acceptable image quality even with the necessarily low pediatric doses.

Acknowledgments

We thank the staff of the Children's Hospital of Michigan PET Center for their help in performing all studies. These studies were supported in part by the Mental Illness Research Association (MIRA) of Michigan, to which we are grateful.

References

Chakraborty, P. K., Mangner, T. J., Chugani, D. C., Muzik, O., and Chugani, H. T. (1996). A high-yield and simplified procedure for the synthesis of [^{11}C]alpha-methyl-L-tryptophan. *Nucl. Med. Biol.* 23: 1005–1008.

Chugani, D. C., Muzik, O., Chakraborty, P. K., Mangner, T. J., and Chugani, H. T. (1998). Human brain serotonin synthesis capacity measured in vivo with [^{11}C]alpha-methyl-tryptophan. *Synapse* 28: 33–43.

Cohen, Z., Tsuiki, K., Takada, A., Beaudet, A., Diksic, M., and Hamel, E. (1995). In vivo-synthesized radioactively labelled alpha-methyl serotonin as a selective tracer for visualization of brain serotonin neurons. *Synapse* 21: 21–28.

Diksic, M., Nagahiro, S., Sourkes, T. L., and Yamamoto, Y. L. (1990). A new method to measure brain serotonin synthesis in vivo. I. Theory and basic data for a biological model. *J. Cereb. Blood Flow Metab.* 9: 1–12.

Herholz, K. (1988). Non-stationary spatial filtering and accelerated curve fitting for parametric imaging with dynamic PET. *Eur. J. Nucl. Med.* 14: 477–484.

ICRP Publication 53 (1988). Radiation dose to patients from radiopharmaceutical. *In* "Annals of the International Commission on Radiological Protection," pp. 15–20. Pergamon. New York.

Missala, K., and Sourkes, T. L. (1988). Functional cerebral activity of an analogue of serotonin formed in situ. *Neurochem. Int.* 12: 209–214.

Muzik, O., Behrendt, D., Mangner, T. J., and Chugani, H. T. (1994). Design of a pediatric protocol for quantitative brain FDG studies with PET not requiring invasive blood sampling. *J. Nucl. Med.* 35: 104P (abstr.).

Muzik, O., Chugani, D. C., Chakraborty, P. K., Mangner, T. J., and Chugani, H. T. (1997a). Analysis of [^{11}C]alpha-methyl-tryptophan kinetics for the estimation of serotonin synthesis rate in vivo. *J. Cereb. Blood Flow Metab.* 17: 659–669.

Muzik, O., Shen, C., and Chugani, D. C. (1997b). Improved positioning method to obtain the left ventricular input function for quantitative PET brain studies using anatomical information from the transmission scan. *J. Nucl. Med.* 38: 94P (abstr.).

Nishizawa, S., Benkelfat, C., Young, S. N., Leyton, M., Mzengeza, S., DeMontigny, C., Blier, P., and Diksic, M. (1997). Differences between males and females in rates of serotonin synthesis in human brain. *Proc. Natl. Acad. Sci. U.S.A.* 94: 5308–5313.

Shoaf, S. E., and Schmall, B. (1996). Pharmacokinetics of alpha-methyl-L-tryptophan in rhesus monkeys and calculation of the lumped constant for estimating the rate of serotonin synthesis. *J. Pharmacol. Exp. Ther.* 277: 219–224.

CHAPTER

31

Brain Mapping of the Effects of Aging on Histamine H₁ Receptors in Humans: A PET Study with [¹¹C]Doxepin

MAKOTO HIGUCHI,[*] MASATOSHI ITOH,[†] KAZUHIKO YANAI,[‡] NOBUYUKI OKAMURA,[*]
ATSUSHI YAMAKI,[§] TATSUO IDO,[†] HIROYUKI ARAI,[*] and HIDETADA SASAKI[*]

[*] Department of Geriatric Medicine
Tohoku University School of Medicine
1-1 Seiryo-machi, Aoba-ku, Sendai 980-77, Japan
[†] Cyclotron and Radioisotope Center
Tohoku University
Aoba, Aramaki, Aoba-ku, Sendai 980-77, Japan
[‡] Department of Pharmacology I
Tohoku University School of Medicine
2-1 Seiryo-machi, Aoba-ku, Sendai 980-77, Japan
[§] Laboratory of Neuroinformation Science
Tohoku Gakuin University
Sendai, Japan

The binding of [¹¹C]doxepin to histamine H₁ receptors in the living human brain (using positron emission tomography (PET) was measured) and the relationship between age and the kinetic parameters of tracer binding was analyzed. Ten normal subjects underwent a dynamic PET scan after intravenous administration of [¹¹C]doxepin. A new analytical method to produce parametric images for tracer delivery (R_0, a target-to-reference ratio of cerebral blood flow), receptor binding (k_3 and k_4, i.e., rate constants for association with and dissociation from the receptors, respectively), and binding potential (BP = k_3/k_4) was applied using dynamic images and a standard plasma curve. The parametric images were spatially normalized, smoothed, and analyzed statistically by statistical parametric mapping (SPM96) software. Correlation between age and R_0, k_3, k_4, or BP was estimated on a voxel-by-voxel basis. k_3 and BP values showed an age-related decline with a significant correlation in areas that have been considered important sites for neuromodulation by histamine neurons, whereas k_4 was unchanged with aging. These results demonstrate the usefulness of this method to assess receptor binding in physiological or pathological conditions.

I. INTRODUCTION

The histamine neurons in human brain originate in the hypothalamus, one of the important sites in the limbic system. There is a wide distribution of the neuronal fibers and histamine H₁ receptors in the neocortex. These wide areas of distribution may be involved in cortical arousal. The density of histamine H₁ receptors is especially high in cingulate cortex, prefrontal cortex, amygdala, and uncal cortex. These areas play important roles in the limbic system. It is possible that the cognitive decline and disturbance of memory, attention, and sleep observed in the elderly result in part from age-related alterations in histaminergic neurotransmission.

There is an age-related decline of histamine H₁ receptor binding in the human brain, as demonstrated in a previous study with positron emission tomography (PET) and graphical analysis (Logan et al., 1990) based on regions of interest (ROI) (Yanai et al., 1992). Graphical analysis, however, may be affected by regional changes in cerebral blood flow and the ROI-based analysis requires a priori information about the location of structures of interest. Furthermore, graphical analysis requires a time-radioactivity curve of arterial plasma corrected for metabolites of the tracer. There are several problems in the use of measured arterial plasma data, including a relatively low signal-to-noise ratio in plasma data at late time points of the PET study, the difficulty of HPLC measurement of metabolites, and the risk of arterial cannulation.

The objectives of this study are

1. To investigate *in vivo* histamine H_1 receptors in the human brain with PET and [^{11}C]doxepin, a selective radioligand for histamine H_1 receptors.
2. To apply a novel analytical method to estimate the kinetic parameters of the tracer and to generate parametric images in a reasonably short time without individual plasma data.
3. To elucidate brain areas showing age-related changes in histamine H_1 receptor binding on a voxel-by-voxel basis using parametric images and statistical parametric mapping (SPM) software.

II. MATERIALS AND METHODS

A. Subjects

Normal volunteers studied were eight males and two females whose ages ranged from 20 to 80 years (mean ± SD: 43 ± 23.8 years).

B. Theory for a New Analytical Method

A new method applied here, approximate kinetic imaging (AKI), is able to generate parametric images of k_3, the rate constant for association of the tracer with the H_1 receptor; k_4, the rate constant for dissociation of the tracer from the H_1 receptor; BP, the binding potential of the tracer (k_3/k_4); and R_0, a target-to-reference ratio of rate constants for the tracer delivery (K_1 or k_2). A reference ROI was placed on the cerebellum, which is considered to contain a negligibly small amount of H_1 receptor (Kanba and Richelson, 1984).

R_0 is estimated by the following equation:

$$R_0 = \frac{\int_0^{t_2} C_t^s(s)\,ds \cdot \int_0^{t_1} ds \int_0^s C_t^r(u)\,du - \int_0^{t_1} C_t^s(s)\,ds \cdot \int_0^{t_2} ds \int_0^s C_t^r(u)\,du}{\int_0^{t_2} C_t^r(s)\,ds \cdot \int_0^{t_1} ds \int_0^s C_t^r(u)\,du - \int_0^{t_1} C_t^r(s)\,ds \cdot \int_0^{t_2} ds \int_0^s C_t^r(u)\,du}, \quad (1)$$

where t_1 and t_2 are times in the early phase of the scan (3 and 5 min post-tracer injection, respectively, in this study) and $C_t^s(t)$ and $C_t^r(t)$ are the radioactivities measured by PET at time t in the specific voxel and the reference ROI, respectively. In this equation, R_0 is determined by assuming that at early times the amount of tracer being associated with the receptors is much greater than the amount being disassociated. The estimation of k_3 and k_4 is achieved as follows

$$k_3 = \frac{\int_0^{t_3} C_b^s(s)\,ds \cdot \int_0^{t_4} ds \int_0^s C_b^s(u)\,du - \int_0^{t_4} C_b^s(s)\,ds \cdot \int_0^{t_3} ds \int_0^s C_b^s(u)\,du}{\int_0^{t_3} ds \int_0^s C_f^s(u)\,du \cdot \int_0^{t_4} ds \int_0^s C_b^s(u)\,du - \int_0^{t_4} ds \int_0^s C_f^s(u)\,du \cdot \int_0^{t_3} ds \int_0^s C_b^s(u)\,du} \quad (2a)$$

$$k_4 = \frac{\int_0^{t_3} C_b^s(s)\,ds \cdot \int_0^{t_4} ds \int_0^s C_f^s(u)\,du - \int_0^{t_4} C_b^s(s)\,ds \cdot \int_0^{t_3} ds \int_0^s C_f^s(u)\,du}{\int_0^{t_3} ds \int_0^s C_f^s(u)\,du \cdot \int_0^{t_4} ds \int_0^s C_b^s(u)\,du - \int_0^{t_4} ds \int_0^s C_f^s(u)\,dU \cdot \int_0^{t_3} ds \int_0^s C_b^s(u)\,du} \quad (2b)$$

with $C_f^s(t) = R(t) \cdot C_t^r(t)$, $C_b^s(t) = C_t^s(t) - R(t) \cdot C_t^r(t)$, where t_3 and t_4 are times in the late phase of the scan (70 and 90 min post-tracer injection, respectively, in this study), $C_f^s(t)$ and $C_b^s(t)$ are the concentrations of free and specifically bound tracer in the specific voxel, respectively, and $R(t)$ is a ratio of the radioactivities in the free and plasma compartments between the specific voxel and the reference ROI. The function $R(t)$ is employed to take into account the regional difference in tracer delivery

$$R(t) = \frac{(1-CBV)R_0 K_1' e^{-R_0 k_2' t} \otimes C_p(t) + CBV \cdot C_p(t)}{(1-CBV) K_1' e^{-k_2' t} \otimes C_p(t) + CBV \cdot C_p(t)}, \quad (3)$$

in which CBV is a fraction of the cerebrovascular volume (usually fixed to 0.05), K_1' and k_2' are transfer rate constants between tissue and plasma in the reference ROI, $C_p(t)$ is the plasma radioactivity at time t, and $\otimes$ denotes convolution. The use of the function $R(t)$ is a simple approach to improve on a direct substitution of $C_t^r(r)$ for $C_f^s(t)$, although $R(t)$ might include k_3 and k_4 in a more rigorous definition.

C. PET Measurement and Image Analysis

PET scans were obtained using the ECAT PT931/04-12 scanner (CTI Inc, Knoxville, TN) after intravenous injection of 10–15 mCi of [^{11}C]doxepin. Seven PET slices were produced simultaneously, and dynamic images were collected from 0 to 90 min after injection. Arterial blood samples were obtained, and an individual metabolite-corrected plasma time–radioactivity curve was generated for use in the graphical analysis (Logan *et al.*, 1990). A cerebellar ROI was placed and a cerebellar time–radioactivity curve was obtained. Standard values of K'_1 and k'_2 were determined by averaging individual values of K'_1 and k'_2 estimated by a two-compartment analysis of the cerebellar ROI data using individual $C_p(t)$ curves. A standard $C_p(t)$ curve was generated by averaging data using individual $C_p(t)$ curves corrected for metabolites. The "AKI" method was applied to generate parametric images of k_3, k_4, BP, and R_0, using the cerebellar time–radioactivity curve, the standard values of K'_1 and k'_2, and the standard curve of $C_p(t)$. This method provided a parametric image data base of normal subjects for analysis without blood sampling. In an ROI-based analysis, the use of a standard $C_p(t)$ curve and standard values of K'_1 and k'_2 has been demonstrated to produce R_0, k_3, and k_4 values that are similar to those obtained using individual $C_p(t)$, K'_1, and k'_2 (Higuchi et al., 1998). Parametric images of BP were also constructed by the graphical method, where an individual $C_p(t)$ was employed to estimate the volume of distribution (V_d) in each voxel; the V_d in the cerebellar ROI was used for the determination of the BP value in each voxel.

D. Statistical Analysis

Statistical parametric mapping (SPM96) software (Friston et al., 1994, 1995) was used for spatial normalization, smoothing, and statistical analyses of the parametric images. Because the field of view of the PET scanner covered only 54 mm in the axial direction, an eight-parameter affine transformation was used to match the individual R_0 images to the regional CBF template, which conformed to the standard anatomical space (Talairach and Tournoux, 1988). The estimated parameters for the normalization of the R_0 images also were applied to the spatial normalization of the k_3, k_4, and BP (by the "AKI" and graphical analyses) images. Following the spatial normalization step, the images

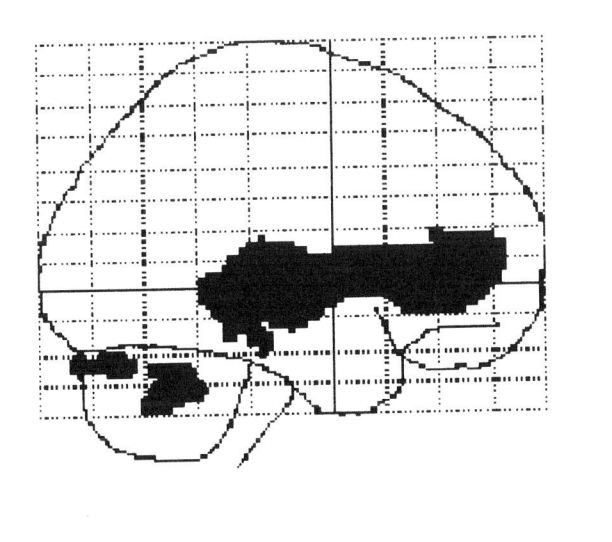

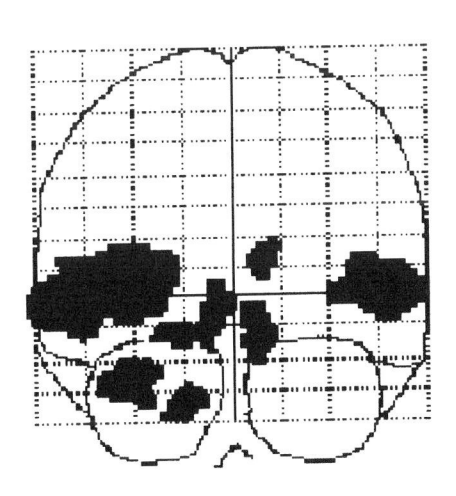

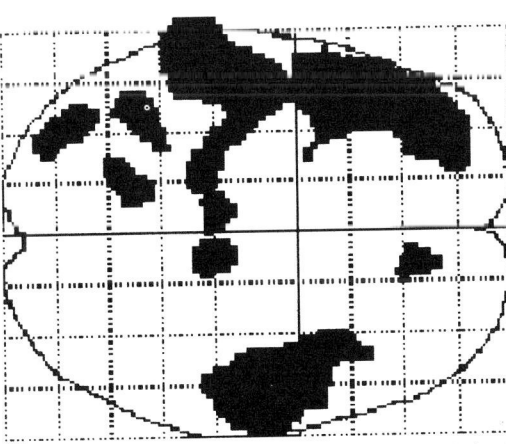

FIGURE 1 SPM{Z} map for the negative correlation between age and BP estimated by the AKI method. The map is thresholded at $p = 0.0001$ (uncorrected) and is displayed as orthogonal projections.

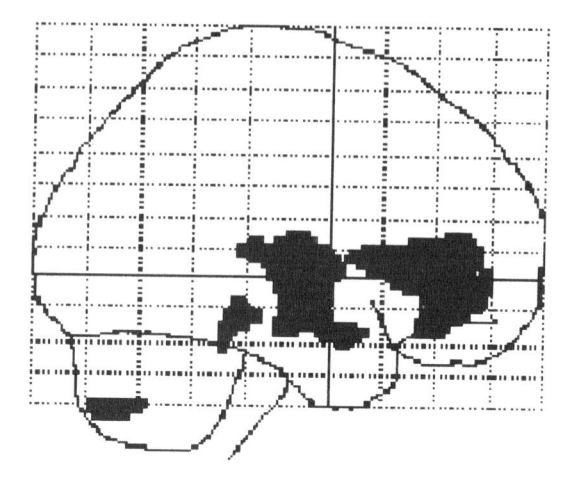

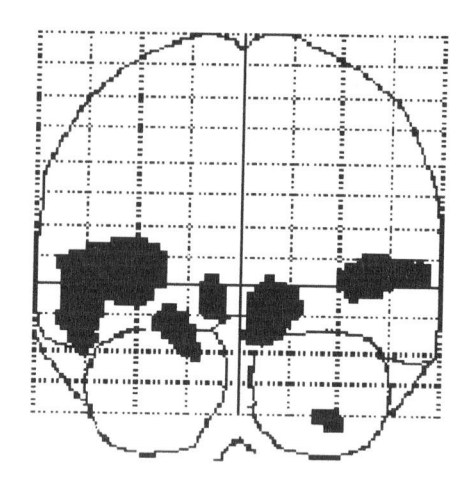

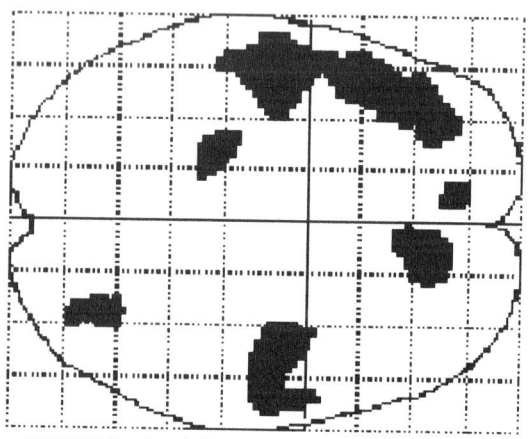

FIGURE 2 SPM{Z} map for the negative correlation between age and k_3 estimated by the AKI method. The map is thresholded at $p = 0.001$ (uncorrected).

were smoothed using an isotropic Gaussian kernel with full width at half maximum of 16 mm.

Statistical analysis on a voxel-by-voxel basis was performed to determine correlations between age and values of k_3, k_4, BP, or R_0. Analyses of k_3, k_4, and BP were carried out without any global normalization, and the analysis for R_0 was performed with an analysis of covariance (ANCOVA) correction for the global value. SPM{t} maps were transformed to SPM{Z} maps, and the distribution of Z values was evaluated.

III. RESULTS AND DISCUSSION

The distribution of Z values for the negative correlation between subject age and the BP value estimated by the "AKI" method is depicted in Fig. 1. The SPM{Z} maps are thresholded at $p = 0.0001$ and are displayed as orthogonal projections. Histamine H_1 receptor binding declined with age, with a close correlation, especially in the temporal cortices bilaterally, the left prefrontal cortex, the right anterior cingulate cortex, and the left parahippocampal cortex.

There was an age-related decline of k_3 with a significant correlation ($p < 0.001$) particularly in the temporal cortices bilaterally, the left prefrontal cortex, the anterior cingulate cortices, and the left parahippocampal cortex (Fig. 2). The pattern of k_3 reduction with increasing age was similar to the pattern of the BP reduction. Figure 3 shows the relationship between age and k_3 in the right anterior cingulate cortex, where a good linear correlation was observed.

No age-related decline of k_4 was found. There were a few small areas presenting an age-related increase of k_4, with a significant correlation ($p < 0.001$) (Fig. 4). Accordingly, it is likely that k_4 remained unchanged

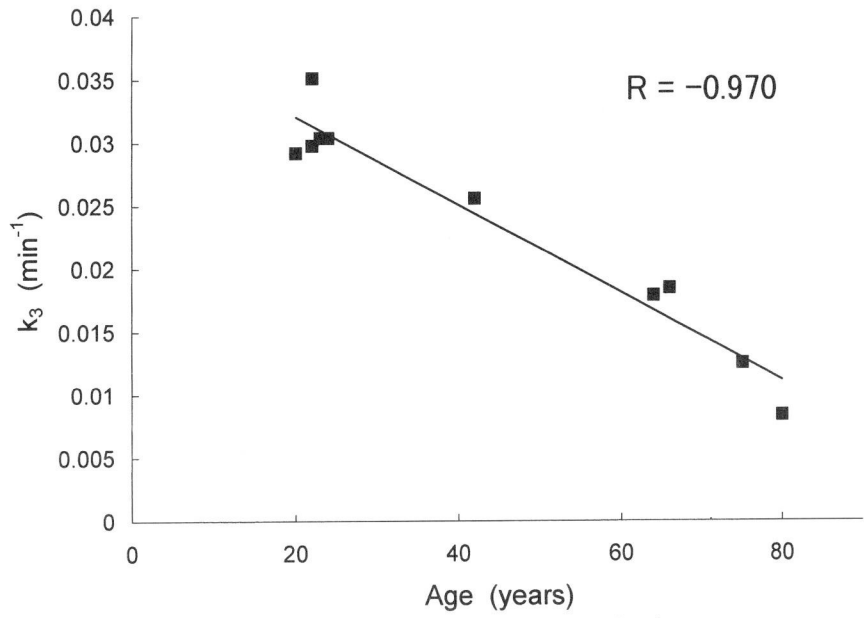

FIGURE 3 Plot of k_3 versus age in the right anterior cingulate cortex.

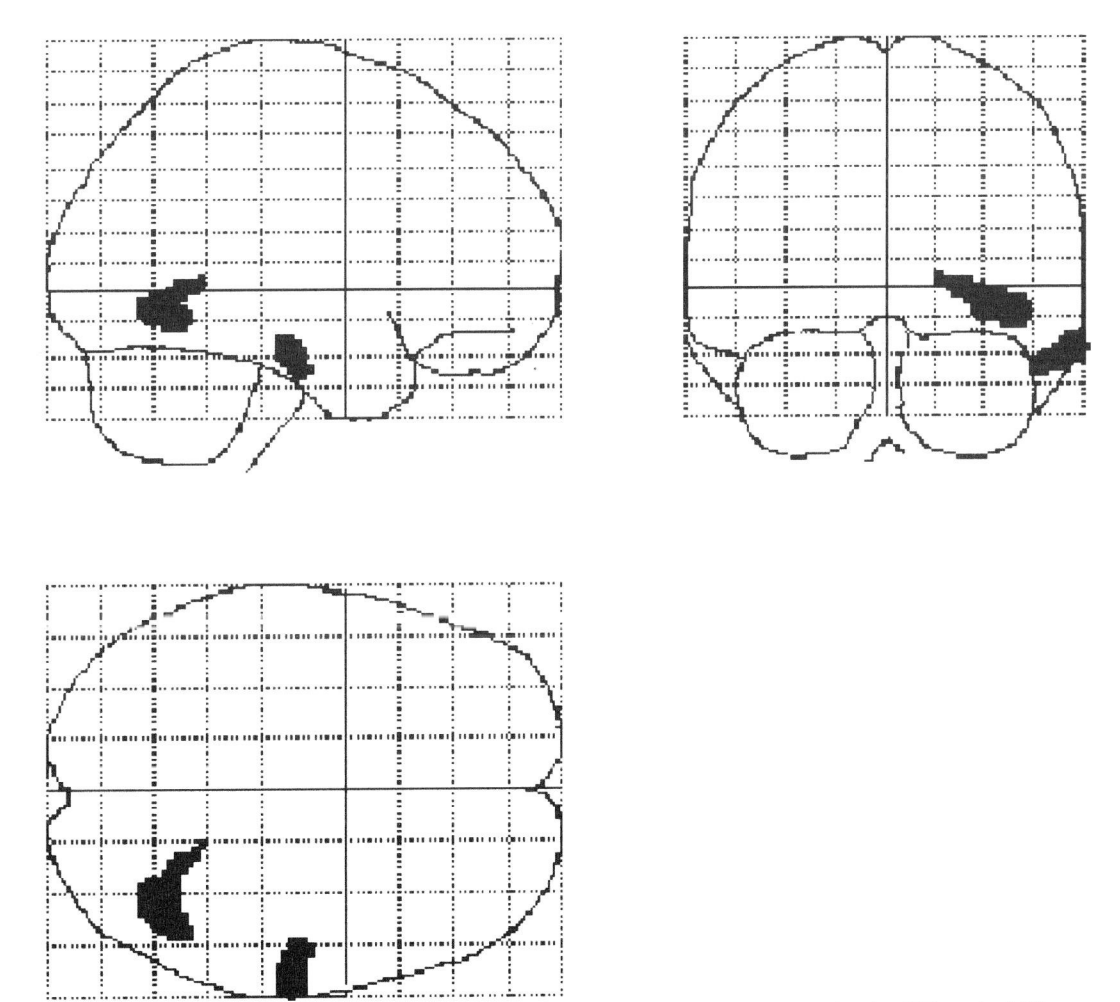

FIGURE 4 SPM{Z} map for the positive correlation between age and k_4 estimates by the AKI method. The map is thresholded at $p = 0.001$ (uncorrected).

during aging and that the BP (k_3/k_4) was reduced with increasing age primarily due to the age-related decline of k_3.

The value of R_0, considered proportional to rCBF, presented an age-related decline with a significant correlation ($p < 0.001$) in the bilateral temporal cortices (Fig. 5). These areas, however, contained the Sylvian fissures, suggesting that the decline of the R_0 value may be in part due to cortical atrophy and the consequent partial volume effect with increasing age. Although the changes of k_3 and BP in the bilateral temporal cortices may also include effects of cortical atrophy, the decreased k_3 and BP in the prefrontal cortex, the anterior cingulate cortex, and the parahippocampal cortex were not caused by age-related morphological changes, as no changes in k_4 or R_0 during aging were observed in these areas.

Figure 6 shows areas with a significant negative correlation ($p < 0.001$) between age and BP as estimated by graphical analysis. The pattern of BP reduction was slightly different from the pattern obtained by the AKI method. In the comparison analysis, the BP value estimated by graphical analysis was lower than that estimated by the AKI method in voxels with a low value of R_0 (i.e., low rCBF) (data not shown). The BP value estimated by graphical analysis was occasionally negative. Therefore, the calculation of BP by graphical analysis may be affected by regional differences in CBF.

The time required for the calculations by the AKI method was comparable to the time for the utilization of graphical analysis.

Histamine H_1 receptor binding showed an age-related decline, with close correlations between age and binding in the temporal cortex, the prefrontal cortex, the anterior cingulate cortex, and the parahippocampal cortex. These areas have a relatively high density of histamine H_1 receptors and are important

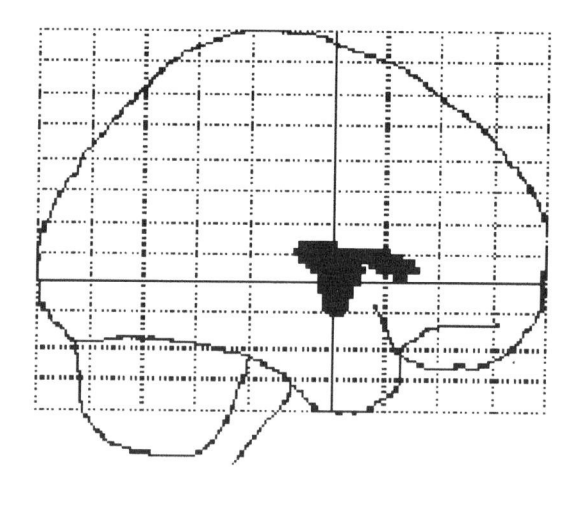

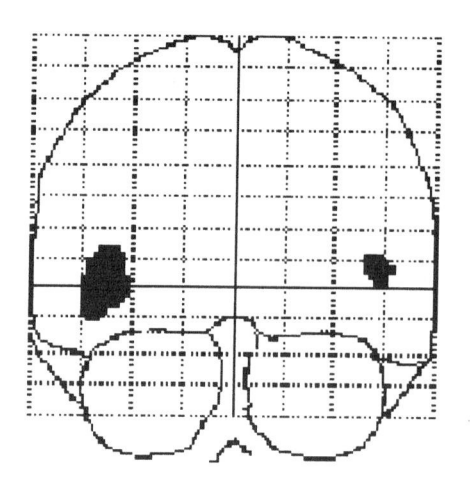

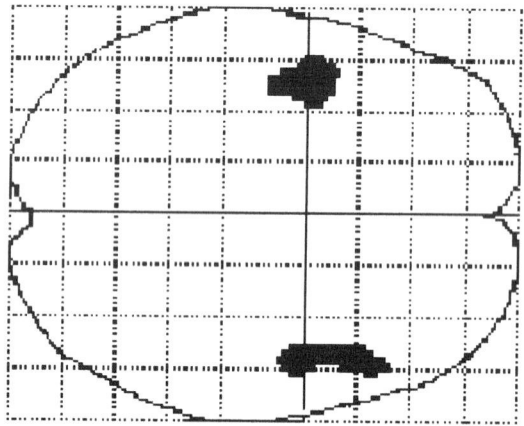

FIGURE 5 SPM{Z} map for the positive correlation between age and R_0 estimated by the AKI method. The map is thresholded at $p = 0.001$ (uncorrected).

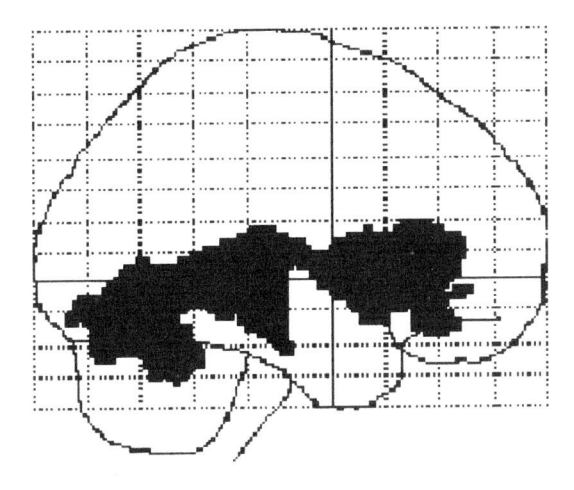

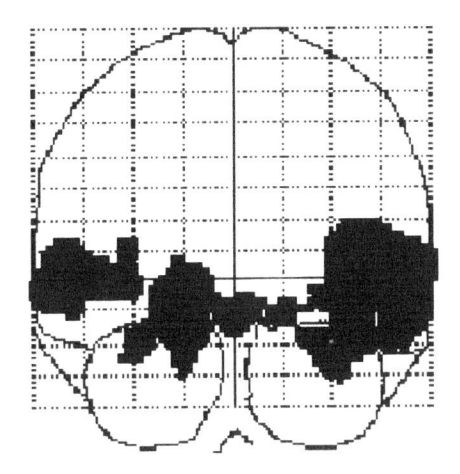

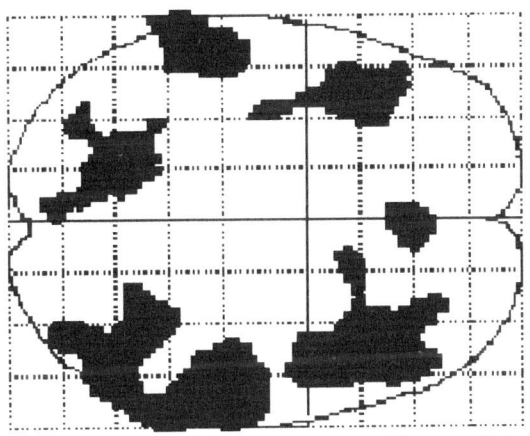

FIGURE 6 SPM{Z} map for the positive correlation between age and BP value estimated by graphical analysis. The map is thresholded at $p = 0.001$ (uncorrected).

sites for neuromodulation by central histamine neurons. The results of this study indicate that the reduction of histamine H_1 receptor binding during aging may be closely associated with the age-related decline of various physiological functions such as sleep and wakefulness, attention, cognition, and memory, which are in part regulated by histaminergic neurotransmission.

The new analytic method applied here does not require arterial blood sampling, provided standard values for the transfer rate constants between plasma and tissue in the reference region and a standard plasma time–radioactivity curve are available. This method is useful in studies of diseases with reduced rCBF because it allows a separation of measures of receptor binding (k_3, k_4, and BP) from tracer delivery (R_0). Accordingly, it will provide further information about the pathogenesis of dementia, including Azheimer's disease. It is applicable to image analyses using other radioligands when a region devoid of receptors is present. Furthermore, this study has demonstrated that the application of SPM to PET measurements with a radioligand will help in understanding alterations in receptor binding under diverse physiological and pathological conditions.

Acknowledgments

We thank S. Watanuki and M. Miyake for their technical assistance in the PET studies. We are also grateful to Professor J. J. Frost for helpful suggestions.

References

Friston, K. J., Worsley, K. J., Frackowiak, R. S. J., Mazziotta, J. C., and Evans, A. C. (1994). Assessing the significance of focal activations using their spatial extent. *Hum. Brain Mapp.* **1**: 214–220.

Friston, K. J., Holmes, A. P., Worsley, K. J., Poline, J. P., Frith, C. D., and Frackowiak, R. S. J. (1995). Statistical parametric maps in functional imaging: A general linear approach. *Hum. Brain Mapp.* **2**: 189–210.

Higuchi, M. *et al.* (1988). In preparation.

Kanba, S., and Richelson, E. (1984). Histamine H_1 receptors in human brain labelled with [^{3}H]doxepin. *Brain Res.* **304**: 1–7.

Logan, J., Fowler, J. S., Volkow, N. D., Wolf, A. P., Dewey, S. L., Schlyer, D. J., MacGregor, R. R., Hitzemann, R., Bendriem, B., Gatley, S. J., and Christman, D. R. (1990). Graphical analysis of reversible radioligand binding from time-activity measurements applied to [N-^{11}C-methyl]-(-)-cocaine PET studies in human subjects. *J. Cereb. Blood Flow Metab.* **10**: 740–747.

Talairach, J., and Tournoux, P. (1988). "A Co-Planar Stereotaxic Atlas of a Human Brain." Thieme, Stuttgart.

Yanai, K., Watanabe, T., Meguro, K., Yokoyama, H., Sato, I., Sasano, H., Ito, M., Iwata, R., Takahashi, T., and Ido, T. (1992). Age-dependent decrease in histamine H_1 receptor in human brains revealed by PET. *NeuroReport* **3**: 433–436.

SECTION IV

STATISTICAL ANALYSIS

CHAPTER

32

PET Analysis Using a Variance Stabilizing Transform[1]

URS E. RUTTIMANN, DANIEL RIO, ROBERT R. RAWLINGS, PAUL ANDREASON, and DANIEL W. HOMMER

Section of Brain Electrophysiology and Imaging
Laboratory of Clinical Studies
National Institute on Alcohol Abuse and Alcoholism
National Institutes of Health
Bethesda, Maryland 20892

Two random-field analysis methods based on either z-field or t-field statistics are currently in use, differing, respectively, by whether a spatially homogeneous error variance is assumed. A third method is proposed, which estimates and applies a variance stabilizing transform (VST) prior to estimating the z-field. This study investigates the feasibility and performance of this new method relative to the two currently used methods and an approximation method that assumes separate homogeneous variance within gray matter and white matter/cerebrospinal fluid regions, which were segmented using coregistered magnetic resonance images. Intersubject mean standard deviation (SD) images were formed from coregistered [^{18}F]fluorodeoxyglucose positron emission tomography scans of 32 subjects. Various polynomial functions of the mean were fitted to the plot of pixel means versus pixel SDs to construct a VST. Residual error analysis found a simple power function adequate, resulting in a VST $y \propto x^{1-b}$, with $b = 0.246$. The relative performance of the four methods in the analysis of drug-induced changes in regional cerebral metabolism of glucose was assessed in 13 normal subjects. The t-field method detected $< 1\%$ of the pixels identified in the z-field as statistically significant ($p = 0.1$ per volume). VST and segmentation methods detected similar fractions each (47.5 and 56.0%), validating the transformation concept. The number of pixels identified in z-statistic images was considered inflated because the global SD underestimated the regional SDs in gray matter, where all significant pixels were located. Hence, the VST corrects for overly optimistic results produced by z statistics without requiring segmentation, whereas t statistics incur an unacceptably large loss of statistical power for typical study sizes.

I. INTRODUCTION

Hypothesis-driven inference requires specification of a data model, which is typically parametric, however, the high dimensionality of data usually necessitates the use of sequential univariate testing procedures. Data models that provide means to make corrections for multiple testing commonly characterize the error variance by a continuous random field with zero mean and known smoothness (Worsley *et al.*, 1996). Depending on assumptions regarding the spatial homogeneity of this error variance, two different analytic routes are possible. The assumption of a spatially homogeneous variance permits pooling of the local error estimates over all pixels to yield an estimator with degrees of freedom high enough that the error variance can be regarded as known. Hence, a homogeneous Gaussian random field model is appropriate, yielding z-statistic images for inference testing. If the assumption of spatial homogeneity is considered untenable, the variance must be estimated at each individual pixel, leading to t-statistic images where the degrees of freedom depend on the number of scans. There is considerable debate

---

[1] Transcripts of the BRAINPET97 discussion of this chapter can be found in Section VIII.

regarding the tenability of a homogeneous field model (Friston, 1995). The assumption of homogeneity produces a stable variance estimate, but can introduce bias in the calculation of z-statistic image if the local variances deviate substantially from the pooled estimate. In contrast, t-statistic images provide unbiased, but possibly unstable inferences (i.e., low power) because the degrees of freedom are typically low. As a possible solution to this dilemma, the following procedure is proposed. Estimate and apply a global image transformation to positron emission tomography (PET) data that makes the local variance to some extent independent of the local mean (variance stabilizing transform). This renders the assumption of spatial homogeneity more acceptable and an analysis based on z-statistic images less objectionable.

II. MATERIALS AND METHODS

A. Subjects

The database totaled 32 subjects, comprising 13 normal male volunteers and 19 abstinent male alcoholics. Images from all subjects were used to find a suitable variance stabilizing transform. The 13 normal subjects served to investigate the impact of using such a variance stabilizing transform on estimating drug-induced changes in regional cerebral metabolism of glucose (rCMRglc).

B. Scanning Procedure

Two sequential [^{18}F]fluorodeoxyglucose (FDG) (PET) scans were obtained from each subject with the Scanditronix PC1024-7B scanner, one before (3 mCi) and one after (5 mCi) administration of 0.08 mg/kg m-chlorophenylpiperazine (mCPP). The scanner produced by filtered backprojection a volume stack of 21 axial emission images, yielding matrices of 128 × 128 pixels with a size of 2 × 2 mm. The in-plane resolution was approximately 7.5 mm (FWHM), and the nominal slice separation was 3.6 mm, with 11 mm axial resolution (FWHM). In addition to the PET images, 11 subjects had full volumetric coronal (124 slices, 256 × 256 pixels) magnetic resonance (MR) brain scans acquired on a GE-Signa 1.5-T scanner.

C. Spatial Registration

The set of PET images from one of the normal volunteers served as the reference three-dimensional (3D) brain template for the purpose of intersubject registration. Registration was accomplished by a fully automated method, implementing 3D affine transformations, with reslicing into the reference axial image planes accomplished by tricubic spline interpolation (Thévenaz et al., 1997). Registration of the reference PET image set to the MR scan of the same subject was accomplished by a modified version of an algorithm (Besl and McKay, 1992) that minimizes displacement vectors of nearest points on a set of surfaces to be matched.

D. Variance Stabilizing Transform

If the intersubject standard deviation (SD) of the pixel intensity X is not constant, but a function of the pixel mean μ, e.g., SD = $f(\mu)$, then it may be possible to apply a transformation $g(X)$ such that SD[$g(X)$] is approximately independent of the mean of $g(X)$. Two problems are associated with this approach: (i) what is $f(\mu)$, and (ii) given $f(\mu)$, what is $g(X)$?

A solution to (i) was attained by fitting a few "reasonable" model functions $f_i(\mu)$ to the observed means and corresponding standard deviations of pixels located within the intracranial space. Estimates of the parameters specified by various model functions were obtained by a Marquardt–Levenberg nonlinear least-squares estimation procedure (Marquardt, 1963). The image sample consisted of all 32 coregistered predrug scans, which were randomly assigned into two equal-sized groups (group 1 and group 2). For each group, average and standard deviation images were computed, providing two independent sets of parameter estimates for each model function. The fit of the different models was assessed by cross-validation, i.e., the residual errors in each group were evaluated by using the parameter estimates derived from the other group.

An answer to (ii) is straightforward, once $f(\mu)$ is known in analytical form. The variance of a random variable X subjected to the nonlinear transform $g(X)$ can be approximated by the first-order terms of a Taylor series expansion of $g(X)$ about the mean $\mu = E(X)$,

$$g(X) = g(\mu) + \frac{dg}{dX}\bigg|_{\mu} (X - \mu) + \cdots. \quad (1)$$

Using the relation Var$(a + bX) = b^2$ Var(X), where $b = dg/dX|_{\mu}$, it follows from Eq. (1) that Var[$g(X)$ ≈ $[dg(X)/dX]_{\mu}^2$ Var(X). If this variance is to be a constant k^2, then with $\sqrt{\text{Var}(X)} = f(X)$, it is required that $dg(X)/dX = k/f(X)$ be approximately true, and hence, apart from an additive constant

$$g(X) \propto \int \frac{dX}{f(X)}. \quad (2)$$

Equation (2) was solved analytically for a set of prototype functions $f_i(X)$.

E. Statistical Analysis

It was of interest to identify in a sample of $N = 13$ normal subjects, brain regions where the cerebral metabolic rate of glucose (CMRglc) changed in response to the administration of mCPP. To that end, each subject's predrug image was subtracted from the postdrug image, the difference images smoothed by a 2D Gaussian filter to yield an in-plane resolution with FWHM ≈ 10.3 mm, and then averaged across the subjects. The results were analyzed by the random field methods developed by Worsley et al., (1996), except that the analysis was performed without division of data by the global CMRglc. Under the assumption of a stationary Gaussian random field (z field), i.e., a spatially homogeneous error variance, the pixel variances can be pooled to yield a more stable estimator. The probability that a local minimum in a 2D z field exceeds a threshold t is approximately

$$P(z_{max} \geq t) \approx \#resels \frac{4 \ln 2}{(2\pi)^{3/2}} t e^{-t^2/2} \qquad (3)$$

where

$$\#resels = \frac{Area}{FWHM^2}. \qquad (4)$$

Area is the size of the search region considered, and FWHM is the (known) smoothness of the Gaussian field.

If the error variance is considered to be location dependent, a t-field model is appropriate and the probability that a local maximum exceeds t is

$$P(t_{max} \geq t) \sim \#resels \frac{4 \ln 2}{(2\pi)^{3/2}} \frac{\Gamma\left(\frac{\eta+1}{2}\right)}{\left(\frac{\eta}{2}\right)^{1/2} \Gamma\left(\frac{\eta}{2}\right)}$$

$$\times t \left(1 + \frac{t^2}{\eta}\right)^{-(\eta-1)/2}, \qquad (5)$$

where $\eta = N - 1$ are the degrees of freedom, and N in our case is given by the number of image pairs. Corresponding expressions for 3D fields and for small search regions have been developed (Worsley et al., 1996), but are not of concern here.

The variance stabilizing transform was applied to the predrug and postdrug images of each subject to yield a more ideal z field. To assess the validity of this alternative method, a fourth methodology was employed that incorporated prior information about the variance inhomogeneity by a dichotomy. Based on the assumption of a higher variance in gray matter than in ventricular or white matter regions, segmentation (Momenan et al., 1997) into two complementary subregions was performed with the aid of the coregistered MR volume. Assuming a homogeneous variance within each of these two subregions, two separate analyses based on Eq. (3) were performed. For conciseness of the discussion, methods using Eqs. (3) or (5) directly or applying a variance stabilizing transform or image segmentation before using Eq. (3) are labeled M1, M2, M3, and M4, respectively.

III. RESULTS AND DISCUSSION

A. Variance Stabilizing Transform

Figure 1 shows a scatter plot of the observed standard deviations versus the means at corresponding pixel locations, accumulated from all intracranial pixels of the group 1 images. Superimposed are the fitted curves based on three proposed model functions. Qualitatively, the fits are very similar, except at either end of the data range, where the scatter is greatest. Group 2 data produced very similar results. Table 1 summarizes the results of the fits in terms of the residual standard error of the regression, both for parameter resubstitution and cross-validation. The heuristics for the selection of the fit functions were as follows: $f_1(x)$ represented the simplest power function of the mean; in $f_2(x)$ the total imaging noise variance was considered a mixture of white noise (a) and Poisson noise (b); and $f_3(x)$ constituted the next higher complexity by including second order powers of the mean. Although the residual errors observed with parameter resubstitution in group 2 seem to indicate that more complex models yielded somewhat better fits, it is clear that the simplest model provided an adequate data description. More importantly, the cross-validation results with $f_1(x)$ produced the smallest (group 1) and second smallest (group 2) residual errors. The difference between the two residual error estimates is partly a consequence of the correlation of the pixel means used as the regressor, causing underestimation of the resubstitution residual error. This correlation arises from the spatial correlation among neighboring pixels, which was kept to a minimum by employing the unfiltered scans in the estimation process.

Equation (2) shows that by selecting $f_1(x)$, only parameter b is relevant, yielding the stabilizing transform $g(x) \propto x^{1-b}$, which would also have resulted by

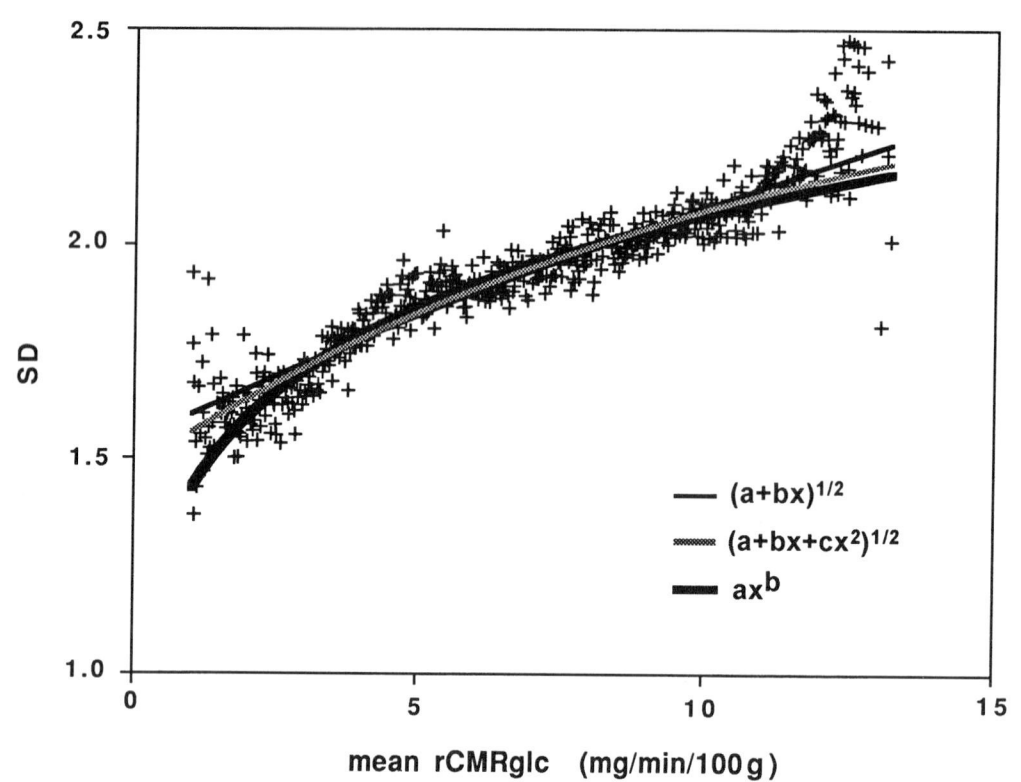

FIGURE 1 Scatter plot of intersubject standard deviations vs means at corresponding pixels, with regression curves for three fit functions superimposed. Group 1 data: 16 subjects, 15 slices each.

using the familiar Box–Cox transform procedure (Box and Cox, 1964). The estimate derived from the pooled group 1 and group 2 sample was $b = 0.246$. Note that for ideal Poisson image noise, often assumed in PET modeling, this parameter would amount to $b = 0.5$, which clearly does not characterize the observed noise.

Figure 2 shows for a slice at the level of the dorsal thalamus the means (a) and SDs before (b) and (c) application of the transform $f_1(x)$ to group 1 data. The association of larger SDs in Fig. 2b at locations with higher means has been effectively removed in Fig. 2c. Note that as a result of the more uniform SDs in Fig. 2c, misregistration errors become more prominent, as is particularly evident in the region near the posterior interhemispheric space, where intersubject anatomical variability associated with left/right cerebral asymmetry was greatest. However, despite the visible regional SD variations displayed in Fig. 2b, a formal χ^2 test (Worsley et al., 1996) did not reject variance homogeneity, typifying the generally low statistical power associated with tests for distributions.

B. Comparison of Methods

Figure 3 displays the coregistered MR image for the standard shown in Fig. 2a, with pixels overlaid in white where significant ($p = 0.1$/volume) increases in rCMRglc were detected by M1, M3, or M4 (no significant activations were detected in that slice by M2). In general, M1 identified more activated pixels. In this particular slice, bilateral responses in the thalamus and the left insular area were common to all methods, whereas the activations in the left middle frontal gyrus and the left caudate nucleus showed some inconsistencies among the methods.

TABLE 1 Residual Standard Errors of Regression for the Fit Functions [SD = f_i(mean)]

$f_i(x)$	Group 1	Group 2
ax^b	4.92[a]	4.87
	15.58[b]	15.45
$\sqrt{a + bx}$	5.02	4.41
	15.80	15.47
$\sqrt{a + bx + cx^2}$	4.92	4.40
	15.76	15.40

[a] Resubstitution.
[b] Cross-validation.

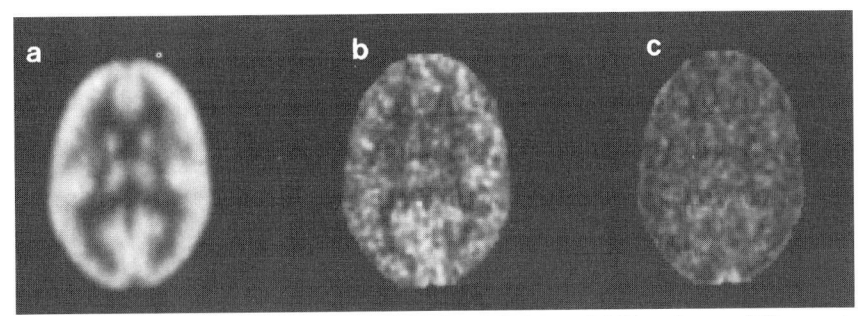

FIGURE 2 Axial PET slice at the level of the dorsal thalamus. (a) Mean image of 13 normal subjects before drug administration; (b) intersubject standard deviations in the difference image between postdrug and predrug scans before and (c) after application of an estimated variance stabilizing transform.

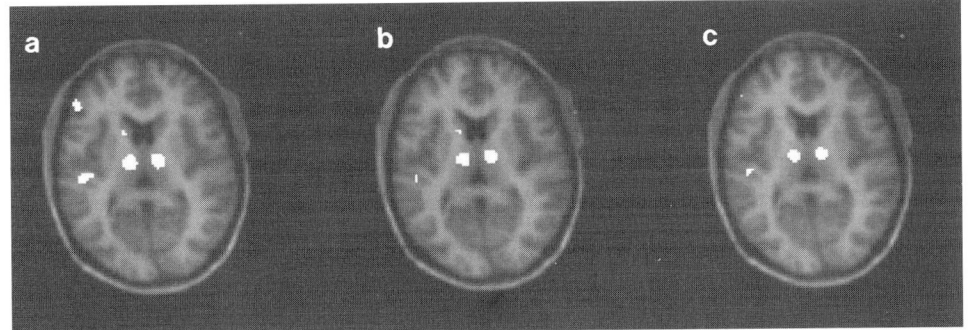

FIGURE 3 Reconstruction of a coregistered axial MR image of the reference subject at the level of the dorsal thalamus. Superimposed in *white* are regions where a significant increase in rCMRglc after administration of mCPP was detected by (a) a z-field model without transformation (M1), (b) a z-field model after application of a variance stabilizing transform (M3), and (c) a model employing two separate z fields in gray matter and white matter/ventricular regions (M4).

A quantitative comparison of the four thresholding methods is shown in Table 2. Most striking in the comparison of M1 versus M2 is the much higher threshold required for the *t*-field method (5) with a sample size of $N = 13$ than for the z-field method (3). Therefore, a relatively large loss of statistical power is incurred by using a local rather than a global estimate of the standard deviation. As a result, despite the higher global maximum in the *t* field, less than 1% of the pixels detected by M1 were found significant by M2. The preprocessing by a variance stabilizing transform performed in M3 decreased z_{max} by a small amount; however, the main effect was a reduced number of detected pixels (47.5%) as compared to M1. M4, which incorporated gray/white segmentation, produced results comparable to those of M3 (56.0%). Because method M4 may be considered a dichotomous approximation to the continuous transform applied in M3, the corresponding variance stabilization is likely to be less perfect. Although the slightly larger number of pixels detected by M4 may be fortuitous, the similarity of the results produced by M3 and M4 does lend credibility to the variance stabilizing approach. Based on this mutual validation, the number of pixels detected by M1 must be regarded as inflated. This interpretation is supported by the observation that all detected pixels were located in gray matter regions, where

TABLE 2 Comparison of the Four Thresholding Methods

Method[a]	Threshold[b]	Significant voxels n	%	z_{max} or t_{max}
M1	4.18	336	100	6.17
M2	7.27	2	0.6	7.46
M3	4.18	160	47.6	5.88
M4	4.18	188	56.0	5.74

[a] M1, z field without transformation; M2, *t* field; M3, z field after transformation; and M4, separate z fields in gray and white/CSF regions.

[b] $p = 0.1$ per volume: 8 slices, $p = 0.0125$ per slice; 830 2D resels with 10.3 mm FWHM.

generally larger local SDs prevailed than in white matter or CSF areas (Fig. 2b). Consequently, the pooled SD used in M1 underestimated the underlying true SDs in gray matter regions and produced z values that were inflated relative to those produced by M3 or M4. However, M3 is preferred over M4 because it does not require coregistration to an MR image to accomplish segmentation, procedures that are both labor intensive and fraught with their own problems.

Many arguments can be advanced to discredit the assumption of a spatially homogeneous variance; however, there is usually insufficient power provided by commonly used sample sizes to formally reject it, as was the case in this analysis. The issue is not whether a model is correct in the exact sense (it never is!) but whether its use provides more benefits in terms of power and stability than is lost by the introduction of bias due to an incorrect assumption. From this point of view, M2 incurs such a huge loss of statistical power for typically employed sample sizes to render it practically unusable, whereas M1 introduces a small bias with consequent loss of specificity. Hence, if there is a concern regarding a possible bias introduced by the adoption of a z-field model, the use of a t-field model is not a viable solution, whereas the removal of variance inhomogeneity by a suitable transform prior to applying a z-field model is an efficient alternative.

References

Besl, P. J., and McKay, N. D. (1992). A method for registration of 3-D shapes. *IEEE Trans. Pattern Anal. Mach. Intell.* **14**: 239–256.

Box, G. E. P., and Cox. D. R. (1964). Analysis of transformations. *J. Roy. Stat. Soc., Ser. B* **26**: 211–252.

Friston, K. (1995). Statistical parametric mapping: Ontology and current issues. *J. Cereb. Blood Flow Metab.* **15**: 361–370.

Marquardt, D. W. (1963). An algorithm for least-squares estimation of nonlinear parameters. *SIAM J. Appl. Math.* **11**: 431–441.

Momenan, R., Hommer, D., Rawlings, R., Ruttimann, U. E., Kehrich, M., and Rio, D. (1997). Intensity adaptive segmentation of single echo T1-weighted magnetic resonance images. *Hum. Brain Mapp.* 5:194–205.

Thévenaz, P., Ruttimann, U., and Unser, M. (1998). Multiresolution image registration without landmarks. *IEEE Trans. Image Processing* **7**: 27–41.

Worsley, K. J., Marrett, S., Neelin, P., Vandal, A. C., Friston, K. J., and Evans, A. C. (1996). A unified statistical approach for determining significant signals in images of cerebral activation. *Hum. Brain Mapp.* **4**: 58–73.

CHAPTER 33

Error and *t* Images Depend on ANOVA Design and Anatomical Standardization in PET Activation Analysis[1]

MICHIO SENDA, KENJI ISHII, KEIICHI ODA, NORIHIRO SADATO,* RYUTA KAWASHIMA,[†]
IWAO KANNO,[‡] HINAKO TOYAMA, and ITARU TATSUMI

*Tokyo Metropolitan Institute of Gerontology
as Multicenter Project with * Fukui Medical School
[†] Tohoku University and
[‡] Akita Institute for Brain and Blood Vessels
35-2 Sakaecho Itabashi
Tokyo 173, Japan*

In a positron emission tomography activation analysis with task replications within subject, a number of analysis of variance (ANOVA) designs are applicable with different definitions of t and error. The characteristics of t (and z) maps and error images and how they depend on the ANOVA design and on the anatomical standardization method have been investigated. Six subjects underwent measurement of regional cerebral blood flow with [^{15}O]water under resting and while thinking of verbs associated with auditorily presented nouns, three times for each. The images were anatomically standardized with LINear, SPM95, or HBA. ANOVA was performed pixel by pixel to compute t statistics for the task main effect (verb vs rest) in four different ANOVA designs: (i) two way (subject and task) (2W), (ii) two way with interaction (2WI), (iii) two way with interaction, except that the "subject" was considered a random factor (2WI-RF), and (iv) three way (subject, task, and replication). The left frontal cortex extending from Broca's area to the premotor cortex was activated by the verb generation. The foci localization in the z images depended both on the anatomical standardization method and on the ANOVA design, and the variation ranged from 1 to 3 cm. SPM tended to present a higher peak z than LIN and HBA. The z images of 2W and 2WI looked alike, but 2WI-RF and 3W each presented a different z map within the activated area. The peak z score by 2WI-RF was lower than the others. The error images for 2W, 2WI, and 3W were heterogeneous, being high in gray and low in white.

I. INTRODUCTION

In the positron emission tomography (PET) activation analysis on a group of subjects, analysis of variance (ANOVA) or related techniques such as analysis of covariance (ANCOVA) are usually employed to create statistical maps of activation (*t* or *z* map) after anatomical standardization of the brain images of each subject for intersubject averaging.

In the case of task replications within subjects, in which regional cerebral blood flow (rCBF) is measured twice or more under the same condition (task) within subject, a number of ANOVA designs are applicable with different definitions of *t* and error. The majority of previous reports of PET activation with task replications seem to use a simple two-way design without looking into how the result depends on the ANOVA design. Some investigators do not even describe specifically which ANOVA design they use. Woods *et al.* (1996) proposed three-way ANOVA for PET activation analysis with task replications and pointed out theoretically that it can reveal interactions between subject and replication as well as between task and replication. The error term of ANOVA reflects the variations that are not explained by the ANOVA model. There has been a

---

[1] Transcripts of the BRAINPET97 discussion of this chapter can be found in Section VIII.

controversy as to whether the error is heterogeneous across the brain or can be treated as homogeneous. Although numerous reports have been published of PET activation studies, very few have addressed this issue or presented the error images. The homogeneity of error may depend both on the task paradigm and on the ANOVA design.

The present study analyzed PET activation data on the verb generation paradigm using four types of ANOVA designs. This chapter looked at the z maps and error images as well as variation components and how they depend on the ANOVA design. The effect of the anatomical standardization method has also been investigated by using three different methods on identical data. The study has been conducted as part of the Japanese Multicenter PET Project.

II. MATERIALS AND METHODS

A. Data Acquisition

Data were acquired in four institutes in Japan: Tokyo Metropolitan Institute of Gerontology (Tokyo), Fukui Medical School (Fukui), Tohoku University (Sendai), and Research Institute of Brain and Blood Vessels-Akita (Akita). Tokyo and Sendai used a two-dimensional (2D) PET scanner (HEADTOME-IV, Shimadzu), whereas a 3D scanner was used in Akita (HEAD-TOME-V) and in Fukui (GE-ADVANCE). In each institute, six young right-handed normal male subjects (ages 19–29) were recruited for a PET activation study with the covert verb generation paradigm. rCBF was measured with bolus or slow bolus intravenous injection of [^{15}O]water under the resting condition (REST) and while the subject tried to think of as many verbs as possible that are associated with auditorily presented nouns (VERB), three times for each in the alternating order (RVRVRV or VRVRVR). The radioactivity images were acquired for 60 sec (Fukui and Sendai), 90 sec (Akita), or 120 sec (Tokyo), starting at injection (Tokyo), 10 sec after injection (Akita), or at the arrival of radioactivity in the brain (Fukui and Sendai). Photon attenuation was corrected with measured transmission data. The images were converted into CBF images with the PET-ARG method (Herscovitch *et al.*, 1983) using measured arterial time course (Tokyo) or standard input function of the institute (Akita, Fukui, Sendai). T1-weighted sagittal and transaxial magnetic resonance (MR) images of each subject were also acquired for anatomical reference. The protocol has been approved by the ethics committee or its equivalent in each institute.

B. Anatomical Standardization

First, the PET images were registered to the subject's own MR images by Babak Ardekani using his method (Ardekani *et al.*, 1995). The result of the registration procedure was inspected and accepted by one of the authors (K.I.), an expert of brain PET and MR interpretation. The images were then anatomically standardized by three methods: Linear (LIN), SPM, and HBA. In LIN, anterior commissure (AC) and posterior commissure (PC) were identified and the brain size was measured on the MR images. PET images were then reoriented to the AC–PC line and were scaled to Talairach's atlas (Talairach and Szikla, 1967) with a different scaling factor for each axis (Senda *et al.*, 1994). In SPM (version 95), the AC–PC line was automatically estimated from PET without using the MR images, and the PET images were transformed into Talairach's space (Talairach and Tournoux, 1988) using a 12-parameter affine transformation and a 6-parameter 3D quadratic deformation (Friston *et al.*, 1995). In HBA, the computerized brain atlas (Roland *et al.*, 1994) was processed interactively with multiple linear and nonlinear transformations so that the brain contour, ventricles and major sulci defined in the atlas would match the structures identified in each subject's MR images, and the PET images that had been registered to the MR images were transformed into the atlas space with the reverse transformations.

C. Statistical Analysis

First, rCBF values were normalized by the global CBF value to account for global CBF fluctuations. Then the images were smoothed with a 3D Gaussian filter of 10 mm FWHM. For each set of six subject data from the same PET center, ANOVA was performed pixel by pixel with four different designs to compute the F statistics for the task main effect (VERB vs REST) and its square root ($= t$). To make the comparison between t statistics with different degrees of freedom (df) easier, the t value was transformed into equivalent z scores of the normal distribution. The mean difference of the task effect (ΔCBF) was also computed and mapped.

Let X_{ijk} denote the rCBF for subject i ($= 1$–6), task j ($=$ rest, verb), and replication k ($= 1$–3). The ANOVA designs employed in the present study are:

1. $X_{ijk} = M + S_i + T_j + Error$ (two-way ANOVA, 2W) $df = 29$.
2. $X_{ijk} = M + S_i + T_j + (ST)_{ij} + Error$ (two-way ANOVA with interaction, 2WI), $df = 24$.
3. Same as just described, except that the "subject" was considered a random factor and mean square

for (ST) instead of mean square for $Error$(MSE) was used to compute F (2WI-RF), $df = 5$.

4. $X_{ijk} = M + S_i + T_j + (ST)_{ij} + R_k + (SR)_{ik} + (TR)_{jk} + Error$ (three-way ANOVA, 3W), $df = 10$.

where M is the intercept; S, T, and R are the main effect of subject, task, and replication, respectively; and (ST), (SR), and (TR) are the interaction effects of two of the three factors.

In addition to the t statistics for the task effect (T), the mean square for the subject effect (S), the subject*task interaction (ST), and the $Error$ were computed and their square roots (RMS) were mapped to evaluate the components of variation.

III. RESULTS AND DISCUSSION

In general, the VERB vs REST z maps showed an activation more or less in the left frontal cortex extending from the inferior frontal gyrus (Broca's area) upward to the premotor cortex as well as in the supplementary motor area (SMA) and in the anterior cingulate gyrus. This activation was consistent among the centers and was observed invariably no matter which anatomical standardization method was used (Fig. 1). However, the exact localization of foci peak in the z images depended on the anatomical standardization method, and the variation ranged over 1–3 cm (Table 1). This variation exceeds the difference in the definition of structures in the atlas on which each method is based. As for the peak values, SPM tended to present higher z scores than LIN and HBA.

The z images created by the four different ANOVA designs were compared. The z images of 2W and 2WI derived from the identical data through the same transformation method looked alike, but 3W presented somewhat different distribution within the activated area. However, 2WI-RF showed lower z values with substantially different z distribution (Fig. 2 and Table 1).

The peak z value of the foci in the present study was comparable among 2W, 2WI, and 3W. As the

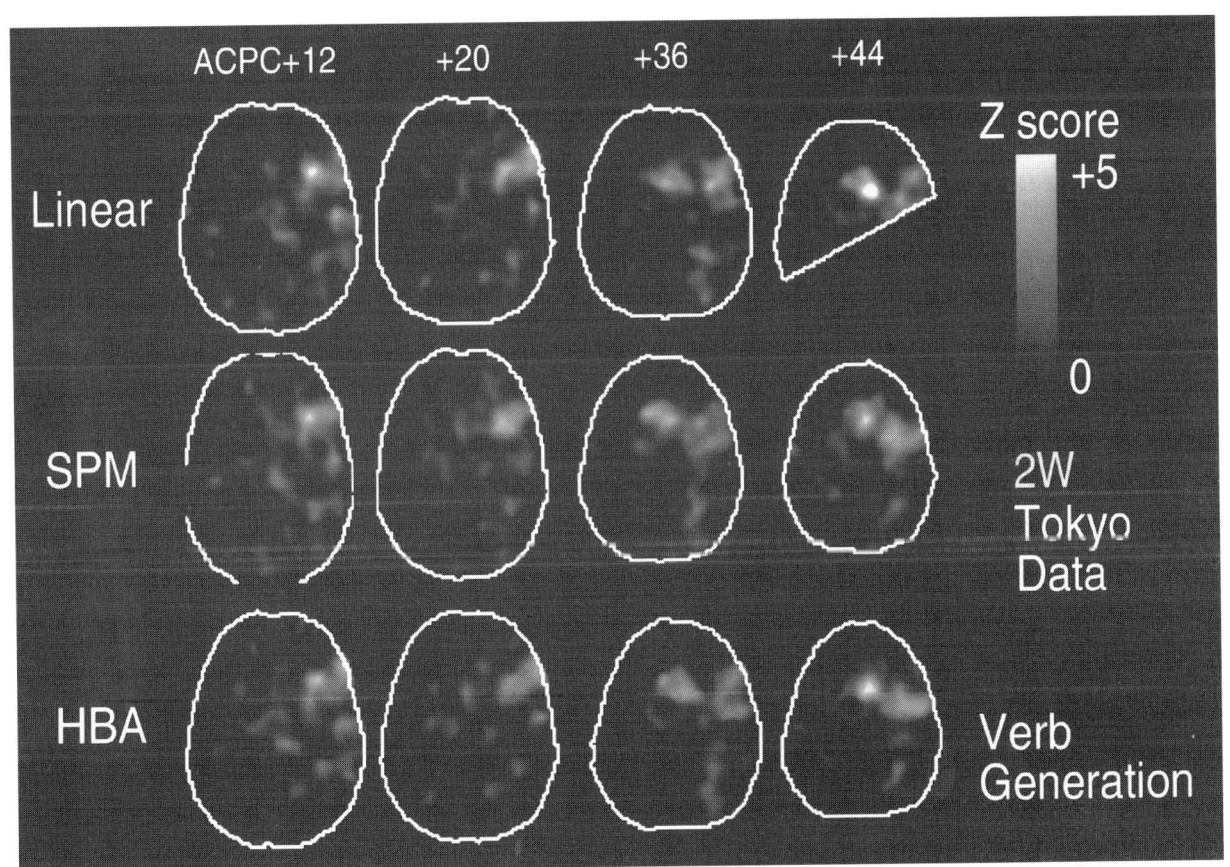

FIGURE 1 z maps of verb genreation vs rest based on two-way ANOVA design computed from data of six subjects acquired in Tokyo and anatomically standardized with three different methods (Linear, SPM, HBA; see text for details).

TABLE 1 Dependence of Peak z Score and Its Localization in and Around Broca's Area on ANOVA Design and Anatomical Standardization[a]

ANOVA Design	LIN x	y	z	Peak	SPM x	y	z	Peak	HBA x	y	z	Peak
2W	−31	23	12	4.34	−30	18	12	4.79	−53	43	10	4.65
2WI	−53	15	4	4.86	−30	8	24	5.77	−33	35	10	4.86
2WI-RF	−59	19	4	3.73	−26	30	8	3.51	−35	33	−6	4.32
3W	−43	25	20	4.56	−36	6	20	5.50	−39	25	18	5.08

[a] Based on data from six subjects acquired in Tokyo. Coordinates (x, y, z) are millimeters from AC, with positive directions toward right, anterior, and superior. Peak values are in z score units. See text for details of ANOVA designs (2W, 2WI, 2WI-RF, 3W) and anatomical standardization methods (LIN, SPM, HBA).

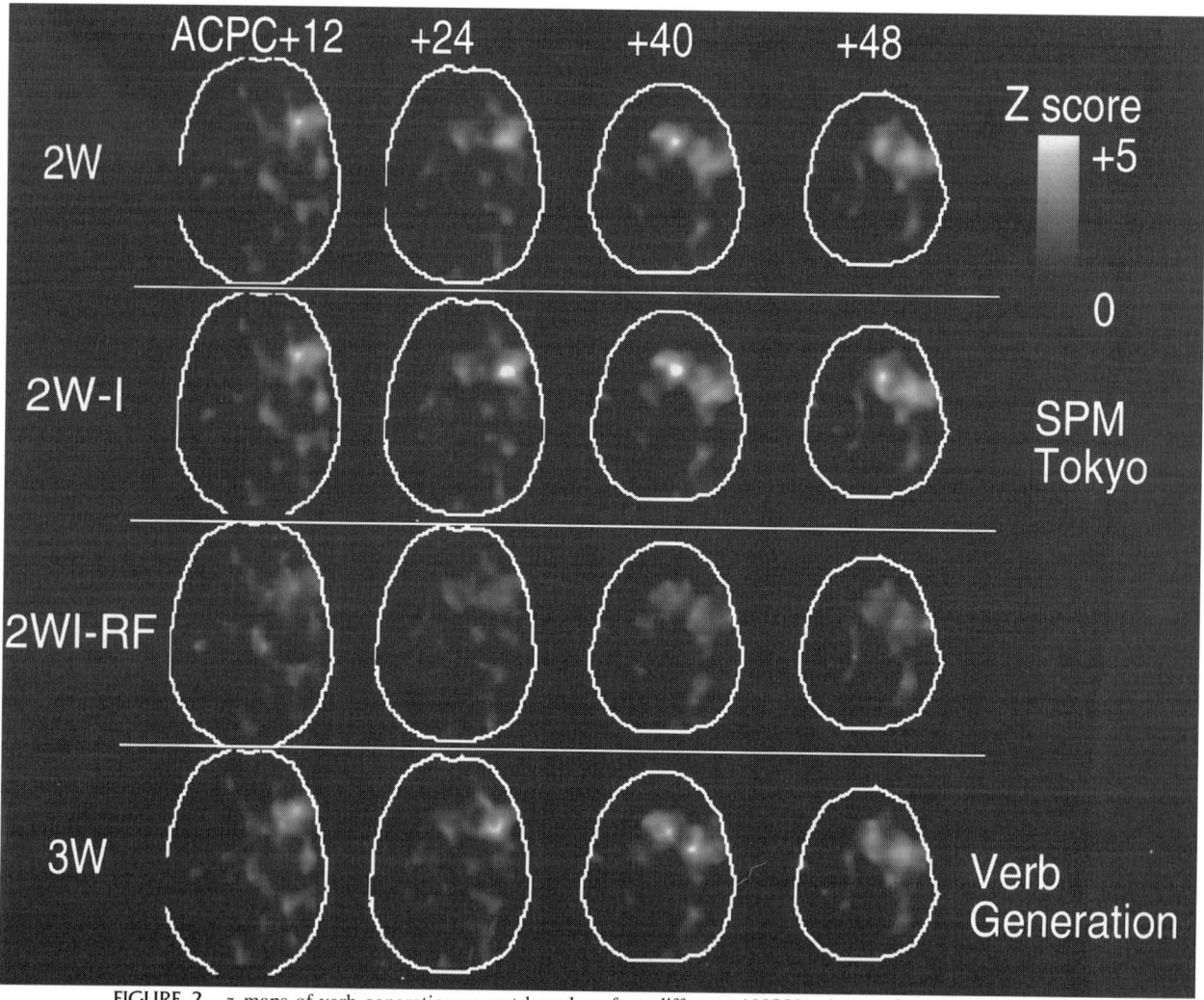

FIGURE 2 z maps of verb generation vs rest based on four different ANOVA designs (2W, 2WI, 2WI-RF, 3W; see text for details) computed from data of six subjects acquired in Tokyo and anatomically standardized with SPM.

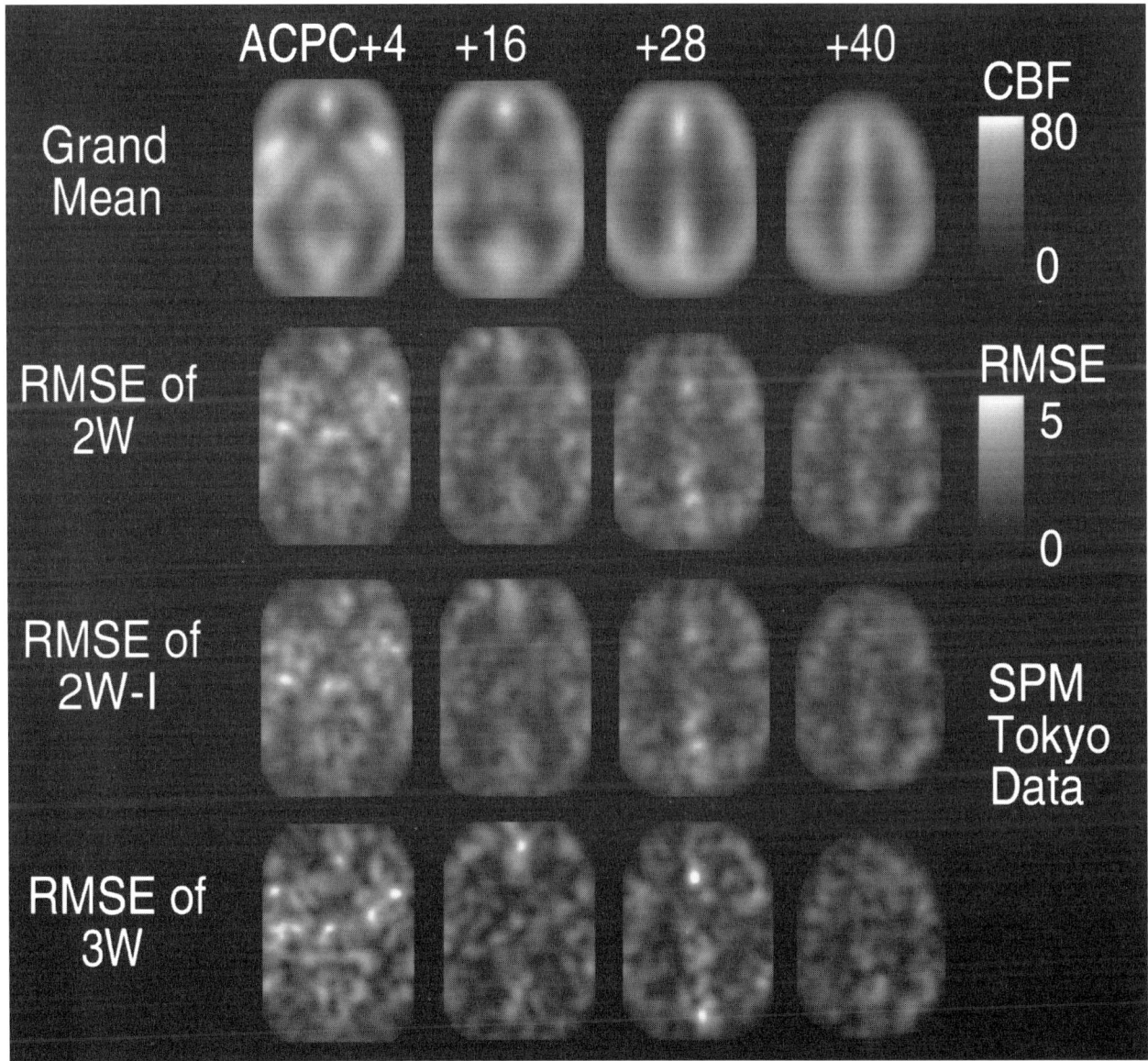

FIGURE 3 Mean CBF images and root mean square for error (RMSE) based on three different ANOVA designs (2W, 2WI, 3W; see text for details) computed from data of six subjects acquired in Tokyo and anatomically standardized with SPM.

ANOVA model becomes complex, the root mean square for error (RMSE) decreases, but the degrees of freedom decrease simultaneously, which may result in similar peak values. In this sense, 3W provided no more sensitive detection of the activation foci than 2W or 2WI.

The "subject" in the ANOVA model is theoretically considered a random factor because the group of six subjects is a random sample from the population. Our interest lies not in the difference between specific subjects, but in the intersubject variation as a whole. Therefore, the 2WI-RF design is a theoretically correct model. In the present study, however, 2WI-RF showed lower z scores than the other designs, suggesting that this approach may be less sensitive in detecting activation foci for a small number of subjects.

The dependence of anatomical standardization on the foci localization was addressed previously (Senda *et al.*, 1994), in which an identical data set on vibrotactile stimulation of the fingers using three different methods of anatomical standardization was transformed and a variation of 1 cm in vibrotactile foci localization was found. The activation by the vibrotactile stimulation is narrower than the present verb generation, the latter having two or more peaks in and around Broca's area. The relative height of the peaks is affected by the anatomical standardization and ANOVA design. This chapter described the localization with the highest peak, which may have caused the variation to be much higher than the previous study.

The error images (RMSE for 2W, 2WI, 3W) were not homogeneous across the brain, being high in gray and low in white, possibly due to higher rCBF fluctuation in the gray matter. Within the gray, the central structures tended to show higher RMSE than the cortical rim (Fig. 3), probably due to higher statistical noise in the center than in the periphery. The result suggests that one should be cautious about the assumption of homogeneous error across the entire brain. It may be possible to segment the brain into regions having similar RMSE and to compute the pooled variance for each segmented region.

The RMS for the subject effect (S), reflecting intersubject mismatch in gray matter distribution, was high on the brain contour and gray–white borders, which was most pronounced in LIN but was still notable in SPM and HBA, especially in the cingulate cortex. This indicates that the gray matter structure is not properly normalized anatomically with the LIN method and that the normalization is still imperfect even with SPM or HBA, especially in the cingulate cortex. The RMS for the subject*task interaction (ST), reflecting individual variation in activation, had inconsistent hot spots in Broca, left prefrontal, cingulate, and visual cortex, depending on the data set.

References

Ardekani, B. A., Braun, M., Hutton, B. F., Kanno, I., and Iida, H. (1995). A fully automated multimodality image registration algorithm. *J. Comput. Assist. Tomogr.* **19**: 615–623.

Friston, K. J., Ashburner, J., Frith, C. D., Heather, J. D., and Frackowiak, R. S. J. (1995). Spatial registration and normalization of images. *Hum. Brain Mapp.* **3**: 165–189.

Herscovitch, P., Markham, J., and Raichle, M. E. (1983). Brain blood flow measured with intravenous $H_2^{15}O$. I. Theory and error analysis. *J. Nucl. Med.* **24**: 782–789.

Roland, P. E., Graufelds, C. J., Wahlin, J., Ingelman, L., Andersson, M., Ledberg, A., Pedersen, J., Akerman, S., Dabringhaus, A., and Zilles, K. (1994). Human brain atlas: For high resolution functional and anatomical mapping. *Hum. Brain Mapp.* **1**: 173–184.

Senda, M., Kanno, I., Yonekura, Y., Fujita, H., Ishii, K., Lyshkow, H., Miura, S., Oda, K., Sadato, N., and Toyama, H. (1994). Comparison of anatomical standardization methods regarding the sensorimotor foci localization and between-subject variation in $H_2^{15}O$ PET activation, a three-center collaboration study. *Ann. Nucl. Med.* **8**: 201–207.

Talairach, J., and Szikla, G. (1967). "Atlas of Stereotaxic Anatomy of Telencephalon." Masson, Paris.

Talairach, J., and Tournoux, P. (1988). "Co-Planar Stereotaxic Atlas of the Human Brain." Thieme, Stuttgart.

Woods, R. P., Iacoboni, M., Grafton, S. T., and Mazziotta, J. C. (1996). Improved analysis of functional activation studies involving within-subject replications using a three-way ANOVA model. *In* "Quantification of Brain Function Using PET" (R. Myers, V. Cunningham, D. Bailey, and T. Jones, eds.), pp. 353–358. Academic Press, San Diego, CA.

CHAPTER

34

Calculation of the Probability That an Activation Site Has Occurred by Chance

JOHN R. VOTAW,* SCOTT. T. GRAFTON,*,† and JOHN M. HOFFMAN*,‡

Departments of * Radiology and ‡ Neurology
Emory University
Atlanta, Georgia 30322

The application of functional brain mapping requires estimating the probability that an identified activation site has occurred by chance. This chapter presents a technique, without making assumptions about the local or global variance, for calculating the probability that an arbitrary group of n pixels with n arbitrary values has occurred by chance. This technique uses locally pooled variance—intermediate between previous techniques that either globally pool variance or do not pool variance at all. Also, the calculation takes into account that the pixel values are not all equal—a site with several pixels that are far above the threshold is recognized as more significant than a site that has all values very near the threshold. Equations are derived from first principles using the n-dimensional normal distribution by transforming from the image space where pixels are correlated to a space where the pixels are independent. The transformation is specified by the eigenvectors that diagonalize the covariance matrix, which is assumed known. Equations for both z and t images are presented. Using these equations, investigators can become more sophisticated in their analysis in order to more accurately determine the statistical significance of their activation results.

I. INTRODUCTION

The application of functional brain mapping has instigated a continual evolution of image analysis techniques designed to identify significant changes in blood flow. It is now routine to perform pixel-by-pixel statistical comparisons (Neter *et al.*, 1990; Friston *et al.*, 1995). This advance has reduced observer bias, but the use of pixel-by-pixel statistic techniques is compromised by the large number of pixels that can be included in an analysis (greater than 100,000).

Friston *et al.* (1990) showed that the number of independent resolving elements in an image space is less than the total number of pixels and that the correction for number of independent elements could be reduced by using images of greater smoothness. This notion was formalized when connected to the theory of random Gaussian fields (Friston *et al.*, 1991; Worsley *et al.*, 1992). Friston and Worsley then developed a series of algorithms that used a defined statistical threshold to determine the probability of an activation of any size occurring in a given volume. Worsley (1994) refined this estimate by considering the geometry of the search volume. To maximize sensitivity with these techniques, the user maximized image smoothness and minimized the search volume.

Intuitively, these techniques have a limit in that the size of an activation does not provide additional statistical certainty. This latter effect was partially addressed in a revised probability estimate by Friston *et al.* (1994). With this application, optimal smoothing depends on the size of the activation site of interest and often must be found by iteration. Application of this algorithm has led to major gains in assessing statistical certainty and reducing the likelihood of false positives in positron emission tomography imaging. A weakness of this technique is that it does not account for differences of statistical certainty for each pixel located in an activation site above a particular threshold. Thus, an activation site with pixels that are equal to a particular threshold is treated identically to an activation site where many pixels are far above the threshold. This chapter presents a technique for assessing statistical certainty using both the number of pixels above a threshold of interest and the magnitude of each pixel.

II. METHODS

A. Correlated Gaussian Images

Each pixel is assumed to be from a normal distribution and is correlated with other pixels. The probability that n pixels have values between x and $x + dx$ is given by the n-dimensional normal probability distribution

$$P(x)\,d^n x = \frac{1}{(2\pi)^{n/2}|\text{cov}|^{1/2}} e^{-x\,\text{cov}^{-1}\,x^T/2}\,d^n x, \quad (1)$$

where x is an n-dimensional vector of the pixel values, cov^{-1} is the inverse of a square matrix with cov_{ij} = the covariance between pixels i and j, and T indicates vector transpose. The probability that each of the n pixels has a value greater than a corresponding cutoff value, a_i, is

$$P(x_i > a_i) = \int_{a_1}^{\infty} dx_1 \int_{a_2}^{\infty} dx_2 \cdots \int_{a_n}^{\infty} dx_n\, P(x). \quad (2)$$

The technique for solving Eq. (2) is illustrated in two dimensions in Fig. 1. The probability that pixel 1 is greater than a_1 and pixel 2 is greater than a_2 is the integral of the function under the shaded region. A pure rotation of the function gives Fig. 1B. A change of variables so that the isocontours are circular is illustrated in Fig. 1C. The circular contours indicate that the transformed function is radially symmetric. Hence, the integral is greatly simplified. However, the limits of integration are no longer easily defined. These geometric steps are the equivalent of first diagonalizing cov^{-1} and then making a change of variables so that all elements in the diagonalized matrix are equal.

After these transformations, the integral is of the form

$$P(x_i > a_i) = \frac{1}{(2\pi)^{n/2}} \int_{r_0}^{\infty} \int_{f_1(r)}^{f_2(r)} e^{-r^2/2} r^{n-1}\, d\Omega\, dr, \quad (3)$$

where $f_1(r)$ and $f_2(r)$ are functions that define the transformed limits of integration at radius r. This integral can be solved numerically as a summation in one dimension:

$$P(x_i > a_i) = \frac{\Delta r}{(2\pi)^{n/2}} \sum_{j=0}^{\infty} e^{-r_j^2/2} r_j^{n-1} \Omega(r_j), \quad (4)$$

where $r_j = r_0 + (j + 0.5)\Delta r$ and $r_j^{n-1}\Omega(r_j)$ is the portion of the (n-dimensional) surface area of the (n-dimensional) sphere of radius r_j that is bounded by the limits of integration. Because the kernel falls off exponentially fast, the summation can be accurately evaluated in relatively few steps.

The aforementioned intuitive argument is made rigorous as follows. Let S be a similarity transformation matrix for cov^{-1} such that $S^T S$ is the unit matrix, then Eq. (1) becomes

$$\begin{aligned} P(x)\,d^n x &= \frac{1}{(2\pi)^{n/2}|\text{cov}|^{1/2}} e^{-xS^T S\,\text{cov}^{-1}\,S^T S x^T/2}\,d^n x \\ &= \frac{1}{(2\pi)^{n/2}|\text{cov}|^{1/2}} e^{-xS^T E S x^T/2}\,d^n x, \quad (5) \end{aligned}$$

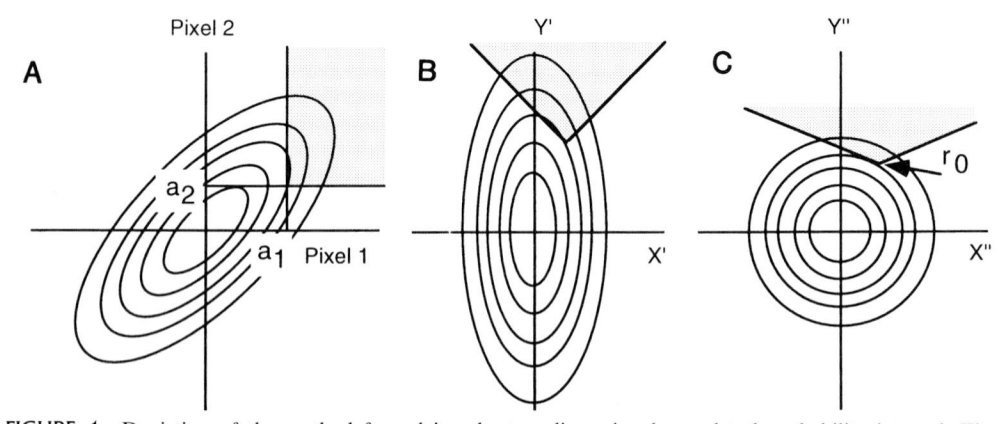

FIGURE 1 Depiction of the method for solving the two-dimensional correlated probability integral. The elliptical contours represent the isoprobabilities [Eq. (1)] that the two pixels have given values. The maximal contour is at the origin, and each successive contour outward represents a smaller value. The probability that pixel 1 is greater than a_1 and pixel 2 is concurrently greater than a_2 is obtained by integrating the function under the shaded region. (A) The problem as stated in Cartesian coordinates. (B) The function has been rotated and (C) The coordinate axes have been scaled so that the isocontours are circular.

where E is the diagonal matrix of eigenvalues. Now let $y_i = \sqrt{E_j}(Sx^T)_j$, where j indicates the jth element in the vector. Because the similarity transformation is a pure rotation, it does not change the differential volume element (the Jacobian of the rotation transformation is unity). The scaling operation reduces the volume element by the square root of the product of the eigenvalues, but the determinate, |cov|, is the product of the inverse of the eigenvalues of cov^{-1}, so

$$P(x)\,d^n x = \frac{\prod_{i=1}^{n}\sqrt{\frac{1}{E_i}}}{(2\pi)^{n/2}|\text{cov}|^{1/2}}e^{-yy^T/2}\,d^n y$$

$$= \frac{1}{(2\pi)^{n/2}}e^{-yy^T/2}\,d^n y$$

$$= \frac{e^{-r^2/2}}{(2\pi)^{n/2}}r^{n-1}\,d\Omega\,dr. \quad (6)$$

In the last step, the function has been transformed to polar coordinates where $d\Omega$ represents the angular component of the differential volume. This justifies Eq. (4). The transformation has mapped the space spanned by the pixel values into a space where the separate y_i are mutually independent.

The remaining task is to evaluate $r_j^{n-1}\Omega(r_j)$. In the original space, the integration region was bounded by a set of hyperplanes that intersect to form a set of n rays that originate at a common point and extend to $+\infty$ parallel to one of the n axes. Any $n-1$ of these rays define a hyperplane that is one of the limits of integration. Because all transformations are linear, each of the new rays (and hyperplanes) can be determined by finding the ray that connects any two transformed points from the original ray.

The hypersurface area is evaluated by Monte-Carlo integration as shown in Fig. 2. First, the intersection of each of the limit rays and the sphere of radius r_j is determined. To improve the efficiency of calculation, these points are rotated as a group to the polar axis and the polar angle, θ_1, of the point furthest away from the axis is determined. Next, the point on the intersection of the sphere of radius r_j and each of the transformed limit planes that is closest to the symmetry axis and its polar angle, θ_2, is determined. The desired area is greater than the hyper area of the polar cap of angle θ_2 but less than the hyper area of the polar cap of angle θ_1. Random points are chosen on the sphere that fall between θ_1 and θ_2. Each point that is above each of the transformed limit of integration planes is in the integration region. The desired quantity is the area of the cap of angle θ_2 plus the area of the band between

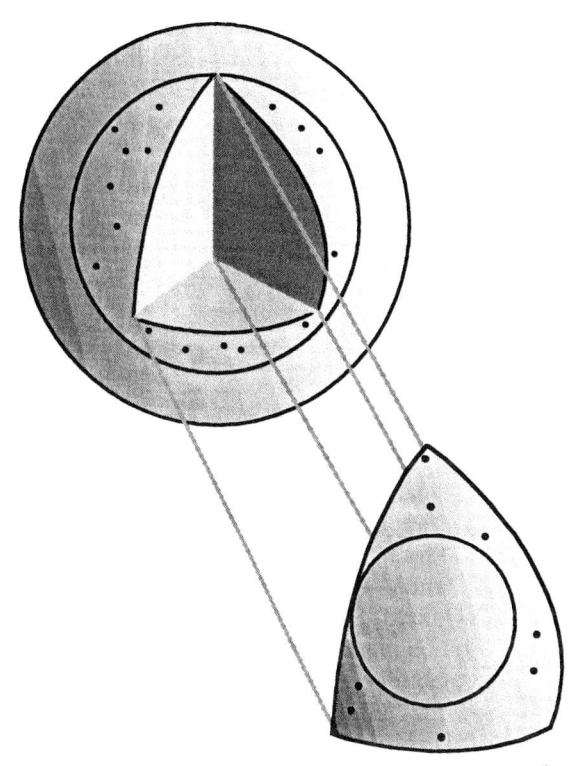

FIGURE 2 Illustration of the technique used to calculate $r_i^{n-1}\Omega(r_i)$, the surface area of the wedge that has been removed from the sphere, when $n = 3$. The wedge is defined by the transformed limits of integration. The area of the wedge is bounded by the area of the small circle on the wedge and the large circle on the sphere, both of which are analytically calculable. Random points are distributed between the two circles to determine what fraction of the area between the circles belongs to the wedge. This area plus the area of the small circle is the desired quantity.

θ_1 and θ_2 multiplied by the fraction of random points that are within the integration region (Press *et al.*, 1988).

B. Correlated *t*-Statistic Images

In most cases, the covariance matrix must be estimated from the physics of the system and data, and a statistic similar to the student's *t*-statistic should be used in the analysis. It is assumed here that the covariance between pixels is introduced by the point spread function of the scanner (which is known).

In the following, it is assumed that data were collected as paired data and that significant differences between the conditions are desired. The image pixel values are not independent and so cannot be used directly to determine the variance in data. However, after the transformation following Eq. (5), the transformed pixel values, y_i, are independent. To slightly increase the degrees of freedom, the variance is pooled across the region of interest:

$$s^2 = \frac{\sum_{j=1}^{n} \sum_{i=1}^{n_{\text{rep}}} (y_{i,j} - \bar{y}_{.,j})^2}{n(n_{\text{rep}} - 1)}, \quad (7)$$

where n_{rep} is the number of replications of each condition, $\bar{y}_{.,j}$ is the average across all replications of pixel j, and $y_{i,j}$ is the transformed value of the ith replication of pixel j.

The appropriate t statistic for pixel j is

$$t_j = \frac{\bar{x}_{.,j}}{\left(\frac{s}{\sqrt{n_{\text{rep}}}}\right)} = \frac{\frac{\bar{x}_{.,j}}{\sigma/\sqrt{n_{\text{rep}}}}}{\sqrt{\frac{s^2}{\sigma^2}}} = \frac{Z_j}{\sqrt{\frac{V}{n(n_{\text{rep}} - 1)}}}, \quad (8)$$

where σ is the true population variance of radioactivity uptake in a pixel in the brain (as opposed to the measured variance seen in an image pixel), $Z_j = \bar{x}_{.,j}/(\sigma/\sqrt{n_{\text{rep}}})$ is a normal variable, and $V = n(n_{\text{rep}} - 1)(s^2/\sigma^2)$ is a χ^2 variable with $v = n(n_{\text{rep}} - 1)$ degrees of freedom. The Z_j are correlated due to the scanner point spread function.

The probability distribution of t_j can be derived in an analogous way to the standard deviation of the t statistic (see Appendix):

$$P(t; n, v) \, dt = \frac{\Gamma((n+v)/2)}{(\pi v)^{n/2} \Gamma(v/2)}$$
$$\times \left(1 + \frac{t \, \text{cov}^{-1} t^T}{v}\right)^{-(n+v)/2} dt. \quad (9)$$

Here cov^{-1} is due to the scanner response only and has unit diagonal elements.

The desired probability is found by integrating this function analogous to Eq. (2). Using the same transformation as following Eq. (5) and transforming to radial coordinates, the probability integral can be estimated from the summation

$$P(t_i > a_i; n, v) = \frac{\Gamma\left(\frac{n+v}{2}\right) \Delta\tau}{(\pi v)^{n/2} \Gamma\left(\frac{v}{2}\right)}$$
$$\times \sum_{i=0}^{\infty} \left(1 + \frac{\tau_i^2}{v}\right)^{-(n+v)/2} \tau_i^{n-1} \Omega(\tau_i), \quad (10)$$

where τ is the radial component of t expressed in polar coordinates after the rotation and scaling operations [described after Eq. (5)] and $\tau^{n-1} d\Omega$ is the portion of the hypersphere of radius τ that is within the limits of integration.

C. Determination of the Covariance in Data

The covariance matrix is the autocorrelation of the point response function multiplied by the true brain pixel variance σ^2. It can be shown that the autocorrelation of a normalized gaussian with FWHM = α is a normalized gaussian with FWHM = $\sqrt{2}\alpha$. In this case, the elements of the covariance matrix are

$$\text{cov}_{ij} = \sigma^2 e^{-2\ln(2)(r_{ij}/\text{FWHM})^2}, \quad (11)$$

where r_{ij} is the distance between pixels i and j.

D. Implementation Notes

When the size of the pixels is small compared to the resolution of the scanner, the covariance matrix is very nearly singular. In cases of large n and small pixel sizes, the largest eigenvalues can be more than 10 orders of magnitude greater than the smallest. The very large eigenvalues indicate that the corresponding directions are very highly correlated with another direction and hence unimportant to the solution. Essentially, directions corresponding to these eigenvalues can be ignored and the dimensionality of the problem reduced without altering the accuracy of the solution. For an illustrated example, see Fig. 1. Large eigenvalues cause values along the corresponding eigenvector to be greatly expanded. In Fig. 1, the eigenvalue corresponding to the x'' direction is greater than the y'' direction. The greater the eigenvalue, the flatter the integration region will become. For a very large eigenvalue, the problem becomes separable into integrals over each direction. For the problem illustrated in Fig. 1, the integral over x'' will be essentially from $-\infty$ to $+\infty$ when the corresponding eigenvalue is made much larger. Because the probability function is normalized, the integral over x'' in this case is unity and the problem is reduced to a one-dimensional integral (in y'').

The error introduced by eliminating one of the dimensions is related to how fast values on the limit planes change in the other dimensions as one moves out along the eliminated dimension. This corresponds to the slope of the lines demarcating the integration region in Fig. 1C, which is the ratio of the $y'':x''$

eigenvalues. In practice, no difference was seen in the calculations when dimensions corresponding to eigenvalues more than a factor of 100 greater than the smallest eigenvalue were eliminated.

Inversion of the covariance matrix is not necessary as a similarity transformation that diagonalizes the inverse of the covariance matrix will also diagonalize the covariance matrix. Therefore, the covariance matrix can be diagonalized and then the inverse is affected by taking the inverse of the eigenvalues. In practice, the largest eigenvalues and corresponding eigenvectors of the cov matrix are found using the power method accelerated with Chebyshev polynomials (Chatelin, 1993).

Eliminating dimensions in this way greatly increases the speed of the calculation for two reasons: (1) it reduces the number of calculations and (2) it makes the Monte-Carlo integration more efficient. The accuracy of a Monte-Carlo calculation is determined by the number of random points that fall within the integration region. When one of the dimensions is greatly expanded, this number falls off rapidly. Consider Fig. 2. If one of the eigenvectors were much greater than the other two, then the wedge to be integrated would be narrow and extend from the equator on one side to the equator on the other side. In this case, θ_1 would be relatively small and θ_2 essentially $\pi/4$. If this were the case, nearly all of the random points (between θ_1 and θ_2) would fall outside of the integration region, requiring that many more random points be used. Typically, when a region of 50–500 pixels is being considered, the problem can be solved in 5–15 dimensions using 1000 random points per radial step in approximately 50 sec on a SUN sparc 10 computer.

III. RESULTS

The equations were tested by simulation using the normally distributed random number generator in IDL. Images of normal deviates were created and then convolved with a smoothing kernel representing the point spread function of the scanner. The first tests were designed to test the assumptions involved in creating the t statistics. Images were averaged to determine Z, and s^2 was calculated for 100,000 separate simulations and these variables were verified to follow normal and χ^2 distributions, respectively.

Figure 3 shows the result of a test of the t-statistic calculation. Three runs of a program that simulates 10,000 activation experiments is presented. Each study had three repetitions of two conditions (activation and rest). In each, six images were created by filling each pixel with an independent normally distributed (mean = 0, variance = 2) random number and then smoothing with the point spread function. The activation images were averaged and subtracted from the average rest image and the nine element t-statistic vector for a 3×3 ROI was formed as in Eqs. (7) and (8). The smallest component of the t vector was recorded for

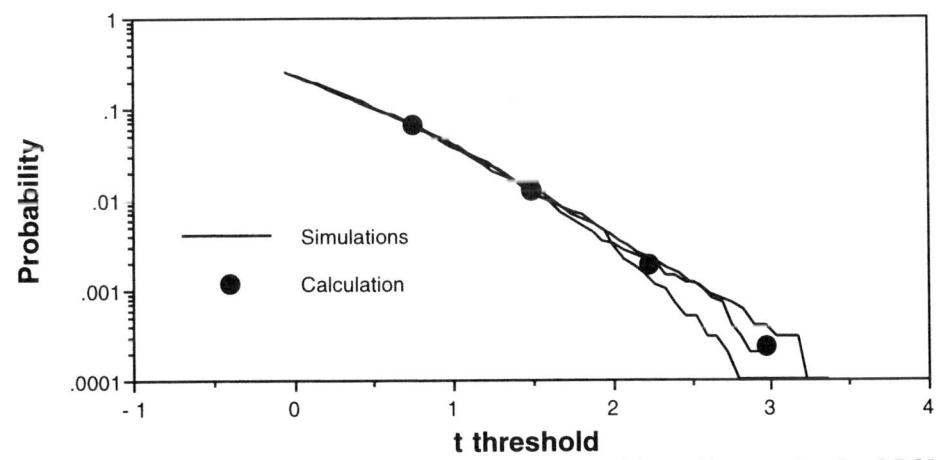

FIGURE 3 Simulation to determine the probability of obtaining a false-positive event in a 3×3 ROI in a dual condition triple replicate activation study. The probability is that all nine t values in the ROI will exceed the t threshold. Values calculated with Eq. (10) are superimposed. (Note: The assumptions used in this simulation are very specific and hence the results should not be applied to a general activation experiment.)

each experiment. The values were converted to a false-positive probability distribution and plotted in Fig. 3. The probability is that all of the nine components of the t vector are greater that the t threshold. The dots represent the calculated values according to Eq. (10). The calculation kept eigenvalues up to a factor of 100 greater than the minimum eigenvalue. In this nine-dimensional problem, the solution was calculated in five dimensions.

IV. DISCUSSION

This methodology allows calculation of the probability that a general region of interest (ROI) is activated by chance. The definition of an activated ROI may include different values for each pixel in the ROI. When a constant threshold is used, "narrow focal activations are most powerfully detected by high thresholds, whereas broader, more diffuse activations are best detected by low thresholds" (Friston *et al.*, 1994). For maximal power in the analysis, the threshold has to be matched to the size and magnitude of the ROI. By taking into account the individual pixel values, the probability that the ROI could have been activated by chance is more accurately calculated.

Equation (11) gives the formula for the covariance between pixels if the point spread function is a Gaussian. If the point spread function is not Gaussian, then the formula for the covariance needs to be modified but all other results remain unchanged.

As described, this methodology is applicable to experiments where the ROI is defined *a priori*. The false-positive rate when considering multiple comparisons is not presented in this work.

The approach taken here contains fewer assumptions than previous works. Variance is not required to be constant across the image, the point spread function is not required to be Gaussian, and the individual pixel values are taken into account. It has been shown that variance is increased at the center of the head (Votaw, 1996). When this is the case, analysis methods that assume uniform variance and use a constant threshold are forced to compromise between missing some activated regions at the edge of the head and accepting a greater false-positive rate at the center. Here, variance is pooled across the ROI. This has the advantage of increasing the degrees of freedom over pixel-by-pixel approaches, yet does not make global assumptions about the variance.

Calculation of the false-positive rate is more accurate when making fewer assumptions. In certain cases, this will permit reporting significant results that would not otherwise be identified. For example, an ROI that contains pixel values that vary widely can now be seen as more significant than when considering only that each of the pixels exceeds a lower threshold value. Using these equations, investigators can become more sophisticated in their analysis in order to more accurately determine the statistical significance of their activation results.

APPENDIX

Derivation of the multidimensional t-statistic probability function follows very closely the derivation of the standard t statistic found in statistic texts (see, e.g., Walpole and Myers, 1989, pp. 200–222). In sampling from a normal population, $\bar{Z}_i$ and s^2 are independent, and consequently so are Z and V in Eq. (8). Therefore, the joint probability of Z [Eq. (1)] and V (χ^2 distribution) is given by the product of their distribution

$$P(Z,V;v)\,dZ\,dV = \frac{e^{-Z\,\text{cov}^{-1}\,Z^T/2}}{(2\pi)^{n/2}} \frac{1}{2^{v/2}\Gamma(v/2)}$$
$$\times V^{v/2-1}e^{-V/2}\,dZ\,dV. \quad (A1)$$

Define the transformation $u = V$ and $t_i = Z_i/\sqrt{V/v}$. The Jacobian of the transformation is $(u/v)^{n/2}$, where n is the dimension of Z (number of pixels being considered) and, after some rearranging,

$$g(t,u;v,n)\,dt\,du = \frac{1}{(2\pi)^{n/2}2^{v/2}\Gamma(v/2)}\left(\frac{u}{v}\right)^{n/2}$$
$$\times e^{-u(1+t\,\text{cov}^{-1}\,t^T/v)/2}u^{v/2-1}\,du\,dt. \quad (A2)$$

The t distribution is found by integrating over u

$$P(t;n,v)\,dt = \int_0^\infty g(t,u;n,v)\,du\,dt$$
$$= \frac{dt}{(2\pi v)^{n/2}2^{v/2}\Gamma(v/2)}$$
$$\times \int_0^\infty e^{-u(1+t\,\text{cov}^{-1}\,t^T/v)/2}u^{n/2+v/2-1}\,du. \quad (A3)$$

Use the substitution, $w = u(1 + t \, \text{cov}^{-1} \, t^T/v)/2$, and noting that t is a constant in this integral, the integral becomes

$$P(t; n, v) \, dt = \frac{dt}{(2\pi v)^{n/2} 2^{v/2} \Gamma(v/2)}$$

$$\times \left(\frac{2}{1 + \frac{t \, \text{cov}^{-1} \, t^T}{v}} \right)^{(n+v)/2}$$

$$\times \int_0^\infty w^{n/2 + v/2 - 1} e^{-w} \, dw$$

$$= \frac{\Gamma\left(\frac{n+v}{2}\right)}{(\pi v)^{n/2} \Gamma(v/2)}$$

$$\times \left(1 + \frac{t \, \text{cov}^{-1} \, t^T}{v} \right)^{-(n+v)/2} dt, \quad (A4)$$

which is Eq. (9). If only one pixel is considered, $n = 1$, $\text{cov}^{-1} = 1$, and Eq. (A4) is seen to be the standard student t distribution.

References

Chatelin, F. (1993), "Eigenvalues of Matrices." Wiley, New York.

Friston, K. J., Frith, C. D., Liddle, P. F., Dolan, R. J., Lammertsma, A. A., and Frackowiak, R. S. J. (1990). The relationship between global and local changes in PET scans. *J. Cereb. Blood Flow Metab.* **10**: 458–466.

Friston, K. J., Frith, C. D., Liddle, P. F., and Frackowiak, R. S. J. (1991). Comparing functional (PET) images: The assessment of significant change. *J. Cereb. Blood Flow Metab.* **11**: 690–699.

Friston, K. J., Worsley, K. J., Frackowiak, R. S. J., Mazziotta, J. C., and Evans, A. C. (1994). Assessing the significance of focal activations using their spatial extent. *Hum. Brain Mapp.* **1**: 210–220.

Friston, K., Holms, A. P., Worsley, K. J., Poline, J. B., Frith, C. D., and Frackowiak, R. S. J. (1995). Statistical parametric maps in functional imaging: A general linear approach. *Hum. Brain Mapp.* **2**: 189–210.

Neter, J., Wasserman, W., and Kutner, M. H. (1990). "Applied Linear Statistical Models." Irwin, Burr Ridge, IL.

Press, W., Flannery, B. P., Teukolsky, S. A., and Vetterling, W. T. (1988). "Numerical Recipes in C. The Art of Scientific Computing." Cambridge University Press, New York.

Votaw, J. R. (1996). Signal-to-Noise ratio in neuro activation PET studies. *IEEE Trans. Med. Imag.* **15**(2): 197–205.

Walpole, R. E., and Myers, R. H. (1989). "Probability and Statistics for Engineers and Scientists." Macmillan, New York.

Worsley, K. J. (1994). Local maxima and the expected euler characteristic of excursion sets of Chi-square, F and t fields. *Adv. Appl. Probab.* **26**: 13–42.

Worsley, K. J., Evans, A. C., Marrett, S., and Neeling, P. (1992). A three-dimensional statistical analysis for CBF activation studies in human brain. *J. Cereb. Blood Flow Metab.* **12**: 900–918.

CHAPTER 35

Multifiltering Signal Detection and Statistical Power in Brain Activation Studies

JOHN DARRELL VAN HORN, TIMOTHY M. ELLMORE, JOHN L. HOLT, GIUSEPPE ESPOSITO, and KAREN FAITH BERMAN

Unit on PET, Clinical Brain Disorders Branch
National Institute of Mental Health
National Institutes of Health
Bethesda, Maryland 20892

Varying levels of image smoothing in relation to experimental effect size was assessed by computing voxel-based statistical power estimates on brain positron emission tomography (PET) activation data measured in subjects during the Wisconsin card sorting task. Forty normal subjects underwent regional cerebral blood flow measurements using [^{15}O]water PET. In general, the amount of smoothing necessary to achieve maximal power varied widely, but systematically, across the brain, tending to follow known neuroanatomical constraints, such as the juxtaposition of functionally reactive brain tissue and vasculature as well as other natural boundaries. However, statistical power was more readily improved as sample size increased in all brain areas. Variations in the optimum filter size across the brain suggest that neuroanatomical structure should be carefully considered in the selection of filter size, but also indicate that the best way to improve the reliability of findings is with increased sample size.

I. INTRODUCTION

Current analytic methods for functional neuroimaging data sets typically employ a single, low-pass Gaussian filter to adjust for intersubject variability in gyral anatomy, to minimize effects due to subtle errors in image registration, and to improve signal-to-noise ratios. Typical filter sizes range from 10 to 20 mm in the x, y, and z dimensions, but no "rule" exists to guide in the *a priori* selection of the best filter size. Previous studies, however, have shown that the ability to identify an embedded signal is dependent on the full width at half-maximum (FWHM) of the filter size that is used (Poline and Mazoyer, 1991). Furthermore, the filter size needed to maximize significance and spatial extent has the same area as that being sought (Worsley *et al.*, 1996). A multifiltering approach (Poline and Mazoyer, 1994) has been found to enhance the detection of activation foci by constructing the test-statistic map from only the most significant of tests conducted over a range of filter widths. However, mapping only the most statistically significant voxels does not provide sufficient information on the probable reproducibility of a result, which requires statistical power calculations.

Statistical power is defined as one minus the probability of accepting the null hypothesis (H_O:) when H_O: is indeed false (the type II error). Sufficient power (e.g., $\geq 90\%$) helps ensure that with a particular sample size, one can reliably identify significant effects existing in data when they are indeed present (Cohen, 1977). Therefore, statistical power is the long-term probability, for a given effect size estimate, significance threshold, and sample size, of correctly rejecting H_O: (Cohen, 1992). Statistical power is an important consideration in experiments attempting to detect the 3–5% neurophysiological changes reported in some cognitive activation experiments (Poline and Mazoyer, 1993; Paulesu *et al.*, 1995). This chapter applies a multifiltering method to the computation of voxel-based statistical power estimates on neuroimaging data. This approach offers a unique method for investigating how to optimize the reliability of signal detection.

II. MATERIALS AND METHODS

A. Subjects and Cognitive Conditions

Forty normal subjects were taken from a sample of normal volunteers recruited for a study examining the effects of aging and cognition (Esposito *et al.*, 1998). Each subject was scanned under two separate conditions, the Wisconsin card sorting test (WCST) and a matched sensorimotor control task (WCSC). The design of the cognitive paradigm and specifics of task administration are described in more detail elsewhere (Berman *et al.*, 1995). All subjects provided written consent to a protocol approved by the NIMH Human Studies Review Board and the NIH Radiation Safety Committee; were screened for major medical illnesses, psychiatric disorders, and substance abuse; and were instructed to abstain from the use of nicotine and caffeine for 4 hr prior to positron emission tomography (PET) scanning.

B. PET Scans

PET scans were performed on a Scanditronix PC2048-15B brain tomograph (15 contiguous slices; spatial resolution 6–6.5 mm in-plane and axially). Regional cerebral blood flow (rCBF) measurements were performed following an intravenous bolus of approximately 42 mCi of [^{15}O]water 1 min after initiation of the cognitive task. Head movements were limited with a thermoplastic mask individually fitted to each subject. The arterial input function was measured via an automated arterial blood sampling and counting system (Daube-Witherspoon *et al.*, 1992), and absolute rCBF (cc/min/100 g) was calculated with a pixel-by-pixel least-squares method (Koeppe *et al.*, 1985).

C. Image Registration and Statistical Analysis

The two scans for each subject were registered using automated image registration (AIR) (Woods *et al.*, 1992), and the realigned images were spatially normalized to the stereotactic atlas of Talairach and Tournoux (1988) using the Statistical Parametric Mapping '95 (SPM95) package (Friston *et al.*, 1995). The images were smoothed using a series of isotropic Gaussian filters as described later. Absolute rCBF was normalized as a ratio to the global mean (i.e., proportional scaling) and analyzed via linear contrasts. The *t*-test values in the resulting statistical map were squared to produce an *F*-statistic map with each voxel having one numerator degree of freedom.

D. Image Smoothing

In order to investigate the influence of image smoothing on voxel-based statistical effect size and power, a series of isotropic Gaussian filters having FWHM of 0, 5, 10, 15, and 20 mm were applied to data, and power estimation was performed using each level of filter size. Power plots for each of several specific brain locations detailed later were constructed to inspect power as a function of smoothing filter and sample size. Additionally, the filter widths producing maximal effect size values in the $N = 40$ sample were mapped into Talairach space at various power level thresholds, thereby providing a reference map of the amount of smoothing required in each specific brain location to maximize effect size and power.

E. Statistical Power Calculations

The effect size magnitudes on the total $N = 40$ sample were used to approximate population level standardized differences at each voxel. A voxel-wise power estimation method developed by Van Horn *et al.* (1997) was applied to the activation effect size estimates and weighted by sample size values for 5, 10, 15, 20, 25, and 40 to produce expected statistical and power maps for these sample sizes at each of the aforementioned filter widths. Sufficient power to correctly reject Ho: was defined as 80%.

Additionally, the three most significant voxels in the activation map of the $N = 40$ at a filter FWHM of 10 mm were considered individually; these voxels corresponded to the left inferior temporal lobe (BA 37), the left inferior parietal lobule (BA 40), and the right inferior frontal gyrus (44/46).

III. RESULTS

As an example of overall WCS activation, the general pattern of results when data were smoothed with a 10-mm filter is presented in Fig. 1 (uncorrected $p < 0.01$). However, at nearly all brain locations, voxel-based effect sizes varied greatly as a function of spatial smoothing (Fig. 2), considerably modulated voxel-based statistical power, and appeared to follow known neuroanatomical constraints such as the juxtaposition of functionally reactive brain tissue and areas containing large cerebral vasculature (Fig. 3, see color insert). In many cortical areas, smaller filter sizes were required in order to maximize power, whereas smoothing with larger filters resulted in decreased power (e.g., left middle frontal gyrus). In other areas, such as the poste-

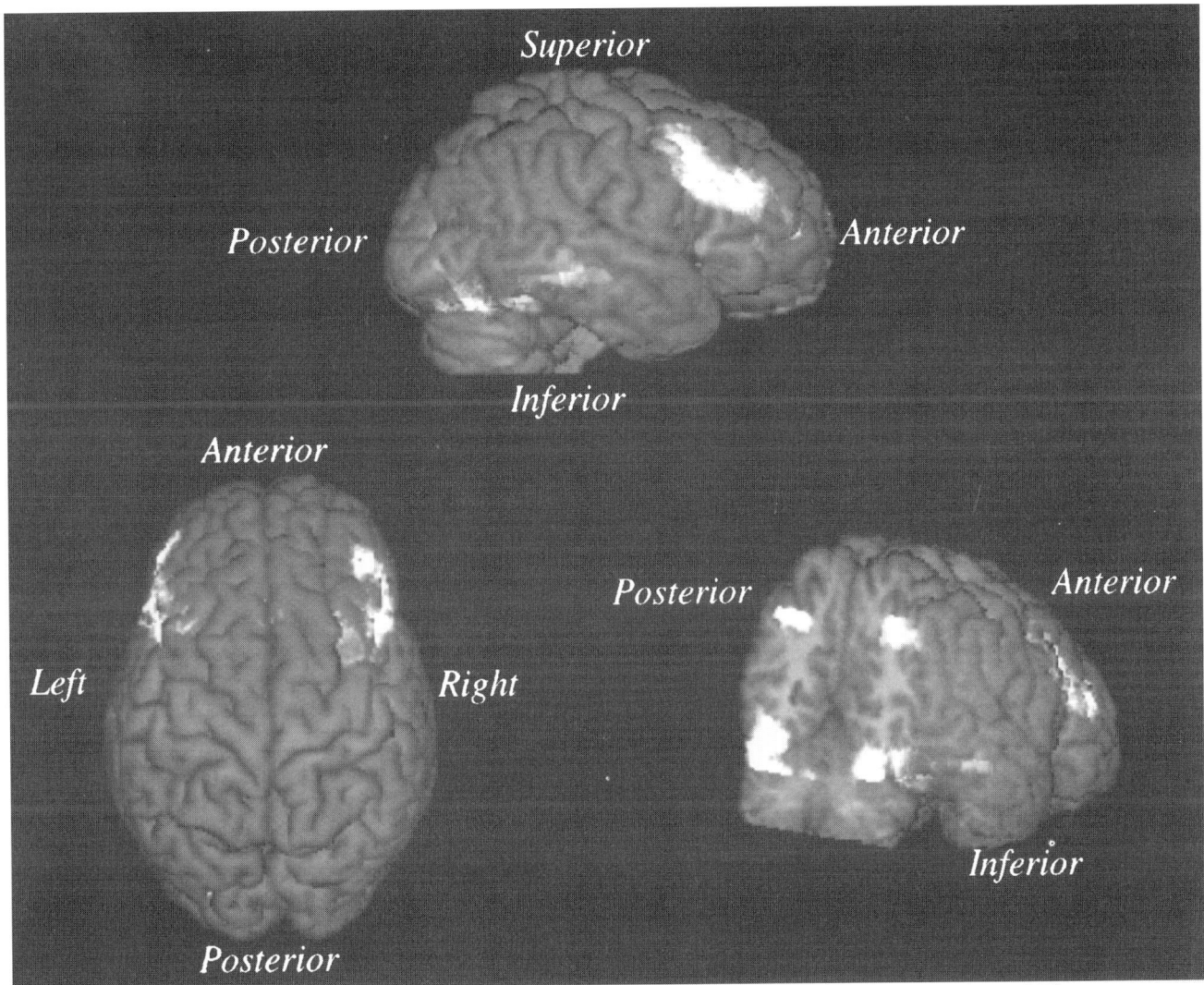

35. Filtering and Power in Brain Activation Studies   239

FIGURE 1 Three views of the significant regional cerebral blood flow activation (uncorrected $p < 0.01$) during the Wisconsin card sorting test ($N = 40$). Results were superimposed onto a representative sagital volume MRI that was also warped to Talairach space using SPM95.

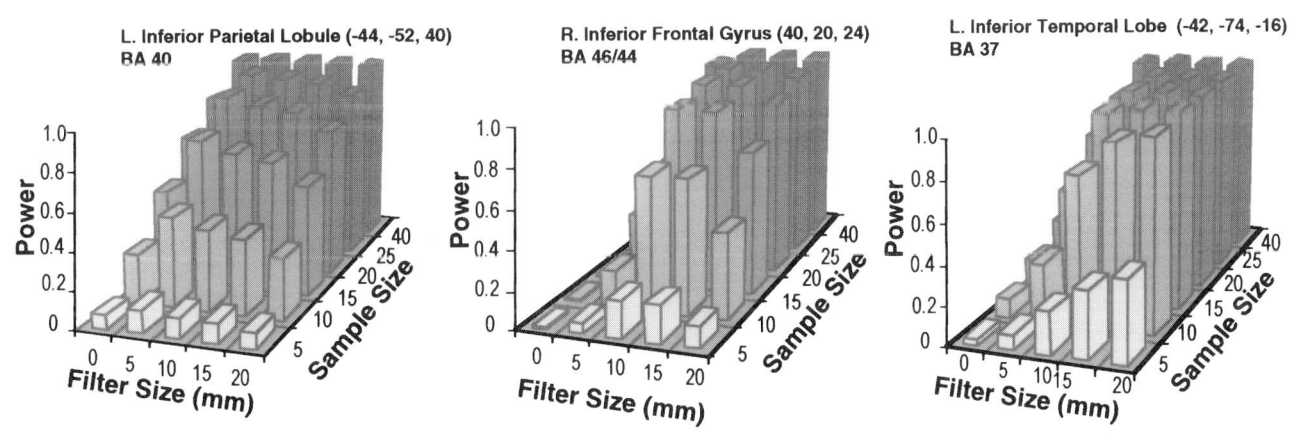

FIGURE 2 Statistical power as a function of sample size and image smoothing filter width (0–20 mm) for several brain locations found to be the most highly activated (uncorrected $p < 0.01$) in the $N = 40$ analysis based on the 10-mm filter size.

rior inferior temporal lobe, maximal power estimates were observed only with the largest smoothing filter (20 mm). The largest filter size also produced maximal power in many white matter areas, in this case likely reflecting maximal spatial averaging with truly reactive gray matter structures. Thresholding at several levels of power, based on this sample size, showed a decreasing pattern of brain areas having sufficient power, but, in particular, indicated that primarily larger filter sizes (10–20 mm) were necessary to produce maximal effect sizes having above 80% power. However, the improvement in power related to image smoothing was always small compared to the increase in power gained by increases in sample size. For instance, in the right inferior frontal gyrus with $N = 5$, the improvement in power using a 15-mm filter above that when no filtering was applied was only 17.7%, whereas using an even larger filter of 20 mm at $N = 5$ only provides a 10% improvement over no smoothing at all. By comparison, a 50% increase in power was achieved by the increasing the sample size from 5 to 10 and holding the 15-mm filter size constant.

IV. DISCUSSION

Previous implementations of multifiltering have concentrated principally on improving only the statistical significance of activation (e.g., Poline and Mazoyer, 1994). The present investigation varied smoothing filter to assess its influence on the statistical power of activation across the brain. These data show that image smoothing greatly modulates the statistical effect size and hence directly influences the estimated probability of replication. These basic results agree well with previous observations in the literature regarding the matched filter theorem (Worsley *et al.*, 1996) and the "scale-space" search for significance. However, variations in the filter width required to maximize effect size and power across brain voxels indicate that the neuroanatomical structure and location underlying hypothesized activations should be carefully considered when choosing an *a priori* filter size to apply to image data. More focal structures within nonfunctionally related areas (e.g., large cerebral vasculature) or areas proximal to the natural boundaries of the brain (e.g., the lateral sulci) may require the use of smaller filter widths in order to avoid the inclusion of nonbrain areas in the resulting smoothed voxel value. Conversely, areas of large, neuroanatomically uniform extent (e.g., cerebellum) may permit the use of larger filter widths to ensure the best signal reproducibility.

Nevertheless, image smoothing alone is unlikely to appreciably improve the statistical power of results in studies having small sample sizes. Excessive smoothing in an attempt to increase significance, in some instances, is likely to be deleterious to already weak power. Therefore, increasing study sample size remains the most dependable strategy by which to improve statistical power for detecting experimentally induced effects. Still, methods for adaptive filtering of functional neuroanatomy based on structural neuroanatomical constraints should be investigated further as they offer an additional approach for improving the ability to detect subtle neurophysiological effect sizes.

References

Berman, K. F., Ostrem, J. L., Randolph, C., Gold, J., Goldberg, T. E., Coppola, R., Carson, R. E., Herscovitch, P., and Weinberger, D. R. (1995). Physiological activation of a cortical network during performance of the Wisconsin Card Sorting test: A Positron Emission Tomography study. *Neuropsychologia* **8L** 1027–1046.

Cohen, J. (1977). "Statistical Power Analysis for the Behavioral Sciences." Academic Press, New York.

Cohen, J. (1992). A power primer. *Psychol. Bull.* **112**: 155–159.

Daube-Witherspoon, M. E., Chou, K. S., Green, S. L., Carson, R. E., and Herscovitch, P. (1992). Factors affecting dispersion correction for continued blood withdrawal and counting systems. *J. Nucl. Med.* **33**: 1010.

Esposito, G., Kirkby, B. S., Van Horn, J. D., Ellmore, T. E., Weinberger, D. R., and Berman, K. F. (1998). Context-dependent, neural system-specific neurophysiological concomitants of aging: Mapping PET correlates during cognitive activation. Submitted for publication.

Friston, K. J., Holmes, A. P., Worsley, K. J., Poline, J. P., Frith, C. D., and Frackowiak, R. S. J. (1995). Statistical parametric maps in functional imaging: A general linear model approach. *Hum. Brain Mapp.* **2**:189–210.

Koeppe, R. A., Holden, J. E., and Ip, W. R. (1985). Performance comparison of parameter estimation techniques for the quantification of local cerebral blood flow by dynamic positron computed tomography. *J. Cereb. Blood Flow Metab.* **5**: 224–234.

Paulesu, E., Connelly, A., Frith, C. D., Friston, K. J., Heather, J., Myers, R., Gadian, D. G., and Frackowiak, R. S. J. (1995). Functional MR imaging correlations with positron emission tomography. *Neuroimag. Clin. North Amer.* **5**: 207–225.

Poline, J. B., and Mazoyer, B. (1991). Towards an individual analysis of brain activation PET studies: Optimal filtering of difference images. *J. Cereb. Blood Flow Metab.* **11**(Suppl. 2): S564.

Poline, J. B., and Mazoyer, B. (1993). Analysis of individual positron emission tomography activation maps by detection of high signal-to-noise-ratio pixel clusters. *J. Cereb. Blood Flow Metab.* **13**: 425–437.

Poline, J. B. and Mazoyer, B. (1994). Enhanced detection in brain activation maps using a multifiltering technique. *J. Cereb. Blood Flow Metab.* **14**: 639–642.

Talairach, J., and Tournoux, P. (1988). "Co-Planar Stereotaxis Atlas of the Human Brain." Thieme, Stuttgart.

Van Horn, J. D., Ellmore, T. E., Esposito, G., Weinberger, D. R., and Berman, K. F. (1997). Voxel-based statistical power in functional neuroimaging studies. *NeuroImage* **5**: S464.

Woods, R. P., Cherry, S. R., and Mazziotta, J. C. (1992). Rapid automated algorithm for aligning and reslicing PET images. *J. Comput. Assist. Tomogr.* **16**: 620–633.

Worsley, K. J., Marrett, S., Neelin, P., and Evans, A. C. (1996). Searching scale space for activation in PET images. *Hum. Brain Mapp.* **4**: 74–90.

CHAPTER 36

Measuring Activation Pattern Reproducibility Using Resampling Techniques[1]

S. C. STROTHER,*,†,‡ K. REHM,* N. LANGE,¶ J. R. ANDERSON,† K. A. SCHAPER,† L. K. HANSEN,§ and D. A. ROTTENBERG*,†,‡

*Departments of * Radiology and ‡Neurology*
University of Minnesota, and
† PET Imaging Center
Veterans Affairs Medical Center
Minneapolis, Minnesota 55417
¶ Brain Imaging Center
Harvard Medical School and McLean Hospital
Boston, Massachusetts 02178
§ Department of Mathematical Modeling
Technical University of Denmark
Lyngby, Denmark DK-2800

This study presents two "similarity measures" of activation pattern reproducibility that do not require a choice of model-dependent noise definitions and thresholds. Both measures are derived from scatter plots that provide a visual comparison of the signal levels across all pairs of Talairach voxels for pairs of activation images. The first measure is the Pearson product-moment correlation of the scatter plot. The second measure is the ratio of the "signal" and "noise" histogram widths for all Talairach voxel signal levels projected onto the major and minor axes, respectively, from a principal components analysis of the scatter plot. Empirical population histograms of these similarity measures for many pair-wise image comparisons of independent groups of subjects were generated using statistical resampling techniques. The resulting "reproducibility histograms" were used to examine the impact of four different data analysis models on groups of 1, 4, 8, and 12 subjects for [^{15}O]water positron emission tomography scans of a simple motor task. For each group size, activation images from models using single voxel noise estimates were less reproducible than images from models using multivoxel noise estimation procedures. This occurs be-cause single voxel noise estimates produce a lower "activation signal-to-noise ratio" compared to multivoxel estimates when measured using the model-independent noise scale of the second similarity measure. These data demonstrate that the number of activated voxels found using thresholding procedures depends strongly on group size and the particular data analysis model used and that models using single voxel noise estimation procedures have suboptimal activation signal-to-noise ratios.

I. INTRODUCTION

A central problem in brain activation experiments is the reproducibility of the resulting functional activation patterns if the experiment were to be repeated with an independent group of subjects. This is an important issue that is influenced by data analysis model biases, resulting from normalization procedures, definitions of noise, and the setting of activation thresholds, and by the small sample sizes that are typically analyzed (e.g., Strother et al., 1997; Vitouch and Glück, 1997). Earlier work has examined the reproducibility of suprathreshold, spatially localized activation foci (Grabowski et al., 1996; Andreasen et al., 1996). To address activation pattern reproducibility while avoiding model-

---

[1] Transcripts of the BRAINPET97 discussion of this chapter can be found in Section VIII.

dependent noise definitions and thresholding biases, the use of scatter plots has been proposed (Strother *et al.*, 1997). With these plots, signal-level reproducibility is visualized in all pairs of Talairach-aligned voxels for pairs of activation images from independent groups of subjects. The activation image reproducibility of such plots may then be summarized using pattern similarity measures. The aims of this work were threefold: (1) to quantitatively measure the impact of different models and group sizes on activation image reproducibility; (2) to achieve Aim 1 without model biases using scatter plots and resampling techniques to generate empirical histograms of activation image similarity measures; and (3) to use scatter plots to define "activation signal" and "orthogonal noise" distributions and to compare the ratios of "activation signal-to-noise ratios" across models using a model-independent noise scale.

II. MATERIALS AND METHODS

[^{15}O]water PET scans were acquired from 28 normal right-handed subjects using the sequential finger-to-thumb opposition task and scanning procedures—eight alternating baseline and activation scans/subject—described in Strother *et al.* (1995). Pairs of independent groups of 1, 4, 8, and 12 subjects were randomly drawn with replacement from this 28-subject pool (2, 8, 16, and 24 subjects/draw, respectively). All 378 unique subject pairs were drawn, and groups of 8, 16, and 24 subjects were randomly drawn with replacement 500 times. For each group size, from 1 to 12 subjects, each pair of independent groups was analyzed using four models: (i) removal of blocked subject effects and scan-mean covariates using a general linear model (GLM) followed by evaluation of baseline activation contrasts using voxel-based independent t tests with single voxel variance estimates to generate a statistical parametric image (GLM/SPI; Friston *et al.*, 1995); (ii and iii) removal of global effects by ratio normalization (i.e., voxels divided by scan means) followed by voxel-based paired t tests with single voxel and pooled variance estimates [ratio-normalized, single voxel-paired t test (RN/SVPt); ratio-normalized, pooled voxel-paired t test (RN/PVPt); Fox *et al.*, 1988; Worsley *et al.*, 1992]; and (iv) removal of a mean image and global scaling effects using a scaled subprofile model with principal components analysis (SSM/PCA; Moeller and Strother, 1991) with discriminant analysis of the resulting eigenvectors of scan weights using a canonical variates analysis (CVA; Mardia *et al.*, 1979) of the baseline activation group separation (SSM/CVA)—this is a Fisher's linear discriminant analysis with only two groups (Ardekani *et al.*, this volume, Chapter 38).

For each model and group size, scatter plots were generated for all pairs of independent groups. Histograms of the correlation coefficients (r) from these scatter plots, i.e., r histograms as a function of model and group size, yielded empirical distributions of activation image similarities. In addition, for each model, a PCA of the scatter plots for all pairs of 12-subject groups was performed; all Talairach voxel values were projected onto the major and minor PCA axes to form histograms of the "reproducible activation signal" and "orthogonal noise" distributions, respectively. Both PCA axes were scaled by the standard deviation of the orthogonal noise distribution, and the origin along the reproducible activation signal axis was estimated by projecting the origin of the scatter plot axes onto the signal axis parallel to the noise axis. For each scatter plot, the 99% confidence intervals (CIs) of the PCA-derived signal and noise histograms were computed and histograms of the activation signal-to-noise ratio (signal 99% CI)/(noise 99% CI) were generated for all four models.

III. RESULTS AND DISCUSSION

Figure 1 displays scatter plots for independent pairs of groups of 1, 4, 8, and 12 subjects analyzed with the GLM/SPI model. As expected the plots become more "cigar" shaped (i.e., reproducible) with increasing group size, a trend seen for all the other models. The shaded regions of the lower-right panel in Fig. 1 demonstrate that for groups of 12 subjects many of the voxels that lie above a simple threshold (e.g., $t = 5$) in one group lie below that threshold in another independent group. This result illustrates the relatively modest power of such analyses in groups of up to 12 subjects. Grabowski *et al.* (1996) have commented on this point, which is discussed in more detail in Strother *et al.* (1997) and is analyzed more formally in Van Horn *et al.* (this volume, Chapter 35) as a function of image smoothing.

In Fig. 2, the average or median r value increases with group size for all four models. Moreover, the spread of r distributions in Fig. 2, which provides a measure of pattern dissimilarity for disjoint groups of equal size, decreases with group size. For a given group size, there is little difference in the r distributions between the two models using single voxel noise estimates (RN/SVPt and GLM/SPI; dashed profiles) or between the two models using multivoxel noise estimates (RN/PVPt and SSM/CVA; solid profiles). Nevertheless, for all group sizes the single voxel noise models have consistently lower r distributions than the

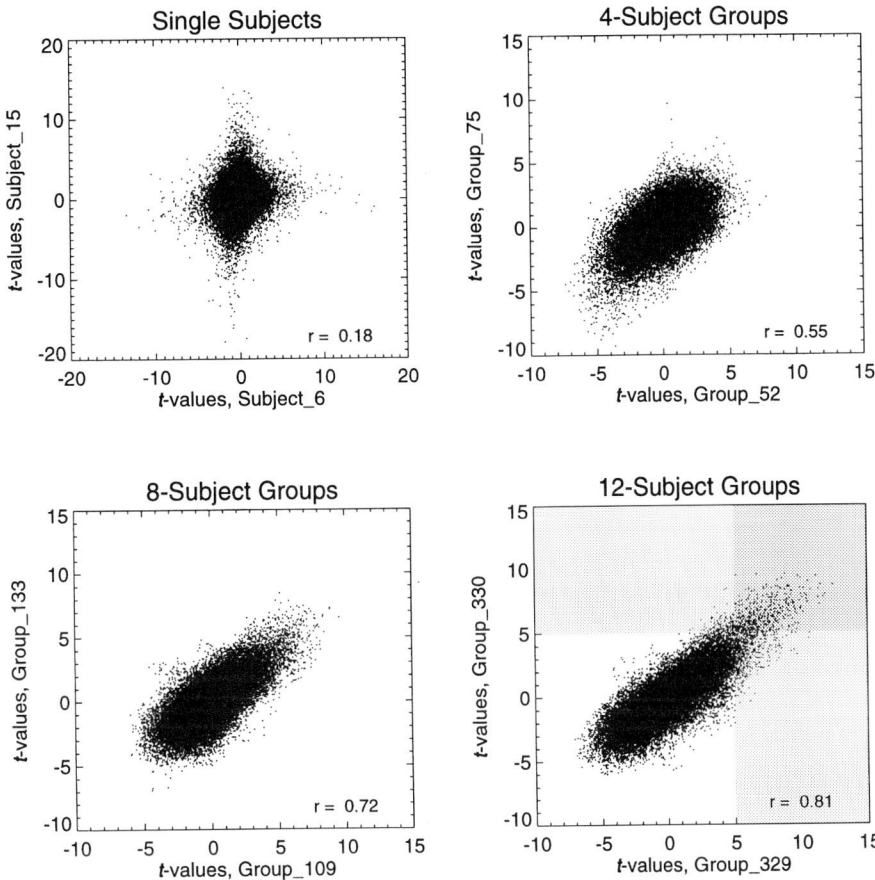

FIGURE 1 Scatter plots of all pairs (1 pair/point) of Talairach voxel t values for pairs of activation images derived from pairs of independent groups of 1, 4, 8, and 12 subjects. For each group, activation images were produced by removal of blocked subject effects and scan mean covariates using a general linear model (GLM) followed by evaluation of baseline activation contrasts using voxel-based unpaired t tests with single voxel variance estimates to generate a statistical parametric image (GLM/SPI, see Section II). Correlation coefficients (r) are displayed for each scatter plot. In the lower-right panel the shaded light-gray rectangles define the areas contributing to $P(t_{\text{Image2}} > 5 \mid t_{\text{Image1}} < 5)$ and $P(t_{\text{Image1}} > 5 \mid t_{\text{Image2}} < 5)$, and the darker-gray square is the area contributing to $P(t_{\text{Image2}} > 5 \mid t_{\text{Image1}} > 5)$ for both activation images from the independent groups of 12 subjects.

models using multivoxel estimates to stabilize the noise. This suggests that in some as yet to be understood sense, the single voxel noise models are less able to extract stable activation signals from the true noise distributions; however, there are unknown bias variance trade-offs influencing these results. Reproducibility measures must be interpreted with the caveat that the signal being reproduced has some unknown, model-dependent bias relative to the unknown true activation signal for the population.

Figure 3 demonstrates the second similarity measure based on "reproducible activation signal" and "orthogonal noise" distributions defined on the PCA axes of scatter plots. The left-hand panel illustrates a comparison of the activation images from the same independent groups of 12 subjects analyzed with the GLM/SPI model. The process of projecting voxel values of the scatter plot onto the PCA axes and rescaling by the noise-axis standard deviation is designed to create signal and noise distributions on a common scale that may be compared across models. After rescaling (right-hand panel), the noise distribution (dashed line) is seen to be very close to a standard normal curve (thin solid line). A similar result is obtained for all 500 draws of 24 subjects for all four models. The extraction of signal and noise distributions from the PCA axes in Fig. 3 may be thought of as a spatially distributed, multivariate, split t test for the 24

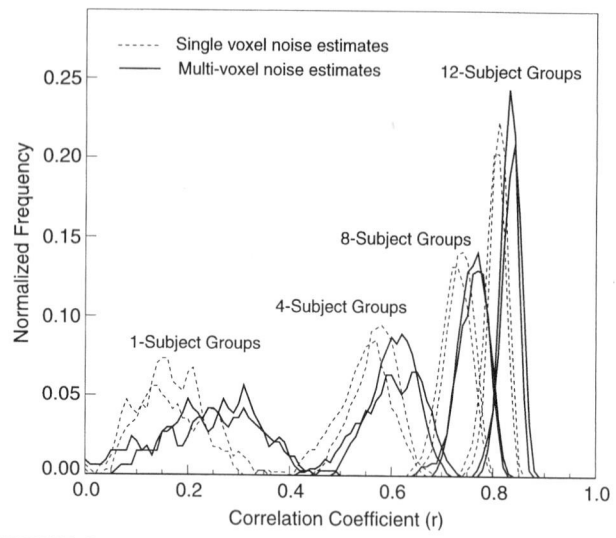

FIGURE 2 Correlation coefficient (r) histograms (normalized by number of activation image pairs per histogram; 378–500—see Section II) for pairs of activation images from resampled pairs of independent groups of 1, 4, 8, and 12 subjects. For each resampled pair of independent groups, the correlated activation images were produced by two models using single voxel noise estimates (RN/SVPt and GLM/SPI—see Section II; dashed lines) and two models using multivoxel noise estimates (RN/PVPt and SSM/CVA—see Section II; solid lines).

subjects comprising the two independent groups being compared. As such it has the robustness advantages of the univariate split t test to deviations from normality of the population distribution from which the sample was drawn (K. Worsley, personal communication).

Figure 4 displays histograms of the ratios of the 99% confidence intervals of the "reproducible activation signal" and "orthogonal noise" distributions (see Fig. 3) for all 500 draws of 24 subjects for all four models. Higher values on the abscissa reflect more extended signal tails on the major PCA axis relative to a standard-normal noise scale, independent of the particular model that produced the original activation images. The shift toward higher activation signal-to-noise ratios for multivoxel noise estimates in Fig. 4 demonstrates that the shifts in r distributions seen in Fig. 2 reflect a real increase in reproducible, activation signal-to-noise levels. In addition, the extended tail of the SSM/CVA distribution in Fig. 4 demonstrates that there are subgroups of subjects that are uniquely well fitted by the SSM/CVA model, although for most pairs of 12-subject groups SSM/CVA and RN/PVPt perform with similar activation signal-to-noise levels.

This chapter introduced activation pattern similarity measures together with resampling techniques to generate histograms that measure the population-based reproducibility of activation patterns. Using these tools

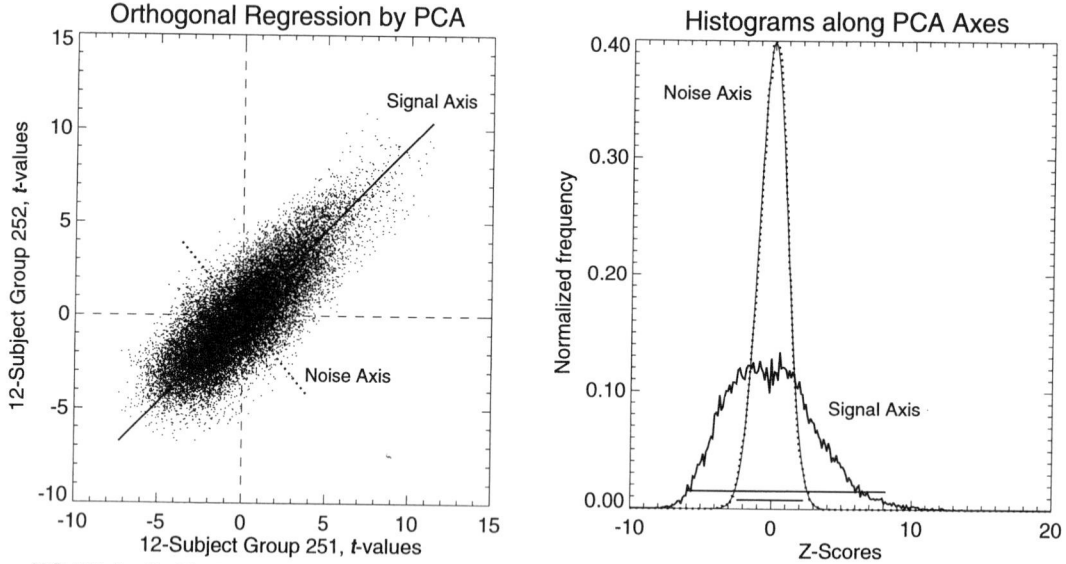

FIGURE 3 (Left) "Reproducible activation signal" and "orthogonal noise" axes are plotted as major (solid line) and minor (dashed line) axes, respectively (from a principal components analysis, PCA), on the scatter plot of the activation images from a pair of independent groups of 12 subjects analyzed with GLM/SPI (see Section II). (Right) Normalized signal (solid line) and noise (dotted line) histograms of all Talairach voxel t values projected onto the PCA axes are plotted, after rescaling both axes by the standard deviation of the noise-axis histogram to obtain standardized units (i.e., Z scores); the noise histogram has been over plotted with the theoretical shape of a standard normal density function (thin solid line). The horizontal solid lines at the base of the signal and noise histograms in the right panel depict the respective 99% confidence intervals.

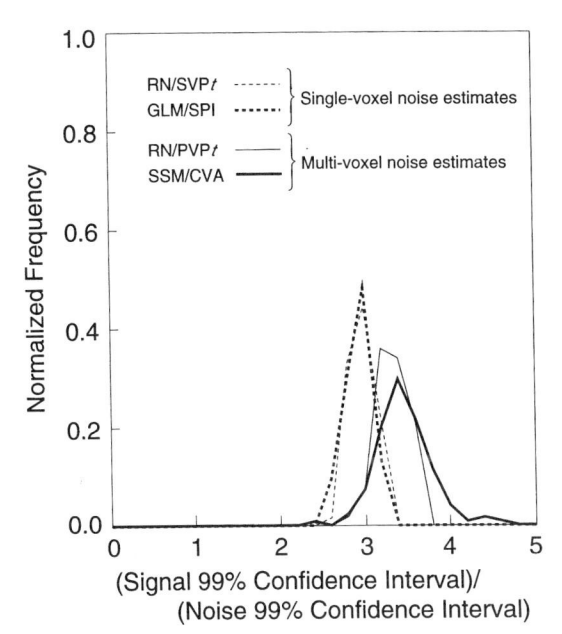

FIGURE 4 Histograms of the activation signal-to-noise ratios defined by the 99% confidence intervals of the "reproducible activation signal" and "orthogonal noise" distributions (see Fig. 3) from scatter plots for all 500 random draws of 24 subjects (see Section II). For each 24-subject draw, scatter plots were generated from pairs of activation images by analyzing the two independent groups of 12 subjects with each of four models: ratio-normalized, single voxel-paired t test with single voxel variance estimates, RN/SVPt; general linear model preprocessing, single voxel-independent t tests with single voxel variance estimates, GLM/SPI; ratio-normalized, single voxel-paired t test with pooled voxel variance estimates, RN/PVPt; and scaled subprofile model preprocessing, canonical variates analysis of resulting eigenvectors, SSM/CVA.

to compare a number of data analysis models demonstrated the somewhat surprising result that models using single voxel noise estimates are suboptimal because their reproducible "activation signal-to-noise ratios" are smaller than those of models using multivoxel noise estimation procedures.

Acknowledgments

This work has been funded in part by Human Brain Project Grants DA09246 and MH57180 from the National Institutes of Health. Our thanks to Dana Daly and David Bonar for data collection and preliminary data analysis. We acknowledge many stimulating discussions on multivariate data analysis and model "generalizability" with Claus Svarer, Benny Lautrup, and Niels Mørch.

References

Andreasen, N. C., Arndt, S., Cizadlo, T., O'Leary, D. S., Watkins, G. L., Boles Ponto, L. L., and Hichwa, R. D. (1996). Sample size and statistical power in [^{15}O]H$_2$O studies of human cognition. *J. Cereb. Blood Flow Metab.* **16**: 804–816.

Fox, P. T., Mintun, M. A., Reiman, E. M., and Raichle, M. E. (1988). Enhanced detection of focal brain responses using intersubject averaging and change-distribution analysis of subtracted PET images. *J. Cereb. Blood Flow Metab.* **8**: 642–653.

Friston, K. J., Holmes, A. P., Worsley, K. J., Poline, J.-P., Frith, C. D., and Frackowiak, R. S. J. (1995). Statistical parametric maps in functional imaging: A general approach. *Hum. Brain Mapp.* **2**: 189–210.

Grabowski, T. J., Frank, R. J., Brown, C. K., Damasio, H., Ponto, L. L. B., Watkins, G. L., and Hichwa, R. D. (1996). Reliability of PET activation across statistical methods, subject groups, and sample sizes. *Hum. Brain Mapp.* **4**: 23–46.

Mardia, K. V., Kent, J. T., and Bibby, J. M. (1979). "Multivariate Analysis." Academic Press, London.

Moeller, J. R., and Strother, S. C. (1991). A regional covariance approach to the analysis of functional patterns in positron emission tomographic data. *J. Cereb. Blood Flow Metab.* **11**: A121–A135.

Strother, S. C., Anderson, J. R., Schaper, K. A., Sidtis, J. J., Liow, J.-S., Woods, R. P., and Rottenberg, D. A. (1995). Principal component analysis and the subprofile scaling model compared to intersubject averaging and statistical parametric mapping: I. "Functional connectivity" of the human motor system studied with [^{15}O] PET. *J. Cereb. Blood Flow Metab.* **15**: 738–775.

Strother, S. C., Lange, N., Anderson, J. R., Schaper, K. A., Rehm, K., Hansen, L. K., and Rottenberg, D. A. (1997). Activation pattern reproducibility: Measuring the effects of group size and data analysis models. *Hum. Brain Mapp.* **5**: 312–316.

Vitouch, O., and Glück, J. (1997). "Small group PETing:" Sample sizes in brain mapping research. *Hum. Brain Mapp.* **5**: 74–77.

Worsley, K. J., Evans, A. C., Marrett, S., and Neelin, P. (1992). A three-dimensional statistical analysis for CBF activation studies in human brain. *J. Cereb. Blood Flow Metab.* **12**: 900–918.

CHAPTER 37

Reproducibility of Regional Metabolic Covariance Patterns: Comparison of Four Populations

D. EIDELBERG,* J. R. MOELLER,[†] V. DHAWAN,* A. ANTONINI,* L. MORAN,* J. MISSIMER,[‡] and K. L. LEENDERS[‡]

*Department of Neurology
North Shore University Hospital
Manhasset, New York 11030 and
New York University School of Medicine
New York, New York 10016
[†] Department of Psychiatry
Columbia College of Physicians and Surgeons
New York, New York 10032
[‡] Paul Scherrer Institute, Villigen, Switzerland

In a previous [^{18}F]fluorodeoxyglucose positron emission tomography study, regional metabolic data from a combined group of Parkinson's disease (PD) patients and normal controls (N) using the scaled subprofile model (SSM) and principal component analysis (PCA) were analyzed. Using this method, a unique pattern of regional metabolic covariation was identified that accurately discriminated patients from controls (Eidelberg et al., 1994). In order to assess the reproducibility of this pattern as a potential marker for PD, its topography was compared with that of the covariance patterns identified by SSM analysis of three other independent populations of PD patients and controls studied in different laboratories with different tomographs. The following patient populations were studied: Group A (original cohort: 22 PD, 20 N; FWHM 7.5 mm); group B (18 PD, 20 N, FWHM 4.2 mm); group C (15 PD, 15 N, FWHM 8.5 mm); and group D (14 PD, 10 N; FWHM 12 mm). Region weights for the PD-related SSM pattern identified in group A were correlated with corresponding region weights for the group B, C, and D PD-related SSM patterns. Subject scores for the original group A pattern were computed for every subject (Eidelberg et al., 1995a) in each of the study populations. Values for PD and N were compared by discriminant analysis. The PD-related pattern identified in group A was highly correlated with the patterns identified in the other populations ($R^2 > 0.61$, $p < 0.01$). Computed subject scores for this pattern accurately discriminated PD from N in all four populations with a sensitivity of 75–85% ($p < 0.001$). The PD-related SSM covariance pattern reported previously is highly reproducible across patient populations and tomographs. The expression of this pattern in individual subjects can accurately discriminate patients from controls in multiple populations studied with different tomographs.

I. INTRODUCTION

The scaled subprofile model (SSM; Moeller et al., 1987; Moeller and Strother, 1991; Alexander and Moeller, 1994) approach has been used in the study of Parkinson's disease (PD), a common akinetic-rigid movement disorder. In previous studies, a specific pattern of regional metabolic covariation associated with PD was identified (Eidelberg et al., 1990, 1994). This covariance pattern was characterized by relative hypermetabolism of the lentiform nucleus and thalamus associated with metabolic decreases in the primary and association motor cortices. Subject scores for this pattern, representing its expression in individual patients, were found to correlate significantly with independent measures of clinical disability as measured by the unified Parkinson's disease rating scale (UPDRS; Fahn et al., 1984). Additionally, these subject scores correlated with presynaptic nigrostriatal dopaminergic function as measured by striatal uptake of [^{18}F]DOPA (Eidelberg *et al.*, 1990, 1995a) and with changes in motor performance following stereotoxic pallidotomy (Eidelberg *et al.*, 1996).

Another technique of computing the expression of this network in individual subjects has been developed. This algorithm is referred to as topographic profile rating (TPR; Eidelberg *et al.*, 1995a; Moeller *et al.*, 1996). This approach has been used to compute subject scores for the original PD-associated pattern prospectively on an individual scan basis. Indeed, in PD these computed measures of network expression were found to be highly predictive of independent measures of clinical disability. In parallel studies, it was also found that a related covariance pattern can differentiate early stage PD from normals with an accuracy comparable to [^{18}F]DOPA/positron emission tomography (PET) (Eidelberg *et al.*, 1990, 1995b). Moreover, using the TPR algorithm, subject scores for this pattern accurately distinguished drug-responsive from drug-resistant parkinsonians at the earliest stages of disease (Eidelberg *et al.*, 1995b). These preliminary findings suggested that quantifying the expression of regional metabolic covariance patterns can assist in the early differential diagnosis of parkinsonism and can provide an objective measure of disease progression.

The current study assessed the reproducibility of SSM covariance patterns associated with PD by comparing the topographies of patterns identified in four independent patient populations scanned in different laboratories using PET tomographs of varying resolution and sensitivity.

II. MATERIALS

A. Subject Groups

The following four populations were studied with quantitative [^{18}F]fluorodeoxyglucose (FDG) and PET:

Group A: Twenty-two PD patients and 20 normal volunteers were scanned using the SuperPETT3000 tomograph at North Shore University Hospital, Manhasset, New York (7.5 mm FWHM; Robeson *et al.*, 1993). Clinical and metabolic data from this group were reported previously (Eidelberg *et al.*, 1994).

Group B: Eighteen PD patients and 20 normal volunteers were scanned using the General Electric Advance Tomograph at North Shore University Hospital (4.3 mm FWHM; DeGrado *et al.*, 1994).

Group C: Fifteen PD patients and 15 normal volunteers were scanned using the ECAT 933-16 tomograph at Paul Scherrer Institute, Villigen, Switzerland (8.5 mm FWHM).

Group D: Fourteen PD patients and 10 normal volunteers were scanned using the PC4600 tomograph at Memorial Hospital, New York, New York (12 mm FWHM; Kearfott and Carroll, 1984). Clinical and metabolic data from this group were reported previously (Eidelberg *et al.*, 1990).

III. METHODS

A. Pattern Reproducibility

Similar quantitative FDG/PET methods were used to calculate global and regional rates of glucose metabolism (GMR and rCMRGlc, respectively) in all patients and control groups (Eidelberg *et al.*, 1990, 1994; Antonini *et al.*, 1996). rCMRGlc values for 30 standardized regions of interest (ROIs) were computed in each subject as described elsewhere (Eidelberg *et al.*, 1994, 1997). In each population, SSM was used to analyze combined rCMRGlc data from PD patients and normal controls (Moeller and Strother, 1991; Alexander and Moeller, 1994). In each population, this analysis was used to identify a disease-related regional metabolic covariance pattern such that its expression in individual subjects (subject scores) accurately discriminated PD patients from controls. The discriminant analysis was restricted to subject scores for either the first or the second SSM principal component patterns or to a linear combination of both. SSM patterns were considered disease related if their subject scores discriminated the patients from controls at $p < 0.001$ (F test according to Wilks' λ). Region weights for the PD-related SSM pattern identified in group A (topographic profile 1 in Eidelberg *et al.*, 1994; Fig. 1) were correlated with corresponding region weights for the PD-related covariance patterns identified in the SSM analyses of groups B, C, and D.

B. Prospective Discrimination

In subsequent analyses, the group A pattern was used as a prospective marker for between-group discrimination on an individual scan basis. This was accomplished using TPR (Eidelberg *et al.*, 1995a; Moeller *et al.*, 1996). Region weights for the group A pattern were projected into rCMRGlc data from each member of groups B, C, and D to compute individual subject scores for this pattern. This computational procedure was performed in an automated blinded fashion (Eidelberg *et al.*, 1995b). In each population, computed subject scores for normal and PD cohorts were compared using discriminant analysis (F test according to Wilks' λ).

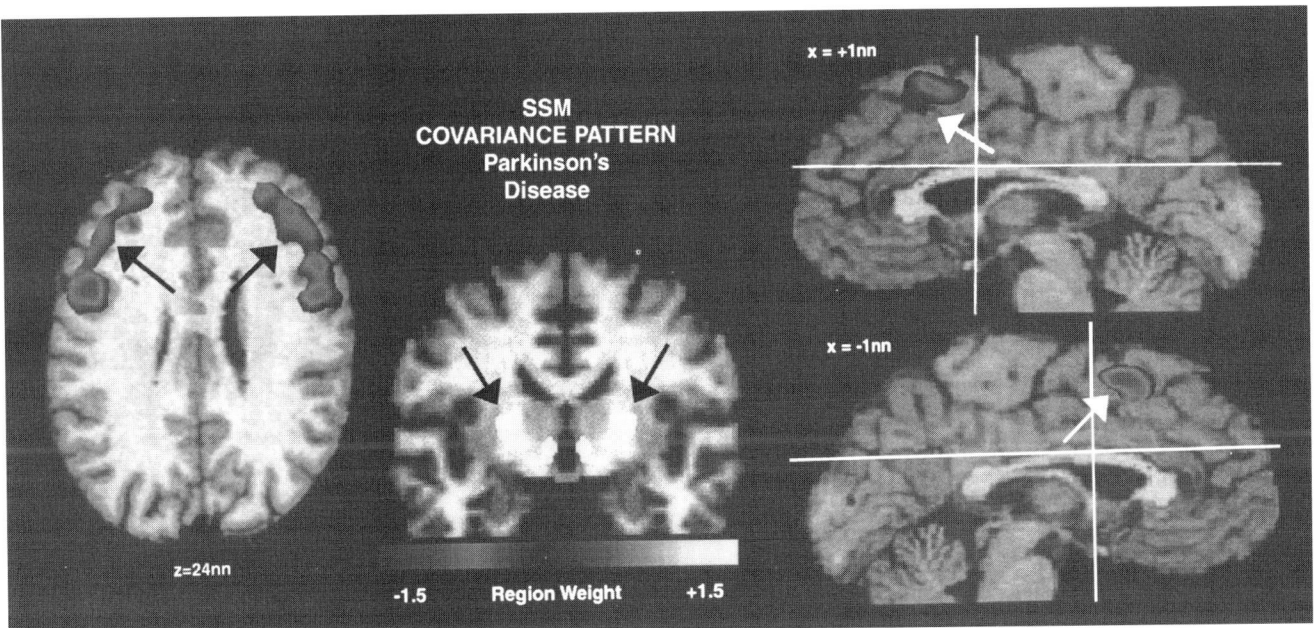

FIGURE 1 Display of the regional metabolic covariance pattern associated with Parkinson's disease identified using the scaled subprofile model (SSM) (Eidelberg *et al.*, 1994). Region weights have been overlaid on standard Talairach-transformed magnetic resonance imaging sections. The pattern is characterized by relative lentiform and thalamic hypermetabolism covarying negatively with bilateral metabolic reductions in motor and premotor regions and in the supplementary motor area.

IV. RESULTS

A. Pattern Reproducibility

In all four populations, SSM analysis of rCMRGlc data from PD patients and controls disclosed PD-related covariance patterns of similar topography (Fig. 2). The patterns were characterized by relative hypermetabolism of the basal ganglia and thalamus covarying with metabolic decreases in lateral frontal and paracentral motor areas and in the parietooccipital association cortex. Region weights for the patterns identified

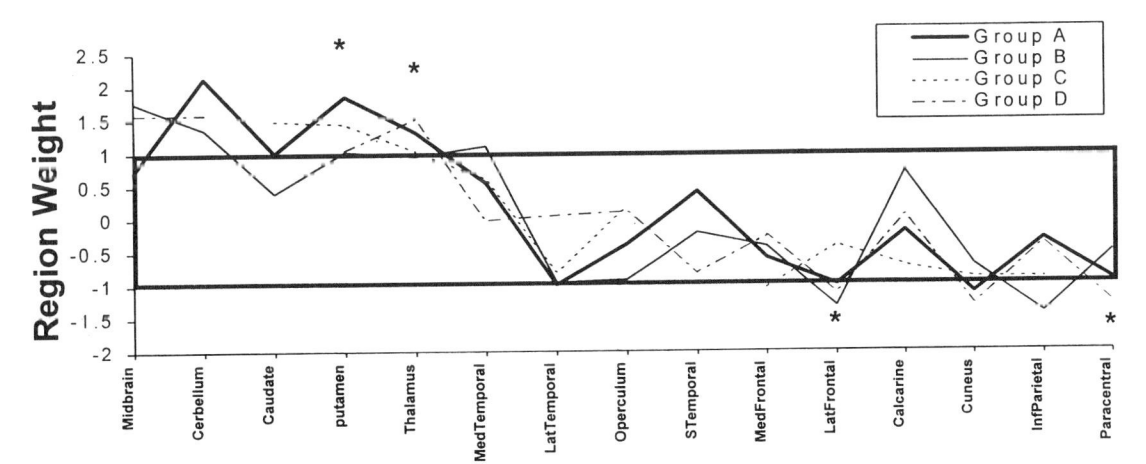

FIGURE 2 Region weights for the metabolic covariance patterns associated with Parkinson's disease. These patterns were identified in separate SSM analyses of four populations of patients and controls. Regions having weights with absolute values ≥ 1 (those outside box) contributed significantly to the pattern topography, accounting for more than 50% of the variance in normalized metabolic data ($p < 0.01$, corrected for multiple ROIs). Pattern-related increases in lentiform-thalamic metabolism covaried with relative metabolic decreases in motor cortical regions (asterisk).

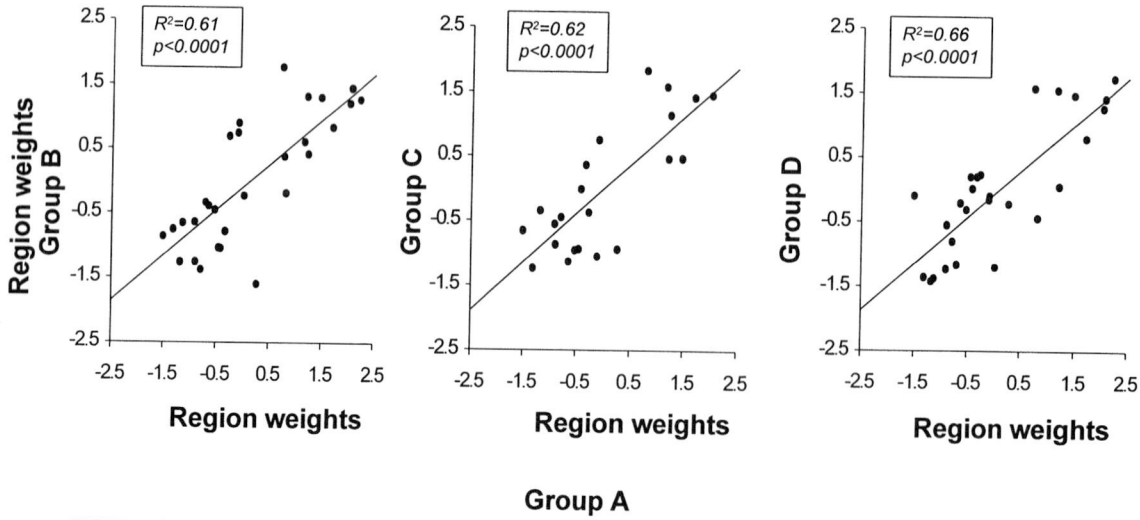

FIGURE 3 Correlations between region weights for the Parkinson's disease-related covariance pattern identified originally in group A (Eidelberg *et al.*, 1994; Fig. 1) and the corresponding regional values for the Parkinson's disease-related patterns identified in groups B, C, and D. These patterns were identified in scaled subprofile model analyses of regional metabolic data sets composed of both patients and normals (see text). Significant correlations were evident between group A region weights and those for groups B, C, and D ($p < 0.0001$).

in the analysis of groups B, C, and D were highly correlated with SSM pattern region weights from the original group A analysis ($R^2 > 0.61$, $p < 0.003$; Fig. 3).

B. Prospective Discrimination

Subject scores for the group A pattern were computed for each member of groups B, C, and D using TPR. These computed subject scores accurately discriminated PD patients from controls in all four populations studied ($p < 0.001$; Fig. 4).

V. DISCUSSION

This study demonstrates that disease-related SSM covariance patterns are highly reproducible across pa-

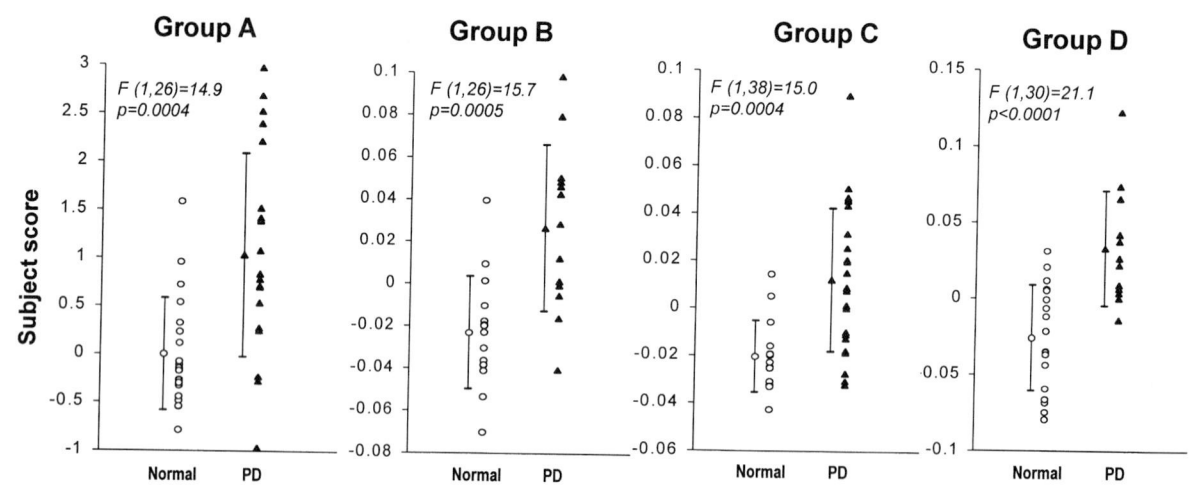

FIGURE 4 Discriminant function analysis of subject scores for the group A Parkinson's disease-related pattern (Fig. 1) computed individually for each patient (▲) and control subject (○) in groups A, B, C, and D (see text). Group A pattern subject scores discriminated PD patients from normals with a comparable degree of accuracy ($p < 0.001$) in all four populations studied.

tient populations and tomographs. The pattern previously reported in the group A analysis (Eidelberg et al., 1994) was used as a benchmark against which to compare the corresponding SSM patterns identified in the other populations. PD-related patterns identified by SSM analysis were topographically similar in the populations studied, with common region weight variability of approximately 60%. Thus, the disease-related topography extracted in SSM analyses of rCMRGlc data from PD patients and controls is stable despite differences in instrumentation and image quantification across centers (Eidelberg et al., 1994; Moeller et al., 1996). Indeed, these technical effects are of similar magnitude in the correlations between the group A pattern region weights and the corresponding pattern region weights identified in the analysis of groups B, C, and D. These findings extend those reported previously (Eidelberg et al., 1994) in which significant region weight correlations were identified for the PD-related patterns identified in groups A and D. The current study demonstrates the reproducibility of the group A pattern in two additional populations, including one scanned in a high-resolution tomograph (group B). The topographic stability of PD-related covariance patterns across populations supports the notion of their biological relationship to the inherent functional anatomy of parkinsonism (Eidelberg et al., 1994, 1996, 1997).

The utility of the group A pattern as a potential metabolic imaging marker for the diagnosis of PD was also tested (Eidelberg et al., 1995a, b). By projecting region weights for this pattern into rCMRGlc data from each PD patient and control subject, it was possible to compute its expression on an individual scan basis. Indeed, the resulting subject scores computed by TPR discriminated PD patients from normal controls in all four populations studied. However, although the overall accuracy of discrimination was comparable across groups, the degree of overlap between patients and controls was somewhat variable. This has been attributed to the inclusion of different numbers of early stage patients in each group. In early stage disease, subject scores for bihemispheric SSM patterns may fall into the normal range, although discriminant accuracy increases with disease progression (Eidelberg et al., 1994). In contrast, at its earliest stages, PD is associated with clinical and metabolic asymmetries that are most evident with SSM contrast analysis, i.e., that utilizing left–right *differences* in rCMRGlc (Eidelberg et al., 1990, 1995b, c). Indeed, subject scores for these patterns are often elevated with early stage disease and may be employed for differential diagnosis at or near the time of clinical onset (Eidelberg et al., 1995b). The reproducibility of SSM contrast patterns in groups A and D has already been demonstrated (Eidelberg et al., 1994). Additional FDG/PET studies of independent patient cohorts restricted to early stage PD will be needed to assess the reproducibility of these SSM contrast topographies, as well as the discriminant accuracy of their corresponding subject scores. Additionally, further validation experiments across multiple populations and PET instruments are required to assess the reproducibility of the SSM patterns associated with the normal aging process (Moeller et al., 1996; Moeller and Eidelberg, 1997).

In summary, this study demonstrates the high degree of reproducibility of PD-related SSM covariance patterns identified in FDG/PET data obtained in the resting state. The topography of these patterns is consistent with abnormal patterns of functional connectivity evident in both experimental animal models and human subjects (Eidelberg et al., 1994, 1996, 1997). Moreover, these patterns may be used prospectively across populations and PET instruments for purposes of differential diagnosis and the objective assessment of disease progression (see Eidelberg et al., 1995a).

Acknowledgments

This work was supported by NIH NS RO1-35069 and by generous grants from the National Parkinson Foundation and the Parkinson Disease Foundation. David Eidelberg is a Cotzias Fellow of the American Parkinson Disease Association. We acknowledge the important contributions of Dr. Phoebe Spetsieris and Dr. Thomas Chaly to this work. We thank Mr. Claude Margouleff for help with the PET studies, Dr. Robert Dahl and Mr. Ralph Matacchieri for cyclotron support, and Ms. Lauren Moran for manuscript preparation.

References

Alexander, G. E., and Moeller, J. R. (1994). Application of the scaled subprofile model to functional imaging in neuropsychiatric disorders: A principal component approach to modeling brain function in disease. *Hum. Brain Mapp.* **2**: 1–16.

Antonini, A., Leenders, K. L., Spiegel, R., Meier, D., Vontobel, P., Weigell-Weber, M., Sanchez-Pernaute, R., de Yebenez, J. G., Boesiger, P., Weindl, A., and Maguire, R. P. (1996). Striatal glucose metabolism and dopamine D_2 receptor binding in asymptomatic gene carriers and patients with Huntington's disease. *Brain* **119**: 2085–2095.

De Grado, T. R., Turkington, T. G., Williams, J. J., Stearns, C. W., Hoffman, J. M., and Coleman, R. E. (1994). Performance characteristics of a whole-body PET scanner. *J. Nucl. Med.* **35**: 1398–1406.

Eidelberg, D., Moeller, J. R., Dhawan, V., Sidtis, J. J., Ginos, J. Z., Strother, S. C., Cedarbaum, J., Greene, P., Fahn, S., and Rottenberg, D. A. (1990). The metabolic anatomy of Parkinson's disease: Complementary ^{18}F-fluorodeoxyglucose and ^{18}F-fluorodopa positron emission tomography studies. *Movement Disord.* **5**: 203–213.

Eidelberg, D., Moeller, J. R., Dhawan, V., Spetsieris, P., Takikawa, S., Ishikawa, T., Chaly, T., Robeson, W., Margouleff, D., Przedborski, S., and Fahn, S. (1994). The metabolic topography of parkinsonism. *J. Cereb. Blood Flow Metab.* **14**: 783–801.

Eidelberg, D., Moeller, J. R., Ishikawa, T., Dhawan, V., Spetsieris, P., Chaly, T., Robeson, W., Dahl, J. R., and Margouleff, D. (1995a). The assessment of disease severity in parkinsonism with ^{18}F-fluorodeoxyglucose and positron emission tomography. *J. Nucl. Med.* **36**: 378–383.

Eidelberg, D., Moeller, J. R., Ishikawa, T., Dhawan, V., Spetsieris, P., Chaly, T., Belakhlef, A., Mandel, F., Przedborski, S., and Fahn, S. (1995b). Early differential diagnosis of Parkinson's disease with ^{18}F-fluorodeoxyglucose and positron emission tomography. *Neurology* **45**: 1995–2004.

Eidelberg, D., Moeller, J. R., Ishikawa, T., Dhawan, V., Spetsieris, P., Przedborski, S., and Fahn, S. (1995c). The metabolic topography of idiopathic torsion dystonia. *Brain* **118**: 1473–1484.

Eidelberg, D., Moeller, J. R., Ishikawa, T., Dhawan, V., Spetsieris, P., Silbersweig, D., Stern, E., Woods, R., Fazzini, E., Dogali, M., and Beric, A. (1996). Regional metabolic correlates of surgical outcome following unilateral pallidotomy for Parkinson's disease. *Ann. Neurol.* **39**: 450–459.

Eidelberg, D., Moeller, J. R., Kazumata, K., Antonini, A., Sterio, D., Dhawan, V., Spetsieris, P., Alterman, R., Kelly, P. J., Dogali, M., Fazzini, M., Fazzini, E., and Beric, A. (1997). Metabolic correlates of pallidal neuronal activity in Parkinson's disease. *Brain* **120**: 1315–1324.

Fahn, S., Elton, R. L., and the UPDRS Development Committee. (1984). Unified Parkinson disease rating scale. In "Recent Developments in Parkinson's Disease" (S. Fahn, C. D. Marsden, D. Calne, and M. Goldstein, eds.), Vol. 2. pp. 293–304. Macmillan, Florham Park, NJ.

Kearfott, K. J., and Carroll, L. R. (1984). Evaluation of the performance characteristics of the PC4600 positron emission tomograph. *J. Comput. Assist. Tomogr.* **8**: 502–513.

Moeller, J. R., and Eidelberg, D. (1997). Divergent expression of regional metabolic topographies in Parkinson's disease and normal aging. *Brain* **120**: 2197–2206.

Moeller, J. R., and Strother, S. C. (1991). A regional covariance approach to the analysis of functional patterns in positron emission tomographic data. *J. Cereb. Blood Flow Metab.* **11**: A121–A135.

Moeller, J. R., Strother, S. C., Sidtis, J. J., and Rottenberg, D. A. (1987). The scaled subprofile model: A statistical approach to the analysis of functional patterns in positron emission tomographic data. *J. Cereb. Blood Flow Metab.* **7**: 649–658.

Moeller, J. R., Ishikawa, T., Dhawan, V., Spetsieris, P., Mandel, F., Alexander, G. E., Grady, C., Schapiro, M. B., and Eidelberg, D. (1996). The metabolic topography of normal aging. *J. Cereb. Blood Flow Metab.* **16**(3): 385–398.

Robeson, W., Dhawan, V., Babchyck, B., Takikawa, S., Zanzi, I., Margouleff, D., and Eidelberg, D. (1993). SuperPETT 3000 time-of-flight tomograph: Optimization of factors affecting quantification. *IEEE Trans. Nucl. Sci.* **40**: 135–142.

CHAPTER 38

On the Detection of Activation Patterns Using Principal Components Analysis

B. A. ARDEKANI,[*] S. C. STROTHER,[†] J. R. ANDERSON,[†] I. LAW,[‡] O. B. PAULSON,[‡] I. KANNO,[*] and D. A. ROTTENBERG[†]

[*] Department of Radiology and Nuclear Medicine
Research Institute for Brain and Blood Vessels
Akita, Japan
[†] PET Imaging Service
Veterans Affairs Medical Center
Minneapolis, Minnesota 55417
[‡] Department of Neurology, Rigshospital
Copenhagen, Denmark

Principal components analysis (PCA) of images obtained from positron emission tomography (PET) activation studies reveals an inter- and intrasubject subspace in data. The activation pattern is usually contained in the first component of the intrasubject subspace. However, this observation alone is not always sufficient to define the activated regions because (a) the activation pattern may not lie entirely on a single principal component (PC) but may be spread across several components (this is particularly true when the number of subjects increases and/or multicenter data are used) and (b) it is difficult to apply conventional parametric models in order to assess the statistical significance of the resulting activation image. This chapter demonstrates that these difficulties can be overcome by (a) using the Fisher's linear discriminant analysis (FLDA) to obtain an activation pattern as a linear combination of all PCs and (b) applying a nonparametric statistical method in order to test the significance of activation. Multicenter [^{15}O]water PET scans were collected at three different centers in Japan, Denmark, and the United States. Each center scanned three right-handed subjects while performing a sequential finger-to-thumb opposition task. Four "baseline" and four "activation" scans were obtained from each subject. Data were analyzed using PCA. The activation pattern was extracted using FLDA. The statistical significance of activated regions was tested using a nonparametric method. The results were compared with pooled variance t-statistic images for which significance levels were obtained using a Gaussian random field model.

I. INTRODUCTION

Principal components analysis (PCA) of images obtained from brain positron emission tomography (PET) activation studies provides important insights into the underlying covariance structure of data. Using PCA, Strother *et al.* (1995) revealed an inter- and intrasubject subspace in data and demonstrated that the activation pattern is usually contained in the first component of the intrasubject subspace. However, this observation alone is not always sufficient to define an activation pattern because (a) the activation pattern may not lie entirely on a single principal component (PC) but may be spread across several components (this is particularly true when the number of subjects increases and/or multicenter data are used) and (b) it is difficult to apply conventional parametric models in order to assess the statistical significance of the resulting activation image.

A method for detecting activated regions in PET functional activation studies has been developed. The method combines PCA, Fisher's linear discriminant analysis (FLDA), and nonparametric statistical significance testing. Application of PCA to PET studies has been proposed by several researchers (Clark *et al.*, 1985; Moeller and Strother, 1991; Friston *et al.*, 1993). The use of FLDA to construct an activation component as a linear combination of several PCs has been suggested by Strother *et al.* (1995). The application of nonparametric statistical testing methods to PET studies has been proposed by Holmes *et al.* (1996).

II. MATERIALS AND METHODS

A. Data Acquisition

[^{15}O]water PET scans were collected at three different centers in Akita, Copenhagen, and Minneapolis. Each center scanned three right-handed subjects. Four "baseline" and four "activation" scans were obtained from each subject, resulting in a total of $N = 72$ volumes of image data. Activation scans were obtained while the subjects performed a left-handed sequential finger-to-thumb opposition task at a rate of 1 Hz. Each scan followed a bolus injection of approximately 15 mCi of [^{15}O]water. The finger opposition task started at the beginning of the injection and continued for 60 sec. Data acquisition began when the activity near the brain reached a threshold level and continued for 90 sec. Baseline and activation scans were acquired alternatively (BABABABA) with an interscan interval of approximately 8 min.

B. Image Preprocessing

In order to correct for possible head movements during the scanning session, the eight scans taken from each subject were registered (intrasubject registration) using six-parameter rigid-body linear transformations. Following registration, the eight scans were averaged to produce an average image for each subject.

In order to reduce anatomical variability between subjects, the average images of the nine subjects were nonlinearly deformed to match a standard PET atlas in Talairach space. The anatomic standardization method used has been developed by Minoshima *et al.* (1994). Briefly, several hundred landmarks are automatically identified on a subject's average image. These landmarks are then registered *precisely* to a set of corresponding landmarks that have been previously manually defined on the atlas, while deforming the image as a whole to a state with minimum "bending energy."

Using the combination of the linear transformation obtained from intrasubject registration and the nonlinear transformation obtained from intersubject registration, each of the 72 original scans was resliced to match the standard atlas. The transformed images were then smoothed using a $3 \times 3 \times 3$ boxcar kernel. The resulting volumes were of matrix size $128 \times 128 \times 60$ and voxel size $2.25 \times 2.25 \times 2.25$ mm^3.

The standardized images were averaged and thresholded to find the intracerebral volume elements (voxels).

The number of intracerebral voxels identified was $P = 125{,}750$. In order to account for global differences in activity between scans, the voxels in each of the 72 volumes were divided by their average activity over all intracerebral voxels. In addition, after normalization, the average of all 72 scans was computed and subtracted from each volume.

C. Principal Component Analysis

Let each scan, after having been processed as described earlier, be represented by a $P \times 1$ matrix (vector) $\mathbf{x}_i$ $(i = 1, 2, \ldots, N)$, where P is the number of intracerebral voxels and N is the number of scans. Thus, each scan $\mathbf{x}_i$ can be considered as a point in a P-dimensional space. This space is referred to as the *image space*. Because there are N scans and their mean is subtracted, they must lie in an $(N-1)$-dimensional subspace of the image space. This subspace is referred to as the *feature space*. The feature space can be represented by a set of $N - 1$ basis vectors, and each of the N scans can be represented by a set of $N - 1$ numbers representing their coordinates with respect to these basis vectors. Because $(N - 1) \ll P$, this represents a substantial data compression. Furthermore, given the coordinates of a scan in the feature space, its corresponding point can be found in the image space and vice versa. This means that linear operations on images can be performed (much faster) in the feature space, saving substantial computing time when many operations have to be performed on images. Results are then easily translated back to the image space.

Many different sets of basis vectors can be identified for the feature space. One that has the optimum representation property can be obtained using singular value decomposition (SVD). The *optimum representation property* means that if the number of features is reduced to $M < (N - 1)$ by discarding some information, discarding a particular set of $N - 1 - M$ basis vectors obtained by SVD results in the smallest expected residual error (Therrien, 1989).

Let the $N \times P$ data matrix $\mathbf{X}$ be defined as

$$\mathbf{X} = [\mathbf{x}_1 \quad \mathbf{x}_2 \quad \cdots \quad \mathbf{x}_N]^T. \tag{1}$$

The SVD of $\mathbf{X}$ is given by

$$\mathbf{X} = \mathbf{U}\mathbf{\Lambda}^{1/2}\mathbf{V}^T, \tag{2}$$

where $\mathbf{\Lambda}$ is an $(N-1) \times (N-1)$ diagonal matrix with its diagonal elements equal to the $N - 1$ nonzero eigenvalues of $\mathbf{X}$ arranged in descending order, $\mathbf{U}$ is an $N \times (N - 1)$ orthogonal matrix with its columns equal to the $N - 1$ eigenvectors corresponding to the $N - 1$

nonzero eigenvalues of **X**, and **V**^{T} is an $(N-1) \times P$ orthogonal matrix. The $N-1$ rows of the matrix **V**^{T} represent the optimum representation basis vectors for the feature space. Efficient implementation of Eq. (2) using PCA is also known as Q analysis in the statistical literature (Jackson, 1991) and has been used in PET data analysis by Clark *et al.* (1985) and others (Strother *et al.*, 1995). This approach has been summarized by Weaver (1995). In this method, **U** and **Λ** are obtained by PCA of the $N \times N$ symmetric matrix **XX**^{T} as

$$\mathbf{XX}^T = \mathbf{U\Lambda U}^T. \quad (3)$$

Then, **V**^{T} is obtained from

$$\mathbf{V}^T = \mathbf{\Lambda}^{-1/2}\mathbf{U}^T\mathbf{X}. \quad (4)$$

Now define the $N \times (N-1)$ matrix **Z** as

$$\mathbf{Z} = \mathbf{U\Lambda}^{1/2} = [\mathbf{z}_1 \; \mathbf{z}_2 \; \cdots \; \mathbf{z}_N]^T. \quad (5)$$

By substituting Eq. (5) into Eq. (2), it can be seen that each image $\mathbf{x}_i$ in the image space corresponds to a unique $(N-1)$-dimensional vector $\mathbf{z}_i$ in the feature space. All the information about the large image matrix **X** has been compressed into the much smaller matrix **Z**. Given $\mathbf{z}_i$, the corresponding $\mathbf{x}_i = \mathbf{Vz}_i$ can always be found.

D. Fisher's Linear Discriminant Analysis

An activation image that is obtained by linear operations on image vectors $\mathbf{x}_i$ can be represented by a point or equivalently a vector in the P-dimensional image space. Alternatively, the corresponding point in the feature space can be found by performing linear operations on the feature vectors $\mathbf{z}_i$. Also, since scaling by a constant factor does not change the information contained in an activation pattern, the activation pattern can be represented by a line or *direction* in the feature space. The question arises then that, for a given set of images represented by points $\mathbf{z}_i$ ($i = 1, 2, \ldots, N$) in the feature space, which direction in the feature space represents the "optimum" activation pattern? This chapter employs the optimality criterion suggested by Fisher (1936).

Consider a direction **a** in the feature space. Let y_i denote the magnitude of the projection of $\mathbf{z}_i$ in the direction of **a**, i.e., $y_i = \mathbf{a}^T\mathbf{z}_i$. Let N_1 and N_2 represent the number of baseline and activation scans, respectively ($N = N_1 + N_2$). Denote the baseline scans by $\mathbf{z}_i^1$ ($i = 1, \ldots, N_1$) and the activation scans by $\mathbf{z}_i^2$ ($i = 1, \ldots, N_2$). The numbers y_i can be divided into two groups: y_i^1 and y_i^2, corresponding to the projections of $\mathbf{z}_i^1$ and $\mathbf{z}_i^2$ in the direction of **a**, respectively. Fisher's optimality criterion is the ratio of the between-groups sum of squares to the within-groups sum of squares given as

$$\frac{N_1(y_{\cdot}^1 - y_{\cdot\cdot})^2 + N_2(y_{\cdot}^2 - y_{\cdot\cdot})^2}{\sum_{i=1}^{N_1}(y_i^1 - y_{\cdot}^1)^2 + \sum_{i=1}^{N_2}(y_i^2 - y_{\cdot}^2)^2}, \quad (6)$$

where a "·" replacing an index denotes the average of the variable over that index. Now define the $N_1 \times (N-1)$ matrix $\mathbf{Z}_1$ as

$$\mathbf{Z}_1 = \begin{bmatrix} \mathbf{z}_1^1 & \mathbf{z}_2^1 & \cdots & \mathbf{z}_{N_1}^1 \end{bmatrix}^T, \quad (7)$$

and let $\mathbf{z}_2$ denote a similar matrix for the activation scans. The direction **a**, which maximizes Eq. (6), is given by Mardia *et al.* (1979) as

$$\mathbf{a} = \mathbf{W}^{-1}\mathbf{d}, \quad (8)$$

where $\mathbf{d} = \mathbf{z}_{\cdot}^2 - \mathbf{z}_{\cdot}^1$,

$$\mathbf{W} = \mathbf{Z}_1^T\mathbf{H}_1\mathbf{Z}_1 + \mathbf{Z}_2^T\mathbf{H}_2\mathbf{Z}_2, \quad (9)$$

and $\mathbf{H}_1$ and $\mathbf{H}_2$ are centering matrices of appropriate sizes.

E. Nonparametric Testing

After obtaining the activation pattern **a** from FLDA in the feature space, an activation image, **A**, can be computed as $\mathbf{A} = \mathbf{Va}$. This image is then examined to locate statistically significant effects. Holmes *et al.* (1996) proposed a nonparametric testing method for this purpose. This chapter applies similar ideas but uses different methods for randomization and voxel classification.

Randomization is achieved by assigning labels "BA" (baseline–activation) or "AB" (activation–baseline) to each pair of consecutive scans from each subject with equal probability. Hence, if there were $N/2$ pairs of scans, then there would be $2^{N/2}$ possible randomizations. In this procedure, a large number of randomizations, K, are generated. For each randomization, an activation image $\mathbf{A}^k$ ($k = 1, 2, \ldots, K$) is found using FLDA. The elements of vectors $\mathbf{A}^k$ are arranged in descending order $\mathbf{A}_j^k$ ($j = 1, 2, \ldots, P$), i.e.,

$$\mathbf{A}_1^k \geq \mathbf{A}_2^k \geq \cdots \geq \mathbf{A}_P^k. \quad (10)$$

Furthermore, an activation image, **A**, is computed for the actual baseline–activation labeling, which is known from the experimental protocol and ordered similarly ($\mathbf{A}_1 \geq \mathbf{A}_2 \geq \cdots \geq \mathbf{A}_P$). Activated voxels are then detected as follows:

1. Set $j = 1$.
2. Given a significance level p (e.g., 0.01), find a threshold T_j such that only $p \times 100\%$ of $\mathbf{A}_j^k$ values are greater than T_j.
3. If $\mathbf{A}_j > T_j$, declare voxel j to be activated, increase j by 1, and go to step 2 or else stop and declare the remaining voxels as nonactivated.

It may be proved that this test has strong control over type I error.

III. RESULTS AND DISCUSSION

PCA of the 72 PET scans showed that the feature space can be divided into two subspaces. The first eight components comprised the intersubject space. The remaining components formed the intrasubject space. Figure 1a shows the coordinates of the 72 scans with respect to the first dimension of the intersubject space. This dimension clearly captures intersubject variations.

It appears that this component also contains some intercenter variations. The first 12 baseline–activation scan pairs are from Akita, the next 12 pairs are from Copenhagen, and the last 12 are from Minneapolis. The first dimension of the intrasubject space, the ninth component overall, clearly represents an activation component (Fig. 1b). However, as noted by Strother et al. (1995), this result may not always be obtained when the number of subjects increases or when different models are used for global activity normalization. This was the motivation for using FLDA in order to extract an activation pattern.

When performing FLDA, it is not necessary to use all $N - 1 = 71$ components of the feature space. The higher components usually capture noise and their removal improves the results by filtering the noise as well as increasing computational efficiency. Therefore, the dimensions of the feature space were reduced by only retaining the first 18 components and discarding the others. Representation of the feature space in terms of the basis vectors found by SVD ensures that reducing the dimension of the feature space in this way results in the minimum expected residual error. FLDA yields the activation pattern given by Eq. (8). It is the product of the inverse of the within-groups sum of squares and products matrix $\mathbf{W}$ with the vector $\mathbf{d}$, which is the difference between the means of the activation and baseline pattern vectors. Elements of vectors $\mathbf{d}$ and $\mathbf{a}$ are plotted in Fig. 2. The activation pattern $\mathbf{a}$ is very close to the mean difference $\mathbf{d}$ with the exception that the contributions from the first few components have been attenuated. This is advantageous because the first few components represent the intersubject/intercenter differences and contain the largest variations.

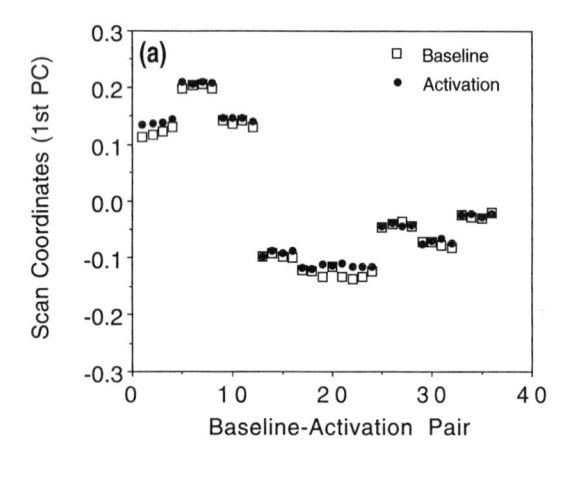

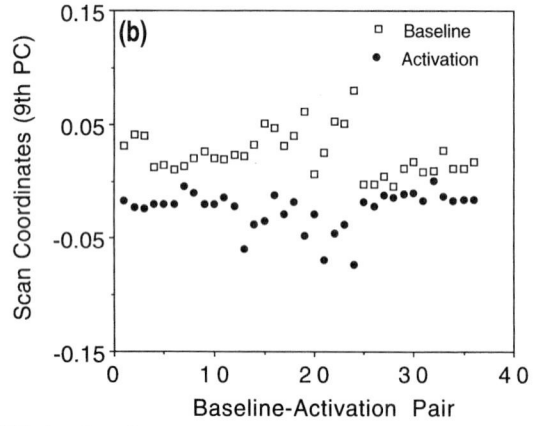

FIGURE 1 Coordinates of the 36 baseline–activation scan pairs with respect to the first PC (a) and the ninth PC (b).

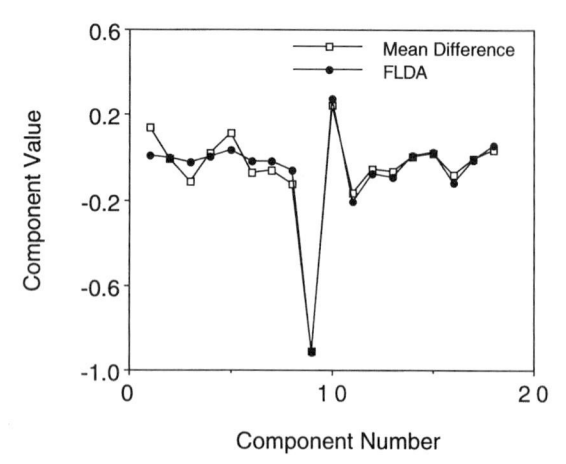

FIGURE 2 Plot of the activation pattern vector $\mathbf{a}$ detected using FLDA versus the activation pattern vector $\mathbf{d}$, the difference between the means of the activation and baseline pattern vectors.

In order to perform the nonparametric significance testing procedure, 2000 randomizations were generated. After generation of a randomization k, FLDA was performed and an activation pattern was computed from Eq. (8). The activation pattern was then translated into the image space by computing the product $\mathbf{A}^k = \mathbf{Va}$. The P elements of $\mathbf{A}^k$ were then sorted into a descending order ($\mathbf{A}_1^k \geq \mathbf{A}_2^k \geq \cdots \geq \mathbf{A}_P^k$) and stored. A program written in C language running on an SGI Indy R5000 workstation took approximately 11 sec per iteration.

After completion of the 2000 iterations, the threshold levels T_j corresponding to $p = 0.01$ were found. The results are plotted in Fig. 3 for the largest 25,000 voxels. The corresponding $\mathbf{A}_j$ of the activation image obtained using the actual baseline–activation labeling is also plotted in Fig. 3. The threshold for significant activation at $p = 0.01$ level is the point where the two plots cross. In this case, the threshold level is 390. At this level, 19,308 voxels were declared activated. The activated voxels are shown in Fig. 4a (see color insert), superimposed on a standard MR image in Talairach space. The results obtained using the software package developed by Minoshima *et al.* (1993) are shown in Fig. 4b (see color insert). These are pooled variance *t*-statistics images with the same significance level of $p = 0.01$ obtained using a Gaussian random field model (Worsley *et al.*, 1992). The locations of detected activation regions are the same in both methods. The sizes of the regions differ because of the heavy smoothing used in the latter method.

In conclusion, this chapter introduced a method for detecting activated regions in PET functional activation studies. The method combines PCA, FLDA, and nonparametric statistical significance testing.

Acknowledgments

The authors thank Dr. Satoshi Minoshima for making his software package available. This research was funded in part by the Human Brain Project Grant P20-MH57180 from the United States National Institutes of Health. B.A.A. is supported by a STA Postdoctoral Research Fellowship from the Japan Science and Technology Agency.

References

Clark, C., Carson, R., Kessler, R., Margolin, R., Buchsbaum, M., DeLisi, L., King, C., and Cohen, R. (1985). Alternative statistical models for the examination of clinical positron emission tomography/fluorodeoxyglucose data. *J. Cereb. Blood Flow Metab.* **5**: 142–150.

Fisher, R. A. (1936). The use of multiple measurements in taxonomic problems. *Ann. Engen.* (*London*) **7**: 179–188.

Friston, K. J., Frith, C. D., Liddle, P. F., and Frackowiak, R. S. J. (1993). Functional connectivity: The principal-component analysis of large (PET) data sets. *J. Cereb. Blood Flow Metab.* **13**: 5–14.

Holmes, A. P. Blair, R. C., Watson, J. D. G., and Ford, I. (1996). Nonparametric analysis of statistic images from functional mapping experiments. *J. Cereb. Blood Flow Metab.* **16**: 7–22.

Jackson, J. E. (1991). "A User's Guide to Principal Components." Wiley, New York.

Mardia, K. V., Kent, J. T., and Bibby, J. M. (1979). "Multivariate Analysis." Academic Press, San Diego, CA.

Minoshima, S., S. Koeppe, R. A., Fessler, J. A., Mintun, M. A., Berger, K. L., Taylor, S. F., and Kuhl, D. E. (1993). Integrated and automated data analysis method for neuronal activation studies using ^{15}O-water PET. *In* "Quantification of Brain Function. Tracer Kinetics and Image Analysis in Brain PET" (K. Uemura, N. A. Lassen, T. Jones, and I. Kanno, eds.), pp. 409–415. Elsevier, Amsterdam and New York.

Minoshima, S., Koeppe, R. A., Frey, K. A., and Kuhl, D. E. (1994). Anatomic standardization: Linear scaling and nonlinear warping of functional brain images. *J. Nucl. Med.* **35**: 1528–1537.

Moeller, J. R., and Strother, S. C. (1991). A regional covariance approach to the analysis of functional patterns in positron emission tomographic data. *J. Cereb. Blood Flow Metab.* **11**: A121–A135.

Strother, S. C., Anderson, J. R., Schaper, K. A., Sidtis, J. J., and Rottenberg, D. A. (1995). Linear models of orthogonal subspaces and networks from functional activation PET studies of the human brain. *In* "Information Processing in Medical Imaging" (Y. Bizais, C. Barillot, and R. Di Paola, eds.), pp. 299–310. Kluwer Academic Publishers, Dordrecht, The Netherlands.

Therrien, C. W. (1989). "Decision Estimation and Classification." Wiley, New York.

Weaver, J. B. (1995). Efficient calculation of principal components of imaging data. *J. Cereb. Blood Flow Metab.* **15**: 892–894.

Worsley, K. J., Evans, A. C., Marrett, S., and Neelin, P. (1992). A three-dimensional statistical analysis for CBF activation studies in human brain. *J. Cereb. Blood Flow Metab.* **12**: 900–918.

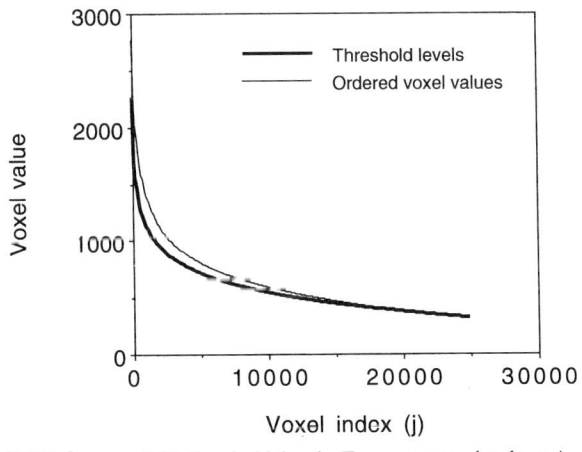

FIGURE 3 $p = 0.01$ threshold levels T_j versus voxel values $\mathbf{A}_j$ of the activation image plotted in descending order. The first crossing point of the two graphs indicates the threshold level that must be applied to the activation image in order to detect the activated voxels.

SECTION V

TRACER DEVELOPMENT

CHAPTER 39

One for All or One for Each? Matching Radiotracers and Regional Brain Pharmacokinetics[1]

MICHAEL R. KILBOURN, THINH B. NGUYEN, SCOTT E. SNYDER, and ROBERT A. KOEPPE

*Division of Nuclear Medicine
Department of Internal Medicine
University of Michigan Medical School
Ann Arbor, Michigan 48109*

In the development of new radiotracers for brain imaging, it is almost always considered that a single radiotracer will be useful for quantification of a biochemical process throughout the brain. It has been demonstrated that when there are large regional differences in concentrations of binding sites or enzymes, a single radiotracer may not be optimal for use in all regions. In the example of substrates for brain acetylcholinesterase, quantification of enzyme hydrolysis rates in a region of high enzyme concentrations such as the striatum requires a substrate with a slow reaction rate; for the cortex, a region of low enzyme concentrations, a substrate with a high hydrolysis rate is best. By this means of adjusting the reaction rate, proper radiotracer pharmacokinetics are obtained in both brain regions, such that the enzymatic hydrolysis rate for the enzyme can be clearly separated from the effects of substrate delivery to the tissues. This concept can and should be extended to the quantification of other biochemical processes, such as binding to receptors or transporters, when there are very large regional differences in the concentrations of targeted sites.

I. INTRODUCTION

A large number of radiolabeled compounds have been synthesized as potential *in vivo* tracers of biochemical processes in the mammalian brain. A topic not previously addressed has been whether a single radiotracer, even one with excellent *in vivo* pharmacokinetic and metabolic properties, is appropriate for quantification of a biochemical process that exhibits a very large dynamic range between different brain regions. This problem occurs in studies of receptor or transporter radioligand binding (e.g., striatal- vs cortical-binding sites for dopamine receptors or transporters) as well as enzyme-binding sites or rates [e.g., striatal vs cortical acetylcholinesterase (AChE)] where regional values of binding site densities or enzyme activities can vary from 30- to 100-fold (or maybe more).

Why is this of concern? Admittedly, much of the primary interest in radiotracer binding often involves either (a) regions of high concentrations of the target site, such as dopamine receptors or transporters in the striatum, or (b) regions of low concentrations, such as AChE in the cortex. However, there are often lingering, unanswered questions involving the physiological importance of changes that occur in other regions of the brain. For example, do changes in dopamine transporters in the cortex (Wang *et al.*, 1995) have physiological or medical importance? Also, are there changes in striatal cholinergic neurotransmission that are relevant to behavioral or motor dysfunctions in Alzheimer's disease (Boissiere *et al.*, 1997)?

This chapter examines the concept that quantification of biochemical measures in regions with disparate concentrations of a specific target (enzyme, receptor, or transporter) may be more easily and better accomplished by utilization of two or more radiotracers tailored to the characteristics of each region of interest.

---

[1] Transcripts of the BRAINPET97 discussion of this chapter can be found in Section VIII.

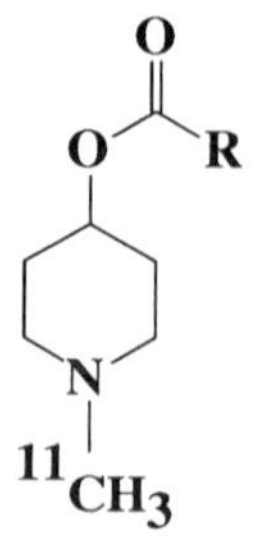

FIGURE 1 Structures of carbon-11-labeled piperidinyl esters used as *in vivo* acetylcholinesterase substrates.

II. MATERIALS AND METHODS

A. Synthesis of Carbon-11-Labeled Esters

The esters shown in Fig. 1 were prepared in no-carrier-added form by the alkylation of the appropriate N-desmethyl precursors, using [^{11}C]methyl triflate and the procedure reported previously (Bormans *et al.*, 1996).

B. PET Imaging Studies

Imaging studies of the time-dependent distribution of radioactivity in the brain after bolus intravenous radiotracer injection were done in a single nemistrina macaque monkey, using a PCT 4600A scanner and procedures reported previously (Kilbourn *et al.*, 1996). The kinetic rate constant for *in vivo* hydrolysis of the esters, k_3, was estimated using an analysis technique based on the shape of the tissue time–radioactivity curves for various brain regions (Frey *et al.*, 1997).

III. RESULTS AND DISCUSSION

As the field of radiotracer development for *in vivo* imaging of brain biochemistry has matured, the necessary pharmacokinetic characteristics of a radiotracer have become better appreciated. When the goal is quantification of a particular biochemical process (e.g., receptor or transporter binding, enzyme rate) and not simply imaging of the distribution of sites throughout the brain, criteria for validation of a new radiotracer become more stringent. A particular problem common to many radiotracers is delivery dependence. Simply put, if the rate associated with the mechanism of localization (binding to a specific site or enzymatic action) is much greater than the rate of delivery to the tissue, then essentially every radiotracer molecule delivered is retained and the distribution reflects the delivery rate, not the biochemical process of interest.

This can be appreciated by examining a series of simulated tissue time–activity curves. For an example of an enzyme substrate that is irreversibly trapped (so $k_4 = 0$), with a set of fixed values for rates of entry and egress from the tissue ($K_1 = 0.5$ ml/g/min, $k_2 = 0.5$ liter/min), varying the forward reaction rate k_3 (where k_3 is the enzyme reaction rate times the concentration of enzymes) over three decades of values produces the family of curves shown in Fig. 2. It is very easy to appreciate the differences between the curves for lower k_3 values, which show considerable washout of the radioactivity over time, from the curves generated using high k_3 values, which simply peak at the early initial time and stay constant or even increase. These curves in Fig. 2 also demonstrate the loss of sensitivity at higher k_3 values. Between any two curves the true k_3 values differ by a factor of approximately threefold; this difference is fairly well maintained between the curves at the slower reaction rates, but when k_3 becomes very high, the differences between curves diminish and thus sensitivity to the higher rates of reaction is lost.

Of course, it is not at all as simple as that. The delivery dependence of a tracer is a function of the relative rates of the four kinetic parameters (K_1, k_2, k_3, and k_4). In the past, these discussions have largely considered all regions of the brain as equal, i.e., radiotracers with proper kinetics in one region of the brain are assumed to have the proper behavior in all regions of the brain. Studies conducted here demonstrate that this may not be necessarily correct. When there are very large differences in kinetic rates for the biochemical process being targeted, the ratios of rates for the delivery and binding (or enzyme rate) steps change dramatically. Whereas the delivery rate of the radiotracer does not, for the most part, vary substantially across brain regions, at least for gray matter structures, for an enzyme such as acetylcholinesterase the human brain ratio of enzyme activities between high (striatum) and low (cortex) regions can be 50- to 100-fold.

An Example: Acetylcholinesterase Substrates

As an example of the need for regionally specific radiotracers, this chapter examines the characteristics of radiolabeled esters that serve as substrates for the enzyme acetylcholinesterase. The authors and others have developed such esters for imaging of enzymatic activity in animal and human brain (Irie *et al.*, 1995, 1996; Bormans *et al.*, 1996; Kilbourn *et al.*, 1996). Human studies have been reported with two of the esters shown in Fig. 1: N-[^{11}C]methylpiperidinyl acetate ([^{11}C]AMP; Nagatsuka *et al.*, this volume, Chapter 59) and N-[^{11}C]methylpiperidinyl propionate ([^{11}C]PMP;

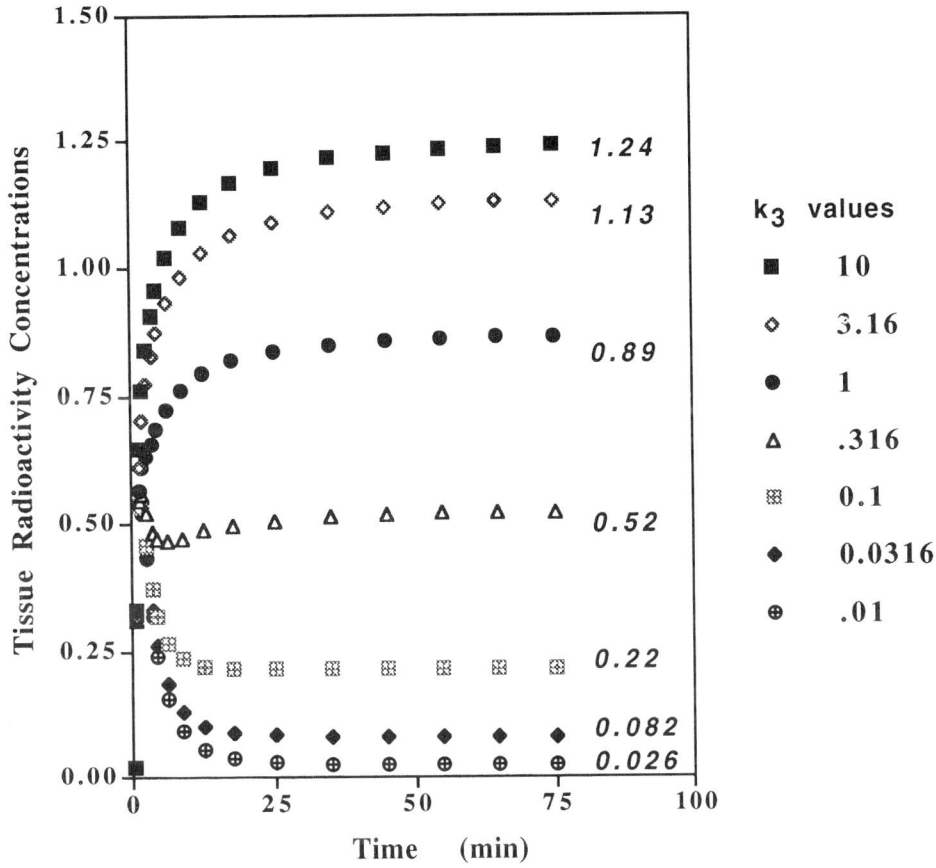

FIGURE 2 Simulated time–activity curves for radiotracer trapping by irreversible enzyme action. Values for K_1 and k_2 were fixed; k_4 equals 0; and k_3 values varied from 0.1 to 10 min^{-1}. Values in italics are tissue concentrations for the final time point of the curve.

Kuhl et al., 1996). Both of these esters provide good measures of AChE enzymatic activity in cortical regions of the human brain (Koeppe et al., 1997; Nagatsuka et al., this volume, Chapter 59), although they provide absolute values that differ by threefold (frontal cortex k_3 values: AMP, 0.077 ± 0.009 min^{-1}; PMP, 0.027 ± 0.002 min^{-1}). As the in vivo estimates of enzymatic activity (k_3) represent the product of the number of enzymes and the hydrolysis rate for the ester, this threefold difference should—and does—reflect the inherent reactivity of the two esters toward AChE, which has been determined to be approximately fourfold using in vitro assays (Irie et al., 1996).

In regions of higher enzyme concentrations, however, both [^{11}C]AMP and [^{11}C]PMP become much less suitable due to greater delivery dependence; [^{11}C]AMP, with the inherent higher reactivity, becomes less useful at lower enzyme concentrations than [^{11}C]PMP. For the striatum, as the region of highest AChE concentration, estimation of k_3 using either radiotracer, particularly [^{11}C]AMP, becomes more difficult. A different radiotracer may be needed for AChE activity measures in the striatum.

Through the synthesis of a variety of radiolabeled esters that serve as in vivo substrates for AChE, the authors have examined whether the concept of multiple, region-specific radiotracers can in fact be implemented. Three of the esters ([^{11}C]AMP, [^{11}C]PMP, and N-[^{11}C]methylpiperidinyl (isobutyrate[^{11}C]iBMP) were studied in a monkey brain using PET imaging of the regional pharmacokinetics following peripheral administration. Tissue time–radioactivity curves for the striatum (top) and cortex (bottom) are shown in Fig. 3. These kinetic curves make it much easier to see the important differences among the three tracers, which can be summarized as follows:

a. In the striatum, both [^{11}C]AMP and [^{11}C]PMP are highly extracted and nearly completely retained. For [^{11}C]iBMP, however, only approximately 50% of the peak activity is irreversibly retained.

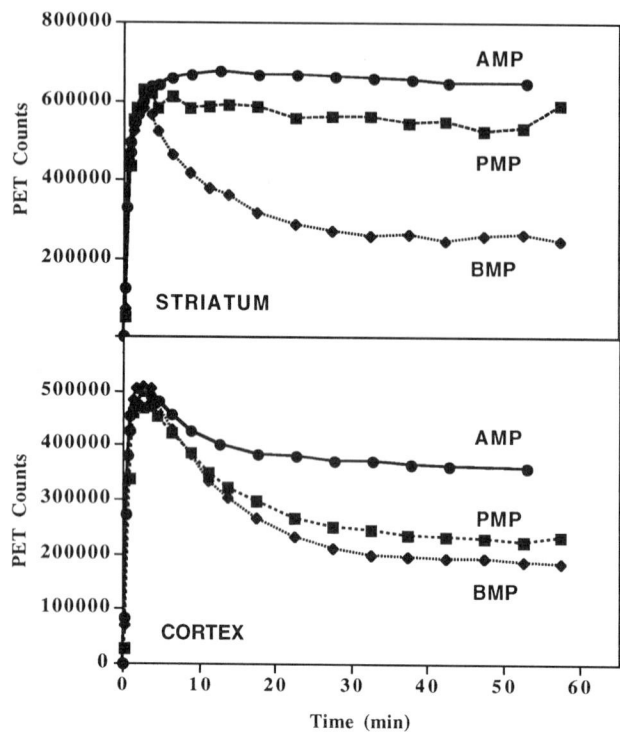

FIGURE 3 Time–activity curves for [¹¹C]AMP, [¹¹C]PMP, and [¹¹C]iBMP in striatum (top) and cortex (bottom) of the monkey brain. Data have been normalized to a common injected dose of radiotracer.

TABLE 1 Kinetic Rate Constants (k_3) for *in vivo* Hydrolysis of Radiolabeled Esters in Monkey Cortex and Striatum

Ester[a]	Control (k_3, min^{-1})	1.0 mg/kg THA (k_3, min^{-1})	Change (%)
[¹¹C]AMP, cortex	0.0854	n.d.[b]	
[¹¹C]AMP, striatum	0.1175	n.d.	
[¹¹C]PMP, cortex	0.0453 ± 0.003	0.0185 ± 0.002	−59
[¹¹C]PMP, striatum	0.0751 ± 0.008	0.0841 ± 0.010	+11
[¹¹C]iBMP, cortex	0.0327 ± 0.001	0.0192 ± 0.004	−42
[¹¹C]iBMP, striatum	0.0445 ± 0.002	0.0198 ± 0.005	−56

[a] *N*-[¹¹C]methylpiperidinyl acetate ([¹¹C]AMP), *N*-[¹¹C]methylpiperidinyl propionate ([¹¹C]PMP), and *N*-[¹¹C]methylpiperidinyl isobutyrate ([¹¹C]iBMP). For [¹¹C]PMP and [¹¹C]iBMP, paired studies were done with an intervening injection of 1.0 mg/kg of tetrahydroaminoacridine (THA, tacrine). Where appropriate, data are shown as mean ±SD.
[b] Not determined.

b. In the cortex, all three tracers show significant washout of radioactivity, with [¹¹C]PMP and [¹¹C]iBMP exhibiting losses of greater than 50% of the peak activities.

All of these kinetic curves were then analyzed by a shape analysis method that estimates the kinetic constant k_3 (forward rate constant for enzyme action times the concentration of enzymes) without the need for a plasma input function or use of a reference region devoid of enzyme (Frey *et al.*, 1997). The calculated values for each ester in cortex and striatum are shown in Table 1. Because of the delivery limitation of [¹¹C]AMP and [¹¹C]PMP in the striatum, the k_3 values calculated for that region are poor estimates of the enzyme rate; it is likely the true hydrolysis rate in the striatum is actually much higher. Cortical values for k_3, however, show a very nice dependence on ester structure and a simply remarkable (too good to be true?) correlation ($r^2 = 1.00$, correlation not shown) with the *in vitro* rates of hydrolysis of the three esters by rat brain cortical cholinesterases (Irie *et al.*, 1996).

As perhaps a more pressing question, can two different tracers be used to measure *changes* in the same enzyme in different regions of the brain? To test this, paired tests ([¹¹C]PMP/[¹¹C]PMP and [¹¹C]iBMP/[¹¹C]iBMP) were performed with an injection of 1 mg/kg tetrahydroaminoacridine (THA, tacrine) between the two scans. If permeation of the drug is fairly uniform throughout the brain, then one would expect similar inhibition of total enzyme activity in all regions of the brain, irrespective of the concentration of enzyme. This was exactly what was observed. In the cortex, measurement of AChE using [¹¹C]PMP showed a 60% decrease; in the striatum, using [¹¹C]iBMP as the radiolabeled substrate, the decrease of AChE activity is 56%! It has thus been demonstrated that the same extent of pharmacological inhibition of the enzyme, in two regions of the brain, can be measured using two different radiotracers.

Is this necessary or would one radiotracer have sufficed to measure AChE activity throughout the brain? [¹¹C]PMP appears unsatisfactory in the monkey striatum for pharmacokinetic reasons. As shown in Table 1, the dose of tacrine failed to reduce the k_3 determined for [¹¹C]PMP in the striatum (in reality, neither k_3 value is likely to be the correct value, and the increase measured is meaningless). [¹¹C]iBMP fails in the cortex for pharmacological reasons: Irie *et al.* (1996) have clearly shown that [¹¹C]iBMP has poor selectivity for AChE and is readily hydrolyzed by butyrylcholinesterase (BuChE). Thus, in the monkey cortex, [¹¹C]iBMP is likely measuring a mixture of enzymatic activities. In humans, there is a significant proportion of BuChE compared to AChE (actually more enzyme molecules for BuChE and an enzymatic activity one-half that of AChE) (Atack *et al.*, 1986; Brimjoin and Hammond, 1988). Although similar data are lacking for monkey cortex [a dissociation between AChE and BuChE has been noted (Graybiel and Ragsdale, 1982)], data are consistent with a mixture of AChE and BuChE in the monkey cortex. The effect of tacrine on [^{11}C]iBMP hydrolysis in the cortex is thus difficult to interpret; tacrine, although marketed as an acetylcholinesterase inhibitor, is actually very potent (and maybe even better) as an inhibitor of butyrylcholinesterase (Freeman and Dawson, 1991). Results show that the dose of tacrine administered to the monkey inhibited [^{11}C]iBMP hydrolysis by cortical cholinesterases, but whether this was AChE or BuChE or a mixture is unknown. In the striatum, where AChE dominates (at least in humans and probably in monkeys), the contribution of BuChE to the hydrolysis of any of the radiolabeled substrate esters is minimal.

IV. CONCLUSIONS

This chapter used acetylcholinesterase as an example, as there is very clearly a large (> 50-fold) difference in enzyme concentrations in different regions of the human brain (Atack et al., 1986). Measurements of altered enzyme activity in such different regions require separate radiotracers with appropriate kinetics for the concentration of enzymes in each region.

Can and should this concept of multiple radiotracers be extended to other systems? The authors believe it should and can provide another example. For the measurement of the vesicular monoamine transporter, the authors have developed and validated a new radiotracer, α-(+)-[^{11}C]dihydrotetrabenazine (DTBZ), for use in the caudate and putamen of the mammalian brain, including humans. This radiotracer shows excellent pharmacokinetic behavior in that region, with a good signal-to-noise ratio that allows quantification of the numbers of binding sites in normal and diseased brain (Koeppe et al., 1996). It would be of interest to extend the use of DTBZ to other brain regions, particularly the cortex where there are clinical questions regarding the monoaminergic projections. However, use of DTBZ in the cortex is limited by the very low concentrations of vesicular monoamine transporters present there; although there are clearly detectable binding sites in the cortex, the signal-to-noise ratio is quite poor and quantification suffers. In this example, the ratio between regions of high (striatum) and low (cortex) concentrations (*in vitro* B_{max} values) is about twenty. From studies done with AChE substrates, the authors thus conclude that to study the vesicular monoamine transporter in the cortex, one would need a radiotracer with a higher binding potential in that region, which would (in practical terms) translate into a radiotracer with a higher binding affinity. Research into that area is underway.

Acknowledgment

This work was supported by grants from the National Institutes of Health (NS 24896 and T32-CA09015) and the Department of Energy (DE-FG021-87ER60651).

References

Atack, J. R., Perry, E. K., Bonham, J. R., Candy, J. M., and Perry, R. H. (1986). Molecular forms of acetylcholinesterase and butyrylcholinesterase in the aged human central nervous system. *J. Neurochem.* **47**: 263–277.

Boissiere, F., Faucheux, B., Agid, Y., and Hirsch, E. C. (1997). Choline acetyltransferase mRNA expression in the striatal neurons of patients with Alzheimer's disease. *Neurosci. Lett.* **225**: 169–172.

Bormans, G., Sherman, P., Snyder, S. E., and Kilbourn, M. R. (1996). Synthesis of carbon-11 and fluorine-18 labeled 1-methyl-4-piperidyl-4-fluorobenzoate and their biodistribution in mice. *Nucl. Med. Biol.* **23**: 513–517.

Brimjoin, S., and Hammond, P. (1988). Butyrylcholinesterase in human brain and acetylcholinesterase in human plasma: Trace enzymes measured by two-site immunoassay. *J. Neurochem.* **51**: 1227–1231.

Freeman, S. E., and Dawson, R. M. (1991). Tacrine: A pharmacological review. *Prog. Neurobiol.* **36**: 257–277.

Frey, K. A., Koeppe, R. A., Kilbourn, M. R., Snyder, S. E., and Kuhl, D. E. (1997). PET quantification of cortical acetylcholinesterase inhibition in monkey and human. *J. Nucl. Med.* **38**: 146P.

Graybiel, A. M., and Ragsdale, C. W., Jr. (1982). Pseudocholinesterase staining in the primary visual pathway of the macaque monkey. *Nature* (*London*) **299**: 439–442.

Irie, T., Fukushi, K., Akimoto, Y., Tamagami, H., and Nozaki, T. (1995). Design and evaluation of radioactive acetylcholine analogs for mapping brain acetylcholinesterase (AChE) in vivo. *Nucl. Med. Biol.* **21**: 801–808.

Irie, T., Fukushi, K., Namba, H., Iyo, M., Tamagami, H., Nagatsuka, S., and Ikota, N. (1996). Brain acetylcholinesterase activity: Validation of a PET tracer in a rat model of Alzheimer's disease. *J. Nucl. Med.* **37**: 649–655.

Kilbourn, M. R., Snyder, S. E., Sherman, P. S., and Kuhl, D. E. (1996). In vivo studies of acetylcholinesterase activity using a labeled substrate, N-[^{11}C]methylpiperidin-4-yl propionate. *Synapse* **22**: 123–131.

Koeppe, R. A., Frey, K. A., Vander Borght, T. M., Karlamangla, A., Jewett, D. M., Lee, L. C., Kilbourn, M. R., and Kuhl, D. E. (1996). Kinetic evaluation of [^{11}C]dihydrotetrabenazine by dynamic PET: Measurement of the vesicular monoamine transporter. *J. Cereb. Blood Flow Metab.* **16**: 1288–1299.

Koeppe, R. A., Frey, K. A., Snyder, S. E., Kilbourn, M. R., and Kuhl, D. E. (1997). Kinetic analysis alternatives for assessing AChE activity with N-[C-11]methylpiperidinyl propionate (PMP): To constrain or not to constrain? *J. Nucl. Med.* **38**: 198P.

Kuhl, D. E., Koeppe, R. A., Snyder, S. E., Minoshima, S., Frey, K. A., and Kilbourn, M. R. (1996). Mapping acetylcholinesterase in human brain using PET and N-[C-11]methylpiperidinyl propionate. *J. Nucl. Med.* **37**: 21P.

Wang, G. J., Volkow, N. D., Fowler, J. S., Ding, Y. S., Logan, J., Gatley, S. J., MacGregor, R. R., and Wolf, A. P. (1995). Comparison of two PET radioligands for imaging extrastriatal dopamine transporters in human brain. *Life Sci.* **57**: 187–191.

CHAPTER 40

Use of Information Technology in the Search for New PET Tracers

A. J. ABRUNHOSA,[*,†] F. BRADY,[*] S. K. LUTHRA,[*] H. MORRIS,[*] J. J. DE LIMA,[†] and T. JONES[*]

*MRC Cyclotron Unit
Hammersmith Hospital
London W12 0NN, United Kingdom
†IBILI-Faculdade de Medicina
Celas, 3000 Coimbra, Portugal*

The true potential of positron emission tomography (PET) for studying various receptor systems and biochemical pathways has barely been tapped due to lack of suitable radiotracers. To overcome this, a means of rationalizing the selection of molecules for development as PET radiotracers is needed. Huge libraries of compounds exist within the pharmaceutical industry as well as in the public domain. The problem is to develop a way of "filtering" or interrogating these huge compound data bases to find lead molecules for PET. Information technology offers the potential to integrate and make most efficient use of the information available. We have set out to create an information technology resource for PET science. At the core of this is a PET chemistry data base, which already contains detailed information on more than 500 radiolabeled compounds. The main function of this resource is to serve as a base from which we can explore the application of computational chemistry methods to the selection and development of new and improved PET tracers. An overview of computational methods currently used to establish relationships between chemical structures and biological activity is presented. From this, an approach that utilizes components from a number of computational methods that could provide a rational approach to PET tracer selection is outlined.

I. INTRODUCTION

Positron emission tomography (PET) is a unique molecular imaging technique with the ability of generating parametric images that depict specific molecular pathways or molecular interactions within the body. Underpinning this multidisciplinary science is the identification of molecules to act as specific and selective tracers of the biomolecular events that are of interest. The true potential of PET has barely been tapped due to a dearth of suitable candidate radiotracers possessing acceptable characteristics for PET studies. This problem is compounded by the fact that to date, there is little consensus as to what "optimum" characteristics are required for successful radiotracers for particular applications (Kilbourn, 1996). A major problem is that compounds that possess biological activity and show promising pharmacological properties *in vitro* often fail *in vivo* as PET tracers. This is in part due to the fact that although compounds with high affinity and good selectivity are selected for radiolabeling, they often exhibit unsuitable kinetics, high nonspecific binding, and interference from radiolabeled metabolites *in vivo*, which precludes their use as PET tracers. As a result, although more than 500 compounds have been labeled with carbon-11 or fluorine-18 in the past decade as potential radiotracers, only about two dozen compounds have reached routine use for clinical PET studies in laboratories worldwide. The problem is to devise a more rational way of selecting potential candidates for development as PET tracers.

In general, the source for many of the compounds developed for PET is the pharmaceutical industry. The increasing use of combinatorial chemistry, creating real and virtual libraries of compounds in the search for new drugs, increases daily the number of compounds available. Although the vast majority of compounds in these data bases will not possess biological properties

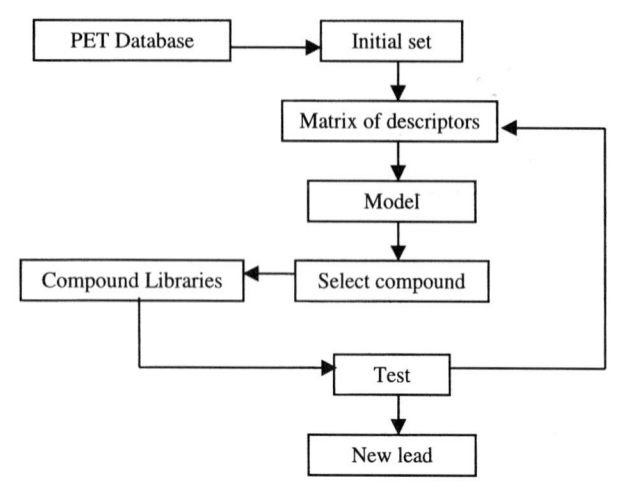

FIGURE 1 An information technology-based rational approach to the selection of lead compounds for PET.

making them suitable as drug candidates, some may have the characteristics required of PET tracers. Additionally, vast numbers of compounds can be found in various organic databases (e.g., Beilstein ca. 7 million compounds). The challenge is to develop a way of "filtering" or interrogating these huge compound data bases to find lead molecules that might have application in PET.

Developments in information technology (IT) offer the potential to integrate and make more efficient use of the information available (Weinstein et al., 1997). We have set out to create an information technology resource for PET science. By integrating this resource with molecular modeling and computational chemistry components, we intend to create sets of descriptors for PET tracers that could then be used to interrogate external compound libraries. This should help provide a rational approach to selection and optimization of lead compounds (Fig. 1).

II. METHODS

A. Information Technology

The need to store chemical information in digital format has led to the development of new ways of representing chemical formulas, structures, and properties. This has made possible the development of searchable data bases. In addition to Chemical Abstracts, other chemistry data bases currently available to the academic community include, e.g., for synthetic organic chemistry, ISIS/RLX (Fletcher et al., 1996) and Beilstein, which lists more than 10 million reactions; for inorganic and organometallic chemistry, GMELIN; for protecting group chemistry, ISIS/SPG. Structural data bases include the Cambridge Crystallographic Data Centre (Allen et al., 1979; Allen and Kennard, 1993). Spectroscopic (SpecInfo) and bibliographic sources are also available.

Until recently, access to most of these data bases was only possible using X-windows systems with proprietary client–server architectures. With the advent of the Internet, most of these are now accessible through user-friendly, web-based interfaces and, in some cases, using web browsers. They are thus very easily integrated in the computer desktop environment, making it possible to incorporate chemical structures in word processing documents, spreadsheets (e.g., Accord for Excel), and data base management systems (e.g., Accord for Access).

This has led to the development of new languages for representing chemistry (Casher and Rzepa, 1995). Structures are stored in data bases in the form of molecular fingerprints, a bitwise representation of the different substructures that are contained in a molecule. An ASCII version of a molecular structure, named a SMILES string, was developed to be used in electronic mail (e-mail), and a series of Multimedia Internet Mail Extensions (MIME types) allowed the interchange of chemical entities in the Internet. Another very important advance was the development of a Chemistry Mark-up Language (CML), a form of HyperText Markup Language (HTML) that allows two- and three-dimensional chemical structures to be visualized and manipulated in a web browser (e.g., Netscape, Mosaic).

B. Computational Chemistry

Computational chemistry implements algorithms that utilize concepts of theoretical chemistry for the prediction of chemical properties of molecules (Hopfinger, 1985). An extension of this approach is used in medicinal chemistry in an attempt to establish structure–activity relationships (SARs) between a certain biological effect or pharmacological characteristic of a family of compounds and their physicochemical characteristics (Sun and Cohen, 1993). When the relationship assumes the form of an equation, it is referred to as a quantitative structure–activity relationship (QSAR).

The physicochemical characteristics (usually called descriptors) used in classical QSAR fall into three main categories: (i) hydrophobicity, (ii) electronic effect, and (iii) steric effects.

i. Hydrophobicity plays a major role in the interactions between drugs and their targets but is also very important for the ability of the drug to cross the blood–brain barrier and its accumulation in hydrophilic compartments. The partition coefficient, usually measured between octanol and water, has been a reliable measure of the hydrophobicity of a molecule, and its logarithm ($\log P$) has been shown to be usually highly correlated with biological activity. The substituent hydrophobicity constant (π) is a measure of how hydrophobic a certain group is in a molecule as compared to hydrogen and is present in many of the classical QSAR equations.

ii. The electronic effects try to take into consideration hydrophilic or more polar parts of the molecules. The Hammet substituent constant (σ) is a measure of the electron donating/withdrawing effects of a substituent in a ring, taking into consideration possible interactions with other ring substituents.

iii. Molar refractivity (MR) is a parameter proposed to account for the difference between volumes occupied by different substituents.

C. Molecular Modeling

Numerous molecular graphics packages are available that allow the user to view and manipulate molecular structures, providing insight into how the structure of the molecule might be related to its chemical or biological behavior (e.g., MSI, Trypos, Daylight). Many of these packages implement routines that attempt to solve the energy equations for the structures and display the result in the form of a new optimized structure (Cohen *et al.*, 1990). These calculations can be made by (1) classical mechanics and by (2) quantum mechanics.

1. In the algorithms that use the classical mechanics model (e.g., MM1, MM2) the steric energy of the molecule is calculated as the sum of the individual contributions of several calculated force fields. The most common are bond stretching, angle bending, angle torsion, and other interactions such as van der Waals forces and hydrogen bonds. The empirically derived equations provide reasonable approximations and are relatively fast.

2. The more accurate methods use quantum mechanics. Although the solution of the Schrödinger equation would provide the ultimate solution for the calculation of all the chemical properties, its direct calculation is not a trivial task. Instead some semiempirical methods (e.g., AM1, PM3) give reasonably accurate results and are not very time-consuming. However, they are still much more demanding in terms of computational power so their calculation is usually beyond the scope of a personal computer.

Both methods compute many theoretically derived properties, such as electron density; orbitals (e.g., HOMO, LUMO); electrostatic potential; partial charge; and bond orders and bond strains for entire molecules, chemical groups, or individual atoms.

These parameters can be used for QSAR studies, providing a much more accurate set of descriptors that can also take into account structural information (3D QSAR). The challenge is to decide from this great variety of properties which are the best descriptors. Powerful statistical methods are used in this process. Partial least squares is, by far, the most common, sometimes with very good correlation factors ($p > 0.9$). The prediction ability of the proposed model can be tested by cross-validation, and a good choice of the initial set of compounds is critical. The methods are not restricted to linear regression techniques (Maddalena and Johnston, 1995; Gálvez *et al.*, 1996; Mestres *et al.*, 1997); very promising results have been achieved with cluster analysis, principal component analysis, neural networks, and even genetic algorithms.

The resulting equations can then used to build a "pharmacophore model." This consists of an invariant molecular skeleton with some active areas where a change in chemical properties is predicted to have a positive or negative effect on the chemical or biological activity of the compound. This model can then be used to predict the activity of a proposed compound or to screen a library of structures in the search for new ones.

An alternative approach is the use of structure-based drug design (SBDD) strategies (Kuntz, 1992). The structures of proteins that act as targets for many of the drugs in use today are known from crystallographic or nuclear magnetic resonance studies. The Protein Databank at Brookhaven lists more than 5000 of these structures, and new structures are being solved at a rate of about 1 per day (Bernstein *et al.*, 1977). It is possible, using computer docking studies, to predict how molecules will bind in the active site and which ones will be the most effective in doing so (Stoddard and Koshland, 1993). It is also possible to build molecules that will perfectly "fit" into the active site and establish the most effective interactions with it. This strategy of *de novo* SBDD can be very useful in the design of antagonists.

Another very promising technique, when the structure of the active site is not known, is comparative molecular fields analysis (CoMFA). In this variant of 3D QSAR, the optimized structures of several compounds that are known to bind to the same active site are aligned in the three-dimensional space by the invariant parts of the molecules (Waller and Marshall, 1993). Based on the molecular volumes, a 3D surface is generated and probed for steric (van der Waals) and electrostatic (coulombic) fields, usually with an sp^3 hybridized carbon with a $+1$ charge. Statistical analysis will then generate a 3D contour map where the changes of any of the fields are associated with changes in the activity of the compounds. This technique has proven to give better statistical results than classical QSAR (Carroll *et al.*, 1991).

III. RESULTS AND DISCUSSION

In order to take into account all of the facets of PET science that might be relevant to radiotracer development it was necessary to create an information technology resource devoted to this area. The first step in the development of this resource has been to create a chemistry data base with literature reported PET tracers that will form the core of a PET science resource. The task of creating such a resource was not a trivial one. The information required is dispersed across a broad range of disciplines. From a purely chemical point of view, information is required on radioisotope production, labeling strategies, precursor synthesis, radiochemistry, automation, quality control, and metabolite analysis. Additionally, information on the pharmacological and biochemical properties of PET radiotracers is also essential.

Some information in these areas already exists in digital form and can be accessed in a variety of data bases by the academic community. One difficulty is that most data are stored in incompatible formats and are accessed using a great variety of platforms. These issues have been resolved by using a common format in the design of a dedicated PET chemistry data base that is also compatible with existing bioinformatics resources. The technology that has been used to develop this data base has allowed us to merge several types of information, including chemical and structural data (2D and 3D), in a single document, while retaining compatibility with the applications that created them. This technology has also allowed us to make the PET chemistry data base available on the Intranet of the authors' PET center.

The PET chemistry data base is radiotracer centered. It has been created by compiling information in four major areas (chemistry, radiochemistry, biological, and others) as shown in Fig. 2. All of this is complemented by a structured reference data base.

Currently, the data base contains information on ca. 500 compounds that have been labeled with carbon-11 and is available on the Intranet of the authors' PET center. Navigation is facilitated by a chemical guide and a pharmacological guide. At a basic level, this resource will provide the radiosynthetic chemist with a readily accessible tool for helping devise synthetic routes to radiolabeling new molecules, as well as easy and comprehensive access to the PET chemistry litera-

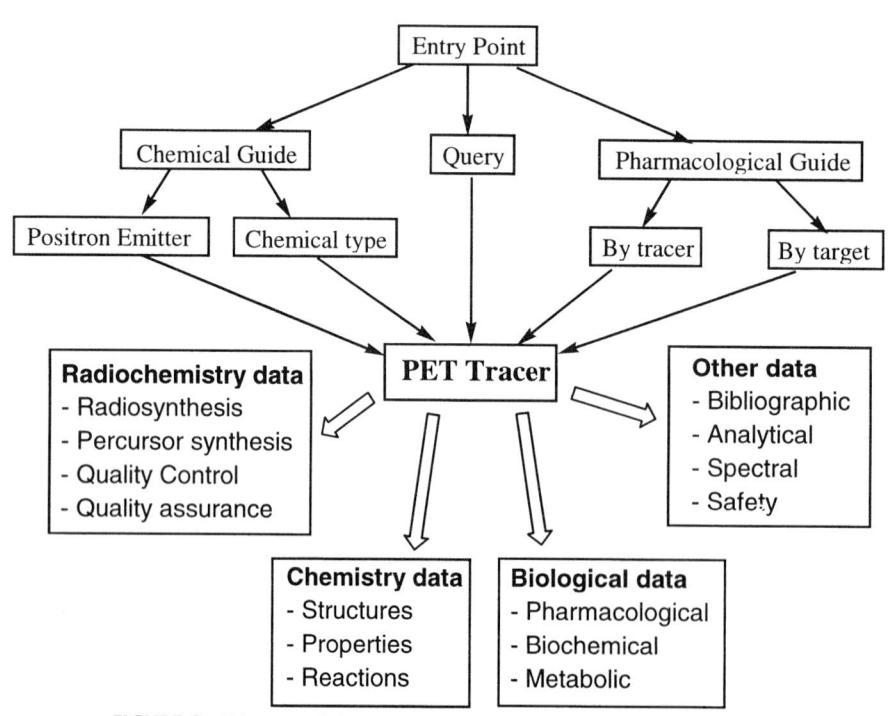

FIGURE 2 Diagram of the organization of the PET chemistry data base.

ture coupled with links to the relevant wider chemistry and literature.

The main function of this resource is to serve as a base from which we can explore the application of computational chemistry methods to the selection and development of new and improved PET tracers. One approach currently being investigated is using a series of analogs, having selectivity for a particular target site, that have been radiolabeled and evaluated as PET tracers. Molecular modeling techniques are being used to provide us with a large number of descriptors for this series of compounds. The goal is to establish a relationship between some of these descriptors and biological activity. The main problem faced here is of a statistical nature. The matrix of descriptors has many columns containing the numerous physicochemical properties that can be measured or derived from a single compound but only a few rows containing the radiolabeled analogs. The challenge of the approach is to obtain QSAR equations with good statistical significance. QSAR equations will then be used to predict how different structural alterations and labeling strategies might affect the performance of a molecule as a PET tracer. In the cases where the QSAR equations can be converted into pharmacophore models, it might be possible to use them to screen virtual or real compound libraries for molecules that would fit the desired profile, providing us with a rational approach to the selection of new lead compounds for PET tracers.

The successful implementation of the strategy described depends on the clear and careful definition of the desired characteristics of a tracer for a particular application. This can only be achieved by the close integration of a number of areas in PET science. The PET IT resource provides a platform for such an integrated approach. When complete, this resource will provide a global overview to PET science and will facilitate the flux of information relevant to development of this area. The PET IT resource could also provide a window of opportunity for dialogue with the pharmaceutical industry.

Acknowledgment

This work was supported by a grant from the Portuguese Government: JNICT-PRAXIS XXI/BD/4563/94.

References

Allen, F. H., and Kennard, O. (1993). 3D search and research using the Cambridge structural database. *Chem. Des. Autom. News* **8**(1): 31–37.

Allen, F. H., Bellard, S., Brice, M. D., Cartwright, B. A., Doubleday, A., Higgs, H., Hummelink, T., Hummelink-Peters, B. G., Kennard, O., Motherwell, W. D. S., Rodgers, J. R., and Watson, D. G. (1979). The Cambridge crystallographic data centre: Computer-based search, retrieval, analysis and display of information. *Acta Crystallogr., Sect. B* **B35**: 2331–2339.

"Beilstein Cross Fireplus Reactions and Gmelin." Beilstein Informationssysteme GmbH. Germany.

Bernstein, F. C., Koetzle, T. F., Williams, G. J. B., Meyer, E. F. Jr., Brice, M. D., Rodgers, J. R., Kennard, O., Shimanouchi, T., and Tasumi, M. (1977). The Protein Data Bank: A computer-based archival file for macromolecular structures. *J. Mol. Biol.* **112**: 535–542.

Carroll, F. I., Gao, Y., Rahman, M. A., Abraham, P., Parham, K., Lewin, A. H., Boja, W., and Kuhar, M. J. (1991). Synthesis, ligand binding, QSAR, and CoMFA study of 3β-(p-Substituted phenyl)tropane-2β-carboxylic acid methyl esters. *J. Med. Chem.* **34**: 2719–2725.

Casher, O., and Rzepa, H. S. (1995). Chemical collaboratories using World-Wide Web servers and EyeChem-based viewers. *J. Mol. Graphics* **13**(5): 268–270.

Cohen, N. C., Blaney, J. M., Humblet, C., Gund, P., and Barry, D. C. (1990). Molecular modeling software and models for medicinal chemistry. *J. Med. Chem.* **33**: 883–894.

Fletcher, D. A., McMeeking, R. F., and Parkin, D. (1996). The United Kingdom Chemical Database Service. *J. Chem. Inf. Comput. Sci.* **36**: 746–749.

Gálvez, J., García-Domenech, R., Alapont, C. deG., and Julián-Ortiz, J. V. (1996). Pharmacological distribution diagrams: A tool for *de novo* drug design. *J. Mol. Graphics* **14**(5): 272–276.

Hopfinger, A. J. (1985). Computer-assisted drug design. *J. Med. Chem.* **28**: 1134–1139.

Kilbourn, M. R. (1996). *In vivo-in vitro* correlations: An example from vesicular monoamine transporters. *In* "Quantification of Brain Function Using PET" (R. Myers, V. Cunningham, D., Bailey, and T. Jones, eds.), pp. 3–8. Academic Press, San Diego, CA.

Kuntz, I. D. (1992). Structure-based strategies for drug design and discovery. *Science* **257**: 1078–1082.

Maddalena, D. J., and Johnston, G. A. R. (1995). Prediction of receptor properties and binding-affinity of ligands to benzodiazepine/GABA(A) receptors using artifical neural networks. *J. Med. Chem.* **38**(4): 715–724.

Mestres, J., Rohrer, D. C., and Maggiora, G. M. (1997). A molecular field-based similarity approach to pharmacophore identification. *J. Am. Chem. Soc.* **213**: 92–comp.

Stoddard, B. L., and Koshland, D. E., Jr. (1993). Molecular recognition analyzed by docking simulations: The aspartate receptor and isocitrate dehydrogenase from *Escherichia coli*. *Proc. Natl. Acad. Sci. U.S.A.* **90**: 1146:1153.

Sun, E., and Cohen, F. E. (1993). Computer-assisted drug discovery —a review. *Gene* **137**: 127–132.

Waller, C. L., and Marshall, G. R. (1993). Three-dimensional quantitative structure-activity relationship of angiotesin-converting enzyme and thermolysin inhibitors. II. A comparison of CoMFA models incorporating molecular orbital fields and desolvation free energies based on active-analog and complementary-receptor-field alignment rules. *J. Med. Chem.* **36**: 2390–2403.

Weinstein, J. N., Myers, T. J., O'Connor, P. M., Friend, S. H., Fornace, A. J., Jr., Kohn, K. W., Fojo, T., Bates, S. E., Rubinstein, L. V., Anderson, N. L., Buolamwini, J. K., van Osdol, W. W., Monks, A. P., Scudiero, D. A., Sausville, E. A., Zaharevitz, D. W., Bunow, B., Viswanadhan, V. N., Johnson, G. S., Wittes, R. E., and Paull, K. D. (1997). An information-intensive approach to the molecular pharmacology of cancer. *Science* **27**: 343–349.

CHAPTER 41

Statistical Power Analysis of *in Vivo* Studies in Rat Brain Using PET Radiotracers[1]

D. HUSSEY, J. N. DASILVA, E. GREENWALD, K. CHEUNG, S. KAPUR, A. A. WILSON, and S. HOULE

Vivian M. Rakoff PET Centre
Clarke Institute of Psychiatry and University of Toronto
Toronto, Ontario, M5T 1R8 Canada

The objective of this study is to analyze the statistical power of in vivo rat competition studies performed in the developmental process of new radioligands for positron emission tomography (PET). These calculations are important for determining the number of animals necessary to detect a significant change of radiotracer uptake for a control versus a treated group of rats. The analysis is based on a series of studies carried out in rats with the D_1 agonist [^{11}C]SKF 82957 and the D_1 antagonist [^{11}C]SCH 23390. The average variance of the percentage injected dose of tracer per gram of brain tissue at 45 min postinjection in different brain regions was calculated for three groups of rats. The power to detect a change due to drug treatment on the radioligand uptake in each region was calculated for different numbers of rats. The average coefficient of variance for all three groups showed similar results, ranging from 8 to 23%. This study provides an empirical basis for selecting the number of rats necessary to detect a significant change in tracer uptake. Using the striatum as an example, total blocking studies may significantly ($p < 0.05$) be determined with a small number of rats (90% change with n = 4 rats). Smaller changes require larger number of rats (25% change with n = 8 rats) to obtain the same statistical power. These results suggest that these power calculations can be used for any in vivo rat studies with PET radioligands having similar coefficients of variance.

I. INTRODUCTION

The power of a statistical test is the probability that it will yield statistically significant results when a change exists in the population in question (Cohen, 1988). Prior to injection of positron emission tomography (PET) radiotracers into humans, new radiotracers are usually evaluated *in vivo* in animals to determine their regional brain uptake and pharmacological binding profile. Binding specificity and selectivity of a new radioligand are determined by treating groups of animals with selective drugs. Radioligands can also be evaluated in groups of animals treated with drugs known to change the density of a given receptor and/or the concentration of endogenous neurotransmitter. The *a priori* probability of achieving this significance is usually not examined in rat *in vivo* studies performed in the developmental process of new radioligands for PET.

The possibility of detecting a difference in radioligand binding between the treated group and the control group merely due to noise is described as Type I error (false positive). A Type II error (false negative) occurs when no significant change in binding is found, when there was one present (Strainer, 1990). In general, Type I error is reduced for most studies by setting significance levels such that results by chance only are very small (conventionally $p < 0.05$). Techniques used to reduce Type II errors are less understood and are examined in this chapter. The ability to estimate variance in the measurement and thus calculate estimates of the power and sample size for these types of experi-

---

[1] Transcripts of the BRAINPET97 discussion of this chapter can be found in Section VIII.

ments can help in reducing and understanding Type II error (Kapur et al., 1995). The goals of this study are to analyze the statistical power of three *in vivo* PET radioligand studies in rats using postsynaptic dopamine D_1 radioligands ([^{11}C]SKF 82957 and [^{11}C]SCH 23390) and also to determine if using the striatum-to-cerebellum ratio will change these power estimates. The striatum-to-cerebellum ratio provides an index of the bound to free ratio, as the cerebellum contains negligible amounts of D_1 receptors.

II. MATERIALS AND METHODS

A. Biodistribution Studies

Animal experiments were conducted in accordance with the recommendations of the Canadian Council on Animal Care and with the approval from the Animal Care Committee at the Clarke Institute of Psychiatry. Male Sprague–Dawley rats (200–260 g at start of study, up to 450 g at end of study for chronic group) were obtained from the Charles River Breeding Farm (Montreal, Canada) and were given food and water *ad libitum*. Three groups were analyzed: the first used the D_1 agonist [^{11}C]SKF 82957 (DaSilva et al., 1996) following 21 days of saline injections plus 3 days of washout and will be referred to as chronic ($n = 28$); the second also used [^{11}C]SKF 82957 but with no pretreatment ($n = 15$); and the third used the D_1 antagonist [^{11}C]SCH 23390 (Halldin et al., 1986) following chronic treatment with saline for 21 days plus 3 days washout ($n = 12$). Tracer injections were given via lateral tail vein (previously vasodilated in a warm 45°C water bath) while animals were lightly restrained in restraining boxes. The rats were sacrificed by decapitation 45 min after radiotracer injection. Blood samples were taken immediately after each animal was sacrificed, and brains were rapidly removed and dissected. The brain regions collected included striatum, hippocampus, cerebellum, frontal cortex, olfactory tubercles, hypothalamus, and thalamus. All tissues were washed in saline, blotted dry, weighed, and counted in a gamma counter along with aliquots of the injected solution for reference and conversion back into nCi/g. The rat tails and the syringes used for injection were measured in a dose calibrator (Capintec) and residual radioactivity contained within them was used to correct for actual injected dose. Radioactivity levels were expressed as percentage injected dose per gram (%ID/g) of tissue and as percentage injected dose per gram of tissue per body weight of rat (%ID/g/BW).

B. Power Estimation

Estimates of baseline variance were calculated for the seven brain regions collected using %ID/g. The baseline variance was also calculated for the three groups of rats using the striatum-to-cerebellum ratios. The results for each group discussed is a compilation of several individual studies carried out over several months and combined for this analysis to yield a larger sample size for variance estimation. Bilateral structures were found not to have a left-to-right difference and were combined as one region. The power to detect a given change depends on the statistical test being employed. *In vivo* competition studies generally result in a comparison between %ID/g means for all brain structures in the treated group as compared to the control group. Literature does not always give a clear directional hypothesis, and the number of rats is desired to be small. Under these conditions the following formula can be used to calculate the power:

$$z(1 - \beta) = \frac{\Delta}{\sigma}\sqrt{\frac{n}{2}} - z(1 - \alpha),$$

where σ^2 is the estimated variance for the measurement, n is the number of rats, α is the probability of Type I error, and Δ is the expected effect size (percentage increase or decrease in %ID/g) (Rosner, 1990). The power in percentage is given by $(1 - \beta) \times 100\%$. This equation assumes that the variance and number of rats for both control and treated groups will be the same. This chapter evaluates the power associated with changes in %ID/g and striatum-to-cerebellum ratios of 5, 12, 25, 50, and 90% with 4–15 rats.

III. RESULTS

The regional brain distribution obtained for the three groups is summarized in Table 1. Regions rich in D_1 receptors, such as the striatum and olfactory tubercles, showed higher radioligand uptake as compared to cerebellum (devoid of D_1 receptors). The range of the coefficients of variance associated with each of the three experiments was similar, ranging from 8 to 23%. Table 2 gives the estimates of statistical power for various levels of effect size (increase or decrease of %ID/g from 5 to 90%). Figure 1 shows the power curves of the striatum for all three groups for all effect sizes.

TABLE 1 Regional Brain Uptake (%ID/g) of [¹¹C]SKF 82957 and [¹¹C]SCH 23390 in Rats 45 min Postinjection[a]

Region	[¹¹C]SKF chronic Mean ± SD	% SD	[¹¹C]SKF no treatment Mean ± SD	% SD	[¹¹C]SCH 23390 Mean ± SD	% SD
Hypothalamus	0.10 ± 0.02	22	0.15 ± 0.03	20	0.30 ± 0.03	11
Olfactory Tubercles	0.29 ± 0.05	18	0.38 ± 0.04	11	1.75 ± 0.26	15
Cerebellum	0.06 ± 0.01	17	0.08 ± 0.01	17	0.17 ± 0.02	11
Hippocampus	0.10 ± 0.02	17	0.14 ± 0.03	23	0.37 ± 0.05	14
Striatum	0.29 ± 0.04	14	0.47 ± 0.10	21	1.92 ± 0.19	10
Thalamus	0.11 ± 0.02	17	0.17 ± 0.02	10	0.42 ± 0.04	8
Front cortex	0.11 ± 0.02	19	0.16 ± 0.03	18	0.44 ± 0.05	12
Ratio S/C	4.92 ± 0.65	13	5.51 ± 1.03	19	11.25 ± 1.64	15
Average (seven regions)		18		17		12

[a] Mean of %ID/g ± SD of control rats; $n = 28$ for [¹¹C]SKF 82957 chronic, $n = 15$ for [¹¹C]SKF 82957 no treatment, and $n = 12$ for [¹¹C]SCH 23390.

TABLE 2 Estimates of Power for All Effect Sizes ($n = 4$ to 15) in the Cerebellum and Striatum and for the Striatum-to-Cerebellum Ratio[a]

% change[b]	n	[¹¹C]SKF 82957 chronic Cerebellum	Striatum	S/C[c]	[¹¹C]SKF 82957 no treatment Cerebellum	Striatum	S/C	[¹¹C]SCH 23390 Cerebellum	Striatum	S/C
5.0	4	6	7	8	6	5	6	9	11	7
12.0	4	17	23	25	18	12	15	33	41	21
25.0	4	54	73	76	57	38	48	88	95	68
50.0	4	98	100	100	99	91	97	100	100	100
90.0	4	100	100	100	100	100	100	100	100	100
5.0	8	8	11	11	9	7	8	14	17	10
12.0	8	29	41	44	31	20	25	57	68	38
25.0	8	83	95	97	86	65	77	99	100	93
50.0	8	100	100	100	100	100	100	100	100	100
90.0	8	100	100	100	100	100	100	100	100	100
5.0	12	10	13	13	10	8	9	17	20	12
12.0	12	35	49	53	37	24	30	67	78	45
25.0	12	90	98	99	92	75	85	100	100	97
50.0	12	100	100	100	100	100	100	100	100	100
90.0	12	100	100	100	100	100	100	100	100	100
5.0	15	12	17	18	13	9	11	23	28	15
12.0	15	48	66	70	51	34	42	83	92	62
25.0	15	98	100	100	99	90	96	100	100	100
50.0	15	100	100	100	100	100	100	100	100	100
90.0	15	100	100	100	100	100	100	100	100	100

[a] All power values at $p < 0.05$.
[b] Absolute percentage change from mean.
[c] Ratio of striatum to cerebellum.

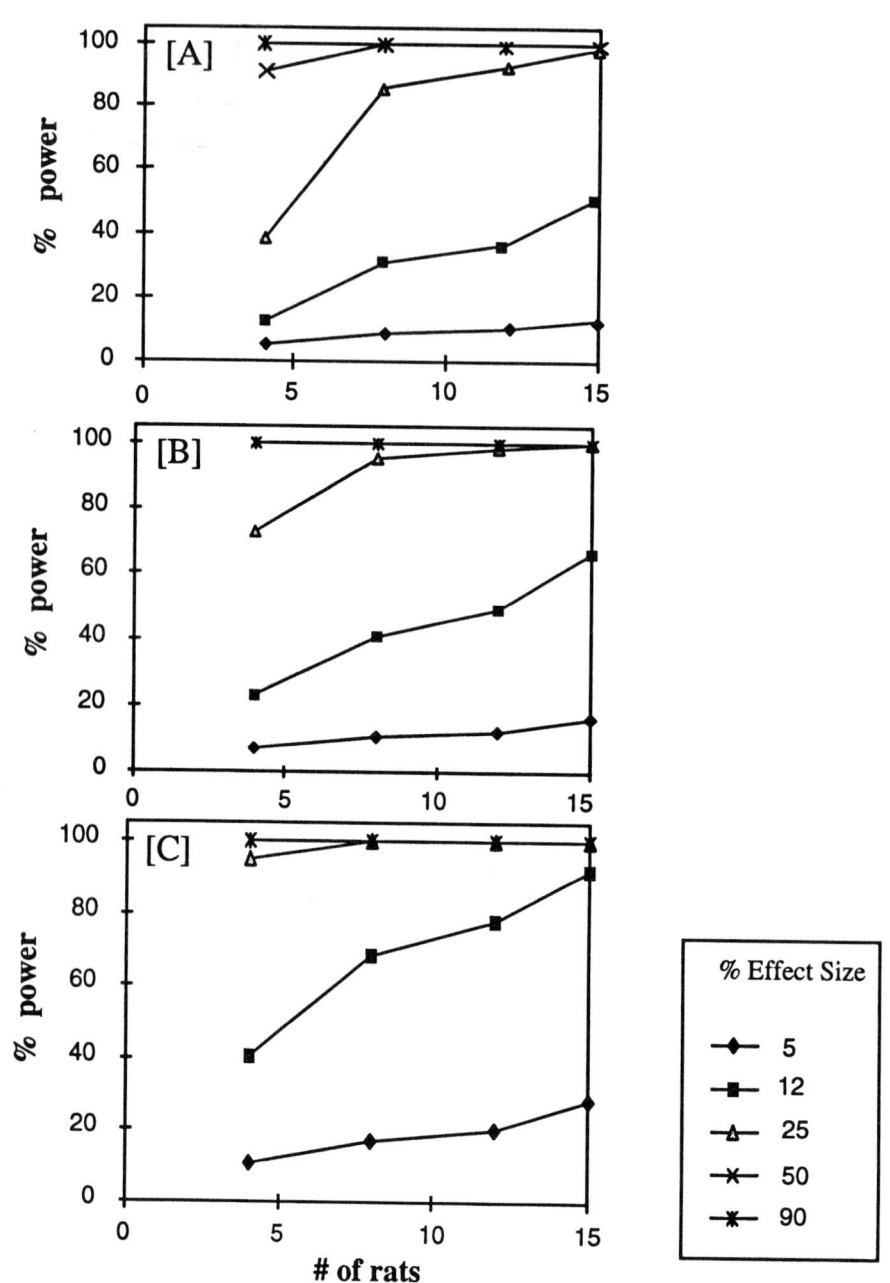

FIGURE 1 Power curves for [^{11}C]SKF 82957 no treatment (A), [^{11}C]SKF 82957 chronic saline treatment (B), and [^{11}C]SCH 23390 (C), all at $p < 0.05$.

IV. DISCUSSION

Since typical rat *in vivo* competition studies often achieve large radioligand binding differences between control and treated groups, the power of these studies is high and significance can readily be achieved with low numbers of rats (Pauwels *et al.*, 1994; Shine *et al.*, 1997). The issue of statistical power is more important in studies where the radioligand binding differences between control and treated groups are smaller (10–50%) than those achieved with total blocking studies.

This study provides empirical estimates of the power for typical rat *in vivo* competition studies with PET radiotracers for effects between 5 and 90%. The average coefficient of variance for the three groups in all brain regions was 15%, which is similar to studies reported by other groups (Hume *et al.*, 1994). As expected, the power to detect large differences in binding is high (> 95% power for effects of 50 to 90%) even with small numbers of rats ($n = 4$). However, small changes (e.g., 5%) are difficult to detect even with large numbers of rats (< 20% power with $n = 15$ rats) (see Table 2). This suggests that experiments that lie in between these values would benefit from these power estimates prior to determining the number of rats necessary to obtain significant results.

No practical difference was seen between the two radiotracers examined, [^{11}C]SKF 82957 and [^{11}C]SCH 23390, with average coefficients of variance for all regions of 17 and 12%, respectively. If it is assumed that a power of 80% or more is the minimum criteria for proceeding with an experiment, then all three groups showed similar trends. For instance, given a significance level of $p < 0.05$, a change of 25% or more in effect size can be significantly reached with $n = 8$ rats. Effect sizes below 25% should be examined closely, as the power at these levels is small (see Fig. 1). By using the same criteria (80% threshold for significance), no advantage was observed in the power by using the striatum-to-cerebellum ratios as compared to the striatum binding values between treated and control groups. Since many factors contribute to the overall estimate of variance in these type of experiments (e.g., rat body weight and dissection and injection techniques, as well as instrumentation quality controls), these results should strictly apply to experiments using similar procedures as outlined earlier. Nonetheless, these results suggest that these power calculations can be used for any *in vivo* rat studies having similar coefficients of variance.

Acknowledgments

The authors thank John Chambers, Celia Valente, David Wilson, and Corey Jones for their help.

References

Cohen, J. (1988). "Statistical Power Analysis for the Behavioral Sciences," 2nd ed. Erlbaum, Hillsdale, NJ.

DaSilva, J. N., Wilson, A. A., Valente, C., Hussey, D., Wilson, D., and Houle, S. (1996). *In vivo* binding of [^{11}C]SKF 75670 and [^{11}C]SKF 82957 in rat brain: Two dopamine D-1 receptor agonist ligands. *Life Sci.* **58**: 1661–1670.

Halldin, C., Stone-Elander, S., Farde, L., Ehrin, E., Fasth, K-J., Langstrom, B., and Sedvall, G. (1986). Preparation of ^{11}C-Labelled SCH 23390 for the *in vivo* study of dopamine D-1 receptors using positron emission tomography. *Appl. Radiat. Isot.* **37**: 1039–1043.

Hume, S. P., Ashworth, S., Juffry-Opacka, J., Ahier, R. G., Lammertsma, A. A., Pike, V. W., Cliffe, I. A., Fletcher, A., and White, A. C. (1994). Evaluation of [O-*methyl*-^{3}H]WAY-100635 as an *in vivo* radioligand for 5-HT$_{1A}$ receptor in rat brain. *Eur. J. Pharmacol.* **271**: 515–523.

Kapur, S., Hussey, D., Wilson, D., and Houle, S. (1995). The statistical power of [^{15}O]-water PET activation studies of cognitive processes. *Nucl. Med. Commun.* **16**: 779–784.

Pauwels, T., Depthy, S., Goldman, S., Monclus, M., and Luxen, A. (1994). Effect of catechol-O-methyl transferase inhibition on peripheral and central metabolism of 6-[^{18}F] Fluoro-1-DOPA. *Eur. J. Pharmacol.* **257**: 53–58.

Rosner, B. (1990). "Fundamentals of Biostatistics," 3rd ed. PWS-Kent Co., Boston.

Shiue, C. Y., Shiue, G. G., Mozley, P. D., Kung, M. P., Zhuang, Z. P., Kim, H. J., and Kung, H. F. (1997). P-[^{18}F]-MPPF: A potential radioligand for PET studies of 5-HT$_{1A}$ receptors in humans. *Synapse* **25**: 147–154.

Streiner, D. L (1990). Sample size and power in psychiatric research. *Can. J. Psychiatry* **35**: 616–620.

CHAPTER 42

A Human Liver Model of Metabolism as a Tool in the Identification of Potential PET Radiotracers[1]

A. A. WILSON,* T. INABA,† N. FISCHER,† J. N. DASILVA,*,† and S. HOULE*

Clarke Institute of Psychiatry PET Centre
†*Department of Pharmacology*
University of Toronto
Toronto, Ontario, M5T 1R8 Canada

The potent 5-HT$_{1A}$ antagonist WAY 100635 is metabolized in vivo by human liver. The metabolite, WAY 100634, readily crosses the blood–brain barrier and may confound positron emission tomography (PET) studies of [^{11}C]WAY 100635. WAY 100635 and three analogs were studied by an in vitro model of human liver metabolism using HPLC to analyze the reaction mixtures. The objective of the study was to determine whether the analogs were likely to suffer the same metabolic fate as WAY 100635 in vivo, an outcome that would limit their potential as PET radiotracers for imaging 5-HT$_{1A}$ receptors. In human liver cytosolic media, WAY 100635 was quantitatively metabolized to WAY 100634 (amide hydrolysis), but none of the three analogs were affected. In human liver microsomal media, WAY 100635 was again metabolized predominantly to WAY 100634 with some more polar products as well. All three analogs were also metabolized to some extent in the microsomal preparations, but none followed the major pathway of WAY 100635, i.e., no metabolite from simple amide hydrolysis was detected. The metabolic products from microsomal incubation were all more polar (by reverse-phase HPLC) than the anticipated product of amide hydrolysis. In vitro screening of potential PET ligands using human liver preparations can provide useful information in helping decide which ligands merit further study.

I. INTRODUCTION

A successful radiotracer for positron emission tomography (PET) or single photon emission computed tomography (SPECT) imaging of neuroreceptors must meet several criteria. A critical property for a potential imaging radiotracer is a lack of labeled metabolites that can cross the blood–brain barrier. At best, the presence of labeled metabolites in the brain contributes to nonspecific binding and the unbound or free signal. Even more damning is the scenario whereby the labeled metabolites bind to specific sites, either those targeted or others. Thus, information on the potential human metabolism of a putative radiotracer would be extremely useful at an early stage of development.

A growing technique in the drug development industry is to study the metabolism of new compounds at an early stage of development in vitro by using liver reaction systems to study the conversion of drug substrates to primary and secondary metabolites (Miners et al., 1994). This technique was applied in the development of radiotracers for imaging the 5-HT$_{1A}$ receptor system by PET.

WAY 100635 is a potent, selective antagonist of the 5-HT$_{1A}$ receptor. Small animal studies and PET studies with nonhuman primates and humans using [O-methoxy-^{11}C]WAY 100635 showed good brain uptake, appropriate regional brain distribution, and appropriate pharmacology for a 5-HT$_{1A}$ receptor imaging radiotracer (Fletcher et al., 1994; Mathis et al., 1994; Pike et al., 1994). Unfortunately, in humans, WAY 100635 is metabolized rapidly by hepatocytes to WAY 100634,

---

[1] Transcripts of the BRAINPET97 discussion of this chapter can be found in Section VIII.

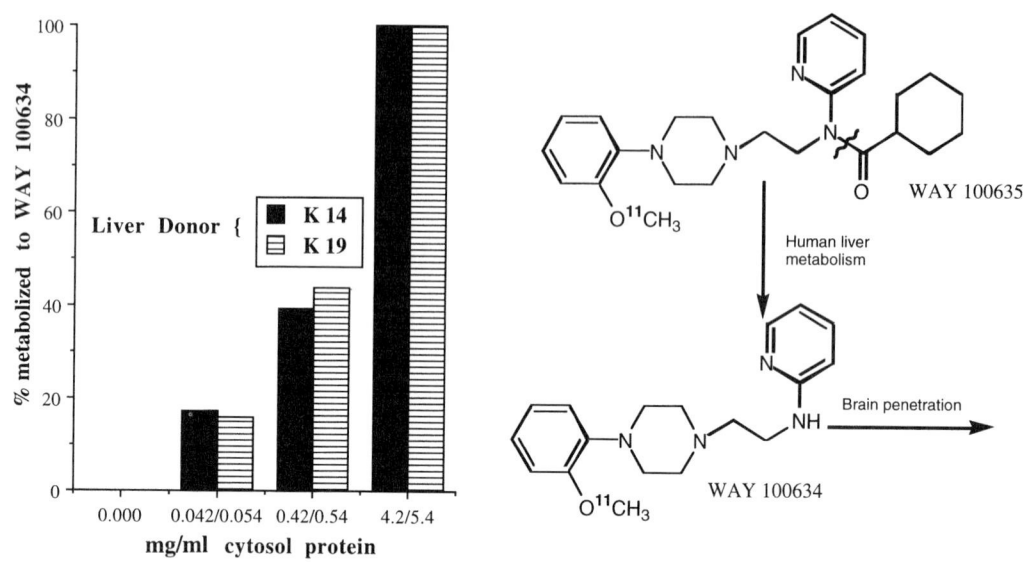

FIGURE 1 In humans, WAY 100635 is metabolized rapidly in liver by amide cleavage to give WAY 100634, which readily crosses the blood–brain barrier. In human liver cytosolic media from two donors (K14 and K19), WAY 100635 is also metabolized to WAY 100634.

which readily crosses the blood–brain barrier (Fig. 1). The presence of this labeled metabolite reduces signal-to-noise ratios and makes quantification difficult. This metabolic pathway is entirely absent in rats and is found only to a small extent in nonhuman primates (Houle et al., 1996; Osman et al., 1996).

Since the metabolism of WAY 100635 to WAY 100634 (amide hydrolysis) was species specific, it was speculated that it may be compound specific as well, i.e., structurally similar analogs might retain affinity for the 5-HT$_{1A}$ receptor while not suffering the same metabolic fate. A variety of analogs was synthesized and screened by in vitro competitive binding assays for 5-HT$_{1A}$ affinity. Three candidates, CPA, BPA, and CPC 222 (Fig. 2), were chosen for metabolism studies based on their high affinities.

II. MATERIALS AND METHODS

Analysis of the mixtures by HPLC was performed with an in-line UV (254 nm) detector in series with a Berthold LB 407A radioactivity detector. Peak areas were measured using Hewlett-Packard 3396 recording integrators. WAY 100635, CPA, BPA, and CPC 222 were prepared by reaction of the appropriate acid chloride with the corresponding secondary amine (Zhuang et al., 1994). All were purified by flash chromatography and stored as hydrochloride salts. WAY 100634 (precursor to WAY 100635 and CPC 222) was prepared by literature methods (Pike et al., 1994).

4-(2-Aminoethyl)-1-(2-methoxyphenyl)piperazine (precursor to CPA and BPA) was prepared by reaction of (2-methoxyphenyl)piperazine with bromoacetonitrile followed by reduction with lithium aluminum hydride (Perrone et al., 1995). Radiolabeling of CPA, BPA, and CPC 222 with ^{11}C was carried out by reaction of the appropriate hydroxyphenyl precursor with [^{11}C]iodomethane either in dimethylformamide solution or by solid-state methods as described for [^{11}C]WAY 100635 (Wilson et al., 1996).

Human Liver Metabolism Modeling

Microsomes and cytosol were prepared from renal transplant donor livers stored in a human liver bank maintained at the University of Toronto (Campbell et al., 1987). Incubations were carried out as described previously (Inaba et al., 1988). Briefly, 10 μl of an aqueous solution of ligand (final concentration 10–100 μM) was added to an appropriate volume of 0.1 M phosphate buffer (pH 7.4) at 0°C and (for microsomal protein) 50 μl of cofactor solution added (10 mM NADPH in 50 mM MgCl$_2$). The liver protein solution of a predetermined volume was then added, giving a final volume of 500 μl, and the incubation was initiated by warming to 37°C with shaking. Reactions were terminated by the addition of acetonitrile (500 μl) and centrifugation (1000 g, 2 min) to deposit protein. Samples were stored at −10°C until they were analyzed. Analyses were performed by reverse-phase HPLC using a UV detector (254 nm) and/or a Berthold LB

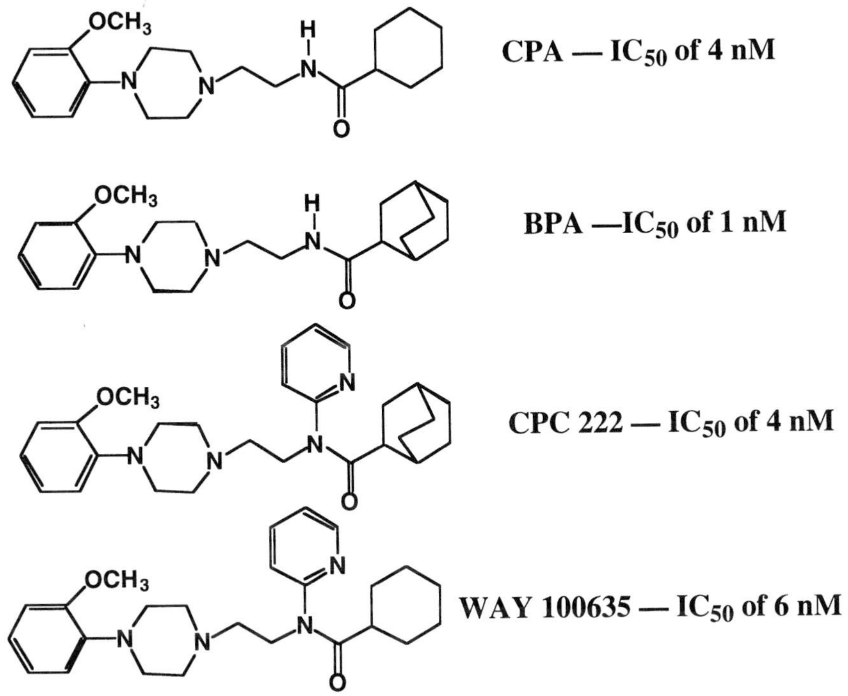

FIGURE 2 WAY 100635 and analogs: Structures and binding affinities for the 5-HT$_{1A}$ receptor.

507A radiation detector. Assays were varied with regard to the concentrations of liver protein and substrate and duration of incubation. Control studies, without protein, were done to confirm that no spontaneous decomposition took place under the conditions of the assays. Identification of the putative metabolites was achieved by coinjection with authentic material using four different HPLC columns (Alltech C$_8$ Econosil, Waters C$_{18}$ Novapak, Waters C$_{18}$ μBondapak, and Alltech C$_{18}$ Econosil).

III. RESULTS AND DISCUSSION

The liver model was first tested using WAY 100635 as substrate. Incubation of the drug with increasing concentrations of human liver cytosol protein produced a corresponding increase in the amount of WAY 100634 formed. Generation of the amide hydrolysis product was quantitative with no other metabolic products being detected by the HPLC analysis system. Results from two different liver donors (K14 and K19) were in close accord (see Fig. 1). Under similar cytosolic incubation conditions, none of the three new candidate ligands, CPA, BPA, or CPC 222, were metabolized, even when WAY 100635 was transformed completely to WAY 100634 (Fig. 3).

In microsomal media (with NADPH cofactor), WAY 100635 was again rapidly metabolized. In addition to WAY 100634, other more polar metabolic products were observed by HPLC analysis. The three test compounds were also metabolized in human liver microsomal media, albeit somewhat more slowly than WAY 100635 (Fig. 4). However, no product of simple amide hydrolysis was observed by HPLC analysis, i.e., no WAY 100634 was detected as a metabolic product from CPC 222 and no 4-(2-aminoethyl)-1-(2-methoxyphenyl)piperazine was generated from CPA or BPA. All metabolic products observed by reverse-phase HPLC eluted before the product of amide cleavage, i.e., they were more polar than WAY 100634 or 4-(2-aminoethyl)-1-(2-methoxyphenyl)piperazine. One of the test compounds (CPC 222) was labeled with ^{11}C and incubated with microsomes in the same manner as unlabeled CPC 222. The pattern of the radiochromatogram closely matched that found using "cold" CPC 222.

These results suggested that none of the three candidate ligands would suffer from the same metabolic problems as WAY 100635 and indicated that further studies were merited. One of them (CPC 222) is currently undergoing human PET trials after displaying

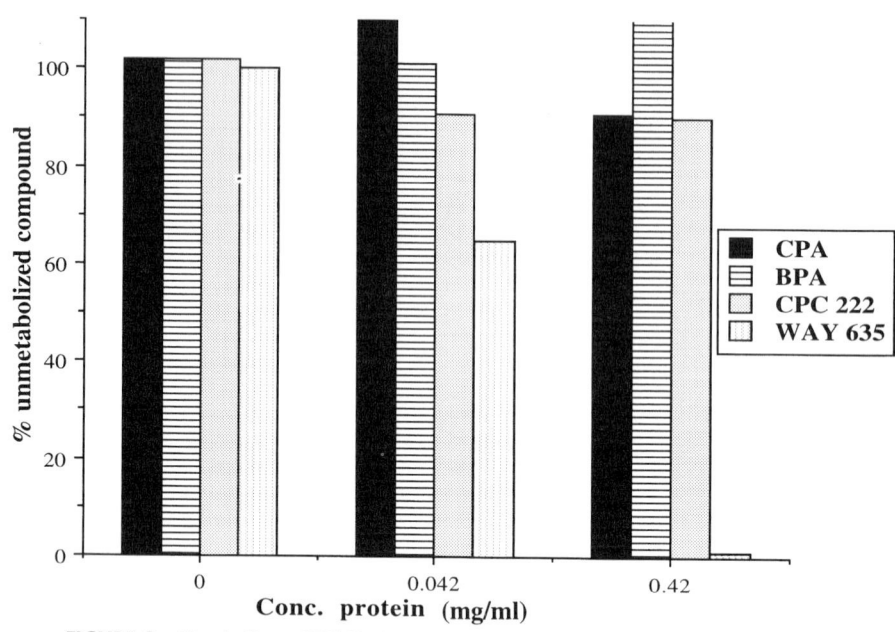

FIGURE 3   Metabolism of WAY 100635 and analogs in human liver cytosolic media.

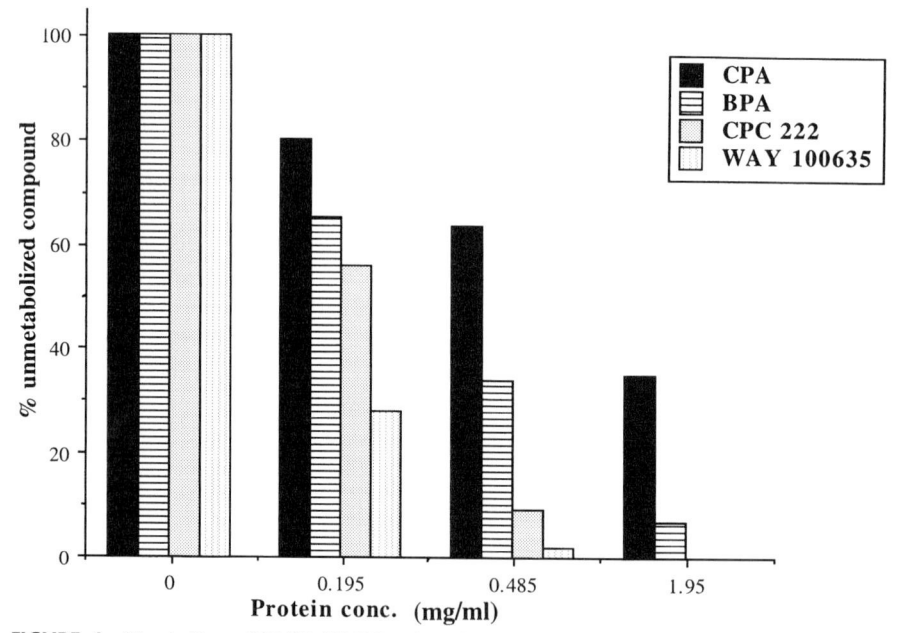

FIGURE 4   Metabolism of WAY 100635 and analogs by human liver microsomes (+NADPH).

promising biodistribution and pharmacological results in rats.

In addition to providing early information on potential metabolism problems in humans, screening candidate PET ligands through a liver metabolism model has other potential benefits. Comparing "hot" and "cold" results also provides useful information on the fate of the label, for example. The information obtained in determining the best HPLC conditions for metabolite analysis can be transferred to the analysis of human plasma for metabolites upon initiation of human trials. In the present case, only minor adjustments from the liver assay were required in designing the HPLC assay of [^{11}C]CPC 222 and metabolites in human plasma. Other potential applications include obtaining sufficient quantities of metabolites for characterization by spectroscopic methods. Finally, the technique can also provide a source of labeled metabolites that can be administered to small animals to test for blood–brain barrier penetration and tissue localization. This would provide indirect evidence on the confounding ability of metabolites of a given potential radioligand in human studies.

References

Campbell, M. E., Grant, D. G., Inaba, T., and Kalow, W. (1987). Biotransformation of caffeine, paraxanthine, theophylline and theobromine by polycyclic aromatic hydrocarbon inducible cytochrome P-450 in human liver microsomes. *Drug Metab. Dispos.* **15**: 237–249.

Fletcher, A., Bill, D. J., Cliffe, I. A., Forster, E. A., and Reilly, Y. (1994). A pharmacological profile of WAY 100635, a potent and highly selective 5-HT$_{1A}$ receptor antagonist. *Br. J. Pharmacol.* **112**: 91P.

Houle, S. H., DaSilva, J. D., and Wilson, A. A. (1996). Evaluation of brain uptake of the metabolite [*O-methyl*-C-11]WAY 100634 in humans during PET scanning with the 5-HT$_{1A}$ radioligand [*O-methyl*-C-11]WAY 100635. *Can. Assoc. Nucl. Med. meet.*, Quebec City.

Inaba, T., Tait, A., Nakano, M., Mahon, W. A., and Kalow, W. (1988). Metabolism of diazepam *in vitro* by human liver: Independent variability of N-demethylation and C$_3$-hydroxylation. *Drug Metab. Dispos.* **16**: 605–608.

Mathis, C. A., Simpson, N. R., Mahmood, K., Kinahan, P. E., and Mintun, M. A. (1994). [C-11]WAY 100635: A radioligand for imaging 5-HT(1A) receptors with positron emission tomography. *Life Sci.* **55**: 403–407.

Miners, J. O., Veronese, M. E., and Birkett, D. J. (1994). *In vitro* approaches for the prediction of human drug metabolism. *Annu. Rep. Med. Chem.* **29**: 307–316.

Osman, S., Lundkvist, C., Pike, V. W., Halldin, C., McCarron, J. A., Swahn, C. G., Ginovart, N., Luthra, S. K., Bench, C. J., Grasby, P. M., Wikstrom, H., Barf, T., Cliffe, I. A., Fletcher, A., and Farde, L. (1996). Characterization of the radioactive metabolites of the 5-HT$_{1A}$ receptor radioligand, [*O-methyl*-C-11]WAY-100635, in monkey and human plasma by HPLC: Comparison of the behaviour of an identified radioactive metabolite with parent radioligand in monkey using PET. *Nucl. Med. Biol.* **23**: 627–634.

Perrone, R., Berardi, F., Colabufo, N. A., Leopoldo, M., Tortorella, V., Fiorentini, F., Olgiati, V., Ghiglieri, A., and Govoni, S. (1995). High affinity and selectivity on 5-HT(1A) receptor of 1-aryl-4-[1-tetralin)alkyl]piperazines. *J. Med. Chem.* **38**: 942–949.

Pike, V. W., McCarron, J. A., Hume, S. P., Ashworth, S., Opacka-Juffry, J., Osman, S., Lammertsma, A. A., Poole, K. G., Fletcher, A., White, A.C., and Cliffe, I. A. (1994). Pre-clinical development of a radioligand for studies of central 5-HT$_{1A}$ receptors *in vivo*—[^{11}C]WAY-100635. *Med. Chem. Res.* **5**: 208–227.

Wilson, A. A., DaSilva, J. N., and Houle, S. (1996). Solid-phase radiosynthesis of [^{11}C]WAY 100635. *J. Labelled Compd. Radiopharm.* **38**: 149–154.

Zhuang, Z.-P., Kung, M.-P., Chumpradit, S., Mu, M., and Kung, H. F. (1994). Derivatives of 4-(2′-Methoxyphenyl)-1-[2′-(N-2″-pyridinyl-p-iodobenzamido)ethylpiperazine (p-MPPI) as 5-HT$_{1A}$ ligands. *J. Med. Chem.* **37**: 4572–4575.

CHAPTER 43

HPLC Analysis of the Metabolism of 6-[^{18}F]Fluoro-L-DOPA in the Brain of Neonatal Pigs

G. VORWIEGER, P. BRUST, R. BERGMANN, R. BAUER,* B. WALTER,* F. FÜCHTNER,
J. STEINBACH, and B. JOHANNSEN

Institut für Bioanorganische und Radiopharmazeutische Chemie, Forschungszentrum Rossendorf
PF 51 01 19, D-01314 Dresden, Germany
* Institut für Pathophysiologie
Friedrich-Schiller-Universität
07740 Jena, Germany

Little information on dopamine metabolism in the immature brain is presently available. Therefore, positron emission tomography studies with newborn piglets are currently in progress. For validation, HPLC analysis of brain samples at different time points after injection of 6-[^{18}F]fluoro-L-DOPA (FDOPA) was performed. This study presents an optimized analysis of FDOPA metabolites, characterized by good separation performance, short processing, minimal sample dilution, and recovery correction. In all brain regions studied (caudate, putamen, frontal cortex, midbrain, cerebellum), six distinct metabolite peaks of FDOPA were reproducibly found, which were identified on the basis of their retention times with respect to the peaks of unlabeled standards. At early time points of tracer circulation (< 10 min) the maximum peaks in the various regions were observed for fluorodopamine and for FDOPAC. Later (at 50 min) 6-fluorohomovanillic acid and 3-O-methyl-FDOPA contributed most to the total activity measured. The initial rate of FDOPA metabolism was highest in striatum and lowest in frontal cortex and cerebellum. Preliminary results differ from those obtained in adult rats and monkeys, suggesting developmental differences. The knowledge of the patterns of the aromatic amino acid decarboxylase (AADC) activity in different brain regions of newborn pigs may have implications for the understanding of the neurodevelopmental effects of perinatal oxygen deprivation.

I. INTRODUCTION

Dopamine (DA) is a major neurotransmitter in the mammalian central nervous system that has been implicated in the etiology of neurological and psychiatric diseases (Agid, 1991; Davis et al., 1991). Disturbances in its metabolism during early postnatal development are expected to contribute to the etiology of epilepsy (Bergamasco et al., 1984) and other neuropsychiatric disorders (Volpe, 1987; Hill, 1991). The synthesis of this neurotransmitter depends on the activity of the enzymes tyrosine hydroxylase (EC 1.14.16.2) and aromatic L-amino acid decarboxylase (AADC; EC 4.1.1.28). Although tyrosine hydroxylase is expected to be the rate-limiting and therefore regulated enzyme (Nagatsu et al., 1964; Udenfriend et al., 1966), increasing experimental evidence has been obtained that AADC is regulated as well (Zhu et al., 1992; Young et al., 1993; Cumming et al., 1994; Cho et al., 1996). Studies on the regulation of dopamine synthesis during early postnatal development are important. Under certain pathophysiological circumstances, such as asphyxia during birth, changes in dopamine metabolism occur (Pastuszko et al., 1993; Huang et al., 1994). Changes in endogenous dopamine, which were found during asphyxia, may be involved in the regulation of AADC activity (Cho et al., 1996). Positron emission tomography (PET) imaging with 6-[^{18}F]fluoro-L-DOPA (FDOPA) has been used in humans to measure the activity of the AADC (Gjedde *et al.*, 1991; Huang *et al.*, 1991; Melega *et al.*, 1991). However, because data allowing insights in the regulation of DA synthesis during this life period are difficult to obtain in humans, the authors therefore developed a model to measure the metabolism of FDOPA in newborn piglets. This study describes analytical methods for identification of FDOPA metabolites in the pig brain.

II. MATERIALS AND METHODS

A. Animal Preparation

Male newborn piglets (age, 2 to 6 days; weight, 1.8 to 3.2 kg) anesthetized with 0.5% isoflurane (70% N_2O, 30% O_2) were injected with 40 MBq/kg FDOPA (6.3 ± 3.7 GBq/mmol). Blood gases, blood glucose and lactate, blood pressure, electroencephalogram, and electrocardiogram were monitored. After euthanasia at 4, 8, or 50 min postinjection (pi), the brain was rapidly exposed, dissected, and mechanically disintegrated (see later).

B. Synthesis of FDOPA

FDOPA was synthesized according to Namavari *et al.* (1992) in a modified procedure. [^{18}F]F$_2$ produced by the ^{20}Ne(d, α)^{18}F reaction was absorbed in a Vigreux absorption column containing a precursor solution of 60 mg N-formyl-3,4-di-boc-6-(trimethylstannyl)-L-DOPA ethyl ester in 15 ml trichlorofluoromethane. Thereafter, 2.5 ml 48% HBr was added, and the temperature was increased to 70°C. After evaporation of the remaining solvent with N_2 carrier gas, the temperature was maintained at 130°C for hydrolysis (5 min). The reaction mixture was then filtered through a 0.22-μm filter, cooled down for 2 min, and injected onto the HPLC column (RP-18, LiChrosorb, 7 μm, 250 × 10 mm, guard column: LiChrospher RP-18, 5–20 μm, 30 × 9 mm). FDOPA was eluted with a retention time of about 11 min. The eluent (105.6 mM of CH_3COONa and CH_3COOH, pH 4.7, isotonic, prepared with aqua ad injectabilia) was sterilized by filtration. The total preparation time after end of bombardment was about 50 min; the yield was between 30 and 35% (corrected for decay).

C. Sample Preparation and HPLC Analysis

The whole preanalytic procedure was performed in one single 12-ml polypropylene test tube. Each brain

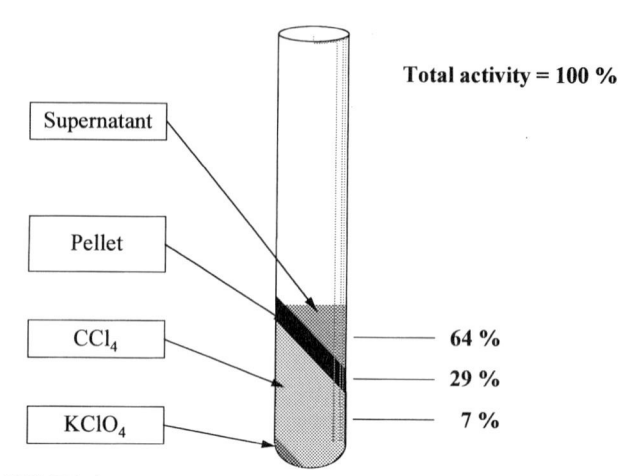

FIGURE 1 Activity distribution among the different phases in the sample preparation tube.

tissue sample was immediately submerged in 2.5 ml of the solvent CCl_4 that had been precooled to −25°C. After difference weighing of the tube in a Dewar container, 0.5 volume parts (referring to the obtained tissue weight) of a precooled solution containing 2.4 M $HClO_4$ and 0.8 M Na_2HPO_4 were added for deproteinization and protection against oxidation of catechols. The phosphate buffering adjusts the pH of the deproteinization solution to 2.1 and the pH of the finally processed sample to 2.6–3.0, thus preventing hydrolysis. Disintegration of the sample was achieved by a 1-min shearing force mixing with a ULTRA TURRAX at 24,000 rpm, whereby the tube underwent slight vertical motions. The resulting emulsion remained stable for at least 1 hr. At this stage the total tissue radioactivity was measured in a well counter (COBRA II, Canberra-Packard, energy window 450–1500 keV, counting time 0.5 min). An aliquot of 2 M KH_2PO_4 solution (referring to the deproteination solution) containing 4.5 mM octyl sulfate was then added. After vigorous shaking the mixture was separated by centrifugation (7 min at 16,000 g and 0°C) into two solid (pellet and potassium perchlorate) and two liquid phases (CCl_4 and a clear supernatant) (Fig. 1). Because the solid protein phase forms a barrier against the CCl_4, the concentrated tissue extract could be removed without loss.

Analysis was carried out by gradient HPLC in the following configuration: quarternary gradient pump, autosampler (0.5 ml sample loop; injection volume 360 μl), UV detector (λ = 280 nm), all parts of the Hewlet–Packard 1050 system, and a flow scintillation analyzer (150 TR, Canberra Packard) with a PET flow cell (100 μl volume; energy window: 400–1500 keV). The analytes were separated on a C18 reverse-phase column (250 × 4 mm, LiChrosorb, 7 μm) fitted with a guard column (4 × 4 mm, LiChrospher 100 RP-18, 5 μm) at a temperature of 30°C. A binary gradient was chosen at a flow rate of 1.5 ml/min (start: 0% mobile phase B; 4 min: 25% B; 10 min: 80% B; 10.1 min: 100% B; 11.2 min: 100% B; 12.8 min: 0% B; 16 min: 0% B; total method time: 16 min). Mobile phase A consisted of 70 mM KH$_2$PO$_4$, 1.5 mM sodium octyl sulfate, and 0.1 mM EDTA, adjusted to pH 3.4 with H$_3$PO$_4$. Mobile phase B was prepared by adding two parts (v/v) acetonitrile (MeCN) to one part "A." The system was reequilibrated every 30 samples with water and MeCN. Concomitantly, the guard columns were renewed and the flow direction reversed. Thus, the system proved to remain stable for at least 500 samples.

The peaks of the radiochromatograms were identified on the basis of their retention times with respect to the peaks of unlabeled standards, together with the plausibility of the succession of metabolite kinetics. The recovery of the preanalytic procedure was calculated by the following procedure: from samples with a roughly balanced content of all the six metabolites (i.e., 8-min samples), the activity distribution among the three main phases was determined and related to the total activity measured. For the specific recovery calculation of each single metabolite, the pellets of the original procedure were treated like a common brain sample. Then, the absolute activities of each single peak were calculated from the sample activity injected into the HPLC and the area percentage output of the software, both from the original sample and from the pellet reextract. The specific recovery, then, is the ratio of the absolute peak activity in the original supernatant to the sum of reextract and original peak. Loss by partition in CCl$_4$ was considered to be equal for all metabolites.

The recovery of the HPLC system was estimated by a set of three experiments: (A) comparison of the cumulative activity of all eluted fractions of one run with the injected activity, (B) collection of the waste liquid during the system reequilibrating rinses with water and MeCN, respectively (the rinse volume in both cases was > 50 column volumes; the collection was continued until the measured activity of the fractions reached background level again), and (C) before rinsing: measuring the activity of the whole guard column. The activities obtained in experiments (B) and (C) were compared to the sum of all samples' activity from one animal (about 10 samples) applied to the HPLC system and given in percentage.

III. RESULTS AND DISCUSSION

A. Chromatography and Validation

This study presents an optimized analysis of FDOPA metabolites in the brain of newborn piglets that is characterized by good separation performance, short processing, and high sensitivity. The first feature contributing to the high sensitivity is the use of CCl$_4$ as disintegration medium. The second feature is the application of the solubility properties of alkali perchlorates. In addition to efficient cooling, the CCl$_4$ allows the mechanical homogenization of the minimally diluted samples. The further addition of concentrated KH$_2$PO$_4$ solution results in the precipitation of most perchlorate. Thus the sample matrix is adapted to the mobile-phase composition of the following HPLC without resolubilization of the protein.

The radioactivity distribution among the three main phases in the preparation tube (Fig. 1) was estimated in samples and showed a roughly balanced content of all the six metabolites (8 min pi). The loss of radioactivity due to the extraction procedure, which was mainly caused by adsorption at the precipitate, was about 36% and similar to values described previously (Firnau et al., 1988; Cumming et al., 1988).

In a similar experiment, the pellets of five samples were pooled and handled like brain tissue. The reextraction of this pellet sample was repeated twice. The distribution ratio of activity between the pellet and the supernatant remained stable (pellet: 27 and 30% of total activity). The activity found in the solvent of the two reextract samples dropped down to 1–2%. For the calculation of the specific recovery of each of the six metabolites (see Section II) the chromatograms of the brain sample and of the reextracted pellet were compared. The values obtained were 76% for FDOPA, 77% for 3-O-methyl-FDOPA (3-O-M-FDOPA), 59% for L-3,4-dihydroxy-6-[^{18}F]fluorophenylacetic acid (FDOPAC), 58% for 6-[^{18}F]fluorohomovanillic acid (FHVA), 66% for [^{18}F]fluorodopamine (FDA), and 72% for 3-methoxy-6-[^{18}F]fluorotyramine (FMT). Obviously, the metabolites do not partition equally between the aqueous phase and the pellet as has been shown by Firnau et al. (1988) for plasma samples using another method. The differential recovery of the various metabolites has been mostly ignored or not mentioned by other authors in later studies on FDOPA metabolism. In fact, it leads to the need for correction of the metabolite ratios for procedures based on precipitation steps if they are used for the calculation of rate constants of FDOPA metabolism.

Recovery and memory effects were also checked for the chromatographic system. The cumulative activity eluted was compared to the total injected activity, and good agreement was observed. The relative eluted activity was $104 \pm 1\%$ ($n = 3$). Correspondingly, nearly no activity (0.17%) was found on the precolumn (measured as a whole before washing). The cumulative activity measured in the washing fluid after reequilibration (more than 50 column volumes) was 1.6% in water and 0.4% in MeCN, respectively.

The third important feature that contributes to the high sensitivity of the method is the combined use of the ion pair reagent octyl sulfate and gradient chromatography. Figure 2 shows a chromatogram of a brain sample overlaid with the gradient used. For calculation of mobile-phase composition, which is responsible for the respective metabolite elution, a correction has been made for the dwell and void volume (sum: 4.28 ml), causing a shift of the overlaid gradient time of 2.85 min referred to the programmed method gradient. A complete baseline separation of the six metabolite peaks of FDOPA in brain was only achieved by the addition of octyl sulfate to the mobile phase. FDOPAC, and in particular FDA and FMT, showed a considerably increased retention on the column and concomitantly an improved peak symmetry. The same kind of HPLC analysis can be applied to plasma samples (data not shown). Among the six metabolites identified in brain, four were also found in high amounts in plasma. However, no FDA and only low amounts of FMT were detected. Apart from this, one major peak and additional small peaks were found in plasma. They remain to be identified.

Although such an application for FDOPA analysis has been described previously by Cumming *et al.* (1987), the combined use of ion pair and gradient chromatography is generally not recommended (Snyder *et al.*, 1988) because of the long reequilibration needed. In the study of Cumming *et al.* (1987), the time just for the gradient ramp was about 40 min. Other authors have used ion pairs without the gradient, resulting in similar run times (Melega *et al.*, 1990, 1991). Compared to these previous results, the present method allows a considerably higher sample throughput.

Table 1 shows the actual mobile-phase composition at the arrival of the elution volume together with the retention times of the FDOPA metabolites compared to those of nonfluorinated standards. An increase in the retention time due to fluorination occurs for all metabolites, as has been described previously (Cumming *et al.*, 1987). The nonfluorinated standards were used to identify the fluorinated metabolites. In separate experiments the standards were added to the supernatants after the extraction procedure, and the

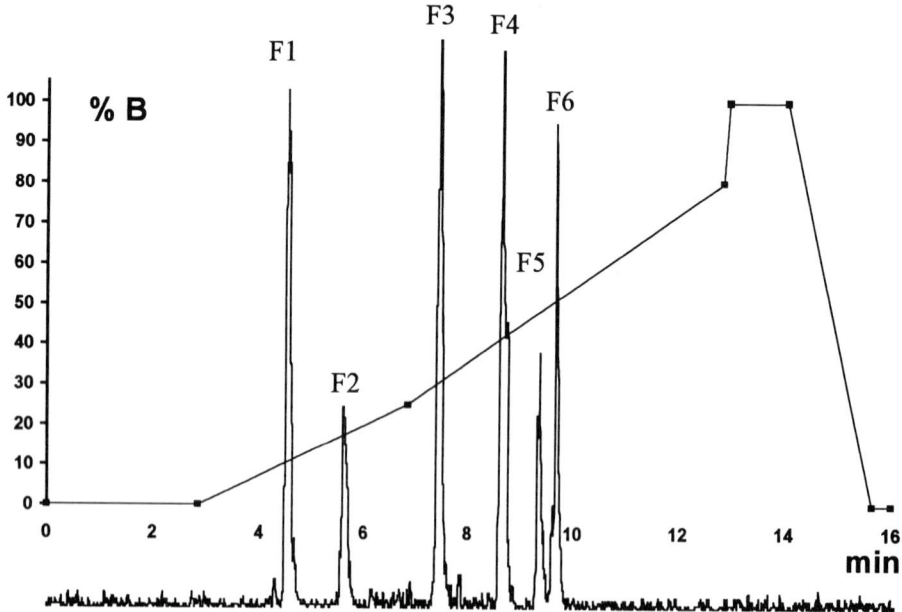

FIGURE 2 A typical trace of the extracted brain metabolites together with the applied binary gradient. The gradient shown (square symbols) is corrected by the dwell volume of the chromatographic system and by the column void volume. % B; percentage of mobile-phase "B" in the binary gradient (for composition, see Section II, C). The ordinate of the trace (cpm equivalents) is not shown. F1, FDOPA; F2, 3-*O*-M-FDOPA; F3, FDOPAC; F4, FHVA; F5, FDA; F6, FMT.

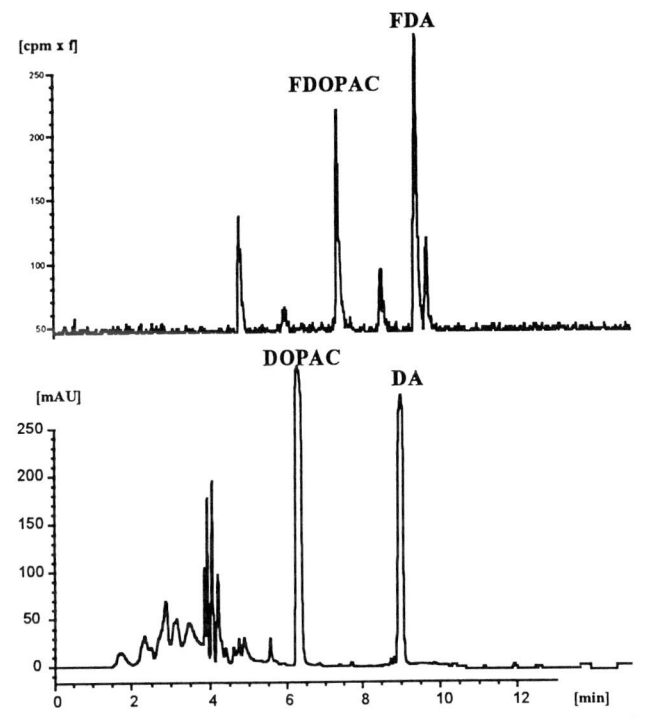

FIGURE 3 Chromatograms of a common brain sample from the UV detector (280 nm, lower trace) and from the radioactivity monitor (upper trace). The deproteination solution was spiked with the standards DOPAC and dopamine.

chromatograms of the UV detector and the radioactivity detector were compared. Figure 3 shows an example of such an experiment for DOPAC and dopamine standards. Because there is a negligible void volume between the two detectors, no correction for the time difference was necessary for the flow rate used (1.5 ml/min).

B. FDOPA Metabolism in Newborn Pigs

Five different brain regions were prepared in the described manner. The caudate nucleus and the putamen were chosen as the main targets of the nigrostriatal pathway where the highest rate of dopamine synthesis may be expected. Furthermore, the midbrain containing dopamine cell body regions, as well as the frontal cortex and the cerebellum as regions with low dopaminergic innervation, were included. In all regions, six distinct peaks were reproducibly found. The corresponding percentage of total ^{18}F activity is shown in Table 2. The very rapid FDOPA metabolism in brain tissue (FHVA, the product of as many as four enzyme reactions, reaches 15% of total activity even at 4 min pi), as also found in rats by Cumming et al. (1987, 1995), is confirmed for the pig at a rather early ontogenetic stage. The results in this study differ from studies performed in adult rats where negligible amounts of the decarboxylated metabolites were seen in the cerebellum (Cumming et al., 1987, 1988). In another study with [^{3}H]DOPA as the tracer, AADC activity was found to be diminished by 95% in the cortex compared to the striatum (Cumming et al., 1995). Interestingly, none of the detected peaks corresponds to sulfoconjugated metabolites, as was also found in rat studies with FDOPA (Cumming et al., 1988, 1994), whereas unfluorinated conjugated DOPAC and HVA have been found in the rat brain (Dedek et al., 1979; Freeman et al., 1993). It is known that these sulfoconjugated metabolites of FDOPA are present in plasma (Cumming et al., 1987; Melega et al., 1991). In human plasma, but not in rats, the sulfoconjugated form is the dominant pool of many catechols and their metabolites, including DOPA (Weicker, 1988), the mesenteric organs being the main sulfoconjugating source (Eisenhofer et al., 1996). Thus the occurrence of sulfoconjugated FDOPA metabolites in plasma does not allow one to draw conclusions regarding brain sulfoconjugation in pigs. Data from this study suggest that the rate constant of elimination of these compounds from the brain is much higher than the rate constant of their formation in the brain. Alternatively, these compounds may undergo rapid hydrolysis during the extraction procedure and therefore remain undetected.

TABLE 1 Mobile-Phase Composition at the Arrival of the Elution Volume, Together with Retention Times of FDOPA Metabolites Compared to Those of Nonfluorinated Standards

Substance	FDOPA	3-O-M-FDOPA	FDOPAC	FHVA	FDA	FMT
Retention time (min)	4.48	5.68	7.35	8.78	9.40	9.60
Elution at [% B]	10.20	17.71	29.58	42.40	48.38	50.21
Substance	DOPA	3-O-M-DOPA	DOPAC	HVA	Dopamine	MT
Retention time (min)	4.16	5.45	6.27	7.98	8.99	9.21
Elution at [% B]	8.19	16.25	21.37	35.35	44.71	46.54

TABLE 2 Relative Amounts (Means ± SD % Total Activity, $n = 3$) of FDOPA and Its Metabolites in Pig Brain at 4 min pi

Brain region	FDOPA	3-O-M-FDOPA	FDOPAC	FHVA	FDA	FMT
Cerebellum	28.2 ± 2.9	8.4 ± 0.8	19.7 ± 1.5	14.6 ± 1.6	16.4 ± 7.1	12.0 ± 1.6
Midbrain	10.8 ± 1.6	10.1 ± 2.4	29.9 ± 7.2	18.2 ± 4.2	18.5 ± 6.1	12.4 ± 2.7
Frontal cortex	26.4 ± 4.8	7.1 ± 0.1	28.4 ± 5.3	12.7 ± 1.6	12.3 ± 5.0	13.2 ± 2.9
Caudate nucleus	17.7 ± 0.9	7.8 ± 1.0	31.1 ± 4.0	13.3 ± 2.9	16.8 ± 5.6	13.2 ± 1.1
Putamen	12.5 ± 1.8	7.3 ± 1.8	28.6 ± 5.9	16.1 ± 2.2	21.3 ± 5.1	14.2 ± 4.0

In accordance with the functional anatomy the five brain regions studied may be classified into three different metabolic types (Fig. 4). In regions belonging to the nigrostriatal dopaminergic system (caudate nucleus, putamen) a rapid turnover of FDOPA occurs. An enrichment of FDA was distinctly detected at 50 min pi combined with an increase of the total [^{18}F] concentration. Peak concentrations were observed for FDA between 4 and 8 min, for FDOPAC at 8 min, and for FVHA and 3-O-M-FDOPA at 50 min pi. A second metabolic type is found in the cerebellum and frontal cortex, often used as negative control regions during FDOPA PET studies. These regions with the lowest level of dopaminergic innervation contain just about 7% of the AADC activity of the striatum in humans (Lloyd et al., 1975). In the occipital cortex and cerebellum of rats the formation of dopamine, DOPAC, and HVA from exogenous L-DOPA was 70–80% less than in striatum (Brannan et al., 1996). A slow turnover of FDOPA is found in these regions. Nearly no FDA is left at 50 min. In contrast to the caudate/putamen, no vesicular storage of dopamine is expected in these regions. The third metabolic type is represented by the midbrain, which contains the neurons of the substantia nigra. As reported by others, it contains similar baseline levels of AADC as the striatum, whereas the increase of enzyme activity elicited by phorbol ester administration is more pronounced in striatum than in midbrain (Young et al.. 1994). Here, the FDA content at 50 min has already dropped down to trace levels

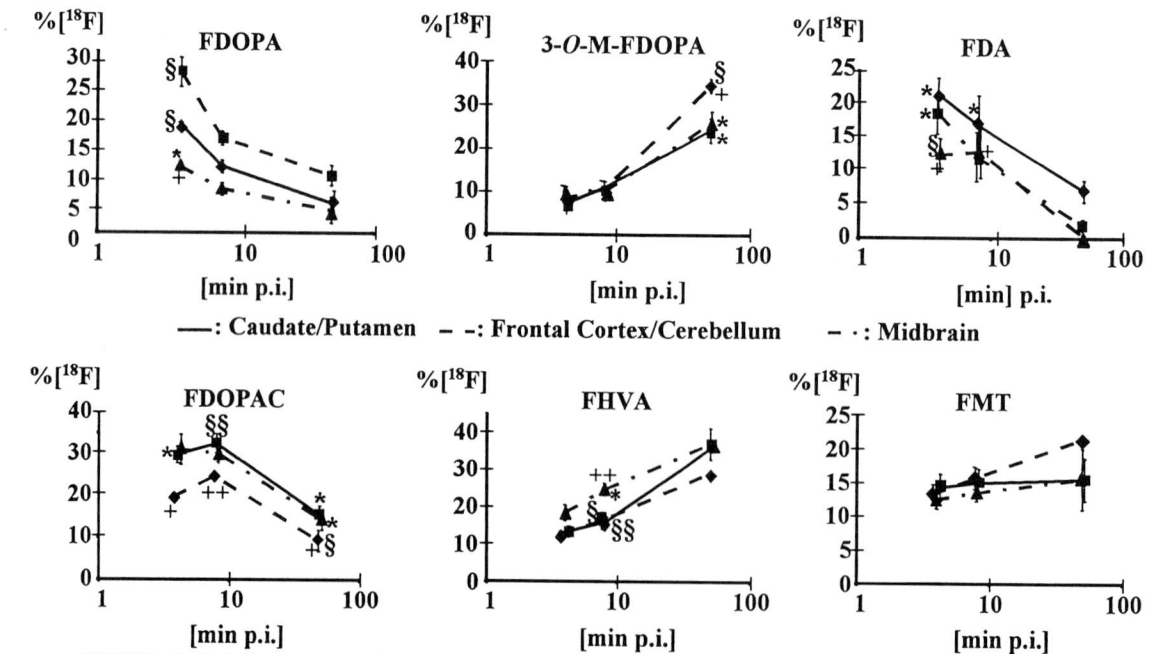

FIGURE 4 Relative amounts (means ± SD % total activity, $n = 3$) of FDOPA, 3-O-M-FDOPA, FDA, FDOPAC, FHVA, and FMT in pig brain as a function of time in three groups of brain regions. *$p < 0.05$ toward frontal cortex/cerebellum; +$p < 0.05$, ++$p < 0.01$ toward caudate/putamen; §$p < 0.05$, §§$p < 0.01$ toward midbrain.

again. However, a rapid turnover of FDOPA occurs. Generally, differences between the various brain regions are much smaller than expected from studies performed in adult animals, which excludes the possibility of using a reference region for compartment analysis of FDOPA PET data as is possible in adult humans (Huang et al., 1991; Hoshi et al., 1993). Obviously, a high "nonspecific" AADC activity most likely not related to dopaminergic neurons is present in the cortex and the cerebellum. The biological significance of such a suggested enzyme activity is unclear. It may be related to the synthesis of trace amines that may serve as neurotransmitters as well (Young et al., 1994). Also, DOPA has been suggested as a neurotransmitter. In this case, AADC may serve as the enzyme limiting the action of it.

It has been described that endothelial cells also contain AADC. Part of the metabolites measured in this study may, therefore, be due to brain endothelium. However, the specific activity of the enzyme in rat brain tissue and brain microvessels was similar (about 5 nmol/mg protein/hr; Hardebo et al., 1980). In microvessels from pigs and humans, similar enzyme activities were measured as in rats. Because of the low mass contribution of brain endothelium to the total brain tissue mass, the authors therefore conclude that the influence of the endothelium on the metabolite analysis is negligible.

Another explanation for the presence of decarboxylated metabolites in the cerebellum could be an immature blood–brain barrier (BBB) that allows blood metabolites to pass into the brain. This may be excluded, however, because of the faster occurrence of all six metabolites in all brain regions compared to plasma regardless of metabolite lipophilicity and charge. In addition, in separate studies using $^{99m}TcO_4$ as the tracer, the authors could demonstrate the tightness of the BBB in newborn piglets (Bergmann, R. and Brust, P.; unpublished data). Therefore, the authors conclude that a considerable amount of AADC activity is present in newborn pigs, not only in the regions of nigrostriatal system, but also in regions with an expected low dopaminergic innervation—a fact that has to be considered in the analysis of FDOPA PET data obtained in newborn piglets.

Acknowledgments

We are grateful to Dr. P. Cumming for helpful discussions during the preparation of the manuscript. The excellent technical assistance of Mrs. R. Scholz, Mrs. R. Lücke, and Mrs. R. Herrlich is gratefully acknowledged. This work was supported in part by the Saxon ministry of Science and Art, 7541.82-FZR/309, and by the Thuringian State Ministry of Science, Research and Arts, Grant 3/95-13.

References

Agid, Y. (1991). Parkinson's disease: Pathophysiology. *Lancet* **337**: 1321–1324.

Bergamasco, B., Benna, P., Ferrero, P., and Gavinelli, R. (1984). Neonatal hypoxia and epileptic risk: A prospective study. *Epilepsia* **25**: 131–146.

Brannan, T., Prikhojan, A., and Yahr, M. D. (1996). An in vivo comparison of the capacity of striatal versus extrastriatal brain regions to form dopamine from exogenously administered L-DOPA. *J. Neural Transm.* **103**: 1287–1294

Cho, S., Duchemin, A. M., Neff, N. H., and Hadjiconstantinou, M. (1996). Modulation of tyrosine hydroxylase and aromatic l-amino acid decarboxylase after inhibiting monoamine oxidase-A. *Eur. J. Pharmacol.* **314**: 51–59.

Cumming, P., Boyes, B. E., Martin, W. R. W., Adam, M., Grierson, J., Ruth, T., and McGeer, E. G. (1987). The metabolism of [18F]6-fluoro-L-3,4-dihydroxyphenylalanine in the hooded rat. *J. Neurochem.* **48**: 601–608.

Cumming, P., Häusser, M., Martin, W. W. R., Grierson, J., Adam, M. J., Ruth, T. J., and McGeer, E. G. (1988). Kinetics of in vitro decarboxylation and the in vivo metabolism of 2-[18F]-and 6-[18F]fluorodopa in the hooded rat. *Biochem. Pharmacol.* **37**: 247–250.

Cumming, P., Kuwabara, H., and Gjedde, A. (1994). A kinetic analysis of 6-[18F]fluoro-L-dihydroxyphenylalanine metabolism in the rat. *J. Neurochem.* **63**: 1675–1682.

Cumming, P., Kuwabara, H., Ase, A., and Gjedde, A. (1995). Regulation of DOPA decarboxylase activity in brain of living rat. *J. Neurochem.* **65**: 1381–1390.

Davis, K. L., Kahn, R. S., Ko, G., and Davidson, M. (1991). Dopamine in schizophrenia: a review and reconceptualization. *Amer. J. Psychiatry* **148**: 1474–1486

Dedek, J., Baumes, R., Tien-Duc, N., Gomeni, R., and Korf, J. (1979). Turnover of free and conjugated (sulphonyloxy) dihydroxyphenylacetic acid and homovanillic acid in rat striatum. *J. Neurochem.* **33**: 687–695.

Eisenhofer, G., Aneman, A., Hooper, D., Rundqvist, B., and Friberg, P. (1996). Mesenteric organ production, hepatic metabolism, and renal elimination of norepinephrine and its metabolites in humans. *J. Neurochem.* **66**: 1565–1573.

Firnau, G., Sood, S., Chirakal, R., Nahmias, C., and Garnett, E. S. (1988). Metabolites of 6-[18F]fluoro-L-dopa in human blood. *J. Nucl. Med.* **29**: 363–369.

Freeman, K., Lin, P., Lin, L., and Blank, C. L. (1993). Monoamines and metabolites in the brain, In "High Performance Liquid Chromatography in Neuroscience Research" (R. B. Holman, A. J. Cross, and M. H. Joseph, eds.), pp. 27–58. Wiley, Chichester.

Gjedde, A., Reith, J., Dyve, S., Léger, G., Guttman, M., Diksic, M., Evans, A. C., and Kuwabara, H. (1991). DOPA decarboxylase activity of the living human brain. *Proc. Natl. Acad. Sci. U.S.A.* **88**: 2721–2725.

Hardebo, J. E., Emson, P. C., Falck, B., Owman, C., and Rosengren, E. (1980). Enzymes related to monoamine transmitter metabolism in brain microvessels. *J. Neurochem.* **35**: 1388–1393.

Hill, A. (1991). Current concepts of hypoxic-ischemic cerebral injury in the term newborn. *Pediatr. Neurol.* **7**: 317–325.

Hoshi, H., Kuwabara, H., Léger, G., Cumming, P., Guttman, M., and Gjedde, A. (1993). 6-[18F]fluoroL-DOPA metabolism in living human brain: A comparison of six analytical methods. *J. Cereb. Blood Flow Metab.* **13**: 57–69.

Huang, C. C., Lajevardi, N. S., Tammela, O., Pastuszko, A., Delivoria-Papadopoulos, M., and Wilson, D. F. (1994). Relationship of extracellular dopamine in striatum of newborn piglets to cortical oxygen pressure. *Neurochem. Res.* **19**: 649–655.

Huang, S.-C., Yu, D.-C., Barrio, J. R., Grafton, S., Melega, W. P., Hoffman, J. M., Satyamurthy, N., Mazziotta, J. C., and Phelps, M. E. (1991). Kinetics and modeling of L-6-[18F]fluoro-DOPA in human-positron emission tomographic studies. *J. Cereb. Blood Flow Metab.* **11**: 898–913.

Lloyd, K. G., Davidson, L., and Hornykiewicz, O. (1975). The neurochemistry of Parkinson's disease: Effect of L-dopa therapy. *J. Pharmacol. Exp. Ther.* **195**: 453–464.

Melega, W. P., Luxen, A., Perlmutter, M. M., Nissenson, C. H. K., Phelps, M. E., and Barrio, J. R. (1990). Comparative in vivo metabolism of 6-[18F]fluoro-L-DOPA and [3H]L-DOPA in rats. *Biochem. Pharmacol.* **39**: 1853–1860

Melega, W. P., Grafton, S. T., Huang, S.-C., Satyamurthy, N., Phelps, M. E., and Barrio, J. R. (1991). L-6[18F]fluoro-DOPA metabolism in monkeys and humans: Biochemical parameters for the formulation of tracer kinetic models with positron emission tomography. *J. Cereb. Blood Flow Metab.* **11**: 890–897.

Nagatsu, T., Levitt, M., and Udenfriend, S. (1964). Tyrosine hydroxylase—the initial step in norepinephrine biosynthesis. *J. Biol. Chem.* **239**: 2910–2917.

Namavari, M., Bishop, A., Satyamurthy, N., Bida, G., and Barrio, J. R. (1992). Regioselective radiodestannylation with [^{18}F]F$_2$ and [^{18}F]CH$_3$COOF: A high yield synthesis of 6-[^{18}F]fluoro-L-dopa. *Appl. Radiat. Isot.* **43**: 989–996.

Pastuszko, A., Saadat-Lajevardi, N., Chen, J., Tammela, O., Wilson, D. F., and Delivoria-Papadopoulos, M. (1993). Effects of graded levels of tissue oxygen pressure on dopamine metabolism in the striatum of newborn piglets. *J. Neurochem.* **60**: 161–166.

Snyder, L. R., Glajch, J. L., and Kirkland, J. J. (1988). "Practical HPLC Method Development." Wiley, New York.

Udenfriend, S., Zaltzman-Nirenberg, P., Gordon, R., and Spector, S. (1966). Evaluation of the biochemical effects produced in vivo by inhibitors of the three enzymes involved in norepinephrine biosynthesis. *Mol. Pharmacol.* **2**: 95–105.

Volpe, J. J. (1987). "Neurology of the Newborn." Saunders, Philadelphia.

Weicker, H. (1988). Determination of free and sulfoconjugated catecholamines in plasma and urine by high-performance liquid chromatography. *Int. J. Sports Med.* **9**(Suppl.): 68–75.

Young, E. A., Neff, N. H., and Hadjiconstantinou, M. (1993). Evidence for cyclic AMP-mediated increase of aromatic L-amino acid decarboxylase activity in the striatum and midbrain. *J. Neurochem.* **30**: 2331–2333.

Young, E. A., Neff, N. H., and Hadjiconstantinou, M. (1994). Phorbol ester administration transiently increases aromatic L-amino acid decarboxylase activity of the mouse striatum and midbrain. *J. Neurochem.* **63**: 694–697.

Zhu, M. Y., Juorio, A. V., Paterson, I. A., and Boulton, A. A. (1992). Regulation of aromatic L-amino acid decarboxylase by dopamine receptors in the rat brain. *J. Neurochem.* **58**: 637–641.

CHAPTER

44

Characterization of the Radiolabeled Metabolites of [^{18}F]Altanserin: Implications for Kinetic Modeling

BRIAN LOPRESTI,* DANIEL HOLT,* N. SCOTT MASON,* YIYUN HUANG,*
JAMES RUSZKIEWICZ,* JENNIFER PEREVUZNIK,* JULIE PRICE,* GWENN SMITH,*,†
JAMES DAVIS,* and CHESTER MATHIS*

*Departments of * Radiology and † Psychiatry*
University of Pittsburgh School of Medicine
Pittsburgh, Pennsylvania 15213

Altered serotonergic neurotransmission has been implicated in many neuropsychiatric illnesses, and evidence supporting a role for 5-HT$_{2A}$ receptors in these disorders has motivated the development of a 5-HT$_{2A}$-selective radiotracer. [^{18}F]Altanserin has demonstrated many favorable characteristics for in vivo imaging of 5-HT$_{2A}$ receptors, including reversibility, saturability, and high specific binding. However, previous studies indicated that radiolabeled metabolites of [^{18}F]altanserin may cross the blood–brain barrier (BBB). The goal of this work was to identify and characterize the radiolabeled metabolites of [^{18}F]altanserin in rat plasma and verify BBB penetration. Identification of the metabolites of [^{18}F]altanserin was aided by reference to those of ketanserin, a close structural analog. Two metabolites of [^{18}F]altanserin were identified: [^{18}F]altanserinol and [^{18}F]4-(4-fluorobenzoyl)piperidine ([^{18}F]4-FBP). After [^{18}F]altanserin injection, the radioactivity in rat plasma was separated using HPLC methods and was compared to authentic standards of altanserinol and 4-FBP. Radiosynthetic methods were also developed that yielded [^{18}F]altanserinol and [^{18}F]4-FBP, which were injected into rats to study their brain uptake and regional brain distribution. These two putative metabolites were found to cross the BBB and resulted in uniformly distributed brain radioactivity concentrations that were unaffected by pretreatment with 5-HT$_{2A}$ antagonists. Injection of [^{18}F]4-FBP resulted in threefold greater brain radioactivity concentrations than [^{18}F]altanserinol. [^{18}F]4-FBP was further metabolized in rats, and its secondary metabolite also crossed the BBB, but exhibited no specific binding to 5-HT$_{2A}$ receptors. Hence, these metabolites contributed only to uniformly distributed, nonspecific binding throughout the rat brain. The in vivo specific binding of [^{18}F]altanserin in positron emission tomography studies is quantifiable, as the contribution of these metabolites to the signal from the brain can be estimated and corrections performed if necessary.

I. INTRODUCTION

Serotonergic dysfunction has been associated with many neuropsychiatric and neurodegenerative disorders, including major depression, schizophrenia, and Alzheimer's disease (Meltzer, 1991; Kahn et al., 1993; Cross et al., 1984). Serotonergic abnormalities in these disorders have been implicated by postmortem brain studies of 5-HT$_{2A}$ receptors and indirectly by the therapeutic efficacy of drugs (e.g., atypical neuroleptics) that block the 5-HT$_{2A}$ receptor. This evidence strongly supports the need to develop a radioligand for *in vivo* positron emission tomography (PET) imaging studies of 5-HT$_{2A}$ receptors. Several 5-HT$_2$ radioligands have been evaluated and have demonstrated less than ideal properties with respect to *in vivo* nonspecific binding, selectivity, affinity, metabolism, or biodistribution (reviewed by Frost, 1990, and Pike, 1995). The radioligand [^{18}F]altanserin has been shown to exhibit many suitable *in vivo* properties for PET studies (Lemaire et al., 1991).

Altanserin is an antagonist with high affinity for 5-HT$_{2A}$ receptors (K_i = 0.13 nM; Leysen, 1990) and relatively low affinities for other binding sites such as 5-HT$_{2C}$ receptors (K_i = 40 nM), α_1 receptors (K_i =

4.6 nM), and D$_2$ receptors (K_i = 62 nM) (Leysen, 1990; 1991). The brain uptake of [^{18}F]altanserin in rats was rapid, and tissue radioactivity concentration ratios for frontal cortex/cerebellum were 10.3 ± 0.6 at 120 min postinjection (Lemaire et al., 1991). Studies in human subjects demonstrated binding saturability and reversibility and frontal cortex/cerebellum ratios of 2.6 at 60 min postinjection (Sadzot et al., 1995). The specific binding of [^{18}F]altanserin in human brain was shown to be highly correlated with the known distribution of 5-HT$_{2A}$ receptors, and the test–retest variability of [^{18}F]altanserin PET studies in young human controls was within 10% (Smith et al., in review).

Kinetic modeling of [^{18}F]altanserin PET studies in humans demonstrated the need for a three-compartment model to fit cerebellar data, consistent with a slow nonspecific component (Biver et al., 1994). Another analysis method was proposed that accounted for blood–brain barrier (BBB) passage of radiolabeled metabolites and used the total metabolites as a second input function to estimate a distribution volume for a tissue metabolite compartment (Mintun et al., 1996). Both modeling approaches provided acceptable fits to tissue data, but experimental confirmation of the brain uptake of radiolabeled metabolites was needed to confirm the validity of the metabolite model.

The radiolabeled metabolites of [^{18}F]altanserin were expected to be similar to ketanserin, as altanserin is a closely related structural analog (Meuldermans et al., 1984). Consequently, [^{18}F]altanserinol and [^{18}F]4-(4-fluorobenzoyl)piperidine ([^{18}F]4-FBP) were predicted to be the principal metabolites of [^{18}F]altanserin (Fig. 1). Both ^{18}F-labeled and unlabeled (nonradioactive) standards of these two putative metabolites and [^{18}F]labeled were synthesized (Mason et al., 1997). The unlabeled standards were used to confirm the presence of the radiolabeled metabolites in rat plasma following the injection of [^{18}F]altanserin, as well as to determine the extent of the conversion of [^{18}F]altanserinol to [^{18}F]altanserin. In addition, the brain uptake and regional brain distribution of ^{18}F-labeled altanserinol and 4-FBP as well as a third metabolite ([^{18}F]4-FBP-met) were examined in rats.

II. METHODS

A. Radiosynthesis

[^{18}F]Altanserin was prepared according to the method of Lemaire et al. (1991), and [^{18}F]altanserinol and [^{18}F]4-FBP were synthesized according to the methods of Mason et al. (1997). All radiolabeled compounds were greater than 90% chemically pure and greater than 95% radiochemically pure. The specific activities of [^{18}F]altanserin, [^{18}F]altanserinol, and [^{18}F]4-FBP were greater than 1.5 Ci/μmol.

B. Plasma Metabolite Analysis

Plasma metabolites of [^{18}F]altanserin in rats were determined from heparinized arterial whole blood (2 ml) that was centrifuged for 2 min at 13,000 g. The resulting plasma supernatant (500 μl) was added to an

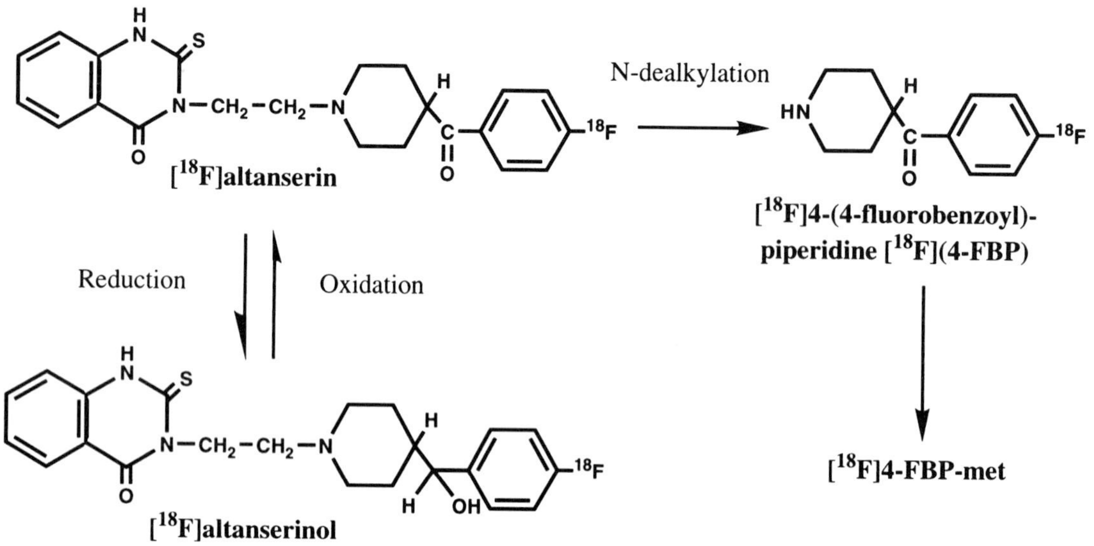

FIGURE 1 Hypothesized metabolism of [^{18}F]altanserin showing three radiolabeled metabolites. [^{18}F]4-FBP-met, whose structure is unknown, was observed in high concentrations in rat plasma following injection of [^{18}F]4-FBP and likely results from further metabolism of [^{18}F]4-FBP.

aqueous solution of acetic acid (10 ml, 50 mM) containing cold altanserin (5.6 μg) and eluted through two C18 Sep-Pak Light cartridges (Waters) connected in series. The Sep-Paks were washed with an aqueous triethylamine (TEA) solution (5 ml, 0.1%) followed by deionized water (5 ml). The Sep-Paks were reversed and eluted with methanol (MeOH, 1.5 ml). The eluent was filtered (Millipore HV syringe filter), and an equal volume of deionized water was added to the filtrate and vortexed (2 min). Aliquots of this sample were analyzed by a step-gradient HPLC system (Waters C18 Symmetry column eluted from 0 to 30 min with 15% MeCN/85% buffer (0.294 M acetic acid/0.0283 M ammonium acetate, pH 5.2) and from 31 to 60 min with 25% MeCN/75% buffer at 2.0 ml/min). This system resolved the metabolites of [^{18}F]altanserin retained by the Sep-Pak and eluted [^{18}F]altanserin in about 50 min. HPLC elution fractions (2 ml) were collected and counted in a gamma counter (Packard 5010). The recoveries of radioactivity from the Sep-Paks and HPLC column were calculated for each plasma sample and were used to determine the relative contributions of [^{18}F]altanserin, [^{18}F]altanserinol, [^{18}F]4-FBP, and [^{18}F]FBP-met to the total radioactivity of each sample.

C. Identification of Metabolites

Authentic cold (nonradiolabeled) standards of altanserin, altanserinol, and 4-FBP were used to confirm the identities of the radiolabeled plasma metabolites separated by HPLC analysis of rat plasma following the injection of [^{18}F]altanserin, [^{18}F]altanserinol, and [^{18}F]4-FBP.

D. Investigation of [^{18}F]Altanserin Regeneration from [^{18}F]Altanserinol

To investigate the potential metabolic conversion of [^{18}F]altanserinol to [^{18}F]altanserin (Van Peer et al., 1986), rats were injected with [^{18}F]altanserinol and plasma samples were analyzed using an isocratic HPLC system (Waters C18 Symmetry column eluted with a 25% MeCN/75% buffer A) in a manner similar to that described above.

E. Brain Uptake of Radioactivity

Brain radioactivity concentrations were measured following the tail vein injection of [^{18}F]altanserinol, [^{18}F]4-FBP, or [^{18}F]FBP-met. Rats were killed, and their brains were removed and dissected to provide samples of frontal cortex, parietal cortex, occipital cortex, striatum, hypothalamus, hippocampus, and cerebellum. The remainder of the brain was collected to provide whole brain uptake values. All samples were counted, weighed, and the percent injected dose per gram (%ID/g) was calculated. Following a 5-min pretreatment of the potent 5-HT$_{2A}$ antagonist SR 46349B (3 mg/kg) (Rinaldi-Carmona et al., 1994), brain radioactivity concentrations 60 mins after injection of [^{18}F]altanserinol and [^{18}F]4-FBP were determined in a similar manner.

III. RESULTS

A. Plasma Metabolite Analysis

Following the injection of [^{18}F]altanserin, three radiolabeled metabolites were found in the plasma of rats. The time course of these metabolites is shown in Table 1. A representative chromatogram of the coinjection of a plasma sample and the authentic standards of 4-FBP, altanserinol, and altanserin is shown in Fig. 2.

B. Identification of Metabolites

Following injection of [^{18}F]altanserin in rats, radiolabeled metabolites were identified that coeluted with authentic standards of 4-FBP and altanserinol. An unidentified metabolite was also observed, and it was

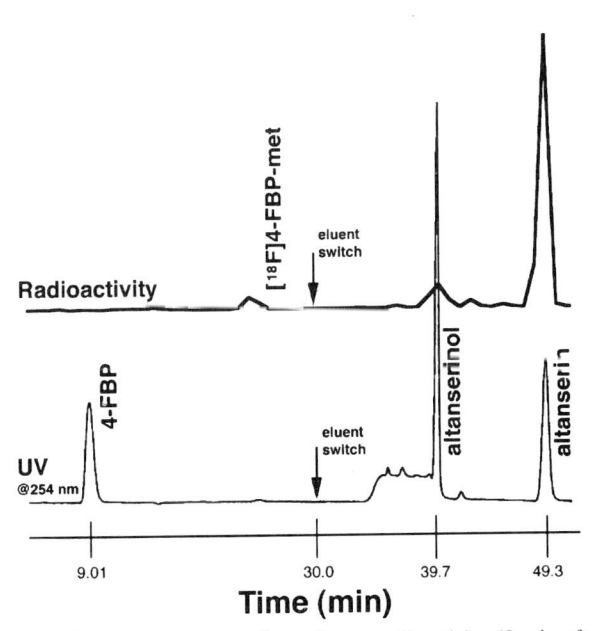

FIGURE 2 Chromatogram of rat plasma radioactivity 60 min after the injection of [^{18}F]altanserin. The retention times of coinjected cold standard of 4-FBP, altanserinol, and altanserin are indicated. The arrows indicate the point at which the mobile phase was switched from 15% acetonitrile/85% buffer A to 25% acetonitrile/75% buffer A. Note the absence of [^{18}F]4-FBP in rat plasma and the presence of a more lipophilic radiolabeled metabolite ([^{18}F]FBP-met) that elutes much later than [^{18}F]4-FBP.

TABLE 1 Time Course of [^{18}F]Altanserin and [^{18}F]Labeled Metabolites in the Plasma of Rats ($n = 3$)[a]

Time (min)	% of total radioactivity in plasma sample			
	[^{18}F]Altanserin	[^{18}F]Altanserinol	[^{18}F]4-FBP-met	[^{18}F]4-FBP
2	87.0 ± 2.0	7.4 ± 0.5	< 0.5	< 0.5
30	71.0 ± 9.8	6.9 ± 2.8	2.2 ± 0.3	1.0 ± 0.2
60	73.5 ± 2.9	8.7 ± 2.6	3.4 ± 0.9	< 0.5
120	58.7 ± 2.7	8.7 ± 3.1	14.7 ± 10.0	~ 1.0

[a] The contribution of [^{18}F]altanserin, [^{18}F]altanserinol, [^{18}F]4-FBP-met, and [^{18}F]4-FBP is given as a percentage of the total radioactivity in the plasma sample. The difference between the sum of these four components and 100% was a result of radioactivity not retained on or eluted from the Sep-Pak cartridges.

found to be more lipophilic than [^{18}F]4-FBP. This same metabolite was present in high concentrations in rat plasma following the injection of [^{18}F]4-FBP, which indicated that it likely resulted from the further metabolism of [^{18}F]4-FBP and was therefore termed [^{18}F]4-FBP-met. [^{18}F]4-FBP-met was also observed in the plasma and urine of baboons and was isolated from baboon urine and injected into rats to determine its brain uptake characteristics.

C. Investigation of [^{18}F]Altanserin Regeneration from [^{18}F]Altanserinol

Injection of [^{18}F]altanserinol in rats resulted in a plasma radioactivity peak that coeluted with a cold standard of altanserin and indicated that [^{18}F]altanserin was oxidatively generated from [^{18}F]altanserinol. [^{18}F]Altanserin accounted for 16 and 35% of the total plasma radioactivity at 30 and 60 min, respectively, following [^{18}F]altanserinol injection.

D. Brain Uptake of Radioactivity

Assays of whole brain radioactivity concentrations in the rat indicated significant blood–brain barrier penetration by [^{18}F]altanserinol, [^{18}F]4-FBP, and [^{18}F]-FBP-met. Regional and whole brain radioactivity concentrations (%ID/g) after the injection of [^{18}F]altanserin, [^{18}F]altanserinol, [^{18}F]4-FBP, and [^{18}F]4-FBP-met are shown in Fig. 3. Regional brain radioactivity concentrations (%ID/g) 60 min after injection of [^{18}F]altanserinol or [^{18}F]4-FBP with and without preadministration of 3 mg/kg of SR 46349B are compared in Fig. 4 to those following an injection of [^{18}F]altanserin in untreated rats.

IV. DISCUSSION

These results confirm brain uptake of radioactivity in the rat following the injection of three radiolabeled metabolites of [^{18}F]altanserin: [^{18}F]4-FBP, [^{18}F]4-FBP-met, and [^{18}F]altanserinol. Although [^{18}F]altanserinol is more lipophilic than either [^{18}F]4-FBP or [^{18}F]4-FBP-met, brain radioactivity concentrations after

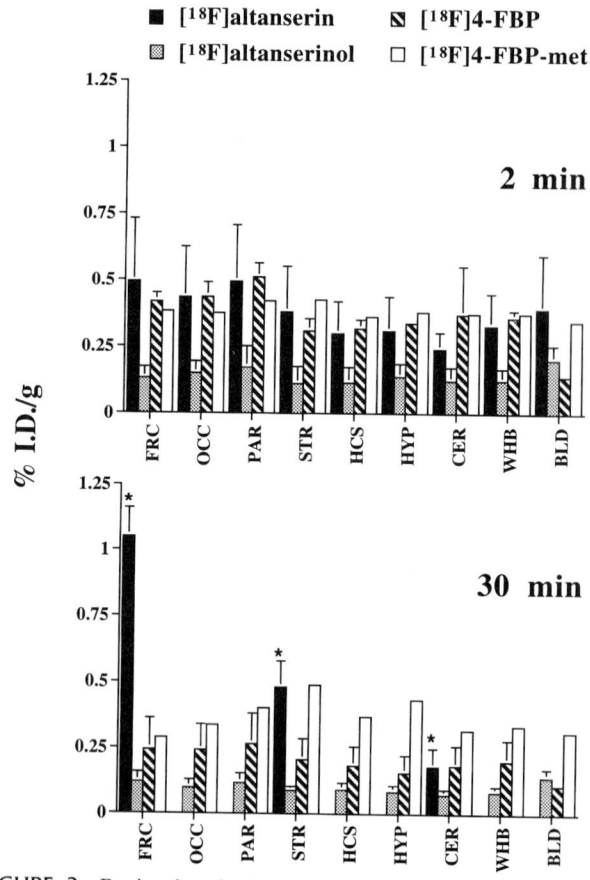

FIGURE 3 Regional and whole rat brain radioactivity concentrations expressed as percentage injected dose per gram (%ID/g) at 2 and 30 min after injection of [^{18}F]altanserin ($n = 3$, $n = 4$), [^{18}F]altanserinol ($n = 4$, $n = 5$), [^{18}F]4-FBP ($n = 2$, $n = 2$), and [^{18}F]4-FBP-met ($n = 1$, $n = 1$). FRC; frontal cortex; OCC; occipital cortex; PAR; parietal cortex; STR; striatum; HCS; hippocampus; HYP; hypothalamus; CER; cerebellum; WHB; whole brain; BLD; blood. Asterisks indicate data from Lemaire et al. (1991).

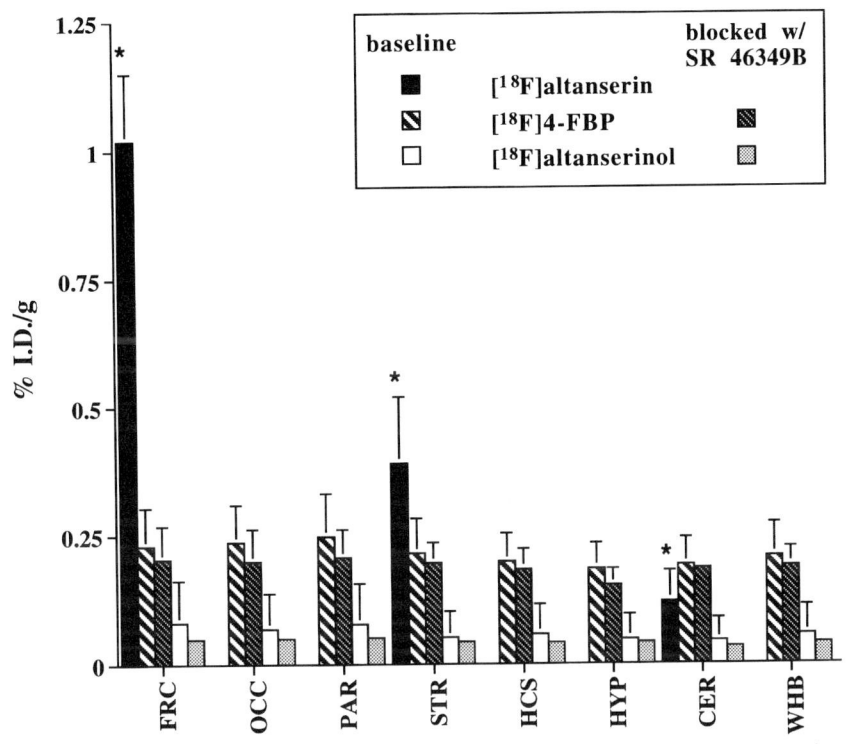

FIGURE 4 Regional and whole rat brain radioactivity concentrations 60 min after injection of [^{18}F]4-FBP or [^{18}F]altanserinol both at baseline and after preadministration of 3 mg/kg of the potent 5-HT$_{2A}$ antagonist SR 46349B as compared to regional rat brain radioactivity concentrations following [^{18}F]altanserin injection in untreated rats. For a list of abbreviations; see the legend to Fig. 3. Asterisks indicate data from Lemaire *et al.* (1991).

[^{18}F]altanserinol injection were three- to fourfold lower than those following injection of either [^{18}F]4-FBP or [^{18}F]4-FBP-met.

The presence of [^{18}F]altanserin in rat plasma following the injection of [^{18}F]altanserinol indicated that an oxidative metabolic pathway was active in rats. However, this pathway is not likely a major contributor to specifically bound radiotracer following the injection of [^{18}F]altanserin, as plasma concentrations of [^{18}F]altanserinol were low compared to those of the unmetabolized parent at all time points (Table 1). Furthermore, preadministration with SR 46349B did not result in significantly lower regional brain radioactivity concentrations in tissue with high 5-HT$_{2A}$ receptor density following an injection of [^{18}F]altanserinol as compared to regional brain radioactivity concentrations in untreated rats (Fig. 4).

Brain radioactivity concentrations at 2 min were severalfold higher for [^{18}F]4-FBP than [^{18}F]altanserinol. However, [^{18}F]4-FBP was rapidly metabolized to [^{18}F]4-FBP-met and accounted for < 1.0% of the total plasma radioactivity in rats at all time points following the injection of [^{18}F]altanserin, whereas [^{18}F]4-FBP-met comprised up to ~ 15% @ 120 min. [^{18}F]4-FBP-met also showed high brain uptake and exhibited slower tissue clearance than either [^{18}F]altanserinol or [^{18}F]4-FBP. Low plasma concentrations of [^{18}F]4-FBP indicated that the high brain uptake of radioactivity observed following the injection of [^{18}F]4-FBP may be partly attributable to the high brain uptake of its metabolite, [^{18}F]4-FBP-met.

In summary, three metabolites of [^{18}F]altanserin enter rat brain tissue. Brain uptake of [^{18}F]4-FBP did not represent a substantial contribution to the total uptake of radioactivity in brain tissue following the injection of [^{18}F]altanserin due to its low plasma concentration. Of greater importance in the rat were the brain uptakes of [^{18}F]4-FBP-met, which demonstrated good BBB penetration and slow tissue clearance, and [^{18}F]altanserinol, which demonstrated relatively high plasma concentrations but lower BBB penetrations. Assessment of regional brain radioactivity concentrations after injection of [^{18}F]altanserinol, [^{18}F]4-FBP, or [^{18}F]4-FBP-met demonstrated that radioactivity was distributed in a manner consistent with only nonspecific binding. Consequently, *in vivo* quantitative measurements of the specific binding of [^{18}F]altanserin using PET should be achievable.

Acknowledgment

This work was supported in part by NIH grants NS 22899, MH 49936, MH 52247, and MH 57078.

References

Biver, F., Goldman, S., Luxen, A., Monclus, M., Forestini, M., Mendlewicz, J., and Lostra, F. (1994). Multi-compartmental study of [F-18]altanserin binding to brain 5-HT$_2$ receptors in human using positron emission tomography. *Eur. J. Nucl. Med.* **21**: 937–946.

Cross, A. J., Crow, T. J., Ferrier, I. N., Johnson, J. A., Bloom, S. R., and Corsellis, J. A. N. (1984). Serotonin receptor changes in dementia of the Alzheimer type. *J. Neurochem.* **43**: 1574–1581.

Frost, J. J. (1990). Imaging the serotonergic system by positron emission tomography. *Ann. N.Y. Acad. Sci.* **600**: 272–278.

Kahn, R., Davidson, M., Siever, L., Gabriel, S., Apter, S., and Davis, K. (1993). Serotonin function and treatment response to clozapine in schizophrenic patients. *Am. J. Psychiatry* **150**: 1337–1342.

Lemaire, C., Cantineau, R., Guillaume, M., Plenevaux, A., and Christiaens, L. (1991). Fluorine-18-Altanserin: A radioligand for the study of serotonin receptors with PET: Radiolabeling and *in vivo* bioligical behavior in rats. *J. Nucl. Med.* **32**: 2266–2272.

Leysen, J. E. (1990). Gaps and peculiarities in 5-HT$_2$ receptor studies. *Neuropsychopharmacology* **3**: 361–369.

Leysen, J. E. (1991). Use of 5-HT receptor agonists and antagonists for the characterization of their respective binding sites. *In* "Drugs as Tools in Neurotransmitter Research. Neuromethods." (A. B. Boulton, G. B. Baker, and A. V. Jurio, eds.), Vol. 12, pp. 299–349. Humana Press, Clifton, NJ.

Mason, N. S., Huang, Y., Holt, D. P., Perevuznik, J. J., Lopresti, B. J., and Mathis, C. A. (1997). Synthesis of two radiolabeled metabolites of [F-18]altanserin:[F-18]3-[2-[4-(4-fluorophenylmethanol)-1-piperidinyl]ethyl]-2,3-dihydro-2-thioxo-4-(1H)-quinazolinone and [F-18]4-(4-fluorobenzoyl)piperidine. *Int. Symp. Radiopharm. Chem.*, 1997, Uppsala, Sweden. *J. Labeled Comp. Radiopharm.* **40**: 161–162.

Meltzer, H., (1991). Serotonergic dysfunction in depression. *Br. J. Psychiatry* **8**: 25–31.

Meuldermans, W., Hendrickx, J., Lauwers, W., Swysen, R., Hurkmans, F., Knaeps, F., Woestenborghs, R., and Heykants, J. (1984). Excretion and biotransformation of ketanserin after oral and intravenous administration in rats and dogs. *Drug Metab. Dispos.* **12**: 772–781.

Mintun, M. A., Price, J. C., Smith, G. S., Lopresti, B., Hartman, L., Simpson, N., and Mathis, C. A. (1996). Quantitative 5-HT$_{2A}$ receptor imaging in man using [F-18]altanserin: A new model accounting for labeled metabolites. *J. Nucl. Med.* **37**: 109P.

Pike, V. W. (1995). Radioligands for PET studies of central 5-HT receptors and reputake sites-current status. *Nucl. Med. Biol.* **22**: 1011–1018.

Rinaldi-Carmona, M., Congy, C., Pointeau, P., Vidal, H., Breliere, J. C., and Le Fur, G. (1994). Identification of binding sites for SR 46349B, a 5-HT$_2$ receptor antagonist, in rodent brain. *Life Sci.* **54**: 119–127.

Sadzot, B., Lemaire, C., Maquet, P., Salmon, E., Plenevaux, A., Degueldre, C., Hermanne, J. P., Guillaume, M., Cantineau, R., Comar, D., and Franck, G. (1995). Serotonin 5HT$_2$ receptor imaging in the human brain using positron emission tomography and a new radioligand, [F-18]altanserin: Results in young normal controls. *J. Cereb. Blood Flow Metab.* **15**: 767–797.

Smith, G. S., Price, J. C., Mathis, C. A., Lopresti, B. J., Holt, D. P., Mason, N. S., Simpson, N. R., Huang, Y., Sweet, R. A., Cidis-Meltzer, C., Sashin, D. Test-Retest variability of serotonin 5-HT$_{2A}$ receptor binding measured with positron emission tomography and [F-18]altanserin in the human brain. In Review.

Van Peer, A., Woestenborghs, L., Embrechts, L., and Heykants, J. (1986). Pharmacokinetic approach to equilibrium between ketanserin and ketanserin-ol. *Eur. J. Clin. Pharmacol.* **31**: 339–342.

CHAPTER 45

Preliminary Evaluation of the Glycine Site Antagonists [¹¹C]L 703,717 and [³H]MDL 105,519 as Putative PET Ligands for Central NMDA Receptors: *In Vivo* Studies in Rats

J. OPACKA-JUFFRY, H. MORRIS, S. ASHWORTH, S. OSMAN, E. HIRANI, A. M. MACLEOD,*
S. K. LUTHRA, and S. P. HUME

PET Methodology Group
MRC Cyclotron Unit
Hammersmith Hospital
London W12 0HS
* Merck Sharp & Dohme Research Laboratories
Terlings Park
Harlow CM20 2QR, England

Two selective antagonists of the glycine site of the N-methyl-D-aspartate (NMDA) receptor—[³H]MDL 105,519, (Z)-2(phenyl)-3-(4,6-dichloroindol-3-yl-2-carboxylic acid) propenoic acid, and [¹¹C]L703,717, 3-substituted 4-hydroxyquinolin-2-(1H)-one—were evaluated in rats as potential in vivo ligands for positron emission tomography (PET). Labeling of L703,717 was carried out by reaction of the phenolic precursor with [¹¹C]iodomethane, giving 98% radiochemical purity and a specific activity of 30–40 GBq/μmol at the end of synthesis. Following intravenous injection of [³H]MDL 105,519 or [¹¹C]L 703,717, radioactivity in plasma was measured and uptake characteristics were assessed in brain within a time period relevant to PET scanning (up to 60 min). Discrete brain regions, such as frontal cortex, entorhinal cortex, striatum, hypothalamus, thalamus, hippocampus, colliculi, medulla oblongata, and cerebellum, were sampled and the temporal distribution of radioactivity analyzed. Additionally, in vitro plasma protein binding of both radioligands was assessed, and the brain uptake index of [³H]MDL 105,519 was measured to evaluate the first-pass extraction. Both [³H]MDL 105,519 and [¹¹C]L 703,717 had a very poor extraction in the rat brain, and consequently, a very low regional biodistribution, in the range of 0.01–0.1% injected dose per gram of tissue, during the time studied. This was consistent with their very high (> 99.8%) in vitro binding to plasma proteins and a pronounced in vivo retention of the radioactivity in plasma. Neither of the glycine site antagonists studied show potential for further development as in vivo PET ligands.

I. INTRODUCTION

Glutamate is the main excitatory neurotransmitter in the mammalian central nervous system, and the N-methyl-D-aspartate (NMDA) receptor is one of its recognition sites. The NMDA site is complex, its activity being dependent on Ca^{2+} and Mg^{2+} and modulated by a number of compounds, including polyamines and glycine (Lipton, 1993). NMDA receptors play a role in memory and learning processes and, in addition, appear to be involved in neuronal death following anoxia, stroke, hypoglycemia, and traumatic brain injury (Choi, 1988). They are likely to be implicated in the mechanisms of neurodegenerative disorders, such as Huntington's disease and Parkinson's disease (DiFiglia, 1990; Zuddas et al., 1992). Although there is a strong clinical interest in positron emission tomography (PET) imaging of NMDA receptors in the brain, suitable radioligands are not available at present.

This study reports the *in vivo* evaluation of two radiolabeled selective antagonists of the glycine site of the NMDA receptor: [³H]MDL 105,519 [(Z)-2(phenyl)-3-(4,6-dichloroindol-3-yl-2-carboxylic acid)propenoic acid, *in vitro* K_d = 3.77 nM (Siegel et al., 1995)] and

[¹¹C]L703,717 [3-substituted 4-hydroxyquinolin-2-(1H)-one, IC$_{50}$ = 4.5 nM (Kulagowski et al., 1994)]. Both compounds have been reported to be pharmacologically active in vivo (Kulagowski et al., 1994; Baron et al., 1995).

II. MATERIAL AND METHODS

A. Radioligands

[³H]MDL 105,519 (Amersham, UK; sp. act. 77 Ci/mmol, radioactive concentration 70.2 mCi/ml, radiochemical purity 97.7%) was stored at −20°C. An aliquot of ethanol stock solution, evaporated under N$_2$ to halve its volume, was diluted ~10 times with 0.9% saline immediately before administration to rats.

L703,717 (7-chloro-4-hydroxy-3-(3-(4-methoxybenzyl)phenyl)2-(1H)-quinolone) was radiolabeled with carbon-11 by reaction of the O-desmethyl precursor (7-chloro-4-hydroxy-3-(3-(4-hydroxybenzyl)phenyl)2-(1H)-quinolone) (0.5 mg) with [¹¹C]iodomethane in DMF (200 μl) containing K$_2$CO$_3$ (0.9 mg) and Kryptofix 2.2.2 (3 mg). The reaction mixture was heated at 95°C for 5 min. The product [¹¹C]L703,717 was isolated using semipreparative HPLC (μBondapak C18, 7.8 × 300 mm column, 70:30 MeOH/0.07 M ammonium formate mobile phase, 2.5 ml/min, 254 nm). [¹¹C]L703,717 eluted at 11–12 min and the O-desmethyl precursor eluted at 7–8 min. The mobile phase was removed and the product formulated in 20% ethanol/0.9% NaCl. [¹¹C]703,717 was produced in ~45 min from end of bombardment (EOB), with radiochemical purity > 98% and a specific activity of 30–40 GBq/μmol at end of synthesis. The overall radiochemical yield was ~25% from [¹¹C]CO$_2$ at EOB. Full details of the radiosynthesis are reported elsewhere (Morris et al., 1997).

B. Animals and Radiotracer Administration

The study was carried out by licensed investigators in accordance with Home Office "Guidance on the Animals (Scientific Procedures) Act 1986." Male Sprague–Dawley rats (270–320 g) were used, awake, but in restrainers. Each rat was injected with either ~10 MBq of [¹¹C]L703,717 or ~300 kBq of [³H]MDL 105,519 (0.25–0.3 ml) via a catheterized tail vein. The specific activity of [¹¹C]L703,717 at time of injection varied from 8 to 13 GMq/μmol. The amount of stable L703,717 coinjected was 3.7 ± 0.2 nmol/kg.

C. Pretreatment

Designated rats were given 3-nitropropionic acid (3NP, RBI) at a dose of 30 mg/kg ip, 3 hr before radioligand injection. 3NP is known to cause NMDA receptor activation in the striatum and cortex when administered acutely to rats, as demonstrated by an increase in [³H]MK-801 binding (Wüllner et al., 1994). Nonradioactive MDL 105,519 was kindly provided by Dr. B. Baron, at the Hoechst Marion Roussel Inc. Research Institute, Cincinnati, Ohio. It was administered to rats at an iv dose of either 10 or 70 mg/kg, 5 min before radioligand injection. In order to facilitate bleeding, rats were heparinized (200 units/0.2 ml/rat) 10–15 min before radioligand injection.

D. Blood and Tissue Sampling

Blood samples (five to six per rat) were taken via a catheterized tail artery at times ranging between 10 sec and 45 min for [¹¹C]L703,717 or 60 min for [³H]MDL 105,519 and were centrifuged to obtain plasma. Rats were killed by an iv overdose of sodium pentobarbitone (Euthatal) at 2, 5, 10, 20, 30, and 45 min after [¹¹C]L703,717 or at 2, 5, 10, 20, 30, and 60 min after [³H]MDL 105,519 injection. Rats pretreated with 3NP were killed at 10, 20, or 30 min after radiotracer. Immediately after sacrifice, the brains were removed rapidly and dissected. Eight brain tissues were sampled: entorhinal cortex, frontal cortex, striatum, hypothalamus combined with thalamus, hippocampus, colliculi (inferior + superior), medulla oblongata, and cerebellum. The radioactivity content of brain and blood samples was measured using liquid scintillation or gamma counting, and data were expressed as a percentage of the total radioactivity injected per rat: (cpm per g tissue/cpm injected) × 100. Because of high levels of radioactivity in blood and low uptake of the radioligand to the brain, brain data were corrected for blood radioactivity using an average blood volume of 1% of the brain tissue weight (Cremer and Seville, 1983).

E. Brain Uptake Index

The brain uptake index (BUI) of [³H]MDL 105,519 was measured according to the method of Oldendorf (1970). BUI is the percentage ratio of the first-pass extraction of a test radioligand relative to the known first-pass extraction of a reference substance. [¹⁴C]Iodoantipyrine ([¹⁴C]IAP, Amersham) was used as a reference substance with an unrestricted (100%) extraction to the brain. A mixture of [¹⁴C]IAP and [³H]MDL 105,519 was diluted either in Krebs buffer (pH 7.4) or

80% rat plasma to obtain final concentrations of 10 and 30 µCi/ml, respectively. A 200-µl bolus was injected into the right carotid artery of deeply anesthetized rats, and the animals were guillotined 6 sec later. The brains were removed rapidly and the radioactivity in the right hemisphere was measured using liquid scintillation counting. BUI was calculated as a ratio:

[(^{3}H/^{14}C) dpm in brain/(^{3}H/^{14}C) dpm in injectate]
× 100.

F. *In Vitro* Binding to Plasma Proteins

Solutions of [^{3}H]MDL 105,519 or [^{11}C]L703,717 in saline were mixed with human or rat blood and incubated at 37°C for 10 min. Centrifuged plasma samples were filtered using Amicon MPS-1 filter units (MW cutoff ~ 30,000). Aliquots of the blood, plasma, and filtrate were counted using liquid scintillation or gamma counting.

III. RESULTS AND DISCUSSION

A. Blood and Plasma

Blood and plasma time–radioactivity curves of [^{3}H]MDL 105,519 and [^{11}C]L703,717 are shown in Figs. 1A and 1B, respectively. Both radioligands had essentially similar profiles, where an initial rapid clearance of radioactivity within the first 5 min after radioligand injection was followed by a slow decrease. Plasma radioactivity per gram was retained at a high level of ~ 1% injected dose at the final time points. The striking retention of the radioligands in plasma *in vivo* was consistent with their very high plasma protein binding *in vitro*, calculated as 99.9 and 99.8% for human and rat plasma, respectively.

Pretreatment with stable MDL 105,519 at the higher dose of 70 mg/kg led to a considerable reduction in the radioactivity uptake in plasma (not shown), and therefore the lower dose of 10 mg/kg was used to assess the "specific signal" in brain. 3NP pretreatment did not affect the radioactivity blood/plasma profiles of either radioligand (not shown).

B. Brain

Both [^{3}H]MDL 105,519 and [^{11}C]L703,717 had very poor extraction in the brain and, consequently, a very low regional biodistribution. All the brain regions sampled showed a very low initial uptake of radioactivity per gram, in the range of 0.01–0.08% injected dose during the time studied (Figs. 2 and 3). The restricted uptake of the radioligands was consistent with their very high affinity to plasma proteins and a minimal brain barrier permeability. The BUI (%) values for [^{3}H]MDL 105,519 were 7.1 ± 2.6 for buffer injectate and 2.9 ± 0.7 for the plasma injectate (mean SD, from $n = 4$).

The apparently heterogeneous uptake was similar for each ligand (linear correlation coefficient = 0.753; $p < 0.05$) and reflected regional differences in blood flow (i.e., delivery of the radioligands), as listed in Cremer and Seville (1983). For each ligand, the regional loss of radioactivity correlated with the initial extraction and (for MDL 105,519) there was no significant effect of pretreatment with a blocking dose of stable compound. The final radioactivity content per gram (at 45 or 60 min after injection) was 0.005–0.015% of that injected, remaining approximately two orders of magnitude lower than the blood radioactivity levels.

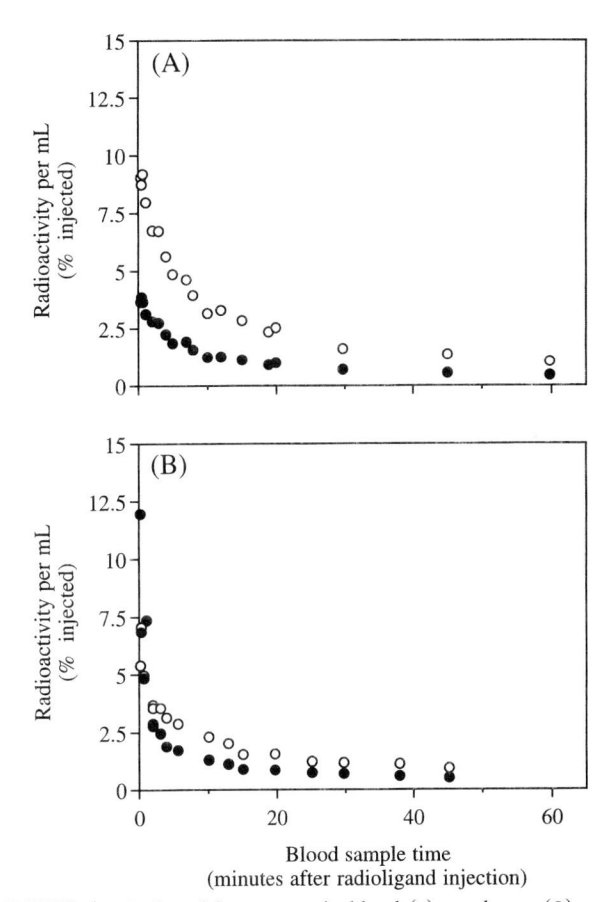

FIGURE 1 Radioactivity content in blood (●) or plasma (○) as a function of time after iv injection of [^{3}H]MDL 105,519 (A) or [^{11}C]L703,717 (B).

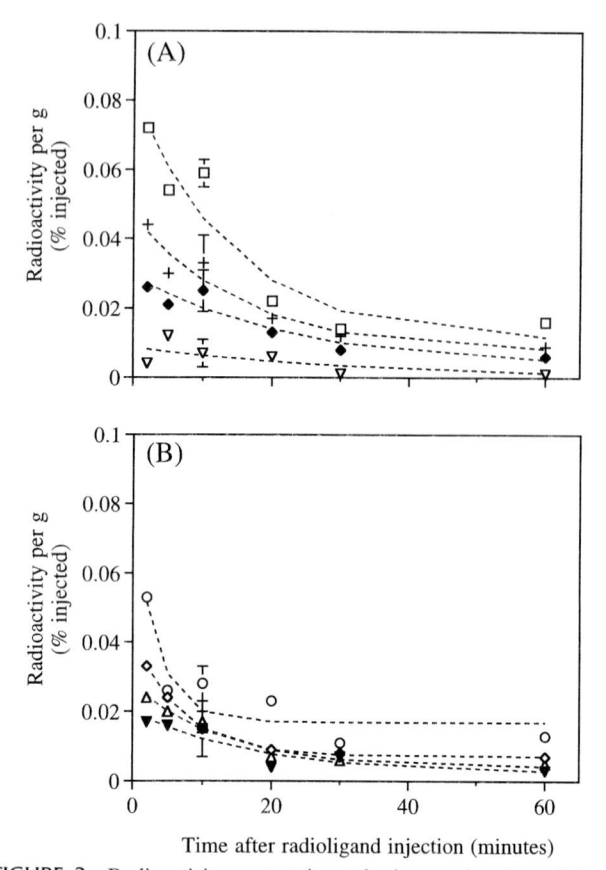

FIGURE 2 Radioactivity content in rat brain as a function of time after iv injection of [^{3}H]MDL 105,519. For clarity, the tissues have been separated into two arbitrary groups (A and B) and the data sets linked by including simple mathematical fits (dashed lines are exponential + constant). Data points are from single animals or a group of three, where the standard deviation has been included to represent the expected error on the individual measurements. Tissues are entorhinal cortex (□), hypothalamus + thalamus (◆), striatum (▽), and cerebellum (+) in A, and frontal cortex (▲), hippocampus (◇), colliculi (○), and medulla (△) in B.

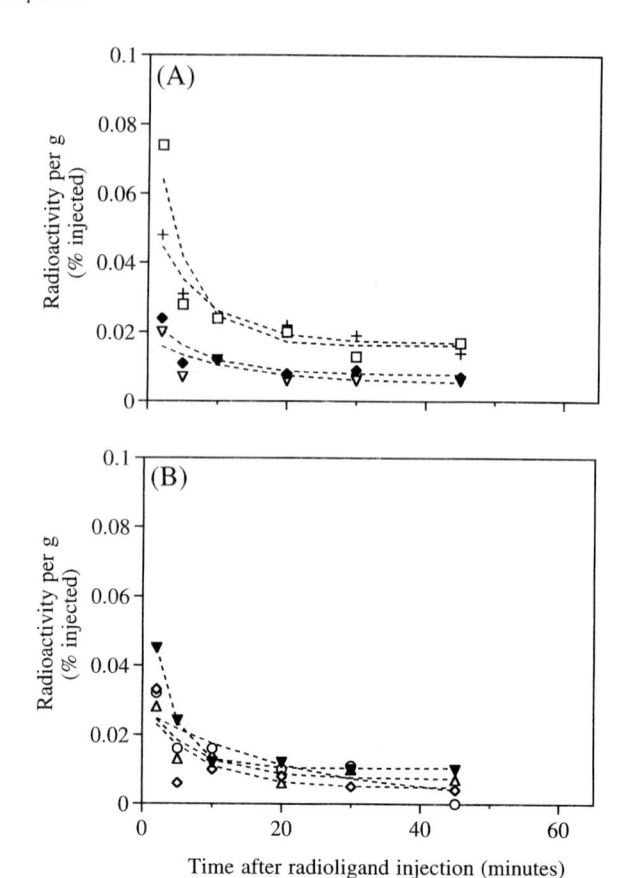

FIGURE 3 Radioactivity content in rat brain as a function of time after injection of [^{11}C]L703,717. Symbols and the fitted curves are explained in the legend to Fig. 2.

Both [^{3}H]MDL 105,519- and [^{11}C]L703,717-related radioactivity contents in most of the brain regions studied, including the striatum and cortex, appeared to be increased as a result of 3NP acute treatment when rats were sacrificed at 20 or 30 min after radioligand injection (Table 1). This may imply that radioligand binding is enchanced under the conditions of NMDA

TABLE 1 Radioactivity Content (% Injected) per Gram Tissue in Rats with or without 3NP Pretreatment (30 mg/kg ip)[a]

Tissue	[^{3}H]MDL 105,519 Control	[^{3}H]MDL 105,519 3NP pretreated	[^{11}C]L703,717 Control	[^{11}C]L703,717 3NP pretreated
Blood	0.87 ± 0.16	0.85	0.78 ± 0.13	0.84 ± 0.27
Plasma	2.15 ± 0.49	2.27	1.37 ± 0.27	1.54 ± 0.23
Frontal cortex	0.013 ± 0.000	0.018	0.011 ± 0.001	0.017 ± 0.003
Striatum	0.011 ± 0.005	0.019	0.006 ± 0.001	0.011 ± 0.002

[a] Killed at 20 or 30 min after iv injection of either [^{3}H]MDL 105,519 or [^{11}C]L703,717 (mean values with SD from two to three animals per treatment, 20 and 30 min data combined, with the exception of data from [^{3}H]MDL 105,519 with 3NP pretreatment, where a single rat was killed at 30 min after radioligand injection).

receptor activation. However, considering the very low extraction of both radioligands in the brain, this finding alone does not modify the final conclusion that neither [^{3}H]MDL 105,519 nor [^{11}C]L703,717 is a suitable candidate for further development as *in vivo* PET ligands.

Acknowledgment

We thank Professor David Brooks for his continued support.

References

Baron, B. M., Siegel, B. W., Harrison, B. L., Kehne, J. H., Mc-Closkey, T. C., Schmidt, C. J., Taylor, V. L., Fadayel, G. M., Murawsky,
M. K., van Giersbergen, P. L. M., White, H. S., and Salituro, F. G. (1995). Pharmacological characterization of MDL 105,519, a systematically active antagonist of the NMDA receptor-associated glycine recognition site. *Soc. Neurosci. Abstr.* **21**: 1106.

Choi, D. W. (1988). Gutamate neurotoxicity and diseases of the nervous system. *Neuron* **1**: 623–634.

Cremer, J. E., and Seville, M. P. (1983). Regional brain blood flow, blood volume and haematocrit values in the adult rat. *J. Cereb. Blood Flow Metab.* **3**: 254–256.

DiFiglia, M. (1990), Excitotoxic injury of the neostriatum: a model for Huntington's disease. *Trends Neurosci.* **13**: 286–289.

Kulagowski, J. J., Baker, R., Curtis, N. R., Leeson, P. D., Mawer, I. M., Moseley, A. M., Ridgill, M. P., Rosley, M. Stansfield, I., Foster, A. C., Grimwood, S., Hill, R. G., Kemp, J. A., Marshall, G. R., Saywell, K., and Tricklebank, M. D. (1994). 3′-(Arylmethyl)- and 3′-(aryloxy)-3-phenyl-4-hydroxyquinolin-2-(1H)-ones: Orally active antagonists of the glycine site on the NMDA receptor. *J. Med. Chem.* **37**: 1402–1405.

Lipton, S. A. (1993). Prospects for clinically tolerated NMDA antagonists: Open channel blockers and alternative states of nitric oxide. *Trends Neurosci.* **16**: 527–532.

Morris, H. J., Luthra, S. K., Brown, D. J., MacLeod, A. M., and Brady, F. (1997). Synthesis of [^{11}C]L 703,717, a potential ligand for PET studies of the glycine site of the NMDA receptor. *J. Labelled Compd. Radiopharm.* **40**: 640–641.

Oldendorf, W. H. (1970). Measurement of brain uptake of radiolabeled substance using a tritiated water internal standard. *Brain Res.* **24**: 372–376.

Siegel, B. W., Baron, B. M., Harrison, B. L., Gross, R. S., Hawes, C., and Towers, P. (1995). [^{3}H]MDL 105,519, a high affinity radioligand for NMDA receptor-associated glycine recognition site. *Soc. Neurosci. Abstr.* **21**: 1106.

Wüllner, U., Young, A. B., Penney, J. B., and Beal, M. F. (1994). 3-Nitropropionic acid toxicity in the striatum, *J. Neurochem.* **63**: 1772–1781.

Zuddas, A., Oberto, G., Vaglini, F., Fascetti, F., and Corsini, G. U. (1992). MK-801 prevents 1-methyl-4-phenyl-1,2,3,6-tetrahydropyridine-induced parkinsonism in primates. *J. Neurochem.* **59**: 733–739.

SECTION VI

PARAMETER ESTIMATION

CHAPTER

46

Evaluation of the Contribution of Protocol Design to Model Parameter Uncertainty

JULIAN C. MATTHEWS,[*,†] VINCENT J. CUNNINGHAM,[‡] and PAT M. PRICE[*]

*PET Oncology Group and ‡Methodology Group
MRC Cyclotron Unit
Hammersmith Hospital, London, United Kingdom
† Wolfson Brain Imaging Centre
Addenbrookes Hospital
Cambridge, United Kingdom

The accuracy of parameters from the dynamic modeling of positron emission tomography (PET) data is influenced by the propagation of statistical uncertainty through the kinetic analysis. The design of PET acquisition protocols is critical in ensuring that the uncertainty does not preclude interpretation of the parameter values. Two methods have been investigated to calculate sensitivity indexes, quantifying the relative propagation of uncertainty from PET data to parameters, as a function of parameters. The first method is for scans where data are binned into discrete time frames, and the second continuous method is for list mode data. Both methods are dependent on a PET data noise model that can be used to incorporate camera characteristics via a noise equivalent counts (NEC) noise model. Monte Carlo simulations for a single-compartment model were conducted to validate the analysis. The two methods were used to assess a variety of blood flow protocols, including both intravenous (iv) [^{15}O]water studies and inhaled [^{15}O]CO$_2$ studies. For the protocols investigated, only minor reductions in the uncertainty of blood flow are possible by using shorter time frames. The longer protocols were better for estimating the fractional volume of tissue occupied by water (V_D). Shorter [^{15}O]water iv boluses were better than the longer inhalation of [^{15}O]CO$_2$ for high flow regions. The methods developed provide a systematic way of evaluating most aspects of scan protocol design. These methods will be of particular importance when limitations such as camera dead time and frame duration restrictions demand the reassessment and comparison of tracer administration protocols.

I. INTRODUCTION

Quantitative values such as cerebral blood flow (CBF) and the cerebral metabolic rate of glucose (CMRGlc) are commonly measured using positron emission tomography (PET). These values are often calculated using dynamic data sets and relating the kinetics of the label in tissue to the kinetics in arterial blood using kinetic models. Model parameters that best predict the observed kinetics can be estimated from which physiological values of interest can be calculated.

Because of the counting nature of PET data and the ill-conditioned reconstruction, PET images are inherently noisy. Such noise will manifest itself as uncertainties in the estimated model parameters, which, if too great, may preclude their usefulness. Exactly how the noise maps through to parameter uncertainty depends on a number of factors, including (1) the true values of the model parameters (assuming the model is appropriate) and (2) the acquisition protocol (the length and frequency of data collection and the schedule and amount of radioactivity injected).

Typically the design of these protocol variables is done intuitively and by trial and error, with systematic methods of analysis rare. The result can be protocols that either are inadequate for measuring a desired value or overly conservative, making acquisition, processing, and modeling laborious. Methods based on Fisher information matrices (FIMs) have been used previously to design protocols for [^{18}F]fluorodeoxyglucose (FDG) (Ho-Shun et al., 1996; Li et al., 1996) and receptor ligand models (Delforge et al., 1989; 1990; Millet et al., 1996). Normally this consists of calculating a D-optimal (the determinant of the inverse of the FIM) protocol given a model and typical model parameters. This approach fails to take account that only a small subset or combination of parameters may be desired, and that protocols are normally required that are good (but not necessarily optimal) for a *range* of parameters. For instance, for assessing the viability of damaged brain, CBF values over a wide range of values are required.

In this chapter the sensitivity of model parameters to protocol design is assessed by partitioning linear models into single-compartment models. Covariance matrices of the model parameters (the inverse of the FIM) are then calculated for the case when data are binned into discrete time frames and for the case when continuous data collection is possible via list mode acquisition. The methods are then applied to Kety–Schmidt blood flow modeling.

II. MATERIALS AND METHODS

A. Theory

The sensitivity of parameters in a *single*-compartment model is investigated for both discrete time frame data and for continuous list mode data. As well as the direct use with list mode data, the later analysis provides a target value that cannot be improved by altering the duration of time frames. The reason for concentrating on single-component models is that linear models can be expressed as

$$y(t) = \sum_i \gamma_i e^{-\beta t} \otimes u(t), \qquad (1)$$

where $y(t)$ is the measurement, $u(t)$ is the input function, and γ_i and β_i are rate constants of inflow and outflow, partitioning complex models into linear combinations of single-compartment models (Fig. 1), and is similar to the model used in spectral analysis (Cunningham and Jones, 1993). The parameters β_i express the speed of dynamics in each component. If protocols are too short or do not have fine enough sampling, then they will be insufficient to measure the parameters associated with very slow and fast dynamics. Similarly, the identification of fast kinetic parameters can be precluded by insufficient fast dynamics in the input function. Hence, if an idea of the speed of the dynamics associated with a particular compound is known,

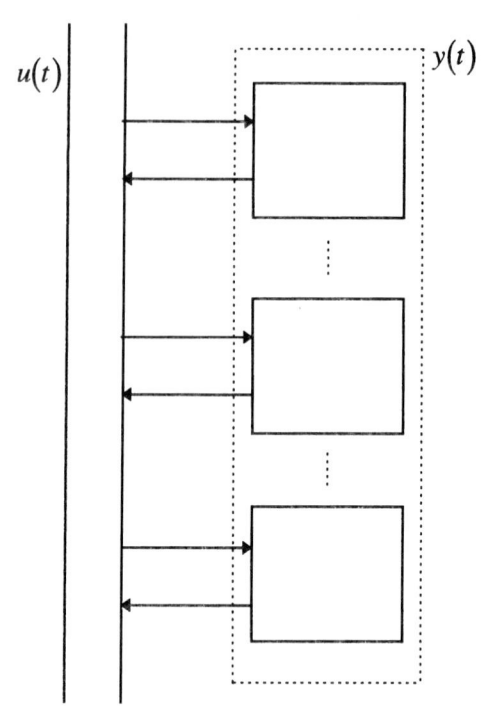

FIGURE 1 Partitioning of multicompartment models into single compartments.

then protocols can be designed that prevent excessive sensitivity of any individual single-compartment parameter. For single-compartment models such as the Kety–Schmidt blood flow model, the analysis is quantitative as well as qualitative.

1. Normalization

The statistical variations of PET measurement differ over time and need to be corrected for either (a) transforming the space of PET measurements to a new equal variance space and using conventional measures of topology or (b) defining a metric which in turn can be used to define notions of distance

$$\|a\| = \sqrt{\langle a, a \rangle} \qquad (2)$$

and geometry

$$\theta(a, b) = \cos^{-1}\left(\frac{\langle a, b \rangle}{\|a\| \|b\|}\right).[1] \qquad (3)$$

Both these methods are equivalent and can be performed so that perturbations have a unit variance probability density function with distance.

For the discrete case, if y is the vector of mean concentration during predefined time frames (i.e., $y_i =$

---

[1] $\theta(\underline{a}, \underline{b})$ is the angle between the vector $\underline{a}$ and $\underline{b}$.

$\int_{t_i}^{t_{i+1}} y(t)dt/(t_{i+1} - t_i))$, then either the transformation

$$\underline{z} = \mathbf{D}\underline{y}, \quad \text{where } \mathbf{D}^T\mathbf{D} = \text{Cov}(\underline{y})^{-1}{}^2 \quad (4)$$

or the metric

$$\langle \underline{a}, \underline{b} \rangle = \underline{a}^T \text{Cov}(\underline{y})^{-1} \underline{b} \quad (5)$$

are equivalent. Typically, PET data frames are not correlated and

$$\text{Cov}(y_i) \propto \frac{N_i^2}{T_i \text{NEC}_i}, \quad (6)$$

where N_i is the mean number of true counts per second, NEC_i is the noise equivalent counts per second (Strother et al., 1990; Bailey et al., 1991), and T_i is the duration of the frame for each time frame.

For the continous case, if $y(t)$ is the mean concentration of activity during a scan of length T, then either the transformation

$$z(t) = \int_0^T \text{Cov}[y(t), y(s)]^{-1/2} y(s) \, ds \quad (7)$$

or the metric

$$\langle a(t), b(t) \rangle = \int_0^T \int_0^T a(t_1) \text{Cov}[y(t_1), y(t_2)]^{-1} \\ \times b(t_2) \, dt_1 \, dt_2 \quad (8)$$

are equivalent. For typical list mode PET data,

$$\text{Cov}[y(t_1), y(t_2)] \propto \frac{N^2(t)}{\text{NEC}(t)} \quad (9)$$

if $t = t_1 = t_2$ and 0 otherwise.

2. First Method

For a single-compartment model,

$$y_i = \frac{1}{t_{i+1} - t_i} \int_{t_i}^{t_{i+1}} k_1 e^{-\beta t} \otimes u(t) \, dt, \quad (10)$$

where the parameters k_1 and β describe a two-dimensional surface in the multidimensional space of measurements (Fig. 2). For noiseless data with the model and input function correct, the measured value will lie on this surface, with noise causing perturbations away from the true value. Subsequent parameter estimation will calculate model parameters associated with the point on the surface closest (as previously defined) to the perturbed measurements. For small perturbations, perturbations perpendicular to the surface will not

---

[2] $\text{Cov}(\underline{y})$ is the covariance of $\underline{y}$.

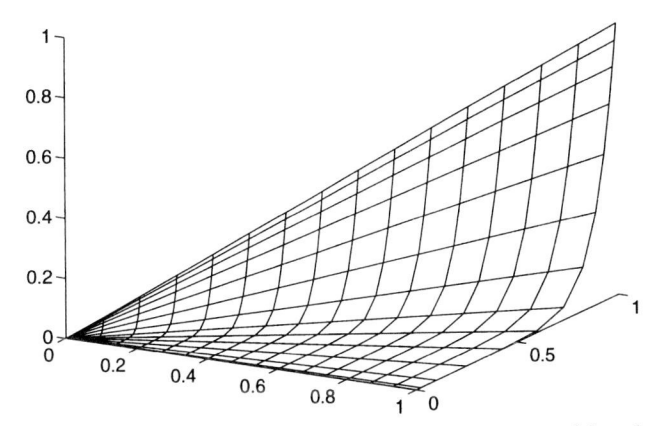

FIGURE 2 Surface described by single-compartment model and input function.

affect the estimated parameters, and perturbations within the surface can be characterized by relative changes in magnitude ($\Delta r/r$) of the measurement vector and changes in angle ($\Delta \theta$). The relationship of those changes to relative changes in the parameters ($\Delta k_1/k_1$ and $\Delta \beta/\beta$) can be calculated numerically as

$$\begin{pmatrix} \Delta r/r \\ \Delta \theta \end{pmatrix} = \begin{pmatrix} 1 & a(\beta) \\ 0 & b(\beta) \end{pmatrix} \begin{pmatrix} \Delta k_1/k_1 \\ \Delta \beta/\beta \end{pmatrix}, \quad (11)$$

where the functions a and b are not dependent on k_1 and are calculated by varying β by small amounts from predefined values and examining how the magnitude and angle of the predicted measurements change.

The inverse of this relationships can be calculated, taking into account radioactive decay (with $k_2 = \beta - \lambda$, where λ is the decay constant), resulting in

$$\begin{pmatrix} \Delta k_1/k_1 \\ \Delta k_2/k_2 \end{pmatrix} = \frac{1}{r} \begin{pmatrix} 1 & -a(k_2 + \lambda)/b(k_2 + \lambda) \\ 0 & \frac{1 + \lambda/k_2}{b(k_2 + \lambda)} \end{pmatrix} \begin{pmatrix} \Delta r \\ \Delta x \end{pmatrix}, \quad (12)$$

where Δr and Δx are absolute changes within the surface in the direction of the measurement vector and perpendicular to it. Hence the covariance matrix of relative changes in k_1 and k_2 can be calculated, which in turn can be used to calculate the coefficient of variation of parameters:

$$\frac{\sigma_{k_1}}{k_1} = \frac{1}{r} \sqrt{1 + \frac{a^2(k_2 + \lambda)}{b^2(k_2 + \lambda)}} \quad (13)$$

and

$$\frac{\sigma_{k_2}}{k_2} = \frac{1}{r}\left|\frac{(1+\lambda/k_2)}{b(k_2+\lambda)}\right|, \quad (14)$$

where $r = \|y\|$. These estimates of the parameter sensitivity to variations in PET data can be calculated from the following three sensitivity indexes:

$$S_{k_1} = \sqrt{1 + \frac{a^2(k_2+\lambda)}{b^2(k_2+\lambda)}}, \quad (15)$$

$$S_{k_2} = \left|\frac{(1+\lambda/k_2)}{b(k_2+\lambda)}\right|, \quad (16)$$

and

$$S_r = \frac{1}{r} = \frac{1}{k_1\|e^{-(k_2+\lambda)t} \otimes u(t)\|}. \quad (17)$$

S_{k_1} and S_{k_2} are dependent on all the protocol design variables apart from the radioactivity administered when the NEC versus true count rate relationship is linear. S_{k_1} also has a value greater than 1 and both S_{k_1} and S_{k_2} are not dependent on k_1. S_r is mainly dependent on the dose injected and the length on the scan.

3. Flow Model

These calculations can be extended to the Kety–Schmidt model, calculating

$$S_F = \sqrt{1 + \frac{a^2(F/V_D+\lambda)}{b^2(F/V_D+\lambda)}}, \quad (18)$$

$$S_{V_D} = \sqrt{1 + \left(\frac{1+a(F/V_D+\lambda)+\frac{\lambda V_D}{F}}{b(F/V_D+\lambda)}\right)^2}, \quad (19)$$

and

$$S_r = \frac{V_D^{-1}}{\|F/V_D e^{-(F/V_D+\lambda)t} \otimes u(t)\|}, \quad (20)$$

where F is the regional blood flow per unit volume of tissue and V_D is the fractional volume of tissue that the labeled water occupies.

4. Second Method

For the continous case, dynamic models and the differences between the measurements and the predicted values can be expressed in state space notation:

$$\dot{x} = f(x, u, \alpha), \quad (21)$$
$$y = g(x, u, \alpha), \quad (22)$$

and

$$J = \tfrac{1}{2}\int_0^T [y(t) - \hat{y}(t)]^T \operatorname{Cov}[y(t)]^{-1}[y(t) - \hat{y}(t)]\,dt, \quad (23)$$

where x is the vector of state space variables, $\hat{y}$ is the measurement, u is the input function, α is the vector of model parameters, and J is the least-squared cost. For a single-compartment model,

$$f(x, u, \alpha) = k_1 u - (k_2 + \lambda)x, \quad (24)$$
$$g(x, u, \alpha) = x, \quad (25)$$

and x is the concentration within the compartment.

With such notation the parameter covariance matrix can be calculated by solving

$$\Lambda(t) = \int_0^t e^{f_x(t)(t-s)} f_\alpha(s)\,ds\,^3; \quad (26)$$

$$L(t) = g_\alpha(t) + g_x(t)\Lambda(t); \quad (27)$$

$$\operatorname{Cov}(\alpha) = \left[\int_0^T L^T(t) \operatorname{Cov}[y(t)]^{-1} L(t)\,dt\right]^{-1}. \quad (28)$$

From the covariance matrix, parameter sensitivity indexes can be calculated:

$$S_{\alpha_i} = \frac{r\sigma_{\alpha_i}}{\alpha_i}; \quad (29)$$

$$S_r = \frac{1}{r}, \quad (30)$$

where

$$r = \int_0^T y^T(t) \operatorname{Cov}[y(t)]^{-1} y(t)\,dt, \quad (31)$$

σ_{α_i} is the standard deviation of the ith parameter, and α_i is the true parameter value.

These equations can be used to evaluate the parameters in single-compartment models, as well as more complex models. As with the discrete case the sensitivity index for a single-compartment model does not depend on k_1 as f_x, f_α, g_x, and g_α are all independent of k_1. Sensitivity indexes for the Kety–Schmidt model parameters can be calculated either from S_{k_1} and S_{k_2} or directly from the equations described earlier.

---

[3] f_x refers to the partial derivative of f with respect to x.

B. Monte Carlo Simulations

The calculations for the discrete case were validated using Monte Carlo simulations, with $k_1 = k_2$. One hundred different values of k_2 were chosen varying over a wide range of kinetic speeds. For each value of k_2, 100 noisy measurement vectors were generated, from which the parameters of a single-compartment model were estimated using nonlinear fitting (Nelder–Mead simplex method). For these calculations, an input function and noise characteristics were taken from a [^{15}O]water brain study, with labeled water administered continuously over 60 sec. Only small perturbations were used ($\sigma_y = 1\%$ of the maximum concentration at this value).[4] From the simulations, mean parameter values and sensitivity indexes were estimated.

C. Blood Flow Protocol Assessment

Four protocols previously used to calculate blood flow using dynamic PET acquisition and Kety–Schmidt modeling were compared. For two of the protocols, [^{15}O]water was injected into a vein, and for the other two protocols, [^{15}O]CO$_2$ gas was inhaled for a duration of 3.5 min. For each type of study two protocols were examined, a short scan and a scan with extended duration (Table 1). Typical input functions and noise characteristics were taken from patient data [^{15}O]water head scans and a [^{15}O]CO$_2$ scan of the abdomen). Sensitivity indexes from the four protocols were calculated using both the discrete method and the continous data method. For the continous case, indexes were calculated for several scan durations and the values extrapolated to a scan of infinite duration. These extrapolated continuous values are only dependent on the administration of the radiochemical and not on the collection of PET data.

III. RESULTS

A. Monte Carlo Simulations

Figure 3 shows estimated (via the Monte Carlo simulations) values of the parameters and sensitivity indexes, together with the true parameter values and calculated indexes. Both the estimated parameter values and the sensitivity indexes of k_1 and k_2 were very similar to the calculated values. The small differences noted at very fast dynamics are probably due to numerical errors in the evaluation of the convolution operations or the nonlinearity of the model.

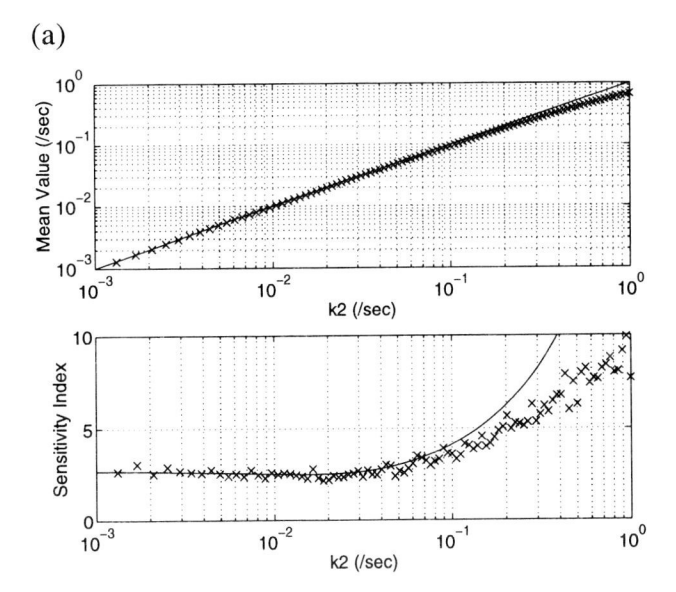

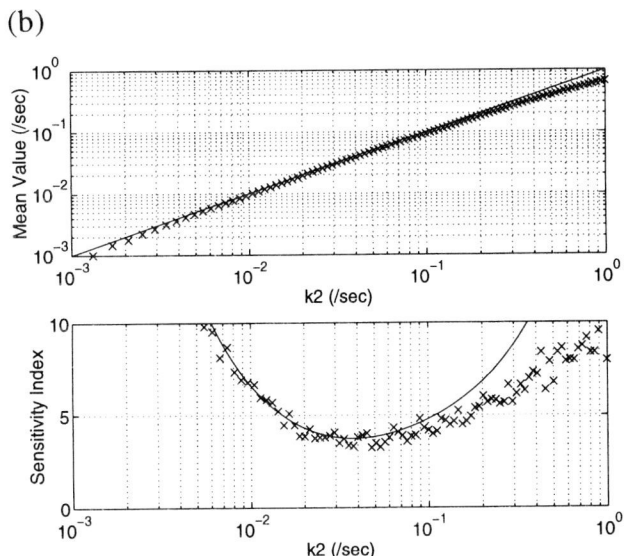

FIGURE 3 Comparison of parameter values and sensitivity indices calculated using the discrete method (solid lines) and estimated from Monte Carlo simulations ("x") for (a) k_1 and (b) k_2.

TABLE 1 Dynamic blood flow protocols

Radiochemical	Frame durations
[^{15}O]water	12 × 5 sec; 12 × 10 sec
[^{15}O]water	12 × 5 sec; 12 × 10 sec; 10 × 15 sec
[^{15}O]CO$_2$	6 × 5 sec; 6 × 10 sec; 6 × 20 sec; 2 × 10 sec; 2 × 20 sec; 4 × 30 sec; 3 × 60 sec
[^{15}O]CO$_2$	5 × 6 sec; 4 × 15 sec; 4 × 30 sec; 3 × 5 sec; 3 × 15 sec; 4 × 40 sec

---

[4] σ_y is the standard deviation of y.

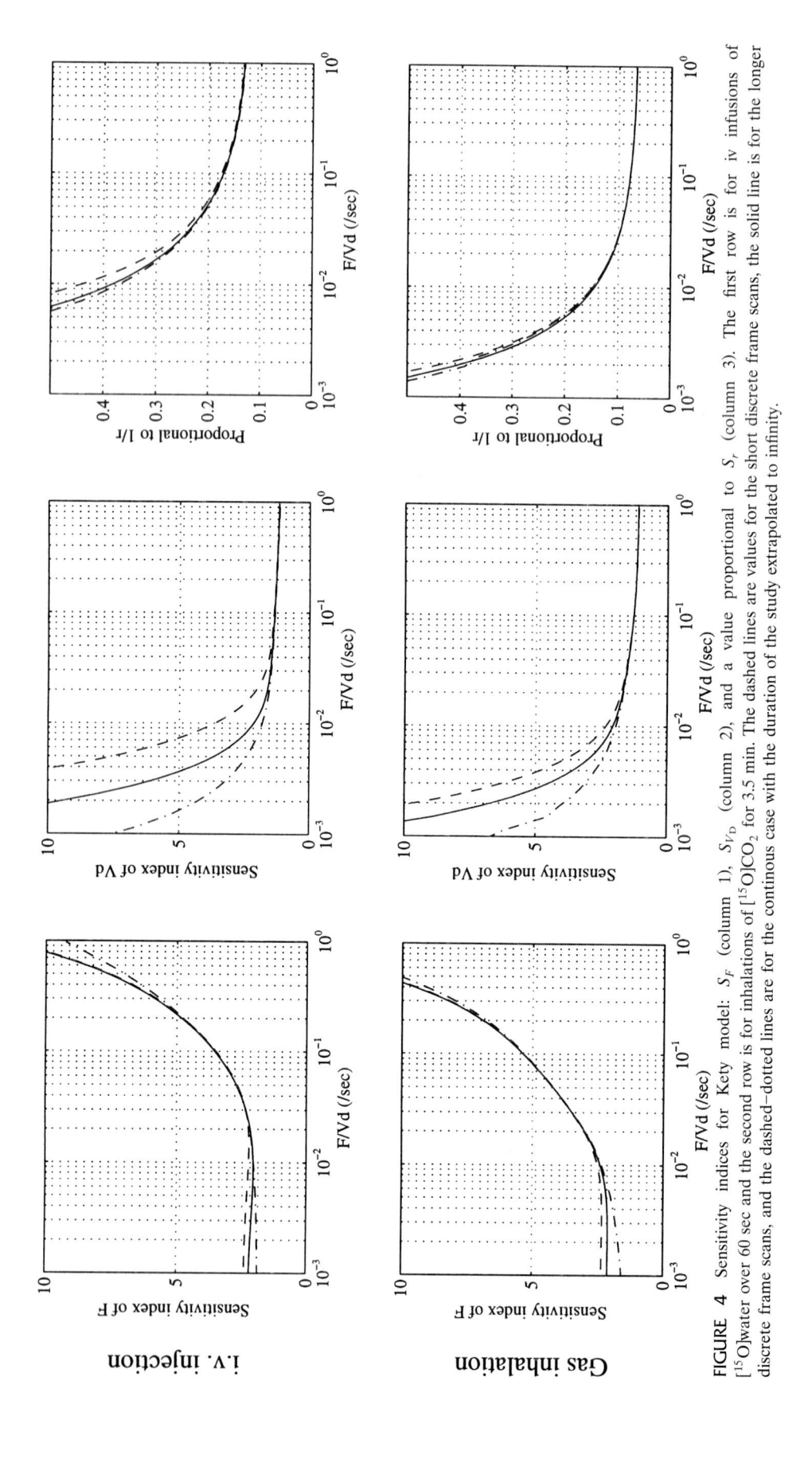

FIGURE 4 Sensitivity indices for Kety model: S_F (column 1), S_{V_D} (column 2), and a value proportional to S_r (column 3). The first row is for iv infusions of [^{15}O]water over 60 sec and the second row is for inhalations of [^{15}O]CO$_2$ for 3.5 min. The dashed lines are values for the short discrete frame scans, the solid line is for the longer discrete frame scans, and the dashed–dotted lines are for the continous case with the duration of the study extrapolated to infinity.

B. Blood Flow Protocol Assessment

Graphs of the sensitivity indexes for the four protocols and the continous case are shown in Fig. 4. The graphs of $S_r = 1/r$ are with $V_D = 1$, but are only proportional to the true value as a noise model is used which is proportional and not equal to the variance on the PET measurements. Hence direct comparison of the S_r graphs between the [^{15}O]water and [^{15}O]CO$_2$ studies cannot be made.

For physiologically realistic values of F and V_D ($F/V_D < 0.02$), sensitivity values differ very little with frame duration and length of scan, although the 60-sec injection protocols appear slightly better for high flow values. This is presumably due to the lack of the fast dynamics in the input function for the inhalation studies. The degradation in the indices with discrete finite duration data sets is minimal. For low values of flow, the scans of longer duration appear better for estimating V_D, with scans of at least 5 and 10 min desirable for 60-sec [^{15}O]water and 3.5-min [^{15}O]CO$_2$ scans. The limitation of estimating V_D for very low values of flow is a fundamental limitation imposed by the halflife of ^{15}O. Very little differences are apparent between the discrete frame and continuous values of S_r, with less sensitivity for high flow values.

IV. DISCUSSION

For single-compartment models, the sensitivity of model parameters as a function of the parameters was calculated and provides a systematic method for assessing most aspects of protocol design. For this simple case, the sensitivity indexes (apart from S_r) just dependent on a single parameter. For multicompartment models, this is no longer the case, with the sensitivity dependent on the speed of the kinetics and the relative contribution of these kinetics to the measurements. This makes it very difficult to design adequate protocols for the expected range of parameters. However, protocols good at estimating parameters for the individual compartments are almost certainly good at estimating combinations of parameters. This assumption can be used to design an administration schedule and time frames that are capable of measuring the range of expected kinetics with minimal uncertainty. However, quantitative results for the single-compartment case cannot be extended, with the identification of model parameters with distinct kinetics less susceptible to noise than when kinetic speeds are subtly different.

This chapter examined four blood flow protocols and all were found to be reasonably good at estimating the desired parameters, with data sampling more than sufficient and camera characteristics unimportant as NEC versus count rate was approximately linear for the cases studied. The methods will be of greater importance when hardware or logistic restrictions prevent frequent kinetic sampling and camera characteristics such as dead time demand the careful design of the quantity and schedule of the radioactivity administered. In particular, in the future: (1) protocols for estimating cerebral blood flow while minimizing the number of frames acquired will be investigated, which will have advantages for the rapid processing of 3D data sets; and (2) protocols for quantitative blood flow measurements on the ECAT 966 camera will be designed, where due to the high sensitivity of the camera, characteristics such as dead time are significant.

Acknowledgments

The authors thank Cathryn Brock and Paula Wells for providing PET blood flow data and many others at the Cyclotron Unit, Hammersmith Hospital, for their contribution to the design and acquisition of these data. This work was also partly funded by the Cancer Research Campaign.

References

Bailey, D. L., Jones, T., Spinks, T. J., Gilardi, M. C., and Townsend, D. W. (1991). Noise equivalent count measurements in a neuro-PET scanner with retractable septa. *IEEE Trans. Med. Imag.* **10**: 256–260.

Cunningham, V. J., and Jones, T. (1993). Spectral analysis of dynamic PET studies, *J. Cereb. Blood Flow Metab.* **13**: 15–23.

Delforge, J., Syrota, A., and Mazoyer, B. M. (1989). Experimental design optimisation: Theory and application to estimation of receptor model parameters using dynamic positron emission tomography. *Phys. Med. Biol.* **34**: 419–435.

Delforge, J., Syrota, A., and Mazoyer, B. M. (1990). Identifiability analysis and parameter identification of and in vivo ligand receptor model from PET data. *IEEE Trans. Biomed. Eng.* **37**: 653–661.

Ho-Shun, K., Feng, D., Hawkins, R. A., Meikle, S., Fulham, M. J., and Li, X. (1996). Optimized sampling and parameter estimation for quantification in whole body PET. *IEEE Trans. Biomed. Eng.* **43**: 1021–1027.

Li, X., Feng, D., and Cehn, K. (1996). Optimal image sampling schedule: A new effective way to reduce dynamic image storage space and functional image processing time. *IEEE Trans. Med. Imag.* **15**: 710–718.

Millet, P., Delforge, J., Pappata, S., Syrota, A., and Cinotti, L. (1996). Error analysis on parameter estimates in the ligand-receptor model: Application to parameter imaging using PET data. *Phys. Med. Biol.* **41**: 2739–2756.

Strother, S. C., Casey, M. E., and Hoffman, E. J. (1990). Measuring PET scanner sensitivity: Relating countrates to image signal-to-noise ratios using noise equivalent counts. *IEEE Trans. Nucl. Sci.* **37**: 728–788.

CHAPTER 47

An Assessment of Optimal Image Sampling Schedule Design in Dynamic PET–FDG Studies

D. HO,* D. FENG,* and L. C. WU[†]

*Biomedical and Multimedia Information Technology Group
Department of Computer Science
The University of Sydney, Sydney, New South Wales 2006, Australia
[†]National PET/Cyclotron Center
Taipei Veterans General Hospital
Taiwan

In dynamic positron emission tomography (PET), the reliability of sampled images is directly influenced by the sampling schedules used to acquire data. As sampling intervals increase in duration, signal-to-noise ratios improve as more counts are recorded. However, in order to obtain quantitative information from the dynamic process, a minimum number of image samples are required. Optimal image sampling schedule (OISS) design addresses this issue and may be used to provide the minimum number of image samples required for successful quantitation. This chapter assesses OISS using dynamic PET data obtained from clinical fluorodeoxyglucose studies. Image data were first collected according to a conventional sampling schedule that consists of 22 images. Optimally sampled data were obtained using a novel technique to reorganize and combine the 22 images into 4 images. Subsequently, rate constant parameters estimated from the originally sampled and optimally resampled data were compared.

I. INTRODUCTION

Parametric imaging using positron emission tomography (PET) permits the estimation of biomedical system parameters on an image-wide or pixel-by-pixel level. In general, the determination of these parameters usually requires measurements obtained from a predefined sampling schedule. The choice of this sampling schedule is fundamental in that it determines the statistical accuracy and reliability of parameters estimated from clinical studies (Carson *et al.*, 1983; Cobelli *et al.*, 1985; D'Argenio, 1981). Optimal image sampling schedule (OISS) design plays an important role as it can be used to increase the cost effectiveness of a clinical procedure through improved experimental design, make efficient use of limited resources, and allow maximum information to be extracted from the clinical study under investigation (Carson *et al.*, 1983; Li *et al.*, 1996).

The application of OISS has been proven theoretically and tested by computer simulation in reducing large volumes of image data obtained from PET dynamic studies without adversely affecting functional image quality or causing loss of useful information (Li *et al.*, 1996). For example, image data acquired from a typical PET [^{18}F]fluorodeoxyglucose (FDG) dynamic study may be reduced by more than 80% using OISS. This chapter assesses OISS using PET dynamic data obtained from clinical FDG studies based on a conventional sampling schedule that consists of 22 images. To obtain optimally sampled data, a novel technique to reorganize and combine the 22 images into an optimally sampled group of images was used. OISS parameter estimates were compared to estimates obtained from originally sampled dynamic data to validate whether the same quality of results was achieved.

II. MATERIALS AND METHODS

A. Theory

The well-known four-parameter FDG model was used for tracer kinetic modeling to extract quantitative information from acquired PET dynamic data (Huang *et al.*, 1980). This model has been extensively investigated and discussed (Huang *et al.*, 1980; Phelps *et al.*, 1979). In general, for the extraction of model parameters, observed PET measurement data are usually fitted to the FDG model by means of a nonlinear least-squares (NLS) procedure.

Measurements acquired using PET are recorded in terms of accumulated samples over the scanning interval. Conventionally, measurements for the total concentration of FDG in tissue $c_i^*(t)$ are represented by the average value over each scan interval at its mid-scan time (Hawkins *et al.*, 1986). However, errors can be propagated during model fitting due to the effect of averaging and the use of mid-scan times. This is particularly the case with OISS PET dynamic data as sampling intervals are lengthened. To avoid the introduction of these errors, the accumulated measurements $\int c_i^*(t)\,dt$ are directly employed during NLS model fitting. The residual sum of squares cost function used during model fitting is given by

$$\Phi(p) = \sum_{k=1}^{k=N} \{z(t_k) - \mu(t_k|p)\}^2,$$

where p is the vector of FDG model parameters ($k_1^* - k_4^*$) to be estimated, $z(t_k)$ represents the accumulated PET measurements during scan interval k, and $\mu(t_k|p) = \int c_i^*(t)dt$, the FDG model, integrated over the kth interval. The Levenberg–Marquardt optimization algorithm was used to minimize the cost function and obtain parameter estimates (Marquardt, 1963).

B. FDG–PET Data Acquisition

For each patient, a bolus of FDG tracer in the amount of 200–400 MBq (about 0.5 mg) was injected intravenously, and arterial blood samples were taken concurrent to the injection of FDG. Samples of blood (2–3 ml) were taken at 8 × 15-sec intervals for the first 2 min, then at 2.5, 3, 3.5, 7, 10, 15, 20, 30, 60, 90, and 120 min. The samples were immediately placed on ice and plasma subsequently separated for the determination of FDG and glucose concentration. The PET scanning schedule in all studies was 10 × 12-sec scans, 2 × 0.5-min scans, 2 × 1- min scans, 1 × 1.5-min scan, 1 × 3.5-min scan, 2 × 5-min scans, 1 × 10-min scan, and 3 × 30-min scans, for a total study duration of 120 min.

An eight-ring, 15-slice PET scanner (General Electric/Scanditronix PC4096-15WB) was employed to measure regional distribution of radioactivity in the brain. For each PET dynamic data set, cross-sectional images (in a 128 × 128 matrix with a 2-mm pixel size) were acquired. Transmission scans were utilized to correct PET data for attenuation. Data were also decay corrected to the time of injection and scatter corrected. The filtered backprojection algorithm applying a Hanning filter with a filter width of 4.2 mm (corresponding to a cutoff frequency of 0.48 Nyquist) was used to reconstruct PET image data. Transaxial and axial spatial resolutions of 6.5 mm full width at half maximum in the center of the field of view were achieved.

C. Resampling of PET Temporal Frame Data

In order to obtain optimally sampled data it is necessary to resample the original 22 image sample data. It has previously been shown that for the four-parameter FDG model, 4 image samples are sufficient to obtain parameter estimates of equivalent statistical accuracy and reliability in relation to the conventional technique that requires 22 image samples (Li *et al.*, 1996). The scanning intervals of these image samples are 2.733, 12.950, 61.383, and 42.934 min, for a total study duration of 120 min.

The application of the novel technique to obtain optimally sampled data from the 22 image samples may require both combination and interpolation procedures. In the simplest instance where only the combination procedure is required, adjacent image frames in the original sampled data are summed to provide OISS image data. For example, if original image frame data were sampled using the sampling schedule given by $[[t_0,t_1],[t_1,t_2],\ldots,[t_9,t_{10}]]$, and OISS is defined by $[[t_0,t_2],[t_2,t_5],[t_5,t_{10}]]$, it would simply be a matter of summing accumulated radioactivity counts for the first two original image frames to obtain the first OISS image frame, the next three original image frames to obtain the second OISS frame, and so on.

When OISS sampling intervals overlap intervals for original sampled data, i.e., when either the start or the end sampling times for OISS lie within the sampling intervals for the original sampled data, an interpolation procedure is necessary as accumulated radioactivity counts data are not directly available in adjacent frames. For example, if the OISS is defined by $[[t'_0,t'_1],[t'_1,t'_2],[t'_2,t'_3]]$ (where $t'_0 \equiv t_0$, $t'_3 \equiv t_{10}$, t'_1 lies in between $[t_1,t_2]$ and t'_2 in between $[t_4,t_5]$), interpolation is used to obtain data within these intervals. Subsequently, adjacent image frames and interpolated frame data are combined to obtain OISS image data.

Three distinct cases need to be considered when the interpolation procedure is used, as shown in Fig. 1.

Case 1

Interpolation is used to obtain data from the last image frame in which the OISS sampling interval overlaps. In this case, resampled data are obtained by

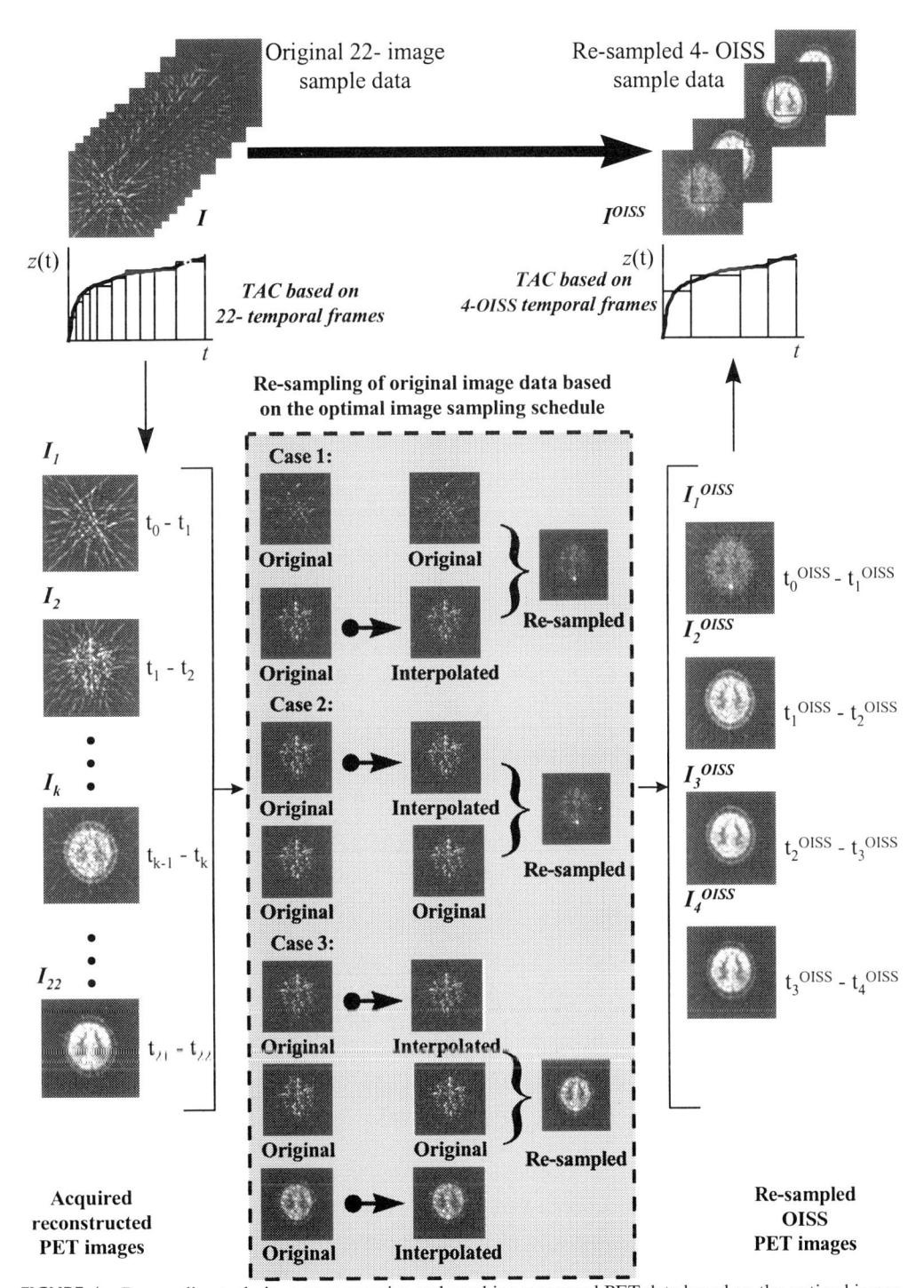

FIGURE 1 Resampling technique to reorganize and combine measured PET data based on the optimal image sampling schedule (OISS). (Left) Acquired reconstructed PET images. (Center) Three interpolation cases to consider. (Right) Resampled OISS PET images.

interpolating the value of each pixel in the overlapping image frame of the original data. The equation governing this operation is given by

$$\text{pixel}^{\text{resample}}([t_i, t'_i]) = \text{pixel}^{\text{original}}([t_i, t_{i+1}]) \frac{t'_i - t_i}{t_{i+1} - t_i}.$$

Case 2

Interpolation is used to obtain data from the first image frame in which the OISS sampling interval overlaps. The equation governing this operation is given by

$$\text{pixel}^{\text{resample}}([t'_i, t_i]) = \text{pixel}^{\text{original}}([t_{i-1}, t_i]) \frac{t_i - t'_i}{t_i - t_{i-1}}.$$

Case 3

Interpolation is used for both the first and the last image frames in which the OISS sampling intervals overlap. In this case, both Case 1 and Case 2 are necessary.

In this chapter linear interpolation was used to resample the 22 image sample PET–FDG dynamic data based on OISS design (Li *et al.*, 1996). Although spline interpolation is more accurate, errors due to linear interpolation were found to be negligible. In Fig. 1 the novel resampling technique is illustrated. Original sampled image data obtained from a PET dynamic study are shown on the left side of Fig. 1. Using the novel technique described by the three dis-

TABLE 1 Estimated Parameters Obtained from Original 22 and OISS 4 Image Sample Data for Three FDG–PET Dynamic Studies

Tissue region of interest	Original sampled 22 image frame data					OISS resampled 4 image frame data				
	k_1^*	k_2^*	k_3^*	k_4^*	CMRGlc	k_1^*	k_2^*	k_3^*	k_4^*	CMRGlc
Clinical study 1										
Gray ROI01	0.08831	0.22317	0.05103	0.00672	6.06859	0.11553	0.37260	0.06768	0.00732	7.22520
Gray ROI02	0.10846	0.28966	0.05446	0.00720	6.09837	0.12706	0.37790	0.06080	0.00717	7.06470
Gray ROI03 (left frontal)	0.12229	0.39008	0.08370	0.00961	8.78890	0.13948	0.48597	0.08895	0.00894	8.77920
Gray ROI04 (visual cortex)	0.15575	0.41586	0.06306	0.00779	8.34350	0178385	0.50132	0.06778	0.00751	8.42340
Gray ROI05 (left temporal)	0.09199	0.27040	0.05490	0.00642	6.31570	0.09935	0.31261	0.05912	0.00647	6.42820
Gray ROI06 (right temporal)	0.08886	0.30605	0.05842	0.01036	5.79440	0.09870	0.37060	0.06319	0.00986	5.84980
White ROI07	0.03346	0.17363	0.06407	0.01364	3.66900	0.04558	0.33748	0.09119	0.01273	3.94520
White ROI08	0.03340	0.15239	0.05115	0.01113	3.41470	0.03809	0.20763	0.06248	0.01097	3.58410
White ROI09 (left hemisphere)	0.04681	0.15897	0.03589	0.00467	3.50750	0.05269	0.20927	0.04569	0.00604	3.84100
White ROI10	0.018161	0.27424	0.06098	0.00848	6.03970	0.09750	0.37907	0.07050	0.00828	6.22060
Clinical study 2										
Gray ROI01	0.09952	0.21683	0.05117	0.00007	6.60930	0.11640	0.29958	0.06450	0.00171	7.17180
Gray ROI02	0.13406	0.37541	0.06782	0.00554	7.13440	0.12312	0.31077	0.06042	0.00551	6.96990
Gray ROI03 (right frontal)	0.11909	0.35697	0.09652	0.00873	8.81570	0.11035	0.29465	0.08611	0.00900	8.67980
Gray ROI04 (visual cortex)	0.10395	0.21997	0.05166	0.00758	6.87590	0.11361	0.26685	0.05947	0.00814	7.20090
Gray ROI05 (right parietal)	0.10065	0.20879	0.04958	0.00325	6.71750	0.12231	0.31518	0.06633	0.00486	7.39610
White ROI06	0.07745	0.25762	0.04360	0.00607	3.89880	0.08093	0.28383	0.04784	0.00681	4.06010
White ROI07	0.09548	0.24766	0.04973	0.00280	5.55330	0.10719	0.31065	0.05844	0.00391	5.90240
Clinical study 3										
Gray ROI01	0.11352	0.45495	0.11117	0.01142	7.97710	0.11347	0.45222	0.10915	0.01108	7.89490
Gray ROI02	0.10751	0.32688	0.07519	0.00991	7.19490	0.12075	0.40498	0.08256	0.00970	7.31680
Gray ROI03 (frontal lobe)	0.11239	0.34826	0.09378	0.01053	8.53180	0.11968	0.39407	0.09866	0.01025	8.57520
Gray ROI04 (visual cortex)	0.12242	0.35506	0.07072	0.00919	7.27570	0.13147	0.39978	0.07341	0.00890	7.29850
Gray ROI05 (right parietal)	0.08486	0.21746	0.04518	0.00876	5.22370	0.09625	0.27958	0.05314	0.00912	5.50110
Gray ROI06 (left parietal)	0.06152	0.16906	0.03723	0.01110	3.97290	0.07475	0.25441	0.05077	0.01201	4.44960
White ROI07	0.02661	0.09836	0.04198	0.01035	2.84790	0.03638	0.25130	0.09563	0.01308	3.58860
White ROI08	0.02841	0.16886	0.07227	0.01254	3.04720	0.03974	0.36540	0.11160	0.01213	3.32740
White ROI09 (right hemisphere)	0.03730	0.09939	0.03148	0.00769	3.21090	0.04142	0.13871	0.04704	0.01035	3.75390

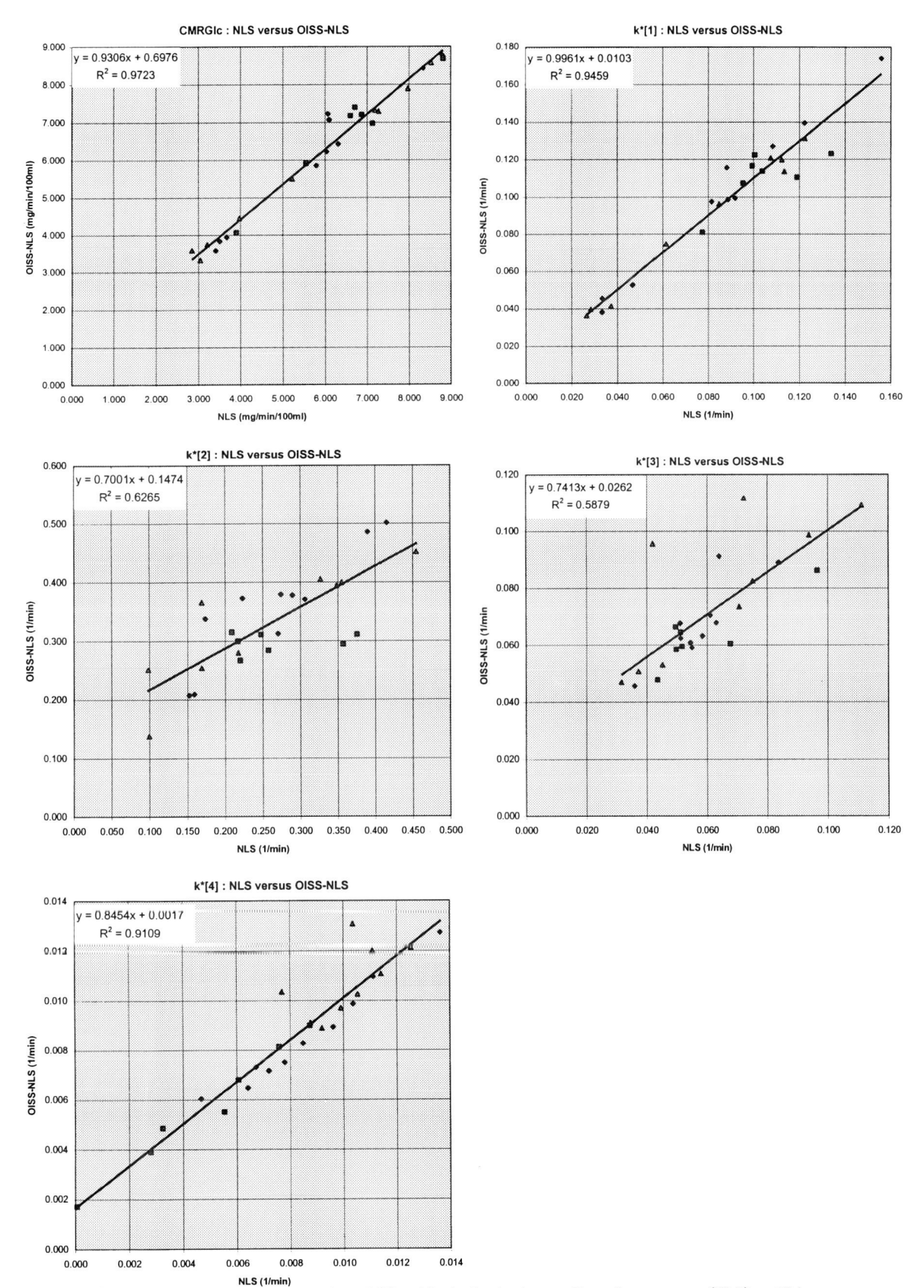

FIGURE 2 Correlation for CMRGlc and $k_1^* - k_4^*$ obtained using nonlinear least squares (NLS) on 22 image data sets and OISS–NLS on 4 image data sets.

tinct cases, a reduced set of images is obtained based on OISS as shown on the right side of Fig. 1.

III. RESULTS AND DISCUSSION

To obtain tracer kinetics of FDG, a total of 26 regions of interest (ROI) of varying sizes were drawn on original sampled 22 image data and OISS 4 image data for the three subjects. Subsequently, rate constant parameters were estimated and the physiological parameter for the metabolic rate of glucose (CMRGlc) was calculated. In Table 1, the rate constants and physiological parameter obtained from the 26 ROIs are shown. The parameter estimates for gray and white matter are consistent for both original and OISS sampled data. The mean and standard deviation of CMR-Glc estimates in gray and white matters obtained from original and OISS data were, respectively: for original image data, 7.01 ± 1.88 and 3.35 ± 1.03 ml/min/100 g; and for OISS image data, 7.38 ± 1.69 and 3.90 ± 0.97 ml/min/100 g. The slight discrepancy between these estimates is due to the loss of finely sampled dynamic information through the application of the resampling technique.

In Fig. 2, estimates for CMRGlc and $k_1^* - k_4^*$ obtained from 22 image and OISS 4 image data and their lines of correlation are shown. Results obtained from CMRGlc, k_1^*, and k_4^* using optimally sampled data were in excellent agreement with those using original sampled data. Correlation coefficients $r = 0.9723$ for CMRGLc, $r = 0.9459$ for k_1^*, and $r = 0.9109$ for k_4^* were achieved. Results for k_2^* and k_3^* achieved correlation coefficients of $r = 0.6265$ and $r = 0.5879$, respectively. From these results, it is clear that OISS resampled data are able to provide enough detailed information for accurately and reliably estimating CMRGlc and the rate constants k_1^* and k_4^*. However, rate constants k_2^* and k_3^* obtained from OISS resampled data, although reliable, are slightly overestimated. These results are consistent with previous results using sensitivity analysis approaches (Huang *et al.*, 1980).

In this study, it took NLS 10–20 iterations on average to meet the 0.1% convergence criteria using original 22 image sample data. For OISS resampled data, NLS took significantly less computational time. On average, NLS required 5–7 iterations for fitting of ROI data to the FDG model using OISS 4 image data. In general, during each iteration of NLS, five subiterations were required.

Although the application of the resampling technique was to reconstructed PET images, it is possible to apply this technique directly to projection data, which would reduce the number of images to be reconstructed. For example, for the FDG model studied, only 4 images would require reconstruction based on the OISS design in comparison to 22 images based on the conventional sampling schedule. Furthermore, as fewer image samples and hence measurements are obtained, storage space and data processing times are reduced greatly.

References

Carson, E. R., Cobelli, C., and Finkelstein, L. (1983). "The Mathematical Modeling of Metabolic and Endocrine Systems: Model Formulation, Identification and Validation." Wiley, New York.

Cobelli, C., Ruggeri, A., DiStefano, J. J., III, and Landaw, E. M. (1985). Optimal design of multi-output sampling schedules—software and applications to endocrine-metabolic and pharmacokinetic models. *IEEE Trans. Biomed. Eng.* **32**: 249–256.

D'Argenio, D. Z. (1981). Optimal sampling times for pharmacokinetic experiments. *J. Pharmacokinet. Biopharm.* **9**: 739–756.

Hawkins, R. A., Phelps, M. E., and Huang, S. C. (1986). Effects of temporal sampling, glucose metabolic rates, and disruptions of the blood-brain barrier on the FDG model with and without a vascular compartment: Studies in human brain tumors with PET. *J. Cereb. Blood Flow Metab.* **6**: 170–183.

Huang, S. C., Phelps, M. E., Hoffman, E. J., Sideris, K., Selin, C., and Kuhl, D. E. (1980). Non-invasive determination of local cerebral metabolic rate of glucose in man. *Am. J. Physiol.* **238**: E69–E82.

Li, X., Feng, D., and Chen, K. (1996). Optimal image sampling schedule: A new effective way to reduce dynamic image storage space and functional image processing time. *IEEE Trans. Med. Imag.* **15**: 710–718.

Marquardt, D. W. (1963). An algorithm for least squares estimation of non-linear parameters. *J. Soc. Ind. Appl. Math.* **2**: 431–441.

Phelps, M. E., Huang, S. C., Hoffman, E., Selin, C., Sokoloff, L., and Kuhl, D. (1979). Tomographical measurement of local cerebral glucose metabolic rate in humans with (F-18) 2-fluoro-2-deoxy-D-glucose: Validation of method. *Ann. Neurol.* **6**: 371–388.

# CHAPTER 48

Simultaneous Extraction of Physiological and Input Function Parameters from PET Measurements[1]

DAGAN FENG,* KOON-PONG WONG,*,† CHI-MING WU,† and WAN-CHI SIU†

*Biomedical and Multimedia Information Technology Group
Basser Department of Computer Science
The University of Sydney, Sydney, New South Wales 2006, Australia
†Department of Electronic Engineering
The Hong Kong Polytechnic University
Hung Hom, Kowloon, Hong Kong

A new cascaded modeling approach for the quantification on a regional or pixel basis of rate constants, plasma input function parameters, and regional cerebral metabolic rate for glucose (rCMRGlu) with PET and [^{18}F]fluoro-2-deoxy-D-glucose is presented. The proposed approach is based on the assumption that all tissue kinetics are driven by the same input function. This assumption is made by all conventional tracer kinetic modeling techniques. Therefore, information about this input function and also the tissue kinetic model have been embedded in the output tissue curves that can be obtained from drawing regions of interest or from a dixel of dynamic PET images, and the input function can be recovered from the PET images if the input function model is properly accounted for. Simulation studies were carried out to evaluate the method. Results show that the proposed method is able to quantify the parameters in both the tissue model and the input function. Good agreement was found between the generated input curve and the fitted curves obtained from the proposed method. The proposed method may prove an alternative to the gold standard of radial arterial blood sampling and dynamic PET data acquisition.

I. INTRODUCTION

The use of positron emission tomography (PET) with [^{18}F]fluoro-2-deoxy-D-glucose (FDG) to determine the regional cerebral metabolic rate for glucose (rCMRGlu) is now generally accepted. The three-compartment tracer kinetic model, originally developed by Sokoloff et al. (1977) to measure rCMRGlu in the albino rat with [^{14}C]2-deoxy-D-glucose (DG), was later applied in humans by Reivich et al. (1979), Phelps et al. (1979), and Huang et al. (1980), respectively, after the synthesis of FDG (Ido et al., 1978). Since then, the FDG–PET technique has been used to study resting rCMRGlu in normal subjects and in various neuropathological conditions (Mazziotta and Phelps, 1986; Grafton and Mazziotta, 1992). To measure the rCMRGlu quantitatively using the FDG–PET technique, the plasma time–activity curve (PTAC), i.e., the input function, is usually required. The PTAC is generally obtained from continuous blood sampling at the radial artery or arterialized vein (Phelps et al., 1979). However, arterial blood sampling is invasive in nature, and the subsequent blood collection and processing is time-consuming, requires extra staff, and increases the radiation exposure to clinical personnel.

Several efforts have been devoted to reducing the invasiveness of blood sampling. The arterial–venous (a-v) blood sampling method using a heated limb has been accepted as an alternative to radial artery blood sampling in FDG–PET studies, as the discomfort of arterial puncture can be removed (Phelps et al., 1979; Huang et al., 1980). However, it requires prolonged warming of the water bath to ensure adequate capillary-venule shunting and needs extra procedures for calibrating the counting equipment and counting the blood samples.

---

[1] Transcripts of the BRAINPET97 discussion of this chapter can be found in Section VIII.

In this study, a cascaded modeling approach is proposed for acquiring the input function together with the tissue kinetic model parameters from dynamic neurologic PET images. With this method, two venous blood samples are taken near the end of the uptake period for calibrating and scaling the blood curve and the plasma glucose concentration. Identifiability and statistical reliability of this method were tested rigorously by computer simulation, and the results are discussed.

II. MATERIALS AND METHODS

A. PTAC Model

As mentioned previously, quantification of kinetic model parameters by PET requires the measurement of the PTAC as the input function. In the simulation study, a fourth-order PTAC model proposed by Feng et al. (1993), which can be described by a sum of exponentials with a pair of repeated eigenvalues, was used for generating the arterial input function. The mathematical expression for this PTAC model without the delay factor is given by

$$c_p^*(t) = (A_1 t - A_2 - A_3)e^{\lambda_1 t} + A_2 e^{\lambda_2 t} + A_3 e^{\lambda_3 t}, \quad (1)$$

where λ_1, λ_2, and λ_3 (in min^{-1}) are the eigenvalues of the model and A_1, A_2, and A_3 (in μCi/ml) represent the coefficients of the model.

B. Tracer Kinetic Model

In this study, a three-compartment model proposed by Sokoloff et al. (1977) with modification by Phelps et al. (1979) and Huang et al. (1980) was used to describe the tracer kinetics of FDG. The FDG model consists of three compartments: (1) the blood pool region representing the concentration of FDG in the plasma [$c_p^*(t)$], (2) the concentration of free FDG in tissue [$c_e^*(t)$], and (3) the concentration of FDG-6-phosphate in tissue [$c_m^*(t)$] (Phelps et al., 1979; Huang et al., 1980). The rCMRGlu corresponding to a local region is related to the rate constants (k_1^*–k_4^*) by

$$\text{rCMRGlu} = \frac{c_p}{LC} \frac{k_1^* k_3^*}{k_2^* + k_3^*}, \quad (2)$$

where LC denotes the lumped constant, which accounts for the differences between FDG and glucose in transportation and phosphorylation, and c_p denotes the cold glucose concentration in plasma (Phelps et al., 1979; Huang et al., 1980). A combined parameter $K = k_1^* k_3^* / k_2^* + k_3^*$, which is proportional to rCMRGlu, is defined to simplify the discussion.

C. Method

In dynamic PET studies, various tissue time–activity curves (TTACs) can be obtained by placing different regions of interest (ROIs) on reconstructed PET images. It may be assumed that the TTACs are driven by the same input function (PTAC) and they are equal to the convolution results of the PTAC and the impulse response functions (IRFs) of the tissue kinetic model in the corresponding ROIs, as shown in Fig. 1. Therefore, the authors hypothesize that the input function and IRF parameters can be estimated simultaneously using *two* or *more* TTACs at a time, as all the TTACs share the same input function based on the just described assumptions.

Computer simulations were performed to evaluate the identifiability and statistical reliability of this approach. In the PTAC function, the last exponential described by A_3 and λ_3 in Eq. (1) will dominate the tail of the blood curve, and these parameters can be determined by fitting two or more venous samples that are sampled near the end of the uptake period. In the computer simulation, these two parameters were kept constant to emulate the act of taking two venous blood samples. The rate constant parameters (k_1^*–k_4^*) in all the available IRFs plus the input function parameters ($A_{1,2}$, $\lambda_{1,2}$) were estimated simultaneously by minimizing the following cost function using the Marquardt algorithm (Marquardt, 1963).

$$\Phi(\theta) = \sum_{i=1}^{m} \sum_{j=1}^{n} w_{ij} \big[\big(c_p^*(A_1, \lambda_1, A_2, \lambda_2; t) \\ \otimes h_i(k_1^*, k_2^*, k_3^*, k_4^*; t) \big) \\ \otimes \delta(t - t_j') - c_{T_i}^*(t_j') \big]^2, \quad (3)$$

where m is the total number of available IRFs, n is the number of samples for each TTAC, h_i is the IRF of the ith ROI, $c_p^*(t)$ is the PTAC function, $\delta(t - t_j')$ is a Dirac delta function shifted by t_j' time unit, $\otimes$ is the convolution integral operator, and w_{ij} is the weighting used during minimization to reflect the relative weight for each data point. The w_{ij} were inversely proportional to the assumed measurement noise variance $\sigma_i^2(t_j')$ as given by

$$\sigma_i^2(t_j') = \frac{\alpha \times c_{T_i}^*(t_j')}{\Delta t_j} = \frac{1}{w_{ij}}, \quad (4)$$

where α is the proportionality constant that deter-

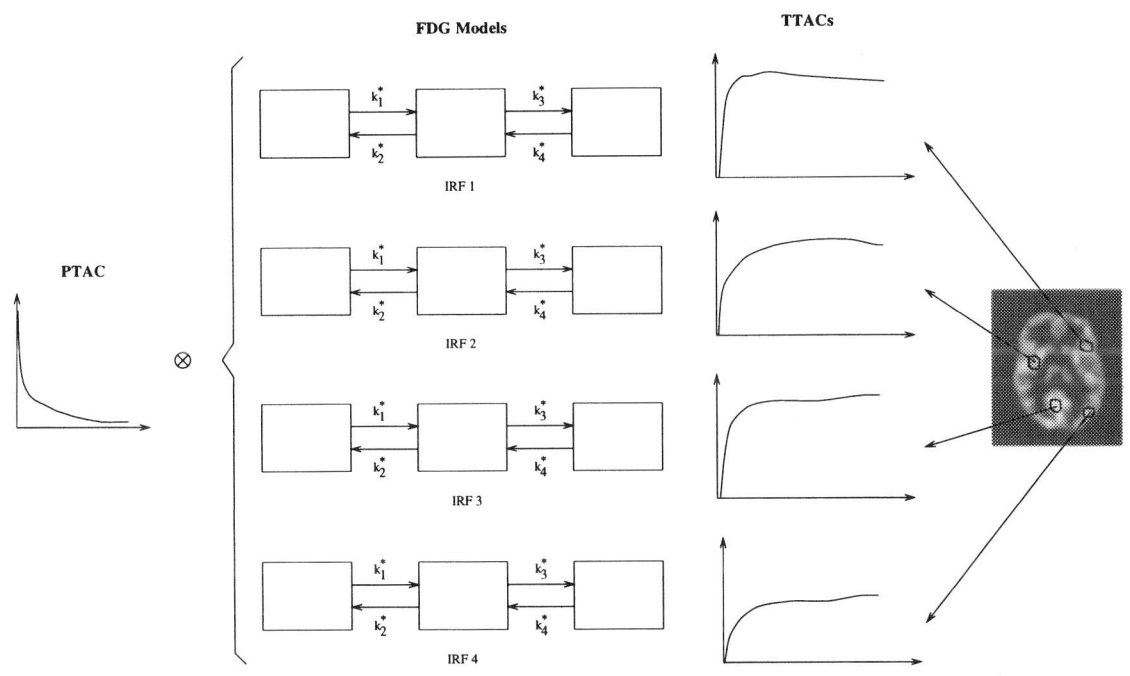

FIGURE 1 The proposed modeling approach. Tissue TACs obtained from ROIs drawn on the PET images are used to fit the convolution of the plasma TAC with the impulse response functions (IRFs) of the tissues, where $\otimes$ is the convolution integration operator. From the fitted rate constant parameters in the FDG model, the physiological parameters corresponding to the local regions can be obtained.

mines the noise level in the measurement and $c^*_{T_i}(t'_j)$ is the PET measurement of the total radionuclide concentration in tissue, which is the average count over the length of the scanning interval $[t_{j-1}, t_j]$ at the mid-scan time t'_j.

D. Simulation Study

The proposed method was evaluated with simulated neurologic FDG-PET studies. The arterial input blood curve (PTAC), representing a rapid bolus administration of FDG tracer, was generated according to Eq. (1) with $A_1 = 851.1225$ μCi/ml, $A_2 = 21.8798$ μCi/ml, $A_3 = 20.8113$ μCi/ml, $\lambda_1 = -4.1339$ min^{-1}, $\lambda_2 = -0.1191$ min^{-1}, and $\lambda_3 = -0.0104$ min^{-1} (Feng et al., 1993); the blood samples were "collected" at 0.25, 0.5, 0.75, 1, 1.25, 1.5, 1.75, 2, 2.5, 3, 3.5, 7, 10, 15, 20, 30, 60, 90, and 120 min. Four ROIs were defined, and their corresponding kinetic model parameters are given in Table 1. The simulated tissue curves in all ROIs were generated according to the following sampling schedule: 10 × 12 sec, 2 × 0.5 min, 2 × 1 min, 1 × 1.5 min, 1 × 3.5 min, 2 × 5 min, 1 × 10 min, and 3 × 30 min, and acquired PET data were assumed to be decay-corrected. Poisson noise with variance structure given by Eq. (4), in which the noise level α was set to 0.0 (noiseless), 0.1, 0.5, 1.0, 2.0, and 4.0, was then added to simulate noisy PET measurements. For each noise level, 100 simulation runs were done; only 1 run was done for the noiseless case. Kinetic model and input

TABLE 1 Kinetic Model Parameters in Different ROIs

Structure	k_1^*	k_2^*	k_3^*	k_4^*	K
Gray matter	0.1020	0.1300	0.0620	0.0068	0.0329
White matter	0.0540	0.1090	0.0450	0.0058	0.0158
Whole brain	0.0780	0.1195	0.0535	0.0063	0.0241
An active region	0.1326	0.1300	0.0806	0.0058	0.0507

Kinetic Model Parameters (in min^{-1})

TABLE 2 Estimation of PTAC Function Parameters A_1, λ_1, A_2, and λ_2 for Different Noise Levels

ROIs used	A_1	Bias (%)	λ_1	Bias (%)	A_2	Bias (%)	λ_2	Bias (%)
True value	851.1225	0.00	−4.1339	0.00	21.8798	0.00	−0.1191	0.00
a. 1 run with noise level $\alpha = 0.0$							
2 ROIs	851.1228	0.00	−4.1339	0.00	21.8807	0.00	−0.1191	0.01
3 ROIs	851.1119	0.00	−4.1338	0.00	21.8806	0.00	−0.1191	0.01
4 ROIs	851.1110	0.00	−4.1338	0.00	21.8808	0.00	−0.1191	0.01
b. 100 runs with noise level $\alpha = 0.1$							
2 ROIs	990.0197	16.32	−4.2008	1.62	27.5072	25.72	−0.1386	16.41
3 ROIs	997.3840	17.18	−4.1758	1.02	27.8425	27.25	−0.1333	11.88
4 ROIs	878.4779	3.21	−4.1328	0.03	22.9888	5.07	−0.1254	5.29
c. 100 runs with noise level $\alpha = 0.5$							
2 ROIs	986.9905	15.96	−4.2327	2.39	27.9057	27.54	−0.1436	20.61
3 ROIs	950.3223	11.66	−4.2239	2.18	26.0587	19.10	−0.1352	13.49
4 ROIs	974.2353	14.46	−4.2936	3.86	25.6998	17.46	−0.1363	14.42
d. 100 runs with noise level $\alpha = 1.0$							
2 ROIs	1193.8261	40.26	−4.6519	12.53	30.5554	29.65	−0.1404	17.85
3 ROIs	1037.2994	21.87	−4.3320	4.79	28.3907	29.76	−0.1326	11.30
4 ROIs	969.3476	13.89	−4.3101	4.26	26.8418	22.68	−0.1328	11.52
e. 100 runs with noise level $\alpha = 2.0$							
2 ROIs	1205.0347	41.58	−4.4853	8.50	31.7101	44.93	−0.1449	21.63
3 ROIs	1048.0574	23.14	−4.3463	5.14	29.7934	36.17	−0.1349	13.25
4 ROIs	1019.0909	19.73	−4.4244	7.03	27.6113	26.20	−0.1446	21.43
f. 100 runs with noise level $\alpha = 4.0$							
2 ROIs	1191.1741	39.95	−4.5363	9.74	31.1864	42.53	−0.1306	9.67
3 ROIs	1239.8401	45.67	−4.6044	11.38	29.4040	34.39	−0.1364	14.51
4 ROIs	1135.3922	33.40	−4.5873	10.97	29.8758	36.55	−0.1386	16.34

function parameters were calculated by minimizing the cost function for different numbers of ROIs at each noise level. Statistical criteria such as the mean, standard deviation (SD), coefficient of variation (CV = SD/mean), and absolute bias of the kinetic model and input function parameters were calculated at different noise levels for comparison.

III. RESULTS AND DISCUSSION

Table 2 shows the statistical results of the input function parameters for the different noise levels. As seen from the table, the parameter estimates of the input function obtained from using three and four ROIs are almost equivalent and generally better than those from two ROIs. Although the parameter estimates do not seem close to their true values, the blood curves generated by these parameter estimates fit original blood data very well. Figure 2a shows a plot of the generated blood curve and fitted blood curves at the highest noise level ($\alpha = 4.0$, which corresponds to 50–100% deviations of measurements at the early scanning intervals), whereas Fig. 2b shows the expanded peak region where the time ranges from 0 to 10 min.

Tables 3 and 4 summarize the statistical results of the parameter estimates of IRF in gray matter and white matter, respectively, using different numbers of ROIs at different noise levels. It can be seen that two ROIs already provide sufficient information to quantify the parameters in the IRFs and PTAC as described earlier. As can be seen from the tables, the CVs of the kinetic parameters in IRFs of gray matter and white matter increased as the noise level increased, and the means and CVs calculated from using three ROIs are very close to those from four ROIs. Although using more ROIs may improve the accuracy of parameter

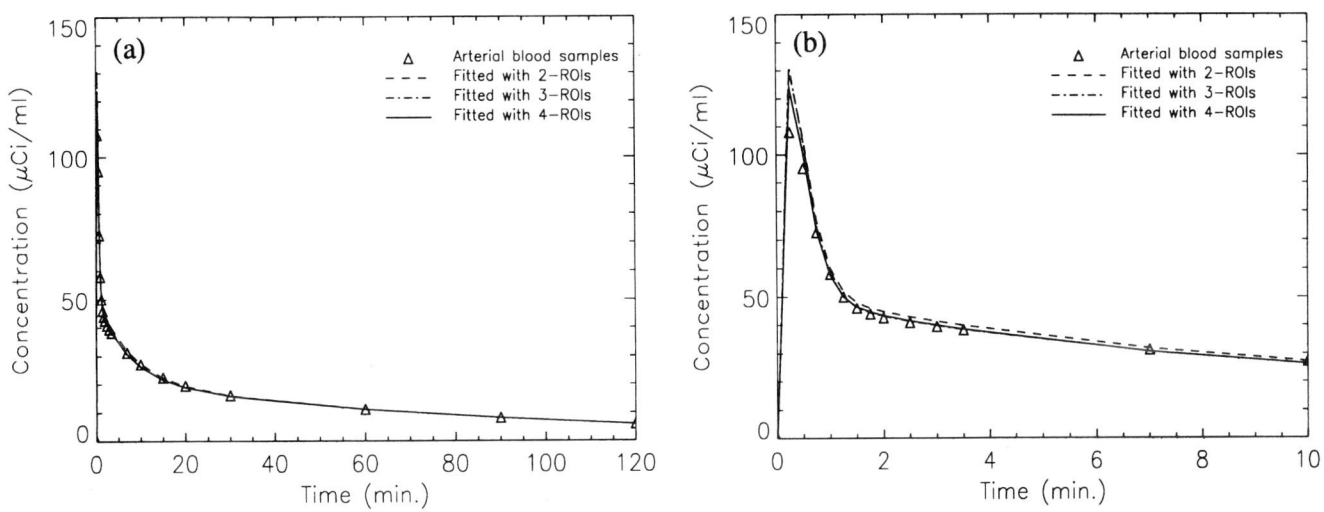

48. Simultaneous Extraction of Parameters in PET

FIGURE 2 (a) The generated blood curve and fitted blood curves obtained using different numbers of ROIs for estimation (noise level $\alpha = 4.0$). (b) Expanded peak regions of (a).

TABLE 3 Estimation Results of Physiological Parameters K and Rate Constants k_1^*–k_4^* in Gray Matter for Different Noise Levels

ROIs used	K	CV (%)	k_1^*	CV (%)	k_2^*	CV (%)	k_3^*	CV (%)	k_4^*	CV (%)
Nominal	0.0329	—	0.1020	—	0.1300	—	0.0620	—	0.0068	—
a. 1 run with noise level $\alpha = 0.0$									
2 ROIs	0.0329	±0.00	0.1020	±0.00	0.1300	±0.00	0.0620	±0.00	0.0068	±0.00
3 ROIs	0.0329	±0.00	0.1020	±0.00	0.1300	±0.00	0.0620	±0.00	0.0068	±0.00
4 ROIs	0.0329	±0.00	0.1020	±0.00	0.1300	±0.00	0.0620	±0.00	0.0068	±0.00
b. 100 runs with noise level $\alpha = 0.1$									
2 ROIs	0.0326	±2.89	0.0955	±24.84	0.1178	±22.54	0.0656	±24.80	0.0074	±18.13
3 ROIs	0.0325	±2.94	0.0930	±20.94	0.1181	±18.85	0.0670	±21.86	0.0073	±14.70
4 ROIs	0.0329	±1.74	0.1022	±18.04	0.1264	±16.95	0.0616	±11.94	0.0070	±11.19
c. 100 runs with noise level $\alpha = 0.5$									
2 ROIs	0.0323	±3.94	0.0956	±27.30	0.1192	±29.59	0.0665	±52.34	0.0072	±18.06
3 ROIs	0.0325	±3.37	0.0988	±23.93	0.1256	±27.67	0.0644	±27.89	0.0070	±16.18
4 ROIs	0.0327	±2.97	0.1004	±26.16	0.1241	±24.86	0.0639	±23.93	0.0072	±19.04
d. 100 runs with noise level $\alpha = 1.0$									
2 ROIs	0.0321	±5.51	0.0980	±36.18	0.1304	±42.73	0.0765	±87.92	0.0074	±28.01
3 ROIs	0.0321	±4.51	0.0956	±21.60	0.1302	±31.66	0.0726	±73.91	0.0071	±20.84
4 ROIs	0.0322	±4.35	0.1006	±28.82	0.1302	±30.05	0.0659	±27.93	0.0070	±24.33
e. 100 runs with noise level $\alpha = 2.0$									
2 ROIs	0.0316	±8.04	0.0927	±29.05	0.1307	±56.90	0.0754	±92.72	0.0072	±24.36
3 ROIs	0.0317	±6.32	0.0972	±34.66	0.1312	±34.57	0.0744	±68.16	0.0070	±27.31
4 ROIs	0.0323	±4.74	0.0999	±27.81	0.1275	±35.56	0.0645	±30.62	0.0070	±23.84
f. 100 runs with noise level $\alpha = 4.0$									
2 ROIs	0.0305	±11.75	0.0935	±27.04	0.1409	±56.20	0.0714	±73.54	0.0064	±31.33
3 ROIs	0.0317	±7.02	0.0943	±24.90	0.1293	±41.55	0.0715	±62.30	0.0071	±33.56
4 ROIs	0.0317	±6.35	0.0999	±39.29	0.1290	±45.32	0.0666	±38.48	0.0070	±29.72

TABLE 4 Estimation Results of Physiological Parameters K and Rate Constants k_1^*–k_4^* in White Matter for Different Noise Levels

ROIs used	K	CV (%)	k_1^*	CV (%)	k_2^*	CV (%)	k_3^*	CV (%)	k_4^*	CV (%)
Nominal	0.0158	—	0.0540	—	0.1090	—	0.0450	—	0.0058	—
a. 1 run with noise level $\alpha = 0.0$
2 ROIs	0.0158	±0.00	0.0540	±0.00	0.1090	±0.00	0.0450	±0.00	0.0058	±0.00
3 ROIs	0.0158	±0.00	0.0540	±0.00	0.1090	±0.00	0.0450	±0.00	0.0058	±0.00
4 ROIs	0.0158	±0.00	0.0540	±0.00	0.1090	±0.00	0.0450	±0.00	0.0058	±0.00
b. 100 runs with noise level $\alpha = 0.1$
2 ROIs	0.0155	±2.56	0.0507	±25.04	0.0975	±25.25	0.0451	±18.10	0.0061	±18.45
3 ROIs	0.0155	±2.49	0.0493	±21.13	0.0974	±21.70	0.0464	±15.54	0.0062	±15.31
4 ROIs	0.0157	±1.57	0.0541	±17.91	0.1061	±18.38	0.0441	±10.19	0.0058	±12.05
c. 100 runs with noise level $\alpha = 0.5$
2 ROIs	0.0154	±3.43	0.0506	±27.08	0.0982	±30.64	0.0449	±22.32	0.0060	±25.07
3 ROIs	0.0155	±2.97	0.0521	±22.99	0.1034	±28.07	0.0450	±19.92	0.0058	±18.45
4 ROIs	0.0155	±3.10	0.0531	±26.03	0.1033	±26.60	0.0445	±19.46	0.0059	±20.30
d. 100 runs with noise level $\alpha = 1.0$
2 ROIs	0.0153	±5.70	0.0517	±34.86	0.1030	±41.15	0.0464	±31.05	0.0062	±37.81
3 ROIs	0.0153	±4.06	0.0503	±21.02	0.1038	±31.14	0.0470	±30.25	0.0058	±27.71
4 ROIs	0.0153	±4.47	0.0534	±28.90	0.1089	±32.87	0.0461	±22.52	0.0058	±24.71
e. 100 runs with noise level $\alpha = 2.0$
2 ROIs	0.0148	±8.68	0.0485	±29.88	0.0972	±45.51	0.0442	±30.28	0.0057	±36.74
3 ROIs	0.0151	±6.88	0.0516	±35.65	0.1055	±38.94	0.0475	±36.44	0.0058	±41.04
4 ROIs	0.0153	±5.39	0.0527	±27.58	0.1042	±34.56	0.0444	±27.54	0.0057	±30.11
f. 100 runs with noise level $\alpha = 4.0$
2 ROIs	0.0144	±12.46	0.0490	±27.39	0.1102	±52.87	0.0464	±45.82	0.0050	±44.46
3 ROIs	0.0151	±7.90	0.0495	±25.41	0.1008	±40.08	0.0458	±29.54	0.0059	±48.36
4 ROIs	0.0317	±8.35	0.0539	±40.74	0.1124	±50.69	0.0469	±40.85	0.0055	±40.81

estimates, the improvement was not significant if more than three ROIs were used. This means that the improvement of accuracy in parameter estimates obtained from using more ROIs is not significant if more ROIs are utilized for estimation, unless the kinetics in the additional regions are very different from those being used. In addition, introducing more ROIs in the calculation will increase the computational complexity. Therefore, there is a trade-off between the computational complexity and the parameter estimation accuracy.

The parameter K is critical in the calculation of rCMRGlu; therefore, it is the major interest in the present study. Figures 3 and 4 show the plots of the CVs and biases of K for gray matter and white matter, respectively, using different number of ROIs at different noise levels. It can be seen that the proposed method can provide very reliable estimates, as the biases and CVs are very small at all noise levels. Even at the highest noise level ($\alpha = 4.0$), the CVs for K in gray matter obtained from using two, three, and four ROIs are 12, 7, and 6%, respectively, and the corresponding biases are 8, 4, and 4%, respectively. As predicted, the CVs and biases of K are largest when two ROIs are used for calculation, and they are generally decreased when more ROIs are employed for estimation. The results of white matter also reveal similar phenomena, and the same arguments apply.

This chapter presented a noninvasive method for extracting the impulse response function and the input function parameters simultaneously by utilizing PET measurements alone. With this method, the input function can be estimated accurately with only two (or more) venous blood samples taken near the end of the uptake period of tracer for calibrating and scaling the blood curve and the plasma glucose concentration. Identifiability and statistical reliability of this method were tested rigorously by computer simulation, and the results demonstrated that this method provides very reliable and robust estimation of rCMRGlu, kinetic model parameters, and input function. Although the feasibility of applying this method in neurologic

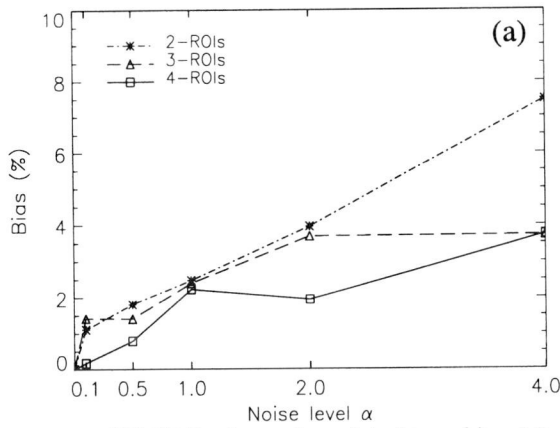

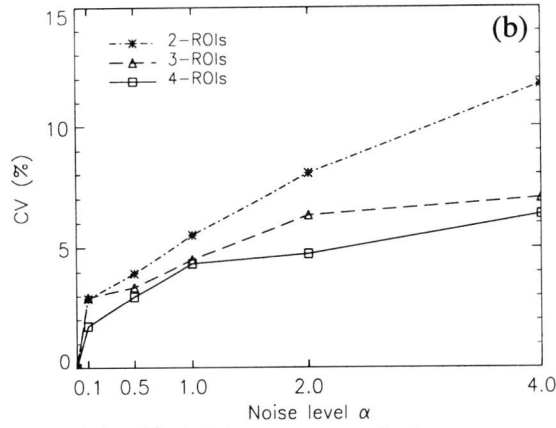

FIGURE 3 Comparison of the biases (a) and the coefficients of variation (b) of K in *gray* matter using two ROIs, three ROIs, and four ROIs in the simulation for different noise levels.

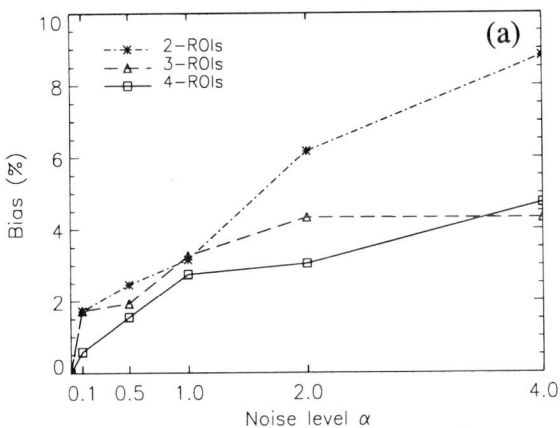

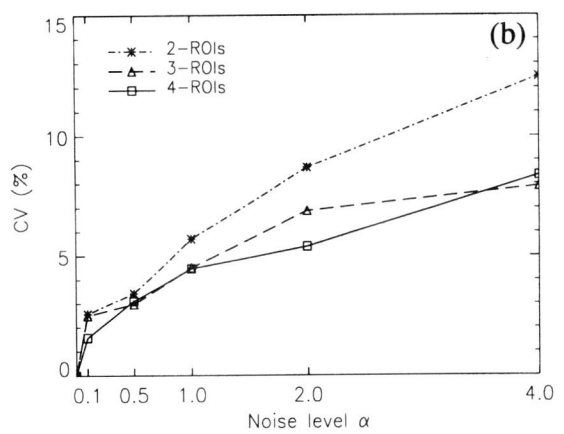

FIGURE 4 Comparison of the biases (a) and the coefficients of variation (b) of K in *white* matter using two ROIs, three ROIs, and four ROIs in the simulation for different noise levels.

FDG–PET was studied by computer simulation only, this method can be applied to the clinical environment and to tracers other than FDG.

References

Feng, D., Huang, S. C., and Wang, X. (1993). Models for computer simulation studies of input functions for tracer kinetic modeling with positron emission tomography. *Int. J. Biomed. Comput.* **32**: 95–110.

Grafton, S. T., and Mazziotta, J. C. (1992). Cerebral pathophysiology evaluated with positron emission tomography. *In* "Diseases of the Nervous System: Clinical Neurobiology" (A. K. Asbury, G. M. McKann, and W. I. McDonald, eds.), pp. 1573–1588. Saunders, Philadelphia.

Huang, S. C., Phelps, M. E., Hoffman, E. J., Siders, K., Selin, C., and Kuhl, D. E. (1980). Noninvasive determination of local cerebral metabolic rate of glucose in man. *Am. J. Physiol.* **238**: E69–E82.

Ido, T., Wan, C. N., Casella, V., Fowler, J. S., Wolf, A. P., Reivich, M., and Kuhl, D. E. (1978). Labeled 2-deoxy-D-glucose analogs: ^{18}F-labeled 2-deoxy-2-fluoro-D-glucose, 2-deoxy-2-fluoro-D-mannose and ^{14}C-2-deoxy-2-fluoro-D-glucose. *J. Labelled Comp. Radiopharm.* **24L** 174–183.

Marquardt, D. W. (1963). An algorithm for least squares estimation of nonlinear parameters. *J. Soc. Ind. Appl. Math.* **2**: 431–441.

Mazziotta, J. C., and Phelps, M. E. (1986). Positron emission tomography studies of the brain. *In* "Positron Emission Tomography and Autoradiography" (M. E. Phelps, J. C. Mazziotta, and H. R. Schelbert, eds.), pp. 493–579. Raven Press, New York.

Phelps, M. E., Huang, S. C., Hoffman, E. J., Selin, C., Sokoloff, L., and Kuhl, D. E. (1979). Tomographic measurement of local cerebral glucose metabolic rate in humans with (F-18)2-fluoro-2-deoxy-D-glucose: Validation of method. *Ann. Neurol.* **6**: 371–388.

Reivich, M., Kuhl, D. E., Wolf, A. P., Greenberg, J. H., Phelps, M. E., Ido, T., Casella, V., Fowler, J. S., Hoffman, E. J., Alavi, A., Som, P., and Sokoloff, L. (1979). The [18]Ffluorodeoxyglucose method for the measurement of local cerebral glucose utilization in man. *Circ. Res.* **44**: 127–137.

Sokoloff, L., Reivich, M., Kennedy, C., DesRosiers, M. H., Patlak, C. S., Pettigrew, K. D., Kakurada, D., and Shinohara, M. (1977). The [14]Cdeoxyglucose method for the measurement of local cerebral glucose utilization: Theory, procedure and normal values in the conscious and anesthetized albino rat. *J. Neurochem.* **28**: 897–916.

CHAPTER 49

Suppression of Noise Artifacts in Spectral Analysis of Dynamic PET Data[1]

VINCENT J. CUNNINGHAM,* ROGER N. GUNN,* HELEN BYRNE,*,†
and JULIAN C. MATTHEWS*,‡

*MRC Cyclotron Unit
Royal Postgraduate Medical School
Hammersmith Hospital
London, United Kingdom
†Department of Mathematics
University of Manchester Institute of Science and Technology
Manchester, United Kingdom
‡Wolfson Brain Imaging Centre
Addenbrookes Hospital
Cambridge, United Kingdom

Spectral analysis is a technique that can be used for the kinetic analysis of dynamic positron emission tomography scans at the voxel level. It is based on the definition of basis functions to describe the expected kinetic behavior of the tracer in the tissue. It is, however, sensitive to noise, particularly when used to derive model parameters such as the total volume of distribution of the tracer. A technique has been developed that uses spectral analysis, but which limits the number of component basis functions that are introduced into the solution sets when fitting data at the voxel level. This significantly reduces the noise sensitivity problem. The technique can utilize an F test or information criteria in testing the suitability of extra component basis functions. In addition, a scheme for spacing the range of basis functions systematically is introduced that is derived from the data collection protocol. The technique is illustrated using both simulated and real data.

I. INTRODUCTION

Spectral analysis provides a general method for the kinetic analysis of dynamic positron emission tomography (PET) data (Cunningham and Jones, 1993; Cunningham et al., 1993; Tadokoro et al., 1993). It is based on the a priori definition of a range of basis functions covering the expected kinetic behavior of the PET tracer in the tissue. Those basis functions whose sum best describes any particular observed time–activity curve of the tracer are selected using a nonnegative least-squares (NNLS) algorithm (Lawson and Hanson, 1974). The method has several advantages in that it is fast and flexible and does not require the prior definition of the number of numerically identifiable components present in the data. This makes it useful at the voxel level. A major disadvantage, however, is that it is sensitive to noise in data. Extra basis functions can be included in the solution set simply to accommodate noise. An approach to this problem based on the inclusion of penalty functions in the NNLS algorithm has already been developed (Cunningham et al., 1993).

[1] Transcripts of the BRAINPET97 discussion of this chapter can be found in Section VIII.

However, in practice, the definition of suitable penalty functions has proved difficult. This chapter describes an alternative approach based on standard statistical tests and information criteria. In addition, some modifications to the method relating to the definition and spacing of the basis functions are introduced.

II. METHODS

A. Spectral Analysis

The method has been described in detail elsewhere (see references in Section I). In brief, for the analysis of dynamic PET data, it is assumed that the impulse response function (IRF) of the tissue can be described by a sum of exponential terms

$$IRF = \sum_{i=1}^{n} \alpha_i \exp[-(\beta_i - \lambda)t],$$ (1)
$$\alpha \geq 0$$

where λ is the isotopic decay constant. This form covers most of the equivalent standard compartmental analyses of total tracer behavior used in the analysis of dynamic PET data.

Working with nondecay-corrected data, basis functions are generated by convolution of the measured input function (usually the time course of the concentration of the parent tracer in arterial plasma) with a range of single exponential terms $[\exp(\beta_i t)]$, followed by integration and averaging over the frame times. The choice of the range of exponential terms β_i is described later. In these implementations, the number of basis functions is set equal to 64 or 100.

The observed time–activity curve to be fitted can then be expressed in terms of a matrix equation:

$$\mathbf{A} \cdot x = b,$$ (2)

where $\mathbf{A}$ is a matrix of weighted basis functions, b is the weighted time–activity curve to be fitted, and x is the solution vector. [The condition number of $\mathbf{A}$ is improved by normalization such that the column vectors in $\mathbf{A}$ are all of unit length, with a corresponding scaling between the x_i and the α_i in Eq. (1).]

A best fit is obtained using the standard nonnegative least-squares algorithm (Lawson and Hanson, 1974). In practice, most of the elements of the solution vector are returned as zero, with a few positive elements identifying the basis functions that contribute to the fit. This is an important general observation relating to the uniqueness of the solution obtained for each fit. Initial inspection of the $\mathbf{A}$ matrix in which the number of basis functions exceeds the number of data points suggests that the solutions might not be unique. This would be the case were it not for the positivity constraint. For basis functions derived from exponential terms, it can be shown that if the number of nonzero peaks in the solution spectrum is less than half the number of data points, then the solution is unique (Gunn, 1996).

B. Range and Spacing of Basis Functions

It is convenient to work with data before correction for the decay of the isotope. A suitable range for β is then from a lower limit of λ to an upper limit of 1 sec^{-1}. A logarithmic spacing of β values between this range was used previously. In this chapter, a spacing equivalent to equal angular spacing of the basis functions across the data space is used, such that

$$\cos(\theta) = a_i \cdot a_j = k,$$ (3)

where θ denotes the angle between columns of the weighted and normalized matrix $\mathbf{A}$ in Eq. (2) and k, the dot product, is a constant for all pairs of adjacent columns (i, j). For a more formal definition of the notions of distance and angle involved in this spacing, see Matthews *et al.* (this volume, Chapter 46). This spacing provides a useful indication of the suitability of the data collection times, as will be illustrated later.

C. Suppression of Noise Artifacts

The basic principle is to limit the number of components introduced into the solution set on the basis of an F test or an information criterion such as Akaike (1974) or Schwarz (1978). As there is not a natural ordering of the variables introduced into the solution set by the NNLS algorithm, the problem is one of determining optimal subsets of increasing order (Lawson and Hanson, 1974). In general this can be computationally prohibitive because of the potentially large number of permutations and because of instability in the solution as the number of parameters in the search is increased. The following approach was therefore developed and implemented.

i. Obtain a fit to data using spectral analysis and the NNLS algorithm to select n components from the set of m basis functions. This provides a best fit, S_n, and the minimum weighted sum of squares, wss_n, compatible with the positivity constraints, and an upper limit, n, on the number of components.

ii. For each case, $k = 1$ to $n - 1$, or until the stopping criterion (below) is reached, search for a positive combination of k basis functions that gives a best fit to S_n using ordinary linear least squares. The corresponding weighted sum of squares of this fit relative to original data, wss_k, is then calculated for the kth case.

iii. The stopping criterion is based on a check for significant improvement of the fit relative to the previous case (e.g., an F test or an information criterion) and a "look ahead" to see if a significant improvement can possibly be obtained in the next case. Because $wss_k > wss_{k+1} \geq = wss_n$ (determined in step i), failure of a test of wss_n against wss_k (based on the introduction of one more component) implies a test of wss_{k+1} against wss_k will fail.

In the current implementation of the method, Hooke and Jeeves' method (Walsh, 1975) was used to carry out the search in step ii. This method was applied to the indexed basis functions. The matrix equation was solved using Householder transformations (Lawson and Hanson, 1974; Golub and Van Loan, 1989). The "look ahead" step improves speed and stability.

D. Simulations

Random sets of data were generated from the sum of two exponentials $[\exp(-0.075t) + \exp(-0.005t)]$ evaluated at specific time points (2, 4, 6, 8, 10, 20, 50, 100, 200, 300, and 500 sec). The standard deviation of data at each point was set equal to its square root multiplied by a scalar common to all the data points. Weights were set equal to the inverse of the resulting variance at each point. Noise was generated at each point by adding the standard deviation multiplied by a random gaussian variate with zero mean and unit variance. One hundred basis functions were generated from single exponential terms evaluated at the time points with $10^{-4} \leq \beta \leq 10^0$.

E. PET Data

Dynamic PET data were taken from a study of a normal male subject (100 kg body weight) who received 362 MBq of the opiate receptor ligand [^{11}C]diprenorphine as an intravenous bolus. The scan was carried out essentially as described by Jones *et al.* (1994) for a tracer-only protocol. Seventeen frames of data were collected (6 × 180 sec, 3 × 240 sec, 4 × 300 sec, 4 × 600 sec). Data were collected in three dimensional mode on a CTI/Siemens 953B PET camera (Spinks *et al.*, 1992). Data were reconstructed with the reprojection algorithm (ramp filter) and a dual window scatter correction applied. The voxel size in the dynamic images was 2.09 × 2.09 × 3.43 mm. The arterial plasma concentration time course of the parent ligand was used as an input function. Sixty-four basis functions were generated as described under Section II, A. The weighting matrix, W, was calculated, assuming that the noise was uniform in the image and uncorrelated between frames, with

W_i = (frame duration)/(true cps) for the ith frame.

III. RESULTS

Simulated noisy data corresponding to the sum of two exponentials as a function of time were generated as described in Section II. Figure 1 shows the distribution of the basis functions used for the analysis of these data. These basis functions were distributed with equal angular separation in the data space. Figure 1 illustrates that the particular set of time points chosen for the simulated data give rise to basis functions that show a regular distribution against the log of the β value in the range $\beta = 10^{-3}$ sec^{-1} to $\beta = 10^0$ sec^{-1}. The two exponential components, with β values equal to 0.005 and .075 sec^{-1}, chosen for simulation purposes are in this range. Figure 1 also illustrates that terms slower than about 10^{-3} sec^{-1} would not be well defined by data at these time points.

Figure 2 illustrates the distribution of the mean size of the basis functions contributing to the fits of 200,000 analyses of random-simulated data. (Individual analyses give a few discrete peaks, each peak consisting of one or two adjacent component basis functions.) Spectral analysis gave a distribution (the broken line in Fig. 2) that clearly identifies the two main components but with noise artifacts. These were distributed across

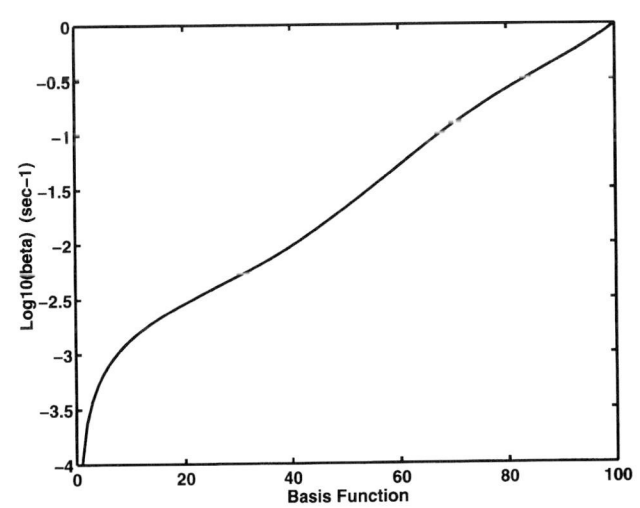

FIGURE 1 Distribution of exponent terms (β) of basis functions with equal angular separation used for the analysis of simulated data. The time points at which the basis functions were evaluated are given in Section II.

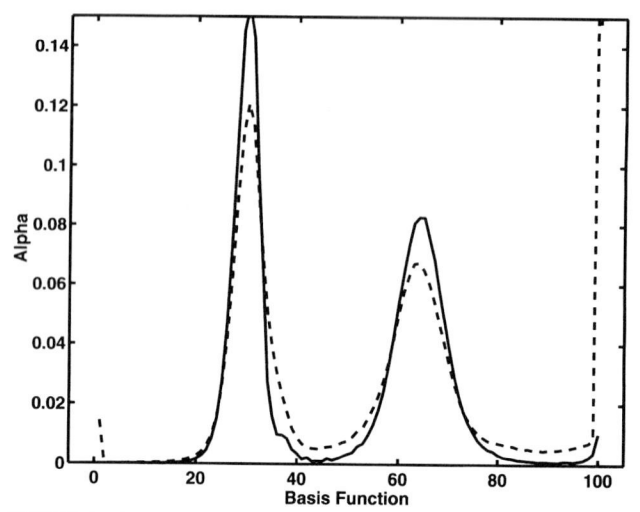

FIGURE 2 Mean distribution of the contribution of basis functions to the fits of 200,000 simulated noisy data sets (see Section II). The error scalar was set to 0.10. The broken line represent spectral analysis and the solid line represents error suppression based on an F test equivalent to $p > 0.01$.

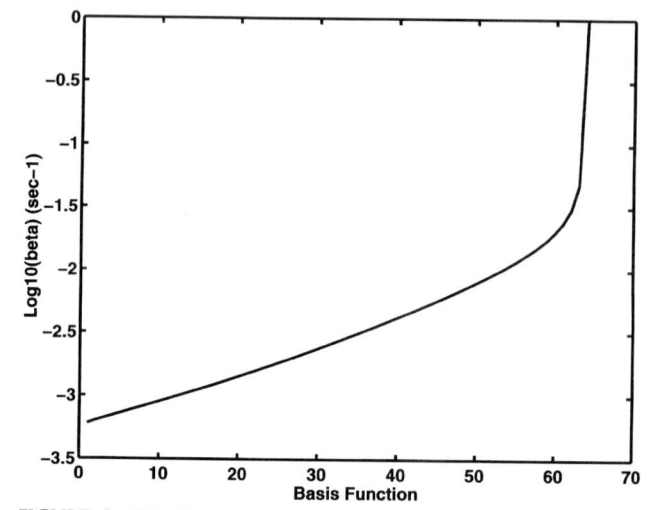

FIGURE 3 Distribution of exponent terms (β) of basis functions with equal angular separation used for the analysis of PET data. The frame times for which the basis functions were evaluated are given in Section II.

the spectrum together with a "pile up" at the ends of the range of basis functions. The sum of the mean values across the spectrum was 2.4. This corresponds to the function extrapolated to zero time, compared with 2.0 in the original noise-free function. Application of the revised algorithm to the same data sets with a stopping criterion based on an F test equivalent to $p < 0.01$ is illustrated by the solid line in Fig. 2. This result illustrates the effect of the new method in suppressing spurious noise components. The sum of the mean values was reduced to 2.05 in this case.

The method was applied to a dynamic PET study of [^{11}C]diprenorphine. Very noisy images of the total volume of distribution of the neuroreceptor ligand were obtained in this study with spectral analysis when the lower limit of the basis functions was set close to the isotopic decay constant of the tracer (i.e., with no limit on the decay-corrected value). The distribution of basis functions for this analysis with equal angular separation in the data space is shown in Fig. 3, which illustrates that the frame protocol (see Section II) gave a regular distribution of basis functions across β values in the range of about 10^{-2} sec^{-1} and slower. This is adequate for tissue diprenorphine-binding kinetics, but faster dynamics (e.g., blood volume effects) with $\beta > 10^{-2}$ sec^{-1} would be poorly defined by this frame protocol.

Figure 4 shows the mean spectral distributions obtained from analyses of this scan, involving about 108,000 voxel time–activity curves, obtained from masked dynamic images. Data are presented to show the relative frequency of occurrence of the basis functions in the solutions. The "pile up" of occurrences at the low and high frequency ends in the spectral analysis (the broken line in Fig. 4) gives rise to noisy extrapolated estimates of, respectively, the total volume of distribution and of delivery (conventionally K_1 in compartmental models).

Application of an F test equivalent to $p < 0.001$ (the solid line in Fig. 4) reduced these noisy components considerably. However, there was also a suppres-

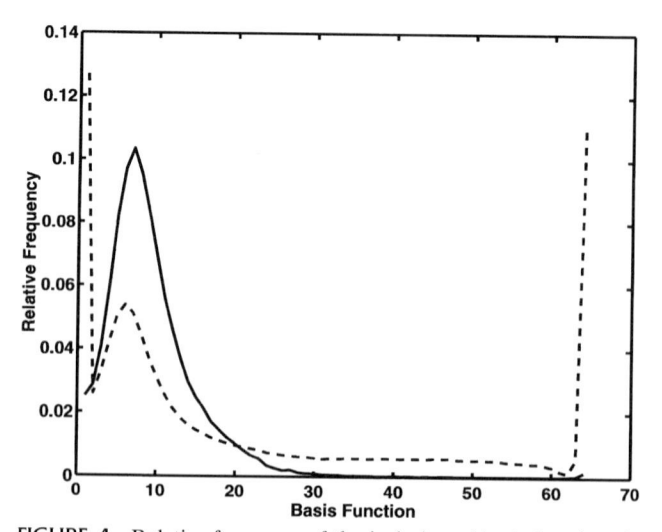

FIGURE 4 Relative frequency of the inclusion of basis functions in the analysis of original (unsmoothed) dynamic PET data consisting of 108,000 voxel time–activity curves, each of 17 frames. The broken line represents spectral analysis and the solid line represents error suppression based on an F test equivalent to $p > 0.001$.

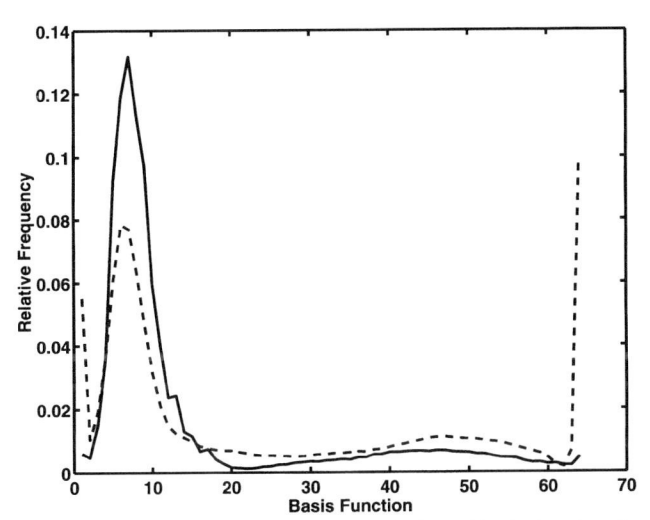

FIGURE 5 Relative frequency of the inclusion of basis functions in the analysis of smoothed dynamic PET data consisting of 108,000 voxel time–activity curves, each of 17 frames. The broken line represents spectral analysis and the solid line represents error suppression based on an F test equivalent to $p > 0.001$.

sion of components introduced in the dynamic range corresponding to basis functions 20–60 (Figs. 3 and 4). (Qualitatively similar effects were seen using other significance levels and the information criteria.) The multicompartmental behavior of [^{11}C]diprenorphine in brain tissue and the presence of regions free of specific binding sites (Jones et al., 1994) suggest that this may involve a loss of true signal among the suppressed noise. This signal was recovered by smoothing data before analysis, as is illustrated in Fig. 5. The original dynamic frames were smoothed with a 3D Gaussian kernel with FWHM of 1 cm. The consequent loss of spatial resolution was accompanied by increased resolution of the kinetic components, which were retained after noise suppression (Fig. 5).

V. DISCUSSION

The method of spectral analysis as originally implemented is sensitive to noise in data. This sensitivity is reflected in a distribution of peaks across the whole spectrum, but is particularly evident when the parameters to be estimated require that data be extrapolated. An example of this is in the estimation of the total volume of distribution of a tracer in the tissue. Although the method is very effective at extracting the tissue IRF as defined within the time course of observed data, estimates of the total volume of distribution require extrapolation of the IRF to infinite time. In some cases, although satisfactory images of the volume of distribution can be obtained at the voxel level [e.g. see the diprenorphine study of Tadokoro et al. (1993)], a "pile up" of peaks at the low frequency end of the spectrum frequently gives noisy images of the volume of distribution. The introduction of penalty functions into the NNLS algorithm can be used to suppress noisy artifacts (Cunningham et al., 1993), but as yet there is no systematic way of generating penalty functions suitable for routine analysis of dynamic PET data. Image quality can be improved by smoothing the dynamic images spatially or by setting the range of basis functions above the decay constant for the isotope. A further alternative is the use of images of the value of the IRF at specific time points in order to compare studies (Tadokoro et al., 1993; Weeks et al., 1997). Early times reflect delivery and late times retention of the tracer, but there is no simple relationship between these values and the underlying pharmacokinetic model. This chapter has presented an approach to the problem that aims to avoid the introduction of spurious components into the solution set and is based on standard statistical tests. The computational feasibility of the method is assisted by the "look ahead" step in the algorithm.

The results presented here have illustrated the application of the technique to noisy dynamic PET data at the voxel level. Any technique for fitting time–activity curves amounts to smoothing data temporally within the constraints imposed by the particular model used. The temporal resolution, as indicated by the number of kinetic parameters or components that can then be identified numerically in the fit, is obviously a function of the noise in data. Spectral analysis can be viewed as an explicitly model-based smoothed deconvolution of the tissue response to the plasma input function. The techniques presented in this chapter have effectively reduced the variance in parameters derived from the model, such as the volume of distribution, by imposing statistically based constraints on the number of components entering into the fits and by reducing spatial resolution in order to gain temporal resolution. The present technique thus provides an extra tool for the exploration and interpretation of dynamic PET data.

References

Akaike, H. (1974). A new look at the statistical model identification. *IEEE Trans. Autom. Control* **19**: 716–723

Cunningham, V. J., and Jones, T. (1993). Spectral analysis of dynamic PET studies. *J. Cereb. Blood Flow Metab.* **13**: 15–23.

Cunningham, V. J., Ashburner, J., Byrne, H., and Jones, T. (1993). Use of spectral analysis to obtain parametric images from dynamic PET studies. *In* "Quantification of Brain Function. Tracer Kinetics and Image Analysis in Brain PET." (K. Uemura, N. A.

Lassen, T. Jones, and I. Kanno, eds.), pp. 101–112. Elsevier, Amsterdam and New York.

Golub, G. H., and Van Loan, C. F. (1989). "Matrix Computations," 2nd ed. Johns Hopkins University Press, Baltimore.

Gunn, R. N. (1996). Mathematical modelling and identifiability applied to positron emission tomography data. Ph.D. Thesis, University of Warwick, Warwick, UK.

Jones, A. K. P., Cunningham, V. J., Sang-Kil, H. K., Takehiko, F., Liyi, Q., Luthra, S. K., Ashburner, J., Osman, S., and Jones, T. (1994). Quantitation of [^{11}C]diprenorphine cerebral kinetics in man acquired by PET using presaturation, pulse-chase and tracer-only protocols. *J. Neurosci. Methods* **51**: 123–134.

Lawson, C. L., and Hanson, R. J. (1974). "Solving Least Squares Problems." Prentice-Hall, Englewood Cliffs, NJ.

Schwarz, G. (1978). Estimating the dimension of a model. *Ann. Stat.* **6**: 461–564.

Spinks, T. J., Jones, T., Bailey, D. L., Townsend, D. W., Grootoonk, S., Bloomfield, P. M., Gilardi, M. C., Casey, M. E., Sipe, B., and Reed, J. (1992) Physical performance of a positron tomograph for brain imaging with retractable septa. *Phys. Med. Biol.* **37**: 1637–1655.

Tadokoro, M., Jones, A. K. P., Cunningham, V. J., Sashin, D., Grootoonk, S., Ashburner, J., and Jones, T. (1993). Parametric images of ^{11}C-diprenorphine binding using spectral analysis of dynamic PET images acquired in 3D. *In* Quantification of Brain Function. Tracer Kinetics and Image Analysis in Brain PET." (K. Uemura, N. A. Lassen, T. Jones, and I. Kanno, eds.), pp. 289–295. Elsevier, Amsterdam and New York.

Walsh, G. R. (1975) "Methods of Optimization." Wiley, London.

Weeks, R. A., Cunningham, V. J., Piccini, P., Waters, S., Harding, A. F., and Brooks, D. J. (1997). [^{11}C]diprenorphine binding in Huntington's disease; A comparison of region of interest analysis with statistical parametric mapping. *J. Cereb. Blood Flow Metab.* **17**, 943–949.

CHAPTER 50

Constraints in Spectral Analysis

DAVID C. REUTENS and MARK ANDERMANN

McConnell Brain Imaging Centre
Montreal Neurological Institute and Department of Neurology and Neurosurgery
McGill University
Montreal, H3A 2B4 Canada

The general method of spectral analysis is more computationally efficient than traditional methods for the estimation of kinetic parameters that use nonlinear least-squares fitting algorithms. Although spectral analysis was originally proposed as a model-independent method, appropriate compartmental models assist the physiological interpretation of the spectral components obtained. For analysis of a three-compartment model with irreversible uptake into a tissue compartment (describing, e.g., the uptake of [^{11}C]L-deprenyl and [^{18}F]fluorodeoxyglucose), this chapter describes modifications of the spectral analysis method that accommodate constraints based on the model and incorporate the physiological constraint of a fixed partition volume $V_e(K_1/k_2)$.

I. INTRODUCTION

The spectral analysis technique described by Cunningham and Jones (1993) possesses several advantages over traditional compartmental modeling methods that use nonlinear least-squares fitting algorithms for the estimation of kinetic parameters. A major advantage is computational efficiency, an important factor in choosing a method for the generation of parametric images. The spectral method expresses the brain tissue time–activity curve [$M_T(T)$] in terms of the convolution of the input function [$C_a(T)$] with a sum of n exponential terms:

$$M_T(T) = A_0 \int_0^T C_a(u)\,du + \sum_{i=1}^{n} A_i \int_0^T e^{B_i(u-T)} C_a(u)\,du + V_0 C_a(T). \tag{1}$$

The possible exponential terms are predetermined to cover an appropriate spectral range and the convolution integrals need only be calculated once. Values for A_i are obtained using a linear optimization algorithm such as the nonnegative least-squares method. In contrast, in nonlinear least-squares fitting algorithms, convolution integrals are calculated in each iteration.

Spectral analysis was originally proposed as a model-independent method with the range of exponential coefficients constrained to reflect the physical properties of the tracer and the attributes of the tissue. However, as noted by Turkheimer *et al.* (1994), the existence of a valid compartmental model assists the physiological interpretation of the components obtained by spectral analysis. This chapter discusses modifications of the general spectral method to take into account compartmental models and physiological constraints.

II. THEORY

The general spectral method can be readily adapted for the analysis of specific compartmental models. For example, in a three-compartment model with irreversible uptake into a tissue compartment (describing, e.g., the uptake of L-[^{11}C]deprenyl or [^{18}F]fluorodeoxyglucose), the solution of the differential equations describing the model yields the following operational equation

$$M_T(T) = A_0 \int_0^T C_a(u)\,du + A_1 \int_0^T e^{B(u-T)} C_a(u)\,du + V_0 C_a(T), \tag{2}$$

where $B = k_2 + k_3$, $A_0 = K_1 k_3/(k_2 + k_3)$, $A_1 = K_1 k_2/(k_2 + k_3)$, and V_0 is the vascular volume. Here, K_1 is the transfer coefficient describing the transport of tracer from plasma to brain, k_2 is the first-order rate constant for tracer transport from brain to plasma, and k_3 is the first-order rate constant for specific and irreversible binding. The number of spectral components (B) may be constrained to a single component by computing the nonnegative least-squares estimates for A_1 and A_0 sequentially for each possible value of B. The set of values for B, A_0, and A_1, which minimizes the sum of squared residual differences between measured and estimated tissue radioactivity, is chosen. This model-constrained method shares with the general spectral method the attribute that convolution integrals are calculated only once.

In using the model-constrained method, it is useful to incorporate additional physiological constraints, such as a fixed value for the partition volume $V_e(K_1/k_2)$. For the three-compartment model with irreversible uptake into a tissue compartment, the following modification allows the partition volume to be fixed. The coefficients in Eq. (2) can be restated in terms of K_1, V_e, and B:

$$A_0 = \frac{K_1 k_3}{k_2 + k_3} = \frac{K_1(B \cdot V_e - K_1)}{B \cdot V_e} = K_1 - \frac{K_1^2}{B \cdot V_e} \quad (3)$$

$$A_1 = \frac{K_1 k_2}{k_2 + k_3} = \frac{K_1^2}{B \cdot V_e}. \quad (4)$$

If V_0 is assumed to be negligible or if a fixed vascular volume correction is used, the operational equation can now be stated as

$$M_T^*(T) = M_T(T) - V_0 C_a(T) = K_1^2 C(T) + K_1 D(T), \quad (5)$$

where

$$C(T) = \int_0^T \frac{[e^{B(u-T)} C_a(u) - C_a(u)]}{B \cdot V_e} du \quad (6)$$

$$D(T) = \int_0^T C_a(u) \, du.$$

Fitting proceeds in two stages. First, the least-squares fit for K_1 is determined for each possible value of B. For each frame in the positron emission tomography (PET) study and each value of B, $C(T)$ and $D(T)$ are calculated. The least-squares estimate for K_1 is, by definition, the value that minimizes

$$\text{Res} = \sum_{i=1}^{N} (K_1^2 C(T_i) + K_1 D(T_i) - M_T^*(T_i))^2, \quad (7)$$

where N is the number of frames in the PET study. By expansion and differentiation, the least-squares estimate for K_1 is thus a real root of the equation

$$0 = K_1^3 \sum_{i=1}^{N} 2 \cdot C^2(T_i) + K_1^2 \sum_{i=1}^{N} 3 \cdot C(T_i) \cdot D(T_i)$$
$$+ K_1 \sum_{i=1}^{N} [D^2(T_i) - 2 \cdot C(T_i) M_T^*(T_i)]$$
$$- \sum_{i=1}^{N} D(T_i) \cdot M_T^*(T_i), \quad (8)$$

which can be obtained analytically (Press et al., 1992). In the second stage, Eq. (7) is computed for each value of B and its associated K_1 value. The pair of B and K_1 values that minimize the residual sum of squares is chosen.

III. METHODS

The V_e-constrained and model-constrained methods were examined using simulated brain time–activity curves generated using Eq. (2) with plasma time–activity data from an actual PET study. Mean noise levels of 5, 10, and 15% Gaussian noise were then added to simulated time–activity data. The constrained methods were compared with nonlinear least-squares fitting both with and without a fixed V_e. For each parameter, the mean and standard deviation of estimates from 100 sets of "noisy" brain time–activity data was calculated. In order to compare the computational efficiency of the three methods of analysis, the CPU time required for the analysis of 800 sets of data was compared. These data were generated with values of K_1 and k_2 of 0.5 ml g^{-1} min^{-1} and 0.167 min^{-1}, respectively, and values of k_3 ranging from 0.05 to 0.75 min^{-1}. In nonlinear least-squares fitting, 0.5 ml g^{-1} min^{-1} was used as the starting value for K_1 and 0.5 min^{-1} was used as the starting value for k_2 and k_3. The analysis was performed using MATLAB (The Mathworks Inc., MA) and a personal computer (Pentium Pro 200MHz CPU; 64 MB RAM).

IV. RESULTS AND DISCUSSION

The results of simulations are shown in Table 1. The accuracy and precision of the V_e-constrained method and of nonlinear least-squares fitting with constrained

TABLE 1 Comparison of Parameter Estimates (Mean ± Standard Deviation) for 100 Time–Activity Curves at Different "Noise" Levels[a]

Noise (%)	True K_1[b]	Estimated K_1 Method A	Method B	Method C	Method D	True k_3[c]	Estimated k_3 Method A	Method B	Method C	Method D
5	0.5	0.50 ± 0.01	0.50 ± 0.01	0.49 ± 0.01	0.49 ± 0.01	0.05	0.050 ± 0.003	0.050 ± 0.001	0.050 ± 0.003	0.048 ± 0.001
5	0.5	0.50 ± 0.01	0.50 ± 0.01	0.49 ± 0.02	0.50 ± 0.01	0.25	0.25 ± 0.04	0.25 ± 0.01	0.25 ± 0.05	0.26 ± 0.02
5	0.5	0.50 ± 0.02	0.50 ± 0.01	0.50 ± 0.02	0.50 ± 0.01	0.45	0.46 ± 0.10	0.45 ± 0.03	0.45 ± 0.13	0.46 ± 0.04
10	0.5	0.49 ± 0.02	0.49 ± 0.01	0.49 ± 0.02	0.49 ± 0.02	0.05	0.049 ± 0.005	0.050 ± 0.002	0.050 ± 0.006	0.049 ± 0.003
10	0.5	0.49 ± 0.02	0.50 ± 0.02	0.49 ± 0.03	0.50 ± 0.02	0.25	0.25 ± 0.09	0.25 ± 0.02	0.24 ± 0.09	0.25 ± 0.03
10	0.5	0.50 ± 0.03	0.50 ± 0.02	0.50 ± 0.04	0.50 ± 0.02	0.45	0.50 ± 0.28	0.46 ± 0.08	0.48 ± 0.24	0.47 ± 0.09
15	0.5	0.49 ± 0.02	0.49 ± 0.02	0.49 ± 0.02	0.49 ± 0.02	0.05	0.050 ± 0.009	0.049 ± 0.004	0.050 ± 0.009	0.050 ± 0.004
15	0.5	0.50 ± 0.05	0.49 ± 0.02	0.50 ± 0.06	0.49 ± 0.03	0.25	0.27 ± 0.15	0.25 ± 0.05	0.27 ± 0.15	0.26 ± 0.05
15	0.5	0.53 ± 0.08	0.49 ± 0.03	0.50 ± 0.05	0.50 ± 0.03	0.45	0.70 ± 0.59	0.47 ± 0.13	0.50 ± 0.33	0.49 ± 0.14

[a] Unconstrained nonlinear least-squares fitting (Method A), nonlinear least-squares fitting with fixed V_e (Method B), the model-constrained method (Method C), and the V_e-constrained method (Method D) are compared.
[b] ml g^{-1} min^{-1}.
[c] min^{-1}.

V_e were similar. Both methods were less susceptible to the effects of noise in the time–activity curve than unconstrained nonlinear least-squares fitting or the model-constrained method. The CPU time required for parameter estimation in 800 data sites was 1819 sec for unconstrained nonlinear-least squares fitting, 778 sec for constrained nonlinear least-squares fitting, 42 sec for the model-constrained method, and 24 sec for the V_e-constrained method.

In summary, the authors proposed methods based on the spectral method for the estimation of kinetic parameters in a three-compartment model with irreversible binding. The methods accommodate model-based constraints as well as constraints on the value of the partition volume, V_e. In addition, they are computationally efficient and should be of value in the generation of parametric images for tracers such as L-[^{11}C]deprenyl.

References

Cunningham, V. J., and Jones, T. (1993). Spectral analysis of dynamic PET studies. *J. Cereb. Blood Flow Metab.* **13**: 15–23.

Press, W. H., Teukolsky, S. A., Vetterling, W. T., and Flannery, B. P. (1992). "Numerical Recipes in C The Art of Scientific Computing," 2nd edition, pp. 183–185. Cambridge University Press, Cambridge, UK.

Turkheimer, F., Moresco, R. M., Lucignani, G., Sokoloff, L., Fazio, F., and Schmidt, K. (1994). The use of spectral analysis to determine regional cerebral glucose utilization with positron emission tomography and [^{18}F]fluorodeoxyglucose: Theory, implementation, and optimization procedures. *J. Cereb. Blood Flow Metab.* **14**: 406–422.

CHAPTER 51

Generalized Linear Least-Squares Modeling Algorithm for Optimally Sampled PET Image Data

D. FENG,* D. HO,* K. K. LAU,[†] and W.-C. SIU[†]

*Biomedical and Multimedia Information Technology Group
Department of Computer Science
The University of Sydney
Sydney, New South Wales 2006, Australia
[†]Department of Electronic Engineering
The Hong Kong Polytechnic University
Hung Hom, Kowloon, Hong Kong

The original generalized linear least-squares (GLLS) algorithm was developed for nonuniformly sampled biomedical system parameter estimation using finely sampled instantaneous measurements. This algorithm is particularly useful in image-wide estimation of multicompartmental model rate constants and physiological parameters with positron emission tomography (PET), as it is statistically reliable and computationally efficient. However, when PET image data are sampled according to an optimal sampling schedule (OSS) in which only a few images are required [e.g., only four images are required for the four-parameter [^{18}F]fluorodeoxyglucose (FDG) model], the direct application of GLLS is not reliable as instantaneous measurements can no longer be approximated by averaging of accumulated measurements over the sampling intervals. This chapter extends GLLS to OSS-GLLS, which deals directly with fewer accumulated measurements using OSS. The theory and implementation of OSS-GLLS are studied extensively in this chapter. A simulation study using the three-compartment, four-parameter FDG model to determine the metabolic rate of glucose and FDG rate constants is presented.

I. INTRODUCTION

Positron emission tomography (PET) is distinguished from several other medical imaging modalities as it holds the promise of accurately quantifying physiological and/or biochemical functions within the body.

With the application of parametric imaging techniques, visual representations of these functions may be obtained. The original generalized linear least-squares (GLLS) algorithm may be used to obtain parametric images as it is computationally fast, tracer and model configuration independent, and can produce estimation as accurate as the standard model-fitting method (Feng et al., 1996). However, when PET data are sampled according to an optimal sampling schedule (OSS) (Huang, 1994; Li et al., 1996) to drastically reduce measurement samples while preserving the amount of information needed for accurate estimation of physiological parameters, the direct application of GLLS is not reliable as instantaneous measurements can no longer be approximated by averaging of accumulated measurements over sampling intervals.

This chapter extends GLLS to OSS-GLLS, which directly deals with fewer accumulated dynamic PET images obtained from OSS. The detailed theory and algorithm of this new technique are formulated and studied extensively. To investigate the statistical reliability and computational efficiency of OSS-GLLS, a case study using the three-compartment, four-parameter [^{18}F]fluorodeoxyglucose (FDG) model is performed. OSS-GLLS using 4 measurement samples was compared to the standard model fitting method, i.e., nonlinear least squares (NLS) using 22 measurement samples and the original GLLS using 22 measurement samples. Various practical implementation aspects of OSS-GLLS are discussed and results presented.

II. MATERIALS AND METHODS

A. Optimal Sampling Schedule Design

In conventional dynamic PET studies, tracer uptake is measured over an interval of time, acquiring as many measurement samples as the instrumentation allowed in relation to acceptable measurement noise (Huang et al., 1980). Subsequently, instantaneous measurement samples are obtained by averaging of accumulated PET counts over sampling intervals (Mazoyer et al., 1986). With an OSS design, the number of measurement samples can be drastically reduced to as few as the number of model parameters while estimation accuracy is maintained (Li et al., 1996). Thus storage space for dynamic PET data as well as processing time for image analysis can be reduced significantly. However, when dynamic PET data are acquired according to an OSS design, instantaneous measurements can no longer be approximated by averaging of accumulated PET counts over the sampling intervals (Hawkins et al., 1986; Mazoyer et al., 1986). Therefore, it is necessary to directly incorporate sampled accumulated PET measurements into the parameter estimation procedure.

B. OSS Generalized Linear Least-Squares Theory

The time course information of a broad class of labeled tracers can be represented by the following nth order differential equation (Feng et al., 1996)

$$\frac{d^n}{dt^n}y(t) + a_1\frac{d^{n-1}}{dt^{n-1}}y(t) + \cdots + a_n y(t)$$

$$= b_1\frac{d^{n-1}}{dt^{n-1}}u(t) + \cdots + b_n u(t), \quad (1)$$

where $y(t)$ denotes measured tracer activity and $u(t)$ the input function. In PET the following conditions also apply: $(d^{n-1}/dt^{n-1})y(0^-) = (d^{n-2}/dt^{n-2})y(0^-) = \cdots = y(0^-) = 0$ and $(d^{n-1}/dt^{n-1})u(0^-) = (d^{n-2}/dt^{n-2})u(0^-) = \cdots = u(0^-) = 0$. Integrating Eq. (1) from 0 to t, $n+1$ times, the authors obtain the linear accumulated measurement form of the differential equation with respect to parameters $a_1,\ldots,a_n$ and $b_1,\ldots,b_n$ as

$$\int_0^t y(t)\,dt = -a_1\int_0^t\int_0^t y(t)\,dt^2 - \cdots a_n\int_0^t\int_0^t\cdots$$

$$\int_0^t y(t)\,dt^{n+1} + b_1\int_0^t\int_0^t u(t)\,dt^2 + \cdots$$

$$+ b_n\int_0^t\int_0^t\cdots\int_0^t u(t)\,dt^{n+1} \quad (2)$$

The general formulation of OSS linear least squares (LLS) is based on Eq. (2), the accumulated PET measurements and input function. Taking measurement samples at optimally sampled times t_i ($i = 1, 2, \ldots, m$, where m is the number of parameters to be estimated) the following set of linear equations is obtained:

$$\mathbf{y} = \mathbf{X}\theta + \xi, \quad (3)$$

where $\mathbf{y} = [\int_0^{t_1} y(t)\,dt, \ldots, \int_0^{t_m} y(t)\,dt]^T$ are optimally sampled accumulated measurements, θ the parameters to be estimated, $\xi = [\xi(t_1)\ldots\xi(t_m)]^T$ the equation errors, and:

$$\mathbf{X} = \begin{bmatrix} \int_0^{t_1}\int_0^{t_1} y(t)\,dt^2, & \ldots, & \int_0^{t_1}\int_0^{t_1}\cdots\int_0^{t_1} y(t)\,dt^{n+1}, & \int_0^{t_1}\int_0^{t_1} u(t)\,dt^2, & \ldots, & \int_0^{t_1}\int_0^{t_1}\cdots\int_0^{t_1} u(t)\,dt^{n+1} \\ \int_0^{t_2}\int_0^{t_2} y(t)\,dt^2, & \ldots, & \int_0^{t_2}\int_0^{t_2}\cdots\int_0^{t_2} y(t)\,dt^{n+1}, & \int_0^{t_2}\int_0^{t_2} u(t)\,dt^2, & \ldots, & \int_0^{t_2}\int_0^{t_2}\cdots\int_0^{t_2} u(t)\,dt^{n+1} \\ \ddots & & \vdots & \vdots & \ddots & \vdots \\ \int_0^{t_m}\int_0^{t_m} y(t)\,dt^2, & \ldots, & \int_0^{t_m}\int_0^{t_m}\cdots\int_0^{t_m} y(t)\,dt^{n+1}, & \int_0^{t_m}\int_0^{t_m} u(t)\,dt^2, & \ldots, & \int_0^{t_m}\int_0^{t_m}\cdots\int_0^{t_m} u(t)\,dt^{n+1} \end{bmatrix}$$

Because $\mathbf{X}$ is a square matrix, the solution to Eq. (3) may simply be given as

$$\hat{\theta}_{\text{OSS-LLS}} = \mathbf{X}^{-1}\mathbf{y}, \quad (4)$$

where $\hat{\theta}_{\text{OSS-LLS}}$ represents estimated parameters obtained from OSS-LLS. For the sake of simplicity, the noise term has been left out in this derivation. Parameter estimates obtained using OSS-LLS are biased due to correlation of equation errors resulting from integration of measurements in $\mathbf{X}$. To overcome this bias and improve on estimation accuracy, GLLS theory may be used (Feng et al., 1996). GLLS improves on estimation accuracy by applying an autoregressive model for the noise process to filter correlated noise.

To formulate OSS-GLLS it is necessary to determine the correlated noise term. It is easy to show that correlated noise in Eq. (4) is given by the following Laplacian term, $s(s^n + a_1 s^{n-1} + \cdots + a_n)E[s]$ where $E[s]$ is the Laplace transform of ξ. Therefore, to filter this correlated noise, an autoregressive filter $G[s] = s(s^n + \hat{a}_1 s^{n-1} + \cdots + \hat{a}_n)$ is defined where $\hat{a}_1, \ldots, \hat{a}_n$ are rough estimates of $a_1, \ldots, a_n$. Dividing the Laplace transform of Eq. (1) by this filter and rearranging the resulting equation, the following equation is obtained:

$$\frac{s^n}{s(s^n + \hat{a}_1 s^{n-1} + \cdots + \hat{a}_n)} Y[s]$$

$$= -\frac{a_1 s^{n-1} + \cdots + a_n}{s(s^n + \hat{a}_1 s^{n-1} + \cdots + \hat{a}_n)} Y[s]$$

$$+ \frac{b_1 s^{n-1} + \cdots + b_n}{s(s^n + \hat{a}_1 s^{n-1} + \cdots + \hat{a}_n)} U[s]$$

$$+ \frac{s(s^n + a_1 s^{n-1} + \cdots + a_n)}{s(s^n + \hat{a}_1 s^{n-1} + \cdots + \hat{a}_n)} E[s], \quad (5)$$

where $Y[s]$ and $U[s]$ are Laplace transforms of $y(t)$ and $u(t)$, respectively. The Laplacian noise term $s(s^n + a_1 s^{n-1} + \cdots + a_n)E[s]$ corresponds to correlated noise introduced by OSS-LLS. Taking the inverse Laplace transform of Eq. (5), the authors obtain

$$\sum_{i=1}^{n} \lambda_i^{n-1} y_i(t)$$

$$- - \sum_{k=1}^{n-1} a_k \sum_{i=1}^{n} \lambda_i^{k-1} y_i(t)$$

$$- a_n \left\{ \frac{1}{(-1)^n \lambda_1 \lambda_2 \ldots \lambda_n} \int_0^t y(t) \, dt + \sum_{i=1}^{n} \frac{1}{\lambda_i} y_i(t) \right\}$$

$$- \sum_{k=1}^{n-1} b_k \sum_{i=1}^{n} \lambda_i^{k-1} u_i(t)$$

$$- b_n \left\{ \frac{1}{(-1)^n \lambda_1 \lambda_2 \ldots \lambda_n} \int_0^t u(t) \, dt + \sum_{i=1}^{n} \frac{1}{\lambda_i} u_i(t) \right\}, \quad (6)$$

where λ_i are distinct nonzero roots of $s^n + \hat{a}_1 s^{n-1} + \cdots + \hat{a}_n$ (the rough estimates $\hat{a}_1, \ldots, \hat{a}_n$ are obtained from OSS-LLS), $y_i(t) = \{y(t) \otimes e^{\lambda_i t}\}/F'(\lambda_i)$, $u_i(t) = \{u(t) \otimes e^{\lambda_i t}\}/F'(\lambda_i)$, and $F'(\lambda_i) = \sum_{i=1}^{n} \{\prod_{j=1, j \neq i}^{n} \{\lambda_i - \lambda_j\}\}^{-1}$ (Feng et al., 1996). As before, the noise term has been left out as it does not have any direct effect on the derivation of OSS-GLLS. In Eq. (6), the terms on the *left-hand side* correspond to filtered PET accumulated measurements, whereas on the *right-hand side*, the filter model output function incorporating the autoregressive model for the noise process is shown. To finally formulate OSS-GLLS, the authors require the convolution term $y(t) \otimes e^{\lambda_i t}$ expressed in $y_i(t)$ in terms of accumulated measurements. Without loss of generality, let us focus on the convolution term $y(t) \otimes e^{\lambda t}$. Integrating this term by parts, its integral form is obtained as

$$y(t) \otimes e^{\lambda t} = \int_0^t y(t) \, dt - \lambda \left\{ \int_0^t y(t) \, dt \right\} \otimes e^{\lambda t}. \quad (7)$$

OSS-GLLS using accumulated measurements is based on Eq. (6) and the integral form of $y_i(t) = (\int_0^t y(t) \, dt - \lambda_i \{\int_0^t y(t) \, dt\} \otimes e^{\lambda_i t})/F'(\lambda_i)$. Digitizing Eq. (6) at optimally sampled times t_i ($i = 1, 2, \ldots, m$) the following equation is obtained:

$$\mathbf{r} = \mathbf{Z}\theta, \quad (8)$$

where $\mathbf{r} = [\sum_{i=1}^{n} \lambda_i^{n-1} y_i(t_1), \ldots, \sum_{i=1}^{n} \lambda_i^{n-1} y_i(t_m)]^T$ are the filtered PET accumulated measurements, θ the parameters to be estimated, and:

$$\mathbf{Z} = \begin{bmatrix} \Gamma_1^1(t_1), & \ldots, & \Gamma_1^{n-1}(t_1), & \Gamma_2(t_1), & \Gamma_3^1(t_1), & \ldots, & \Gamma_3^{n-1}(t_1), & \Gamma_4(t_1) \\ \Gamma_1^1(t_2), & \ldots, & \Gamma_1^{n-1}(t_2), & \Gamma_2(t_2), & \Gamma_3^1(t_2), & \ldots, & \Gamma_3^{n-1}(t_2), & \Gamma_4(t_2) \\ \vdots & \ddots & \vdots & \vdots & \vdots & \ddots & \vdots & \vdots \\ \Gamma_1^1(t_m), & \ldots, & \Gamma_1^{n-1}(t_m), & \Gamma_2(t_m), & \Gamma_3^1(t_m), & \ldots, & \Gamma_3^{n-1}(t_m), & \Gamma_4(t_m) \end{bmatrix}$$

VI. Parameter Estimation

MRGlc physiological parameter estimate

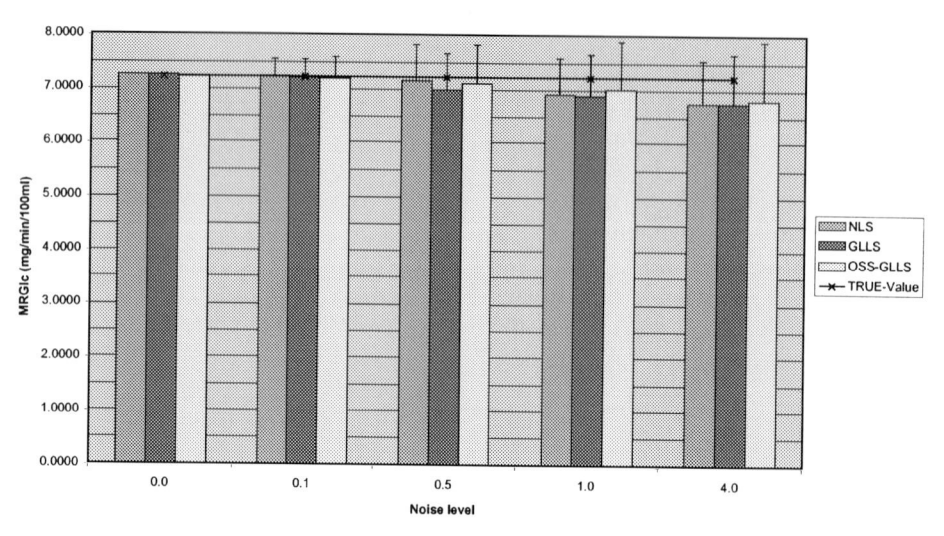

k*[1] rate constant parameter estimate

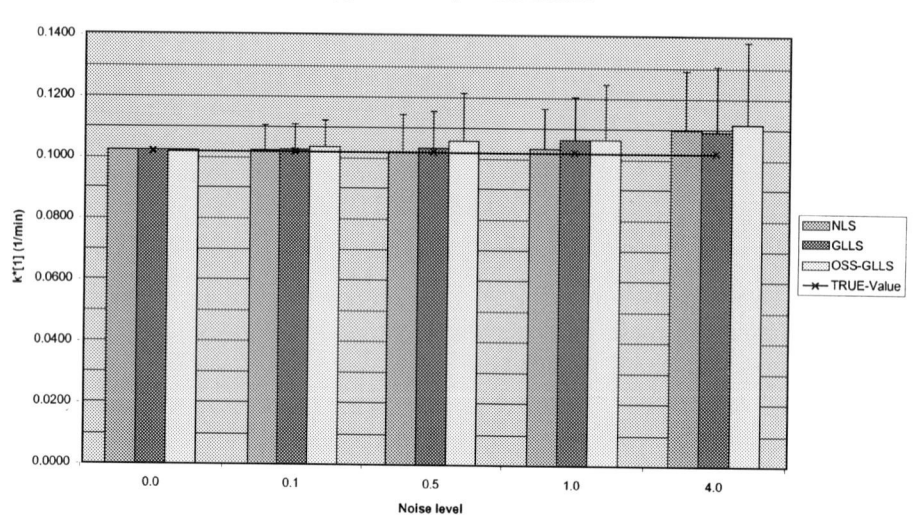

k*[2] rate constant parameter estimate

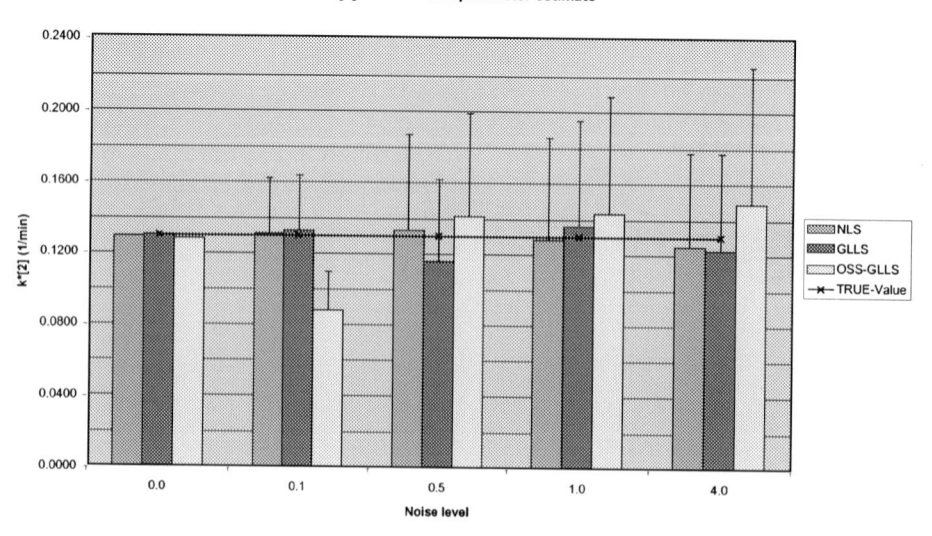

51. GLLS Modeling Algorithm for Optimally Sampled Data

k*[3] rate constant parameter estimate

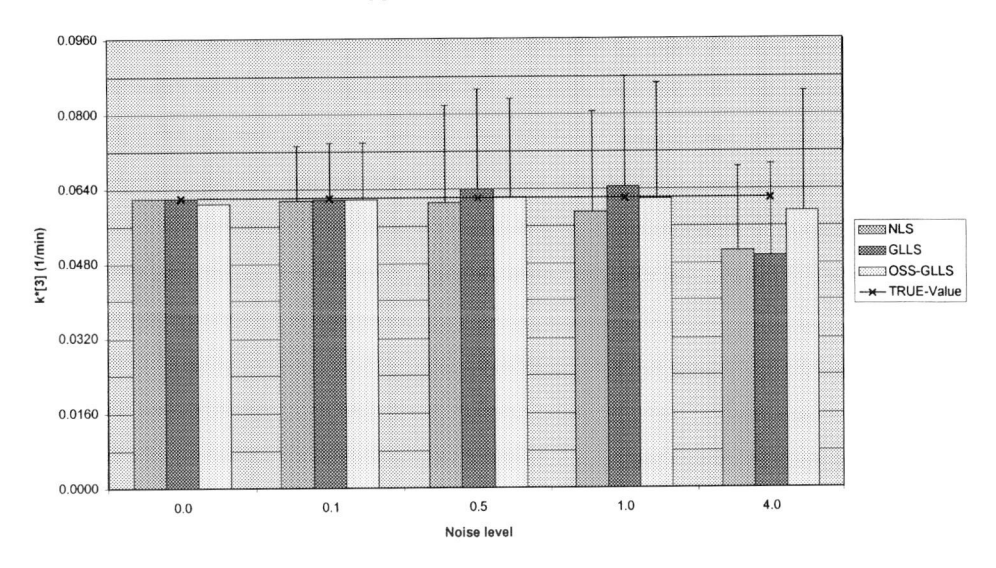

k*[4] rate constant parameter estimate

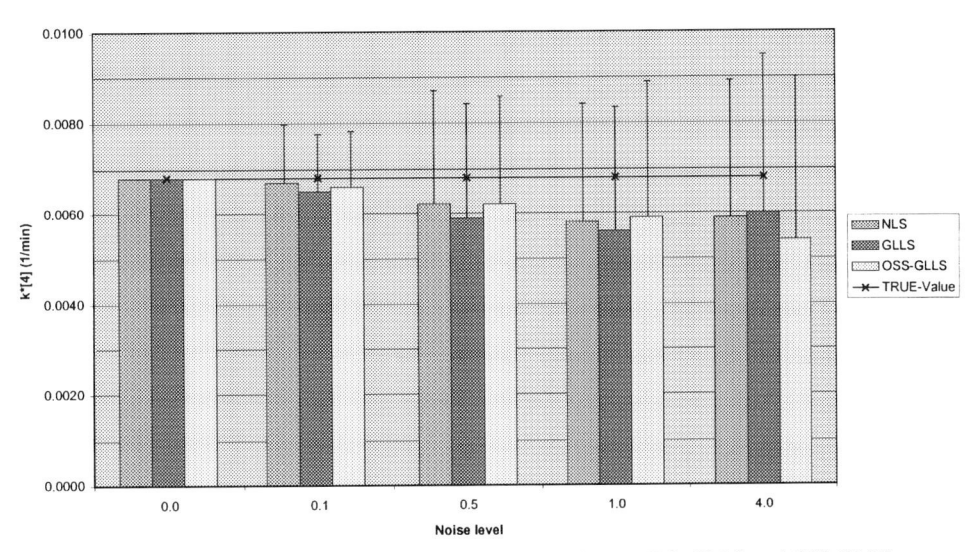

FIGURE 1 Mean and SD parameter estimates obtained from NLS, GLLS, and OSS-GLLS.

with $\Gamma_1^k(t) = \sum_{i=1}^n \lambda_i^{k-1} y_i(t)$, $\Gamma_2(t) = \{(1/[(-1)^n \lambda_1 \lambda_2 \cdots \lambda_n]) \int_0^t y(t)\, dt + \sum_{i=1}^n (1/\lambda_i) y_i(t)\}$, $\Gamma_3^k(t) = \sum_{i=1}^n \lambda_i^{k-1} u_i(t)$ and $\Gamma_4(t) = \{(1/[(-1)^n \lambda_1 \lambda_2 \cdots \lambda_n]) \int_0^t u(t)\, dt + \sum_{i=1}^n (1/\lambda_i) u_i(t)\}$. Solving Eq. (8) for θ yields

$$\hat{\theta}_{\text{OSS-GLLS}} = \mathbf{Z}^{-1} \mathbf{r}, \qquad (9)$$

where $\hat{\theta}_{\text{OSS-GLLS}}$ corresponds to the estimated parameters obtained from OSS-GLLS. OSS-GLLS may be used iteratively until a termination criterion is reached (Feng *et al.*, 1996). However, it has been found that one or two iterations were sufficient to obtain satisfactory estimates.

C. Simulation Study

The FDG model consists of three compartments: (i) FDG in plasma $c_p^*(t)$, (ii) FDG in tissue $c_e^*(t)$, and (iii) FDG-6-phosphate in tissue $c_m^*(t)$. This model has been studied extensively (Huang *et al.*, 1980). The differential equation relating the concentration of radioactivity in the tissues is given by

$$\frac{d^2}{dt^2} c_i^*(t) + P_1 \frac{d}{dt} c_i^*(t) + P_2 c_i^*(t) = P_3 \frac{d}{dt} c_p^*(t) + P_4 c_p^*(t), \qquad (10)$$

where $c_i^*(t) = c_e^*(t) + c_m^*(t)$ is the total FDG concentration in tissue. Utilizing PET, accumulated measurements $\int c_i^*(t)\,dt$ are obtained. The parameters P_1–P_4 correspond to the estimated parameters (a_1, a_2, b_1, and b_2). These parameters are related to FDG rate constant parameters as $P_1 = k_2^* + k_3^* + k_4^*$, $P_2 = k_2^* k_4^*$, $P_3 = k_1^*$, and $P_4 = k_1^*(k_3^* + k_4^*)$. Once P_1–P_4 have been estimated, FDG rate constants can be calculated as $k_1^* = P_3$, $k_2^* = P_1 - P_4/P_3$, $k_3^* = P_1 - k_2^* - k_4^*$, and $k_4^* = P_2/k_2^*$. Subsequently, the glucose metabolic rate (MRGlc) can be determined as $(1/LC)(k_1^* k_3^*)/(k_2^* + k_3^*)c_p$, where LC denotes the lumped constant and c_p the "cold" glucose concentration in plasma (Huang et al., 1980).

A Monte Carlo simulation is performed using the BLD software package (Carson et al., 1983) to compare parameters estimated from OSS-GLLS using 4 measurement samples and GLLS and NLS using 22 measurement samples. The nominal set of values in gray matter are used: $k_1^* = 0.1020$ min^{-1}, $k_2^* = 0.1300$ min^{-1}, $k_3^* = 0.0620$ min^{-1}, $k_4^* = 0.0068$ min^{-1}, $c_p = 91.9$ mg/100 ml, and $LC = 0.418$ (Huang et al., 1980). The input function $c_p^*(t)$ is also obtained from human studies (Feng et al., 1996). Two sampling schedules for the acquisition of PET measurements are used. The sampling times for the 22 measurement sampling schedule are given by 0.2, 0.4, 0.6, 0.8, 1.0, 1.2, 1.4, 1.6, 1.8, 2.0, 2.5, 3.0, 4.0, 5.0, 6.5, 10.0, 15.0, 20.0, 30.0, 60.0, 90.0, and 120.0 min, and for the 4 measurement optimal sampling schedule by 2.733, 15.683, 77.066, and 120.0 min (Huang et al., 1980; Li et al., 1996). Four different levels of noise with a Poisson distribution are added to generated tissue activity data to simulate realistic noisy PET measurements. For each noise level, FDG rate constants, k_1^*–k_4^*, and MRGlc are calculated from 1000 realization. (Note: For noise-free data, only one realization is performed.)

A practical problem that needs to be overcome involves the numerical integration of accumulated PET measurements. Conventionally, trapezoidal integration is used for numerical integration due to nonuniformity of sampled intervals and computational efficiency requirements (Wang and Feng, 1992). When OSS data are used, the number of measurement samples is dramatically reduced while sampling intervals lengthen. Thus, trapezoidal integration which has first-order algebraic accuracy and subsequent numerical integration errors may interfere with modeling of the system and

TABLE 1 Estimated MRGlc Physiological Parameter and FDG Rate Constants Obtained from NLS and Original GLLS Using 22 Measurement Samples and OSS-GLLS Using 4 Measurement Samples

Method	MRGlc	CV$_{\text{MRGlc}}$	k_1^*	CV$_{k_1^*}$	k_2^*	CV$_{k_2^*}$	k_3^*	CV$_{k_3^*}$	k_4^*	CV$_{k_4^*}$
True value	7.2415	—	0.1020	—	0.1300	—	0.0620	—	0.0068	—
a. 1 run with noise level $\alpha = 0.0$
NLS	7.2416	0.0000	0.1020	0.0000	0.1290	0.0000	0.0620	0.0000	0.0068	0.0000
GLLS	7.2415	0.0000	0.1020	0.0000	0.1300	0.0000	0.0620	0.0000	0.0068	0.0000
OSS-GLLS	7.2162	0.0000	0.1019	0.0000	0.1279	0.0000	0.0608	0.0000	0.0068	0.0000
b. 1000 runs with noise level $\alpha = 0.1$
NLS	7.2012	0.0492	0.1021	0.0845	0.1311	0.2351	0.0615	0.1899	0.0067	0.1897
GLLS	7.1921	0.0498	0.1024	0.0844	0.1322	0.2374	0.0617	0.1970	0.0065	0.1906
OSS-GLLS	7.1874	0.0571	0.1033	0.0881	0.0881	0.2441	0.0616	0.1999	0.0066	0.1832
c. 1000 runs with noise level $\alpha = 0.5$
NLS	7.1563	0.0962	0.1017	0.1251	0.1332	0.4001	0.0611	0.3402	0.0062	0.4026
GLLS	6.9842	0.0989	0.1031	0.1208	0.1154	0.4004	0.0638	0.3351	0.0059	0.4261
OSS-GLLS	7.0949	0.1051	0.1054	0.1520	0.1412	0.4080	0.0623	0.3382	0.0062	0.3855
d. 1000 runs with noise level $\alpha = 1.0$
NLS	6.9062	0.1011	0.1031	0.1313	0.1278	0.4491	0.0591	0.3613	0.0058	0.4513
GLLS	6.8867	0.1140	0.1062	0.1333	0.1358	0.4402	0.0642	0.3724	0.0056	0.4883
OSS-GLLS	7.0090	0.1307	0.1063	0.1739	0.1430	0.4626	0.0618	0.4036	0.0059	0.5059
e. 1000 runs with noise level $\alpha = 4.0$
NLS	6.7499	0.1217	0.1096	0.1798	0.1243	0.4292	0.0507	0.3567	0.0059	0.5125
GLLS	6.7525	0.1378	0.1090	0.2001	0.1222	0.4539	0.0496	0.3982	0.0060	0.5845
OSS-GLLS	6.8128	0.1639	0.1115	0.2424	0.1487	0.5177	0.0592	0.4370	0.0054	0.6676

have more influence over the accuracy of estimated parameters. To overcome this problem, the weighted parabola overlapping technique that guarantees integral values of third-order algebraic accuracy within the intervals t_i and t_{i+1}, and second-order accuracy for the first t_0 and last t_m intervals is used (Wang and Feng, 1992).

III. RESULTS AND DISCUSSION

The mean, standard deviation (SD), and coefficient of variation (CV = SD/mean) statistical criteria were used to compare parameter estimates from OSS-GLLS, GLLS, and NLS. In Table 1, five sets of results for MRGlc and rate constants k_1^*–k_4^* are summarized. Results for noise-free data are shown in Table 1a, whereas in Tables 1b–1e results are shown for noisy data. The results demonstrate that OSS-GLLS is able to accurately and reliably quantify the physiological parameter MRGlc, which is consistent with those obtained using NLS or GLLS from the finely sampled 22 measurement PET data. In relation to the FDG rate constant parameters, OSS-GLLS obtained percentage errors less than 21%, which were slightly higher than the NLS and GLLS percentage errors of 18 and 20%, respectively. The mean estimated MRGlc and FDG rate constant parameters and their respective standard deviations for different noise levels are illustrated in Fig. 1.

In the simulation study, NLS took 10–20 iterations on average to meet the 0.1% convergence criteria, whereas for GLLS and OSS-GLLS all iterations were terminated after 2 iterations. The significant improvement in the computational efficiency of GLLS and OSS-GLLS is clearly obvious. Furthermore, with the application of an OSS design, fewer measurement samples are required when OSS-GLLS is used. Hence, this algorithm is most useful for image-wide PET parameter estimation as it is the fastest algorithm of those studied.

References

Carson, R. E., Huang, S. C., and Phelps, M. E. (1983). BLD—a software system for physiological data handling and model analysis. *Proc. 5th Annu. Symp. Comput. Appl. Med. Care*, pp. 502–562.

Feng, D., Huang, S. C., Wang, Z., and Ho, D. (1996). An unbiased parametric imaging algorithm for non-uniformly sampled biomedical system parameter estimation. *IEEE Trans. Med. Imag.* **15**: 512–518.

Hawkins, R. A., Phelps, M. E., and Huang, S. C. (1986). Effects of temporal sampling, glucose metabolic rates, and disruptions of the blood-brain barrier on the FDG model with and without a vascular compartment: Studies in human brain tumors with PET. *J. Cereb. Blood Flow Metab.* **6**: 170–183.

Huang, S. C. (1994). Automated experimental design of scanning sequences for dynamic PET/SPECT studies. *J. Nucl. Med.* **35**: 71P.

Huang, S. C., Phelps, M. E., Hoffman, E. J., Sideris, K., Selin, C., and Kuhl, D. E. (1980). Non-invasive determination of local cerebral metabolic rate of glucose in man. *Am. J. Physiol.* **238**: E69–E82.

Li, X., Feng, D., and Chen, K. (1996). Optimal image sampling schedule: A new effective way to reduce dynamic image storage space and functional image processing time. *IEEE Trans. Med. Imag.* **15**: 710–718.

Mazoyer, B. M., Huesman, R. H., Budinger, T. F., and Knittel, B. L. (1986). Dynamic PET data analysis. *J. Comput. Assist. Tomogr.* **10**: 645–653.

Wang, Z., and Feng, D. (1992). Continuous-time system modeling using the weighted-parabola-overlapping numerical integration method. *Int. J. Syst. Sci.* **23**: 1361–1369.

CHAPTER 52

Parametric Image Generation with Neural Networks

MICHAEL M. GRAHAM,* STEVEN B. GILLISPIE,* MARK MUZI,* and FINBARR O'SULLIVAN[†]

Departments of * Radiology (Nuclear Medicine) and Statistics
[†] University of Washington
Seattle, Washington 98195

An artificial neural network (ANN) is a trainable algorithm that can learn to produce an output appropriate for a given input. Such networks can be applied in a wide variety of pattern recognition tasks, including parameter estimation. The major advantages of using ANNs for parameter estimation are speed and noise tolerance. A two-layer network was used to determine the rate constants and metabolic rate (MR_{fdg}) from dynamic PET images acquired after injection of [^{18}F]fluorodeoxyglucose (FDG). The number of input points was 24, which represented a well-sampled tissue time–activity-curve (TAC) over 60 min after injection of FDG. One thousand noisy training data sets were generated, using the arterial plasma TAC, for each subject (computer time: $\approx$ 1 min). The network was trained with the backpropagation algorithm using the 1000 data sets (time: $\approx$ 15 min). Images were generated using the weights determined by training (time: $\approx$ 7 sec). Times are for a Macintosh 7100/80. The MR_{fdg} images were of high quality (better contrast than the integrated images and less noisy than Patlak images). Parametric images (K_1, k_2, or k_3) were quite noisy. Training the ANN with an average plasma TAC instead of the individual TACs resulted in MR_{fdg} images of equal quality. Preliminary data from four subjects suggest it may be possible to generate quantitatively accurate images by scaling with the count from a single blood sample taken 30 min after injection.

I. INTRODUCTION

Artificial neural networks are a type of computer program that can be trained to recognize patterns. These programs have been trained to estimate the kinetic parameters associated with [^{18}F]fluorodeoxyglucose (FDG) using blood and tissue time–activity curves (TACs) as the input functions (Graham and O'Sullivan, 1996). Because these programs are very fast and are relatively noise tolerant, it is appropriate to examine their use for the generation of parametric images from pixel-wise time–activity curves. However, there are definite limitations to their capabilities. This study examines these limitations as well as the approaches that are necessary to optimize the performance of ANNs for parametric image generation.

A major potential advantage of neural networks for image generation is that quantitative parametric images can be generated very rapidly. This advantage is lost if the network has to be trained separately for each individual patient. This problem was examined by training the network with a set of tissue TACs generated with an average plasma TAC instead of the individual patient's plasma TAC. This generically trained network was used to produce images that were compared to the individually trained network images. In addition, these generically produced images were examined quantitatively to establish the feasibility of scaling the images with the plasma count from a late blood sample.

II. METHODS

One of the most common ANNs is the multilayer perceptron network trained with backpropagation. This is an appropriate ANN for the task of parameter estimation, as the input can be an integral number of values over a wide range and the output is also a number of values over a range. In previous studies (Graham and O'Sullivan, 1996; Graham et al., 1996) the individual inputs were the blood activity and the tissue activity at different times. The outputs were the estimated parameters. In an attempt to simplify the problem for image generation, only the tissue TAC was used as the input function and only a single parameter was used as an output.

The network architecture chosen for this problem is illustrated in Fig. 1. The input layer has 24 nodes. These nodes are assigned values of the tissue TAC. The timing of the images used in generating the tissue TAC were 20-sec intervals × 4, 40-sec intervals × 4, 1-min intervals × 4, 3-min intervals × 4, and 5-min intervals × 8. Each node in the input layer is connected to each node in the next hidden layer. There is a weight associated with each connection. Thus, the input to each node, j, in the hidden layer is $\sum_{i=1}^{24} W_{ij} \cdot A_i$, where W_{ij} is the weight associated with the connection from node i in the input layer to node j in the hidden layer and A_i is the value of node i in the input layer. In a similar fashion, there are a set of weights, W_j, connecting the hidden layer with the output node.

As the weighted sum of all the inputs to a node may be a large number, it is scaled down before producing the output of the node. The scaling function used in this network is $1/(1 + e^{-NET})$, where NET is the sum of all the weights times the values of the nodes in the next higher layer. This function produces an output value for each node in the hidden layer and output layer. Forward calculation of the ANN is relatively simple. The input vector (tissue TAC) is applied to the input layer. Using the weights, values of the hidden and then the output node are calculated.

The network is trained by adjusting the weights until it produces an output as close as possible to the desired output. There are several different ways to adjust these weights. The method used for this effort is called the backpropagation method. Detailed descriptions of this method can be found in almost any text on neural network, such as those by Wasserman (1989) or Freeman and Skapura (1991). Briefly, the training approach is to present a large number of examples to the ANN, including input vectors and target vectors. After each input vector is propagated through the network, the output is compared to the target values. Based on this comparison, an error signal is generated that is used to adjust the weights. This process is done iteratively with many different examples. Gradually, the network performance improves and reaches a state where it is as accurate as it can get, at which point training is stopped. This process may take many hours of computer time.

It is important to present the network with a wide range of examples, as it is likely to perform best if it has seen an example similar to a given data set. Generation of the training sets requires an input function, i.e., a plasma FDG TAC. Plasma TACs from four different patients, along with a mean TAC of all four patients, were used for the input functions. Three of the parameters of the FDG model (K_1, k_2, and k_3) were randomized with uniformly distributed random numbers over the following ranges: K_1, 0.02–0.25; k_2, 0.05–0.40; and k_3, 0.01–0.20. One thousand example data sets were created for training. The four different input functions are shown in Fig. 2 and a sample of the output functions used to train the neural network is shown in Fig. 3.

A. ANN Options

1. Scaling

Based on previous work, the input tissue TACs were scaled to have a range of approximately 0 to 1.0. In addition, the output variables were scaled to values of 0 to 1.0.

2. Number of Nodes

The most obvious option in constructing the ANN is the number of nodes in the hidden layer. There are no definite rules for determining the optimum number. Again, based on previous work, the number of nodes chosen was 7. This seemed to be an optimum number in terms of accuracy of parameter optimization. A

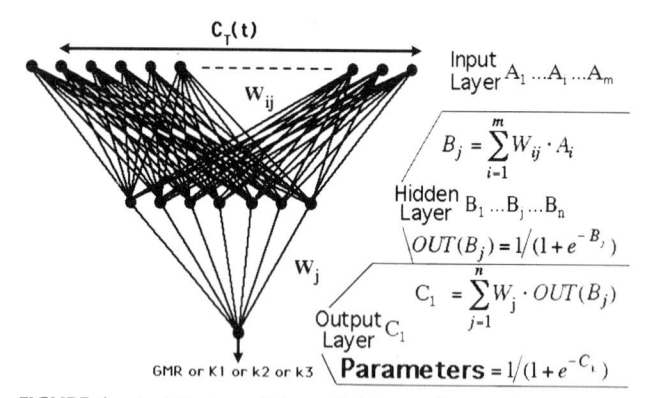

FIGURE 1 Architecture of the artificial neural network used. Tissue time–activity curves (24 points) are used as input vector A. There is a weight (W_{ij}) associated with the connection between each node in the input layer and each node in the hidden layer. The input to the next layer, B, is the sum of the product of the weights times the values of the input nodes. The output of the hidden layer is constrained by the function $1/(1 + e^{-B_j})$ to keep the output in a reasonable range. Similarly, the input to the last layer is the product of W_j times the output from the hidden layer nodes. The single output value (MR_{fdg}, K_1, k_2, or k_3) is calculated in the same way as the output of the hidden layer.

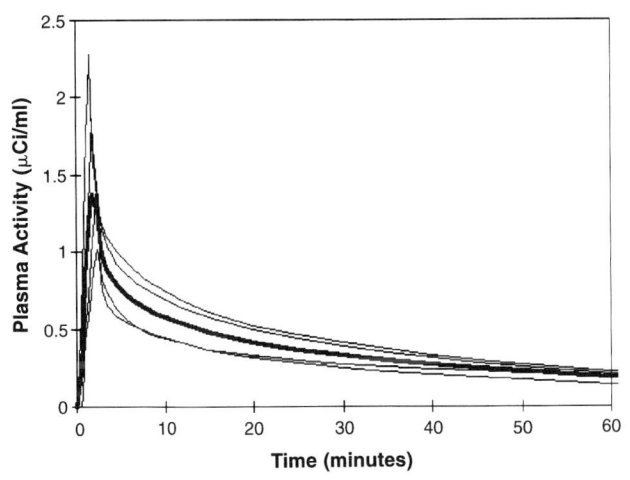

FIGURE 2 Plasma time-activity curves from the four subjects and the average curve used to train the artificial neural networks. The average curve is in bold.

single node was chosen for the output. Previous work showed that greater accuracy is achieved with a single node estimating MR_{fdg}, compared to several nodes estimating various rate constants simultaneously. Some of the following examples depict images of K_1, k_2, or k_3. These were generated with single output node networks trained with target values of K_1, k_2, or k_3.

3. Noise

In generating the training sets from the kinetic model for FDG, considerable noise was added to the output data. The rationale for training with noisy data is that the ANN will ultimately be used to analyze noisy data and it therefore should do better if trained on noisy data. This was shown quite clearly in previous work (Graham and O'Sullivan, 1996).

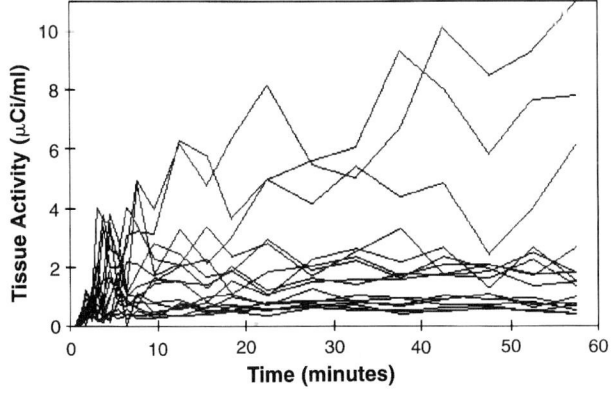

FIGURE 3 Typical noisy tissue TACs used to train the artificial neural network. The parameters K_1, k_2, k_3 were randomly varied. Blood volume was fixed at 0.04 ml^{-1} and k_4 was fixed at 0.007 min^{-1}.

4. Training Procedure

During training the starting weights were randomized between zero and 1.0. Each ANN was trained 10 separate times with different starting weights. The weight sets chosen were the ones that resulted in the best parameter estimates during training. The training data sets consisted of 1000 noisy tissue TACs along with correct parameters. All ANNs were implemented in Pascal, and the programs were run on a Macintosh Power Mac 7100/80. Training times varied from 10 to 40 min.

5. Image Generation

Images were generated by forward operation of the neural network on pixel-wise TACs from a single slice from an FDG PET study. The images were 128 × 128. One set of images was generated with the network trained using the individual patient plasma TACs. The second set was generated with the network trained with the average plasma TAC from the four patients. For comparison, "Patlak" images were also generated using the graphical analysis method (Gjedde, 1982; Patlak et al., 1983) on a pixel-by-pixel basis, using data from 20 to 60 min.

III. RESULTS

Three programs were used. The first generated test data sets, given a plasma TAC input function. It took approximately 1 min to generate the test file of 1000 data sets. The second program trained the neural network using the data sets produced by program 1. For the 24 input nodes/7 hidden nodes/1 output node networks, the training time was approximately 15 min. The third program generated parametric images using the weights determined by program 2. The time required to generate one 128 × 128 byte mode image once the dynamic FDG data set was loaded into memory was 7 sec. All times are for a Power Macintosh 7100/80.

Different sets of weights were obtained by training the ANN with targets of MR_{fdg}, K_1, k_2, or k_3. The performance of these ANNs was evaluated by presenting the network with an independent set of noisy data. The metabolic rate and parameter estimates from this evaluation are shown in Fig. 4. It is clear that the estimates of MR_{fdg} are relatively accurate, whereas those of K_1, k_2, and k_3 are relatively poor. This is reflected in the quality of the parametric images. The FDG metabolic rate images generated with the neural network are of excellent quality. Examples are shown

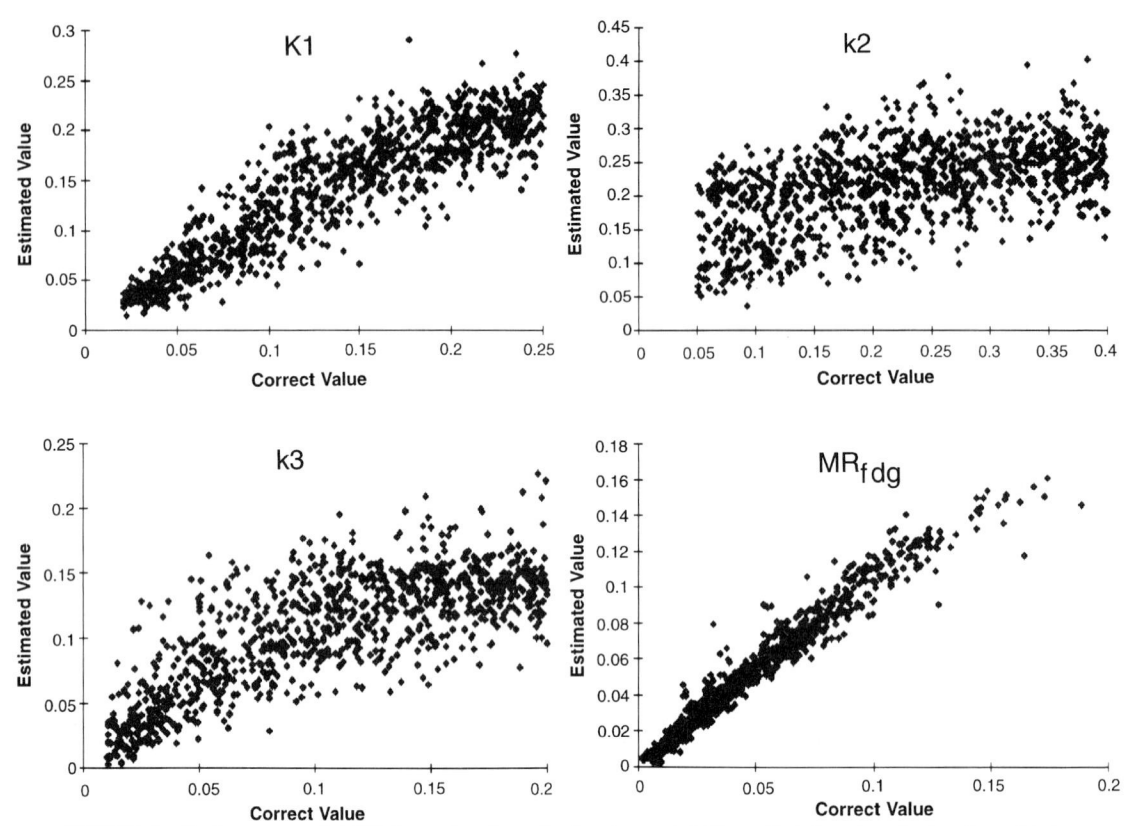

FIGURE 4 Performance of ANN for estimating K_1, k_2, k_3, and MR_{fdg}. Training data had noise added equivalent to 100% standard deviation for a 20-sec image with appropriately less noise for longer images. The network was then tested with a similar but independent data set.

in Fig. 5. The parametric images of K_1, k_2, and k_3 (Fig. 6) are of much lower quality, i.e., much noisier. In general, the neural network images are of higher contrast than the simple integrated images, are less noisy than the Patlak graphical analysis images, and have the advantage that they are calibrated quantitatively so that individual pixel values are in appropriate units to convert to glucose metabolic rate.

Training with data sets created with an average plasma TAC instead of individual plasma TACs resulted in MR_{fdg} parametric images that were essentially identical to the images created using the appropriate individual plasma TACs. Calibration was in error, as the areas under the individual curves differed considerably. Correct calibration factors were determined for each subject, using metabolic rates determined with conventional parameter optimization over the occipital region of each subject. The correct calibration factors were plotted (Fig. 7) against normalized plasma concentrations (patient plasma value divided by average plasma value). The plot for the 30-min time points shows the best correlation, implying that this approach would lead to an error in determining MR_{fdg} of 11% SD. Because four subjects were studied, this is clearly very speculative and will require a larger number of studies to confirm.

IV. DISCUSSION AND CONCLUSION

A trainable artificial neural network performs remarkably well in the task of estimating FDG parameters from noisy pixel-wise tissue TACs, even when the plasma TAC is not known. Although training the network takes a relatively long time, once it is trained it can generate parametric images in only a few seconds. The accuracy of the estimates of MR_{fdg} seems to be better than that achieved with graphical analysis (Giedde, 1982; Patlak et al., 1983) and is also better than more conventional methods of parameter optimization when attempted on a pixel-by-pixel basis.

The major advantages of this approach are that it is very fast, is relatively noise tolerant, and produces quantitative images of metabolic rate. Disadvantages are that it cannot produce reasonable quality rate constant images, presumably because of the extreme noise seen in single pixel TACs, and that the approach requires a relatively large number of dynamic images.

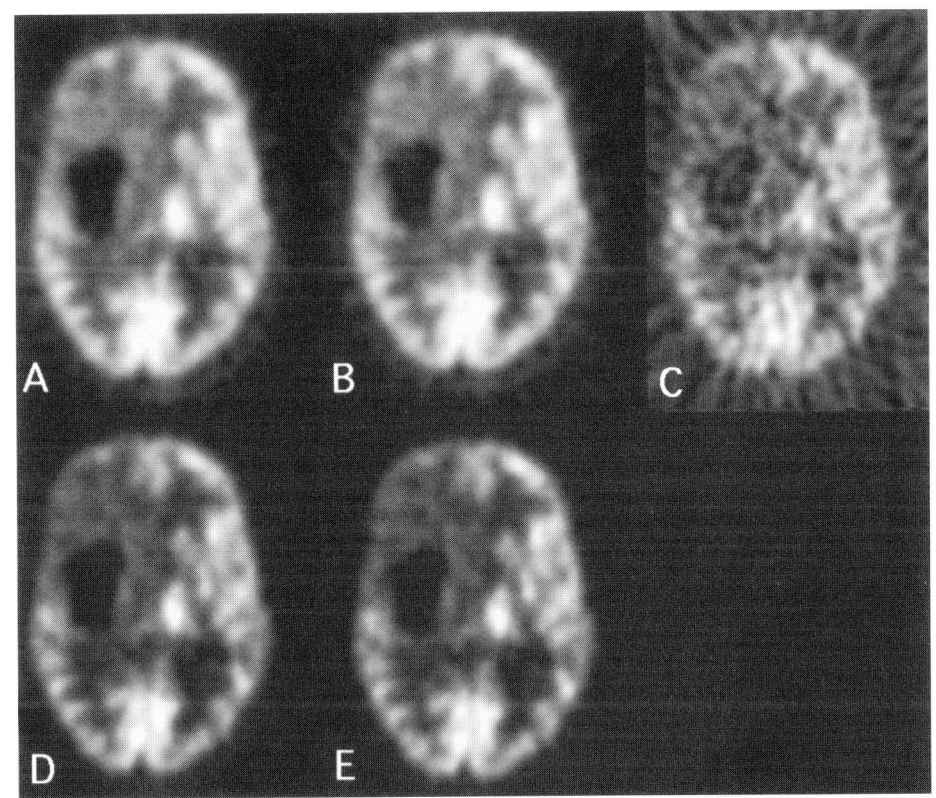

FIGURE 5 Images of the brain from subject 1. (A) Summed image from 30 to 60 min. (B) Summed image from 45 to 60 min. (C) Image generated with the graphical analysis method, using tissue data from 20 to 60 min. (D) Neural network MR$_{fdg}$ image produced using weights derived from training data made with the same subject's plasma TAC. (E) Neural network MR$_{fdg}$ image produced using weights derived from training data made with an average plasma TAC.

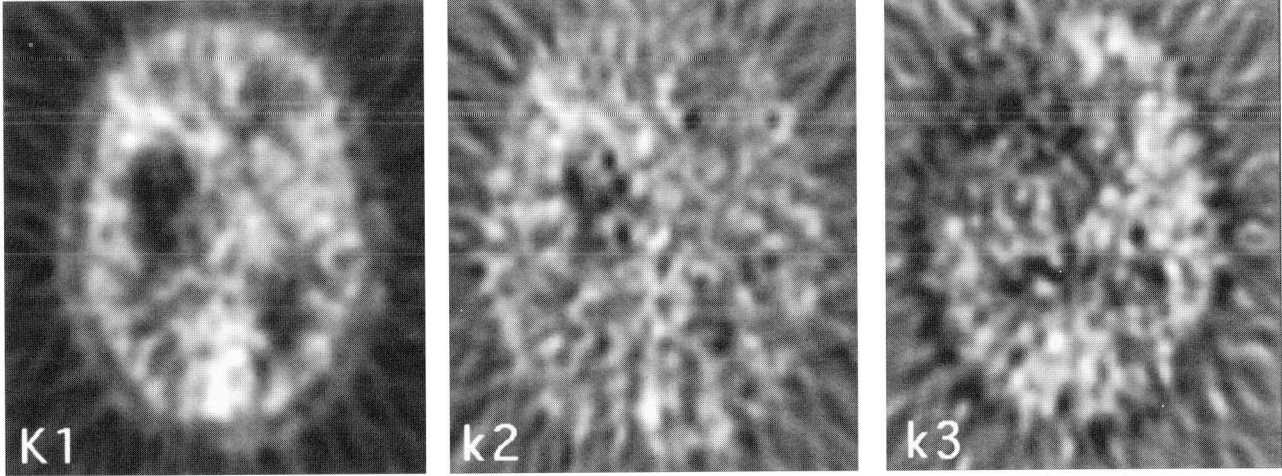

FIGURE 6 Rate constant images generated with three different neural networks for subject 1. Each network was trained separately to estimate the specific rate constant.

30 min samples

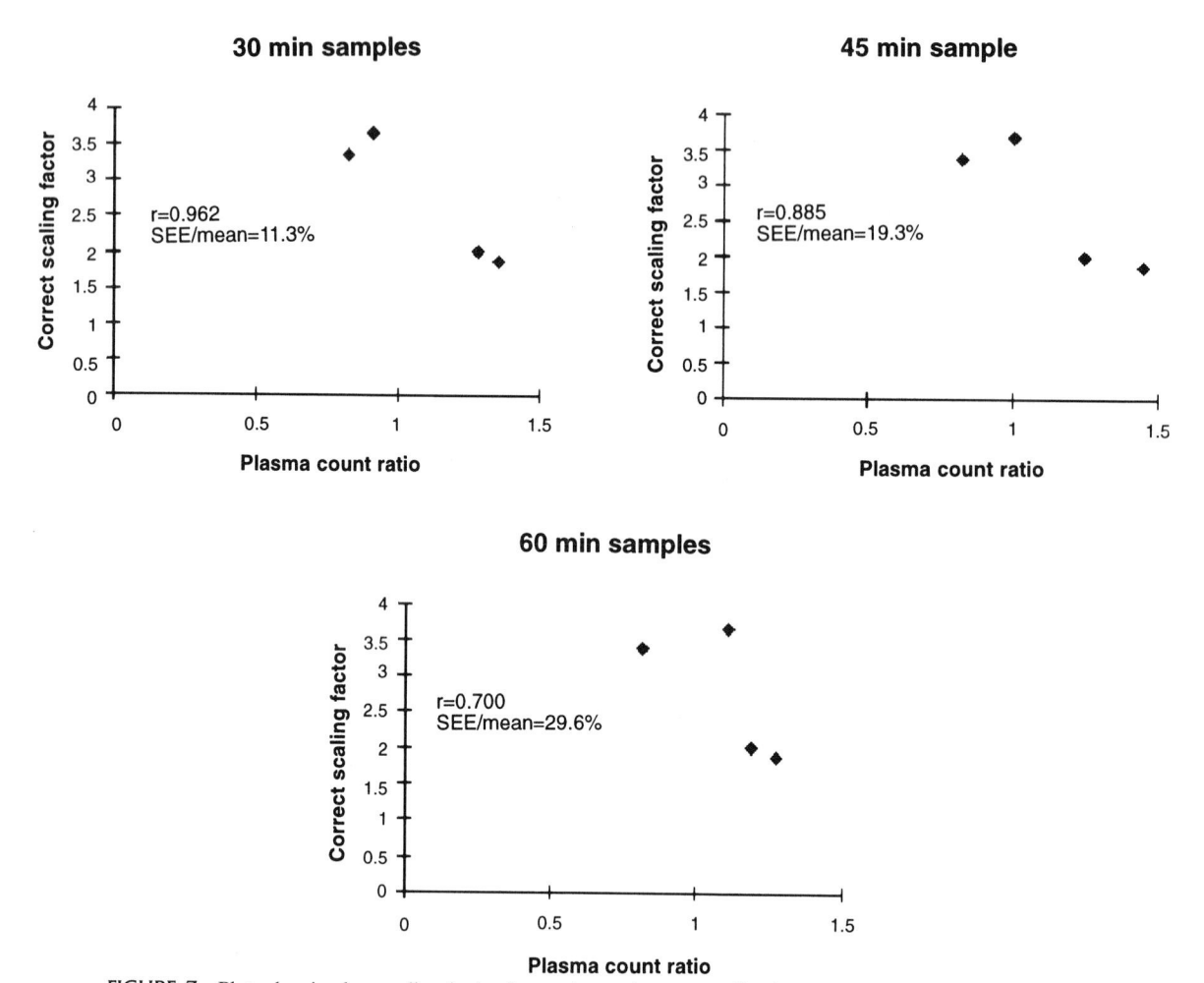

r=0.962
SEE/mean=11.3%

45 min sample

r=0.885
SEE/mean=19.3%

60 min samples

r=0.700
SEE/mean=29.6%

FIGURE 7 Plots showing how well a single plasma time point can predict the correct scaling factor to correct the MR_{fdg} image produced using the average plasma TAC training set. Since the correlation is best and the standard error of the estimate is least for the 30-min sample, this suggests that the count from a blood sample at this time may be used to create quantitatively accurate MR_{fdg} images generated with the generic neural network.

It is possible that the network will work well with considerably fewer images. This is an area that can be explored easily by summing images to create more limited data sets.

Acknowledgment

This work was supported by NIH Grant CA42045.

References

Freeman, J. A., and Skapura, D. M. (1991). "Neural Networks Algorithms, Applications, and Programming Techniques." Addison-Wesley, New York.

Gjedde, A. (1982). Calculation of cerebral glucose phosphorylation from brain uptake of glucose analogs in vivo: A re-examination. *Brain Res. Rev.* **4**: 237–274.

Graham, M. M., and O'Sullivan, F. (1996). Parameter estimation of the fluorodeoxyglucose model with a neural network. *In* "Quantification of Brain Function Using PET" (R. Myers, V. Cunningham, D. Bailey, and T. Jones, eds.), pp. 277–281. Academic Press, San Diego, CA.

Graham, M. M., Gillispie, S. B., Muzi, M., and O'Sullivan, F. (1996). Generation of parametric images with artificial neural networks. *J. Nucl. Med.* **37**: 220P (abstr.).

Patlak, C. S., Blasberg, R. G., and Fenstermacher, J. D. (1983). Graphical evaluation of blood-to-brain transfer constants from multiple-time uptake data. *J. Cereb. Blood Flow Metab.* **3**: 1–7.

Wasserman, P. D. (1989). "Neural Computing Theory and Practice." Van Nostrand-Reinhold, New York.

CHAPTER 53

Kinetic Modeling: Achieving Computer Platform Independence with Java

SYLVAIN HOULE

Vivian M. Rakoff PET Centre
Clarke Institute of Psychiatry and University of Toronto
Toronto, Ontario, M5T 1R8 Canada

Software for kinetic modeling has been written in a variety of computer languages. Portability of these programs between different computer systems has often been hindered by computer system dependencies in the software introduced because of vendor-specific extensions to the programming language or the operating system environment. The new programming language Java allows programs to be written and executed without modification on different computing platforms. Java is an object-oriented language patterned closely after C++, but is considered to be a simpler and improved dialect of that language. This chapter investigates the potential of Java for writing kinetic modeling software for neuroreceptor studies with positron emission tomography.

I. INTRODUCTION

Software for kinetic modeling has been written in a variety of computer languages: Pascal, Fortran, Ratfor, C, and C++. Portability of these programs between different computer systems at the same or different institutions has often been hindered by computer system dependencies in the software introduced because of vendor-specific extensions to the programming language or the operating system environment. The new programming language Java (Sun Microsystems Inc.) allows programs to be written and executed without modification on different computing platforms (Arnold and Gosling, 1996). Java is an object-oriented language patterned closely after C++, but is considered to be a simpler and improved dialect of that language. It has gained rapid popularity and widespread acceptance since its release in mid-1995, particularly for writing Internet applications, known as applets. However, the language can also be used as a general-purpose computer language for stand-alone applications. This chapter investigates the potential of Java for writing kinetic modeling software for neuroreceptor studies with positron emission tomography (PET).

II. MATERIAL AND METHODS

In the original white paper, Sun described Java as "object-oriented, robust, secure, architecture neutral, portable, high performance, interpreted, threaded and dynamic" (Gosling and McGilton, 1995). Of these epithets, portability and architecture neutrality are of particular interest, as they translate, in practice, to the fact that any program written in Java can be run on any computer that supports Java without having to modify the code for different compilers, windowing, or operating systems or, in the words of a Sun trademark, "write-once, run everywhere." These characteristics are highly desirable in the environment of the Internet. These could also be important for researchers, as software written in Java could easily be exchanged not only between different computer platforms within one institution but also with other PET centers. Today, all other languages, ranging from the venerable Fortran to more dedicated software such as Matlab (The Mathworks Inc.), require different versions of the software for each type of computer and operating system (Unix, Microsoft Windows or NT, or Apple Macintosh). Often they are not even compatible among different versions of the same operating system.

The secret behind the computer architecture neutrality of Java is that the programs are not translated into machine code instructions for a particular computer hardware but rather into something known as "byte code" intended to be interpreted by a Java virtual machine (JVM). A byte code interpreter is then written for each computer platform to support the JVM. The result is that platform dependence is eliminated. A Java program can be compiled on any platform and the resultant byte code can then be run on any other platform that implements the JVM.

There are also other features of the language that are appealing. Although Java is an object-oriented programming (OOP) language mainly derived from C just as C++ is, it is a much simpler language than C++. It has retained only the essential elements needed to implement OOP and avoids the complex and often confusing features of C++. Java has uniform data types. For example, an "integer" is always 32 bits long. Having uniform data types eliminates another form of platform dependency that either requires modifying the source code or, worse, shows up as unexpected results. Another major difference in Java is that objects are always handled by reference. There is no pointer arithmetic as in C and C++. Although pointers are a powerful feature, they often lead to programming bugs that are hard to track down. Dynamic memory management is handled automatically by the JVM. When objects in memory are no longer referenced, the memory they occupy is freed by a process known as garbage collection (GC). In C and C++, memory is allocated dynamically with the malloc( ) library call and must be freed by a call to free( ). This process is prone to error and can result in either memory leaks or premature freeing. Memory leaks occurs when an object is not freed after use, and they may eventually cause a program to run out of available memory. Premature freeing occurs when memory is freed while it is still being referenced by one or more pointers. It is a troublesome problem, as it often manifests itself by memory corruption occurring in parts of the program far remote from the original error. Java also does not allow a program to access memory outside of its own environment. By doing so it prevents malicious or unintended access to memory allocated to other programs or to the system.

Java has kept the concept of exceptions popularized by C++. Exceptions are a language feature that greatly simplifies error handling in programs. Exception handling is implemented by a try {...} catch {...} blocks combination. The code within the try block is executed. If an error occurs, the program execution is transferred to the catch block for error handling. Part of the robustness of Java arises from the fact that unhandled exceptions are propagated back through the call hierarchy until the exception is handled. If it is not handled within the program itself, the run-time system will abort the program. The exception mechanism is simpler than other approaches such as using special return values from subroutines to indicate an error (especially since they are often ignored).

The designers of Java have chosen to support a wide range of programming features in class libraries rather than as part of the language itself. By doing so, they maintain the simplicity of the language itself and of the JVM. Rather than using one large class library, Sun provides a set of separate libraries. Each library groups together the classes required for a given functionality. For example, unless you need to access data bases, you do not have to use (or even learn) the data base-related library. These libraries are known as application programming interfaces (APIs). There are two APIs that every Java programmer must become familiar with: the Core API and the AWT (Abstract Window Toolkit). The Core API implements the basic class hierarchy and various utilities (e.g., data structures and I/O). The AWT provides the basic classes needed to use Java in a windowing environment. In keeping with the overall Java philosophy, the AWT is implemented in a way that is independent of the windowing platform. The AWT itself transforms calls from Java methods into equivalent calls to the native windowing system (e.g., X-Window or Microsoft Windows). This approach avoids the need for separate windowing libraries for different windowing environments. At the moment, the AWT creates windows that retain the "feel-and-look" of the native windowing system. The library approach also permits vendors to offer specialized libraries to complement those supplied by Sun.

Java programs can be either applets or applications. Applets are programs specially written to be executed by a web browser. Applications are executed as regular programs outside the web browser environment. As most applets are downloaded over the Internet, they pose a serious security risk to computer systems. Because of this risk, web browsers generally do not allow applets to have access to system resources such as the local file system. Applications do not have such restrictions.

Because of these features, Java appeared ideally suited for kinetic modeling software that would be both easy to write and platform independent. In order to evaluate the suitability of Java for this task, new programs were written in Java and existing Fortran, C, and C++ software was also converted into Java. Software has been written to solve the equations of conventional multicompartment models using the damped SVD variant (Lines and Treitel, 1984) of the Levenberg–Marquardt method to determine the best parameter values. The graphical analysis approach (Patlak and Blasberg, 1985; Logan *et al.*, 1996) and the simplified reference tissue model (Lammertsma and Hume, 1996; Gunn *et al.*, this volume, Chapter 60) have also been implemented. In the process of writing this software, Java software code was ported from older sources such as BLD (Carson *et al.*, 1981) and the numerical recipes in C (Press *et al.*, 1988). For most of this work, Café, Symantec's integrated Java development environment for Windows 95 and NT, and Sun's Java development kit (JDK) for Solaris (Unix) were used. The initial work was done with Java 1.0.2 and, since the beginning of 1997, with Java 1.1. The Java Numerical Library from Visual Numerics was also used.

III. RESULTS AND DISCUSSION

It was found that the basic elements of the language can be learned very quickly, especially by someone with prior experience with C or C++, as the basic syntax is very similar. Object-oriented programming is also much simpler in Java than in C++. Because of all the features available in C++, it may be tempting to try more complex problems only to find oneself facing long debugging sessions. The epitome of both elegance and complexity for scientific programming in C++ can be found in Barton and Nackman (1994). Java does not permit such elaborate intertwining of classes. Because the class constructs are simpler in Java than in C++, one can become proficient in OOP much faster in Java. However, irrespective of the programming language, one must learn how to conceptualize a program in terms of classes and their methods.

Because the Java byte code needs to be interpreted by the JVM, an initial concern was that the programs might run much slower than their C++ counterparts. In general, for tasks requiring human intervention (e.g., mouse or keyboard input), no difference was seen, as the human intervention was the limiting factor. For CPU intensive tasks, Java was slower, as expected. The difference is due not only to the fact that Java is interpreted but also because it carries out more error checking at run time (such as out-of-bound array access). Initial implementations of Java have been reported to be about 10 times slower than an equivalent C++ program. However, new technology, such as just in time (JIT) compiling and other forms of conversion of byte code to native machine code, have greatly reduced the difference (to about 2 times). If the increases in CPU speeds are taken into account, most Java programs run faster than their Fortran or C counterpart did only a few years ago. For kinetic modeling tasks, execution speed was not a problem. In fact, the availability of a simple, robust, and truly portable language more than makes up for any speed difference.

There are, however, some limitations that were encountered in the course of this work. Because most of the interest in Java has been directed from the start toward the WWW, there is a lack of fully functional plotting libraries. The lack of suitable classes for plotting made the task of implementing the graphical techniques more difficult. Also, printing was not possible until release 1.1 of Java and still has major limitations under Microsoft Windows 95 and NT. Given the interest in Java as a general programming language, these two limitations are most likely to be short term.

IV. CONCLUSIONS

Learning to program in Java is helped by prior experience with C++. Because Java is an interpreted language, there is a speed penalty compared to compiled languages such as Fortran, C, or C++. However, execution speed is not a limitation on modern personal computers or workstations. One of the greatest advantages of Java is that it has been designed for writing highly reliable software. Strong typing and explicit declaration of variables and functions, absence of pointers, automatic memory management, and exception handling prevent common programming mistakes associated with pointers and memory allocation in C and C++.

Java is well suited for writing platform neutral kinetic modeling software. These programs run without change on Macintosh, Microsoft Windows (Win95 and NT), and Unix (Solaris and Linux) computer systems. Java has thus the potential of simplifying the sharing of software among PET facilities.

Acknowledgments

The author thanks Adriaan Lammertsma and Roger Gunn for their help regarding the implementation details of the tissue reference model.

References

Arnold, K., and Gosling, J. (1996). "The Java Programming Language." Addison-Wesley, Reading, MA.

Barton, J. J., and Nackman, L. R. (1994). "Scientific and Engineering C++." Addison-Wesley, Reading, MA.

Carson, R., Huang, S. C., and Phelps, M. E. (1981). BLD—A software system for physiological data handling and model analysis. *Proc. 5th Annu. Symp. Comput. Appl. Med. Care*, pp. 562–565.

Gosling, J., and McGilton, H. (1995). "The Java Language Environment," White paper. Sun Microsystems Inc., Mountain View, CA.

Lammertsma, A. A., and Hume, S. P. (1996). Simplified reference tissue model for PET receptor studies. *NeuroImage* 4: 153–158.

Lines, L. R., and Treitel, S. (1984). A review of least squares inversion and its application to geophysical problems. *Geophys. Prospect.* 32: 159–186.

Logan, J., Fowler, J. S., Volkow, N. D., Wang, G. J., Ding, Y. S., and Alexoff, D. (1996). Distribution volume ratio without blood sampling from graphical analysis. *J. Cereb. Blood Flow Metab.* 16: 834–840.

Patlak, C. S., and Blasberg, R. G. (1985). Graphical evaluation of blood-to-brain constants from multiple-time uptake data. Generalizations. *J Cereb. Blood Flow Metab.* 5: 584–590.

Press, W. H., Flannery, B. P., Teukolsky, S. A., and Vetterling, W. T. (1988). "Numerical Recipes in C. The Art of Scientific Computing." Cambridge University Press, Cambridge, UK.

SECTION VII

KINETIC MODELING

CHAPTER 54

Temporally Overlapping Dual-Tracer PET Studies[1]

R. A. KOEPPE, E. P. FICARO, D. M. RAFFEL, S. MINOSHIMA, and M. R. KILBOURN

Division of Nuclear Medicine
University of Michigan
Ann Arbor, Michigan 48109

There has been increased interest in studying multiple neurotransmitter–neuroreceptor systems within the same subject. Currently, examination of two distinct neuropharmacologic measures with positron emission tomography (PET) involves performing two independent scans. This chapter proposes a dual-tracer protocol with a single overlapping PET scan and analysis using a combined compartmental model. Advantages include (1) reduction in scan time by 1.5–2 hr, (2) neuropharmacologic measures would be obtained over nearly the same interval, and (3) interventional protocols involving a pair of dual-tracer scans could be performed in a single session. Simulations were performed for various combinations of three tracers: [^{11}C]flumazenil (FMZ), [^{11}C]dihydrotetrabenazine (DTBZ), and N-[^{11}C]methylpiperidinylpropionate (PMP). Noisy time–activity curves were generated simulating injection separations of 10–30 min. Model parameters were estimated for both tracers simultaneously using the combined model. For FMZ/DTBZ, two parallel two-compartment configurations were used, estimating K_1 and distribution volume (DV) for each ligand. For tracer pairs involving PMP, a parallel two- and three-compartment combination was used, estimating K_1 and DV for FMZ or DTBZ and K_1 and k_3 for PMP. Parameter estimation accuracy is strongly dependent on the choice of tracers, the brain region, the parameter of interest, the injection order, and, obviously, the injection spacing. Examination of the covariance among parameters is important for interpreting model performance. Making specific conclusions that can be generalized to any tracer pair is difficult; however, the use of tracers with rapid kinetics will yield greater success. The authors conclude that dual-tracer single-scan PET is feasible and can be implemented with a number of different radiotracers.

I. INTRODUCTION

The authors' goal is to develop methodology for use with positron emission tomography (PET) that will yield neuropharmacologic information related to two different biochemical systems or two different aspects of the same system of the brain at a single point in time. Currently, if one is interested in examining two distinct neuropharmacologic measures with PET, a scan session would involve injecting the first radioligand, scanning for about 1 hr, waiting for the first ligand to decay, injecting the second radioligand, and then scanning for approximately another hour. There are at least two obvious limitations of such a procedure: (1) the long study duration, making it difficult to scan certain patient groups; and (2) obtaining neuropharmacologic measures at two different points in time separated by about 2 hr when the physiologic/pharmacologic state of the subject may have changed. Furthermore, if one is interested in how two neuropharmacologic systems interact, a research study involving a pharmacologic challenge would require four PET scans, two baseline scans (one for each ligand), and two challenge scans. Such a protocol would require multiple days to perform and, in addition, the radiation dose to the subject may be excessive. A dual-tracer protocol with two radioligands injected 10–30 min apart and overlapping scan-

---

[1] Transcripts of the BRAINPET97 discussion of this chapter can be found in Section VIII.

ning offers a means of addressing these problems. The scan time required for a dual-tracer study would be reduced by 1.5–2 hr. The neuropharmacologic measures estimated from the two radioligands would be obtained over nearly the same time window. The lower doses afforded by three-dimensional (3D) imaging would allow interventional protocols involving a pair of dual-tracer studies (dual-baseline and dual-intervention scans) and could be performed in a single morning or afternoon scan session.

The rationale for a dual-tracer protocol stems from studies that indicate that many synaptic neurochemical indices by themselves may be insufficient to fully characterize neurologic diseases. As an example, multiple markers for striatal dopaminergic synapses are now available, including D1 (measured with [^{11}C]SCH-23390) and D2 ([^{11}C]raclopride; RAC) receptors, the presynaptic dopamine reuptake site ([^{11}C]WIN35,428), the vesicular monoamine transporter ([^{11}C]dihydrotetrabenazine; DTBZ), and DOPA decarboxylase activity ([^{18}F]fluoroDOPA; FDOPA). In diseases involving both the number and the activity of dopaminergic nerve terminals, any or all of these markers have potential applications. Furthermore, there is mounting evidence that endogenous neurotransmitters and neuromodulators may influence *in vivo* measures with some of these agents. Finally, there is interest in examining the effects of pharmacologic treatments or test interventions with these markers. Experimental designs can now be envisioned in which two or more markers are studied under different conditions, permitting the testing of hypotheses regarding altered numbers *and* function of specific neurons.

An example of multitracer dopaminergic characterization is the investigation of altered presynaptic terminals and D2 receptors in unmedicated Parkinson's disease (PD). It has been hypothesized that the observation of reduced DOPA decarboxylase activity in PD may be confounded by compensatory upregulation of enzymatic activity, whereas the density of synaptic vesicles as reflected by VMAT2 binding is a stable marker of presynaptic terminal integrity. This could be directly investigated by performing FDOPA, DTBZ, and RAC studies in unmedicated patients at baseline, followed by repeat characterizations after administering a pharmacologic dose of a D2 agonist drug. If the hypothesis were correct, FDOPA accumulation would be reduced in proportion to reduced RAC binding in the second scans due to D2 receptor-mediated downregulation of DOPA decarboxylase activity. DTBZ binding would remain unchanged. Another example of multitracer protocols possible with existing radioligands is the study of multiple opioid receptors and of possible correlations or distinctions of agonist and antagonist recognition sites using [^{11}C]carfentanil, [^{11}C]diprenorphine, and [^{18}F]cyclofoxy.

While offering these distinct advantages and the potential for yielding information not obtainable with current methodology, dual-tracer protocols present new challenges for both data acquisition (determination of two input functions) and compartmental model/parameter estimation procedures, as well as for the data extraction and statistical analysis steps that follow. This work focuses on the development, implementation, and evaluation of methods for compartmental modeling and parameter estimation of dual-tracer, single-scan PET data.

II. METHODS

A. Background

Typical PET neuropharmacology studies have sufficient statistical quality to estimate from two to at most about five kinetic parameters (Graham, 1985; Carson, 1986; Koeppe, 1990). There will be roughly double the number of parameters to be estimated for a dual-tracer procedure. This approach requires a parallel compartmental model configuration, which accounts for two distinct radiotracer input functions and two sets of rate constants, rather than a single serial configuration. For a parallel configuration, if the two tracers were injected simultaneously it would be impossible, without additional information, to interpret which set of parameter estimates corresponded to which tracer. Thus, the injection of one radiopharmaceutical must be offset relative to the other. Obviously, the further the two injections are separated, the easier it would be to estimate and isolate individual parameters, until at the extreme, one arrived back at the case of performing and analyzing two completely separate scans. However, to minimize the overall study duration and be able to make measurements of different receptor systems nearly simultaneously, one would like to separate the injections by as short a time as possible.

B. Radiotracers

In this initial work, the authors chose to develop and test methods for use with well-characterized radiotracers where the kinetics of single-injection studies had already been investigated. Data simulating the following three tracers were generated: [^{11}C]flumazenil (FMZ), a benzodiazepine receptor antagonist (Koeppe *et al.*, 1991); DTBZ, a ligand for the vesicular monoamine transporter (VMAT2) binding site (Koeppe *et al.*, 1996); and *N*-[¹¹C]methylpiperidinylpropionate (PMP), a substrate for hydrolysis by the enzyme acetylcholinesterase (Koeppe *et al.*, 1997). FMZ and DTBZ can be classified as reversible tracers and have been successfully analyzed using both bolus and continuous infusion protocols. PMP can be classified as an irreversible tracer as no hydrolyzed product can be converted back to authentic PMP ($k_4 = 0$). Both injection orders for all three combinations of tracers were simulated: FMZ:DTBZ, DTBZ:FMZ; FMZ:PMP, PMP:FMZ; and PMP:DTBZ, DTBZ:PMP.

C. Compartmental Modeling and Parameter Estimation

Figure 1 shows two configurations of combined dual-tracer parallel compartmental models. Each model has two input functions and either four or five estimated rate parameters. The combined model for FMZ and DTBZ (top of Fig. 2) consists of two parallel two-compartment two-parameter models, one for each tracer. This configuration is applicable for use with two rapidly equilibrating tracers. The combined model for PMP and DTBZ (bottom of Fig. 2) or PMP and FMZ consists of two parallel models, one again a two-compartment two-parameter model and the other a three-compartment three-parameter model. This configuration is applicable for use with a rapidly equilibrating and an irreversible tracer pair. The parameter k_4 is fixed to zero. Estimates are optimized by nonlinear least-squares analysis. For two reversible tracers, four parameters were estimated, the transport rate constant K_1 and the total distribution volume (DV) for each tracer. For the combined reversible/irreversible model,

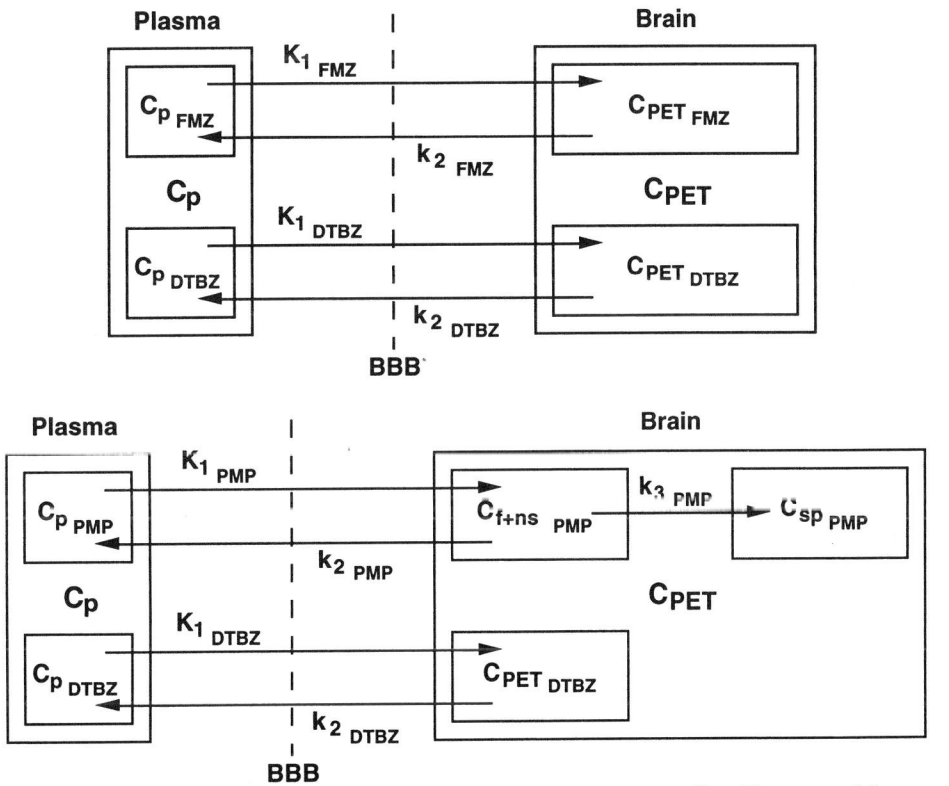

FIGURE 1 Compartmental model configurations for dual-tracer PET studies. The top model was implemented for studies with FMZ and DTBZ. This configuration is applicable to studies involving two rapidly reversible PET radiotracers. Four parameters are estimated, K_1 and DV for both tracers. The bottom model was implemented for studies with PMP and either DTBZ (as illustrated above) or FMZ. This configuration is applicable to studies involving one reversible and one irreversible tracer. Four or five parameters are estimated, K_1 and DV for the reversible tracer, and K_1, k_2, and k_3, or just K_1 and k_3 (assuming a constant fixed value for DV_{f+ns}, the distribution volume of free plus nonspecifically bound tracer) for the irreversible tracer.

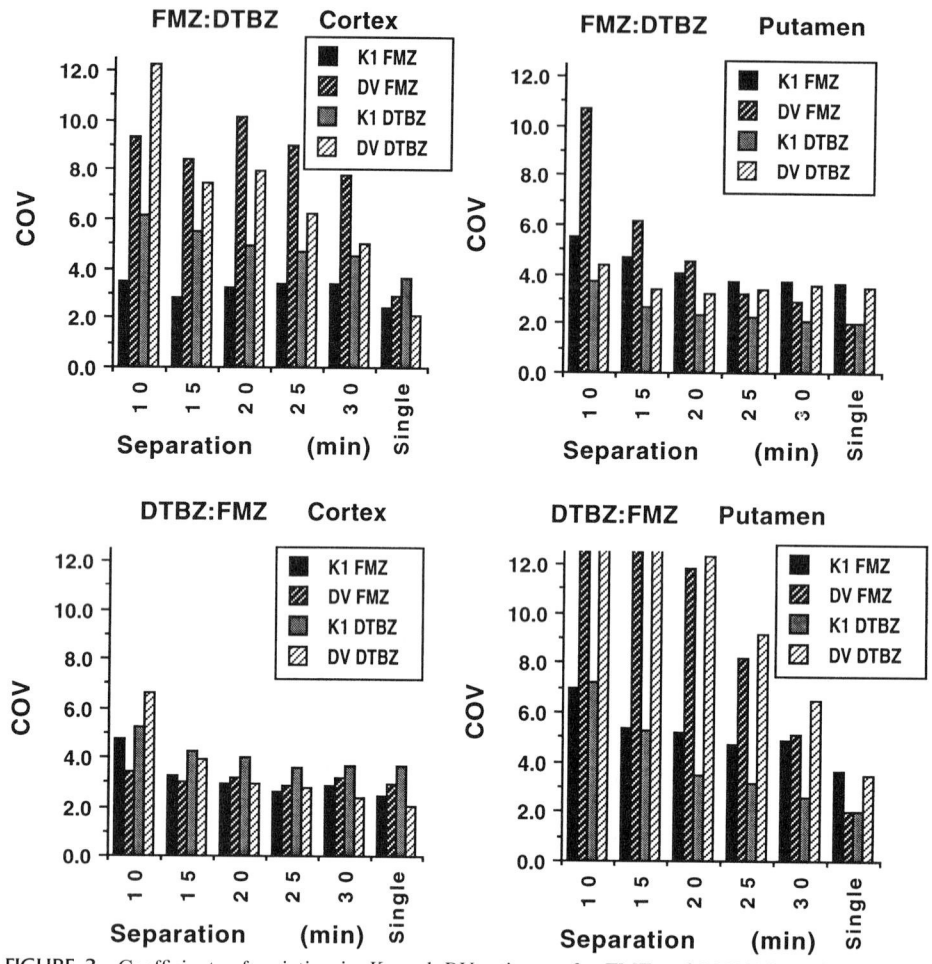

FIGURE 2 Coefficients of variation in K_1 and DV estimates for FMZ and DTBZ from fits to 1024 sets of noisy data. Noise levels approximate those for 8 mCi tracer injections and 3D scanning. Each panel shows COVs for dual-tracer studies with injections separated by 10–30 min as well as COVs for single-injection studies. Results when FMZ is injected first are shown in the top panels, whereas results for DTBZ first are shown below. In general, estimates become less precise with decreasing tracer injection separation. However, this is highly dependent on region, injection order, and parameter. For example, when DTBZ is injected first, cortical estimates are nearly as good as those derived from single scans even down to a 15-min separation. In contrast, estimates of DTBZ DV in putamen are much poorer even at a longer separation of 25 min.

K_1 and DV are estimated for the reversible tracer, whereas K_1, k_2, and k_3 or only K_1 and k_3 (if K_1/k_2 is fixed) are estimated for the irreversible tracer.

D. Simulation Studies

Two types of computer simulations were performed: (1) studies involving simple computer-generated, individual region time–activity curves (TACs) based on kinetic model equations and (2) studies involving a computer-generated phantom based on a digitized Hoffman 3D brain phantom warped into stereotactic brain atlas coordinates (Talairach and Tournoux, 1988). This phantom can account for physical effects involved in the scanning process, including Poisson noise, detector response functions, spatial resolution, attenuation, and random coincidences.

1. Single Region TAC Simulations

Noise-free TACs were generated for each tracer pair from model equations. For FMZ and DTBZ, noise-free curves were calculated using a model consisting of four rate constants: K_1, k_2, k_3, and k_4. Because data were generated assuming a four rate constant two-tissue compartment model, but analyzed with a single-tissue compartment model estimating only K_1 and DV, it was possible to examine the effect of biases caused by simplifications in the model. For PMP, curves were calculated in the same manner, but with $k_4 = 0$ as hydrolysis by AChE is irreversible. To examine the performance of the methods across a range of kinetic behaviors, six different brain regions (cortex, hippocampus, cerebellum, pons, thalamus, and putamen) were simulated using typical values determined from actual studies in humans. The simulated scan sequence started with 30-sec frames followed by progressively longer frames, returning to the shorter 30-sec frames at the time of the second injection, followed again by progressively longer frames for the remainder of the scan. Noise-free dual-tracer data sets were calculated for each region (six total), for both injection orders (two) of each dual-tracer pair (three), at injection separations of 10, 15, 20, 25, and 30 min, creating a total of 180 noise-free data sets. Pseudo-Gaussian noise was added to each noise-free TAC, accounting for both frame duration and radioactive decay, generating 1024 sets of noisy data per condition. The noise level was adjusted to that observed in actual PET VOIs (~ 1 cm^3) for 3D scanning for injected doses of approximately 1, 3, 8, and 20 mCi per tracer. Results reported here are for 8-mCi simulations.

Data were fitted with the true input functions used to generate the noise-free time activity curves and a combined model estimating parameters for both tracers simultaneously. The mean and standard deviation was calculated across the 1024 data sets for each of the 180 conditions, yielding both an estimate of the precision, given by the coefficient of variation ($COV = 100 * SD/mean$), and the bias, given by the difference between the mean and the true parameter value. Performance in terms of both precision and bias in the parameter estimates were compared to those derived from simulations of single-injection single-tracer scans.

2. Computer-Generated 3D Brain Phantom Simulations

Three-dimensional PET data sets of the brain simulating the isotope distribution of each of the three radiopharmaceuticals were produced from the transverse levels of a digitized version of the 3D Hoffman brain phantom (Hoffman *et al.*, 1991). The digitized phantom was then transformed, including nonlinear warping, into the stereotactic coordinate system of Talairach and Tournoux (1988). The brain was divided into 19 regions, identifying various cortical and subcortical structures, cerebellum and brain stem regions, white matter, and cerebrospinal fluid. Temporal sequences of 3D data sets were generated to simulate the spatially varying time course of tracer concentration in the brain by assigning each of the regions a temporally varying radionuclide concentration based on its unique set of kinetic parameters. After assignment of radioactivity values for the tracer distribution, data sets were forward-projected into sinogram space. Poisson noise was generated for each projection ray. Effects of detector response function, decay, scan duration, injected dose, attenuation, and random coincidences were included in the calculation of the noise level for each ray. Following the generation of sinograms, data sets were reconstructed by filtered backprojection. Both VOI and pixel-by-pixel time activity curves were generated from the data sets for subsequent kinetic modeling and parameter estimation as described in the preceding paragraphs.

III. RESULTS AND DISCUSSION

A. Single Region Simulations

The studies for all six combinations of tracer pair and injection order demonstrated that both the precision and the bias in the estimates of model parameters were highly dependent on each of the following factors: (1) individual tracer (FMZ vs DTBZ vs PMP), (2) pairing of tracers (e.g., FMZ and DTBZ vs FMZ and PMP), (3) order of injection for a given pair of tracers (e.g., FMZ followed by PMP vs PMP followed by FMZ), (4) brain region (e.g., cortex vs putamen), (5) parameter (K_1 vs DV vs k_3), and (6) the separation time between the two injections.

Figures 2, 3, and 4 present results in cortex and putamen for both injection orders of one of the tracer pairs (FMZ and DTBZ). Results for other regions and other tracer pairs led to similar conclusions, although specific results of which regions and what parameters could be estimated with the greatest (or least) accuracy varied from condition to condition. The true parameter values for the simulations of Figs. 2–4 were for cortex: $K_{1,FMZ} = 0.3$, $DV_{FMZ} = 7.5$, $K_{1,DTBZ} = 0.4$, $DV_{DTBZ} = 3.8$; and for putamen: $K_{1,FMZ} = 0.36$, $DV_{FMZ} = 3.25$, $K_{1,DTBZ} = 0.48$, $DV_{DTBZ} = 16$. All K_1 and DV values are given in ml g^{-1} min^{-1} and ml g^{-1}, respectively.

Results shown in Fig. 2 demonstrate the effect that the time between injections has on the precision of parameter estimates. At the right of each panel, COVs for single injection studies of FMZ and DTBZ (i.e., nonoverlapping studies) are given for comparison. As expected, estimates of all parameters become less precise when injections are moved closer together. However, the magnitude of this effect varied considerably with region and the order of injections. For the FMZ DTBZ pairing, there was not a particular injection order or separation time between tracer injections that could be considered optimal for all parameters and all regions. The estimate of the K_1 values for both tracers

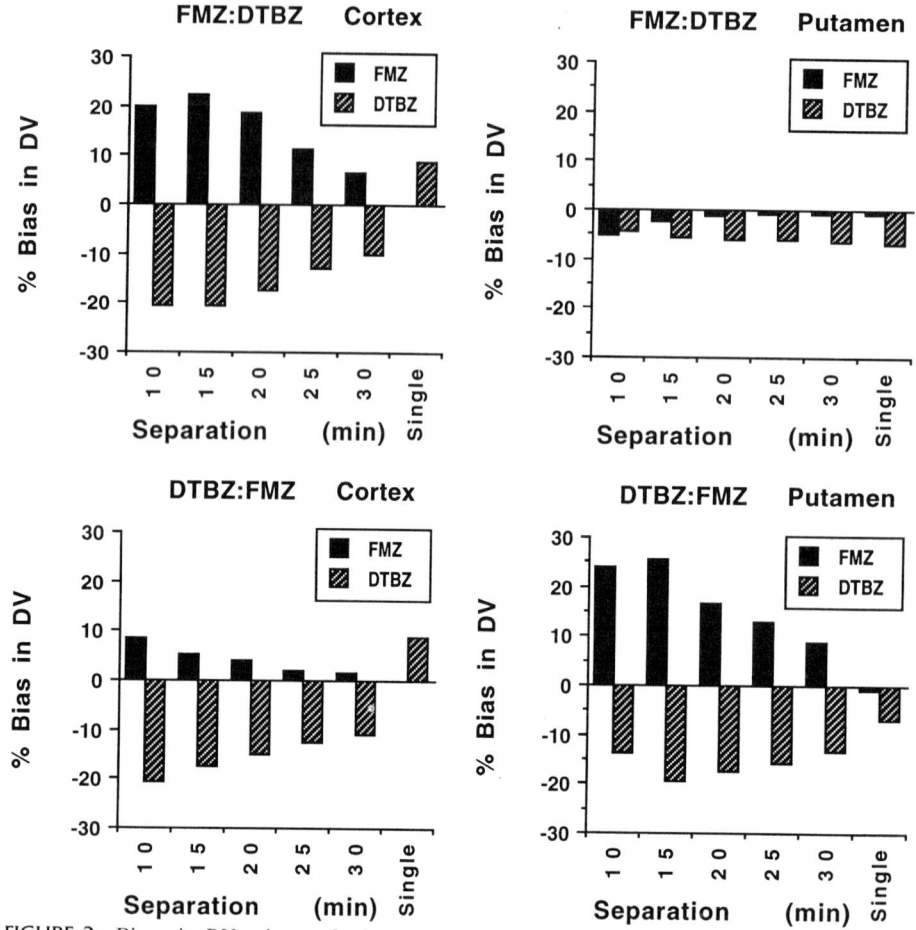

FIGURE 3 Biases in *DV* estimates for FMZ and DTBZ from fits to 1024 sets of noisy data. Results are from the same data sets as in Fig. 2, but are shown only for *DV*. Many of the same conclusions apply here. In general, estimates become more biased with decreasing separation between injections; however, this depends again on both region and injection order. The conditions under which parameters were estimated more precisely (FMZ:DTBZ in putamen; DTBZ:FMZ in cortex) are also the conditions that tend to yield lower biases. In all cases, the change in bias with decreasing injection separation is of opposite sign for the two tracers. For example, when DTBZ is injected first, DTBZ becomes progressively more underestimated, whereas FMZ is overestimated. The covariance between parameters is important, as can be seen in Fig. 4.

suffered very little loss of precision as the injections were moved closer together, for either injection order. Injection offsets of as little as 10–15 min appear acceptable. The *DV* estimates, however, showed a definite interaction between region, more specifically the kinetics of the tracers for a given region, and which tracer should be injected first. If one was primarily interested in the cortex, then DTBZ should be injected first (lower left), whereas if the putamen was of greater interest, FMZ should be injected first (upper right). This same finding occurred for the FMZ and PMP pair (data not shown), but was different for the DTBZ PMP pairing where either order of injection worked quite well in the cortex, whereas neither worked very well in the putamen. For any tracer pairing, estimates were more precise for a specific region when the tracer that exhibits a more rapid clearance from tissue back to blood was injected first. In other words, because K_1 values in gray matter structures are reasonably similar across the brain, the tracer with the lower distribution volume (i.e., higher k_2 values) or the tracer with the lower k_3 value (hence greater back diffusion across the blood brain barrier) should be injected first.

Figure 3 illustrates biases in *DV* estimates for the same simulations shown in Fig. 2. In general, parameter bias increased with shorter separation between injections, but as was seen for precision, the magnitude of the bias depended on both regional kinetics and injection order. When biases of a particular sign occurred in the *DV* of one tracer, corresponding biases

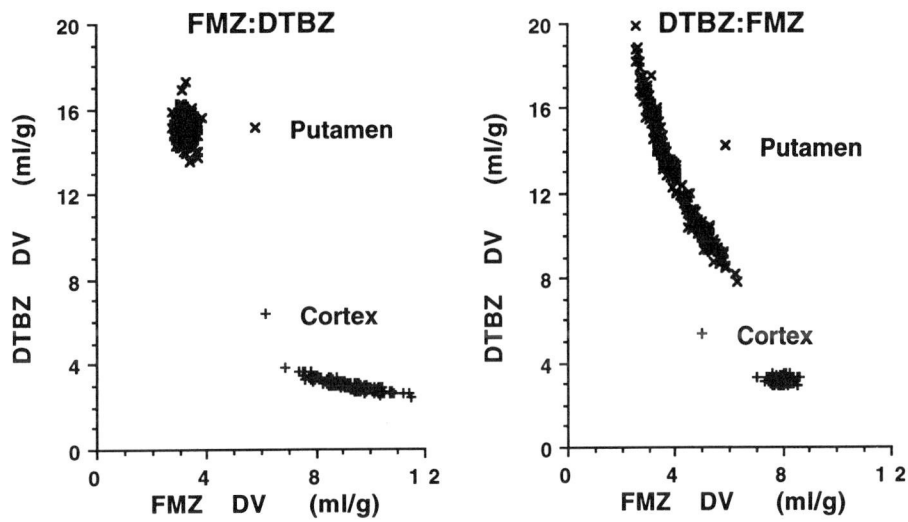

FIGURE 4 Estimates of FMZ and DTBZ *DV* for injections 15 min apart. Note the dependence of the estimates on the order of injection. When FMZ is injected first (left), putamen values are estimated more precisely; however, if DTBZ is injected first (right), cortical estimates are better. In both cases, the region that can be estimated with better precision and less bias is the one where the first tracer injected is the one that exhibits the more rapid kinetic behavior.

of the opposite sign occurred in the *DV* of the other tracer. As has been seen when fitting data from single-injection studies using a two-tissue compartment model (e.g., DTBZ; Koeppe *et al.*, 1996), one can estimate the total distribution volume with more precision than either the free + nonspecific or specific distribution volumes individually. In this case, the sum of the total distribution volumes of the two tracers was estimated with greater accuracy than the *DV* of either tracer separately.

The covariance between FMZ and DTBZ estimates can be seen clearly in Fig. 4. Results are for a 15-min separation between injections. Note that the increases in uncertainty in the estimates of the *DV* of the two tracers in situations where estimation accuracy is poorer (i.e, cortex when FMZ is injected first, and putamen when DTBZ is injected first) were not random but were highly correlated. With this short an interval between injections, and having the tracer with "slower" kinetic behavior injected first, all that can be measured accurately is the sum of the *DV* of the two tracers. There is not sufficient kinetic information to quantify accurately the contribution of each tracer to the total.

Figure 5 compares the time–activity curves derived from simulations using the 3D brain phantom to the noise-free TACs that were used to generate data. Re-

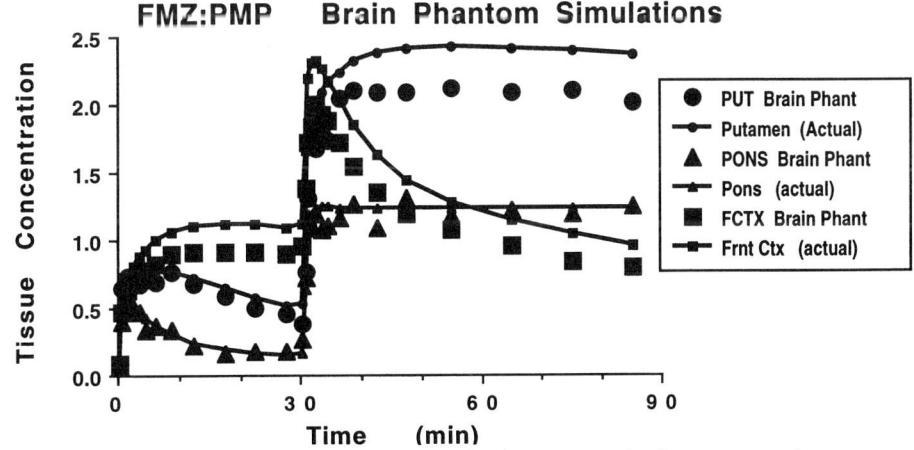

FIGURE 5 Comparison of TACs from the 3D brain phantom to noise-free curves used to generate data. Note the partial volume effects due to the detector response function and filtered back projection.

sults are for FMZ:PMP with a 30-min injection separation. Differences between the curves are due primarily to the detector response function and the smoothing filter used in the filtered-backprojection reconstruction. As one would expect, the partial volume effects were not regionally uniform, but were higher in small, high-activity regions (i.e., putamen) and were lower in regions surrounded by regions of similar or higher activity (i.e., pons, where FMZ binding in nearby regions is higher).

In conclusion, temporally overlapping dual-tracer PET using a single data acquisition sequence is feasible. Under many conditions, the model parameters for two different tracers injected only 10–20 min apart can be estimated nearly as accurately as in standard single-injection studies. It may be difficult to translate the conclusions drawn from the present simulations into general conclusions applicable to any tracer pair, and thus each new pair considered for use in dual-tracer designs must be evaluated independently. However, the first tracer injected should have sufficiently rapid kinetics so that the model parameters describing its *in vivo* behavior are reasonably well determined by the time of the second injection. Future work on dual-tracer scanning will continue with more studies using the computerized 3D brain phantom to explore more fully how tomograph and scanning properties affect these studies, and with the development of the methods needed to rapidly separate the authentic arterial plasma fractions of the two tracers. Dual-tracer methods, such as described here, offer new opportunities for obtaining neuropharmacologic information related to multiple biochemical systems or multiple aspects of the same biochemical system in a single PET scanning session.

Acknowledgment

This work was supported by the Department of Energy Grant DE-FG02-87ER60561.

References

Carson, R. E. (1986). Parameter estimation in positron emission tomography. *In* "Positron Emission Tomography and Autoradiography" (M. E. Phelps, J. C. Mazziotta, and H. R. Schelbert, eds.), pp. 347–390. Raven Press, New York.

Graham, M. M. (1985). Model simplification: Complexity versus reduction. *Circulation* **72**(Suppl. IV): 63–68.

Hoffman, E. J., Cutler, P. D., Digby, W. M., and Mazziotta, J. C. (1991). 3-D phantom to simulate cerebral blood flow and metabolic images for PET. *IEEE Trans. Nucl. Sci.* **NS-37**(2): 616–620.

Koeppe, R. A. (1990). Compartmental modeling alternatives for kinetic analysis of PET neurotransmitter/receptor studies. *In* "Frontiers in Nuclear Medicine: In Vivo Imaging of Neurotransmitter Functions in Brain, Heart, and Tumors" (D. E. Kuhl, ed.), pp. 113–139. American College of Nuclear Physicians, Washington, DC.

Koeppe, R. A., Holthoff, V. A., Frey, K. A., Kilbourn, M. R., and Kuhl, D. E. (1991). Compartmental analysis of [^{11}C]flumazenil kinetics for the estimation of ligand transport rate and receptor distribution using positron emission tomography. *J. Cereb. Blood Flow Metab.* **11**: 735–744.

Koeppe, R. A., Frey, K. A., Vander Borght, T. M., Kilbourn, M. R., Jewett, D. M., Lee, L. C., and Kuhl, D. E. (1996). Kinetic evaluation of [^{11}C]dihydrotetrabenazine by dynamic PET: Measurement of vesicular monoamine transporter. *J. Cereb. Blood Flow Metab.* **16**: 1288–1299.

Koeppe, R. A., Frey, K. A., Snyder, S. E., Kilbourn, M. R., and Kuhl, D. E. (1997). Kinetic analysis alternatives for assessing AChE activity with N-[C-11]methylpiperidinyl propionate (PMP): To constrain or not to constrain? *J. Nucl. Med.* **38**: 198P.

Talairach, J., and Tournoux, P. (1988). "Co-Planar Stereotaxic Atlas of the Human Brain." Thieme, Stuttgart.

CHAPTER 55

Simultaneous Estimation of Perfusion (K_1) and Vascular (V_0) Responses in [^{15}O]Water PET Activation Studies

P.-J. TOUSSAINT and E. MEYER

Positron Imaging Laboratories
McConnell Brain Imaging Centre
Montreal Neurological Institute
McGill University
Montréal, Québec, H3A 2B4 Canada

A two-compartment model for the analysis of [^{15}O]water positron emission tomography (PET) cerebral blood flow studies which, contrary to the Kety one-compartment model, accounts for nonextracted intravascular activity for blood flow tracers whose first-pass extraction fraction is less than unity has been introduced. With this approach, erroneously high blood flow values derived with the Kety model for highly vascularized brain regions are avoided. Furthermore, because the second compartment of the new model, the so-called initial tracer distribution volume, V_0, represents part of the cerebral vascular volume, predominantly the arterial and part of the capillary fraction, the two-compartment analysis offers the possibility of distinguishing changes in tissue perfusion (K_1) from changes in vascular volume (V_0) in response to physiological stimuli. For this purpose, computer simulations were used to investigate the feasibility of simultaneously estimating regional differences rather than the absolute values of K_1 and V_0 between two physiological states from a single dynamic PET study using nonlinear least-squares multiparameter fitting. The fitting parameters were K_1, k_2, V_0, tracer delay Δt, and dispersion time constant τ. Under the assumption that the ratio K_1/k_2 and τ did not change between baseline and activation, it was possible to recover the simulated changes in K_1 and V_0 with an error of less than 6%. In addition, absolute estimates of K_1 were accurate within 5%, whereas the error in V_0 was larger than 20%. As expected, V_0 proved to be extremely sensitive to the "vascular" correction parameters Δt and τ. In conclusion, the two-compartmental analysis provides for a more detailed investigation and interpretation of cerebral activation experiments without the need for additional PET experiments or scan time.

I. INTRODUCTION

The intravenous [^{15}O]H$_2$O method has long been the method of choice for the measurement of cerebral blood flow (K_1) in physiological activation studies in humans by means of positron emission tomography (PET) (Herscovitch *et al.*, 1983; Raichle *et al.*, 1983). Data are usually analyzed on the basis of Kety's one-compartment model (Kety, 1951), which views cerebral tissue, including its vascular components, as a single homogeneous compartment. Building on this model, a two-compartment model with the second compartment, called the initial tracer distribution volume (V_0), accounting for nonextracted intravascular radioactivity (Ohta *et al.*, 1996) has been introduced. The introduction of this second compartment allows a more reliable quantification of tissue perfusion with tracers whose first-pass extraction fraction is less than unity, as is the case for [^{15}O]H$_2$O (Herscovitch *et al.*, 1983; Eichling *et al.*, 1974). In particular, the two-compartment approach avoids the erroneously high blood flow estimates resulting from the Kety model for highly vascularized structures (Ohta *et al.*, 1996). A further potential advantage of this model resides in its capability to distinguish between changes in tissue perfusion (K_1) and in vascular volume (V_0) in response to specific physiological stimuli. For the case of vibrotactile stimulation of the hand, it has already been shown that these responses need not be spatially coincident (Fujita *et al.*, 1997). The authors therefore believe that the two-compartment model allows a more detailed interpretation of cerebral activation experiments. However, the simultaneous quantification of K_1 and V_0 from a single dynamic PET study is beset with a number of difficulties. First, V_0 is expected to depend very critically on the accuracy of the "vascular" correction factors, the tracer delay (Δt), and the dispersion time constant (τ). Second, the five fitting parameters, including K_1, k_2, and V_0, as well as the tracer delay and dispersion correction factors, might not be accurately and simultaneously recovered from experimental PET data. Computer simulations have been used to clarify these concerns. In particular, the accuracy with which changes in K_1 and V_0 between two physiological conditions, rather than absolute values, may be estimated by multiparameter fitting under simplified assumptions has been investigated.

II. METHODS

An ideal bolus type input function, $C_a(t)$, was generated by approximating the arterial blood activity curve from a typical dynamic PET [^{15}O]H$_2$O study to a sum of Γ variates. The measured delayed and dispersed blood curve, $g(t + \Delta t)$, may be described by the convolution of the ideal input function, $C_a(t)$, with a monoexponential dispersion function as $g(t + \Delta t) = C_a(t) * (1/\tau)\exp(-t/\tau)$, where τ is the dispersion time constant (Iida et al., 1986). Deconvolution of this expression yields $C_a(t)$ from $g(t + \Delta t)$ as

$$C_a(t) = g(t + \Delta t) + \tau \frac{dg}{dt}(t + \Delta t). \quad (1)$$

The two-compartment model gives the tissue time–activity curve (Ohta et al., 1996) as

$$C(t) = K_1 \int_0^t C_a(u) e^{-k_2 \cdot (t-u)} du + V_0 \cdot C_a(t). \quad (2)$$

Inserting $C_a(t)$ from Eq. 1,

$$\begin{aligned} C(t) &= \tau \cdot K_1 \cdot g(t + \Delta t) \\ &\quad + (1 - \tau \cdot k_2) \cdot K_1 \\ &\quad \cdot e^{-k_2 \cdot (t + \Delta t)} \int_{\Delta t}^{t + \Delta t} g(u) e^{k_2 \cdot u} du \\ &\quad + V_0 \cdot \left[g(t + \Delta t) + \tau \frac{dg}{dt}(t + \Delta t) \right]. \end{aligned} \quad (3)$$

Tissue curves for a total scan duration of 3 min with frame lengths of 5 sec were simulated for low, average, and high blood flow regions ($K_1 = 25$, 50, and 75 ml/100 g/min, respectively), $K_1/k_2 = 0.9$ ml/g, and $V_0 = 4$ ml/100 g at low K_1 and 2 ml/100 g at average

to high values. Two activation conditions were obtained for each baseline by increasing both K_1 and V_0 by 10% and then by 50%. Twenty data sets were generated with 10% Gaussian random noise superposed.

Nonlinear least-squares fitting was used to estimate sets of K_1, k_2, V_0, and the dispersion time constant τ for the baseline condition while stepping through a range of Δt values, using Eq. (3). The four-parameter set corresponding to the smallest residual sum of squares (RSS) was deemed the best fit. For the two activation conditions, the ratio K_1/k_2, which is closely related to the one-compartment equilibrium tissue–blood partition coefficient for water, and τ were fixed to their baseline fitted values as an approximation in order to reduce the number of fitted parameters from four to two, and the best delay estimate was again obtained by stepping through a range of delays, Δt. The percentage deviations of the estimated percentage increases in K_1 and V_0 from their expected changes of 10 or 50% (ΔK_1 and ΔV_0) were calculated as

$$\left(\frac{\text{estimated \% change}}{\text{expected \% change}} - 1 \right) \times 100.$$

III. RESULTS

Absolute K_1 values were always recovered within ±5% for all baseline and activation conditions with the coefficient of variation (COV) ranging from 0.3 to 1.8% (Fig. 1). As expected, the error for the absolute estimates of V_0 was large (±35% for $K_1 = 75$ ml/100 g/min, Fig. 2). However, the simulated V_0 changes

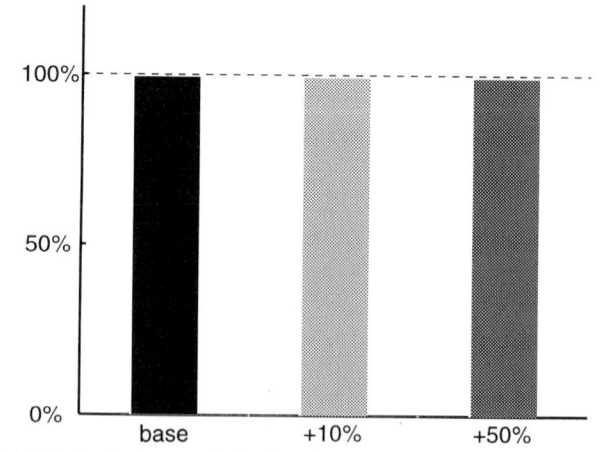

FIGURE 1 Estimates of absolute K_1 expressed as a percentage of their expected values for a baseline four-parameter fit (left bar) and 10 and 50% activation two-parameter fits (middle and right bar, respectively) for baseline $K_1 = 50$ ml/100 g/min. The expected K_1 value is indicated below each bar.

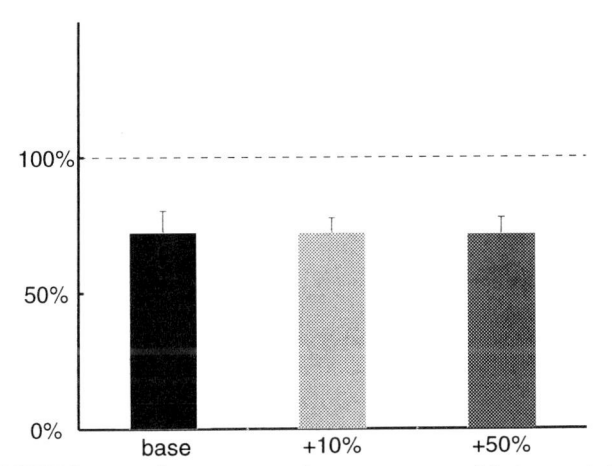

FIGURE 2 V_0 estimates expressed as a percentage of their expected values for a baseline four-parameter fit (left bar) and 10 and 50% activation two-parameter fits (middle and right bar, respectively) for baseline $K_1 = 50$ ml/100 g/min. The corresponding absolute K_1 value is indicated below each bar.

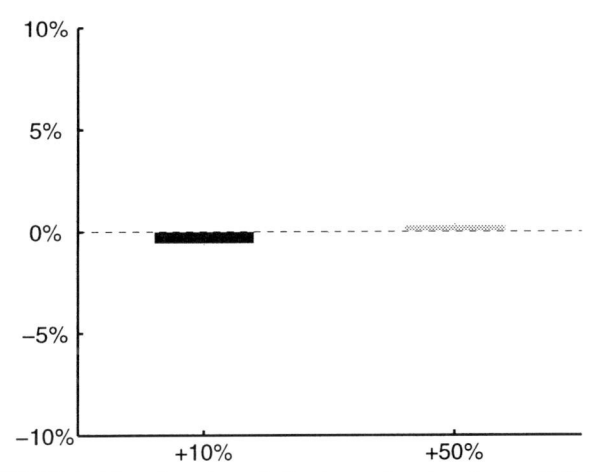

FIGURE 4 Error in estimated K_1 change from baseline to 10 (left) and 50% (right) activation expressed as a percentage of the expected value for baseline $K_1 = 50$ ml/100 g/min. The corresponding "activated" K_1 value is indicated below each bar.

between baseline and activation conditions were recovered with an error (bias) of less than 6% for low and average K_1 (Fig. 3). At high K_1 values, however, the error in ΔV_0 reached 20%. The expected K_1 changes were accurate within ±6% at all levels of K_1 (Fig. 4).

In areas where blood flow is very high, V_0 differences, especially if small, are not recovered accurately. For example, for a baseline K_1 of 100 ml/100 g/min, an error in V_0 change of 42% for a 10% increase was observed, whereas the 50% increase was recovered within 9%. The ΔK_1 values were recovered with 3% for both activation conditions.

IV. DISCUSSION

It has been shown previously that K_1 determinations based on Kety's one-compartment model (Kety, 1951) may be erroneously high, particularly for very vascularized brain regions, because this model does not account for nonextracted intravascular activity from radiotracers with a first-pass extraction fraction of less than unity, such as in the case of [^{15}O]H$_2$O (Ohta et al., 1996). The two-compartment model, which includes a vascular compartment, called the initial tracer distribution volume V_0, corrects this situation. At the same time, this model allows the separation of local tissue perfusion (K_1) from vascular (V_0) responses following physiological activation based on a single dynamic (3 min) [^{15}O]H$_2$O study. However, one more parameter (V_0) has been added to the already plentiful set of unknowns, K_1, k_2, Δt, and τ. Although accurate absolute determination of all five parameters from nonlinear least-squares multiparameter fitting may not be expected, it has been demonstrated that this approach yields accurate estimates of K_1 when Δt, τ, and V_0 are lumped into a single-fit parameter, Δt_{lumped} (Toussaint and Meyer, 1996) whose absolute value, in this context, is meaningless and of no interest. Similarly, using the Kety model, K_1 is accurately estimated from a four-parameter fit of K_1, k_2, τ, and Δt with the latter two parameters often assuming meaningless values (Meyer, 1989). In the present study, with V_0 being a parameter of interest (no lumping), K_1 estimates

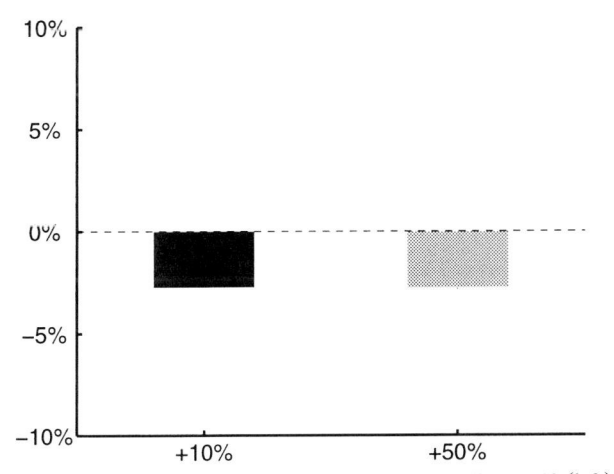

FIGURE 3 Error in estimated V_0 change from baseline to 10 (left) and 50% (right) activation expressed as a percentage of the expected value for baseline $K_1 = 50$ ml/100 g/min. The corresponding "activated" K_1 value is indicated below each bar.

again turned out to be highly reliable, whereas V_0 could not be estimated accurately. It was found that this additional vascular parameter, as to be expected, was extremely sensitive to the "vascular" correction factors τ and Δt. Because, however, changes in K_1 and V_0 rather than their absolute values are of prime interest in a majority of activation experiments, the authors focused their efforts on the estimation of ΔK_1 and ΔV_0. Under the simplifying assumption that the ratio K_1/k_2 and τ did not change between baseline and activation conditions, the authors succeeded in estimating ΔK_1 and ΔV_0 with an error of only about 6%. The assumption of a constant K_1/k_2 ratio is synonymous with the assumption of a constant equilibrium partition coefficient for water, which may well be justified as this parameter depends mainly on the chemical composition of the cerebral tissue under study. Constancy of the dispersion time constant, however, is less warranted as τ will change somewhat due to the variation in blood flow velocity and V_0 between the two conditions.

V. CONCLUSIONS

The simulations described in this chapter demonstrate that the two-compartment model of Ohta *et al.* (1996) allows the simultaneous estimation of changes in tissue perfusion (K_1) and initial vascular tracer distribution volume (V_0) in response to physiological activation from a single [^{15}O]H$_2$O water dynamic PET study by means of nonlinear multiparameter least-squares fitting. This approach also yields accurate absolute values of K_1, but not of V_0. The use of this model allows a more detailed analysis and interpretation of physiological activation studies due to its ability to make a distinction between tissue perfusion and vascular responses as a consequence of physiological stimulation.

Acknowledgments

This work was supported by MRC (Canada) Grant SP-30, the Isaac Walton Killam Fellowship Fund of the Montreal Neurological Institute, and the McDonnell-Pew Program in Cognitive Neuroscience.

References

Eichling, J. O., Raichle, M. E., Grubb, R. L., and Ter-Pogossian, M. M. (1974). Evidence of the limitation of water as a freely diffusible tracer in brain of the rhesus monkey. *Circ. Res.* **35**: 358–364.

Fujita, H., Meyer, E., Reutens, D. C., Kuwabara, H., Evans, A. C., and Gjedde, A. (1997). Cerebral [^{15}O]water clearance in humans determined by positron emission tomography: II. Vascular responses to vibrotactile stimulation. *J. Cereb. Blood Flow Metab.* **17**: 73–79.

Herscovitch, P., Markham, J., and Raichle, M. E. (1983). Brain blood flow measured with intravenous H$_2^{15}$O. I. Theory and error analysis. *J. Nucl. Med.* **24**: 782–789.

Iida, H., Kanno, I., Miura, S., Murakami, M., Takahashi, K., and Uemura, K. (1986). Error analysis of a quantitative cerebral blood flow measurement using O-15 H$_2$O autoradiography and positron emission tomography, with respect to the dispersion of the input function. *J. Cereb. Blood Flow. Metab.* **6**: 536–545.

Kety, S. S. (1951). The theory and application of the exchange of inert gas at the lungs and tissues. *Pharmacol. Rev.* **3**: 1–41.

Meyer, E. (1989). Simultaneous correction for tracer arrival delay and dispersion in CBF measurements by the H$_2^{15}$O autoradiographic method and dynamic PET. *J. Nucl. Med.* **30**: 1069–1078.

Ohta, S., Meyer, E., Fujita, H., Reutens, D. C., Evans, A., and Gjedde, A. (1996). Cerebral [^{15}O]water clearance in humans determined by PET: I. Theory and normal values. *J. Cereb. Blood Flow Metab.* **16**: 765–780.

Raichle, M. E., Martin, W. R. W., Herscovitch, P., Mintun, M. A., and Markham, J. (1983). Brain blood flow measured with intravenous H$_2^{15}$O. II. Implementation and validation. *J. Nucl. Med.* **24**: 790–798.

Toussaint, P.-J., and Meyer, E. (1996). A sensitivity analysis of model parameters in dynamic blood flow studies using H$_2^{15}$O and PET. *In* "Quantification of Brain Function Using PET" (R. Myers, V. J. Cunningham, D. L. Bailey, and T. Jones, eds.), pp. 196–200. Academic Press, San Diego, CA.

CHAPTER 56

Enhancement of the Signal-to-Noise Ratio in [¹⁵O]Water Bolus PET Activation Studies Using a Combined Cold-Bolus/Switched Protocol

JORGE J. MORENO-CANTÚ,*,† CHRISTOPHER J. THOMPSON,*,† ERNST MEYER,*,†
PIERRE FISET,‡ ROBERT J. ZATORRE,*,†,¶ DENISE KLEIN,†,¶
and DAVID C. REUTENS*,†

*Department of Neurology and Neurosurgery
¶ Neuropsychology/Cognitive Neuroscience Unit and † McConnell Brain Imaging Centre
Montreal Neurological Institute
‡ Department of Anaesthesiology
Royal Victoria Hospital
McGill University
Montreal, Quebec, H3A 2B4 Canada

To increase the signal-to-noise ratio (S/N) of [¹⁵O]H₂O bolus positron emission tomography (PET) activation images, a data acquisition protocol that alters the relative distribution of tracer in the uptake and washout phases of the input function was designed and tested. This protocol enhances S/N gains yielded by conventional switched protocols by combining task switching and the use of a large bolus of blood free of tracer (cold bolus). The cold bolus is formed by sequestering blood in the lower limbs with a double cuff prior to tracer injection. The effect of a cold-bolus/switched protocol on the activation signals was first simulated using a compartmental model of the uptake of [¹⁵O]H₂O into the brain. Then, the effectiveness of the protocol was investigated in four volunteers performing a language task. The signal changes introduced when using the cold bolus were analyzed by comparing, across protocols, the magnitude and statistical significance of the resulting activation foci and by comparing changes in the overall image noise. In simulated data, the combined protocol yielded a substantial increase in the activation signal for scan durations greater than 60 sec when compared to the signal yielded by the switched protocol alone. In PET experiments, activation foci obtained using the combined protocol had significantly higher t statistic values than equivalent foci detected using the conventional switched protocol (mean improvement 36%). Analysis of the averaged subtracted images revealed that the improvements in statistical significance of the activation foci were caused by increases in the signal magnitudes and not by decreases in the overall image noise.

I. INTRODUCTION

The signal-to-noise ratio (S/N) of positron emission tomography (PET) activation images can be enhanced by manipulating tracer distribution in order to maximize scan length using a switched protocol (Cherry et al., 1995; Moreno-Cantú et al., 1998). This is achieved by switching task execution from activation to baseline during the activation scans and vice versa during the baseline scans, when tracer concentration in the brain reaches a maximum (Cherry et al., 1995). The effect of task switching is twofold: (a) after execution of the task is switched from activation to baseline, blood flow in the regions activated by the activation task is decreased, thereby reducing tracer clearance, whereas blood flow in regions activated by the baseline task is increased, thus augmenting tracer clearance; and (b) after execution of the task is switched from baseline to activation, the opposite occurs.

This chapter describes a method developed to augment the S/N of activation images obtained with the switched protocol by decreasing tracer dispersion during the uptake stage and by enhancing tracer clearance during the washout period. This is accomplished by combining task switching with (a) a reduction of the blood pool available for tracer mixing during the uptake period and (b) the release of a large bolus of blood, free of tracer, during the washout period. The blood pool reduction is achieved by occluding blood circulation to the lower limbs prior to tracer injection, whereas the cold bolus effect is obtained by releasing the blood previously sequestered in the legs. In a normal adult, the lower limbs contain 10–40% of the total blood volume, depending primarily on body position and limb temperature (Albert, 1971; Rowell, 1986; Henry and Meehan, 1971). The effect of combining a cold bolus and a switched protocol on the activation signals was first simulated using a compartmental model of [^{15}O]H$_2$O uptake in the brain (Herscovitch et al., 1983) and was then tested in a PET experiment designed to identify the neuronal areas activated when subjects perform lexical search and retrieval operations (Klein et al., 1996).

II. METHODS

A. Simulation Studies

The kinetics of uptake of [^{15}O]H$_2$O into brain tissue was simulated using a one-compartment model (Herscovitch et al., 1983). In this model, for a constant cerebral blood flow (F), the radioactivity (M) in the brain at a time $t = T$ after tracer injection is described by

$$M(T) = F \int_0^T C_a(t) \exp\left(\frac{F}{P}(t - T)\right) dt, \quad (1)$$

where C_a is the plasma concentration of tracer and P is the partition coefficient across the blood brain barrier, which is assumed to be 0.9 ml · g^{-1}. Thus, if the regional cerebral blood flow (F_2) changes at time T_s, the radioactivity in the brain at any time T_2 after T_s is given by

$$M(T_2) = M(T_s) \exp\left(\frac{F_2}{P}(T_s - T_2)\right) + F_2 \int_{T_s}^{T_2} C_a(t) \exp\left(\frac{F_2}{P}(t - T_2)\right) dt. \quad (2)$$

1. Input Functions (Fig. 1)

A typical measured input function (Vaffae et al., 1996) was used to simulate tracer kinetics in standard and switched protocols. This input function was then modified to account for the blood volume and tracer concentration changes introduced by the cold-bolus/

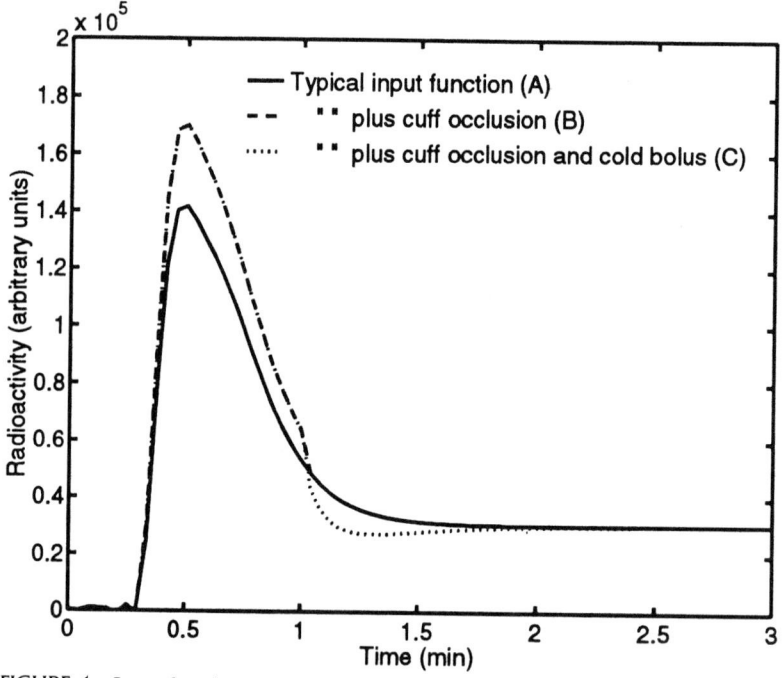

FIGURE 1 Input functions used to simulate tracer uptake in standard, switched, and cold-bolus/switched protocols. (A) Standard and switched protocol simulations used a typical measured input function. (B and C) Cold-bolus/switched protocol simulations employed two input functions obtained by modifying the measured function. Function B accounted for a 60-sec cuff occlusion and instantaneous mixing of the cold bolus. Function C accounted for the 60-sec cuff occlusion plus a realistic time-varying mixing of the cold bolus after cuff deflation.

switched protocol. First, the reduction in the vascular volume caused by the isolation of the vasculature of the legs was simulated by applying a scaling factor to the initial 60 sec of the measured input function to simulate a 20% decrease in blood volume. Second, incomplete mixing of the tracer-free blood released after cuff deflation was simulated. Here, the input function was multiplied element wise by a time-varying scaling factor (R_2) of the form:

$$R_2(t) = 1 \quad t < T_S$$
$$= 1 - \frac{R}{A[\max(R) - \min(R)]} t \geq T_S, \quad (3)$$

where t is the time after injection, T_s is the time when the cuffs were deflated, and R is of the form:

$$R(t) = \left[\frac{\alpha(t - T_s)}{\delta^2} \exp\left(\frac{-(t - T_s)}{\delta} + \beta \right) \right]$$
$$\times \exp(-\lambda(t - T_s)). \quad (4)$$

In the simulations, $\alpha = 625$, $\beta = 1$, $\delta = 20$, $\lambda = 0.1$, and $A = 2$ were used.

2. Time–Activity Curves (Fig. 2)

Values for blood flow used in the simulations were 50 ml · 100 g^{-1} · min^{-1} for mean whole brain flow; 65

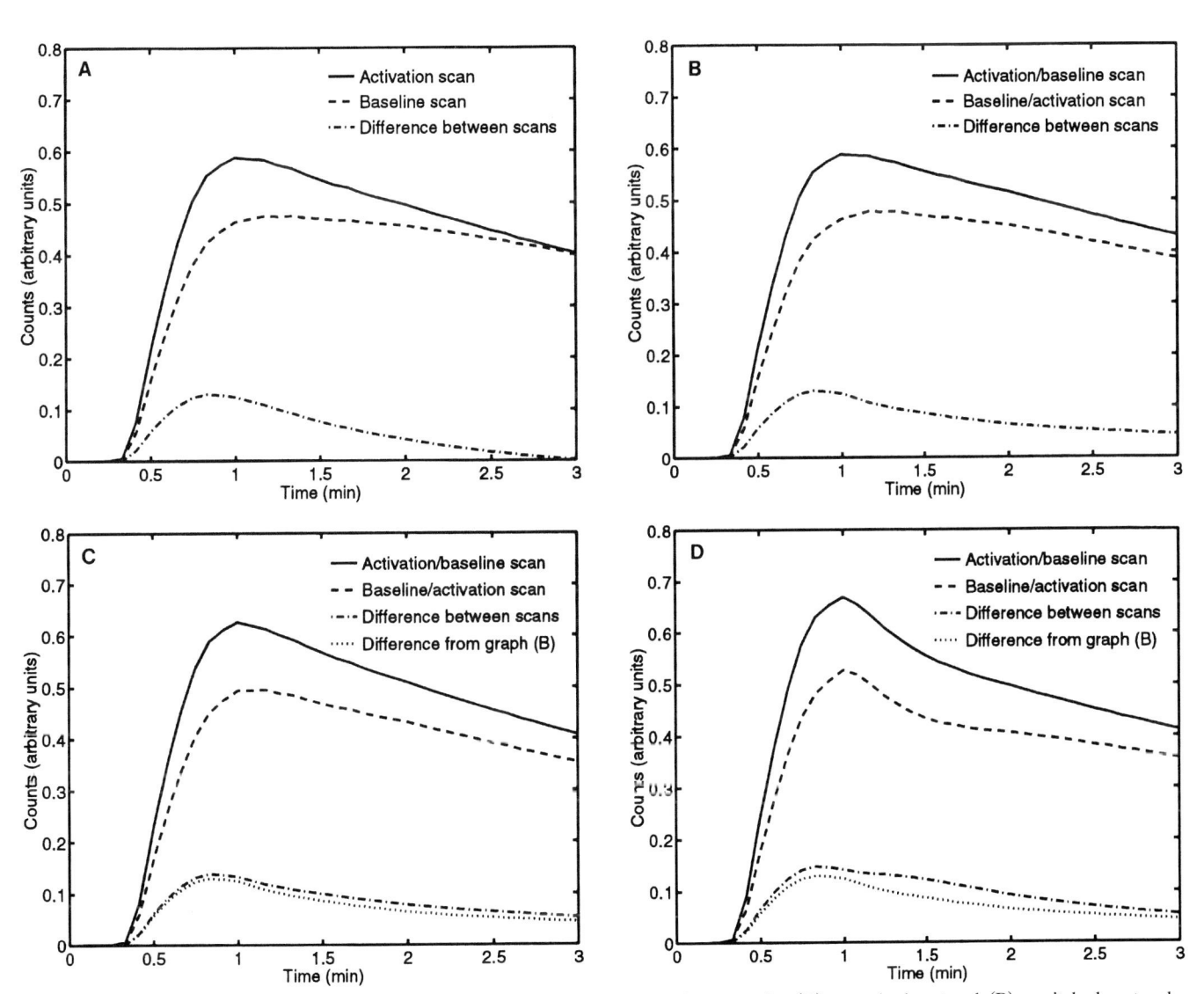

FIGURE 2 Simulated tracer concentration in activated and baseline gray matter in scans using (A) a standard protocol, (B) a switched protocol, (C) a switched protocol with a 60-sec cuff occlusion and instantaneous mixing of the cold bolus, and (D) a switched protocol with a 60-sec cuff occlusion and a realistic time-varying mixing of the cold bolus. The bottom plot in graphs A and B represents the difference in activity concentration between activation and baseline states. Graphs C and D show equivalent plots plus the bottom plot from graph B as reference.

ml · 100 g^{-1} · min^{-1} for flow in gray matter in the baseline state, and 91 ml · 100 g^{-1} · min^{-1} for flow in activated gray matter. Tracer concentration in gray matter was simulated for 3-min scans using standard, switched, and cold-bolus/switched protocols. Time–activity curves were integrated over the simulated scan lengths and normalized by whole brain radioactivity (Fig. 2). The standard activation protocol was simulated using the measured input function, and activated and baseline blood flows which were kept constant for the duration of the scan. The switched protocol was simulated using the measured input function and blood flows corresponding to switching task execution at 60 sec. The combined protocol was simulated for blood flows corresponding to switching tasks and deflating the cuffs at 60 sec using two different input functions. Both functions accounted for the initial decrease in the available blood volume. However, one function simulated instantaneous complete mixing of the cold bolus after cuff deflation whereas the other simulated time-dependent mixing.

3. Signal Magnitude vs Statistical Variability (Fig. 3)

The relative value (V) of the activation signal in a subtracted image with respect to its estimated statistical uncertainty was examined using

$$V = \frac{A - B}{\sqrt{A + B}}, \quad (5)$$

where A and B are the counts in the activation and baseline scans, respectively.

B. PET Experiments

The effect of using a cold-bolus/switched protocol on the S/N of activation images was investigated with PET in four healthy volunteers. All subjects were male, right handed, and native English speakers (mean age = 24 years). Scans were acquired using a Siemens ECAT HR+ PET scanner (Moreno-Cantú et al., 1997).

1. Cognitive Tasks

The neuronal regions activated when generating an element from a given category were studied using an activation and a baseline task. In the activation scans, subjects were presented with an English word every 5 sec describing a common category of objects or concepts. After hearing a word, the subjects were required to find an element belonging to the category just heard and say the element aloud (e.g., hear "flower" → say "rose"; hear "religion" → say "Buddhism"). When the subjects were unable to find a matching element, they said "pass." In the baseline tasks, English words were presented at the same rate as in the activation tasks. After hearing a word, the subjects were required to repeat each word aloud. Subjects were instructed to switch task execution at the presentation of an easily identifiable tone.

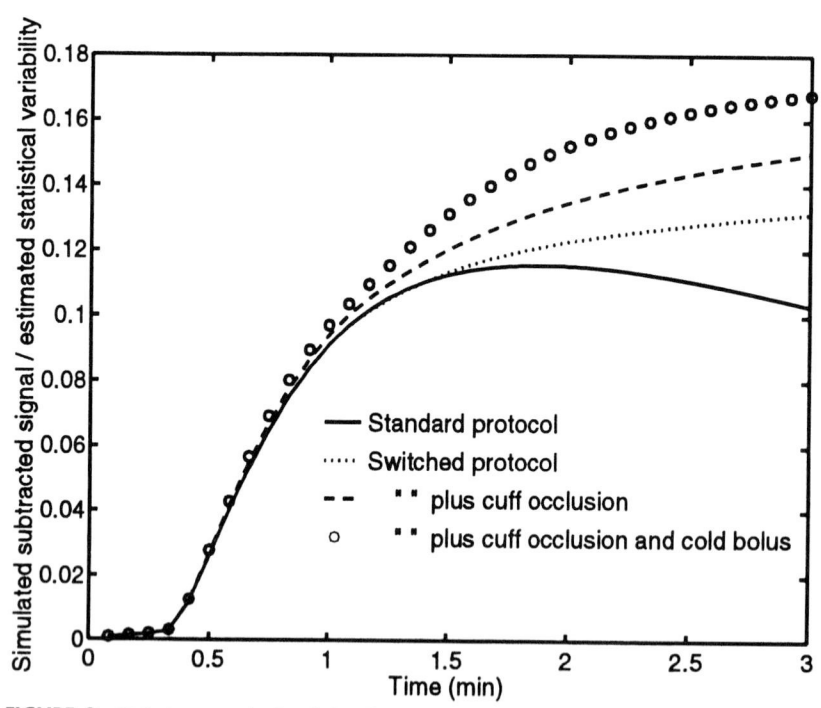

FIGURE 3 Relative magnitude of the simulated subtracted signals with respect to their statistical uncertainty.

2. Data Acquisition

Subjects lay on the scanner bed with their eyes closed in a quiet darkened room. Each subject underwent 12 scans: 3 activation/baseline and 3 baseline/activation scans employing a switched protocol, and 3 activation/baseline and 3 baseline/activation scans using a cold-bolus/switched protocol. The scan order used across subjects was counterbalanced to avoid random bias. All the scans were 3 min long. In each scan, a 3-sec intravenous bolus injection of 15 mCi of [^{15}O]H$_2$O was employed. The categories used while scanning were novel to the subjects in order to minimize practice effects (Raichle et al., 1994). The presentation of stimuli began once the bolus was injected. Data acquisition commenced 14 sec later to account for the delay between tracer injection and arrival in the brain. Task switching occurred approximately 45 sec after data acquisition began (~ 60 sec after injection). In each scan using the combined protocol, a cold bolus was generated by isolating the vasculature of the lower limbs a few seconds prior to the injection of tracer using double cuffs applied to each upper thigh. The cuffs were inflated above arterial pressure, as verified by loss of pulsation in the dorsal pedal artery. The cuffs remained inflated until the subjects switched tasks and were deflated rapidly (~ 1 sec). Subjects also underwent a magnetic resonance (MR) scan to facilitate the identification of the anatomic structures associated with the PET images (Evans et al., 1992). In order to examine the hemodynamic effects of cuff inflation, arterial blood pressure was measured in another group of four subjects, using sphygmomanometry and Korotkoff sounds at the brachial artery, before and 60 sec after cuff inflation.

3. Image Analysis

Individual three-dimensional (3D) scans were reconstructed (Hanning filter: cutoff frequency = 0.3 cycles/sample; image resolution ~ 8 mm), transformed into a stereotaxic coordinate system, and normalized by mean whole brain activity. The images were then grouped in three sets: (a) switched protocol images; (b) cold-bolus/switched protocol images; and (c) an overall set containing all the acquired images. For each set, a t-statistic image was generated for the mean change in normalized activity between baseline and activation scans using a standard deviation pooled across all image voxels (Worsley et al., 1992). Activation foci were identified in the overall set as this group is the largest sample. Corresponding activation foci were then identified in the other two groups and compared in terms of (a) magnitude, (b) statistical significance, and (c) S/N. The statistical significance of the activation foci was determined by using a previously described method based on Gaussian random field theory (Worsley et al., 1992). A focus was deemed statistically significant if the probability of being a false positive was less than 5%.

III. RESULTS AND DISCUSSION

A. Simulation Studies

Results from the simulations suggest that the combined cold-bolus/switched protocol would enhance the signal from averaged subtracted images when compared to matching signals obtained with standard or switched protocols. Simulations that included the reduction in blood volume introduced by lower limb occlusion showed a 14% increase in the subtracted signal compared to the signal from the simulated switched protocol [Fig. 1 (input function B) and Fig. 2C]. When the simulations also included a time-dependent mixing of the cold bolus, the subtracted signal was 27% greater than the signal from the switched protocol [Fig. 1 (input function C) and Fig. 2D].

Subtracted images acquired using the switched protocol yield higher S/N than matching standard protocol images because the difference between activation and baseline states is maintained longer (Figs. 2A and B), allowing longer scanning times (Cherry et al., 1995; Moreno-Cantú et al., 1998). The longer scanning times increase the S/N of the subtracted images by reducing the overall image noise. The signal in switched protocol images is maintained for longer because tracer washout from activated regions is reduced during activation/baseline scans and washout from the activated regions is increased during baseline/activation scans. Cold-bolus/switched protocols extend this concept by manipulating the input function to increase the concentration of tracer delivered during the uptake phase and to decrease tracer concentration during the washout phase. Figure 3 shows the relative magnitude of the simulated accumulated counts (activation−baseline) with respect to their estimated statistical uncertainty. The improvement yielded by the combined protocol is more than twice the one obtained with the switched protocol.

B. PET Experiments

Figure 4 shows mean normalized whole brain radioactivity for actual PET scans using the switched and combined protocols. The dotted vertical line in Fig. 4 indicates the time region where task switch and cuff deflation occurred. Compared to the switched protocol,

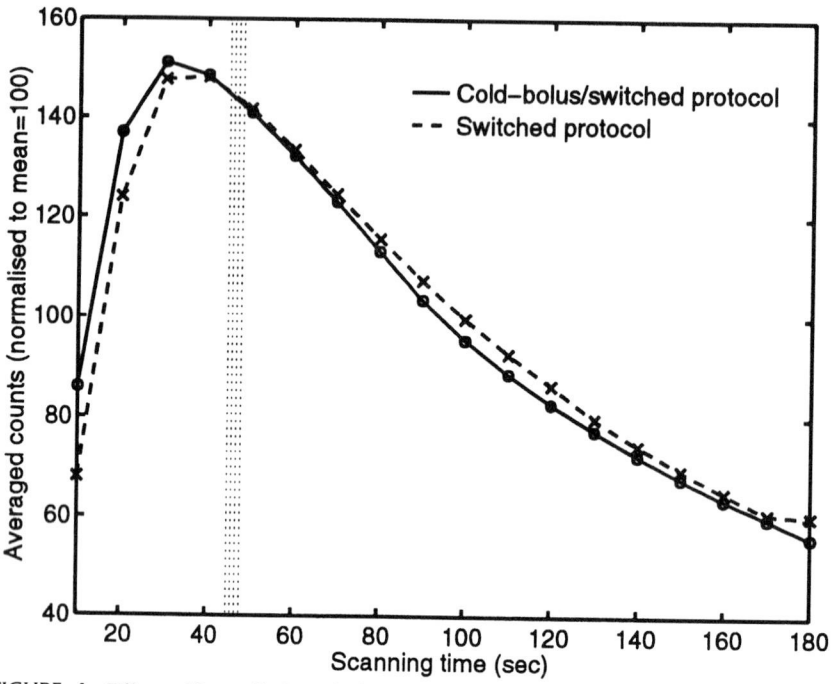

FIGURE 4 Effect of lower limb occlusion and cold bolus release on the observed whole brain counts. At the beginning of a scan, occlusion of blood circulation to the lower limbs results in an increase in counts. After the cuffs are deflated, the released cold bolus causes a time-varying reduction in whole brain counts.

the combined protocol yielded an increase in normalized activity while the cuffs remained inflated (uptake phase). This was probably caused by the reduction in the blood volume available for tracer mixing. After cuff deflation (washout phase), there was a reduction in normalized activity that was likely caused by the release of the cold bolus. As expected, the effect of the cold bolus on tracer concentration was not immediate as the cold bolus does not reach the brain immediately. The effect appeared around 30 sec after cuff deflation (Fig. 4, time = 80 sec) and continued for about 50 sec.

Late in the scans, the difference between total counts from switched and combined protocol scans decreased, since the two protocols are equivalent after the cold bolus is mixed uniformly.

1. Improved Detection of Activation Foci

Table 1 shows the location, magnitude, and significance level of the activation foci identified as statistically significant in the overall group. For all foci, the t statistic was greater with the cold-bolus/switched protocol than with the switched protocol alone. The aver-

TABLE 1 Statistically Significant Activation Foci from Data Averaged across Subjects[a]

Region	Average all scans (pooled SD: 8.54)[b]				Switched protocol scans (pooled SD: 8.27)					Combined protocol scans (pooled SD: 8.76)					%Δt	%Δm
	x	y	z	t	x	y	z	t	m	x	y	z	t	m		
Right cerebellum	38	−78	−32	6.1	38	−80	−32	4.2	12.2	35	−73	−33	6.1	20.0	45.2	63.9
Left insula	−38	20	−5	4.8	−43	22	−3	3.4	9.2	−36	20	−5	4.5	13.0	32.4	41.3
Right insula	31	20	−3	4.8	43	20	0	2.8	7.6	31	20	−3	4.6	13.4	64.3	76.3
Left frontomedial	−1	13	59	6.2	−5	8	69	5.0	13.9	0	13	60	5.2	15.2	4.0	9.4

[a] x, y, z, location in Talaraich space; t, focus t value; m, focus signal magnitude; %Δt and %Δm, relative increases obtained with the combined protocol.

[b] Mean voxel standard deviation in averaged subtracted image.

age increase was 36%. Because the *t* statistic of a focus is proportional to the averaged subtracted signal magnitude (see *m* values in Table 1) and inversely proportional to the overall image noise (see pooled SD in Table 1), the values in Table 1 indicate that the increases in *t* statistic were caused by increases in the signals from the subtracted images (see Δm for signal changes across protocols) and not by changes in the overall image noise (see pooled SD) which remained essentially constant. These observations are in agreement with simulations that predicted an increase in the activation signals from the subtracted images (Figs. 2C and 2D). The overall image noise was not expected to change significantly as the acquisition time for the two protocols tested was the same. Variations in the improvement to the *t* statistic observed among the detected foci were possibly caused by random noise and by differences in the time that brain regions were activated within scans with respect to cold bolus dispersion patterns.

2. Implementation of the Cold-Bolus/Switched Protocol

This protocol was simple to implement and well tolerated by all subjects. The required apparatus is portable and readily available, as it is commonly used for regional anesthesia. Cuff inflation may cause some discomfort (Murphy, 1995; Hagenouw *et al.*, 1986; Sapega *et al.*, 1985); however, this can be minimized by slow inflation of the cuffs. None of the volunteers tested felt that the discomfort interfered with the tasks. No significant risk is associated with the use of the cuffs for the periods required by the combined protocol (Murphy, 1995), and cuff inflation/deflation did not produce significant changes in arterial blood pressure. Use of the cuffs did not generate additional statistically significant foci in the subtracted images.

Acknowledgments

We are grateful to Pierre Ahad for his help in the preparation of the stimuli and to the staff of our brain imaging center for their assistance during the acquisition and analysis of the scans. This work was supported by the Medical Research Council of Canada (SP-30).

References

Albert, S. N. (1971). "Blood Volume and Extracellular Fluid Volume," pp. 15–16. Thomas, Springfield, IL.

Cherry, S. R., Woods, R. P., Doshi, N. K., Banerjee, P. K., and Mazziotta, J. C. (1995). Improved signal-to-noise in PET activation studies using switched paradigms. *J. Nucl. Med.* **36**: 307–314.

Evans, A. C., Marrett, S., Neelin, P., Collins, L., Worsley, K., Dai, W., Milot, S., Meyer, E., and Bub, D. (1992). Anatomical mapping of functional activation in stereotactic coordinate space. *NeuroImage* **1**: 43–53.

Hagenouw, R. R. P. M., Bridenbaugh, P. O., van Egmond, J., and Stuebing, R. (1986). Tourniquet pain: A volunteer study. *Anesthesiology* **65L**: 1175–1180.

Henry, J. P., and Meehan, J. P. (1971). "The Circulation: An Integrative Physiologic Study," pp. 20–21. Yearbook Publ., Chicago.

Herscovitch, P., Markham, J., and Raichle, M. E. (1983). Blood flow measured with intravenous $H_2^{15}O$. I. Theory and error analysis. *J. Nucl. Med.* **24**: 782–789.

Klein, D., Zatorre, R. J., Milner, B., Johnsrude, I. S., Nikelski, J., Meyer, E., and Evans, A. C. (1996). CBF patterns during synonym generation: Group vs. individual study. *NeuroImage* **3**: S444.

Moreno-Cantú, J. J., Thompson, C. J., and Zatorre, R. J. (1997). Evaluation of the ECAT EXACT HR + 3D PET scanner in ^{15}O-water brain activation studies. *In* "1996 IEEE Nuclear Science Symposium Conference Record" (A. DelGuerra, ed.), pp. 1280–1284. IEEE, Piscataway, NJ.

Moreno-Cantú, J. J., Reutens, D. C., Thompson, C. J., Zatorre, R. J., Klein, D., Meyer, E., and Petrides, M. (1998). Signal-enhancing switched protocols to study higher-order cognitive tasks with PET. *J. Nucl. Med.* **39**: 350–356.

Murphy, F. L. (1995). Anaesthesia for orthopaedic surgery. *In* "Wylie and Churchill-Davidson's A Practice of Anaesthesia" (T. E. Healy and P. J. Cohen, eds.), pp. 1254–1255. Edward Arnold, London.

Raichle, M. E., Fiez, V. A., Videen, T. O., MacLeod, A. M., Pardo, J. V., Fox, P. T., and Petersen, S. E. (1994). Practice-related changes in human brain functional anatomy during nonmotor learning. *Cereb. Cortex* **4**(1): 8–26.

Rowell, L. B. (1986). "Human Circulation: Regulation During Stress," p. 118. Oxford University Press, New York.

Sapega, A. A., Heppenstall, R. B., Chance, B., Park, Y. S., and Sokolow, D. (1985). Optimizing tourniquet application and release times in extremity surgery. *J. Bone Joint Surg. Am.* **67**: 303–314.

Vafaee, M., Murase, K., Gjedde, A., and Meyer, E. (1996). Dispersion correction for automatic sampling of O-15-labeled H_2O and red blood cells. *In* "Quantification of Brain Function Using PET" (R. Myers, V. Cunningham, D. Bailey, and T. Jones, eds.), pp. 72–75. Academic Press, San Diego, CA.

Worsley, K. J., Evans, A. C., Marrett, S., and Neelin, P. (1992). A three-dimensional statistical analysis for CBF activation studies in human brain. *J. Cereb. Blood Flow Metab.* **12**: 900–918.

CHAPTER 57

Neutral Amino Acids Influence [^{18}F]Fluorodopa Quantification[1]

P. VONTOBEL, G. KÜNIG, M. BRÜHLMEIER,* I. GÜNTHER, A. ANTONINI, M. PSYLLA, and K. L. LEENDERS

PET Department
Paul Scherrer Institute
Villigen, Switzerland
* Kantonsspital Aarau, Switzerland

[^{18}F]Fluorodopa crosses the blood–brain barrier (BBB) by means of the L-type neutral amino acid transporter, which is nearly saturated under normal conditions. The affinity of the transporter for many other large neutral amino acids (LNAA) in the blood plasma leads to competition effects, by which a high level of one of the LNAA reduces the BBB forward transport of others. This has been shown for [^{18}F]fluorodopa in vivo in humans using positron emission tomography (Leenders et al., 1986) and for two other tracers targeting LNAA BBB transport: [^{11}C]aminocyclohexanecarboxylate (Shulkin et al., 1995) and [^{18}F]fluorophenylalanine (Ito et al., 1995). This chapter shows that despite fasting imposed on healthy control subjects and parkinsonian patients, small variations in selected LNAA plasma concentrations affect [^{18}F]fluorodopa uptake quantification. Variations in putamen [^{18}F]fluorodopa uptake parameters analyzed with the striatum-occipital ratio method or by multiple time graphical analysis based on either [^{18}F]fluorodopa plasma input or occipital cortex input are explained by varying plasma levels of competing amino acids. It is shown that measures based on reference tissue are less sensitive to these effects of competitive inhibition.

I. INTRODUCTION

[^{18}F]Fluorodopa accumulation in brain is determined by blood–brain barrier (BBB) transport and by the activity of DOPA decarboxylase in presynaptic dopaminergic nerve terminals. [^{18}F]Fluorodopa crosses the BBB by means of the L-type neutral amino acid transporter, which is nearly saturated under normal conditions (Oldendorf, 1971; Pardridge, 1977).

Because of its affinity for many natural amino acids in the blood plasma, which compete for the availability of the transporter, elevations in the concentration of one or more large neural amino acids (LNAA) reduce brain uptake of [^{18}F]fluorodopa (Leenders et al., 1986). More recently, the competitive inhibition of BBB transport was shown for [^{18}F]fluorophenylalanine (Ito et al., 1995) and [^{11}C]aminocyclohexanecarboxylate (Shulkin et al., 1995). The adverse effect of transporter saturation on [^{18}F]fluorodopa quantification with the plasma slope-intercept plot was discussed by Hoshi et al. (1993). It was shown in healthy controls that a protein-rich diet ingested 1 hr before tracer administration significantly reduced net [^{18}F]fluorodopa clearance and [^{18}F]fluorodopa blood brain forward transport constants (Kuwabara et al., 1994). This chapter shows that despite fasting imposed on healthy control subjects and patients with Parkinson's disease (PD), small variations in

---

[1] Transcripts of the BRAINPET97 discussion of this chapter can be found in Section VIII.

selected LNAA blood plasma concentrations affect [^{18}F]fluorodopa uptake quantification.

II. MATERIALS AND METHODS

In groups of healthy controls ($n = 11$; mean age 56 ± 9 years) and subjects suffering from PD ($n = 40$) in the same age range with various degrees of disease severity, 2-hr [^{18}F]fluorodopa scans were performed. The tomograph used was a CTI type 933/04-16 positron emission system (PET) system. The scanning procedure and details of the region of interest analysis are described elsewhere (Antonini et al., 1995). Subjects received written instructions not to take a protein-rich meal on the morning of the PET scan. The scan started at about noon for all subjects. Carbidopa (2 mg/kg) was given 1 hr before tracer injection. [^{18}F]Fluorodopa plasma metabolites were analyzed using HPLC and a γ-detector system. The plasma concentration of selected LNAA was measured from a 1-ml sample using HPLC.

The kinetics of [^{18}F]fluorodopa and 3-O-methyl[^{18}F]fluorodopa in the occipital regions with nonspecific tracer uptake were each described with a single reversible tissue compartment model (see Fig. 1). Striatal uptake was quantified using the striatooccipital ratio (SOR) for putamen defined as all counts acquired between 60 and 120 min in putamen divided by the counts acquired in occipital cortex. The slope k_r [1/min] derived from the tissue slope-intercept plot was calculated from 30 to 90 min using the occipital cortex as tissue input. The [^{18}F]fluorodopa uptake rate constant K_{pl} [ml/g/min] derived from the plasma slope-intercept plot was found by linear regression for the time range 30–110 min using [^{18}F]fluorodopa plasma activity input and subtracting occipital cortex activity from striatal activity.

Six neutral amino acid concentrations in plasma were chosen to assess the effect of competitive inhibition on [^{18}F]fluorodopa brain uptake: phenylalanine, leucine, methionine, isoleucine, tyrosin, and valine. The selection criterion was their high affinity for the L-type transporter as measured in rat brain (Smith et al., 1987). The sum of each individual amino acid concentration $C(i)$ divided by its half-saturation constant $K_m(i)$ for the L-type transporter is a measure for their combined inhibition capacity (Smith et al., 1987), i.e., the weighted sum, $\Sigma C(i)/K_m(i)$.

III. RESULTS AND DISCUSSION

A. Characteristics of [^{18}F]Fluorodopa Brain Uptake

Table 1 lists occipital [^{18}F]fluorodopa BBB forward transport constants K_{1d} for the control and PD subjects in order of disease severity. There is no significant difference between group mean K_{1d} values of controls and parkinsonian patients. The striatal uptake measures for putamen (averaged over both brain hemispheres) of PD patients differ significantly from control values. Percentage coefficients of variation (standard

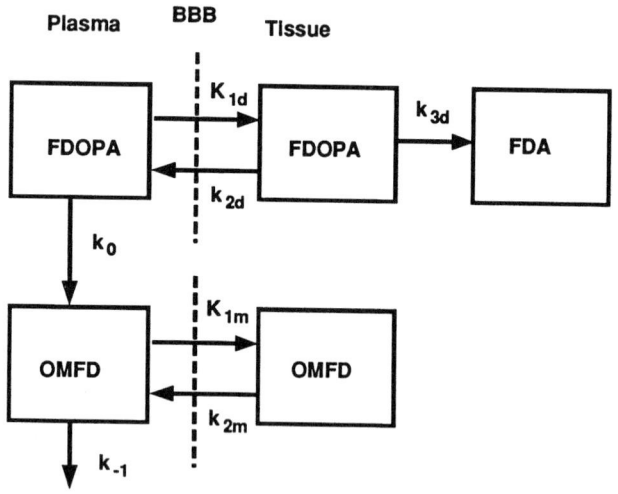

FIGURE 1 Compartment model for [^{18}F]fluorodopa (FDOPA) and 3-O-methyl[^{18}F]fluorodopa (OMFD).

TABLE 1 Average Occipital Cortex [^{18}F]Fluorodopa Forward Transport Constants K_{1d} and Putamen [^{18}F]Fluorodopa Uptake Measures for Healthy Controls and PD Patients

Group	Occipital K_{1d} (ml/g/min)	Putamen K_{pl} (ml/g/min)	K_r (1/min)	SOR
Controls				
Average	0.03785	0.01401	0.00989	2.39
%CV	27	26	7	6
H & Y[a] 1				
Average	0.03800	0.00878	0.00655	1.88
%CV	25	35	28	14
H & Y 2				
Average	0.03397	0.00708	0.00525	1.73
%CV	30	49	34	11
H & Y 3				
Average	0.03702	0.00711	0.00449	1.64
%CV	24	59	59	23
H & Y 4				
Average	0.03107	0.00459	0.00309	1.51
%CV	28	19	28	11

[a] Hoehn and Yahr stage.

TABLE 2 Characteristics of Competing Large Neutral Amino Acids (LNAA)

		Normal range		Control and PD	
LNAA	K_m (μmol/ml)	High (μmol/liter)	Low (μmol/liter)	Average (μmol/liter)	SD (μmol/liter)
Phenylalanine	0.011	80	45	54.0	13.2
Leucine	0.029	215	105	130.1	48.6
Methionine	0.04	45	20	26.7	10.8
Isoleucine	0.056	120	50	71.9	27.9
Tyrosine	0.064	100	40	83.4	41.1
Valine	0.21	325	180	233.8	60.4

deviation/mean * 100) indicate the amount of variation of the [^{18}F]fluorodopa uptake measures.

B. Characteristics of Plasma LNAA Concentrations

Table 2 summarizes the half-saturation constants measured in rat brain (Smith *et al.*, 1987) that were used for analysis together with the range of normal values of each amino acid and group mean values (controls + PD subjects; $n = 51$). Figure 2 shows the distribution of calculated weighted sums of LNAA for each subject together with the normal range. No significant differences in the weighted sum of LNAA were found between control and PD patients. Therefore, correlations with [^{18}F]fluorodopa uptake measures were calculated for controls and PD patients taken together.

C. Correlations of [^{18}F]Fluorodopa Uptake Measures with the Weighted Sum of LNAA

Highly significant negative correlation exists between occipital cortex [^{18}F]fluorodopa BBB forward transport constant K_{1d} and the weighted sum of LNAA ($r = -0.75$, $p < 0.000001$) (see Fig. 3 and Table 3). This shows the effect of inhibition of [^{18}F]fluorodopa transport into brain tissue in the presence of other LNAA, which compete for the same saturated L-type

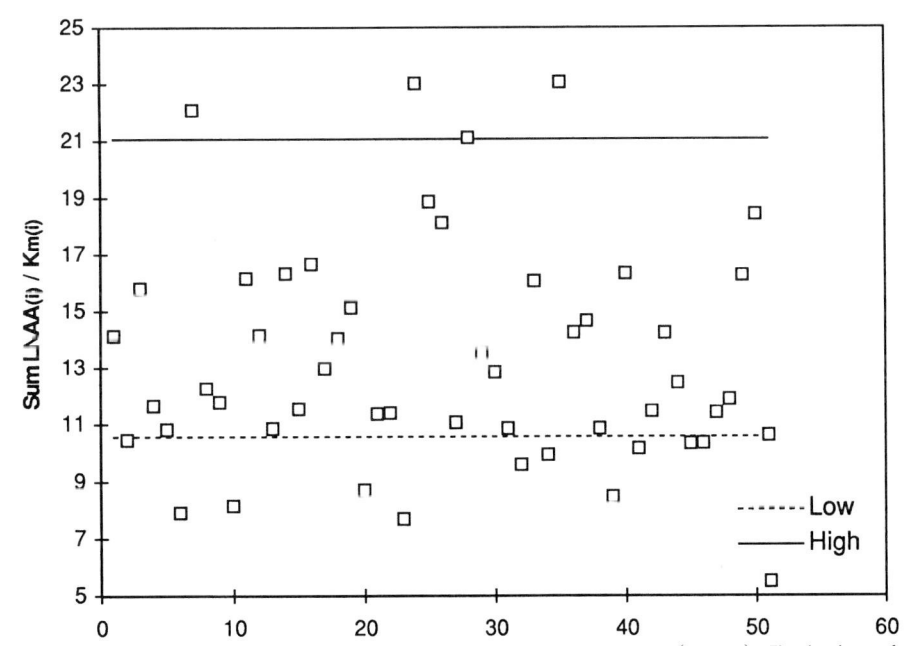

FIGURE 2 Distribution of weighted sum of large neutral amino acids (LNAA). The horizontal lines represent y-axis values calculated from the high and low values of the normal range of large neutral amino acids concentrations listed in Table 2. Patient number is on the abscissa.

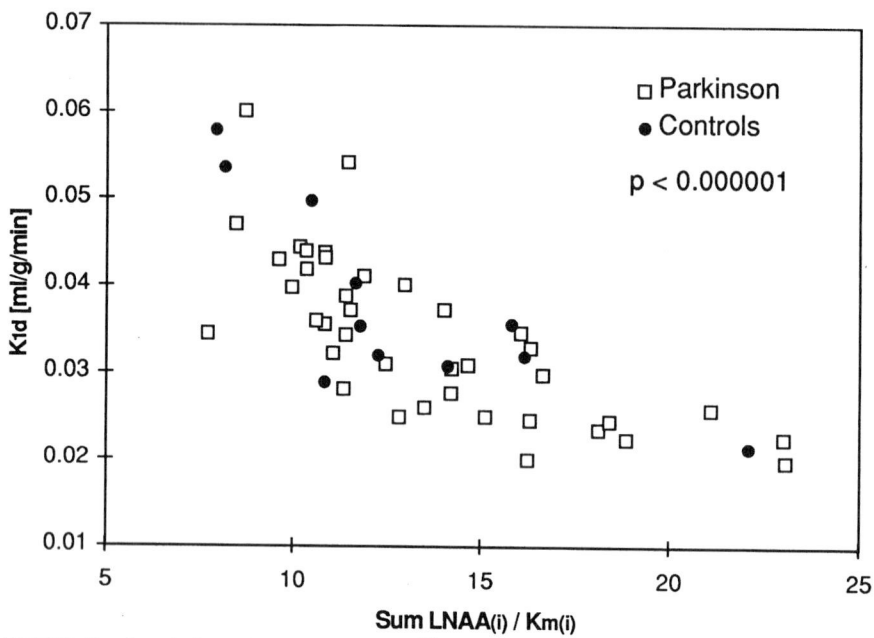

FIGURE 3 Correlation of occipital cortex [^{18}F]fluorodopa forward transport constant K_{1d} values with the weighted sum of large neutral amino acids.

transporter. A similar ($r = -0.44$, $p < 0.002$) negative correlation is seen for the plasma-slope intercept plot-derived net [^{18}F]fluorodopa clearance K_{pl} in striatal regions (see Fig. 4), whereas there is no significant correlation for the striatooccipital ratio calculated for putamen (Fig. 5) and for the slope k_r derived from the tissue slope-intercept plot (Fig. 6). The reason for the different correlations of K_{1d} and K_{pl}, (significant) and k_r and SOR (not significant) is explained by the fact that K_{1d} and K_{pl}, both measure BBB transport: K_{1d} directly and K_{pl} indirectly through the [^{18}F]fluorodopa uptake rate constant, $K_{1d}^* k_{3d}/(k_{2d} + k_{3d})$. In contrast, the occipital tissue reference based SOR and k_r values are relative measures of [^{18}F]fluorodopa accumulation derived from regions suffering from the same degree of competitive inhibition.

TABLE 3 Significance of Correlations between [^{18}F]Fluorodopa Uptake Parameters and Weighted Sum of Large Neutral Amino Acids

	K_{1d}	K_{pl}	k_r	SOR
PD + controls ($n = 51$)	0.000001	0.0020	n.s.[a]	n.s.
PD subjects ($n = 40$)	0.000001	0.0004	n.s.	n.s.
Controls ($n = 11$)	0.004000	n.s.	n.s.	n.s.

[a] Not significant.

D. Method Sensitivity

To assess the sensitivity of the different methods used to discriminate controls from early parkinsonian patients, the authors used the group mean difference of Hoehn and Yahr stage groups and the control groups divided by the standard deviation of the control group (Fig. 7), or counted the percentage of PD group members with values that are more than two control standard deviations from the control group mean value (Fig. 8). Data suggest that methods based on tissue activity input have a higher sensitivity to discriminate early parkinsonian patients from healthy controls than the plasma slope-intercept plot [^{18}F]fluorodopa uptake rate constant K_{pl}.

IV. CONCLUSIONS

Variations in [^{18}F]fluorodopa BBB forward transport rate constants can be explained by competition effects caused by intersubject differences in plasma LNAA levels. The plasma-slope intercept plot-derived uptake rate constant K_{pl}, is sensitive to the blood plasma concentrations of other LNAA, as has been discussed by Hoshi et al. (1993). This leads to higher values for the coefficient of variation of control and patient K_{pl}, group mean values compared to the group mean striatooccipital ratios or the group mean k_r val-

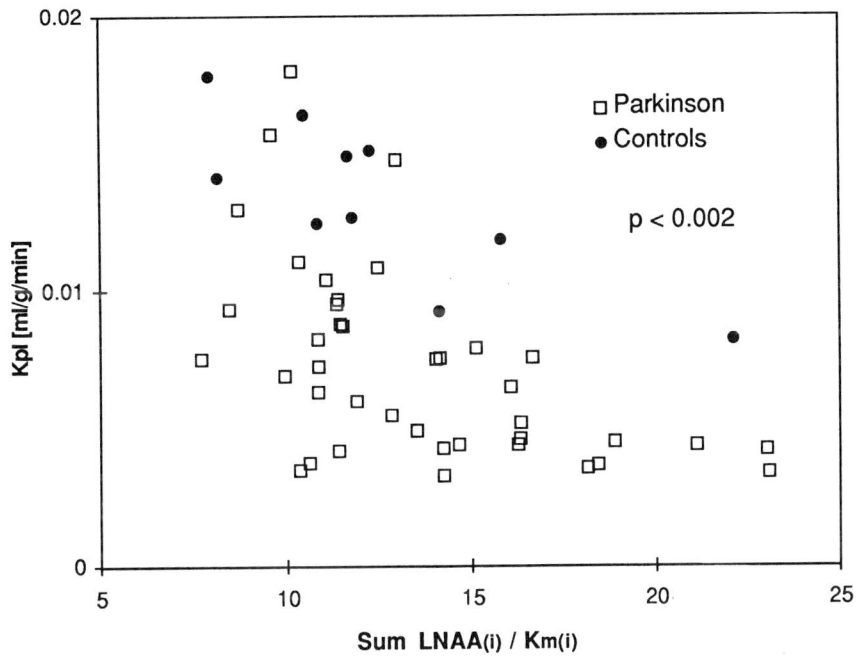

FIGURE 4 Correlation of plasma slope plot K_{pl}, values with the weighted sum of large neutral amino acids.

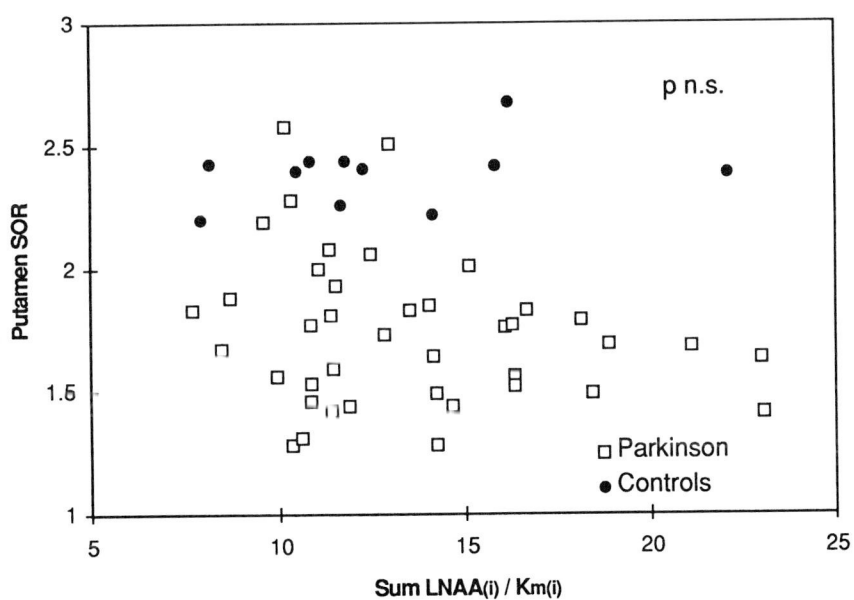

FIGURE 5 Correlation of tissue slope plot k_r, values with the weighted sum of large neutral amino acids.

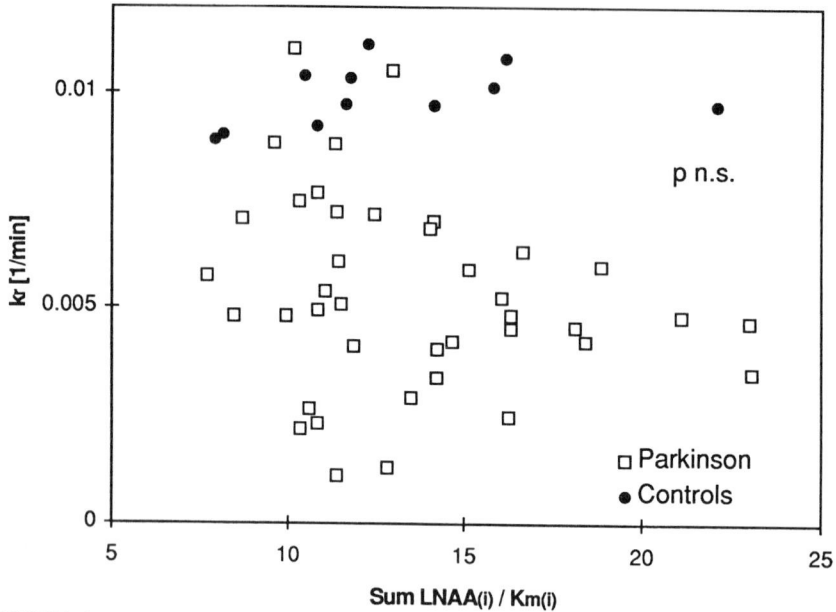

FIGURE 6 Correlation of putamen *SOR* values with the weighted sum of large neutral amino acids.

ues of tissue slope-intercept plots. This is the main reason for a higher sensitivity of the striatooccipital ratio and the tissue slope-intercept plots for discriminating healthy controls from early stage PD patients. More sophisticated kinetic compartment models trying to estimate the decarboxylase activity k_{3d} should not rely on population mean values for BBB forward transport constants unless the plasma level of other LNAA is taken into account.

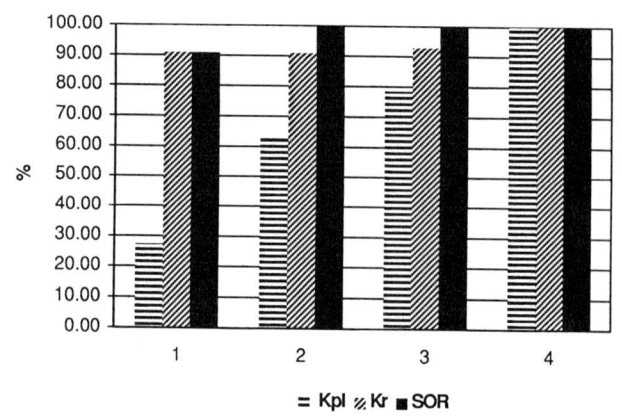

FIGURE 8 Percentage of Hoehn and Yahr group members with mean uptake values < 2 control standard deviations.

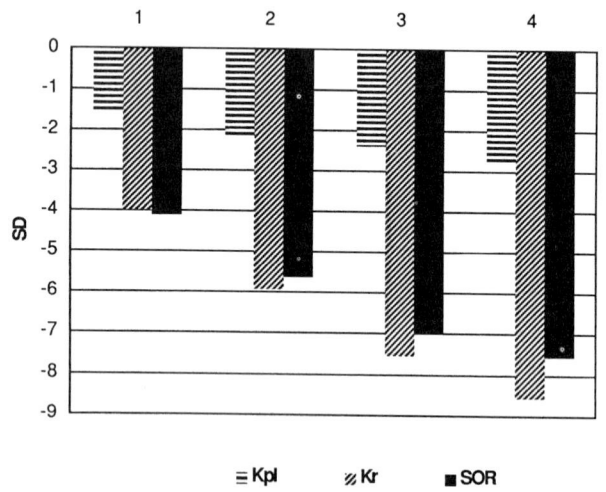

FIGURE 7 Differences of mean putamen [^{18}F]fluorodopa uptake measures divided by control group standard deviations versus Hoehn and Yahr stage.

References

Antonini, A., Vontobel, P., Psylla, M., Günther, I., Maguire, P., Missimer, J., and Leenders, K. L. (1995). Complementary PET studies of the striatal dopaminergic system in Parkinson's disease. *Arch. Neurol. (Chicago)* **52**: 1183–1190.

Hoshi, H., Kuwabara, H., Léger, G., Cumming, P., Guttman, M., and Gjedde, A. (1993). 6-[^{18}F]Fluoro-L-DOPA metabolism in living human brain: A comparison of six analytical methods. *J. Cereb. Blood Flow Metab.* **13**: 57–69.

Ito, H., Hatazawa, J., Murakami, M., Miura, S., Iida, H., Bloomfield, P. M., Kanno, I., Fukuda, H., and Uemura, K. (1995). Aging effect on neutral amino acid transport at the blood-brain barrier measured with L-[2-^{18}F]fluorophenylalanine and PET. *J. Nucl. Med.* **36**: 1232–1237.

Kuwabara, H., Cumming, P., Reutens, D., Diksic, M., Jolly, D., and Gjedde, A. (1994). Protein rich diet suppressed net and unidirectional clearances of 6-[F-18]fluoro-DOPA decarboxylase activity. *J. Nucl. Med.* **35**: 212P.

Leenders, K. L., Poewe, W. H., Palmer, A. J., Brenton, D. P., and Frackowiak, R. S. J. (1986). Inhibition of L-[^{18}F]fluorodopa uptake into human brain by amino acids demonstrated by positron emission tomography. *Ann. Neurol.* **20**: 258–266.

Oldendorf, W. H. (1971). Brain uptake of radiolabeled amino acids, amines and hexoses after arterial injection. *Am. J. Physiol.* **221**: 1629–1639.

Pardridge, W. M. (1977). Kinetics of competitive inhibition of neutral amino acid transport across the blood-brain barrier. *J. Neurochem.* **28**: 103–108.

Shulkin, B. L., Betz, A. L., Koeppe, R. A, and Agranoff, B. W. (1995). Inhibition of neutral amino acid transport across the human blood-brain barrier by phenylalanine. *J. Neurochem.* **64**: 1252–1257.

Smith, Q. R., Momma, S., Aoyagi, M., and Rappoport, S. I. (1987). Kinetics of neutral amino acid transport across the blood-brain barrier. *J. Neurochem.* **49**: 1651–1658.

CHAPTER 58

Imaging of the Dopamine Presynaptic System by PET: 6-[^{18}F]Fluoro-L-DOPA versus 6-[^{18}F]Fluoro-L-*m*-tyrosine[1]

D. J. DOUDET, O. T. DEJESUS,* G. L. Y. CHAN, S. JIVAN, J. E. HOLDEN,* C. ENGLISH,
T. G. AIGNER,[†] and T. J. RUTH

Department of Medicine, Division of Neurology and
Neurodegenerative Disorders Centre and TRIUMF
University of British Columbia
Vancouver, British Columbia, V6T 2B5 Canada
* Department of Medical Physics
University of Wisconsin
Madison, Wisconsin 53706
† National Institute of Mental Health
National Institutes of Health
Bethesda, Maryland 20892

The respective advantages and disadvantages of 6-fluoro-L-DOPA (6FDOPA) and 6-fluoro-L-m-tyrosine (6FMT) labeled with ^{18}F to image the dopamine (DA) presynaptic system in vivo were investigated by positron emission tomography (PET) in normal monkeys and in monkeys with 1-methyl-4-phenyl-1,2,3,6-tetrahydropyridine-induced lesions of the DA nigrostriatal neurons. Peripheral metabolism of the radiotracers was analyzed throughout the course of the PET studies. Using a modification of the extended graphical method of analysis, three measures of central DA function were examined: the rate of uptake of the tracer, the rate of loss of striatal radioactivity, and the effective turnover of DA. 6FMT presented a simpler metabolic profile compared to 6FDOPA, having no peripheral metabolites able to cross the blood–brain barrier. The uptake rates of 6FMT and 6FDOPA (decreased in lesioned animals) were found to be equally sensitive to abnormalities in the DA system. 6FDOPA, but not 6FMT, was able to separate normals from lesioned animals based on their increased striatal radioactivity loss rate and effective DA turnover. 6FMT was found to be an excellent tracer for brief clinical studies. However, 6FDOPA remains a promising tracer for the study of the storage processes of exogenous L-DOPA and of drug therapies aimed at DA preservation.

I. INTRODUCTION

Fluorinated analogs of *m*-tyrosines have been proposed as an alternative to 6-[^{18}F]fluoro-L-DOPA (6FDOPA) to evaluate dopamine (DA) presynaptic function *in vivo* (DeJesus and Mukherjee, 1988; Nahmias *et al.*, 1995; DeJesus *et al.*, 1997). 6FDOPA, an analog of the amino acid L-DOPA used extensively for positron emission tomography (PET) studies of the DA system, is a good substrate for aromatic amino acid decarboxylase (AAAD), as well as for catechol-*O*-methyl-transferase (COMT). In the periphery, the *O*-methylation process is responsible for the formation of 3-*O*-methyl-[^{18}F]fluoro-L-DOPA (3OMFDOPA), a large amino acid that crosses the blood–brain barrier. This complicates considerably the kinetics and interpretation of 6FDOPA data obtained in the brain. L-*m*-Tyrosine is an amino acid that is an excellent substrate for AAAD but not for COMT. Although both 4- and 6-[^{18}F]fluoro-L-*m*-tyrosine (FMT) have been used successfully, the kinetics of 6FMT in brain and plasma make it a better tracer for PET studies (Barrio *et al.*, 1996). The proposed main advan-

[1] Transcripts of the BRAINPET97 discussion of this chapter can be found in Section VIII.

tage of 6FMT over 6FDOPA is its simplified metabolic profile: the peripheral metabolites of 6FMT, 6-[^{18}F]fluoro-m-tyramine (FMA) and sulfated conjugates, do not cross the blood–brain barrier, thus greatly simplifying the modeling of cerebral kinetics.

This study compared the effectiveness of 6FDOPA and 6FMT to separate a group of normal monkeys from a group of monkeys with 1-methyl-4-phenyl-1,2,3,6-tetrahydropyridine (MPTP)-induced lesions of the DA nigrostriatal system. The two tracers were evaluated on their abilities to distinguish between the two groups based on three measures of DA presynaptic function obtained from the standard graphical method of analysis (Patlak and Blasberg, 1985; Holden et al., 1997): the uptake rate constant K_i, the rate of reversibility of tracer trapping, k_{loss}, and an index of effective turnover of DA, k_{loss}/K_i, that reflects DA storage capacity.

II. MATERIAL AND METHODS

Routine synthesis of 6FDOPA was performed (Namavari et al., 1992). 6FMT was synthesized as follows: L-m-tyrosine (36.2 mg; 200 μmol) was dissolved in 12 ml of 1:1 trifluoroacetic acetic (TFA) and glacial-acetic acid. [^{18}F]Acetylhypofluorite was bubbled into the mixture at a flow rate of 120 ml/min. The TFA/glacial acetic acid was removed and the crude product was dissolved in potassium dihydrogen phosphate buffer (0.06 M, 2.5 ml) and purified by HPLC using an Alltech Econosil C-18 column (10 μm, 250 × 25 mm) with an eluant of 4.5% tetrahydrofuran, 0.15% TFA in water at a flow rate of 10 ml/min. The product was detected with a UV detector at 254 nm and a radioactivity detector. Two fractions eluting at 32 and 38 min were determined to be 2-fluoro and 6-fluoro isomers, respectively, by ^{19}F NMR. The 6FMT fraction was collected, evaporated to dryness, and reconstituted with saline. ^{19}F NMR of this fraction showed less than 5% of the 2-fluoro isomer present. Specific activity was approximately 200 mCi/μmol.

Six normal monkeys and six MPTP-lesioned monkeys (*Macaca mulata* and *Macaca fascicularis*) were included in the study. Four out of the six MPTP-lesioned animals had a bilateral lesion, whereas two had a unilateral lesion. Each animal received, on separate occasions within a 4-month period, a PET scan with 6FDOPA and one with 6FMT. Animal preparation and the scanning procedure have been described in detail elsewhere (Doudet et al., 1989). All procedures were approved by the Committee on Animal Care of the University of British Columbia.

The monkey was placed prone in a stereotaxic frame. Thirty-one coronal slices of the head were obtained with a Siemens ECAT 953B-31 scanner (in-plane spatial resolution: 6 mm full width at half maximum; axial resolution: 5 mm) For both the 6FDOPA and 6FMT studies, 8 out of the 12 monkeys received the periperal decarboxylase inhibitor carbidopa (5 mg/kg ip), 40–80 min (58 ± 10 min) before injection of radiotracer. Because of delay in the synthesis of FMT, carbidopa was given 90–150 min prior to injection in 3 animals (110 ± 25 min). To maintain consistency of data, 6FDOPA was also injected late after carbidopa in 3 animals (106 ± 23 min). One animal did not receive carbidopa in either the 6FMT or the 6FDOPA studies.

6FDOPA and 6FMT (respectively, 4.9 ± .43 and 4.7 ± .8 mCi adjusted to 10 ml with saline) were injected intravenously over 1 min with a Harvard injection pump. The PET protocol consisted of six 0.5-min, two 1-min, and one 5-min scans followed by a succession of 10-min scans for a total duration of 200 min. Arterial blood samples were drawn throughout the study to obtain the plasma input function. Metabolite analysis was performed on plasma samples obtained at 2, 5, 10, 15, 30, 60, 90, 120, 150, 180, and 200 min, using a previously described technique, i.e., an alumina extraction method with anion/cation-exchange columns (McLellan et al., 1991). With this method, metabolites of 6FDOPA (^{18}F-labeled fluorodopamine, 6-fluoro-3-methoxytyramine, 6-fluoro-homovanillic acid, 6-fluoro-dihydroxyphenylacetic acid, and sulfated conjugates) and 6FMT [6-fluoro-m-tyramine (FMA) and sulfated conjugates] were retained by the anion and cation columns. 6FDOPA and its main metabolite 3OMFDOPA, both unretained by the columns, were separated by alumina extraction. 6FMT was also unretained by the columns and was contained in the supernatant. Plasma metabolite data were expressed as the fractions of total plasma radioactivity containing 6FDOPA and 6FMT.

Regions of interest (ROIs) were placed over the left and right striatum on four contiguous slices and over an area of nonspecific ^{18}F accumulation in the occipital cortex in two contiguous slices. Time–activity curves were averaged for left striatum, right striatum, and occipital cortex. Left and right striatal data of normal and bilaterally lesioned animals were averaged. Data from the MPTP-injected side of the unilaterally lesioned animals are reported alone.

Conventional graphical analysis was used to calculate the uptake rate constant K_i between 30 and 120 min using plasma 6FDOPA or 6FMT concentration as the input function (Patlak and Blasberg, 1985; Martin et al., 1989). The rate of loss of radioactivity out of the striatum, k_{loss}, was calculated using a modification of the extended graphical analysis of the full time courses acquired between 30 and 200 min. Details of the fitting method and its theoretical basis were described previously (Holden et al., 1997). The ratio k_{loss}/K_i was calculated as the index of effective DA turnover.

Data from normal and MPTP-treated groups were compared using unpaired student t tests. 6FDOPA and the 6FMT data were compared using linear correlation.

III. RESULTS AND DISCUSSION

6FMT has been suggested as an alternative to 6FDOPA because of its simpler metabolic profile and lower rate of peripheral metabolism. As shown in Fig. 1, the fraction of plasma 6FMT remained high over the course of the study, whereas the 6FDOPA fraction decreased rapidly after injection. Because 6FMT has been reported to be an excellent substrate for AAAD, the authors also plotted separately the fractions of 6FMT and 6FDOPA over time in three conditions: without carbidopa administration ($N = 1$), with carbidopa administration 40–80 min prior to tracer injection ($N = 3$), and with carbidopa pretreatment 90–150 min prior to tracer injection ($N = 8$). Data are shown in Fig. 2. Even taking into account the small numbers of subjects in each group, one can note the dramatic decrease in the fraction of 6FMT in the plasma with less and less AAAD inhibition. In contrast, 6FDOPA was less affected by the presence of carbidopa. This suggests that the timing of carbidopa administration is an important factor during studies

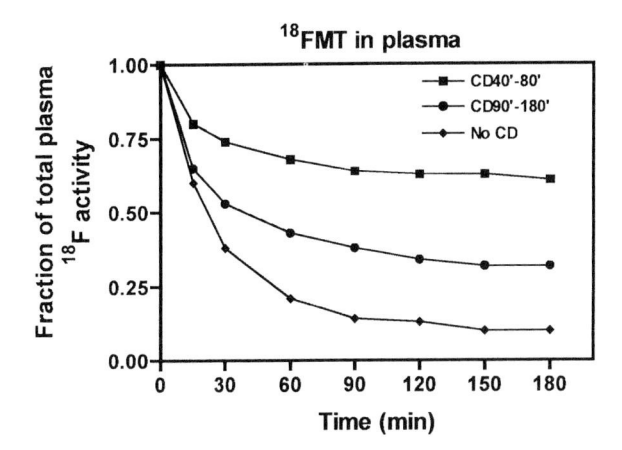

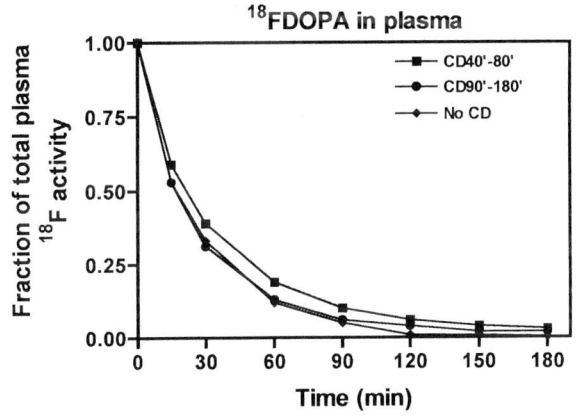

FIGURE 2 Fractions of the total plasma activity made up by 6FMT and 6FDOPA as a function of time and its dependence on the time of administration of carbidopa (CD) 40–80 min before radiotracer injection, 90–180 min before injection, or no carbidopa. 6FMT, an excellent substrate for AAAD, is extremely sensitive to the degree of inhibition of the enzyme. However, 6FDOPA is less sensitive.

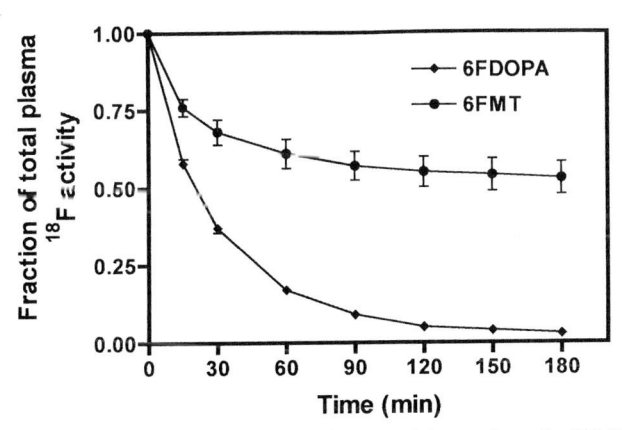

FIGURE 1 Fractions of the total plasma activity made up by 6FMT and 6FDOPA as a function of time. Points represent the mean and SD of plasma data obtained from all the animals that received carbidopa ($N = 11$). Note the high fraction of 6FMT over the course of the study. The relatively large standard deviation is due to the variability in the plasma fraction depending on the time of carbidopa administration.

with 6FMT in patients as the brain counts are directly related to the input function. The existence of a clear contrast between specific and nonspecific regions in 6FMT PET should allow for a better identification of structures, especially in patients with Parkinson's disease when magnetic resonance imaging coregistration is not available. Alternatively, the counting statistics in the brain will be better because in 6FMT studies with carbidopa pretreatment, a reduction in the injected dose could be possible.

As shown in Fig. 3, the uptake rate constant K_i for both 6FMT and 6FDOPA was significantly decreased in MPTP-treated monkeys compared to normal (6FDOPA and 6FMT: $p < 0.0001$). The rate of loss k_{loss} and the effective DA turnover were increased significantly in the MPTP-lesioned striata compared to normal after 6FDOPA injection (respectively, $p = 0.0003$ and $p = 0.0003$), but there was no significant difference in either measurement after 6FMT injection (respectively, $p = 0.697$ and $p = 0.199$). Likewise, there was a significant correlation between 6FDOPA and 6FMT K_i ($p < 0.0001$), but no significant correlation between 6FDOPA and 6FMT k_{loss} ($p = 0.426$) and turnover index k_{loss}/K_i ($p = 0.1$) (Fig. 3).

The factor k_{loss}/K_i which is termed effective DA turnover, is taken to reflect the storage of tracer label following the trapping process. This normalization of the loss rate constant by the uptake rate constant is appropriate because long-term storage depends on both. The ratio is interpreted as a turnover index because it is the inverse of the steady-state distribution volume of trapped label that would accumulate were tracer in plasma held constant.

Thus, for routine, short clinical studies, 6FMT appears to be an attractive tracer. It presents several advantages over 6FDOPA: a simpler metabolic profile, which leads to simplified modeling of cerebral tracer kinetics, and an increased contrast between specific and nonspecific regions. 6FMT appears as sensitive as 6FDOPA to differences in patient populations. The two compounds differ, however, in the behavior of the tracer label following the trapping process. The rate constant that reflects reversibility of the trapping process for 6FMT is small and is apparently not affected by the integrity of the dopaminergic system. Because of the low rate of diffusion of metabolites out of the striatum, the authors were unable to measure accurately the rate of reversibility of 6FMT trapping or to derive a meaningful index of DA turnover. Many studies have shown an increase in DA turnover in the striatum of DA-lesioned animals (Agid *et al.*, 1973). Using 6FDOPA PET, these findings have been confirmed *in vivo*, thus opening the way for further *in vivo* exploration of diseases in which a DA abnormality is suspected. Similarly, the action of DA drugs with potential therapeutic properties may be investigated in more detail. For example, a study of the effects of the central and peripheral COMT inhibitor tolcapone has been reported. Tolcapone had no effect on the 6FDOPA uptake constant but significantly decreased the effective turnover in the striatum (Doudet *et al.*, 1997).

Thus, 6FMT is a powerful tracer that better isolates the decarboxylation step in the trapping process in brief clinical studies aimed at separating patients with clear DA deficits from normal subjects. Because the elimination of its metabolites from striatum is not COMT dependent, however, its behavior following trapping cannot provide insight into the turnover of stored DA made from exogenous L-DOPA. However, it may prove useful to study the role of monoamine oxidase, the other major catabolic enzyme. 6FDOPA remains a valuable tracer in investigating the storage of

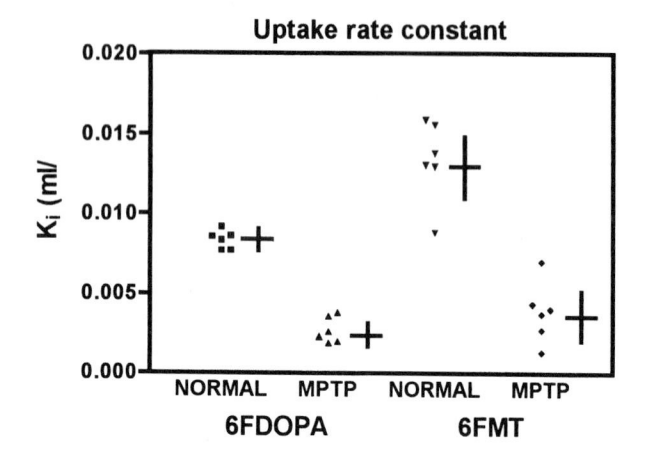

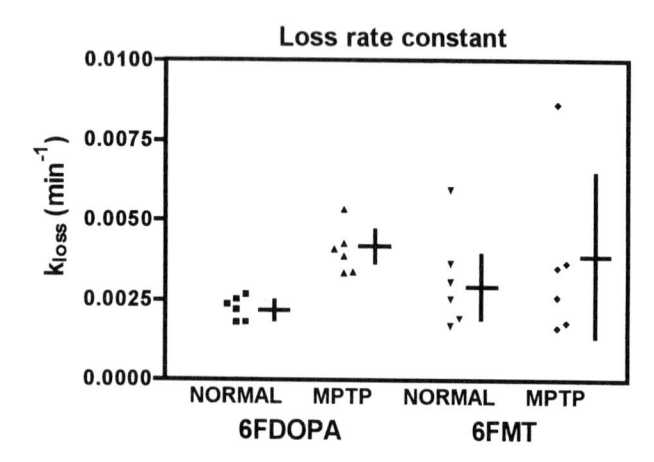

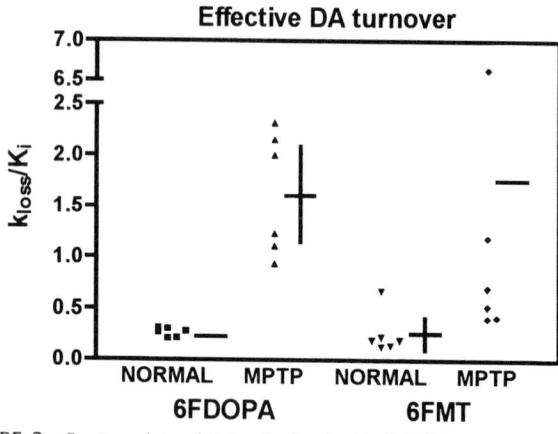

FIGURE 3 Scatter plots of data obtained with 6FDOPA and 6FMT in normal and MPTP-treated monkeys. (Top) Uptake rate constant K_i, (center) loss rate constant k_{loss}, and (bottom) effective DA turnover k_{loss}/K_i. Horizontal and vertical bars represent the mean and standard deviation of data. In the bottom graph, the standard deviation of 6FMT data in MPTP-treated animals has been omitted because, being too large, it would not fit in the graph.

exogenous L-DOPA and in studying drug therapies aimed at DA preservation.

Acknowledgments

This work was supported by the Medical Research Council of Canada and the National Institute of Health, Grants RO1-AG10217 and RO1-NS26621. We thank Merck, Sharp and Dohme, Canada for their gift of the carbidopa. The authors thank the staff of the UBC/TRIUMF PET program for their assistance and contribution to this work.

References

Agid, Y., Javoy, F., and Glowinski, L. (1973). Hyperactivity of remaining dopaminergic neurons after partial destruction of the nigrostriatal dopaminergic system in the rat. *Nature* (*London*) **245**: 150–151.

Barrio, J. R., Huang, S. C., Yu, D. C., Melega, W. P., Quintana, J., Cherry, S. R., Jacobson, A., Namavari, M., Satyamurthy, N., and Phelps, M. E. (1996). Radiofluorinated L-*m*-Tyrosines: new in vivo probes for central dopamine biochemistry. *J. Cereb. Blood Flow Metab.* **16**: 667–678.

DeJesus, O. T., and Mukherjee, J. (1988). Radiobrominated *m*-tyrosine analog as a potential CNS L-DOPA PET tracer. *Biochem. Biophys. Res. Commun.* **150**: 1027–1031.

DeJesus, O. T., Endres, C. J., Shelton, S. E., Nickles, R. J., and Holden, J. E. (1997). Evaluation of fluorinated *m*-tyrosine analogs as PET imaging agents for dopamine nerve terminals: Comparison with 6-fluorodopa. *J. Nucl. Med.* **38**: 630–636.

Doudet, D. J., Miyake, H., Finn, R. T., McLellan, C.A., Aigner, T. G., Wan, R. Q., Adams, H. R., and Cohen, R. M. (1989). 6-^{18}F-L-DOPA imaging of the dopamine neostriatal system in normal and neurologically-normal MPTP-treated rhesus monkeys. *Exp. Brain Res.* **78**: 69–80.

Doudet, D. J., Chan, G. L. Y., Holden, J. E., Morrison, K. S., Dobko, T., Wyatt, R. J., and Ruth, T. J. (1997). Effects of catechol-*O*-methyltransferase inhibition on the rates of uptake and reversibility of 6-fluoro-L-dopa trapping in MPTP-induced parkinsonism in monkeys. *Neuropharmacology* **36**: 363–371.

Holden, J. E., Doudet, D. J., Endres, C. J., Chan, G. L. Y., Morrison, K. S., Vingerhoets, F. J. G., Snow, B. J., Pate, B. D., Sossi, V., Buckley, K. R., and Ruth, T. J. (1997). Graphical analysis of 6-fluoro-L-dopa trapping: Effect of inhibition of catechol-*O*-methyltransferase on the rate of reversibility. *J. Nucl. Med.* **38**: 1568–1574.

Martin, W. R. W., Palmer, M. R., Patlak, C. S., and Calne, D. B. (1989). Nigrostiatal function in humans studied with positron emission tomography. *Ann. Neurol.* **26**: 535–542.

McLellan, C. A., Doudet, D. J., Brucke, T., Aigner, T. G., and Cohen, R. M. (1991). New rapid analysis method demonstrates differences in 6-[^{18}F]Fluoro-L-dopa plasma input curves with and without carbidopa and in hemi-MPTP lesioned monkeys. *Appl. Radiat. Isot.* **42**: 847–854.

Nahmias, C., Wahl, L., Chirakal, R., Firnau, G., and Garnett, E. S. (1995). A probe for intracerebral aromatic amino-acid decarboxylase activity: Distribution and kinetics of [^{18}F]6-fluoro-L-*m*-tyrosine in the human brain. *Movement Disord.* **10**: 298–304.

Namavari, M., Bishop, A., Satyamurthy, N., Bida, G. T., and Barrio, J. R. (1992). Regioselective radiofluorodestannylation with [^{18}F]F2 and [^{18}F]CH3COOF: A high yield synthesis of 6-[^{18}F]fluoro-L-dopa. *Appl. Radiat. Isot.* **43**: 989–996.

Patlak, C. S., and Blasberg, R. G. (1985). Graphical evaluation of blood-to-brain transfer constants from multiple-time uptake data. Generalizations. *J. Cereb. Blood Flow Metab.* **5**: 584–590.

CHAPTER 59

Quantitative Measurement of Acetylcholinesterase Activity in Living Human Brain Using a Radioactive Acetylcholine Analog and Dynamic PET[1]

S. NAGATSUKA,*,† H. NAMBA,*,‡ M. IYO,*,¶ K. FUKUSHI,* H. SHINOTOH,*,§ T. SUHARA,*
Y. SUDO,* K. SUZUKI,* and T. IRIE*

*Division of Clinical Research
National Institute of Radiological Sciences
Chiba 263, Japan
†Tokai Research Laboratories
Daiichi Pure Chemicals Co., Ltd.
Ibaraki 319-11, Japan
‡Division of Neurological Surgery
Chiba Cancer Center Hospital
Chiba 260, Japan
¶National Institute of Mental Health
National Center of Neurology and Psychiatry
Chiba 272, Japan
§Department of Neurology
Chiba University School of Medicine
Chiba 260, Japan

Acetylcholinesterase (AChE) activity in living human brain was characterized by determining the regional metabolic rate constant of a lipophilic acetylcholine analog, N-[^{11}C]methylpiperidyl-4-acetate ([^{11}C]MP4A). The metabolic rate constant is presumed to be proportional to the regional AChE activity in living human brain because of high specificity of [^{11}C]MP4A for human cerebral AChE. Cerebral regional radioactivity data obtained from dynamic positron emission tomography were subjected to kinetic analyses based on a three-compartment model using the input function of arterial unchanged [^{11}C]MP4A. Changes in fitting conditions gave varied rate constants of tracer uptake and washout. However, the metabolic rate constants were rather stable in various fitting conditions. The metabolic rate constants obtained in neocortical regions corresponded well with AChE activity obtained from postmortem studies, suggesting the feasibility of the present method for detection of changes in neocortical AChE activity in living human brain.

I. INTRODUCTION

The Central cholinergic nerve system has been known to be degenerated in the early stage of Alzheimer's disease (AD). The most significant biochemical change obtained from postmortem studies is the decreased activity of two cholinergic enzymes, choline acetyltransferase, and acetylcholinesterase (AChE) (Davies and Maloney, 1976; Perry *et al.*, 1978; Bierer *et al.*, 1995). The reduction of AChE is usually seen in cholinergic degeneration, not only AD, but also in other neurological disorders such as Parkinson's disease with dementia (Ruberg *et al.*, 1986) or in an artificial degeneration by destroying a specific nucleus

[1] Transcripts of the BRAINPET97 discussion of this chapter can be found in Section VIII.

using a neurotoxin (Namba *et al.*, 1991). To achieve *in vivo* detection of cholinergic degeneration, AChE was selected as the target enzyme and several substrate analogs that are metabolically trapped in the brain were designed (Irie *et al.*, 1994). The analogs, characterized by highly lipophilic *N*-methylpiperidine ring, have been evaluated in terms of their metabolic rate and specificity against human cerebral AChE (Irie *et al.*, 1996) to raise a candidate tracer for human positron emission tomography (PET) studies, *N*-[^{11}C]methylpiperidyl-4-acetate ([^{11}C]MP4A). This tracer is highly lipophilic with passage through the brain–blood barrier (BBB) depending on regional cerebral blood flow (rCBF). A portion of [^{11}C]MP4A diffuses back to the block, whereas the rest is specifically hydrolyzed, mostly by AChE, into hydrophilic alcohol, *N*-[^{11}C]methyl-4-piperidinol ([^{11}C]MP4OH), which is less permeable through the BBB and becomes trapped in the brain (Fig. 1). In the present study, the hydrolysis rate constant of [^{11}C]MP4A by AChE (k_3) was estimated quantitatively in the human brain using a three-compartment model analysis of plasma [^{11}C]MP4A and time–radioactivity curves of the brain tissues obtained from dynamic PET.

II. MATERIALS AND METHODS

4-Acetoxypiperidine was reacted with noncarrier-added [^{11}C]methyliodide to form [^{11}C]MP4A, which was then purified with HPLC. Specific radioactivity was higher than 18 GBq/μmol with a radiochemical purity of more than 99%. The final product in physiological saline solution was negative for bacteria and pyrogen tests.

Thirteen healthy volunteers (five men and eight women, ranging 24–80 years old) participated in the study with written informed consent. All the subjects received magnetic resonance (MR) imaging (Gyroscan, Phillips, Netherlands) and quantitative rCBF measurement with *N*-isopropyl-*p*-[^{123}I]iodoamphetamine (IMP) SPECT (Namba *et al.*, 1996) within 1 month before or after the PET study. This study was approved by the Ethics and Radiation Safety Committees of National Institute of Radiological Sciences, Chiba, Japan.

A SIEMENS ECAT EXACT 47 scanner with 24 detector rings, which acquires 47 slices simultaneously, was used. Spatial, transaxial, and axial resolutions were 3.38, 6.0, and 5.4 mm full width at half maximum, respectively. The subject was positioned on a PET scanner bed with eyes covered with an eye mask and the head immobilized by a resinous plate held between teeth. Needles were inserted in the right cubital vein for the tracer injection and in the left radial artery for arterial blood sampling under local anesthesia with 1% xylocaine. After transmission scanning, [^{11}C]MP4A in 5–6 ml physiological saline (555–740 Mbq) was constantly infused for 60 sec using an infusion pump. Serial PET scanning (sampling interval: 10 sec to 3 min) was started right after the rapid increase in PET count rate at the brain transit time (T_b).

Blood sampling was initiated about 6 sec after starting PET scanning. This time delay almost corresponded to the transit time difference between the brain and the blood sampling site. Arterial blood samples were collected in heparinized microtubes containing 0.1 mg physostigmine at designated time points. The plasma samples were then separated by centrifugation. A portion of each plasma sample was extracted with 2 volumes of ethanol and was subjected to thin-layer chromatography to separate [^{11}C]MP4A and [^{11}C]MP4OH. The plasma [^{11}C]MP4A radioactivity was calculated from the percentage of [^{11}C]MP4A fraction determined using a storage phosphor imaging system (Fujix BAS 2000, Fuji Photo Film Co., Ltd, Tokyo, Japan) and the plasma radioactivity was measured with a gamma counter. The plasma time–radioactivity curves of unchanged [^{11}C]MP4A (C_p) and total radioactivity (C_r)

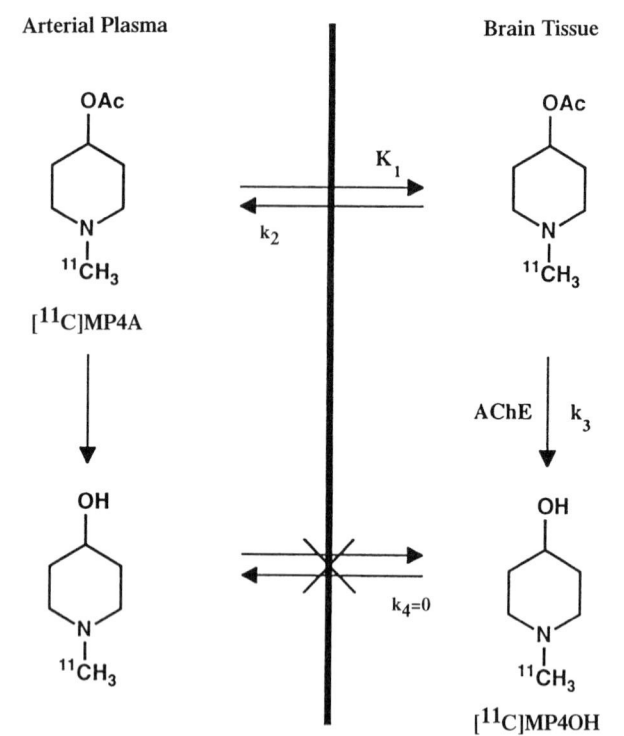

FIGURE 1 The three-compartment model used in the kinetic analysis of the radioactivity distribution.

during and after infusion period (T_i) were obtained by fitting data to the following theoretical function with weight factor of concentration^{-1};

$$C_p \text{ or } C_r = S(1 - e^{-\alpha t})$$

$$(0 \leq t \leq T_i; \text{ during infusion})$$

$$C_p \text{ or } C_r = Ae^{-\alpha(t-T_i)} + Be^{-\beta(t-T_i)} + Ce^{-\gamma(t-T_i)}$$

$$(T_i \leq t; \text{ after infusion}).$$

Regions of interest (ROI) were manually placed on a PET display referring to the corresponding MR images as anatomical landmarks (frontal, temporal, parietal, occipital, sensorimotor cerebral cortices, thalamus, striatum, and cerebellar cortex). The theoretical regional brain time–radioactivity curve is composed of three fractions: unchanged tracer (C_b), hydrolyzed tracer (C_m), and vascular radioactivity (C_v).

$$C_b = \int_0^T K_1 C_p e^{-(k_2+k_3)(T-t)} dt$$

$$C_m = \int_0^T k_3 C_b e^{-k_4(T-t)} dt$$

$$C_v = f_v C_r.$$

Parameter estimates were obtained by the nonlinear least-square method where K_1, k_2, k_3, and k_4 are rate constants shown in Fig. 1 and f_v is the blood volume fraction in cerebral tissue. Because [^{11}C]MP4OH has been shown to pass through the blood–brain barrier very slowly (Irie et al., 1996), it was assumed that the k_4 value was zero in the three-compartment model for this tracer, i.e., the diffusion of the radioactive metabolite through the BBB was negligible during the period of PET study. Corrections of transit time difference and tracer decomposition during the delay period were, if required, achieved during the fitting process.

All calculations were run on a Power Macintosh computer using an operating software written in Visual Basic included in Microsoft Excel Version 5.0b.

III. RESULTS AND DISCUSSION

A. Input Function

Time courses of the [^{11}C]MP4A percentage (Fig. 2A), total, and [^{11}C]MP4A radioactivity (Fig. 2B) in the arterial plasma were obtained after constant infusion for 60 sec. The [^{11}C]MP4A percentage and radioactivity concentration in the whole arterial blood were identical to those in the arterial plasma (data not shown).

[^{11}C]MP4A was rapidly hydrolyzed to a single metabolite, [^{11}C]MP4OH, in the blood. The [^{11}C]MP4A percentage during the infusion period stayed at a constant level. From this level, the halflife of [^{11}C]MP4A in the arterial plasma was calculated as 28.1 ± 5.0 sec ($n = 15$). This halflife of [^{11}C]MP4A *in vivo* was significantly shorter than that obtained from *in vitro* incubation experiments (35.0 ± 2.4 sec, $n = 3$), suggesting that the *in vivo* halflife of [^{11}C]MP4A includes first-pass clearance of the tracer in the lung. The metabolite correction, which is influential in estimating the K_1 value, could be modified to include the transit time difference and the estimated half-life of [^{11}C]MP4A in the blood. However, this additional correction was not applied here because the transit delay at the blood sampling site (about 6 sec) was rather short compared to both of the estimated halflives. Additionally, estimation of k_3 in the kinetic analysis was not affected by tracer decomposition during the delay period.

B. Data Analysis

A three-compartment model composed of an arterial blood compartment and two subcompartments (unchanged tracer and metabolite) in the brain tissue (Sokoloff et al., 1977; Phelps et al., 1979) was applied to the nonlinear least-square analysis. In addition to rate constants governing tracer uptake (K_1), washout (k_2), and metabolism (k_3), the operating software can handle rate constants of metabolite elimination (k_4), cerebral transit time (T_b), and blood volume fraction (f_v) as unknown parameters to be estimated by the kinetic analysis. As described earlier, k_4 was fixed to zero in the present study. T_b and f_v were estimated in a nonweighted analysis. The kinetic analysis gave an almost identical T_b to PET start time and f_v of almost 4%. Therefore, the fixed value of 4% was used as the cerebral blood volume fraction for further analyses.

In noncortical regions (thalamus, striatum, and cerebellar cortex), fitting calculations were often divergent. Therefore, K_1 was fixed to λk_2, where λ was the average value of distribution volume (K_1/k_2) in neocortical regions of each subject.

Two different weight factors, N_p/σ_p^2 and $1/\sigma_c^2$, were applied to the kinetic analysis to compare the results obtained with those obtained from the nonweighted analysis. In these weight factors, N_p, σ_p^2, and σ_c^2 represent number of pixels in a ROI, variance of pixel values in a ROI, and variance of radioactivity counts in a ROI, respectively. The weighted analysis with N_p/σ_p^2 showed a tendency to give slight lower values for K_1, k_2, and k_3 as compared with those obtained from nonweighted analysis. No significant differences were observed in K_1, k_2, and k_3 between nonweighted and weighted analyses when $1/\sigma_c^2$ was used. As the weight factor for further kinetic analyses, the inverse of the variance of radioactivity counts in a ROI was used, which was calculated from radioactivity concentration, volume of the region (ROI size × depth), and tracer decay factor.

C. Neocortical Time–Radioactivity Curve and PET Scanning Duration

Figure 3 shows a typical neocortical time–radioactivity curve obtained from the kinetic analysis dealing with K_1, k_2, k_3, and T_b as unknown parameters. At around 20 min after starting the infusion, plasma [^{11}C]MP4A concentration was almost negligible and cerebral radioactivity concentration reached plateau. The average rCBF of all neocortical regions in all subjects obtained from IMP SPECT was 0.507 ml/min/g. The average neocortical K_1, k_2, and k_3 were 0.524 ml/min/g, 0.124 min^{-1}, and 0.079 min^{-1}, respectively. The average K_1 value obtained from the present kinetic analysis was slightly higher than the average rCBF. This might be partly due to the overestimation of the K_1 value by neglecting the tracer decomposition correction. The distribution volume calculated from the average K_1 and k_2 values was 4.23 ml/g. This rather high distribution volume might be explained by the pH difference between the arterial plasma (pH 7.4) and brain tissue (pH 7.2) and/or the tertiary amine nature of MP4A. The presence of specific binding sites for MP4A is, at present, unclear.

In one subject, PET scanning was conducted for 60 min followed by sequential kinetic analyses with removing the last time frame one by one. Figure 4 shows k_3 values in neocortical regions obtained from sequential analyses to decide the appropriate duration of PET scanning. The minimal scanning duration giving stable k_3 values was around 25 min. In the present study, the scanning duration was 40 min or more.

D. k_3 Values in Various Regions

The regional metabolic rate constant, k_3, is equal to the product of the bimolecular reaction rate constant and the activity of esterases that hydrolyze MP4A. The specificity of MP4A against human cerebral AChE has been confirmed to be 94% (unpublished data). Therefore, k_3 values obtained in the present study are nearly proportional to regional AChE activity. The rank order in the neocortical k_3 values was sensorimotor cortex > temporal cortex > frontal cortex > parietal cortex > occipital cortex (Table 1). AChE activity in the temporal and frontal cortices has been reported to be higher than that in the occipital cortex in postmortem studies (Arai *et al.*, 1984; Atack *et al.*, 1986; Bierer *et al.*, 1995). These results indicate feasibility of the present method for detection of changes in neocortical AChE activity.

The ratios of k_3 values for the temporal cortex:thalamus:cerebellum:striatum were 1:3:13:11, respectively,

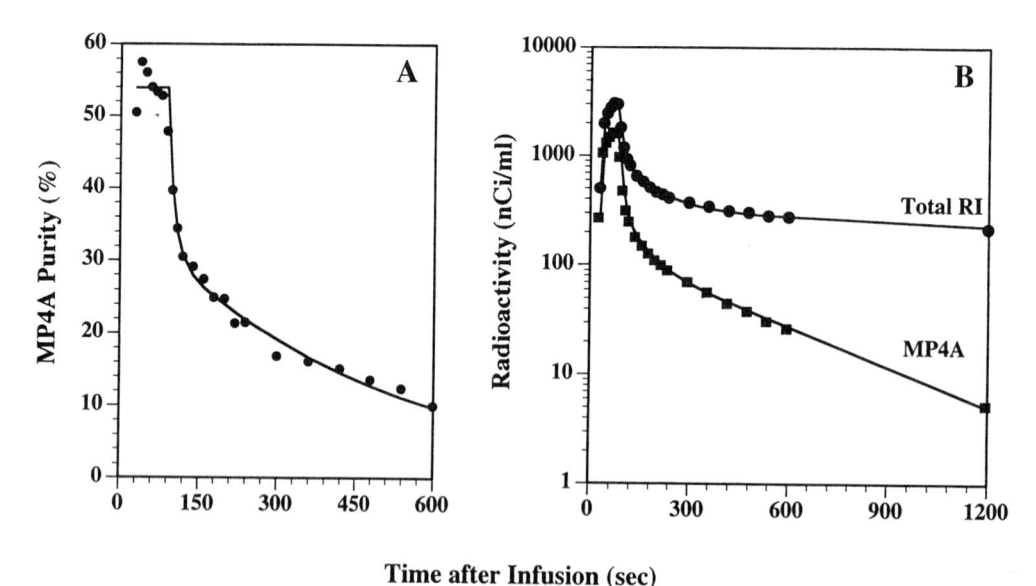

FIGURE 2 Purity of [^{11}C]MP4A (A), total radioactivity and [^{11}C]MP4A concentrations (B) in the arterial plasma.

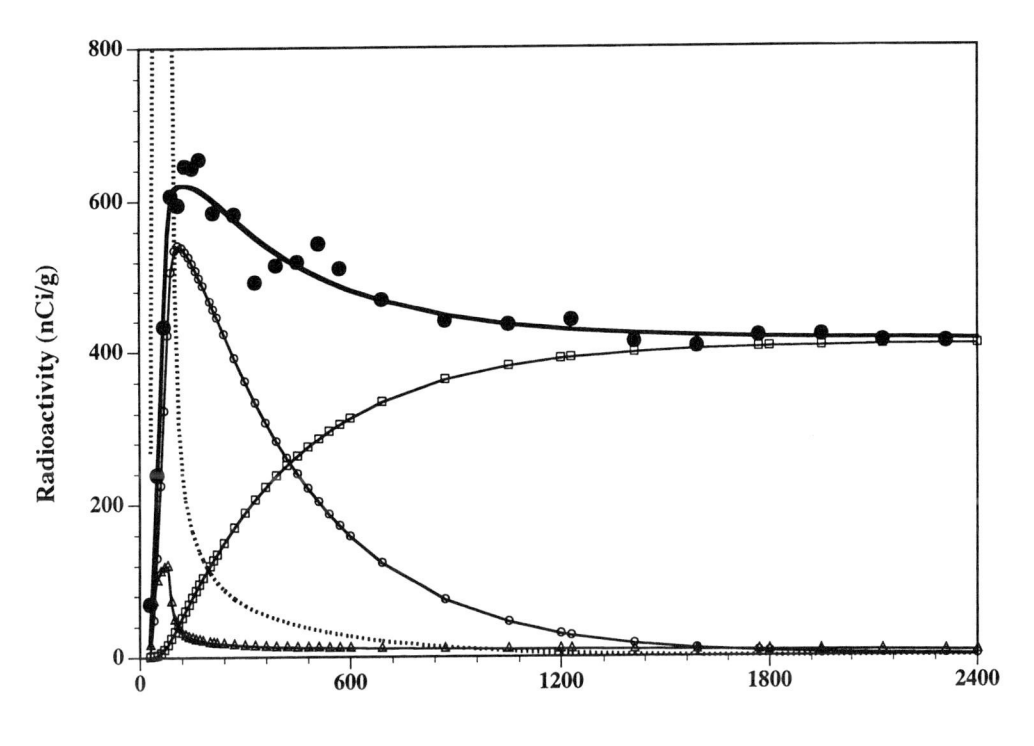

FIGURE 3 Typical time-radioactivity curve obtained from the frontal cortex. Closed circles show observed radioactivity concentration with the bold line showing the theoretical curve obtained by the fitting procedure. The dotted line shows the input function. The symbols (○, □, △) show theoretical concentrations of unchanged [^{11}C]MP4A, metabolized [^{11}C]MP4OH, and vascular radioactivity, respectively.

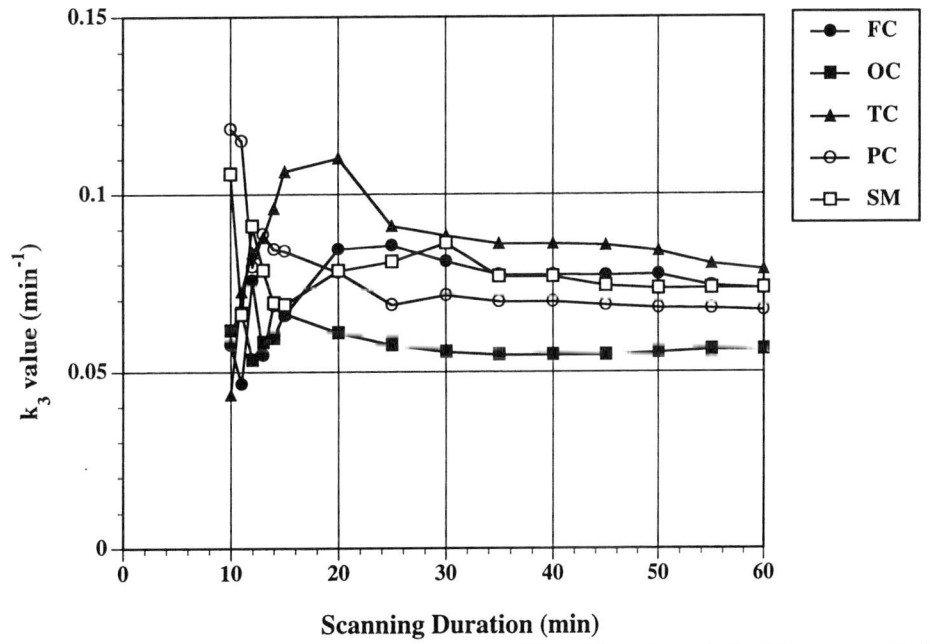

FIGURE 4 Effect of PET scanning duration on the stability of k_3 values obtained from neocortical regions. FC; frontal cortex (●); occipital cortex (■); TC; temporal cortex (▲); PC; parietal cortex (○); and SM; sensorimotor cortex (□).

TABLE 1 k_3 Values (min^{-1}) Obtained from Regions of Living Human Brain

Region	Average ± SD	CV (%)	Relative k_3	AChE
Occipital cortex	0.068 ± 0.010	15.0		
Parietal cortex	0.072 ± 0.010	14.4		
Frontal cortex	0.078 ± 0.009	11.5		
Temporal cortex	0.085 ± 0.010	11.9	1	1
Sensorimotor cortex	0.097 ± 0.013	13.8		
Thalamus	0.285 ± 0.042	14.6	3.3	3
Cerebellum	1.145 ± 0.743	64.9	13.4	8
Striatum	0.951 ± 0.690	72.5	11.2	38

corresponding well with the AChE ratios obtained from the postmortem studies (1:3:8:38), except for striatum. The discrepant result obtained in the striatum is explained by the nature of the tracer (Irie et al., 1996). In the region with very high AChE activity, most of the tracer is hydrolyzed to be accumulated, and washout from the region is negligible. In such a region, rCBF regulating the tracer distribution is the limiting factor of radioactivity accumulation, and AChE activity is no longer the determining factor. Actually, CV values in the cerebellum and the striatum were very high (Table 1), suggesting low reliability of k_3 values in regions with very high AChE activity. Figure 5 shows the effect of the k_3 value on the cerebral time–radioactivity curve. Other parameters, such as K_1, k_2, and the input function, were fixed to those obtained from the frontal cortex of one subject. In this simulation, k_3 was sequentially multiplied by a factor of 2. The condensed simulation curves in the high k_3 area clearly show the saturation of radioactivity accumulation. The k_3 values of striatum and cerebellum were in this saturated area, showing that these values are out of the appropriate range of detection of k_3. In the present kinetic analysis, estimation of k_3 in the thalamus, as well as the striatum and cerebellum, required adding a constraint that K_1 be λk_2. The lack of a significant washout phase in the time–activity curve corresponding to the k_3 value of the thalamus (0.285 min^{-1}) in Fig. 5 might explain the instability of the fitting calculation. Therefore, the upper boundary of the appropriate range seemed to be below the thalamus level. However, a quite decreased radioactivity accumulation was seen in the low k_3 area, suggesting the presence of a lower boundary. A preliminary study to determine the SD of k_3 values showed that the appropriate range of detec-

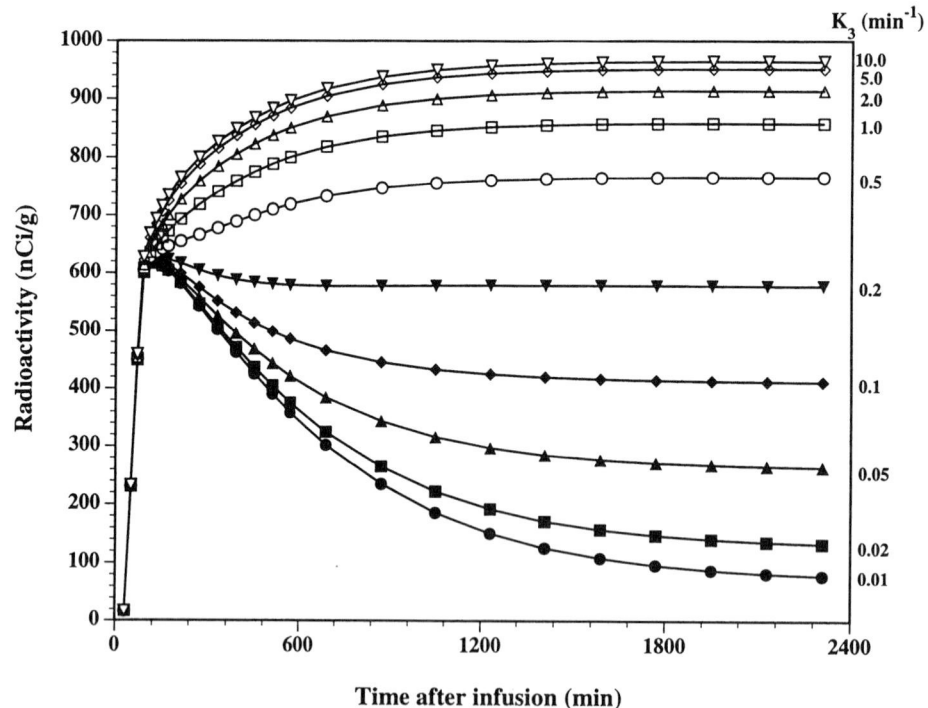

FIGURE 5 Simulated time–radioactivity curves in the frontal cortex. K_1 and k_2 values in this simulation were fixed to 0.446 ml/min/g and 0.140 min^{-1}, respectively. Original data were from frontal cortex of one particular subject shown in Fig. 3.

tion of k_3 was approximately from 0.02 to 0.2 min^{-1}. Neocortical k_3 values determined by the present method ranged from 0.06 to 0.1 min^{-1} in normal subjects and from 0.04 to 0.08 min^{-1} in patients with early AD, respectively (Iyo et al., 1997). Therefore, [^{11}C]MP4A is a very suitable tracer for the determination of neocortical k_3 values reflecting AChE activity *in vivo*.

In conclusion, the present kinetic analysis of dynamic PET data using [^{11}C]MP4A is a promising method to detect changes in neocortical AChE activity in living human brain. Changes in fitting conditions gave varying K_1 and k_2, but rather stable k_3 values could be obtained by the present kinetic analysis. The k_3 values obtained from neocortical regions corresponded well with postmortem AChE activity. The method seems to be applicable not only to early diagnosis of AD, but also to *in vivo* biochemical studies of cholinergic degeneration and/or developing therapeutic drugs.

Acknowledgment

We thank Yasuhiro Wada, Siemens-Asahi Medical Technologies Ltd., for technical support for PET data acquisition.

References

Arai, H., Kosaka, K., Muramoto, O., Moroji, T., and Iizuka, R. (1984). Study of cholinergic neurons in the post-mortem brains from the patients with Alzheimer-type dementia. *Clin. Neurol. (Tokyo)* **24**: 1128–1135.

Atack, J. R., Perry, E. K., Bonham, J. R., Candy, J. M., and Perry, R. H. (1986). Molecular forms of acetylcholinesterase and butyrylcholinesterase in the aged human central nervous system. *J. Neurochem.* **47**: 263–277.

Bierer, L. M., Haroutunian, V., Gabriel, S., Knott, P. J., Carlin, L. S., Purohit, D. P., Perl, D. P., Schmeidler, J., Kanof, P., and Davis, K. L. (1995). Neurochemical correlates of dementia severity in Alzheimer's disease: Relative importance of the cholinergic deficits. *J. Neurochem.* **64**: 749–760.

Davies, P., and Maloney, A. F. J. (1976). Selective loss of central cholinergic neurons in Alzheimer's disease. *Lancet* **2**: 1403.

Irie, T., Fukushi, K., Akimoto, Y., Tamagami, H., and Nozaki, T. (1994). Design and evaluation of radioactive acetylcholine analogs for mapping brain acetylcholinesterase (AChE) *in vivo*. *Nucl. Med. Biol.* **21**: 801–808.

Irie, T., Fukushi, K., Namba, H., Iyo, M., Tamagami, H., Nagatsuka, S., and Ikota, N. (1996). Brain acetylcholinesterase activity: Validation of a PET tracer in a rat model of Alzheimer's disease. *J. Nucl. Med.* **37**: 649–655.

Iyo, M., Namba, H., Fukushi, K., Shinotoh, H., Nagatsuka, S., Suhara, T., Sudo, Y., Suzuki, K., and Irie, T. (1997). Measurement of acetylcholinesterase by positron emission tomography in the brains of healthy controls and patients with Alzheimer's disease. *Lancet* **349**: 1805–1809.

Namba, H., Irie, T., Fukushi, K., Yamasaki, T., Tateno, Y., and Hasegawa, S. (1991). Lesion of the nucleus basalis magnocellularis does not affect cerebral blood flow in rats. *Neurosci. Res.* **12**: 463–467.

Namba, H., Yanagisawa, M., Yui, N., Togawa, T., Kinoshita, F., Iwadate, Y., and Sueyoshi, K. (1996). Quantifying brain tumor blood flow by the microsphere model with N-isopropyl-p-[^{123}I]iodo-amphetamine super-early SPECT. *Ann. Nucl. Med.* **10**: 161–164.

Perry, E. K., Tomlinson, B. E., Blessed, G., Bergmann, K., Gibson, P. H., and Perry, R. H. (1978). Correlation of cholinergic abnormalities with senile plaques and mental test scores in senile dementia. *Br. Med. J.* **2**: 1427–1429.

Phelps, M. E., Huang, S. C., Hoffman, E. J., Selin, C., Sokoloff, L., and Kuhl, D. E. (1979). Tomographic measurement of local cerebral glucose metabolic rate in human with (F-18)2-fluoro-2-deoxy-D-glucose: Validation of method. *Ann. Neurol.* **6**: 371–388.

Ruberg, M., Rieger, F., Villageois, A., Bonnet, A. M., and Agid, Y. (1986). Acetylcholinesterase and butyrylcholinesterase in frontal cortex and cerebrospinal fluid of demented and non-demented patients with Parkinson's disease. *Brain Res.* **362**: 83–91.

Sokoloff, L., Reivich, M., Kennedy, C., Des Rosiers, M. H., Patlak, C. S., Pettigrew, K. D., Sakurada, O., and Shinohara, M. (1977). The [^{14}C]deoxyglucose method for the measurement of local cerebral glucose utilization. *J. Neurochem.* **28**: 897–916.

CHAPTER

60

Parametric Imaging of Ligand–Receptor Interactions Using a Reference Tissue Model and Cluster Analysis

ROGER N. GUNN,* ADRIAAN A. LAMMERTSMA,[†] and VINCENT J. CUNNINGHAM*

*MRC Cyclotron Unit
Hammersmith Hospital
London, W12 0NN, United Kingdom
[†]PET Centre
Free University Hospital
Amsterdam, The Netherlands

A method for quantification of ligand–receptor binding at the voxel level using a simplified reference tissue model and cluster analysis is presented. Cluster analysis is used to extract the reference tissue time course automatically. Parameter estimation of the compartmental structure is then achieved using a basis function approach. This enables generation of parametric images of both delivery and binding without the need for blood sampling. Integration of the "data led" (cluster analysis) and the "structured model" (compartment analysis) provides a powerful tool for the imaging of receptor binding by combining the individual strengths of each method. Data from [^{11}C]raclopride, [^{11}C]SCH 23390, and [^{11}C]WAY 100635 data sets are given.

I. INTRODUCTION

A simplified reference tissue model has been previously derived and applied to region of interest (ROI) dynamic positron emission tomography (PET) data by Lammertsma and Hume (1996). This method has been validated on an ROI basis using nonlinear least-squares methods for [^{11}C]raclopride (D_2 receptor) and [^{11}C]SCH 23390 (D_1 receptor) (Lammertsma et al., 1996; Lammertsma and Hume, 1996).

Subsequently, a basis function method for the implementation of this model at the voxel level, which allows the fast and robust generation of parametric images, has been developed (Gunn et al., 1997). This method is illustrated here using simulated and measured data.

This chapter presents an extension to the method whereby the reference tissue is extracted automatically from dynamic PET data using cluster analysis (Ashburner et al., 1996). Cluster analysis is used as a tissue classification process that discriminates between voxel time–activity curves based on their shape. This process enables automatic delineation of the cerebellum and it is applied to a [^{11}C]WAY 100635 ($5HT_{1A}$ receptor) study. This process enables the combination of a "data-led" technique (cluster analysis) and a "structured model" (compartment analysis) that unite to provide a more powerful tool for parametric imaging. The "data led" approach enables the segmentation of the reference tissue with minimal observer bias compared to the manual definition of a reference ROI. The reference tissue time course is then used as the input to the basis function method for parametric imaging of relative delivery and binding potential.

II. MATERIALS AND METHODS

A. Theory

The simplified reference tissue model (Fig. 1) requires that a radioligand satisfies the following assumptions: (1) a reference region exists that is devoid of specific binding, (ii) labeled metabolites of the parent tracer do not cross the blood–brain barrier, (iii) the degree of nonspecific binding and the volume of distribution of the free/nonspecific compartment is the same in the reference and target tissues, and (iv) the ex-

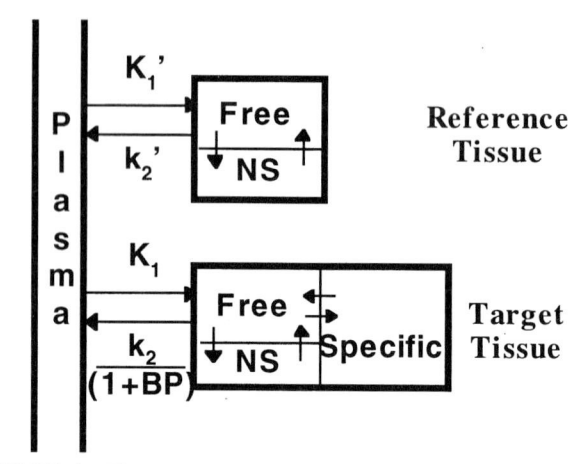

FIGURE 1 The compartmental structure for the reference tissue model.

change between free/nonspecific and specific compartments is rapid such that their composite behavior may be approximated by a single compartment. When these assumptions are met, the following equation may be derived:

$$C_T(t) = R_1 C_R(t) + \left\{ k_2 - \frac{R_1 k_2}{1 + BP} \right\} C_R(t) \otimes e^{-(k_2/(1+BP)+\lambda)t}, \quad (1)$$

where $C_R(t)$ is the concentration time course in the reference tissue, $C_T(t)$ is the concentration time course in the target tissue, R_1 is the ratio of the delivery in the target tissue to the reference tissue, k_2 is the efflux rate constant from the target tissue, BP is the binding potential (cf. Mintun et al., 1984), λ is the physical decay constant of the isotope, and $\otimes$ is the convolution operator.

This model allows for parameter estimates of both relative delivery and binding potential. Relative delivery (R_1) is defined as

$$R_1 = \frac{K_1}{K_1'} = \frac{F(1 - e^{-PS/F})}{F'(1 - e^{-P'S'/F'})}, \quad (2)$$

where F and F' are the blood flows in the target tissue and reference tissue, respectively, and PS and $P'S'$ are the permeability surface area products in the target tissue and the reference tissue, respectively.

Binding potential (BP) is defined as

$$BP = \frac{k_3}{k_4} = \frac{B_{max} f_2}{K_{D_{Tracer}} \left(1 + \sum_i \frac{F_i}{K_{D_i}}\right)}, \quad (3)$$

where B_{max} is the total concentration of specific binding sites, $K_{D_{Tracer}}$ is the equilibrium disassociation constant of the radioligand, f_2 is the "free fraction" of the unbound radioligand in the tissue, and F_i and K_{D_i} are the concentration and equilibrium disassociation constants of competing endogenous ligands.

B. Basis Function Method (BFM)

To enable a fast and robust solution of Eq. (1) at the voxel level, the problem is reformulated in terms of a set of basis functions. Equation (1) can be rewritten as

$$C_T(t) = \theta_1 C_R(t) + \theta_2 C_R(t) \otimes e^{-\theta_3 t}, \quad (4)$$

where $\theta_2 = k_2 - R_1 k_2/(1 + BP)$ and $\theta_3 = k_2/(1 + BP) + \lambda$.

Equation (4) is linear in both θ_1 and θ_2. The nonlinear term θ_3 is accommodated by defining the following set of basis functions:

$$B_i(t) = C_R(t) \otimes e^{-\theta_{3_i} t}. \quad (5)$$

This allows the introduction of realistic parameter bounds on θ_3, which makes the model robust at the voxel level (see Section II,C):

$$\lambda < \theta_3^{min} \leq \theta_{3_i} \leq \theta_3^{max}, \quad i = 1, \ldots, 100, \quad (6)$$

where $\theta_3^{min} \leq k_2^{min}/(1 + BP^{max}) + \lambda$ and $\theta_3^{max} \geq k_2^{max} + \lambda$. For the ligands considered here, 100 discrete values for θ_3 were found to be sufficient ($\theta_3^{min} = 0.001$ sec^{-1}, $\theta_3^{max} = 0.01$ sec^{-1} for [^{11}C]raclopride and [^{11}C]SCH 23390, and $\theta_3^{min} = 0.00075$ sec^{-1}, $\theta_3^{max} = 0.01$ sec^{-1} for [^{11}C]WAY 100635). All convolutions can be performed initially to generate the basis functions B_i, which leads to a fast algorithm. Equation (4) can then be transformed into a linear equation for each basis function:

$$C_T(t) = \theta_1 C_R(t) + \theta_2 B_i(t). \quad (7)$$

Equation (7) is then solved using linear least squares for each basis function in turn. The index i for which the weighted (1/variance) residual sum of squares is minimized is determined by a direct search, and the associated parameter values for this solution are obtained ($\theta_1, \theta_2, \theta_3$). The parameter values for BP, R_1, and k_2 are then easily deduced from the relationships given in Eq. (4).

The sensitivity of the estimated binding potential to the presence of specific binding in the reference tissue was also investigated using noiseless simulated data. These simulation results were compared with an explicit algebraic equation derived from a consideration of the ratio of the expected distribution volumes at equilibrium:

$$BP_{\text{Apparent}} = \frac{BP_{\text{Tissue}} + 1}{BP_{\text{Reference}} + 1} - 1, \qquad (8)$$

where $BP_{\text{Reference}}$ and BP_{Tissue} are the true binding potentials of the ligand in the reference and target tissues, respectively. BP_{Apparent} would be the binding potential estimated in the target tissue.

C. Simulations

A measured [^{11}C]raclopride metabolite corrected input function (Lammertsma et al., 1996) was used to generate perfect reference and target tissue data for typical parameter values ($R_1 = 1.2$, $k_2 = 0.0024$ sec^{-1}, and $BP = 2$). Simulations were performed according to a 1-hr protocol involving 17 frames of data. Normally distributed noise was added to the perfect data. A lower level of noise was added to the reference tissue to approximate the real situation where a large region is used to define it compared to voxel time courses of the target tissue (noise in target tissue = 20 × noise in reference tissue). The noise was scaled such that 1 standard deviation corresponded to the maximum count in each simulated curve. One thousand realizations were produced at each noise level. Data were then analyzed using conventional nonlinear least squares (NLM) (simplex: Nelder and Mead, 1965) and the BFM described earlier.

D. Measured Data

Four human [^{11}C]raclopride studies were acquired after intravenous injection of the radioligand (9.28–9.73 mCi) using an ECAT 953B PET camera (CTI/Siemens, Knoxville, TN) characterized previously by Spinks et al. (1992). Three-dimensional data sets were acquired involving 17 frames (1 × 15, 1 × 5, 1 × 10, 1 × 30, 4 × 60, 3 × 300, 2 × 600, 4 × 300 sec) of data. A convolution subtraction scatter correction was applied (Bailey and Meikle, 1994) and data were reconstructed using a reprojection algorithm (Kinahan and Rogers, 1989), which included a calculated attenuation correction. An ROI template was defined on an integral image for regions of high and low binding and applied to the dynamic data sets producing time–activity curves. The data set was then analyzed on an ROI level using NLM (Lammertsma and Hume, 1996) and on a voxel-by-voxel level using BFM described earlier. A single [^{11}C]SCH 23390 scan was acquired using an identical protocol to the [^{11}C]raclopride scan described previously (injected activity: 10.64 mCi).

E. Automatic Extraction of Reference Tissue

An example of automatic extraction of the reference region for [^{11}C]WAY 100635 is presented using cluster analysis. Cluster analysis was used to segment dynamic PET data, such that the reference region may be extracted purely by the differing shape of time–activity curves in this region. The kinetics of the radioligand in the reference tissue are the fastest (excluding any blood volume effect), allowing the region to be determined by

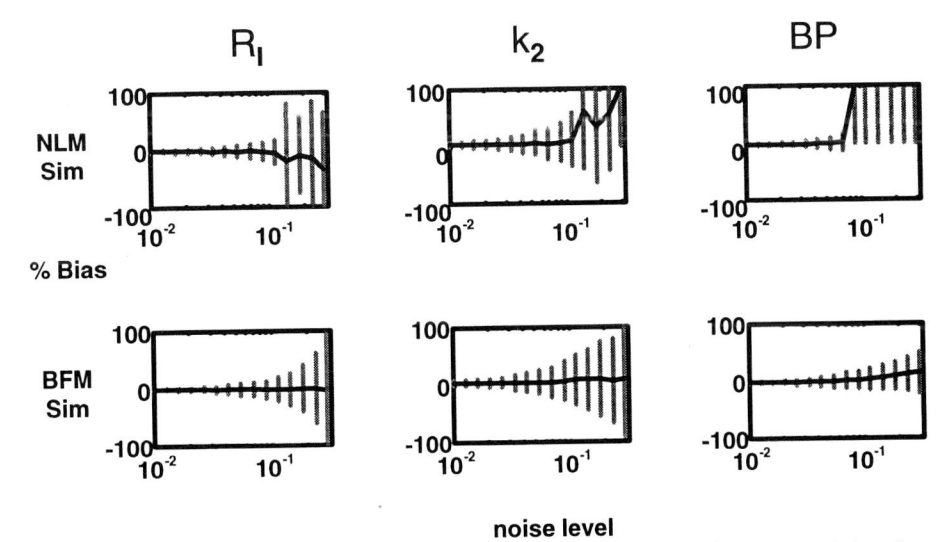

FIGURE 2 Simulation results for the nonlinear least-squares method (NLM) and the basis function method (BFM).

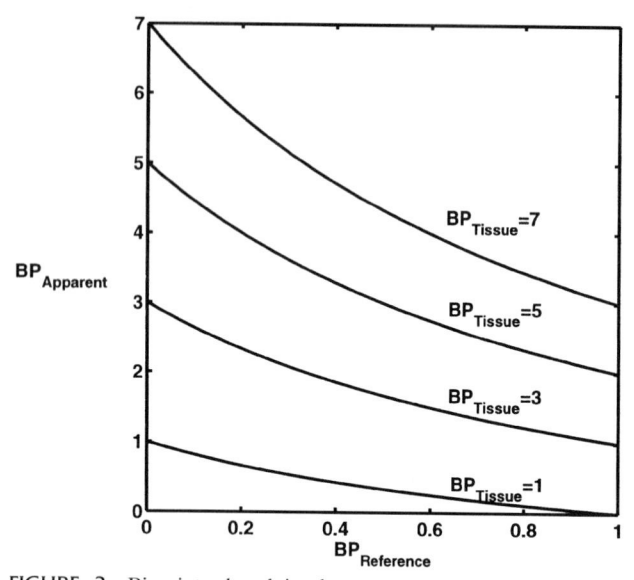

FIGURE 3 Bias introduced in the measure of binding potential when the reference tissue contains some specific binding.

this tissue classification method. A pilot [^{11}C]WAY 100635 study (injected activity: 7.42 mCi) was acquired on the ECAT HR + PET camera (CTI/Siemens). Three-dimensional data acquisition incorporated 21 frames (1 × 30, 1 × 15, 1 × 5, 1 × 10, 1 × 30, 4 × 60, 7 × 300, 5 × 600 sec) of data. A model-based scatter correction was applied (Watson et al., 1996) and data were reconstructed using a reprojection algorithm (Kinahan and Rogers, 1989), which included a segmented attenuation correction (Xu et al., 1994).

Eight cluster images were generated with their associated time–activity curves by applying cluster analysis (Ashburner et al., 1996) to the decay-corrected dynamic data set. These cluster images contain probability values that the corresponding voxel is described purely by the associated cluster–time activity curve. The cluster image corresponding to the cerebellum was identified easily and thresholded at $p > 0.75$, which produced a cerebellum mask. This mask was then applied to dynamic PET data to produce a reference time–activity

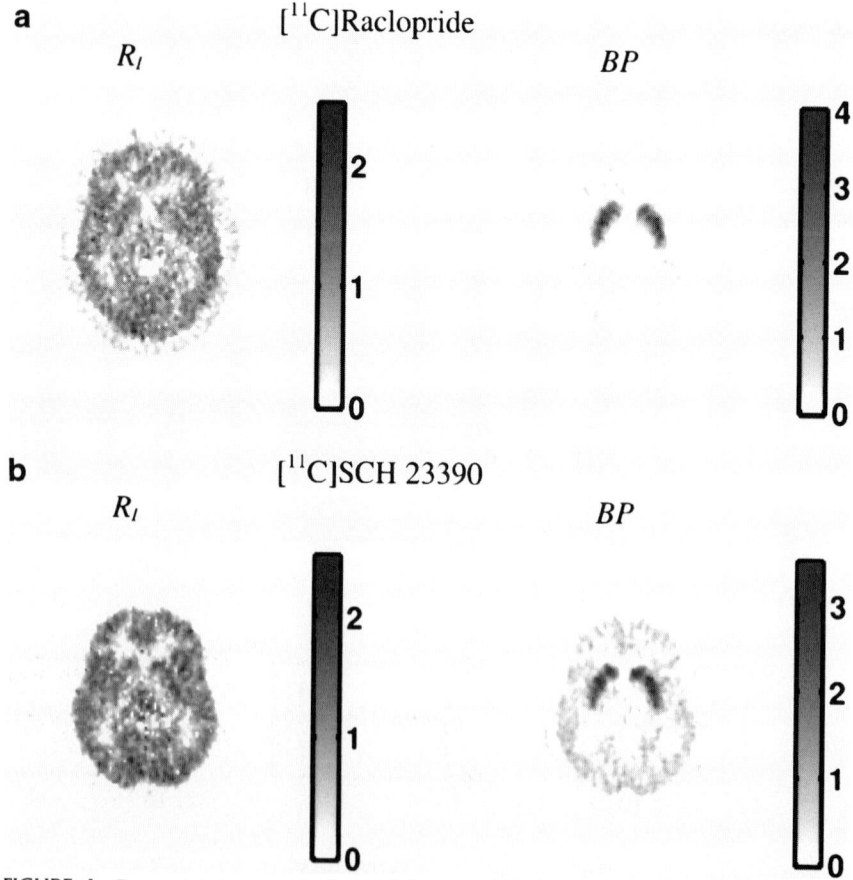

FIGURE 4 Parametric images of relative delivery and binding potential for a midtransaxial plane: (a) [^{11}C]raclopride and (b) [^{11}C]SCH 23390.

curve. Subsequently, parametric images of [¹¹C]WAY 100635 delivery and binding were generated using BFM described earlier.

III. RESULTS AND DISCUSSION

A. Simulations

To investigate the implementation of the simplified reference tissue model at the voxel level, simulated data were used to assess the performance of the NLM and the BFM for a range of noise levels. The mean and standard deviation of the parameter estimates, using BFM and NLM, for 1000 realizations at different noise levels is shown in Fig. 2. At low noise levels (up to 0.1), which are characteristic of ROI time–activity curves, the behavior of the two methods is comparable, but at higher noise levels, which are characteristic of voxel time–activity curves, the NLM method breaks down. The BFM copes well at noise levels considered due to the inclusion of parameter bounds and only a small bias is introduced even at high noise levels.

Assessment of the effect of specific binding in the reference tissue with simulated data agreed with theoretical Eq. (8). A graph illustrating the underestimation of binding potential when the assumption that there is no specific binding on the reference tissue is incorrect is illustrated in Fig. 3.

B. Measured Data

To compare the performance of the two methods it was necessary to apply the NLM to ROI data on which it had previously been validated (Lammertsma and Hume, 1996), whereas the BFM could be applied at the voxel level. Parameter estimates of binding potential were obtained for both methods (the ROIs being applied to the parametric images and mean values calculated) and there was a good correlation between them (parametric = 1.028 ROI − 0.002, $r = 0.999$). Representative parametric images generated using the BFM are shown in Fig. 4.

C. Automatic Extraction of Reference Tissue

Extraction of the cerebellum for the [¹¹C]WAY 100635 study was facilitated using cluster analysis. The temporal characterization of this region was obtained within cluster 2. This cluster also contained some white matter (an area expected to be devoid of specific binding) at lower probability values, but this was removed by thresholding the cluster image at $p > 0.75$. Figure 5 illustrates the segmentation of the cerebellum with

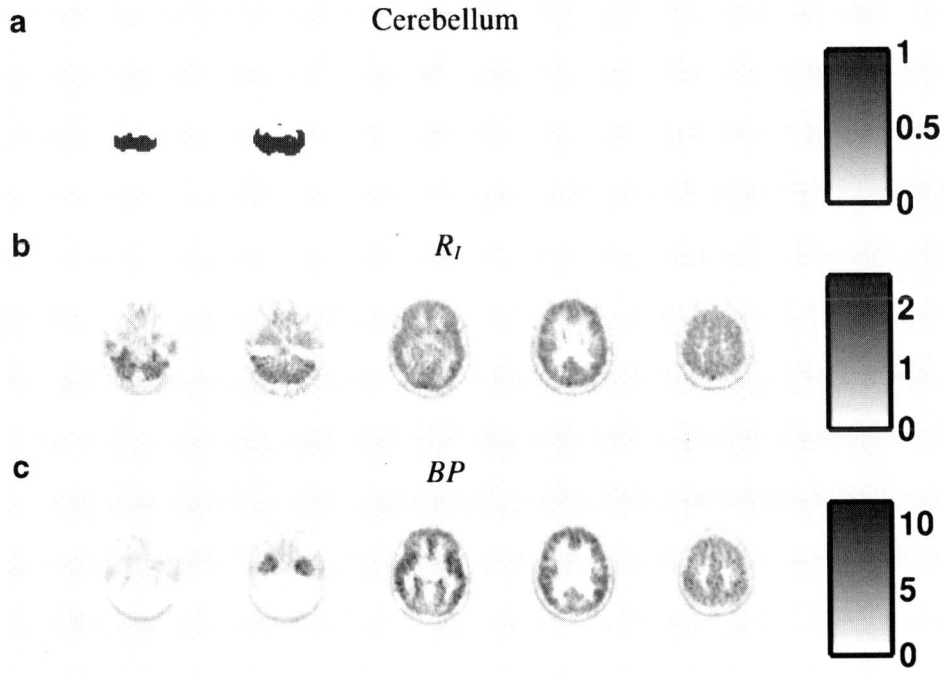

FIGURE 5 Combination of cluster analysis and basis function method for representative planes of a [¹¹C]WAY 100635 study: (a) Cluster analysis segmentation of the cerebellum, (b) parametric images of relative delivery, and (c) parametric images of binding potential.

cluster analysis and the associated R_I and BP images generated with BFM for a selection of transverse planes.

D. Summary

This chapter presents a robust implementation of a reference tissue compartmental model for parametric imaging of ligand binding at the voxel level. The formulation of the problem in terms of a set of basis functions allows for a fast and stable solution. Furthermore, the extension of this approach using cluster analysis to automatically extract the reference tissue is presented. The combination of the "data-led" approach (cluster analysis) and the "structured model" (compartmental analysis) provides a powerful tool for the parametric imaging of PET radioligand binding and delivery.

References

Ashburner, J., Haslam, J., Taylor, C., Cunningham, V. J., and Jones, T. (1996). A cluster analysis approach for the characterization of dynamic PET data. In "Quantification of Brain Function Using PET" (R. Myers, V. Cunningham, D. Bailey, and T. Jones, eds.), pp. 301–306. Academic Press, San Diego, CA.

Bailey, D. L., and Meikle, S. R. (1994). A convolution-subtraction scatter correction method for 3D PET. *Phys. Med. Biol.* **39**: 411–424.

Gunn, R. N., Lammertsma, A. A., Hume, S. P., and Cunningham, V. J. (1997). Parametric imaging of ligand-receptor binding in PET using a simplified reference region model. *NeuroImage* **6**: 279–287.

Kinahan, P. E., and Rogers, J. G. (1989). Analytic 3-D image reconstruction using all detected events. *IEEE Trans. Nucl. Sci.* **36**: 964–968.

Lammertsma, A. A., and Hume, S. P. (1996). Simplified reference tissue model for PET receptor studies. *NeuroImage* **4**: 153–158.

Lammertsma, A. A., Bench, C. J., Hume, S. P., Osman, S., Gunn, K., Brooks, D. J., and Frackowiak, R. S. J. (1996). Comparison of methods for analysis of clinical [^{11}C]raclopride studies. *J. Cereb. Blood Flow Metab.* **16**: 42–52.

Mintun, M. A., Raichle, M. E., Kilbourn, M. R., Wooten, G. F., and Welch, M. J. (1984). A quantitative model for the in vivo assessment of drug binding sites with positron emission tomography. *Ann. Neurol.* **15**: 217–227.

Nelder, J. A., and Mead, R. (1965). A simplex method for function minimization. *Comput. J.* **7**: 308–313.

Spinks, T. J., Jones, T., Bailey, D. L., Townsend, D. W., Grootoonk, S., Bloomfield, P. M., Gilardi, M.-C., Casey, M. E., Sipe, B., and Reed, J. (1992). Physical performance of a positron tomograph for brain imaging with retractable septa. *Phys. Med. Biol.* **37**: 1637–1655.

Watson, C. C., Newport, D., and Casey, M. E. (1996). A single scatter simulation technique for scatter correction in 3D PET. In "Three-Dimensional Image Reconstruction in Radiology and Nuclear Medicine" (P. Grangeat and J.-L. Amans, eds.), pp. 255–268. Kluwer Academic Publishers, Dordrecht, The Netherlands.

Xu, M., Luk, W. K., Cutler, P. D., and Digby, W. M. (1994). Local threshold for segmented attenuation correction of PET imaging of the thorax. *IEEE Trans. Nucl. Sci.* **41**: 1532–1537.

CHAPTER 61

Limitations of Binding Potential as a Measure of Receptor Function: A Two-Point Correction for the Effects of Mass

EVAN D. MORRIS, SVETLANA I. CHEFER, and EDYTHE D. LONDON

Brain Imaging Center
National Institute on Drug Abuse
National Institutes of Health
Baltimore, Maryland 21224

Using simulated time–activity curves, significant effects of mass of ligand on estimates of binding potential (BP) at reasonable mass doses for nonhuman primates have been demonstrated. Compartmental model theory predicts the dependence of BP on mass but this possibility is usually ignored—with justification. However, BP values for nonhuman primates may be more susceptible to artifacts because a typical mass dose per body weight of animal is higher than in a human. Time–activity curves were simulated for [^{11}C]raclopride in rhesus monkeys in order to gauge the likelihood of mass effects in positron emission tomography (PET) studies in which small (e.g., 10%) differences in BP between primate groups who may not have the same average body weight are detected. PET data and simulations in this study are in agreement that small procedural biases in ligand administration between groups could masquerade as changes in BP as large as those being sought. A two-point extrapolation technique based on the observations of Hume et al. (1995) to correct for mass artifacts is introduced and evaluated. Tests of the extrapolation with simulated data show that mass correlation can be reduced or eliminated without adding variability to data. The sensitivity of this correction method to noisy BP estimates favors the use of graphical rather than kinetic calculation of BP. Finally, a discrepancy between the Hume model and predictions regarding the impact of nonspecific binding on estimated BP is explored.

I. INTRODUCTION

Binding potential (BP), first introduced by Mintun et al. (1984), is a commonly used measure of receptor activity that can be derived graphically or kinetically from dynamic positron emission tomography (PET) studies whether or not a plasma curve has been measured (Logan et al., 1990, 1996; Lammertsma et al., 1996). That BP (or the closely related distribution volume ratio) is a fruitful and robust measure is reflected in the large number of literature citations (71) to one paper presenting a popular method for its graphical estimation (Logan et al., 1990). Estimates of BP in PET and single photon emission computed tomography (SPECT) research have been used in lieu of estimating available number of receptors (B'_{max}) when the lack of blood data or convenience dictates. More recently, differences in BP before and after pharmacological perturbation have been used to infer the up- or downregulation of neurotransmitter concentrations in the synapse (Laruelle et al., 1997) or the action of one neurotransmitter system on another (Dewey et al, 1993, 1995). Nevertheless, there are some caveats to keep in mind when using binding potential as an index of receptor number or receptor occupancy.

This chapter sets out the simple theory behind a potentially important limitation on the interpretation of BP arising from its sensitivity to the mass of injected ligand. Through the use of realistic simulations of dynamic PET data with the D_2 antagonist [^{11}C]raclopride, those conditions (e.g., injected mass, subject body weight, receptor density) under which mass might present a significant confounding influence on BP estimates and their interpretation were explored. In addition, a "graphical" method for eliminating any dependence on mass and an evaluation of the proposed method are presented, based on how well it works with simulated time–activity data. Although the methods of BP estimation used here do not require blood curves, these findings are consistent across methods and apply, in theory, to any methods that do not estimate B'_{max} and K_D separately. This simulation study showed that masses of ligand (or conversely, specific activities) that are commonly considered in the PET literature to be immune from mass effects may not be without artifact in all circumstances and that the artifact is exacerbated by low receptor density. In any PET-receptor study, as a minimum, it is important to avoid erroneously attributing an effect of the experimental protocol (e.g., injected mass) to a change in receptor–ligand binding. This is most likely to occur when comparing results across groups of subjects where one population is more prone than the other to small body weight but where only the absolute radioactivity (and not the mass per body weight) has been held constant. Similarly, comparing the BP in regions of high and low BP might be subject to mass artifacts. The study presented here was intended to determine whether the mass of injectate was of concern in attempting to measure age-related changes in D_2 receptor activity in rhesus monkeys with PET.

II. THEORY

A. Definitions and Assumptions

BP is defined as the ratio of available receptor sites to the equilibrium distribution coefficient, B'_{max}/K_D, but is calculated as the ratio of apparent forward binding to dissociation rate constants, k_3/k_4, whenever B'_{max} and K_D cannot be separated. Such is the case in any single-injection PET investigation with a receptor ligand and is certainly true if an input function has not been measured. However, it is important to remember that k_3/k_4 is not identically equal to B'_{max}/K_D. In the conventional three-compartment model (i.e., tissue = free + bound + nonspecific), the mass balance on the bound compartment is

$$dB/dt = k_{on}(B'_{max} - B - B^c)F - k_{off}B, \quad (1)$$

which is exactly equivalent to

$$dB/dt = k_{on}(B'_{max} - B/SA)F - k_{off}B \quad (2)$$

for single-injection experiments (Morris et al., 1996). F and B are the concentration of labeled tracer in free and bound compartments, respectively. B^c is the concentration of unlabeled ligand in the bound state. k_{on} and k_{off} are the true forward and reverse rate constants for the ligand and receptor and SA is the specific activity of the injectate. The steady-state volume of distribution between free and bound compartments is

$$B/F|_{ss} = k_{on}(B'_{max} - B/SA)/k_{off} \quad (3)$$
$$\equiv k_3/k_4 \quad (4)$$

and is equal to the BP provided that k_3 is a constant. In other words, it is necessary that

$$B'_{max} \gg B/SA. \quad (5)$$

B. Model of BP Dependence on Mass

When the specific activity is low enough to violate Eq. (5), then the measured binding potential will decrease with the increasing total (labeled and unlabeled) ligand in the bound state, B/SA. It is reasonable to expect that the average amount of ligand in the bound state over the scanning period is proportional to the amount delivered to the region of interest and thus proportional to the mass of injectate per body weight of subject. An inverse relationship has been proposed by Hume et al. (1995) to relate BP (determined with PET) to the dose (in $\mu g/kg$):

$$BP = {}^{\alpha}B_{max}/({}^{\gamma}K_D + M) + NS \quad (6)$$

to explain the mass–response curves in rats administered raclopride in doses ranging from 0.36 to 360 $\mu g/kg$. ${}^{\alpha}B_{max}$ and ${}^{\gamma}K_D$ are constant terms related by SA to B'_{max} and K_D, respectively. NS is a constant offset term in BP due to nonspecific binding. M is the mass of ligand injected per body weight.

C. Correction Method for Mass Dependence

Inverting the model in Eq. (6) suggests a simple linear correction to BP for mass artifacts. Consider the following two cases.

Case 1

If nonspecific binding is absent, Eq. (6) yields an expression for $1/BP$ that is linear in mass, M:

$$1/BP = M/{}^{\alpha}B_{max} + {}^{\gamma}K_D/{}^{\alpha}B_{max}. \quad (7)$$

In this case, any two BP estimates at different masses should be sufficient to extrapolate $1/BP$ at zero mass.

Case 2

If nonspecific binding of the ligand is present, linearity is preserved as long as $M \ll {}^{\alpha}K_D$. Two BP points should still be sufficient to extrapolate to zero mass provided that mass is small. With only two BP measurements, it will not be possible to fit the nonlinear model [Eq. (6)] for all three unknown parameters: ${}^{\gamma}K_D$, ${}^{\alpha}B_{max}$, and NS. Nevertheless, it is assumed that it is possible to choose two small but different mass values for which the following linearization holds:

$$1/BP = M(^{\alpha}B_{max} + ^{\gamma}K_D \text{ NS})$$
$$+ ^{\gamma}K_D/(^{\alpha}B_{max} + ^{\gamma}K_D \text{ NS}). \quad (8)$$

This expression predicts that two points are again sufficient to find the intercept on the $1/BP$ axis. However, this intercept, $^{\gamma}K_D/(^{\alpha}B_{max} + ^{\gamma}K_D\text{NS})$, retains a contribution from nonspecific binding, NS.

III. METHODS

A. Analysis

Test data sets [composed of a "striatum" and a "cerebellum" time–activity curve (TAC)] were generated by simulating the standard three-compartment model as implemented by Delforge *et al.* (1990). BP was estimated from simulated data using two reference region techniques that do not require a plasma input; rather, the receptor-free "cerebellum" curve was used as the reference. The kinetic method was originally introduced by Cunningham *et al.* (1991) and has been shown to yield BP estimates that correlate closely with the standard three-compartment kinetic analysis using a blood curve (Lammertsma *et al.*, 1996). The graphical method used was introduced by Logan *et al.* (1996) as a bloodless modification of their earlier graphical method for estimating distribution volume ratios. All slopes from the graphical method were based on integrals of data from 30 to 62 min. Because the k_2 value was the same for all simulations, and the B/F ratio was fairly constant after 30 min, the term including a population-average k_2 was neglected as described by the authors.

B. Simulations

The simulation parameters were chosen to produce TACs that resembled experimental [^{11}C]raclopride data from rhesus monkeys, including a partial volume correction (Morris *et al.*, 1998). Model parameters used were: $K_1 = 0.06$ ml/min/g, $k_2 = 0.06$ min^{-1}, $k_{on} = 0.0032$ ml/pmol/min, and $k_{off} = 0.075$ min^{-1}. B'_{max} values for striatum curves were 5, 35, and 100 nM. These receptor densities corresponded to BP levels of 0.21, 1.5, and 4.27, respectively. "Cerebellum" was simulated with its B'_{max} value set to 0. PET output was modeled as the continuous weighted sum of blood, free, bound, and nonspecific compartments integrated over each scan time. The blood volume fraction was fixed at 0.04. The plasma input curve was modeled as a biexponential with an initial 1-min time delay to the peak. Metabolite correction was not applied to the plasma curve. One standard blood curve was used to generate all simulated PET curves. The plasma curve was scaled so that the peak point equaled the total injected activity (10 mCi) normalized by the approximate blood volume of a 10-kg animal (1000 ml). Where nonspecific binding was included, both "striatum" and "cerebellum" were simulated with the same large forward and reverse rate constants for NS binding ($k_5 = k_6 = 1.0$ min^{-1}). Poisson noise was added and the level of noise was chosen to resemble dynamic data acquired on the GE PC4096 plus whole body scanner. Pairs of "cerebellum" and "striatum" were simulated for a large range of injection masses (i.e., 10 mCi, where SA = [20,000 – 0.1] mCi/μmol) for each of three B'_{max} values.

C. Test of Mass–Effect Correction

The simulated data was used to evaluate the correction method proposed earlier. Two-pont extrapolation was tested for both cases (with and without NS binding). Every possible pair of BP estimates, for a given parameter set, were inverted and extrapolated to $1/BP$ at zero mass. The reinverted intercepts were compared to the true BP values to determine bias and against each other to determine variability and the ability of the extrapolation to eliminate correlation of BP with mass.

IV. RESULTS AND DISCUSSION

A. Demonstration of (Experimental) Mass Artifact

Figure 1 displays the correlation between BP and mass that was found after two scans each of nine rhesus monkeys with [^{11}C]raclopride. The animals received between 0.5 and 2 mCi per kg of body weight with SA in the range 1200–1600 mCi/μmol, i.e., 0.1–0.5 μg/kg. The BP values decline with increasing mass as predicted by the theory. Greater mass leads to more bound ligand molecules and greater impact for the term, B/SA relative to B'_{max}. The slope, -2.8 kg/μg, and the intercept, 7.3, translate to a decline of 3.8% in BP for every 0.1 μg/kg added to the mass injected. The BP estimates in Fig. 1 are all from the graphical method.

B. Simulated Mass Artifact

Figure 2 is a plot based on simulated data that is reminiscent of the artifact observed in Fig. 1. These simulations have no nonspecific binding and a B'_{max} of 35 nM (true BP = 1.5). The slope is -0.16 kg/μg and

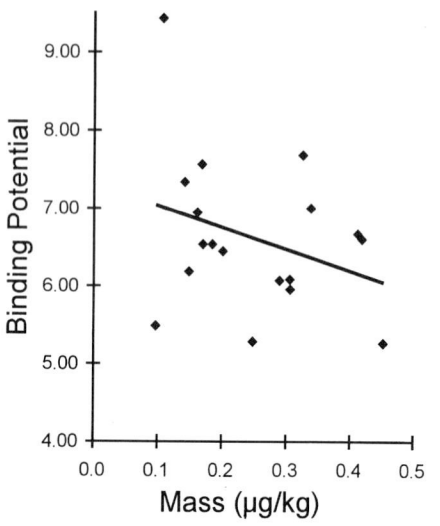

FIGURE 1 Correlation of BP with injected mass—experimental [¹¹C]raclopride data in monkeys. Correlation coefficient = −0.31.

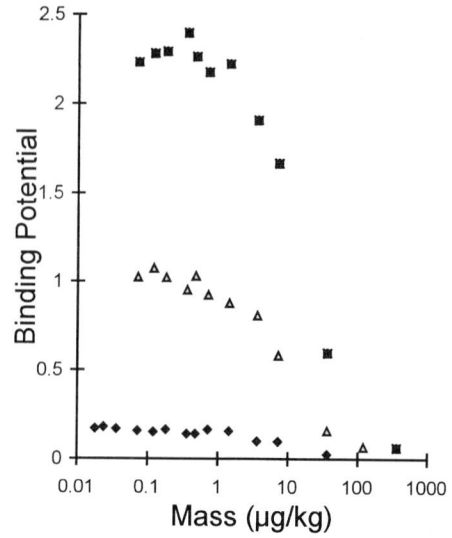

FIGURE 3 Mass–response curves for simulations of $B'_{max} = 100$ (top), 35 (middle), and 5 nM (bottom). Curves appear to obey the model of Hume *et al.* (1995) for NS = 0. The curves, from the top, correspond to BP values of 4.3, 1.5, and 0.21.

the intercept is 1.1, which is a decline of 1.5% in BP for each additional 0.1-μg/kg increase in the mass dose. These BP values were estimated with the graphical method. Within this limited mass range, BP appears to be linearly related to mass. However, if these simulations are extended to lower and higher masses (and to other values of B'_{max}) as shown in Fig. 3, the relationship is hyperbolic as predicted by Hume *et al.* (1995) [Eq. (6)]. Figure 3 displays graphically estimated BP over wide mass ranges of 0.018–360 μg/kg for B'_{max} values of 100, 35, and 5 nM without nonspecific binding. Each curve has a relatively flat ("safe") portion over which the mass effect is negligible and a declining section where the dependence on mass is pronounced. The largest "safe" mass dose varies according to B'_{max}. For a simulated 10-kg monkey with $B'_{max} = 100$ nM, this "safe" dose is between 1.44 and 3.6 μg/kg (probably outside the range required for humans or monkeys). For $B'_{max} = 35$ nM, the largest "safe" mass dose may be as low as 0.12 μg/kg (on the order of what is used in monkeys). For $B'_{max} = 5$ nM, the mass–response curve is noisy, but the largest safe dose appears to be no more than 0.024 μg/kg, within the range of doses administered to monkeys for SPECT (see Laruelle *et al.*, 1997).

C. Comparison of Graphical and Kinetic Methods

Figure 4 shows that the mass dependence of BP is consistent across estimation methods. The kinetic method consistently shows a positive bias whereas the graphical method shows a negative one relative to the true BP value. Although the large variability in the kinetic estimates makes it difficult to identify the largest "safe" dose on the kinetic response curve, it is clear that they consistently overestimate the BP relative to the graphical method. Similar to graphical estimates, the kinetic estimates decline with mass beginning somewhere between 0.1 and 1.0 μg/kg. Graphical data (open squares) shown in Fig. 4 are the same points as the middle curve in Fig. 3 (simulations with $B'_{max} = 35$ nM).

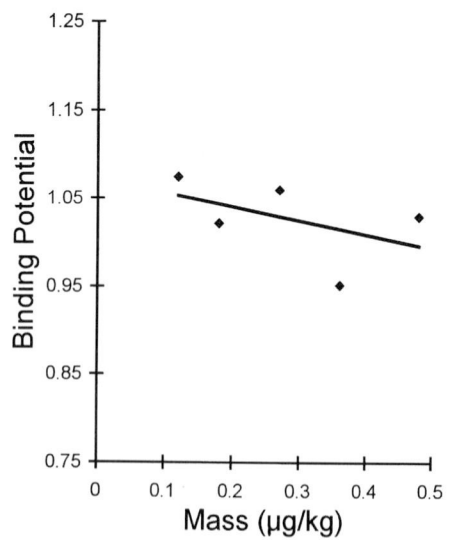

FIGURE 2 Correlation of BP with mass for simulated data. Mass range is 0.1–0.5 μg/kg. Correlation coefficient = −0.47.

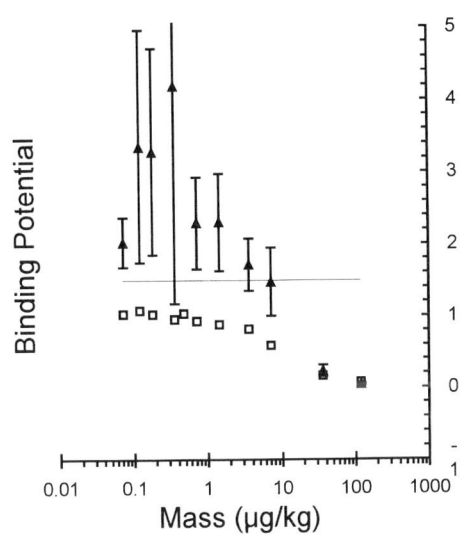

FIGURE 4 Mass–response curves for BP based on kinetic (top curve) and graphical (bottom) estimates. $B'_{max} = 35$ nM in all simulations. NS binding = 0. The solid line represents the value of the true BP = 1.5. Error bars on the kinetic points are based on the covariance matrix of the parameter estimates. No error bars are given for the graphical method because the points on the Logan plot are not independent.

E. How Good is the Mass Correction?

Regardless of biases in the reference region methods, their performance in eliminating the mass effect, demonstrated earlier, on BP can be assessed. Figure 5 displays graphical BP estimates from simulated data for which B'_{max} is 35 nM; the mass range is 0.072–3.6 μg/kg and there is no nonspecific binding. These data correspond to activity doses of 10 mCi and a reasonable range of specific activities (5000–100 mCi/μmol) for [^{11}C]raclopride at time of injection. [*Note*: Carson *et al.* (1997) injected 2–5 mCi of [^{11}C]raclopride into 10-kg monkeys and observed mass effects in their BP estimates with SA of 363 ± 181 mCi/μmol.]

Figure 5a shows the expected decline in BP with increasing mass. These points are a subset of the points on the middle curve in Fig. 3. Figure 5b is a plot of 1/BP vs mass. The points lie roughly on a line with positive slope as predicted by Eq. (7). To correct 1/BP "graphically" (i.e., to locate the 1/BP intercept), any two points on Fig. 5b should suffice. The variability in the intercept values for all these points taken two at a

D. Discrepancies between Models

Biases result from incompatibilities between the reference region techniques and the full three-compartment model. One obvious incompatibility is the presence of a blood component in the PET model. However, even with noiseless simulations and no blood volume fraction, the graphical method gives a low and the kinetic method a high estimate of BP. However, the reference region model is internally consistent. When it is used to both generate data and fit them, there is no bias, at least at moderate error levels, in BP estimates. The graphical technique assumes that the free and bound compartments are in equilibrium. This condition is not strictly satisfied with data presented here. Also, by using least-squares fitting of points that are dependent on the integral of data (i.e., contain cumulative error), the graphical method implicitly imposes a weighting of the points other than the inverse variance weighting prescribed by the unbiased nonlinear least-squares estimator. Thus, *a priori*, it appears that the Logan plot method is a biased estimator of BP, which has been confirmed by Hsu *et al.* (1997). Although the kinetic method properly weights data, it is also quite sensitive to the shape of the cerebellum curve. Thus, when the noisy cerebellum curve is smoothed, the resulting generated input function, $K_1 C_p$, may have a shape that is incompatible with the striatum and give a poor fit.

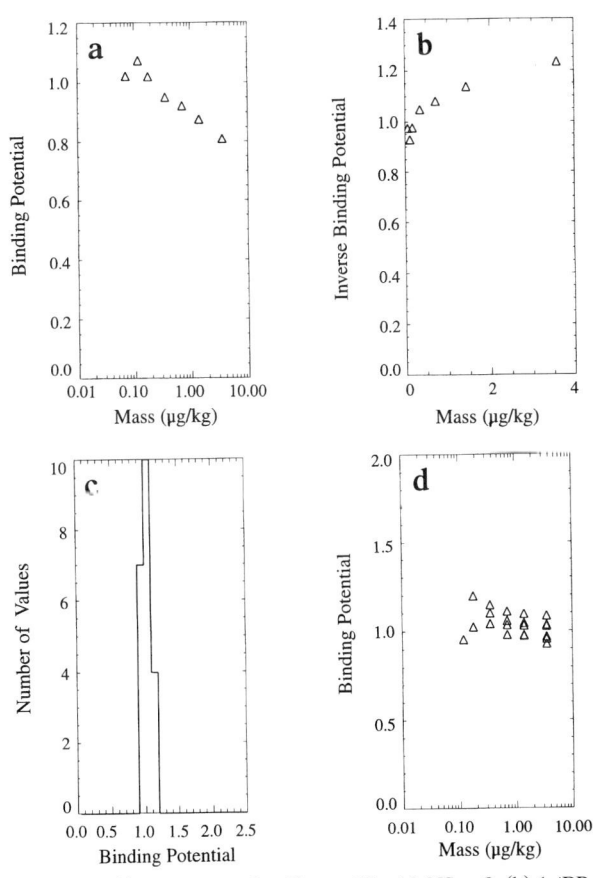

FIGURE 5 (a) BP vs mass for $B'_{max} = 35$ nM, NS = 0, (b) 1/BP vs mass, (c) histogram of corrected BP values, and (d) corrected BP values vs maximum of two masses used in correction.

time is displayed as a histogram of BP values in Fig. 5c. The histogram of corrected BP values, which is binned in 0.1 increments, is narrow and symmetric, reflecting good error properties of the correction. Figure 5d gives each of the corrected BP values plotted against the maximum of the two masses used to generate the corrected value. Regardless of mass correction, the Logan method underestimates the true BP. Prior to correction, the correlation coefficient (r) of BP with mass was -0.88 and the coefficient of variation (COV) was 9.7% in the BP values. Following correction, r was reduced to -0.41 and the COV was 6.6%. Much of the correlation with mass is eliminated without sacrificing precision. Unfortunately, if a similar test is performed with the kinetic estimates of BP, r of -0.59 is improved to -0.082, but the COV is aggravated from 35 to 120%. The poor variability in the correction method reflects the large variability in kinetic estimates themselves.

F. Inclusion of Nonspecific Binding

To this point, the two-point correction has only been tested on parameters that were derived from data that had no nonspecific binding. If fast nonspecific binding is included in simulated data, the mass effect is retained but the amplitude of the mass–response curve is changed. Figure 6 displays three mass–response curves. The top curve is for data in the middle curve of Fig. 3 without nonspecific binding. The bottom curve is for data from the same parameters except that it includes fast nonspecific binding in both the "striatum" and the "cerebellum." The largest "safe" mass dose remains unchanged (compare bottom curve to top), but the presence of nonspecific binding causes the mass-independent portion of the curve to be much lower than in the top curve. This fact, which is even clearer on an inverted plot, is consistent with case 2 of the Hume model in Eq. (8). The intercept is a function of NS. For seven simulations with NS binding and a mass range of 0.072–3.6 μg/kg, r improved dramatically from -0.9 to -0.12 after correction and COV improved slightly from 11.5 to 10.3%.

G. Disagreement with Hume Model

The middle curve in Fig. 6 (solid line) is the mass–response curve that Hume *et al.* (1995) fit to their [^{11}C]raclopride PET data in rats. Although it shows a "safe" range of masses that is comparable to monkey simulations, it reveals a discrepancy with the results described in this chapter. As dictated by the offset term, NS, in Eq. (6), the Hume model predicts that in the presence of NS binding, one will measure a positive BP at least equal to the NS term for all mass doses no matter how large. Simulation data described in this study indicate otherwise. The bottom curve in Fig. 6 clearly goes to zero at a mass dose of 100 μg/kg. What is the reason for this discrepancy? In reviewing the paper by Hume *et al.* (1995), it appears that the model may not have been tested by an adequately large mass range to determine the need for an offset term. Most likely, Hume *et al.* (1995) were constrained by the lethal limits of raclopride. In any case, this finding has implications for the authors' correction method. If the results shown here hold up (i.e., if all mass–response curves for BP go to zero at large masses, even with NS binding), then the prohibition on using a large mass dose to extrapolate to zero is removed (see case 2 in Section II). Of course there may continue to be pharmacological reasons for avoiding large masses of tracer when the ligand, such as raclopride, is a psychoactive substance.

H. Conclusion

Using simulated time–activity curves, this chapter demonstrated significant effects of mass on estimates of binding potential at reasonable mass doses to non-human primates. As PET protocols are often prototyped in monkeys and basic science studies are often performed exclusively in monkeys, it is important not to mistake artifacts of mass for important findings. The same artifacts may occur in humans, most probably in tissues with low receptor density. At a minimum, these simulation results underscore the importance of uniform administration of ligand by mass. A two-pont extrapolation method has been proposed and validated that corrects for the effects of mass based on a model

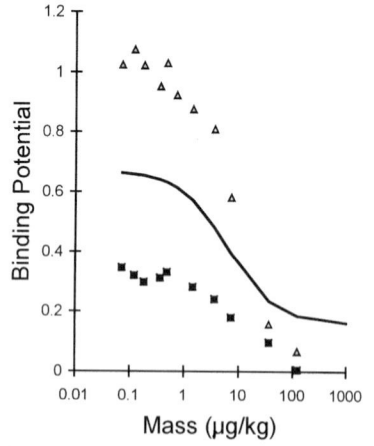

FIGURE 6 Mass–response curves for simulated data with NS = 0 (top), fitted curve from rat data of Hume *et al.* (1995) (middle), and simulations with fast NS binding (bottom).

of BP given by Hume *et al.* (1995). The correction performs better with graphical than with kinetic estimates of BP. Nonspecific binding will bias the corrected value for BP but, contrary to Hume, the authors' simulations suggest that any two estimates of BP, regardless of their injected masses, can be used for the extrapolation.

Acknowledgments

The authors would like to acknowledge their collaborators at Johns Hopkins University PET lab and at the Neuroscience laboratory of the National Institute on Aging for their help in generating the experimental PET data shown in Figure 1 of this paper.

References

Carson, R. E., Breier, A., de Bartolomeis, A., Saunders, R. C., Su, T. P., Schmall, B., Der, M. G., Pickar, D., and Eckelman, W. C. (1997). Quantification of amphetamine-induced changes in [^{11}C]raclopride binding with continuous infusion. *J. Cereb. Blood Flow Metab.* **17**: 437–447.

Cunningham, V. J., Hume, S. P., Price, G. R., Ahier, R. G., Cremer, J. E., and Jones, A. K. P. (1991). Compartmental analysis of diprenorphine binding to opiate receptors in the rat in vivo and its comparison with equilibrium data in vitro. *J. Cereb. Blood Flow Metab.* **11**: 1–9.

Delforge, J., Syrota, A., and Mazoyer, B. M. (1990). Identifiability analysis and parameter estimation of an in vivo ligand-receptor model from PET data. *IEEE Trans. Biomed. Eng.* **37**: 653–661.

Dewey, S. L., Smith, G. S., Logan, J., Brodie, J. D., Simkowitz, P., MacGregor, R., Fowler, J. S., Volkow, N. D., and Wolf, A. P. (1993). Effects of central cholinergic blockade on striatal dopamine release measured with positron emission tomography in normal human subjects. *Proc. Natl. Acad. Sci. U.S.A.* **90**: 11816–11820.

Dewey, S. L., Smith, G. S., Logan, J., Alexoff, D., Ding, Y.-S., King, P., Pappas, N., Brodie, J. D., and Ashby, C. R., Jr. (1995). Serotonergic modulation of striatal dopamine measured with positron emission tomography (PET) and in vivo microdialysis. *J. Neurosci.* **15**: 821–829.

Hsu, H., Alpert, N. M., Christian, B. T., Bonab, A. A., Morris, E. D., and Fischman, A. J. (1997). Noise properties of a graphical assay of receptor binding. *J. Nucl. Med.* **38**: 204P.

Hume, S. P., Opacka-Juffry, J., Myers, R., Ahier, R. G., Ashworth, S., Brooks, D. J., and Lammertsma, A. A. (1995). Effect of L-DOPA and 6-hydroxydopamine lesioning on ^{11}C-Raclopride binding in rat striatum, quantified using PET. *Synapse* **21**: 45–53.

Lammertsma, A. A., Bench, C. J., Hume, S. P., Osman, S., Gunn, K., Brooks, D. J., and Frackowiak, R. S. J. (1996). Comparison of methods of analysis of clinical [^{11}C]raclopride studies. *J. Cereb. Blood Flow Metab.* **16**: 42–52.

Laruelle, M., Iyer, R. N., Al-Tikriti, M. S., Zea-Ponce, Y., Malison, R., Zoghbi, S. S., Baldwin, R. M., Kung, H. F., Charney, D. S., Hoffer, P. B., Innis, R. B., and Bradberry, C. W. (1997). Microdialysis and SPECT measurements of amphetamine-induced dopamine release in non-human primates. *Synapse* **25**: 1–14.

Logan, J., Fowler, J. S., Volkow, N. D., Wolf, A. P., Dewey, S. L., Schlyer, D. J., MacGregor, R. R., Hitzemann, R., Bendriem, B., Gatley, S. J., and Christman, D. R. (1990). Graphical analysis of reversible radioligand binding from time-activity measurements applied to [N-^{11}C-methyl]-(-)-cocaine PET studies in human subjects. *J. Cereb. Blood Flow Metab.* **10**: 740–747.

Logan, J., Fowler, J. S., Volkow, N. D., Wang, G.-J., Ding, Y.-S., and Alexoff, D. L. (1996). Distribution volume ratios without blood sampling from graphical analysis of PET data. *J. Cereb. Blood Flow Metab.* **16**: 834–840.

Mintun, M. A., Raichle, M. E., Kilbourn, M. R., Wooten, G. F., and Welch, M. J. (1984). A quantitative model for the in vivo assessment of drug binding sites with positron emission tomography. *Ann. Neurol.* **15**: 217–227.

Morris, E. D., Fischman, A. J., and Alpert, N. M. (1996). Comparison of two compartmental models for describing receptor ligand kinetics and receptor availability in multiple injection PET studies. *J. Cereb. Blood Flow Metab.* **16**: 841–853.

Morris, E. D., Chefer, S. I., Lane, M. A., Muzic, R. F., Wong, D. F., Dannals, R. F., Matochik, J. A., Bonab, A. A., Villemagne, V. L., Grant, S. J., Ingram, D. K., Roth, G. S., and London, E. D. (1998). Loss of D2 receptor binding with age in rhesus monkeys: Importance of correction for differences in striatal size. Submitted for publication.

CHAPTER 62

Estimation of Nonspecific Binding of [^{18}F]Setoperone, a 5HT$_{2A}$ Receptor PET Radioligand, from Saturation Kinetic Data in Baboon and Human Neocortex

M. C. PETIT-TABOUÉ,[*,†,‡] B. LANDEAU,[*] A. R. YOUNG,[*] P. SCHUMANN,[¶] L. BESRET,[*,§]
M. IBAZIZENE,[§] and J. C. BARON[*,†]

[*] INSERM U320, [†] Cyceron, [‡] Centre Baclesse, [¶] CNRS UMR 6551
University of Caen
[§] CEA DSV/DRM, Caen, France

In most positron emission tomography (PET) radioligand models, it is assumed, but rarely proven, that nonspecific (NS) binding in the target structure is equal to that in the reference structure. [^{18}F]Setoperone is a well-validated and widely used PET radioligand for the study of neocortical 5HT$_{2A}$ receptor (5HT$_{2A}$ R). Single-experiment modeling to quantitate neocortical 5HT$_{2A}$ R binding potential (or k_3/k_4) used the cerebellum as reference structure because it is virtually devoid of 5HT$_{2A}$ R. Previous studies supported the use of cerebellum as reference in both the baboon and the human, but a quantitative estimation of NS binding was not carried out. Thus the aim of this study was to estimate NS binding directly in both species. Four young healthy volunteers were studied before and after 2 weeks of daily doses of an atypical neuroleptic possessing strong 5HT$_{2A}$ R antagonist activity. Presaturation PET experiments were carried out in three baboons 10 min after injection of a large amount of cold setoperone. In both species, visual inspection of the time–activity curves (TACs) indicated full saturation of specific binding in the neocortex without effect on the cerebellum, but neocortical TACs fell below cerebellum TACs after ~ 20 min in humans. Cerebellum TACs were fitted (3-Cpt model), providing the k_5/k_6 ratio and the f_2 fraction. The control neocortical TACs were fitted (4-Cpt model) assuming NS binding similar to cerebellum, providing k_3, k_4, and the k_3/k_4 ratio, whereas treated neocortical TACs could be fitted only with a 3-Cpt model (i.e., $k_3/k_4 = 0$). NS binding for the neocortex in the treated condition, expressed as the f_2 fraction, was significantly larger (i.e., NS binding was smaller) than that estimated for the cerebel-lum in human but not in baboon studies. Modeling of control human studies showed that the use of the cerebellum as reference would result in significant underestimations of the neocortical k_3/k_4 ratio by about 30%. This study suggests that NS binding is uniform in baboons, but is lower in the neocortex compared to the cerebellum in humans. Based on these data however, a systematic correction factor may be applied in future human studies as a convenient way to correct for this uneven distribution in NS binding.

I. INTRODUCTION

In most positron emission tomography (PET) radioligand compartmental models, it is assumed, but rarely proven, that nonspecific (NS) binding in the target structure is equal to that in the reference structure. [^{18}F]Setoperone is a sensitive and widely used PET radioligand for the study of the neocortical 5HT$_{2A}$ receptor (5HT$_{2A}$ R) (Blin et al., 1988, 1990, 1993; Bin and Crouzel, 1992; Chabriat et al., 1995; Fischman et al., 1996). It has been demonstrated previously (Petit-Taboué et al., 1996) that metabolite-corrected, [^{18}F]setoperone single experiment kinetic PET data allows quantification of the binding potential (k_3/k_4) to the neocortical 5HT$_{2A}$ R by means of multicompartmental modeling of cerebellum (3-Cpt model) and neocortex (4-Cpt model). In these studies, the cerebellum (Cb), a structure virtually devoid of 5HT$_{2A}$ R (Schotte *et al.*, 1983), was used as a reference to estimate free and NS binding in the neocortex (NCx). Previous presaturation studies (Blin *et al.*, 1988) supported the use of the cerebellum as reference structure in the baboon. Formal documentation of similar NS binding in neocortex and cerebellum, however, was not provided. Because comparable studies regarding [^{18}F]setoperone in humans are seldom possible because of the risk of pharmacological effects (Mintun *et al.*, 1984), the single available report by Blin *et al.* (1990) provided less definite evidence for uniform NS binding.

The aim of this study was therefore to reevaluate the issue of NS binding of [^{18}F]setoperone in the neocortex of both baboons and humans.

II. MATERIALS AND METHODS

A. PET Experiments

1. Human Studies

After written informed consent, four young healthy volunteers (30.5 ± 5 years, mean ± SD) were studied by PET with [^{18}F]setoperone (synthesized using the method described by Crouzel *et al.*, 1988), before ("control") and after ("challenge") 2 weeks of daily oral treatment with therapeutic doses of an atypical neuroleptic possessing strong 5HT$_{2A}$ R antagonist activity and anticipated to induce substantial occupancy of the neocortical 5HT$_{2A}$ R. Magnetic resonance (MR) images (SPGR sequence) were acquired in each subject using a 1.5-T General Electric device.

After establishing the presence of satisfactory collateral circulation, a thin radial catheter was inserted into the radial artery under xylocaine local anesthesia. The positioning and movement of the patient's head were controlled and prevented by the use of Laitinen's stereotaxic frame (Issal Surgical Instruments, Stockholm). Using a method based on a lateral skull X-ray and derived from that of Fox *et al.* (1985), the subjects were positioned with reference and parallel to the glabella-inion line. In both PET examinations, tracer amounts of [^{18}F]setoperone at high specific radioactivity (SA = 3.92 ± 1.02 Ci/μmol, mean ± SD) were injected intravenously as a bolus (6.73 ± 0.93 mCi, mean ± SD for the eight studies).

2. Baboon Studies

Presaturation ("challenge") experiments, involving the intravenous slow administration of cold setoperone (10 mg) prior to a 5-mCi [^{18}F]setoperone injection, were carried out in three young adult baboons (*Papio anubis*) with the following protocol, chosen for its lack of interference with 5HT$_2$ binding, based on *in vitro* data (Leysen *et al.*, 1981, and personal communication). Initial sedation with the short-acting barbiturate methohexital (20 mg/kg im) was followed by an intravenous injection of etomidate (3 mg/kg), potentiated by clonidine (75 μg iv over 10 min); muscular relaxation was achieved with atracurium (Tacurarium, 0.5 mg/kg). Anesthesia was subsequently maintained with etomidate (0.3 mg/kg/hr, iv infusion), atracurium (0.75 mg/kg/hr, iv infusion), and positive pressure ventilation with a N$_2$O–O mixture (2:1). No significant physiological effect of cold setoperone was observed in these studies.

Percutaneous femoral arterial catheters were inserted for radioactivity measurement, determination of the physiological parameters, and monitoring of arterial pressure. The body temperature was kept within normal limits with heating blankets. The baboons were positioned in order to obtain seven planes parallel to the cantho-meatal line from −27 mm to +45 mm relative to this line. Magnetic resonance (MR) images were acquired in each animal.

3. PET Measurements

PET data were acquired with the seven-slice LETI TTV03 time-of-flight camera (intrinsic resolution: 5.5 × 5.5 × 9 mm, x, y, z). After a transmission scan with ^{68}Ge, the PET acquisition with list mode started from the beginning of the bolus injection (T_0) and lasted 120 min. Data were reconstructed with the following image frame sequences: 9 × 16 sec, 5 × 30 sec, 5 × 60 sec, 6 × 300 sec, and 4 × 1200 sec. Initial (0–2 min) 2-sec head counts were also recorded (Miyazawa *et al.*, 1993) in order to fit the brain-to-arterial shift of [^{18}F]setoperone rise (Mazoyer *et al.*, 1986). PET data were corrected automatically for ^{18}F decay, and scatter was corrected empirically based on the technique of Bendriem *et al.* (1986).

4. Quantification of Arterial Radioactivity and ^{18}F-Labeled Metabolites

Arterial sampling began at the start of radiotracer injection; about 12 samples (∼ 1 ml) were withdrawn in the first minute and about 10 samples (∼ 1 ml) at increasing intervals until the end of the study. Radioactivity in whole blood and total plasma, corrected for ^{18}F decay, was measured in units of nCi/ml by means of a gamma counter (Packard, France) cross-calibrated with the PET camera.

The plasma input function was corrected for the presence of ^{18}F metabolites (Petit-Taboué *et al.*, 1996) as estimated by thin-layer radiochromatography in each subject.

B. PET Data Analysis

1. Time–Activity Curves for Cerebellum and Neocortex

In human experiments, the two PET studies and the coregistered MR images were realigned in 3D using the method of Woods *et al.* (1993). Neocortical ($n = 90$) and cerebellar ($n = 5$) circular regions of interest (ROIs) (14 mm diameter) were defined anatomically on the individual MR images of each subject, resliced according to seven PET planes, with the help of the stereotaxic atlas of the human brain of Talairach and Tournoux (1988).

The radioactivity concentration in these regions (weighted mean of all voxels in each region) was calculated for each sequential scan and plotted versus time, using dedicated software (MIRIAM).

In the baboon studies, 30 circular ROIs (diameter = 1 cm) were defined anatomically on the coregistered T1 MR images following the same procedure, using the PET–MR atlas of Riche *et al.* (1988). Two cerebellar ROIs were placed on the plane CM + 9 mm. Whole cerebral cortex kinetic data were taken as an average of 28 ROIs of the same shape and size on each side of the brain and placed over three adjacent PET planes (CM + 9 mm, CM + 21 mm, and CM + 33 mm).

2. Kinetic Modeling

Kinetic modeling was carried out according to a procedure described previously (Petit-Taboué *et al.*, 1996). In both control and challenge studies, the kinetics of [^{18}F]setoperone in cerebellum and whole neocortex were modeled according to a 3-Cpt and a 4-Cpt model, respectively, whereas a 3-Cpt model was also applied to the neocortex from challenge studies to assess NS binding directly (Fig. 1); the rate constants are K_1 (ml/min/g) and k_2 for transport into and out of free ligand compartment, k_3 and k_4 for binding to and dissociation from the 5HT$_{2A}$ R, and k_5 and k_6 for binding to and dissociation from the NS compartment (k_2–k_6, min^{-1}).

a. Cerebellum

Nonlinear least-squares (NLSQ) 3-Cpt modeling was used to estimate the exchange rates among plasma, free ligand, and NS binding in the cerebellum for both control and challenge studies, V_f, which represents the vascular fraction in the block of tissue under consideration, was determined by fitting. The fraction f_2 calculated as $1/(1 + k_5/k_6)$, represents the fraction of the ligand available for binding to the receptors (Mazoyer, 1991) and, therefore, also the NS binding.

b. Neocortex

Neocortical time–activity curves (TACs) were fitted with NLSQ and a 4-Cpt model in control studies (i.e., taking the cerebellum as reference to estimate NS binding) and with either a 3- or 4-Cpt model in challenge studies. With the 4-Cpt model, the initial starting value for K_1 was that estimated for the cerebellum, whereas values of 0.20 and 0.09 were used for k_3 and k_4, respectively. K_1/k_2, k_5, and k_6 were fixed at cerebellar values. The neocortical binding potential was calculated as the k_3/k_4 ratio (Mintun *et al.*, 1984). With 3-Cpt modeling (challenge studies only), K_1 and k_2, k_5 and k_6 were directly estimated, allowing to directly calculate a f_2 fraction for the neocortex.

3. Ratios

The neocortical to cerebellar ratios, which are assumed to provide a semiquantitative index of the binding potential, were also obtained for control and challenge experiments to calculate occupancies.

III. RESULTS

A. Control Studies

In both species, and as illustrated by Fig. 2A for a human study, both cerebellar and neocortical control TACs showed typical profiles of [^{18}F]setoperone kinetics: a rapid rise of radioactivity followed by a rapid clearance in the cerebellum and a quasi-plateau in the cortex. As in previous studies (Petit-Taboué *et al.*, 1996), and in both species, cerebellum and neocortical TACs were fitted adequately with a 3-Cpt and a 4-Cpt model, respectively. Values of the rate constants, as well as the f_2 fraction and the binding potential estimates, were within the same range as reported previously (Petit-Taboué *et al.*, 1995; 1996).

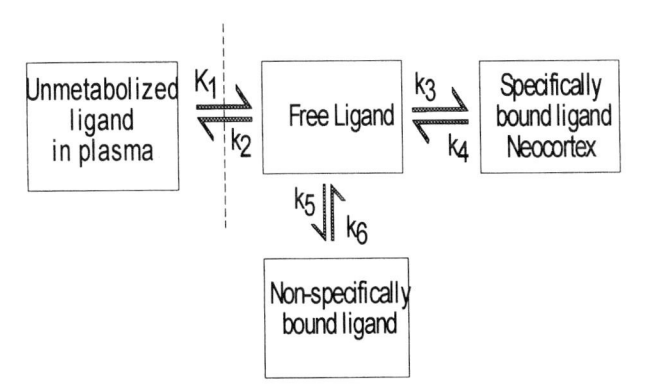

FIGURE 1 Full four-compartment model describing [^{18}F]setoperone behavior in the brain, in which the exchanges among the four compartments are described by rate constants K_1 and k_2 to k_6.

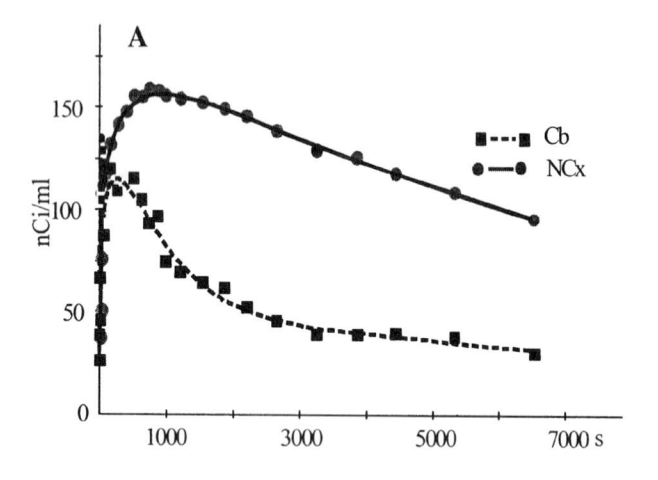

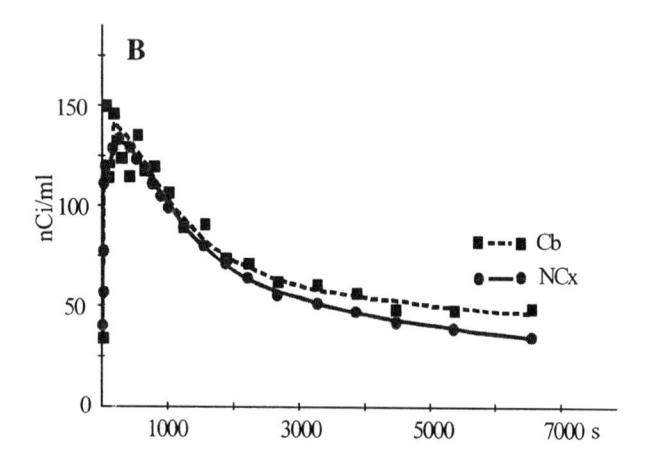

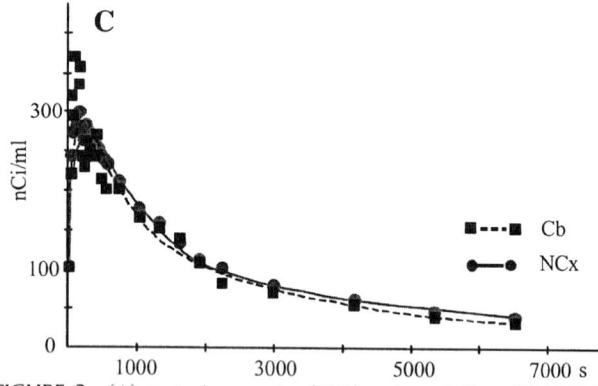

FIGURE 2 (A) control neocortex (NCx) and cerebellum (Cb) TACs in a human subject. (B) Presaturation NCx and Cb TACs in a human subject. (C) Presaturation NCx and Cb TACs in a baboon.

B. Challenge Studies

1. TACs

In both species, the distribution of [^{18}F]setoperone uptake (Figs. 2B and 2C) demonstrated a marked decrease in the cortical accumulation of radioactivity as compared to the control study. In both humans and baboons, no effect of treatment was observed in cerebellum TACs. In humans, neocortical TACs (Fig. 2B) fell progressively below that of the cerebellum, suggesting lower NS binding in the former. Accordingly, receptor occupancy calculated by the ratio method was 112 ± 12% (mean ± SD).

In baboons, however, neocortical TACs were almost exactly superimposable on cerebellum TACs, suggesting full 5HT$_{2A}$ R saturation but with similar NS binding in both structures (Fig. 2C).

Neither the plasma ^{18}F kinetics nor the percentage of metabolism was affected by the treatment.

2. Modeling Results

In both species, neocortical data could be fitted according to a 3-Cpt model (i.e., with a k_3/k_4 ratio equal to 0), but not to a 4-Cpt model, indicating the lack of a specific binding component. Thus, it was possible to evaluate directly K_1, k_2, k_5, and k_6 and, in turn, the f_2 fraction for the neocortex.

In humans, the f_2 fraction for the neocortex was a factor of 1.2 larger than for the cerebellum (0.493 ± 0.047 and 0.575 ± 0.039, respectively, $p < 0.05$ by oaired t-test). Regarding the cerebellum, the f_2 fraction did not differ significantly between the control and the treatment studies (0.452 ± 0.116 and 0.496 ± 0.048, mean ± SD, respectively), confirming the absence of specific binding of [^{18}F]setoperone in this structure.

In baboons, the neocortical f_2 fraction estimated after blocking of the 5HT$_{2A}$ R was not significantly different from that found in the cerebellum (Table 1). These values were in the same range as those reported previously with single experiment studies in the same species and the same modeling approach (Petit-Taboué et al., 1996).

For both species, the free volume of distribution (i.e., K_1/k_2) in treated conditions did not differ between the neocortex and the cerebellum and was similar to that found in the control studies.

IV. DISCUSSION

In both human and baboon experiments, the cerebellum has up until now served as a reference structure to estimate NS binding of [^{18}F]setoperone in the neocortex, be it with NLSQ fitting, Patlak-Logan graphical

TABLE 1 Estimated f_2 Fraction for the Cerebellum and the Neocortex in the Challenge Condition in Baboon Experiments

Subject	Cb	NCx
1	0.435	0.548
2	0.628	0.649
3	0.614	0.572
Mean ± SD	0.559 ± 0.108	0.589 ± 0.053

TABLE 2 k_3/k_4 Ratios for Whole Neocortex

Subjects	k_3/k_4 corrected[a]
1	2.491
2	3.228
3	1.945
4	2.497
Mean ± SD	2.541 ± 0.526

[a] k_3/k_4 ratio for the neocortex using the challenge NCx-fitted rate constants (corrected).

analysis, or simple pseudo-equilibrium (i.e., NCx/Cb ratio) approaches (Chabriat et al., 1995; Petit-Taboué et al., 1996; Fischman et al., 1996).

Previously, the kinetics of [^{18}F]setoperone in the cerebellum of the baboon suggested a lack of specific binding in this structure, which was supported by pretreatment studies with spiperone or ketanserin (Blin et al., 1988). In support of this earlier semiquantitative report, the present study documented that NS binding is similar in cerebellum and neocortex in baboons, and thus that the cerebellum is an adequate reference structure for [^{18}F]setoperone studies in this species.

However, the present findings suggest that this does not equally apply to young healthy human subjects. To estimate the effects of the use of the cerebellum in estimating the binding potential of [^{18}F]setoperone in the neocortex, neocortical TACs from the control condition in human experiments were fitted using as initial K_1, K_1/k_2 ratio, and k_5, k_6 values those found in the treated conditions for the same structure. Compared to these corrected values, the use of the cerebellum as a reference structure entails an underestimation of the k_3/k_4 ratio in the neocortex of about 30% (see Table 2). These results are consistent with the notion that an overestimation of the NS fraction results in an underestimation of B_{max}/K_d (Lassen et al., 1995). Differences in NS fraction between the reference and the target structure have also been suggested with [^{11}C]raclopride (Seeman et al., 1990), with NS binding 35% higher in the striatum compared to the cerebellum, and with [^{18}F]fluoroethylspiperone (Bahn et al., 1989), where the use of the cerebellum as a reference structure led to an underestimation of the rate constant k_3 for the striatum.

Both in humans and in baboons, the volume of distribution (i.e., K_1/k_2) of [^{18}F]setoperone was identical in the cerebellum and in the neocortex, and was not affected by treatment. With other ligands such as [^{18}F]spiperone (Logan et al., 1987), [^{11}C]carfentanil (Frost et al., 1989), and [^{11}C]diprenorphine (Sadzot et al., 1991), an increase of k_2, without a change of K_1, was observed after presaturation.

Such differences in NS binding between the neocortex and the cerebellum are important in calculations of [^{18}F]setoperone of specific binding. Based on data from this study, correcting the Cb by a fraction of 1.2 may be applied systematically in future human studies as a convenient way to correct for this uneven distribution in NS binding of [^{18}F]setoperone.

V. CONCLUSION

This study has validated the use of the cerebellum as a reference structure for estimating [^{18}F]setoperone specific binding in the baboon. Although the lack of specific binding in the cerebellum of human subjects has been documented, this study also suggested a difference in the NS binding fraction in the cerebellum and neocortex, which does not preclude the use of the cerebellum as a reference structure, but needs to be taken into account in future studies. Finally, this study showed that it is important to estimate NS binding directly in at least a few subjects in the development of a clinical PET radioligand.

References

Bahn, M. M., Huang, S. C., Hawkins, R. A., Satyamurthy, N., Hoffman, J. M., Barrio, J. R., Mazziota, J. C., and Phelps, M. E. (1989). Models for in vivo kinetic interactions of dopamine D2-neuroreceptors and 3-(2'-^{18}F fluoroethyl)spiperone examined with positron emission tomography. J. Cereb. Blood Flow Metab. 9: 840–849.

Bendriem, B., Soussaline, F., Campagnolo, R., Verrey, B., Wajnberg, P., and Syrota, A. (1986). A technique for the correction of scattered radiation in a PET system using time of flight information. J. Comput. Assist. Tomogr. 10: 287–295.

Blin, J., and Crouzel, C. (1992). Blood-cerebrospinal fluid and blood-brain barriers imaged by ^{18}F-labeled metabolites of [^{18}F]setoperone studied in humans using positron emission tomography. J. Neurochem. 58: 2303–2310.

Blin, J., Pappata, S., Kiosawa, M., Crouzel, C., and Baron, J. C. (1988). [^{18}F]setoperone: A new affinity ligand for positron emission tomography study of the serotonin-2 receptors in baboon brain in vivo. *Eur. J. Pharmacol.* **147**: 73–82.

Blin, J., Sette, G., Fiorelli, M., Bletry, O., Elghozi, J. L., Crouzel, C., and Baron, J. C. (1990). A method for the in vivo investigation of the serotoninergic 5-HT$_2$ receptors in the human cerebral cortex using positron emission tomography and [^{18}F]labeled setoperone. *J. Neurochem.* **54**: 1744–1754.

Blin, J., Baron, J. C., Dubois, B., Crouzel, C., Fiorelli, M., Attar-Lévy, D., Pillon, B., Fournier, D., Vidailhet, M., and Agid, Y. (1993). Loss of brain 5-HT$_2$ receptors in Alzheimer's disease. *Brain* **116**: 497–510.

Chabriat, H., Tehindrazanarivelo, A., Vera, P., Samson, Y., Pappata, S., Boullais, N., and Bousser, M. G., (1995). 5HT$_2$ receptors in cerebral cortex of migraineurs studied using PET and [^{18}F]fluorosetoperone. *Cephalalgia* (Oslo) **15**: 104–108.

Crouzel, C., Venet, M., Irie, T., Sanz, G., and Boullais, C. (1988). Labelling of serotoninergic ligand with ^{18}F: [^{18}F]setoperone. *J. Labelled Compd. Radiopharm.* **25**: 403–414.

Fischman, A. J., Bonad, A. A., Babich, J. W., Alpert, M., Raucu, S. L., Elmaleh, D. R., Shoup, T. M., Williams, S. A., and Rubin, R. H. (1996). Positron emission tomography analysis of central 5-hydroxytryptamine2 receptor occupancy in healthy volunteers treated with the novel antipsychotic agent ziprasidone. *J. Pharmacol. Exp. Ther.* **279**: 939–947.

Fox, P. T., Perlmutter, J. S., and Raichle, M. E. (1985). A stereotactic method of anatomical localization for positron emission tomography. *J. Comput. Assist. Tomogr.* **9**: 141–153.

Frost, J. J., Douglass, K. H., Mayberg, H. S., Dannals, R. F., Links, J. M., Wilson, A. A., Ravert, H. T., Crozer, W. C., and Wagner, H. N. (1989). Multicompartmental analysis of 11C-carfentanil binding to opiate receptors in humans measured by positron emission tomography. *J. Cereb. Blood Flow Metab.* **9**: 398–409.

Lassen, N. A., Bartenstein, P. A., Lammertsma, A. A., Prevett, M. C., Turton, D. R., Luthra, S. K., Osman, S., Bloomfield, P. M., Jones, T., Patsalos, P. N., O'Connell, M. T., Duncan, J. S., and Vanggaard Andersen, J. (1995). Benzodiazepine receptor quantification in vivo in humans using [^{11}C]flumazenil and PET: Application of the steady-state principle. *J. Cereb. Blood Flow Metab.* **15**: 152–165.

Leysen, J. E., Niemegeers, C. J. E., Van Nueten, J. M., and Laduron, P. M. (1981). [^{3}H]Ketanserin (R 41 468), a selective ^{3}H-ligand for serotonin binding sites. *Mol. Pharmacol.* **21**: 301–314.

Logan, J., Wolf, A. P., Shiue, C. Y., and Foxler, J. S. (1987). Kinetic modeling of receptor-ligand binding applied to positron emission tomography studies with neuroleptic tracers. *J. Neurochem.* **48**: 73–87.

Mazoyer, B. (1991). Investigation of the dopamine system with positron emission tomography: General issues in modelling. *In* "Brain Dopaminergic Systems: Imaging with Positron Tomography" (J. C. Baron, D. Comar, L. Farde, J. L. Martinot, and B. Mazoyer, eds.), pp. 65–83. Kluwer Academic Publishers, Dordrecht, The Netherlands.

Mazoyer, B. M., Huesman, R. H., Budinger, T. F., and Knittel, B. L. (1986). Dynamic PET data analysis. *J. Comput. Assist. Tomogr.* **10**: 646–653.

Mintun, M. A., Raichle, M. E., Kilbourn, M. R., Wooten, G. F., and Welch, M. J. (1984). A quantitative model for the in vivo assessment of drug binding sites with positron emission tomography. *Ann. Neurol.* **15**: 217–227.

Miyazawa, H., Osmont, A., Petit-Taboué, M. C., Tillet, I., Travere, J. M., Young, A. R., Barre, L., MacKenzie, E. T., and Baron, J. C. (1993). Determination of [^{18}F]FDG brain kinetic constants in the anesthetized baboon. *J. Neurosci. Methods* **50**: 263–272.

Petit-Taboué, M. C., Landeau, B., Osmont, A., Barré, L., and Baron, J. C. (1995). Determination of human cortical 5HT$_2$ receptor binding from [^{18}F]setoperone, kinetic PET data: Comparison of different methods. *J. Cereb. Blood Flow Metab.* **15**(Suppl. 1): S641.

Petit-Taboué, M. C., Landeau, B., Osmont, A., Tillet, I., Barré, L., and Baron, J. C. (1996). Estimation of neocortical serotonin-2 receptor binding potential by single-dose fluorine-18-setoperone kinetic PET data analysis. *J. Nucl. Med.* **37**: 95–104.

Riche, D., Hantraye, P., Guibert, B., Naquet, R., Loc'h, C., Mazière, B., and Mazière, M. (1988). Anatomical atlas of the baboon's brain in the orboto-meatal plane used in experimental positron emission tomography. *Brain Res. Bull.* **20**: 283–301.

Sadzot, B., Price, J. C., Mayberg, H. S., Douglass, K. H., Dannals, R. F., Lever, J. R., Ravert, H. T., Wilson, A. A., Wagner, H. N., Feldman, M. A., and Frost, J. J. (1991). Quantification of human opiate receptor concentration and affinity using high and low specific activity 11C-diprenorphine and positron emission tomography. *J. Cereb. Blood Flow Metab.* **11**: 204–219.

Schotte, A., Maloteaux, J. M., and Laudron, P. M. (1983). Characterization and regional distribution of serotonin S2-receptors in human brain. *Brain Res.* **276**: 231–235.

Seeman, P., Niznik, H. B., and Guan, H. C. (1990). Elevation of dopamine D2 receptors in schizophrenia is underestimated by radioactive raclopride. *Arch. Gen. Psychiatry* **47**: 1170–1172.

Talairach, J., and Tournoux, P. "Co-Planar Stereotaxic Atlas of the Human Brain." Thieme, Stuttgart.

Woods, R. P., Mazziotta, J. C., and Cherry, S. R. (1993). MRI-PET registration with automated algorithm. *J. Comput. Assist. Tomogr.* **17**: 536–546.

CHAPTER

63

Estimation of Binding Potential for the 5HT$_2$ Receptor Ligand [^{18}F]Setoperone by a Noninvasive Reference Region Graphical Method

ALI A. BONAB, ALAN J. FISCHMAN, and NATHANIEL M. ALPERT

Division of Nuclear Medicine
Department of Radiology
Massachusetts General Hospital and Harvard Medical School
Boston, Massachusetts 02114

[^{18}F]Setoperone is an important positron emission tomography ligand for measuring concentrations and occupancies of neocortical 5HT$_2$ receptors. In this study, estimates of binding potential (BP, B'_{max}/K_D, k_3/k_4) calculated by nonlinear least-squares fitting (LSF) were compared with results of reference region graphical analysis (RRG). Results of 16 human studies were analyzed by both methods. With the LSF method, cerebellar time–activity curves (TAC's) were fitted to a three-compartment model to derive K_1 (transport from plasma to brain), k_2 (transport rate from brain to plasma), k_5 (association rate for nonspecific binding), and k_6 (dissociation rate for nonspecific binding). K_1/k_2, k_5, and k_6 were fixed at cerebellar estimates, and individual values for k_3 and k_4 were derived by fitting cortical TACs to a four-compartment model. Metabolite-corrected arterial input functions were used in all calculations, and the vascular fraction of brain tissue (V_b) was fixed at 4.5%. In the RRG method, nonspecific binding was assumed to be negligible in the cerebellum and neocortex, and BP was estimated from the cortical/cerebellar distribution volume ratio (DVR). This method does not require blood sampling and uses cerebellar TACs as input functions to calculate BP from cortical TACs. In all cases, BP determined by the RRG method was lower than the results of LSF. Despite this bias, BP determined by the two methods was highly correlated ($r^2 = 0.95$). Compared with LSF, the RRG method has several advantages, including (1) arterial blood sampling and metabolite corrections are not required and (2) the method is readily adaptable to pixel-by-pixel calculations of BP images that can be used for evaluating regional changes in 5HT$_2$ receptor density in clinical applications.

I. INTRODUCTION

Serotonin and its receptors are widely distributed in the central nervous system (CNS), and changes in serotonergic neurotransmission have been associated with a variety of neurological and psychiatric conditions, including schizophrenia, depression, affective disorders, infantile autism, mental retardation, hyperactivity syndromes, eating disorders, sleep disorders, suicidal behavior, Alzheimer's disease, Duchenne muscular dystrophy, Parkinson's disease, Huntington's chorea, and migraine headaches (Azmitia and Whitaker-Azmitia, 1991; Lucki, 1991, 1996; Jimerson et al., 1990; Brewerton, 1995; Mann et al., 1992; Stahl, 1977; Peroutka, 1993; Fuller, 1992). This association between serotonin and psychiatric disease has inspired the development of drugs that inhibit serotonin reuptake and modulate postsynaptic events for treating depression, schizophrenia, eating disorders, and a variety of other conditions (Stahl, 1977).

Traditionally, the primary methods for studying serotonergic effects noninvasively in humans were measurements of 5HT receptors on blood platelets (Pandey et al., 1995; Weizman et al., 1986; Hartig et al., 1985). Although these measurements have yielded significant pathophysiological information and guided the development of a variety of important drugs, they have provided little insights about the specific neural substrates for 5HT effects or the target of drug action. This picture has been changed drastically by the introduction of specific positron emission tomography (PET) and single photon emission computed tomography

(SPECT) tracers for mapping the distribution of 5HT receptors within the CNS.

Several ligands have been proposed for imaging postsynaptic serotonin receptors. In particular, iodo-, bromo-, and N-alkyl derivatives of lysergic acid diethylamide (LSD) have been shown to localize at $5HT_{1c}$ and $5HT_2$ receptors (Lever et al., 1991; Blin et al., 1990). The most promising ligands for PET imaging are ^{76}Br- and ^{11}C-labeled N-methyl-Br-LSD, [^{18}F]setoperone (Blin et al., 1990), and [^{18}F]altanserin (Lemaire et al., 1991).

[^{18}F]Setoperone has a high affinity and selectivity for the $5HT_2$ receptors and has been used successfully for the quantification of $5HT_2$ receptors in the neocortex of nonhuman primates and humans (Petit-Taboué et al., 1996; Fischman et al., 1996). Specific advantages of [^{18}F]setoperone include (1) equilibration, across the blood–brain barrier, is rapid and specific binding to neocortex is reversible (Blin and Crouzel, 1992; Blin et al., 1993); (2) lack of displaceable binding in the cerebellum allows this region to be used as a reference (Blin and Crouzel, 1992; Blin et al., 1993); and (3) metabolites are polar and do not cross the blood–brain barrier. Unfortunately, because [^{18}F]setoperone undergoes extensive in vivo metabolism, quantitative analysis of receptor concentrations requires arterial blood sampling and precise metabolite analysis. This chapter presents a reference region graphical analysis of [^{18}F]setoperone binding potential that does not require blood sampling or metabolite analysis.

II. MATERIALS AND METHODS

A. Radiopharmaceutical Preparation

[^{18}F]Setoperone was prepared by a modification of a method reported previously (Mazière et al., 1988; Fischman et al., 1996). The radiochemical yield was 20–25%, specific activity of the final product was 5000–10,000 mCi/μmol, and chemical and radiochemical purities were greater than 98%.

B. Human Subjects

Eight healthy male volunteers (ages 22–40) participated in the study. Each subject was studied twice: at baseline and after a single 40-mg oral dose of the neuroleptic ziprasidone (Seeger et al., 1993; Seymour et al., 1993; Fischman et al., 1996). The second scan was performed at 4, 8, 12, and 18 hr after the oral dose to detect changes in occupancy levels. The human study protocol was approved by the committees on human studies, pharmacy and radioisotopes of the Massachusetts General Hospital. All subjects gave written informed consent prior to participation in the study.

C. Positron Emission Tomography

Images were acquired with a PC-4096 PET camera (Scanditronix AB, Sweden). The primary imaging parameters of the PC-4096 camera are in-plane and axial resolutions of 6.0 mm FWHM, 15 contiguous slices of 6.5 mm separation, and a sensitivity of ~5000 cps/μCi. All images were reconstructed using a conventional filtered backprojection algorithm to an in-plane resolution of 7 mm FWHM. Attenuation correction was performed from transmission images acquired with a rotating pin source containing ^{68}Ge. All projection data were corrected for nonuniformity of detector response, dead time, random coincidences, and scattered radiation.

Prior to imaging, a venous catheter was placed in the right arm for radiopharmaceutical administration and a radial arterial catheter was placed in the left wrist for blood sampling. The subjects were positioned supine on the imaging bed of the PET camera with their arms extended out of the field of view and their heads immobilized with individually fabricated head holders (Tru Scan Image Inc., Annapolis, MD). Approximately 7 mCi of [^{18}F]setoperone was injected intravenously over 30 sec and serial PET images were acquired. Dynamic image collection was started at the same time as the infusion and images were acquired in 15-sec frames for the first 2.0 min and in 1-min frames for the remaining 88 min. Because of the limited field of view of the PET camera, images were acquired in two bed positions. Arterial blood samples (1 ml) were collected at 15-sec intervals for the first min, at 1.0-min intervals for the next 15 min, and at 5.0-min intervals for the remainder of the study. At 1, 5, 10, 30, 60, and 90 min, 5-ml blood samples were obtained for metabolite analysis.

Specific brain regions were selected for analysis. Three to five circular regions of interest (ROIs) of 10 mm diameter were placed on each side of the frontal, parietal, temporal, and occipital cortex and 10 reference ROIs were placed on each side of the cerebellum. This procedure was repeated for all slices, and the complete set of ROIs was replicated over all time frames. For each frame, the ROIs of like structures were averaged to yield cortical and cerebellar time–activity curves (TACs). The same set of ROIs was used to analyze each scan for a single subject; however, locations were adjusted to compensate for repositioning when necessary.

D. Kinetic Modeling

Figure 1 illustrates the kinetic model that was used to analyze the [^{18}F]setoperone PET data from either the cerebellum or the cerebral cortex. In this model, the tracer is considered to be in one of three states: (1) the free pool, consisting of a ligand that is instantly available to other processes or reactions; (2) a slowly associating and dissociating reversible, nonsaturable, nonspecific binding pool; or (3) bound to the receptor site. The transport rate of ligand from plasma to tissue is K_1 (ml/min/g). Ligand in the free pool may return to plasma, according to the rate constant k_2 (min^{-1}), or enter the nonspecific binding pool according to the rate constant k_5 (min^{-1}), or bind to the receptor site, according to the compound rate constant $k_3 = k_{on} \cdot B'_{max}$, (min^{-1}), where k_{on} is the equilibrium rate constant for association and B'_{max} is the receptor density. The ligand–receptor complex dissociates with a rate constant k_4 (k_{off}, min^{-1}). For cerebellum, k_3 and k_4 were assumed to be zero and this tissue is used as a reference to calculate k_5, k_6, and the ratio K_1/k_2. The binding potential (BP) for [^{18}F]setoperone was calculated as the ratio $k_3/k_4 = B'_{max}/K_D$. Calculation of BP was performed by two methods: nonlinear least-squares fitting (LSF) and reference region graphical analysis (RRG).

E. Nonlinear Least-Squares Fitting

Quantification of the binding potential of cortical 5HT$_2$ receptors for [^{18}F]setoperone was performed by modifications of a method described previously (Petit-Taboué et al., 1996). Individual values for the parameters K_1, k_2, k_5, and k_6 were estimated by fitting the three-compartment model shown in Fig. 1 to cerebellar TACs, using nonlinear least squares and the Marquardt algorithm. To determine binding potential, the parameters K_1/k_2, k_5, and k_6 were fixed at cerebellar values, V_b was fixed at 4.5%, and K_1, k_3, and k_4 were estimated by fitting the cortical TACs to the three-compartment model shown in Fig. 1. The ratio k_3/k_4 was calculated from the individual values of k_3 and k_4.

F. Reference Region Graphical Analysis

The distribution volume ratio (DVR) calculation proposed by Logan and others (Ichise et al., 1995; Logan et al., 1996) is an attractive method for analysis of the kinetics of reversible ligands without blood sampling. The only difficulty with this method is that it requires that nonspecific binding in the cerebellum and neocortex is negligible and can be ignored. To calculate DVR, the integrated tissue radioactivity [$A(t)$] from time zero to T normalized to tissue activity at time T is plotted versus integrated cerebellum time activity data [$CB(t)$], which is also normalized to tissue activity at time T [Eq. (1)]. This plot becomes linear and the asymptotic slope for the cortex data equals DVR. Under these circumstances, BP = DVR − 1.

$$\frac{\int_0^T A(t)\,dt}{A(T)} = \text{DVR}\,\frac{\int_0^T CB(t)\,dt}{A(T)} + \text{int}'. \quad (1)$$

III. RESULTS AND DISCUSSION

Figure 2 shows a representative reference region graphical plot for the cortex with the cerebellum as the reference region. Figure 3 shows examples of LSF

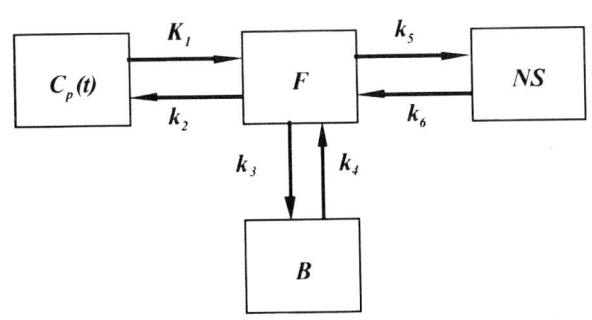

FIGURE 1 Three-compartment model used for analysis of [^{18}F]setoperone.

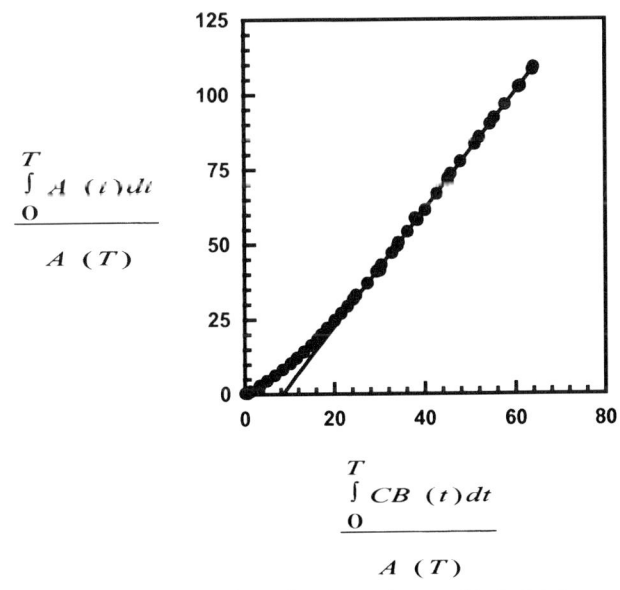

FIGURE 2 Representative reference region graphical plot for cortex with cerebellum as the reference region (slope = DVR).

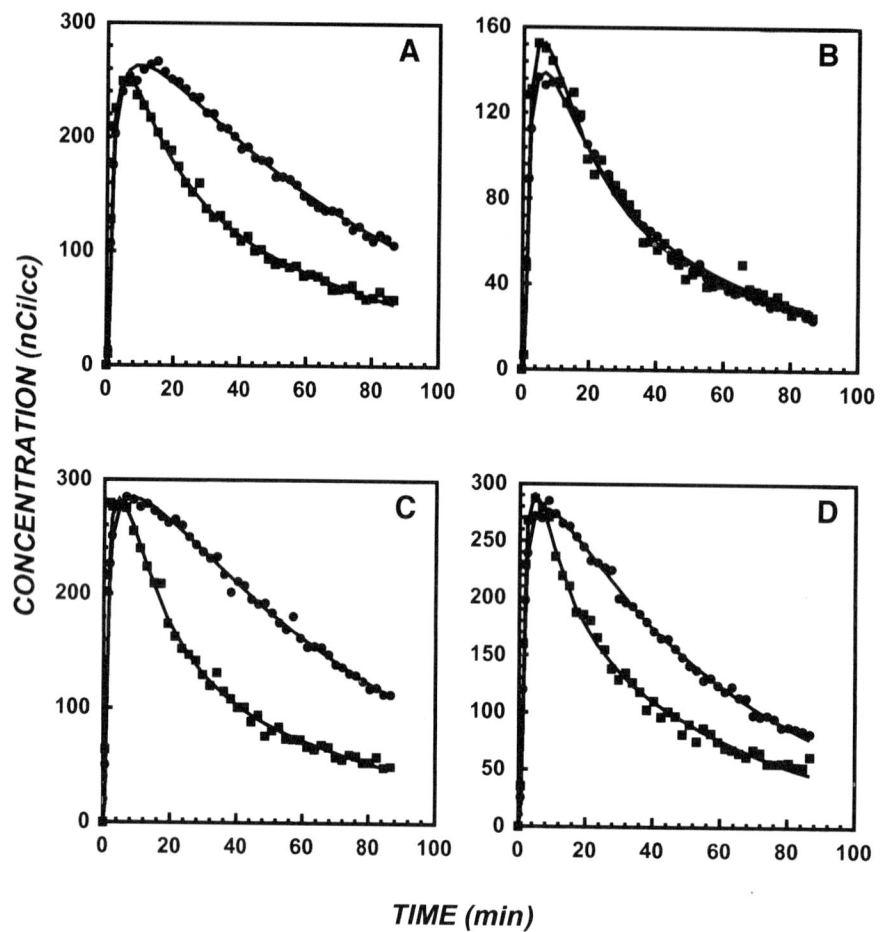

FIGURE 3 Examples of nonlinear least-squares fits of cerebellar and cortical PET data acquired at baseline (A and C) and at 4 (B) and 18 (D) hr after dosing with ziprasidone. Cerebellar and cortical data were fit to the two- and three-compartment models shown in Fig. 1. Cerebral cortical and cerebellar PET data are indicated by circles and squares, respectively.

results. Table 1 shows the individual values for BP derived by both methods of analysis and Fig. 4 shows a plot of BP calculated by fitting verses BP from RRG analysis. As Fig. 4 illustrates, BPs determined by the two methods were highly correlated ($R^2 = 0.95$).

In all cases (Table 1), BP determined by the RRG method was lower than the result obtained by fitting. This is most probably due to the fact that nonspecific binding is neglected in the RRG calculations. This fact is confirmed by the following calculations (Koeppe et al., 1991):

$$DV_{CER} = DV_F + DV_{NS} = DV_F(1 + k_5/k_6) \quad (2)$$

$$DV_{NC} = DV_F + DV_{NS} + DV_S$$
$$= DV_F(1 + k_5/k_6) + DV_S$$
$$= DV_{CER} + DV_F(k_3/k_4) \quad (3)$$

$$DVR = DV_{NC}/DV_{CER} = 1 + \frac{k_3/k_4}{1 + k_5/k_6} \quad (4)$$

$$BP = k_3/k_4 = (DVR - 1)(1 + k_5/k_6). \quad (5)$$

Equation (5) indicates that using the value of (DVR − 1) as BP is biased by the factor $(1 + k_5/k_6)$.

LSF has been used by many authors for the analysis of compartmental models and is considered to be the gold standard. Its advantages include minimum variance unbiased estimation of all the model parameters.

In practice, the LSF model requires accurate determination of the blood curve and metabolite correction. Errors in the input function propagate as errors in the estimation of the model parameters. LSF is impractical for parametric mapping when the number of parameters is greater than 2 or 3.

Compared with LSF, the RRG method has several important advantages, including (1) arterial blood sampling is not necessary; (2) because CNS TACs are used

63. Graphical Analysis of 5HT₂ Receptor-Binding Potential

TABLE 1 BP Estimating Using RRG and Nonlinear Least-Squares Fitting at Baseline (BL) and after Dosing with Ziprasidone in Hours [post (hr)]

No.	Drug status	RRG DVR-1	LSF k_3/k_4
1	BL	0.6961	2.320
1	post (4)	0.027	0.068
2	BL	0.8914	3.053
2	post (4)	0.0776	0.015
3	BL	1.3437	4.445
3	post (8)	0.1167	0.409
4	BL	0.9816	3.004
4	post (8)	0.1252	0.258
5	BL	2.0185	4.731
5	post (12)	0.3124	0.554
6	BL	0.6794	1.940
6	post (12)	0.2014	0.521
7	BL	2.2383	6.057
7	post (18)	1.2532	2.528
8	BL	0.9598	2.805
8	post (18)	0.4933	1.340

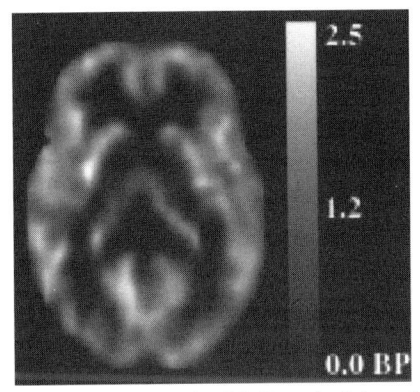

FIGURE 5 BP Image using the RRG method in a healthy volunteer.

in the calculation, metabolite corrections are not required; and (3) the method is readily adaptable to parametric mapping of BP, as illustrated by Fig. 5. These images are useful for detecting regional changes in receptor density or for defining regions of interest for the LSF method.

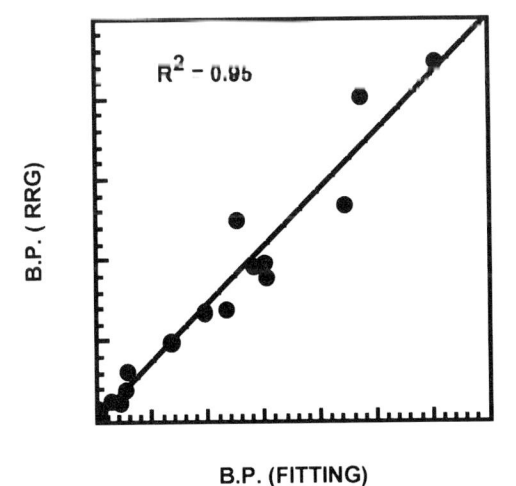

FIGURE 4 Correlation of BP from RRG and nonlinear least-squares fitting methods.

References

Azmitia, E. C., and Whitaker-Azmitia, P. M. (1991). Awakening the sleeping giant: Anatomy and plasticity of the brain serotonergic system. *J. Clin. Psychiatry* **52**(Suppl.): 4–16.

Blin, J., and Crouzel, C. (1992). Blood-cerebrospinal fluid and blood-brain barriers imaged by [¹⁸F] labeled metabolites of ¹⁸F-setoperone studied in humans using positron emission tomography. *J. Neurochem.* **58**: 2303–2310.

Blin, J., Sette, G., Fiorelli, M., Bletry, O., Elghozi, J. L., Crouzel, C., and Baron, J. C. (1990). A method for the in vivo investigation of the serotonergic 5-HT2 receptors in the human cerebral cortex using positron emission tomography and 18F-labeled setoperone. *J. Neurochem.* **54**: 1744–1754.

Blin, J., Baron, J. C., Dubois, B., Crouzel, C., Fiorelli, M., Attar-Levy, D., Pillon, B., Fournier, D., Vidailhe, M., and Agid, Y. (1993). Loss of brain 5-HT₂ receptors in Alzheimer's disease. In vivo assessment with positron emission tomography and [¹⁸F]setoperone. *Brain* **116**: 497–510.

Brewerton, T. D. (1995). Toward a unified theory of serotonin dysregulation in eating and related disorders. *Psychoneuroendocrinology* **20**: 561–590.

Fischman, A. J., Bonab, A. A., Babich, J. W., Alpert, N. M., Rauch, S. L., Elmaleh, D. R., Shoup, T. M., Williams, S. A., and Rubin, R. H. (1996). Positron emission tomographic analysis of central 5HT₂ receptor occupancy in healthy volunteers treated with the novel antipsychotic agent, ziprasidone. *J. Pharmacol. Exp. Ther.* **279**: 939–947.

Fuller, R. W. (1992). Basic advances in serotonin pharmacology. *J. Clin. Psychiatry* **53**(Suppl.): 36–45.

Hartig, P. R., Scheffel, U., Frost, J. J., and Wagner, H. N., Jr. (1985). In vivo binding of 125I-LSD to serotonin 5-HT2 receptors in mouse brain. *Life Sci.* **37**: 657–664.

Ichise, M., Ballinger, J. R., Golan, H., Vines, D., Luong, A., Tsai, S., and Kung, H. F. (1995). SPECT imaging of dopamine D2 receptors in humans with iodine 123-IBF: A practical approach to quantification not requiring blood sampling. *J. Nucl. Med.* **36**: 11P.

Jimerson, D. C., Lesem, M. D., Kaye, W. H., Hegg, A. P., and Brewerton, T. D. (1990). Eating disorders and depression: Is there a serotonin connection? *Biol. Psychiatry* **28**: 443–454.

Koeppe, R. A., Holthoff, V. A., Frey, K. A., Kilbourn, M. R., and Kuhl, D. E. (1991). Compartmental analysis of [11-C]Flumazenil kinetics for the estimation of ligand transport rate and receptor distribution using positron emission tomography. *J. Cereb. Blood Flow Metab.* **11**: 735–744.

Lemaire, C., Cantineau, R., Guillaume, M., Plenevaux, A., and Christiaens, L. (1991). Fluorine-18-altanserin: A radioligand for the study of serotonin receptors with PET: Radiolabeling and in vivo biologic behavior in rats. *J. Nucl. Med.* **32**: 2266–2272.

Lever, J. R., Scheffel, U. A., Musachio, J. L., Stathis, M., and Wagner, H. N., Jr. (1991). Radioiodinated D-(+)-N1-ethyl-2-iodolysergic acid diethylamide: A ligand for in vitro and in vivo studies of serotonin receptors. *Life Sci.* **48**: 73–78.

Logan, J., Fowler, J., Volkow, N. D., Wang, G., Ding, Y., and Alexoff, D. L. (1996). Distribution volume ratio without blood sampling from graphical analysis of PET data. *J. Cereb. Blood Flow Metab.* **16**: 834–840.

Lucki, I. (1991). Behavioral studies of serotonin receptor agonists as antidepressant drugs. *J. Clin. Psychiatry* **52**(Suppl.): 24–31.

Lucki, I. (1996). Serotonin receptor specificity in anxiety disorders. *J. Clin. Psychiatry* **57**(Suppl. 6): 5–10.

Mann, J. J., McBride, P. A., Brown, R. P., Linnoila, M., Leon, A. C., DeMeo, M., Mieczkowski, T., Myers, J. E., and Stanley, M. (1992). Relationships between central and peripheral serotonin indexes in depressed and suicidal psychiatric inpatients. *Arch. Gen. Psychiatry* **49**: 442–446.

Mazière, B., Crouzel, C., Venet, M., Stulzaft, O., Sanz, G., Ottaviani, M., Sejourne, C., Pascal, O., and Bisserbe, J. C. (1988). Synthesis, affinity, and specificity of ^{18}F-setoperone, a potential ligand for in-vivo imaging of cortical serotonin receptors. *Int. Radiat. Appl. Instrum. B* **15**: 463–468.

Pandey, G. N., Pandey, S. C., Dwivedi, Y., Sharma, R. P., Janicak, P. G., and Davis, J. M. (1995). Platelet serotonin-2A receptors: A potential biological marker for suicidal behavior. *Am. J. Psychiatry* **152**: 850–855.

Peroutka, S. J. (1993). 5-Hydroxytryptamine, receptor subtypes and the pharmacology of migraine. *Neurology* **43**(6)(Suppl. 3): S34–S38.

Petit-Taboué, M. C., Landeau, B., Osmont, A., Tillet, I., Barre, L., and Baron, J. C. (1996). Estimation of neocortical serotonin-2-receptor binding potential by single dose ^{18}F-setoperone kinetic PET data analysis. *J. Nucl. Med.* **37**: 95–104.

Seeger, T. F., Schmidt, A. W., Lebel, L. A., Koe, B. K., Zorn, S. H., and Schulz, D. (1993). CP-88,059, a new antipsychotic with mixed dopamine D_2 and serotonin $5HT_2$ antagonist activities. *Proc. Soc. Neurosci.*, p. 1623.

Seymour, P. A., Seeger, T. F., Guanowsky, V., Robinson, G. L., Howard, H., and Heym, J. (1993). Behavioral pharmacology of CP-88,059: A new antipsychotic with both 5HT2 and D2 antagonist activities. *Proc. Soc. Neurosci.*, p. 599.

Stahl, S. M. (1977). The human platelet. A diagnostic and research tool for the study of biogenic amines in psychiatric and neurologic disorders. *Arch. Gen. Psychiatry* **34**: 509–516.

Weizman, A., Carmi, M., Hermesh, H., Shahar, A., Apter, A., Tyano, S., and Rehavi, M. (1986). High-affinity imipramine binding and serotonin uptake in platelets of eight adolescent and ten adult obsessive-compulsive patients. *Am. J. Psychiatry* **143**: 335–339.

CHAPTER 64

[^{18}F]Altanserin PET Studies of Serotonin-2A Binding: Examination of Nonspecific Component

J. C. PRICE, B. LOPRESTI, N. S. MASON, Y. HUANG, D. HOLT, G. S. SMITH,* and C. A. MATHIS

*Departments of Radiology and *Psychiatry*
University of Pittsburgh School of Medicine
Pittsburgh, Pennsylvania 15213

Previous analyses of the regional brain kinetics of [^{18}F]altanserin ([^{18}F]ALT) in humans, using positron emission tomography (PET), indicated the existence of a slow nonspecific binding component. These analyses involved (1) compartmental models that did not account for blood–brain barrier (BBB) passage of radiolabeled metabolites and (2) metabolite models that did. Both analyses yielded regional binding parameters that were highly correlated with the known rank order of human serotonin-2A (5HT$_{2A}$) receptors. In this work, the slow nonspecific binding component was further examined by applying conventional and metabolite methods to PET studies of [^{18}F]ALT and two putative metabolites: [^{18}F]altanserinol ([^{18}F]ALT-ol) and [^{18}F]4-(4-fluorobenzoyl)piperidine ([^{18}F]4-FBP). High specific activity [^{18}F]ALT, [^{18}F]ALT-ol, and [^{18}F]4-FBP PET studies were performed in baboons at baseline and after 5HT$_{2A}$ antagonist pretreatment with ketanserin or SR 46349B (blocking data). Radioligand distribution volumes (DV) were estimated using model-based methods. Following [^{18}F]ALT-ol or [^{18}F]4-FBP injection, the radioactivity concentration in brain increased, indicating that [^{18}F]ALT-ol and [^{18}F]4-FBP (and/or their secondary metabolites) crossed the BBB, yielding cerebellar and frontal DV values that were equivalent to [^{18}F]ALT cerebellar and blocking DV values. Hence, the metabolite-generated brain radioactivity was nonspecifically distributed. Conventional models performed statistically better than metabolite models, although a greater dynamic range in the regional binding parameters was obtained by the latter method. These results support the quantification of [^{18}F]ALT data by standard kinetic methods.

I. INTRODUCTION

Biodistribution studies in rats indicated that [^{18}F]altanserin ([^{18}F]ALT) was a very promising serotonin 5HT$_{2A}$ radiotracer for positron emission tomography (PET) imaging studies of postsynaptic receptors *in vivo* (Lemaire *et al.*, 1991). This promise was based largely on the high affinity ($K_i = 0.13$ nM, 5HT$_{2A}$) and selectivity (90-, 300-, and 400-fold less potent at α_1, D$_2$ and 5HT$_{2C}$, respectively) of [^{18}F]ALT and the possibility of higher specific:nonspecific binding ratios than previously obtained using the 5HT$_{2A}$ radioligands [^{11}C]ketanserin and [^{18}F]setoperone, as summarized by Sadzot *et al.* (1995).

Subsequently, [^{18}F]altanserin PET studies in humans yielded frontal cortex:cerebellum ratios that approached 3 (Biver *et al.*, 1994; Sadzot *et al.*, 1995) and demonstrated the specific displacement of [^{18}F]ALT by ketanserin (Sadzot *et al.*, 1995). It was also shown that conventional compartmental and graphical analyses performed well as quantitative methods for the regional analyses of [^{18}F]ALT kinetics, although cerebellar data required a model with two tissue compartments to account for a slow nonspecific binding component (Biver *et al.*, 1994). These conventional methods do not explicitly account for blood–brain barrier (BBB) passage of radiolabeled metabolites. More recently, a compartmental metabolite model that allowed for radiolabeled metabolites to cross the BBB was shown to describe the regional kinetics of human [^{18}F]ALT PET data (Mintun *et al.*, 1996) and conventional methods provided specific binding parameter test–retest variability of about 10% in receptor-binding regions (Smith *et al.*, 1998). Despite these differences, all kinetic models yielded [^{18}F]ALT specific binding parameters that were highly correlated ($r^2 > 0.9$) with the rank order of 5HT$_{2A}$ receptor densities in human brain (Pazos *et al.*, 1987).

The goal of the present work was to better understand and characterize the slow [^{18}F]ALT nonspecific binding component. Toward this goal, PET studies at baseline and after pretreatment with the 5HT$_{2A}$ antagonists ketanserin or SR 46349B (Rinaldi-Carmona *et al.*, 1992) were performed to investigate the kinetic characteristics of [^{18}F]ALT and two putative metabolites of [^{18}F]ALT in baboons. PET studies were performed following the bolus injection of [^{18}F]ALT, [^{18}F]altanserinol ([^{18}F]ALT-ol), or [^{18}F]4-(4-fluorobenzoyl) piperidine ([^{18}F]4-FBP).

II. METHODS

A. Radiochemistry

[^{18}F]Altanserin was synthesized according to the literature method (Lemaire *et al.*, 1991). Two putative radiolabeled metabolites of [^{18}F]ALT, likely resulting from ketone reduction ([^{18}F]ALT-ol) and N-dealkylation ([^{18}F]4-FBP), were also synthesized (Mason *et al.*, 1997). These two metabolites were identified in baboon plasma following the injection of [^{18}F]ALT, and confirmation of their identities was based on their identical chromatographic properties compared to those of authentic, unlabeled (cold) ALT-ol and 4-FBP. However, it was apparent that 4-FBP was rapidly metabolized into a secondary metabolite which contributed to the polar metabolite component. The definitive characterization of the *in vivo* behavior of all of the radiolabeled metabolites of [^{18}F]ALT in primates is in progress.

B. Data Acquisition

Dynamic PET imaging (20–25 frames) was performed in baboons over 90–120 min after the bolus injection of high specific activity (> 1500 Ci/mmol) [^{18}F]ALT ($n = 7$), [^{18}F]ALT-ol ($n = 1$), or [^{18}F]4-FBP ($n = 1$) using a Siemens-CTI 951R/31 PET scanner in two-dimensional (2D) imaging mode. Four baboons were also studied with [^{18}F]ALT 10 min after ketanserin pretreatment (4 ± 0.7 mg/kg), and two of these baboons underwent an additional [^{18}F]ALT-ol ($n = 1$) or [^{18}F]4-FBP ($n = 1$) study 5 min after SR 46349B pretreatment (1 mg/kg). About 40 arterial blood samples were collected for the determination of the plasma input function, with 20 samples collected in the first 2 min. HPLC was used to determine the contribution of the radiolabeled metabolites in plasma at five or more time points during the study. Plasma protein binding (free fraction) was determined from arterial blood samples using ultrafiltration (Sadzot *et al.*, 1991).

C. Data Analyses

Time–activity curves were generated for the frontal (Frt) and temporal (Tem) cortices, striatum (Str), and cerebellum (Cer) using region-of-interest sampling. Kinetic PET data were analyzed using compartmental modeling methods. All models were compared using the Akaike information criteria (AIC) (Akaike, 1974) with the lowest AIC value indicating the best model fit to data. The general models used in this work are shown diagrammatically in Fig. 1 as two parallel configurations (A and B) that describe the PET measurement in terms of parent radiotracer (A; conventional) and radiolabeled metabolites (B; metabolite). The regional time–activity curves were analyzed using compartmental methods that either included (A + B; metabolite) or did not include (A only; conventional) an additional term to account for radiolabeled metabolites in brain. The conventional method solution (C_{MOD}) was described in terms of a comprehensive four-compartment model (see Koeppe *et al.*, 1991):

$$C_{MOD} = C_2 + C_3 + C_4 + (BV)C_p. \quad (1)$$

Briefly, C_2, C_3, and C_4 correspond to the concentrations of free, specifically bound, and nonspecifically bound drug in brain, respectively. In practice, two-, three-, and four-compartment models (2C, 3C, and 4C) were implemented that were composed of 1 ($C_2 + C_3 + C_4$), 2($C_2 + C_4, C_3$), and 3(C_2, C_3, C_4) tissue compartments (Fig. 1). A blood volume (BV) parameter was included to account for the fractional amount of total plasma radioactivity (C_p) that contributed to the PET measurement. Three conventional model configurations were used to obtain estimates of the distribution volume (DV) of the unmetabolized parent radioligand:

$$2C\ DV = \frac{K_1}{k_2'}, \text{ where } k_2' = \left[\frac{k_2}{1 + k_3/k_4 + k_5/k_6}\right] \quad (2)$$

$$3C\ DV = \frac{K_1}{k_2 f_2}\left[1 + \frac{k_3 f_2}{k_4}\right] \quad (3)$$

$$4C\ DV = \frac{K_1}{k_2}[1 + k_3/k_4 + k_5/k_6], \quad (4)$$

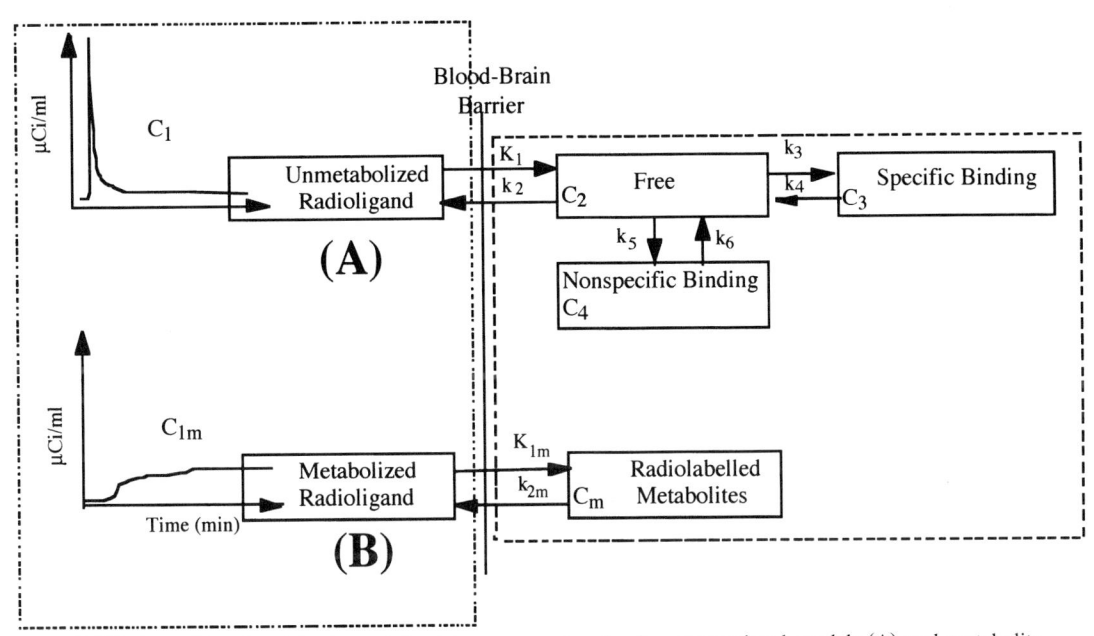

FIGURE 1 PET regional brain–activity data were analyzed using conventional models (A) and metabolite models (A + B). The conventional models provided estimates of the distribution volume (DV) of unmetabolized [^{18}F]ALT. The metabolite models provided an estimate of the distribution volume of all of the radiolabeled metabolites (DV$_{met}$), in addition to the estimate of DV.

where K_1 (ml/min/ml) and k_2 (min^{-1}) are kinetic parameters that represent the bidirectional transport of parent radioligand across the blood–brain barrier, k_3 (min^{-1}) and k_4 (min^{-1}) reflect the association and dissociation rate constants of specific receptor binding, k_5 (min^{-1}) and k_6 (min^{-1}) represent the nonspecific binding parameters, and f_2 is the free fraction of radioligand in tissue. In addition to the conventional estimate of DV, the metabolite models provide an estimate of the volume of distribution of radiolabeled metabolites (DV$_{met}$):

$$\frac{dC_m(t)}{dt} = K_{1m}C_{1m}(t) - k_{2m}C_m(t) \quad (5)$$

$$DV_{met} = \frac{K_{1m}}{k_{2m}}. \quad (6)$$

The metabolite models correspond to two ($C_2 + C_3 + C_4, C_m$) and three ($C_2 + C_4, C_3, C_m$) tissue compartments (Fig. 1). Model Eqs. (5) and (6) allow for the passage of radiolabeled metabolites from plasma (C_{1m}) into brain (C_m), similar to what has been implemented for other PET neuroreceptor studies (Huang et al., 1991; Votaw et al., 1993; Price et al., 1996). The metabolite model fits were performed with all parameters varying freely (2C$_{met}$ and 3C$_{met}$), with K_{1m} fixed to the value of K_1 (2C$_{met-fix}$ and 3C$_{met-fix}$) and with K_1/k_2 and DV$_{met}$ fixed in receptor-binding regions to the cerebellar values (3C$_{met-fix}$) (Mintun et al., 1996). In the context of [^{18}F]ALT and its putative metabolites, [^{18}F]ALT-ol and [^{18}F]4-FBP, the metabolite models made implicit use of the assumption that the metabolite(s) input function(s) was approximated by the total concentration of radioactive metabolites in plasma.

III. RESULTS AND DISCUSSION

A. Plasma Analyses

The average ($n = 8$) amounts of [^{18}F]ALT, the polar metabolites, and [^{18}F]ALT-ol measured in baboon plasma following iv injection of [^{18}F]ALT are shown in Fig. 2. The metabolism of [^{18}F]ALT resulted in ~ 90 and 25% unchanged [^{18}F]ALT at 2 and 90 min, respectively. The polar metabolites comprised ~ 10 and 50% of the plasma radioactivity at 2 and 90 min, whereas [^{18}F]ALT-ol comprised ~ 2 and 12% at these times. Note that the sum of the three components was not equal to 100% at all times and that the "missing radioactivity" was composed of highly polar metabolites that were not retained during the preparation of the sample for HPLC analysis. The plasma-free fraction for [^{18}F]4-FBP was 77%, whereas those for [^{18}F]ALT-ol and [^{18}F]ALT were about 10 and 1%, respectively.

430   VII. Kinetic Modeling

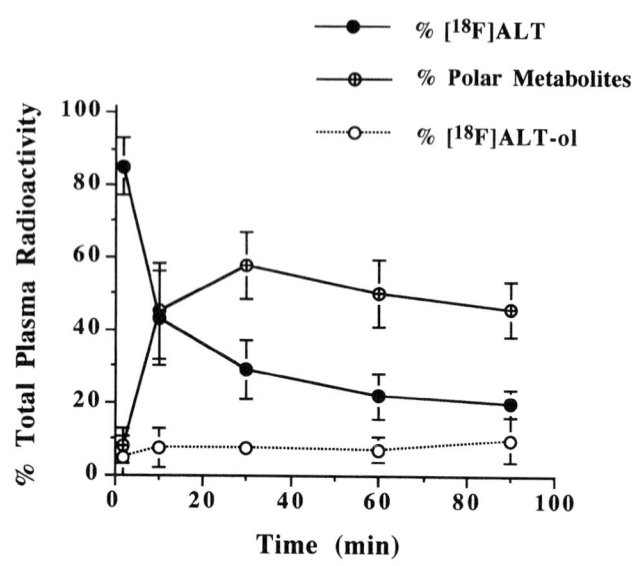

FIGURE 2 The average ($n = 8$) contributions of (unchanged) [^{18}F]ALT, polar metabolites (including [^{18}F]4-FBP), and [^{18}F]ALT-ol components in baboon plasma measured after bolus injection of [^{18}F]ALT (see Lopresti et al., this volume, Chapter 44). The metabolism of [^{18}F]ALT in baboons led to ~ 90 and 25% of unchanged [^{18}F]ALT at 2 and 90 min, respectively. The polar metabolite components were ~ 10 and 50% and [^{18}F]ALT-ol was ~ 2 and 12% at 2 and 90 min, respectively.

B. Data Analyses

In general, the conventional methods provided lower AIC values than the metabolite configurations, although the results obtained with both models were similar (Fig. 3). The 2C$_{met}$ and 3C$_{met}$ results were often associated with large parameter errors, whereas more reliable results were obtained with the constrained metabolite configurations (2C$_{met-fix}$ and 3C$_{met-fix}$). Summaries of the compartmental modeling results are shown in Tables 1A, 1B, and 1C. The coefficient of variation of the mean (CV%) is shown in parentheses.

[^{18}F]ALT cerebellar and regional ketanserin-blocked data were statistically better fit by a 3C model (lowest AIC values). The mean baseline cerebellar 3C DV value was 1.8 with K_1/k_2 and k_3/k_4 values of 1.1 and 0.65, respectively. The regional DV values for the ketanserin-blocked studies lacked receptor density rank order, with mean blocked cerebellar and frontal 3C DV values of 1.9 and 1.8, respectively. Using the 2C$_{met-fix}$ metabolite model, the baseline cerebellar DV and DV$_{met}$ values were 0.92 and 0.90, respectively, indicating that the nonspecific binding component was split between [^{18}F]ALT and its metabolites. It is interesting to note that the baseline [^{18}F]ALT cerebellar and regional blocked DV values, ~ 2, were lower than the occipital and cerebellar 3C DV values for radioligands, such as [^{11}C]diprenorphine (Sadzot et al., 1991) and [^{18}F]cyclofoxy (Carson et al., 1993), which were 3–4.

Because bolus injections of [^{18}F]4-FBP and [^{18}F]ALT-ol yielded fairly homogeneous regional brain distributions at baseline and after blocking, only the cerebellar baseline results are presented in Table 1A with a description of the other regional results to follow. [^{18}F]4-FBP and [^{18}F]ALT-ol data were better fit by a 2C model as compared to the 2C$_{met-fix}$ or 3C models. The 2C$_{met-fix}$ model generally performed well with these data, whereas the conventional 3C model did not, yielding large parameter errors in k_3/k_4 and having problems reaching convergence.

Baseline and SR 46349B-blocked DV values for [^{18}F]4-FBP were 2.0 and 1.8 at baseline and after blocking, respectively, in the same baboon. Mean 2C DV values of 2.2 ± 0.1 and 1.9 ± 0.1 were obtained across regions ($n = 4$) at baseline and after SR 46349B blocking, respectively, In contrast to [^{18}F]4-FBP, the kinetic modeling of regional [^{18}F]ALT-ol data was hindered by its relatively low uptake in brain (Fig. 4). [^{18}F]4-FBP data yielded K_1 values (~ 0.2 ml/min/ml) that were about five times larger than those obtained for [^{18}F]ALT-ol (~ 0.04 ml/min/ml). In addition, the plasma-free fraction for [^{18}F]4-FBP was nearly eight times greater than that for [^{18}F]ALT-ol. Despite low brain uptake, [^{18}F]ALT-ol 2C fits produced cerebellar DV values of 2.1 at baseline and 2.0 after blocking with corresponding frontal DV values of 2.7 and 2.2 in the same baboon. The 2C$_{met-fix}$ metabolite model yielded cerebellar DV and DV$_{met}$ values for [^{18}F]ALT-ol that were similar to those for [^{18}F]ALT. The higher 2C DV value for [^{18}F]ALT-ol in frontal cortex (2.7) compared to its baseline value in cerebellum (2.1) and blocked values in frontal cortex (2.2) and cerebellum (2.0) indicate that [^{18}F]ALT-ol was converted to [^{18}F]ALT *in vivo* in a manner analogous to that reported for ketanserin-ol to ketanserin in rats and humans (Van Peer et al., 1986). This was confirmed by the presence of [^{18}F]ALT in plasma following the injection of [^{18}F]ALT-ol.

Regional DV estimates for [^{18}F]ALT receptor-binding data (Table 1C) were highly correlated with the known rank order of 5HT$_{2A}$ receptors in human brain (Pazos et al., 1987). The conventional 3C and constrained 4C model yielded the lowest AIC values, whereas the 3C$_{met}$ model generally provided equivalent results to the 3C and 4C model values, but large errors in K_{1m}, k_{2m}, and DV$_{met}$ were observed if K_{1m} was not constrained. Conventional model DV estimates for [^{18}F]ALT were always larger than those generated us-

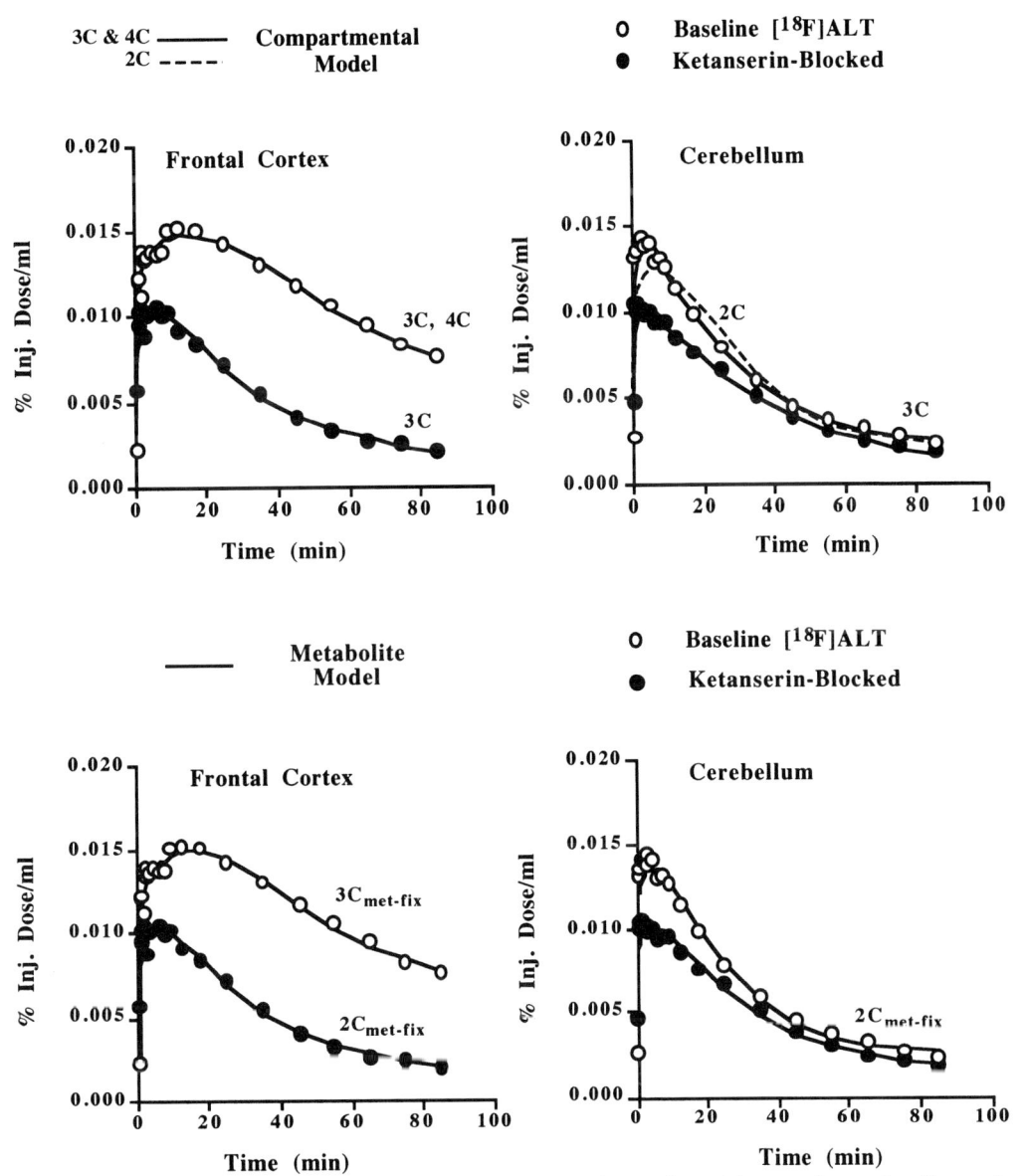

FIGURE 3 Examples of conventional (top) and metabolite (bottom) model curve fits to baseline and ketanserin-blocked [^{18}F]ALT PET data. In general, the conventional models provided better curve fits (lower AIC values) than the metabolite configurations, $2C_{met-fix}$ and $3C_{met-fix}$, although the results were similar for the two methods. For receptor-binding data, the 3C and 4C models yielded the lowest AIC values. For cerebellar and ketanserin-blocked data, the 3C model yielded the lowest AIC values, whereas the 2C model yielded the highest values.

TABLE 1A Cerebellum: [^{18}F]ALT (Baseline and Ketanserin-Blocked), [^{18}F]4-FBP, and [^{18}F]ALT-ol[a]

Cerebellum		[^{18}F]ALT baseline ($n = 7$)	[^{18}F]ALT ketanserin ($n = 4$)	[^{18}F]4-FBP baseline ($n = 1$)	[^{18}F]ALT-ol baseline ($n = 1$)
2C	DV	1.6 (25)	1.7 (22)	2.0	2.1
3C	DV	1.8 (28)	1.9 (18)		
	K_1/k_2[b]	1.1 (23)	1.3 (25)		
	k_3/k_4[b]	0.65 (28)	0.50 (48)		
2C$_{met-fix}$	DV[c]	0.92 (32)	1.1 (29)	1.8	1.0
	DV$_{met}$[c]	0.90 (69)	0.62 (32)	0.28	0.57

[a] The coefficient of variation of the mean (CV%) is shown in parentheses.
[b] Average values of 4C model constraints, see Table 1C.
[c] Average values of 3C$_{met-fix}$ model constraints, see Table 1C.

TABLE 1B Regional [^{18}F]ALT Ketanserin-Blocked Data[a]

Ketanserin pretreatment	($n = 4$)	Frt	Tem	Str
2C	DV	1.9 (24)	2.0 (25)	2.0 (26)
3C	DV	1.8 (38)	2.3 (20)	2.3 (25)
	K_1/k_2	1.4 (18)	1.4 (23)	1.3 (26)
	k_3/k_4	0.27 (130)	0.65 (28)	0.76 (37)
2C$_{met-fix}$	DV	1.3 (34)	1.4 (39)	1.2 (36)
	DV$_{met}$	0.65 (26)	0.69 (29)	0.89 (40)

[a] The coefficient of variation of the mean (CV%) is shown in parentheses.

TABLE 1C Regional Baseline [^{18}F]ALT Data[a]

Baseline	($n = 7$)	Frt	Tem	Str
2C	DV	4.4 (31)	4.2 (28)	2.6 (27)
3C	DV	4.7 (31)	4.5 (29)	2.8 (26)
	K_1/k_2	1.5 (28)	1.5 (40)	1.3 (27)
	k_3/k_4	2.3 (58)	2.5 (68)	1.1 (28)
4C	DV	4.6 (32)	4.4 (29)	2.8 (28)
Fixed	K_1/k_2[b]	1.1	1.1	1.1
	k_3/k_4	2.5 (18)	2.3 (22)	0.84 (16)
Fixed	k_5/k_6[b]	0.65	0.65	0.65
3C$_{met-fix}$	DV	3.8 (28)	3.6 (25)	1.9 (25)
Fixed	K_1/k_2[c]	0.92	0.92	0.92
	k_3/k_4	3.3 (33)	3.1 (40)	1.1 (35)
Fixed	DV$_{met}$[c]	0.90	0.90	0.90

[a] The coefficient of variation of the mean (CV%) is shown in parentheses.
[b] Average of specific parameter constraints from 3C model, see Table 1A.
[c] Average of specific parameter constraints from 2C$_{met-fix}$ model, see Table 1A.

ing the metabolite models as the latter method "removed" the metabolite component (DV$_{met}$) from the regional DV value.

IV. CONCLUSION

Two putative radiolabeled metabolites of [^{18}F]ALT, [^{18}F]4-FBP and [^{18}F]ALT-ol, were injected into baboons and resulted in sustained radioactivity in the brain, indicating that [^{18}F]ALT-ol and [^{18}F]4-FBP (and/or their secondary metabolites) crossed the blood–brain barrier. The conventional compartmental models performed better statistically than those that explicitly allowed for blood–brain barrier passage of metabolites, although a greater dynamic range in the regional binding parameters was obtained using the metabolite methods. Results indicate that the further refinement of metabolite models for [^{18}F]ALT would require more precise constraints than are available at this time. Complex characterization of the radiolabeled metabolites of [^{18}F]ALT and their individual kinetic behaviors in primates remains under investigation. Nonetheless, reduced compartmental models (four parameters) yielded [^{18}F]ALT-binding parameters that were highly correlated ($r^2 > 0.9$) with the rank order of human 5HT$_{2A}$ receptor densities. Overall, the results of this work support the quantification of [^{18}F]ALT data using conventional kinetic methods.

Acknowledgments

The authors thank the UPMC PET facility staff. This work was supported by NIH Grants MH52247, NS22899, and MH49936.

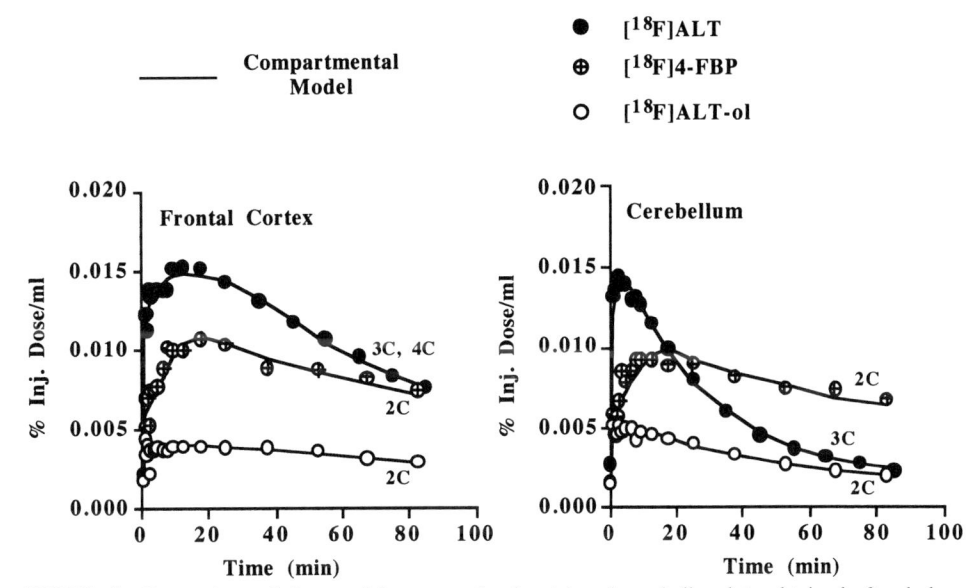

FIGURE 4 Comparison of time-activity curves for frontal and cerebellar data obtained after bolus injection of [^{18}F]ALT, [^{18}F]4-FBP, and [^{18}F]ALT-ol and the corresponding model curve fits. After injection of either [^{18}F]ALT-ol or [^{18}F]4-FBP, radioactivity in the brain increased to levels above that attributed to intravascular radioactivity. Thus, [^{18}F]ALT-ol and [^{18}F]4-FBP (and/or their secondary metabolites) crossed the blood-brain barrier, resulting in [^{18}F]ALT-ol and [^{18}F]4-FBP cerebellar and frontal 2C DV values that were consistent with the [^{18}F]ALT cerebellar and ketanserin-blocked 3C DV values.

References

Akaike, H. (1974). A new look at the statistical model identification. *IEEE Trans. Autom. Control* **AC-19**: 716–723.

Biver, F., Goldman, S., Luxen, A., Monclus, M., Forestini, M., Mendlewicz, J., and Lotstra, F. (1994). Multicompartmental study of fluorine-18 altanserin binding to brain 5HT$_2$ receptors in humans using positron emission tomography. *Eur. J. Nucl. Med.* **21**: 937–946.

Carson, R. E., Channing, M. A., Blasberg, R. G., Dunn, B. B., Cohen, R. M., Rice, K. C., and Herscovitch, P. (1993). Comparison of bolus and infusion methods for receptor quantitation: Application to [^{18}F]-cyclofoxy and positron emission tomography. *J. Cereb. Blood Flow Metab.* **13**: 24–42.

Huang, S.-C., Yu, D.-C., Barrio, J. R., Grafton, S., Melega, W. P., Hoffman, J. M., Satyamurthy, N., Mazziotta, J. C., and Phelps, M. E. (1991). Kinetics and modelings of L-6-[18F]fluoro-DOPA in human positron emission tomographic studies. *J. Cereb. Blood Flow Metab.* **11**: 898–913.

Koeppe, R. A., Holthoff, V. A., Frey, K. A., Kilbourn, M. R., and Kuhl, D. E. (1991). Compartmental analysis of [^{11}C]flumazenil kinetics for the estimation of ligand transport rate and receptor distribution using positron emission tomography. *J. Cereb. Blood Flow Metab.* **11**: 735–744.

Lemaire, C., Cantineau, R., Guillaume, M., Plenevaux, A., and Christiaens, L. (1991). Fluorine-18-altanserin: A radioligand for the study of serotonin receptors with PET: Radiolabeling and *in vivo* biologic behavior in rats. *J. Nucl. Med.* **32**: 2266–2272.

Mason, N. S., Huang, Y., Holt, D. P., Perevuznik, J. J., Lopresti, B., and Mathis, C. A. (1997). Synthesis and characterization of [F-18]4-(4-fluorobenzoyl)piperidine, an [F-18]altanserin metabolite. *J. Nucl. Med.* **38**: 56P.

Mintun, M. A., Price, J. C., Smith, G. S., Lopresti, B., Hartman, L., Simpson, N., and Mathis, C. A. (1996). Quantitative 5-HT$_{2A}$ receptor imaging in man using [F-18]altanserin: A new model accounting for labeled metabolites. *J. Nucl. Med.* **37**: 109P.

Pazos, A., Probst, A., and Palacios, J. M. (1987). Serotonin receptors in the human brain. IV. Autoradiographic mapping of serotonin-2 receptors. *Neuroscience* **21**: 123–139.

Price, J. C., Mathis, C. A., Simpson, N. R., Mahmood, K., and Mintun, M. A. (1996). Kinetic Modeling of serotonin-1A binding in monkeys using ^{11}C WAY 100635 and PET. *In* "Quantification of Brain Function using PET" (R. Myers, V. Cunningham, D. Bailey, and T. Jones, eds.), pp. 257–261. Academic Press, San Diego, CA.

Rinaldi-Carmona, M., Congy, C., Santucci, V., Simiand, J., Gautret, B., Neliat, G., Labeeuw, B., LeFur, G., Soubrie, P., and Breliere, J. C. (1992). Biochemical and pharmacological properties of SR 46349B, a new potent and selective 5-hydroxytryptamine$_2$ receptor antagonist. *J. Pharmacol. Exp. Ther.* **262**: 759–768.

Sadzot, B., Price, J. C., Mayberg, H. S., Douglass, K. H., Dannals, B. F., Lever, J. R., Ravert, H. T., Wilson, A. A., and Wagner, H. N., Jr. (1991). Quantification of human opiate receptor concentration and affinity using high and low specific activity [^{11}C]diprenorphine and positron emission tomography. *J. Cereb. Blood Flow Metab.* **11**: 204–219.

Sadzot, B., Lemaire, C., Maquet, P., Salmon, E., Plenevaux, A., Degueldre, C., Hermanne, J. P., Guillaume, M., Cantineau, R., Comar, D., and Franck, G. (1995). Serotonin 5HT$_2$ receptor imaging in the human brain using positron emission tomography and a new radioligand, [^{18}F]altanserin: Results in young normal controls. *J. Cereb. Blood Flow Metab.* **15**: 787–797.

Smith, G. S., Price, J. C., Mathis, C. A., Lopresti, B. J., Holt, D., Mason, N. S., Simpson, N., Huang, Y., Sweet, R. A., Meltzer, C. C., and Sashin, D. (1998). Test-retest variability of serotonic 5-HT$_{2A}$ receptor binding measured with positron emission tomography (PET) and [^{18}F]altanserin in the human brain. *Synapse*, in press.

Van Peer, A., Woestenborghs, R., Embrechts, L., and Heykants, J. (1986). Pharmacokinetic approach to equilibrium between ketanserin and ketanserin-ol. *Eur. J. Clin. Pharmacol.* **31**: 339–342.

Votaw, J. R., Kessler, R. M., de Paulis, T., Ansari, M. S., Mason, N. S., Schmidt, D. E., Manning, R. G., and Holburn, G. (1993). Quantification of striatal and extrastriatal dopamine D2 receptors using N-allyl-[^{18}F]FPE. *J. Nucl. Med.* **34**: P132.

CHAPTER
65

Characteristics of Neurotransmitter Competition Studies Using Constant Infusion of Tracer[1]

CHRISTOPHER J. ENDRES and RICHARD E. CARSON

Positron Emission Tomography Department
National Institutes of Health
Bethesda, Maryland 20892

Neuroreceptor radioligands have been shown to be sensitive to the level of endogenous neurotransmitter. One method used to study neurotransmitter competition is the bolus + infusion (B/I) method, in which an initial bolus injection is followed by a constant infusion in order to achieve an equilibrium level of tracer in plasma and tissue. Subsequently, stimulus-induced alterations of endogenous ligand concentration cause the tracer level to change from the original equilibrium level. The change in tracer concentration has been assumed to reflect the magnitude of neurotransmitter release. Previously, the conventional receptor compartment model was extended to describe B/I competition studies using [^{11}C]raclopride. This chapter applied the extended model in simulation studies of the B/I method. A wide range of tracer kinetic parameters was used for simulations: K_1 0.1–1.0 ml/min/ml, $V_e = 0.3–6.0$ ml/ml, $k_{on}B_{max} = 0.1–1.0$ min^{-1}, and $k_{off} = 0.02–0.25$ min^{-1}. For tracers within the simulated range, the extended model predicts that the change in tracer concentration measured with the B/I protocol is strongly correlated with the integral of the neurotransmitter pulse. The maximum change in tracer concentration following a neurotransmitter pulse was used as a measure of the sensitivity to displacement by neurotransmitter release. Tracers with binding potential in the range 1–4 were the most sensitive to displacement. Faster tissue-to-blood transport and lower nonspecific binding improved sensitivity. These simulations determine how tracer characteristics affect the sensitivity to neurotransmitter release. However, the accuracy of the extended model and signal-to-noise issues must be considered in determining the ultimate sensitivity of a tracer method.

I. INTRODUCTION

The use of neuroreceptor radioligands has been expanded to study dynamic changes in tracer binding due to competition with endogenous neurotransmitter. The interaction studied most extensively has been the effect of endogenous dopamine on the binding of D_2 receptor ligands (Dewey et al., 1991, 1993; Innis et al., 1992; Volkow et al., 1994). To better characterize the competition between dopamine and [^{11}C]raclopride, bolus + infusion (B/I) studies (Carson et al., 1997) were performed with simultaneous microdialysis (Breier et al., 1997). Using these data, the authors extended the conventional compartment model for reversible ligands to explicitly account for dopamine receptor occupancy (Endres et al., 1997). The conventional model ignores neurotransmitter binding and assumes linear time-invariant tracer kinetics. However, the extended model accounts for the effects of a transient neurotransmitter pulse, causing the tracer binding rate to be time dependent. In [^{11}C]raclopride studies using B/I tracer delivery, the extended model sufficiently described the reduction in tracer concentration following amphetamine-stimulated dopamine release (Endres et al., 1997).

This chapter applies the extended model in simulations of the B/I method. The purpose of this study is to characterize the response to neurotransmitter release over a wide range of tracer kinetics. There are two specific goals for this study. One is to determine an

---

[1] Transcripts of the BRAINPET97 discussion of this chapter can be found in Section VIII.

appropriate interpretation of changes in tracer binding in terms of neurotransmitter release. The second is to determine how kinetic properties of the tracer affect the sensitivity to dynamic changes in neurotransmitter concentration.

II. METHODS

A. The Extended Receptor Model

The extended receptor model is shown in Fig. 1. The conventional compartment model for reversible receptor ligands includes two tissue compartments. One compartment (C_f) represents both free and nonspecific tracer, with the assumption that nonspecifically bound tracer equilibrates rapidly with free tracer. The second compartment (C_b) represents tracer specifically bound to receptor. The model parameters are K_1, the plasma-to-tissue influx constant (ml plasma/min/ml tissue); k_2, the tissue to plasma clearance in the absence of nonspecific binding (min^{-1}); $k_2' = k_2/(1 + NS)$ (NS is the ratio of nonspecific to free tracer at equilibrium); $V_e = K_1/k_2'$, the volume of distribution of free + nonspecific tracer (ml plasma/ml tissue); k_3, the binding rate of tracer to receptors (min^{-1}); $k_3' = k_3/(1 + NS)$; and k_{off}, the dissociation rate of tracer from receptors (min^{-1}). The conventional model assumes time-invariant kinetics and is unable to account for transient changes in kinetics induced by changes in endogenous neurotransmitter.

To account for neurotransmitter competition, the conventional model has been extended to account for receptor occupancy by neurotransmitter explicitly (Endres et al., 1997). The extended model predicts that a time-varying neurotransmitter concentration causes the tracer binding rate k_3 to be time dependent. With the assumption that neurotransmitter binding kinetics are sufficiently rapid to maintain secular equilibrium between free (N_f) and bound (N_b) neurotransmitter, even during a transient neurotransmitter pulse, the tracer binding rate is given by

$$k_3(t) = \frac{k_{on}\left(B_{max} - \dfrac{C_b(t)}{SA}\right)}{1 + \dfrac{N_f(t)}{K_D^N}}, \quad (1)$$

where k_{on} is the bimolecular association rate (nM^{-1} min^{-1}), B_{max} is total receptor concentration (nM), SA is specific activity (mCi/μmol), and K_D^N is the equilibrium dissociation constant for neurotransmitter (nM). It is assumed that tracer and neurotransmitter bind to a single class of receptors with a single affinity.

B. Simulation Studies

There were two objectives for performing simulations with the extended model. One was to determine an appropriate interpretation of the change in tracer concentration in terms of the underlying neurotransmitter pulse. Initial simulations using [^{11}C]raclopride kinetics showed that the change in tracer binding measured with the B/I protocol correlates very well with the integral of the dopamine pulse (Endres et al., 1997). The first goal was to determine if this interpretation can be applied to other tracers. The second goal was to determine how the kinetic characteristics of the tracer affect its sensitivity to neurotransmitter release.

For simulation studies, a wide range of tracer kinetic parameters was tested. The ranges of values chosen were based on the published kinetics of several tracers, including [^{11}C]raclopride (Farde et al., 1989), [^{11}C]flumazenil (Koeppe et al., 1991), and [^{18}F]cyclofoxy (Carson et al., 1993). The parameters were varied as follows: $K_1 = 0.1$–1.0 ml/min/ml, $V_e = 0.3$–6.0 ml/ml, $k_{on}B_{max} = 0.1$–1.0 min^{-1}, and $k_{off} = 0.02$–0.25 min^{-1}. The $k_{on}B_{max}$ product was treated as a single parameter because high specific activity was assumed. Neurotransmitter pulses were modeled as monoexponential curves [$N_f(t) = H \exp(-Rt)$], with $H = (0.375$–$2.5)K_D^N$ and $R = 0.03$–0.15 min^{-1}. These values for H and R were chosen based on values obtained for amphetamine-induced dopamine pulses in rhesus monkeys, as measured with microdialysis (Endres et al., 1997). Overall, three to seven values of each parameter ($K_1, V_e, k_{on}B_{max}, k_{off}, H, R$) were chosen, and time–activity

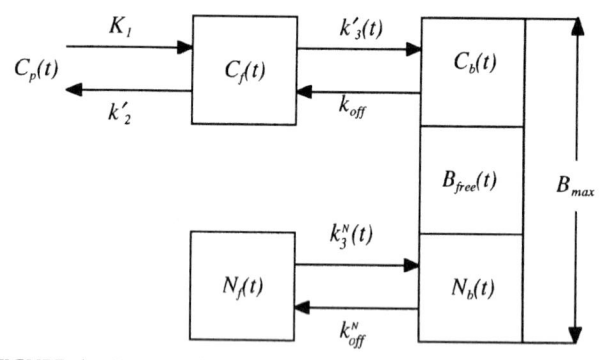

FIGURE 1 An extended compartment model to describe the competition between a reversible receptor-binding tracer and an endogenous neurotransmitter. Terms are defined in the text. The conventional compartment model is represented by tissue compartments for free + nonspecific (C_f) and specifically bound (C_b) tracer. The compartments labeled N_f and N_b represent free and specifically bound neurotransmitter, respectively.

curves were simulated using all possible parameter combinations, giving more than 5000 simulated curves. The plasma input function C_p was set to a constant, i.e., it was assumed that the B/I protocol perfectly attained a constant plasma tracer level. In addition, the initial tissue activity was set to the equilibrium level C_{EQ}, where $C_{EQ} = V_e(1 + k_{on}B_{max}/k_{off})C_p$. Examples of the simulated curves are shown in Fig. 2. The neurotransmitter pulse was initiated, causing the tracer concentration to decrease, achieve a minimum value (C_{MIN}), and then increase toward the original equilibrium level. The sensitivity to the neurotransmitter pulse was computed from the maximum percentage change in concentration ΔC:

$$\Delta C = \frac{C_{EQ} - C_{MIN}}{C_{EQ}} \times 100. \quad (2)$$

In a previous simulation study, the extended model predicated that a change in [^{11}C]raclopride binding measured with the B/I protocol is proportional to the integral of the dopamine pulse (Endres *et al.*, 1997). To determine the generality of this interpretation, the correlation between the change in concentration ΔC and the integral of the neurotransmitter pulse for hypothetical tracers described by the kinetics listed earlier was tested. To understand how particular features affect the change in concentration induced by neurotransmitter release, the relationship between ΔC and tissue-to-blood transport, receptor density, receptor dissociation rate, binding potential, and nonspecific binding was examined. To investigate each feature, simulations were examined to determine the effect of varying selected parameters of interest. For example, to determine the effect of tissue to blood transport on ΔC, the effect of changing k_2 was examined, with the remaining parameters held fixed. Also studied was the effect of changing $k_{on}B_{max}$ to test ΔC vs receptor density, and the effect of changing k_{off} to test ΔC vs receptor dissociation rate or binding potential. To test the effect of nonspecific binding, the authors examined changing $k_2' = k_2/(1 + NS)$ and $k_3' = k_3/(1 + NS)$ simultaneously, such that $k_3'/k_2' = k_3/k_2$.

III. RESULTS AND DISCUSSION

A. Interpretation of ΔC

Over the entire range of parameters tested, there was a very strong correlation between the change in tracer concentration ΔC and the integral of the neurotransmitter pulse. The degree of correlation was somewhat dependent on the integration interval. For example, if the first 30 min of the neurotransmitter pulse was integrated, then the correlation between ΔC and $\int N_f \, dt$ ranged from 0.90 to 0.98. If the first 60 min was used, then the correlation ranged from 0.82 to 1.00. The integration interval that gave the best correlation appeared to correspond to the time at which the tracer was displaced to its minimum value (C_{MIN}). Thus tracers that displace rapidly would have ΔC better correlated with the 30-min integral as opposed to the 60-min integral. Because there was always a large correlation (> 0.80), the extended model predicts that any tracer described by kinetics within the range of these simulations can be used with the B/I paradigm to obtain a measure that can be interpreted as being proportional to the integral of the neurotransmitter pulse. Note that at the highest values of $\int N_f \, dt$, ΔC no longer increases linearly (rolls off) due to saturation.

B. Effects of Tracer Kinetics on Sensitivity to Displacement

The simulations revealed large variability in the sensitivity to displacement of different tracers. Figure 2 shows three simulated curves for theoretical tracers with binding potentials (BP) of 1, 3, and 10. When receptor dissociation rate is fast (BP = 1, dotted curve),

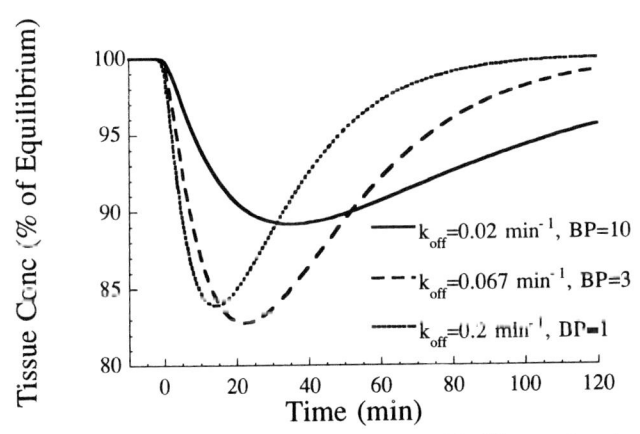

FIGURE 2 Tissue activity profiles for tracers with different dissociation rates (k_{off}) and binding potentials (BP = $k_{on}B_{max}/k_{off}$). All other kinetic parameters are identical [K_1 = 0.16 ml/min/ml, V_e = 0.6 ml/ml, $k_{on}B_{max}$ = 0.2 min^{-1}, $N_f(t) = K_D \exp(-.06t)$]. The tracer with the fastest k_{off} value (dotted line) is quickly displaced and then rapidly recovers as neurotransmitter clears. At a slower k_{off} value (dashed line), the tracer displaces more slowly but shows a larger absolute change in concentration. The larger change is due to a slightly larger specific binding fraction for the slower dissociation rate. At a slower k_{off} value (solid line), the tracer displaces slowly and it takes longer for the tracer concentration to reach a minimum value. At this time, much of the neurotransmitter pulse has already cleared, thus the percentage change in tissue concentration is smaller, despite the larger specific binding fraction.

the tracer is better able to follow the transient neurotransmitter pulse. Thus the tracer returns more quickly to its initial concentration following clearance of neurotransmitter. At a slower dissociation rate (BP = 3, dashed curve), the tracer displaces more slowly but gives a larger change in the total concentration due to the larger specific binding fraction. At an even slower dissociation rate (BP = 10, solid curve) the tracer displaces very slowly and takes a longer time to return to the original concentration following neurotransmitter clearance. Thus, although this tracer has the highest specific binding fraction, it does not show the highest displacement.

For regions of low receptor density, the tracer that had the highest sensitivity had a comparatively slow dissociation rate, whereas a faster dissociation produced larger ΔC for high density regions. As a result of this trend, tracers with optimal sensitivity to displacement had binding potentials of 1 to 4 (Fig. 3). This contrasts with results from simulation of bolus studies where tracers with optimal sensitivity to changes in neurotransmitter concentration had binding potentials of 5 to 10 (Endres and Carson, 1997). The displacement as assessed by ΔC was largest for tracers with rapid dissociation rates at high receptor densities. In addition, it was found that faster tissue-to-blood transport (k_2) and lower nonspecific binding improved sensitivity (data not shown).

C. Other Considerations

The extended model was originally developed using simultaneously acquired [^{11}C]raclopride–PET and dopamine-microdialysis data (Endres et al., 1997). In the time frame of these experiments (90 min), the model sufficiently described the time–activity curves. However, some of the assumptions of the model, such as a single affinity state for agonist binding and rapid equilibration of nonspecific binding, may be inadequate to describe neurotransmitter competition for all tracers. For example, when fitting [^{11}C]raclopride B/I studies to the extended model it was found that the model predicted baseline dopamine receptor occupancy to be 3–15% (Endres et al., 1997), whereas experimental data have shown that inhibition of synaptic dopamine can increase D_2 binding by 30% (Dewey et al., 1995; Laruelle et al., 1996). This inconsistency could be accounted for by including both high and low affinity states for dopamine binding.

These simulations have only considered the tissue signal, i.e., the effect of noise has not been addressed. It is likely that there are cases where a smaller displacement may be measured more reliability than a larger displacement. As an example, from the simulations, it was concluded that faster tissue to blood transport improves sensitivity. However, faster tissue clearance would also be expected to reduce tissue uptake, which could increase the measurement noise. In addition, radiation dose limits and the presence of metabolites will affect signal to noise. The effects of noise and model accuracy need to be considered more fully to evaluate which tracers are most effective for use with the infusion method and to determine whether to use the bolus method or the infusion method to study neurotransmitter competition with a particular tracer.

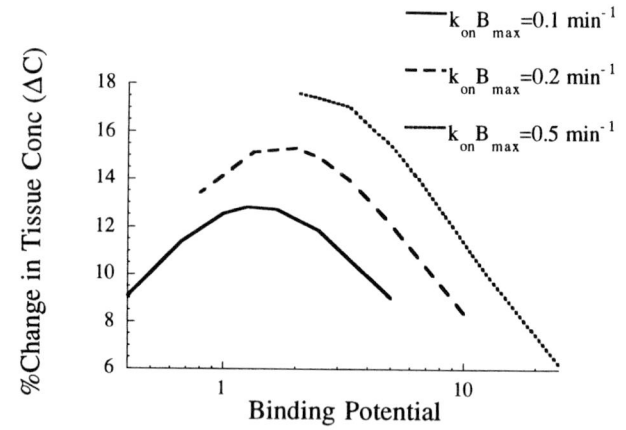

FIGURE 3 The simulated percentage change in tissue concentration (ΔC) plotted vs binding potential ($k_{on}B_{max}/k_{off}$). These curves show that an optimal value for the change in tracer concentration corresponds to a relatively narrow range of binding potentials (1 to 4). Equivalently, tracers with fast receptor dissociation rates (large k_{off}) were more sensitive at high receptor density (large $k_{on}B_{max}$) and tracers with small k_{off} values were most sensitive at low receptor density.

References

Breier, A., Su, T. P., Saunders, R., Carson, R. E., Kolachana, B. S., Debartolomeis, A., Weinberger, D. R., Weisenfeld, N., Malhotra, A. K., Eckleman, W. C., and Pickar, D. (1997). Schizophrenia is associated with elevated amphetamine-induced synaptic dopamine concentrations—evidence from a novel positron emission tomography method. *Proc. Natl. Acad. Sci. U.S.A.* **94**: 2569–2574.

Carson, R. E., Channing, M. A., Blasberg, R. G., Dunn, B. B., Cohen, R. M., Rice, K. C., and Herscovitch, P. (1993). Comparison of bolus and infusion methods for receptor quantitation: application to [^{18}F]cyclofoxy and positron emission tomography. *J. Cereb. Blood Flow Metab.* **13**: 24–42.

Carson, R. E., Breier, A., de Bartolomeis, A., Saunders, R. C., Su, T. P., Schmall, B., Der, M. G., Pickar, D., and Eckleman, W. C. (1997). Quantification of amphetamine-induced dopamine release with [^{11}C]-raclopride and continuous infusion. *J. Cereb. Blood Flow Metab.* **17**: 437–447.

Dewey, S. L., Logan, J., Wolf, A. P., Brodie, J. D., Angrist, B., Fowler, J. S., and Volkow, N. D. (1991). Amphetamine induced decreases in (^{18}F)-N-methylspiroperidol binding in the baboon brain using positron emission tomography (PET). *Synapse* **7**: 324—327.

Dewey, S. L., Smith, G. S., Logan, J., Brodie, J. D., Fowler, J. S., and Wolf, A. P. (1993). Striatal binding of the PET ligand ^{11}C-raclopride is altered by drugs that modify synaptic dopamine levels. *Synapse* **13**: 350–356.

Dewey, S. L., Smith, G. S., Logan, J., Alexoff, D., Ding, Y. S., King, P., Pappas, N., Brodie, J. D., and Ashby, C. R., Jr. (1995). Serotonergic modulation of striatal dopamine measured with positron emission tomography (PET) and in vivo microdialysis. *J. Neurosci.* **15**: 821–829.

Endres, C. J., and Carson, R. E. (1997). Measurement of neurotransmitter release with bolus or infusion delivery of receptor-binding tracers. *J. Nucl. Med.* **38**: 11P.

Endres, C. J., Kolachana, B. S., Saunders, R. C., Su, T., Weinberger, D., Breier, A., Eckelman, W. C., and Carson, R. E. (1997). Kinetic modeling of [^{11}C]raclopride: Combined PET-microdialysis studies. *J. Cereb. Blood Flow Metab.* **17**: 932–942.

Farde, L., Eriksson, L., Blomquist, G., and Halldin, C. (1989). Kinetic analysis of central [^{11}C]raclopride binding to D2-dopamine receptors studied by PET: A comparison to the equilibrium analysis. *J. Cereb. Blood Flow Metab.* **9**: 696–708.

Innis, R. B., Malison, R. T., al-Tikriti, M., Hoffer, P. B., Sybirska, E. H., Seibyl, J. P., Zoghbi, S. S., Baldwin, R. M., Laruelle, M., and Smith, E. O. (1992). Amphetamine-stimulated dopamine release competes in vivo for [^{123}I]IBZM binding to the D2 receptor in nonhuman primates. *Synapse* **10**: 177–184.

Koeppe, R. A., Holthoff, V. A., Frey, K. A., Kilbourn, M. R., and Kuhl, D. E. (1991). Compartmental analysis of [^{11}C]flumazenil kinetics for the estimation of ligand transport rate and receptor distribution using positron emission tomography. *J. Cereb. Blood Flow Metab.* **11**: 735–744.

Laruelle, M., D'Souza, C. D., Zoghbi, S. S., Baldwin, R. M., Charney, D. S., and Innis, R. B. (1996). SPECT measurement of dopamine synaptic concentration in the resting state. *J. Nucl. Med.* **37**: 32P.

Volkow, N. D., Wang, G. J., Fowler, J. S., Logan, J., Schlyer, D., Hitzemann, R., Lieberman, J., Angrist, B., Pappas, N., MacGregor, R., Burr, R., Cooper, T., and Wolf, A. P. (1994). Imaging endogenous dopamine competition with [^{11}C]raclopride in the human brain. *Synapse* **16**: 255–262.

CHAPTER 66

PET Measurement of Endogenous Neurotransmitter Activity Using High and Low Affinity Radiotracers[1]

J. C. PRICE, N. S. MASON, B. LOPRESTI, D. HOLT, N. R. SIMPSON, W. DREVETS,* G. S. SMITH,* and C. A. MATHIS

Departments of Radiology and *Psychiatry
University of Pittsburgh School of Medicine
Pittsburgh, Pennsylvania 15213

Functional imaging of the in vivo competition of dopamine–D_2 radioligands with endogenous dopamine has provided indirect measures of dopamine activity. Binding parameter changes (> 20%) have been observed for [^{11}C]raclopride positron emission tomography studies performed after d-amphetamine administration, and this sensitivity has been partially attributed to its low D_2 receptor affinity. In this work, a d-amphetamine pretreatment/bolus radiotracer injection paradigm was used to examine binding parameter changes for the high and low affinity benzamides, [^{18}F]fallypride and [^{11}C]raclopride; and a preliminary examination of [^{18}F]fallypride extrastriatal D_2 binding at baseline and after haloperidol pretreatment was performed. [^{11}C]raclopride studies were performed in baboons at baseline and after d-amphetamine pretreatment (0.6 mg/kg, n = 3 or 1.0 mg/kg, n = 2). Two baboons had additional [^{18}F]fallypride studies at baseline and after d-amphetamine, and a third baboon was studied at baseline and after haloperidol pretreatment (1 mg/kg). Kinetic analyses provided distribution volume (DV) and DV-derived binding potential (BP_{DV}) estimates. For [^{11}C]raclopride, the average striatal BP_{DV} change was −26 ± 2.6% for the 0.6-mg/kg amphetamine dose, with slightly greater changes of −31 and −39% after the 1.0-mg/kg dose. Lower striatal changes were observed for [^{18}F]fallypride: −14% (0.6 mg/kg) and −29% (1.0 mg/kg). Baseline regional BP_{DV} values were highly correlated ($r^2 > 0.9$) with human D_2 receptor rank order but did not correlate after 1 mg/kg haloperidol pretreatment ($r^2 < 0.1$).

---

[1] Transcripts of the BRAINPET97 discussion of this chapter can be found in Section VIII.

I. INTRODUCTION

Indirect measures of endogenous dopamine activity have been obtained through functional imaging studies of the *in vivo* competition between dopamine–D_2 receptor radioligands and extracellular dopamine. For example, after *d*-amphetamine administration, extracellular dopamine concentrations increase and dopamine competes with the D_2 radioligand for binding to D_2 receptors. In fact, *d*-amphetamine pretreatment (≤ 1 mg/kg) has led to binding parameter decreases of > 20% (relative to baseline) after the bolus injection of the D_2 antagonist [^{11}C]raclopride (Dewey *et al.*, 1993). The sensitivity of [^{11}C]raclopride to extracellular dopamine has been attributed largely to its relatively low affinity for D_2 receptors ($K_D = 1.0$ nM) (Seeman *et al.*, 1990; Hall *et al.*, 1990).

Remarkably, a D_2 antagonist with irreversible *in vivo* binding kinetics and of fivefold higher affinity, [^{18}F]N-methylspiperone ($K_D = 0.2$ nM, Hall *et al.*, 1990), exhibited *in vivo* sensitivity to dopamine in positron emission tomography (PET) studies that were performed following *d*-amphetamine administration (1.0 mg/kg). These studies demonstrated that *in vivo* pharmacokinetic effects such as tissue clearance play an important role in radioligand sensitivity of high affinity radioligands to endogenous dopamine concentrations (Dewey *et al.*, 1991; Logan *et al.*, 1991).

PET imaging studies have also been performed using the D_2 antagonist [^{18}F]fallypride, which has 10- and 30-fold higher affinity ($K_D = 0.03$ nM) than *N*-methyl-[^{18}F]N-methylspiperone and [^{11}C]raclopride (Mukherjee *et al.*, 1995). In rhesus monkeys, [^{18}F]fallypride was displaced by the D$_2$ antagonist haloperidol (K_i = 1 nM, Hall *et al.*, 1990), with a half-time of 18 min (Mukherjee *et al.*, 1995). Preliminary work indicated that [^{18}F]fallypride PET data can be quantified using general compartmental methods with B_{max} estimates that were consistent with *in vitro* binding results (Votaw *et al.*, 1993; Kessler *et al.*, 1993a) and that [^{18}F]fallypride is sensitive to increases in extracellular dopamine concentration (Kessler *et al.*, 1993b; Mukherjee *et al.*, 1996).

The goal of the present work was to examine how the detection of binding parameter changes compares for the high and low affinity benzamides, [^{11}C]raclopride and [^{18}F]fallypride, by performing PET studies in the same baboon at baseline and after *d*-amphetamine administration. In addition, the specificity of extrastriatal D$_2$ receptor binding was investigated by performing [^{18}F]fallypride PET studies at baseline and after a blocking dose of haloperidol (1 mg/kg).

II. METHODS

A. Data Acquisition

The radiosyntheses of [^{11}C]raclopride and [^{18}F]fallypride were performed according to previous methods (Halldin *et al.*, 1991; Mukherjee *et al.*, 1995). High specific activity (> 1500 Ci/mmol) [^{11}C]raclopride (30–40 mCi) and [^{18}F]fallypride (10–15 mCi) PET studies were performed in isoflurane-anesthetized baboons. Radioligand was injected as a slow bolus (30 sec) for baseline and *d*-amphetamine pretreatment studies: [^{11}C]raclopride; 0.6 mg/kg (n = 3) or 1.0 mg/kg (n = 2); and [^{18}F]fallypride; 0.6 mg/kg (n = 1) or 1.0 mg/kg (n = 1). A third baboon was studied at baseline and after haloperidol pretreatment (1 mg/kg) using [^{18}F]fallypride. The *d*-amphetamine was administered by bolus iv injection 5 min prior to radioligand injection (Dewey *et al.*, 1993), and the administration of the haloperidol was performed in the same way.

Dynamic PET data (20–27 frames) were acquired over 90 and 210 min for [^{11}C]raclopride and [^{18}F]fallypride, respectively, using a Siemens-CTI 951R/31 PET scanner (two-dimensional imaging). One animal was also scanned at baseline and again 5 min after haloperidol pretreatment for 210 and 120 min, respectively, using a CTI HR + PET scanner (three-dimensional imaging). Arterial blood samples (35–40) were collected for plasma input function determination with about 20 samples collected in the first 2 min. HPLC was used to determine radiolabeled metabolite concentrations in plasma at multiple (six or more) times throughout the study. The plasma free fraction (f_1) was determined from arterial blood samples collected prior to injection (ultrafiltration) for a small subset of the studies ([^{11}C]raclopride, n = 2; [^{18}F]fallypride, n = 2).

B. Data Analysis

The [^{18}F]fallypride studies were acquired with an average of 20 days between scanning sessions, and the paired studies were aligned using early summed images (0–8 min) and the automated image registration algorithm (Woods *et al.*, 1992). Regions-of-interest were used to generate time–activity curves corresponding to the frontal (Frt) and temporal (Tem) cortices, striatum (Str), thalamus (Thl), and cerebellum (Cer).

Data were analyzed using compartmental modeling and Logan graphical methods. The compartmental model solution (C_{MOD}) is shown in terms of a comprehensive four-compartment configuration that has been well described by Koeppe *et al.* (1991):

$$C_{MOD} = C_2 + C_3 + C_4 + (BV)C_p. \quad (1)$$

Briefly, C_2, C_3, and C_4 correspond to the free, specifically bound, and nonspecifically bound drug in brain, respectively. A blood volume (BV) parameter was included in the model solution and corresponds to a fractional amount of total plasma radioactivity concentration (C_p). In practice, kinetic data were fit using reduced two- and three-compartment models (2C and 3C) and nonlinear least-squares curve fitting. The Logan graphical method (Logan *et al.*, 1990) was routinely implemented with the radioligand distribution volume (DV) obtained from the linear regression slope value with no correction for blood volume. The compartmental model DV and DV-derived binding potential (BP$_{DV}$) values were calculated as

$$2C\ DV = \frac{K_1}{k'_2}, \quad \text{where } k'_2 = \left[\frac{k_2}{1 + k_3/k_4 + k_5/k_6}\right] \quad (2)$$

$$3C\ DV = \frac{K_1}{k_2 f_2}\left[1 + \frac{k_3 f_2}{k_4}\right] \quad (3)$$

$$BP_{DV} = \frac{DV_{ROI}}{DV_{F+NS}} - 1 = \frac{B'_{max} f_2}{K_D}, \quad (4)$$

where K_1 (ml/min/ml) and k_2 (min^{-1}) represent the bidirectional transport of radiotracer across the blood–brain barrier, k_3 (min^{-1}) and k_4 (min^{-1}) reflect the bimolecular association rate to (k_{on}, nM^{-1}min^{-1}) and the unimolecular dissociation rate from (k_{off}, min^{-1}) receptors, respectively, and the nonspecific binding parameters are k_5 (min^{-1}) and k_6 (min^{-1}). B'_{max} (nM) is the concentration of available receptors, K_D (nM) is the equilibrium dissociation constant, and f_2 (ml/ml) corresponds to the free fraction of radioligand in brain tissue, $f_2 = 1/(1 + k_5/k_6)$. The BP$_{DV}$ calculation was performed using a reference region DV value (DV$_{F+NS}$). For D$_2$ receptor studies, it is common to use the cerebellum as the reference region for which specific binding of the radiotracer is assumed to be negligible and the concentrations of free radiotracer and nonspecific binding (F + NS) are assumed to be approximately equal in all regions.

III. RESULTS

A. Plasma Analyses

The concentration of radiolabeled metabolites in plasma differed for the two radioligands. The fraction of unchanged [^{11}C]raclopride corresponded to 70 and 30% at 10 and 60–90 min, respectively. In contrast, [^{18}F]fallypride was metabolized more rapidly with about 50, 18, and 10% at 10, 90, and 150–210 min, respectively. The fraction of unchanged radioligand and the plasma input function (at early and late times) are shown for both radioligands in Figs. 1 and 2. The plasma-free fractions were similar for the two [^{11}C]raclopride baboons (0.10 and 0.09), but different for the two [^{18}F]fallypride baboons (0.19 and 0.08).

B. Data Analyses

The striatal time–activity curves were consistent with the high and low affinities of [^{18}F]fallypride and [^{11}C]raclopride, respectively, whereas the extrastriatal time–activity curves displayed rapid tissue clearance for both radiotracers (Fig. 3). Decreases in the regional uptake of radioligand are visually evident after d-amphetamine relative to baseline, although the decreases in extrastriatal data were less apparent than those observed for striatal data (Fig. 3). [^{11}C]Raclopride data were analyzed using a 3C model (60-min data set). Striatal [^{18}F]fallypride data were fit using 2C and 3C models (210 min), and extrastriatal data were fit using a 3C model (90 min).

Comparisons of the kinetic results for the striatal data that were obtained for the two baboons with comparable [^{11}C]raclopride and [^{18}F]fallypride studies are shown in Table 1. The compartmental results were in good agreement with the Logan graphical results. Relative to baseline, the Logan BP$_{DV}$ values for [^{11}C]raclopride data decreased by 29 and 39% after 0.6- and 1.0-mg/kg doses of d-amphetamine, respectively, with smaller decreases of 14 and 29% for [^{18}F]fallypride data acquired after the same respective doses of d-amphetamine. Examples of Logan-derived BP$_{DV}$ images are shown at baseline and following 1.0 mg/kg d-amphetamine pretreatment for the [^{18}F]fallypride and [^{11}C]raclopride studies performed in the same baboon (Fig. 4).

The [^{11}C]raclopride results described earlier were comparable to those obtained in other baboons that were studied only with [^{11}C]raclopride. A summary of all [^{11}C]raclopride results is shown in Table 2. The average striatal BP$_{DV}$ change for three baboons scanned after 0.6 mg/kg was $-26 \pm 2.6\%$. A trend toward a slightly higher striatal BP$_{DV}$ change is apparent for the [^{11}C]raclopride baboons that received 1.0 mg/kg of amphetamine.

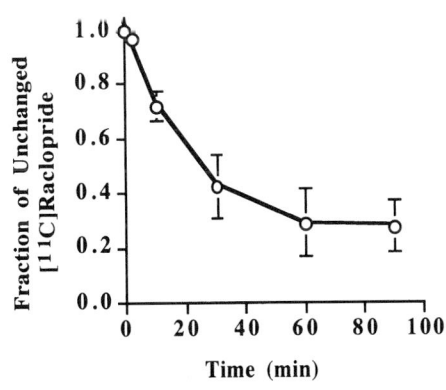

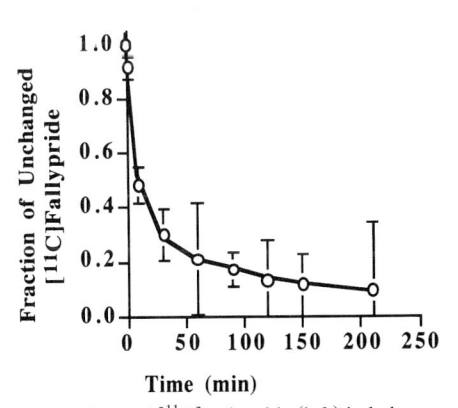

FIGURE 1 The average ($\pm$1SD) value of the fraction of unchanged [^{11}C]raclopride (left) in baboon plasma corresponded to 70 and 30% at 10 and 60–90 min, respectively. [^{18}F]fallypride (right) was metabolized more rapidly with about 50, 18, and 10% unchanged at 10, 90, and 150–210 min, respectively.

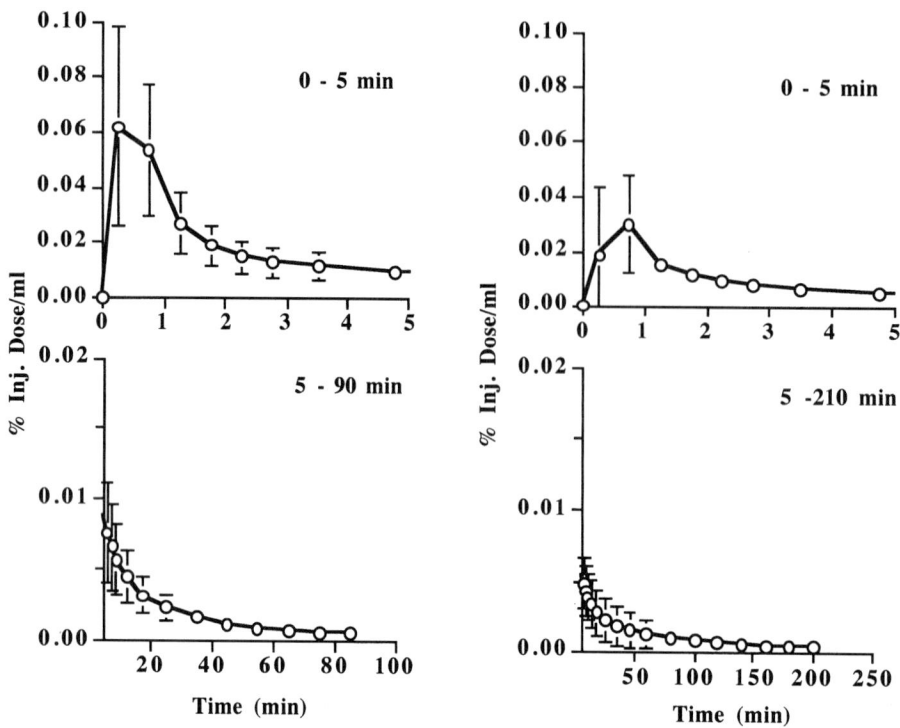

FIGURE 2 Plasma input function average values (±1 SD) at early (top) and late (bottom) study times for [^{11}C]raclopride (left, $n = 4$) and [^{18}F]fallypride (right, $n = 3$). Data indicate a lower percentage injected dose per milliliter in plasma for [^{18}F]fallypride. Plasma clearance was evident for both radioligands throughout the time frame of the PET studies.

The regional BP$_{DV}$ values followed the human D$_2$ receptor specific binding rank order for both radioligands (Str > Thl > Tem > Frt > Cer) and were highly correlated (> 0.9) with the human D$_2$ receptor binding rank order of [^{125}I]epidepride (Kessler et al., 1993a). At baseline, the extrastriatal range of BP$_{DV}$ values from thalamus to frontal cortex was 0.65 to 0.30 for [^{11}C]raclopride and 1.2 to 0.30 for [^{18}F]fallypride. The extrastriatal [^{11}C]raclopride binding parameters generally decreased following d-amphetamine pretreatment with occasional increases observed for the frontal cortex. The extrastriatal [^{18}F]fallypride-binding parameters were generally inconsistent with the postamphetamine striatal reductions as increases in BP$_{DV}$ were more often observed for these regions. The robustness of the extrastriatal-binding parameter measures were investigated very preliminarily by studying a third baboon with [^{18}F]fallypride at baseline and after receptor blockade with haloperidol (1 mg/kg). The baseline [^{18}F]fallypride BP$_{DV}$ values were highly correlated ($r^2 = 1.0$) with human D$_2$ receptor rank order (Kessler et al., 1993a) and uncorrelated after a 1-mg/kg dose of haloperidol ($r^2 = 0.06$).

IV. DISCUSSION

In this work, *in vivo* PET imaging studies in baboons demonstrated that striatal binding parameter changes can be detected for the reversible high affinity benzaminde [^{18}F]fallypride, following d-amphetamine pretreatment doses of 0.6 and 1.0 mg/kg. The magnitude of the change in the striatal [^{18}F]fallypride-binding parameter was lower than that detected in the same baboon using the lower affinity benzamide [^{11}C]raclopride. In addition, there was a trend toward a larger change in the binding parameters for the larger dose of d-amphetamine. Finally, pretreatment with the antagonist haloperidol resulted in [^{18}F]fallypride BP$_{DV}$ values that were not correlated ($r^2 = 0.06$) with the *in vitro* human D$_2$ receptor rank order in contrast to the observation of high correlation at baseline ($r^2 = 1.0$).

The baseline striatal time–activity curves obtained for the two benzamides reflected the different radioligand affinities with peak tissue uptake at 80 and 5 min for [^{18}F]fallypride and [^{11}C]raclopride, respectively. The [^{18}F]fallypride striatal BP$_{DV}$ values were only about a factor of 6 higher than those obtained for [^{11}C]raclo-

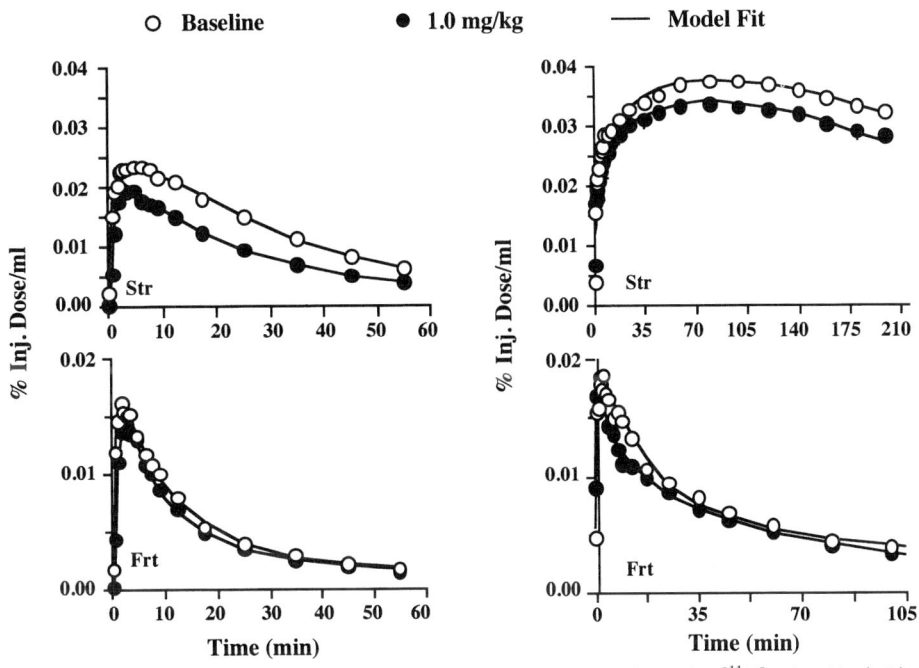

FIGURE 3 Striatal and frontal time–activity curves examples are shown for [11C]raclopride (left) and [18F]fallypride (right) along with the model curve fits to baseline and postamphetamine (1 mg/kg) data. Decreases in the regional uptake of radioligand are visually evident after d-amphetamine relative to baseline, although the decrease for frontal data is less apparent than that observed for striatal data. The kinetic models were applied to 60 min of [11C]raclopride data and to 90 min (extrastriatal) or 210 min (striatal) of [18F]fallypride data.

pride despite the factor of 30 difference between the *in vitro* affinities of these radioligands.

The quantitative results obtained for [11C]raclopride were consistent with a methodological comparison of the d-amphetamine/[11C]raclopride bolus and bolus + constant infusion (B/I) paradigms (administration of d-amphetamine during equilibrium) performed by Carson *et al.* (1997). These studies were performed using rhesus monkeys and a lower d-amphetamine dose of 0.4 mg/kg, with reported changes in the bolus-binding parameters than ranged from 22 to 42%, whereas less of a change was detected for the B/I studies (19 ±

TABLE 1 Comparison of [11C]Raclopride and [18F]Fallypride
(Baseline and change after d-amphetamine)

Striatum		Model		Logan	
		DV	BP$_{DV}$	DV	BP$_{DV}$
Baboon 1	[11C]Raclopride	2.4	3.3	2.5	2.8
(0.6 mg/kg)	% Diff	−25%	−30%	−23%	−29%
	[18F]Fallypride	14	16	13	14
	% Diff	−3.9%	−8.8%	−5.8%	−14%
Baboon 2	[11C]Raclopride	4.1	3.3	4.1	2.8
(1.0 mg/kg)	% Diff	−38%	−45%	−36%	−39%
	[18F]Fallypride	67	20	71	21
	% Diff	−11%	−22%	−12%	−29%

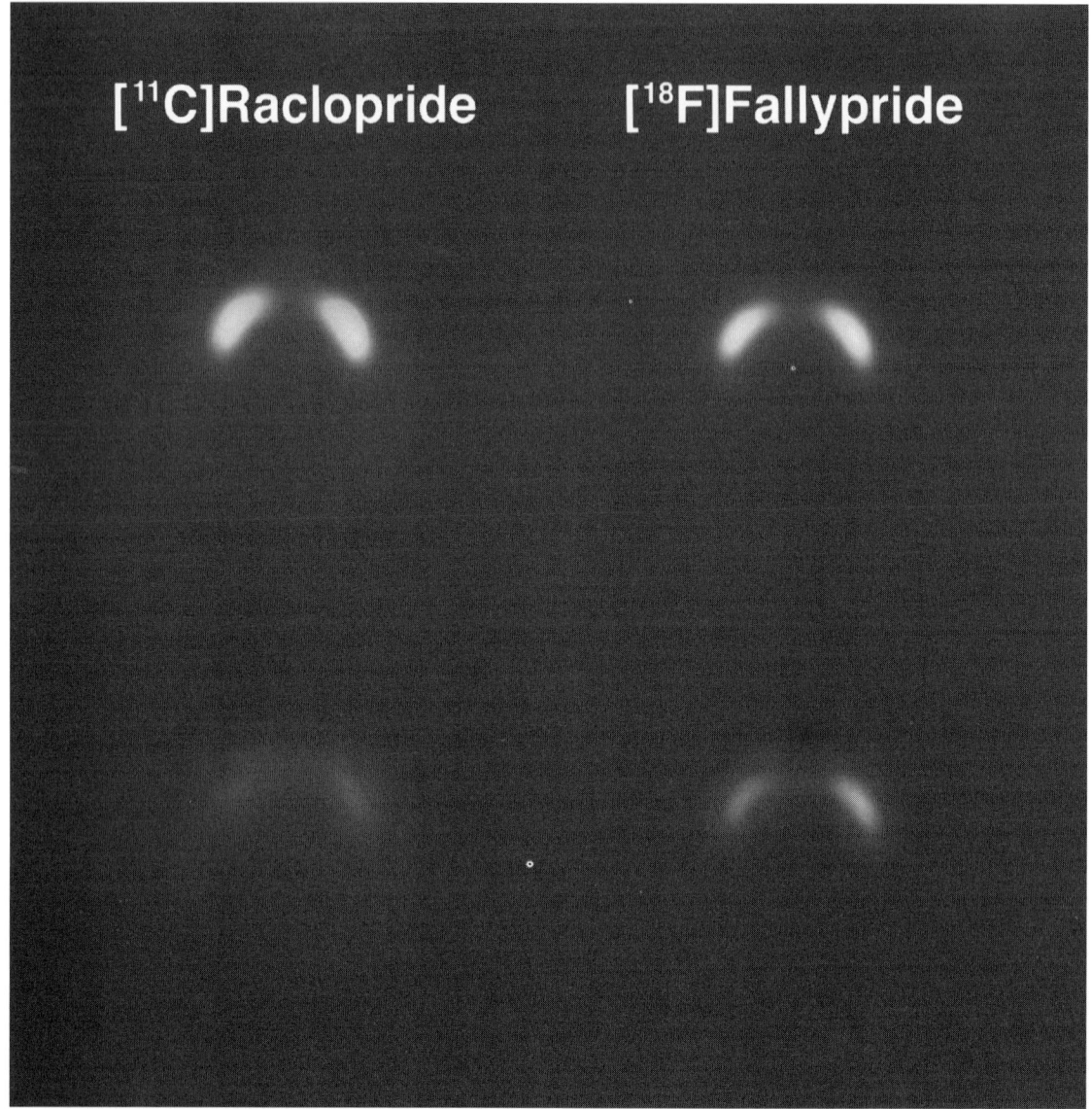

FIGURE 4 Examples of Logan-derived BP_{DV} images are shown for [^{11}C]raclopride (left) and [^{18}F]fallypride (right) at baseline and following 1.0 mg/kg d-amphetamine pretreatment. Decreases in the BP_{DV} images are visually evident after the d-amphetamine dose and correspond to striatal BP_{DV} decreases that are about 10% greater for [^{11}C]raclopride. Images are scaled to a common maximum.

TABLE 2 [^{11}C]Raclopride: 0.6 and 1.0 mg/kg (Baseline and change after d-amphetamine)

	Striatum	Model DV	Model BP_{DV}	Logan DV	Logan BP_{DV}
Baseline	($n = 5$)	3.2 ± 0.70	3.2 ± 0.19	3.2 ± 0.65	2.8 ± 0.32
0.6 mg/kg ($n = 3$)	% Diff	$-15 \pm 11\%$	$-26 \pm 9.5\%$	$-11 \pm 11\%$	$-26 \pm 2.6\%$
1.0 mg/kg ($n = 2$)	% Diff	$-29\%, -38\%$	$-39\%, -45\%$	$-27\%, -36\%$	$-31\%, -39\%$

16%). Similarly, [^{123}I]IBZM, a D$_2$ antagonist with slightly higher affinity than raclopride (K_D = 0.4 nM), has demonstrated *in vivo* sensitivity to extracellular dopamine in humans with binding parameter decreases of about 15% for single photon emission tomography imaging studies using a B/I paradigm and a 0.3-mg/kg dose of *d*-amphetamine (Laruelle *et al.*, 1995).

The quantitative results obtained for [^{18}F]fallypride yielded BP$_{DV}$ values that were consistent with preliminary Logan graphical results obtained in humans (Kessler *et al.*, 1997). In the present work, the DV values for the two [^{18}F]fallypride baboons were different by about a factor of 5, which may be partially attributed to a difference of about 2.5 in the f_1 values with the larger DV associated with the lower value of protein binding (higher free fraction).

Changes were observed in the cerebellar DV values. For [^{11}C]raclopride, the cerebellar DV changes ranged from 0 to 12%, which is similar to the measured test–retest variability reported for [^{11}C]raclopride baboon studies (Dewey *et al.*, 1993). Changes of 6–12% were also observed for the [^{18}F]fallypride cerebellar DV values, which is consistent with the test–retest variability reported for [^{18}F]N-methylspiperone baboon studies (Dewey *et al.*, 1991). The test–retest variability of the [^{18}F]fallypride baboon studies is not yet known.

High affinity D$_2$ radioligands, such as [^{11}C]N-methylspiperone and [^{18}F]N-methylspiperone (K_D = 0.2 nM), exhibit *in vivo* kinetics that are irreversible during the time frame of the PET study with little propensity for displacement by dopamine (K_D = 5 nM; Seeman *et al.*, 1990) or haloperidol (K_i = 1 nM) in contrast to the observed sensitivity of [^{11}C]raclopride for which the *in vitro* affinity (K_D = 1 nM) is five-fold lower. In this work, [^{18}F]fallypride, a radioligand with very high *in vitro* affinity (K_D = 0.03 nM), demonstrated reversible binding kinetics with substantial specific binding parameter reductions (relative to baseline) following *d*-amphetamine administration that were interpreted as an *in vivo* sensitivity to extracellular dopamine. The reductions that were observed for [^{18}F]fallypride were as much as a factor of 2 lower than those observed for [^{11}C]raclopride, in the same baboon. These results are consistent with the finding that *in vivo* sensitivity can be demonstrated for high affinity radioligands due to tissue clearance effects (Logan *et al.*, 1991) and with previous rhesus monkey PET studies that have reported raclopride and haloperidol displacement of [^{18}F]fallypride with clearance half-times of about 12 min (Votaw *et al.*, 1993) and 18 min (Mukherjee *et al.*, 1995). These clearance values correspond to *in vivo* dissociation rates of about 0.06 and 0.02 min^{-1}, respectively, that are on the same order as the *in vivo* k_{off} of 0.10 min^{-1} determined in humans with [^{11}C]raclopride (Farde *et al.*, 1989). Therefore, the differences observed between [^{18}F]fallypride and [^{11}C]raclopride most likely result from properties that are less traditionally associated with high and low affinity ligand comparisons, such as differences in association rate.

V. CONCLUSIONS

Changes in striatal binding parameters can be detected using the reversible and very high affinity benzamide [^{18}F]fallypride, following *d*-amphetamine pretreatment doses of 0.6 and 1.0 mg/kg. The smaller [^{18}F]fallypride binding parameter changes that were observed relative to [^{11}C]raclopride indicate that a threshold of sensitivity may exist for [^{18}F]fallypride below which sensitivity to endogenous dopamine could not be detected. Additional studies are underway to better understand these initial observations.

Acknowledgments

The authors thank the UPMC PET facility staff. This work was supported by NIH Grants MH52247, NS22899, and MH49936 and grants from the NARSAD and Stanley Foundations.

References

Carson, R. E., Breier, A., DeBartolomeis, A., Saunders, R. C., Su, T. P., Schmall, B., Der, M. G., Pickar, D., and Eckelman, W. C. (1997). Quantification of amphetamine-induced changes in [^{11}C] raclopride binding with continuous infusion. *J. Cereb. Blood Flow Metab.* 17: 437–447.

Dewey, S. L., Dewey, S. L., Wolf, A. P., Brodie, J. D., Angrist, B., Fowler, J. S., and Volkow, N. D. (1991). Amphetamine induced decreases in (^{18}F)-N-methylspiroperidol binding in the baboon brain using positron emission tomography (PET). *Synapse* 7: 324–327.

Dewey, S. L., Smith, G. S., Dewey, S. L., Brodie, J. D., Fowler, J. S., and Wolf, A. P. (1993). Striatal binding of the PET ligand ^{11}C-raclopride is altered by drugs that modify synaptic dopamine levels. *Synapse* 13: 350–356.

Farde, L., Eriksson, L., Blomquist, G., and Halldin, C. (1989). Kinetic analysis of central [^{11}C]raclopride binding to D$_2$-dopamine receptors studied by PET—a comparison to the equilibrium analysis. *J. Cereb. Blood Flow Metab.* 9: 696–708.

Hall, H., Wedel., I., Halldin, C., Kopp, J., and Farde, L. (1990). Comparison of the in vitro receptor binding properties of N-[^{3}H]methylspiperone and [^{3}H]raclopride to rat and human brain membranes. *J. Neurochem.* 55: 2048–2057.

Halldin, C., Farde, L., Hogberg, T., Hall, H., Strom, P., Ohlberger, A., and Solin, O. (1991). A comparative PET-study of five carbon-11 or fluorine-18 labelled salicylamides. Preparation and in vitro dopamine D-2 receptor binding. *Nucl. Med. Biol.* 18: 871–881.

Kessler, R. M., Whetsell, W. O., Ansari, M. S., Votaw, J. R., de Paulis, T., Clanton, J. A., Schmidt, D. E., Mason, N. S., and Manning, R. G. (1993a). Identification of extrastriatal dopamine D$_2$ receptors in post mortem human brain with [^{125}I]epidepride. *Brain Res.* 609: 237–243.

Kessler, R. M., Mason, N. S., Votaw, J. R., Ansari, M. S., de Paulis, T., Schmidt, D. E., and Manning, R. G. (1993b). Factors underlying the D-amphetamine induced displacement of dopamine D2 radioligands in rhesus monkeys. *J. Nucl. Med.* **34**: P132.

Kessler, R. M., Mason, N. S., Jones, C., Price, R. C., Manning, R. G., Ansari, M. S., Rieck, R. W., and Shelton, R. (1997). In vivo quantitation of striatal and extrastriatal dopamine D2 receptors in human brain using PET and [^{18}F]N-allyl-5-fluoropropylepidepride. *J. Nucl. Med.* **38**: P289.

Koeppe, R. A., Holthoff, V. A., Frey, K. A., Kilbourn, M. R., and Kuhl, D. E. (1991). Compartmental analysis of [^{11}C]flumazenil kinetics for the estimation of ligand transport rate and receptor distribution using positron emission tomography. *J. Cereb. Blood Flow Metab.* **11**: 735-744.

Laruelle, M., Abi-Dargham, A., van Dyck, C. H., Rosenblatt, W., Zea-Ponce, Y., Zoghbi, S. S., Baldwin, R. M., Chaney, D. S., Hoffer, P. B., Kung, H. F., and Innis, R. B. (1995). SPECT Imaging of striatal dopamine release after amphetamine challenge. *J. Nucl. Med.* **36**: 1182-1190.

Logan, J., Fowler, J. S., Volkow, N. D., Wolf, A. P., Dewey, S. L., Schlyer, D. J., MacGregor, R. R., Hitzemann, R., Bendriem, B., Gatley, S. J., and Christman, D. R. (1990). Graphical analysis of reversible radioligand binding from time-activity measurements applied to [N-^{11}C-methyl]-(-)-cocaine PET studies in human subjects. *J. Cereb. Blood Flow Metab.* **10**: 740-747.

Logan, J., Dewey, S. L., Wolf, A. P., Fowler, J. S., Brodie, J. D., Angrist, B., Volkow, N. D., and Gatley, S. J. (1991). Effects of endogenous dopamine on measures of [^{18}F]-N-methylspiroperidol binding in the basal ganglia: Comparison of simulations and experimental results from PET studies in baboons. *Synapse* **9**: 195-207.

Mukherjee, J., Yang, Z.-Y., Das, M. K., and Brown, T. (1995). Fluorinated benzamide neuroleptics. III. Development of (S)-N-[(1-allyl-2-pyrrolidinyl)methyl]-5-(3-[^{18}F]fluoropropyl)-2, 3-dimethoxybenzamide as an improved dopamine D-2 receptor tracer. *Nucl. Med. Biol.* **22**: 283-296.

Mukherjee, I., Yang, Z.-Y., Brown, T., Roemer, J., and Cooper, M. (1996). Evaluation of extrastriatal dopamine receptors by using ^{18}F-fallypride. *Soc. Neurosci. Abstr.* **22**: P498.

Seeman, P., Niznik, H. B., and Guan, H.-C. (1990). Elevation of dopamine D2 receptors in schizophrenia is underestimated by radioactive raclopride. *Arch. Gen. Psychiatry* **47**: 1170-1172.

Votaw, J. R., Kessler, R. M., de Paulis, T., Ansari, M. S., Mason, N. S., Schmidt, D. E., Manning, R. G., and Holburn, G. (1993). Quantification of striatal and extrastriatal dopamine D2 receptors using N-allyl-[^{18}F]FPE. *J. Nucl. Med.* **34**: P132.

Woods, R. P., Cherry, S. R., and Mazziotta, J. C. (1992). Rapid automated algorithm for aligning and reslicing PET images. *J. Comput. Assist. Tomogr.* **16**(4): 620-633.

CHAPTER 67

Measuring Neurotransmitter Release with PET: Methodological Issues[1]

ALAIN DAGHER,*,† ROGER N. GUNN,† GEOFF LOCKWOOD,† VINCENT J. CUNNINGHAM,†
PAUL M. GRASBY,† and DAVID J. BROOKS†

*McConnell Brain Imaging Centre
Montreal Neurological Institute
Montreal, Quebec, Canada H3A 2B4
†MRC Cyclotron Unit
Hammersmith Hospital
London W12, United Kingdom

The objective of this study is to develop a method of measuring the release of neurotransmitter in the human brain based on the premise that endogenous neurotransmitter competes with a positron emission tomography neuroreceptor tracer for binding at the receptor site. A graphical method was used to estimate the change in distribution volume of the opiate antagonist [^{11}C]diprenorphine in response to stimuli intended to promote the release of opioid peptides in the brain. Six subjects inhaled the anesthetic agent nitrous oxide, which is thought to promote opioid peptide release. There was generalized displacement of the tracer throughout the brain in three out of six subjects. However, simulations showed that changes in cerebral blood flow (CBF) due to the stimulus could act as a confounding factor by causing significant displacement of radiotracer, and that this would be indistinguishable from effects attributed to endogenous transmitter release. Moreover, these studies also suffered from significant motion artifacts despite conventional head restraints. Standard realignment algorithms were applied to correct for motion; however, they introduced systematic errors (false displacements) due to the changing distribution of radioligand during the course of the scan. It was concluded that (1) in a displacement protocol, CBF changes in response to activation may mimic the effects of endogenous transmitter release; and (2) false positives may *also result from slight subject motion. Motion correction algorithms, developed for [^{15}O]water activation studies, cannot be applied to dynamic radioligand images.*

I. INTRODUCTION

The physiological release of neurotransmitter from neurons, a pivotal event in brain function, has so far not been measured in humans. It has been possible to measure neuroreceptor densities with positron emission tomography (PET), but until recently, there has been little success in actually measuring the release of neurotransmitter. Many groups have now measured *in vivo* changes in the uptake of dopamine receptor ligands in response to various pharmacological challenges designed to promote the release of endogenous dopamine (Logan *et al.*, 1991; Young *et al.*, 1991; Dewey *et al.*, 1992, 1993a, b, 1995; Innis *et al.*, 1992; Laruelle *et al.*, 1995; Carson *et al.*, 1997). This technique has provided new insights into the control of dopamine release in schizophrenia (Laruelle *et al.*, 1996; Breier *et al.*, 1997) and Parkinson's disease (Tedroff *et al.*, 1996). However, it would also be interesting to measure neurotransmitter release in response to cognitive or behavioral activations and to generate statistical parametric maps of transmitter release, as is done for cerebral blood flow (CBF) (Friston *et al.*, 1991; Evans *et al.*, 1992).

[1] Transcripts of the BRAINPET97 discussion of this chapter can be found in Section VIII.

The basic premise of neurotransmitter activation studies is that an endogenous neurotransmitter will compete with the PET tracer for binding at the receptor site. In a *displacement* study, the scan starts with the subject in the baseline or normal physiological state. At some point during the scan, an activation is carried out (e.g., a cognitive task, injection of a drug) that stimulates the release of the endogenous neurotransmitter, which displaces the radioligand from its binding sites in the brain. An increase in the washout of neurotransmitter is then measured by PET. The term "displacement" is misleading as an endogenous neurotransmitter does not actively displace the tracer from brain, but favors its washout by reducing the number of available binding sites. The spontaneous dissociation (k_{off}) of ligand from receptor is itself unaffected by the concentration of endogenous transmitter. A displacement study can be accomplished either with a bolus injection of radioligand or with a continuous infusion. In a *competition* study, there are two separate injections of radiotracer (usually in two scanning sessions): one while the patient is in the baseline state and one in the activation state.

II. MATERIALS AND METHODS

A. Theory

The observed time series of PET images of neuroreceptor ligands can be modeled using classical compartmental approaches. For most ligands, a three-compartment model is sufficient to account for the observed activity as a function of time. This is certainly true of the tracers most often used for measuring neurotransmitter release: [^{11}C]raclopride, a dopamine D_2 ligand (Farde *et al.*, 1989), [^{11}C]SCH23390, a dopamine D_1 ligand (Lammertsma and Hume, 1996), and [^{11}C]diprenorphine, a nonselective opioid ligand (Jones *et al.*, 1994).

The concentration of intrasynaptic neurotransmitter determines the number of available binding sites for the radioligand. However, changes in rCBF will also affect the rate of uptake and washout of the radioligand. The rate constants of the compartmental model can be related to blood flow and available receptor sites by the following equations, assuming the radioligand is present in tracer quantities, and neglecting nonspecific tissue binding of tracer (Logan *et al.*, 1990).

$$K_1 = F(1 - e^{-PS_1/F})$$

$$k_2 = \frac{PS_2}{PS_1} F(1 - e^{-PS_1/F})$$

$$k_3 = k_{on}(B_{max} - B_{en})$$

$$k_4 = k_{off},$$

where K_1 (ml g^{-1} min^{-1}) is the plasma to brain transport constant, k_2 (min^{-1}) is the brain to plasma efflux constant, F (ml g^{-1} min^{-1}) is the blood flow per gram of tissue, PS_1 and PS_2 (ml g^{-1} min^{-1}) are the capillary permeability surface area products for blood to brain and brain to blood, respectively, B_{max} (pmol/cc) is the concentration of available receptor sites, B_{en} is the concentration of bound endogenous neurotransmitter, and k_{on} and k_{off} are the association and dissociation constants between ligand and binding site. According to these equations it is evident that a change in endogenous neurotransmitter concentration will only affect k_3. However, most activation paradigms can be expected to have an effect on rCBF, which could act as a confounding factor, as discussed later.

Graphical methods of PET data analysis lend themselves particularly well to the generation of parametric images. They have the advantage of being relatively model independent and computationally simple. Logan *et al.* (1990) have proposed a graphical technique for the analysis of reversible ligands.

By plotting $\int_0^T C_{pet}(t)\,dt/C_{pet}(T)$ against $\int_0^T C_{pl}(t)\,dt/C_{pet}(T)$, one obtains, after a certain time, a straight line with slope

$$DV = \frac{K_1}{k_2}\left(1 + \frac{k_3}{k_4}\right) + V_p,$$

where C_{pl} is the plasma concentration of radiotracer, C_{pet} is the radioactivity concentration in brain measured by PET, $K_d = k_{off}/k_{on}$ is the equilibrium dissociation constant for the tracer, and V_p is the cerebral plasma volume. It is clear that if the plasma volume is neglected and if k_3 is much larger than k_4, then DV, the slope of the Logan plot, is approximately proportional to k_3 and hence to the number of available binding sites. Thus the release of endogenous neurotransmitter can be expected to reduce DV. By using linear regression on a voxel basis for data points before and after activation, one can in principle obtain a ratio of slopes proportional to the relative change in B_{max} and thus generate a *displacement* image. Moreover, by applying a standard formula (Goldstein, 1964) for generating a t statistic from the two linear regressions at each voxel (assuming a constant variance), one can obtain a *t map* of displacement. (Note that B_{max} is used to denote the concentration of available receptor sites, not total receptor sites.)

B. Scanning Protocols

Fifteen healthy male volunteers were scanned with [^{11}C]diprenorphine, a nonselective opiate antagonist. The authors attempted to induce neurotransmitter release either by having the subjects inhale 30% nitrous oxide (N$_2$O, six subjects), which is thought to act as an anesthetic by stimulating the release of endogenous opioid peptides (Quock et al., 1985, 1993; Finck et al., 1995; Guo et al., 1996), or by topical application of capsaicin solution (0.004 M) to the forearm to cause local pain (two subjects). In both paradigms, the activation was started 45 min after injection of tracer and continued for the rest of the scanning period. Seven subjects served as controls and remained in the resting state during the entire scanning period. Following injection of the radiotracer, dynamic frames were acquired on the CTI-953 camera in three-dimensional (3D) mode for 120 min. Arterial plasma samples, corrected for metabolites, were used for determination of the input function. In the nitrous oxide experiments, end-tidal pCO$_2$ and pO$_2$ were measured throughout the experiment. Data were analyzed using the graphical method described earlier to generate statistical displacement maps.

C. Effect of Blood Flow

Most activation paradigms can be expected to change rCBF. This would, in turn, affect the rate constants K_1 and k_2 as defined in the compartmental model described earlier. Thus, blood flow changes will act as a confounding factor, possibly altering the delivery or washout of tracer. It has been assumed that this effect is negligible. Based on the assumption that capillary recruitment does not occur in the brain, most authors have simulated vascular effects by changing flow only, but leaving the capillary PS product (permeability times surface area) constant (Bartenstein et al., 1993; Logan et al., 1994). Bartenstein et al. (1993) simulated the effect of a change in CBF for the tracer [^{11}C]diprenorphine. The effects of even a 10-fold increase in blood flow were shown to be negligible on simulated PET time–activity curves when PS was unchanged.

However, the assumption that changes in local CBF in capillary systems are due only to changes in F, and not in PS, is controversial (Villringer and Dirnagl, 1995). The PS product could change in one of two ways: (1) through capillary recruitment or (2) through capillary dilatation. There is good agreement that the rCBF changes that accompany cerebral activation are due to arteriolar dilatation, which increases blood flow through capillary systems (Villringer and Dirnagl, 1995). However, the consequence of arteriolar dilatation in the capillary beds downstream is uncertain. The current consensus is that recruitment is not significant in the brain, but that dilatation of already perfused capillaries may occur (Duelli and Kuschinsky, 1993; Chen et al., 1995). Even small changes in capillary diameter could be significant as PS depends on the square of the capillary diameter. Chen et al. (1995) have tested the effects on influx rate constants and PS product of regional changes in CBF with a variety of tracers following nicotine administration in rats. They found that slight increases in K_1 and PS could occur, but that these depended on the extraction fraction of the tracer. However, even for tracers with a relatively low K_1 (0.2–0.3 ml g^{-1} min^{-1}), they concluded that a change in PS of about 8% (0–23%) and a change of K_1 of approximately 15% (5–27%) could occur after nicotine injection. The K_1 values for raclopride and diprenorphine fall roughly within this range (Farde et al., 1989; Jones et al., 1994). It is unclear how these findings might affect the interpretability of cognitive activation studies as the rCBF changes induced by nicotine were quite high (30–150%).

In a single human subject, Logan et al. (1994) performed [^{11}C]raclopride scans at rest and during hyperventilation to produce vasoconstriction. They found significant reductions in both K_1 (43%) and DV (18%). Using a visual activation task, Holthoff et al. (1991) found a 20% increase in K_1 for [^{11}C]flumazenil in the occipital cortex. Because the extraction fraction for flumazenil is in the same range as that of raclopride and diprenorphine ($K_1 \cong 0.3$ ml g^{-1} min^{-1}), one must conclude that blood flow changes can also alter the transport characteristics of these tracers.

The effect of blood flow changes on the measurement of neurotransmitter release was simulated by solving the standard three-compartment model using the Runge–Kutta method with a real plasma curve as input function (using Matlab software, The Mathworks Inc., Natick, MA).

III. RESULTS AND DISCUSSION

A. PET Experiments

Statistical parametric images of [^{11}C]diprenorphine displacement were generated as described earlier. In three of the six subjects who received nitrous oxide (N$_2$O) there was diffuse displacement of the labeled diprenorphine, as shown by a statistically significant reduction in DV after the administration of N$_2$O. No such effects were observed in any of the control subjects. The displacement occurred only in areas known to contain moderately high densities of opiate receptors, but not in primary sensorimotor and occipital cortices, which have few opiate receptors. There was, however, homogeneous displacement in all other cortical areas, thalamus, and basal ganglia. Moreover, there were no differences in the effect of nitrous oxide on end-tidal pCO_2 nor was there any correlation between the occurrence of displacement and either the change in pCO_2 or the effect of nitrous oxide on neuropsychological function, as assessed by measuring tapping speed, short-term memory, and reaction times (data not shown).

The displacement of [^{11}C]diprenorphine could have been due either to the release of endogenous opiates or to an increase in CBF. Controversy exists regarding the mechanism of the anesthetic action of nitrous oxide; however, several different studies suggest that it acts by stimulating the release of endogenous opiates (Quock *et al.*, 1985, 1993; Finck *et al.*, 1995; Guo *et al.*, 1996). Experimental evidence also shows that nitrous oxide increases cerebral blood flow, even in the presence of hypocapnea (Field *et al.*, 1993; Reinstrup *et al.*, 1994; Hormann *et al.*, 1997). However, the only PET studies that measured the effect of nitrous oxide on CBF did not show a generalized increase in blood flow (Gyulai *et al.*, 1996, 1997), but rather increased CBF in the anterior cingulate cortex and decreased rCBF in the posterior cingulate, medial temporal, and visual association areas.

B. PET Simulations

Simulations showed that the single scan method employed here could not distinguish between a change in blood flow and the effect of endogenous neurotransmitter release (Fig. 1). The effect on the tissue time–activity curve of a change in K_1 and k_2 (i.e., vascular effect) was indistinguishable from a change in k_3 (i.e., neurotransmitter release). The assumption that vascular effects are negligible in ligand studies can therefore no longer be accepted. Moreover, these simulations demonstrate that the present finding of generalized displacement of [^{11}C]diprenorphine in response to nitrous oxide could be due either to blood flow changes or to the release of opioid peptides.

C. Correcting for Patient Motion

Patient movement during a displacement scan could lead to the detection of false-positive tracer displacements, especially if the movements occur in a systematic way. For example, in a behavioral experiment where the activation paradigm involves performing a task presented on a video monitor, one might expect a consistent change in head position when going from the baseline to the activation state. In [^{15}O]H_2O activation studies, correction for patient motion is achieved with PET to PET coregistration algorithms (Friston *et al.*, 1995; Woods *et al.*, 1992).

In the present studies, even slight patient movement leads to false radioligand displacements (Fig. 2, see color insert). The authors have attempted to apply PET to PET coregistration to the dynamic frames of each scan to correct for this artifact. However, this also creates false ligand displacements. The coregistration algorithms are designed to minimize, in a least-square sense, the difference between two images. However,

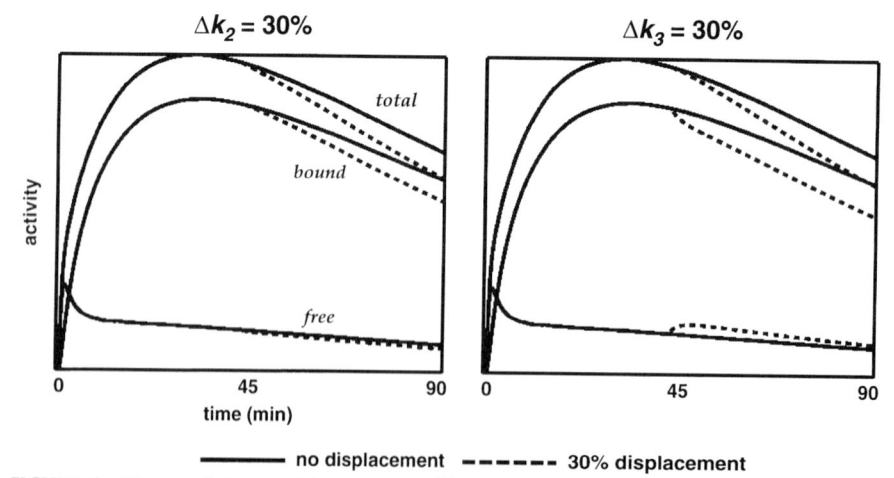

FIGURE 1 Simulated time–activity curves for [^{11}C]diprenorphine. A real plasma input function was used, along with the published values of the four-rate constants, to solve the standard differential equations describing a three-compartment model. The effects of a 30% change in k_2 (left) or k_3 (right) were simulated. As shown, both changes have the same effect on the tissue radioactivity distribution.

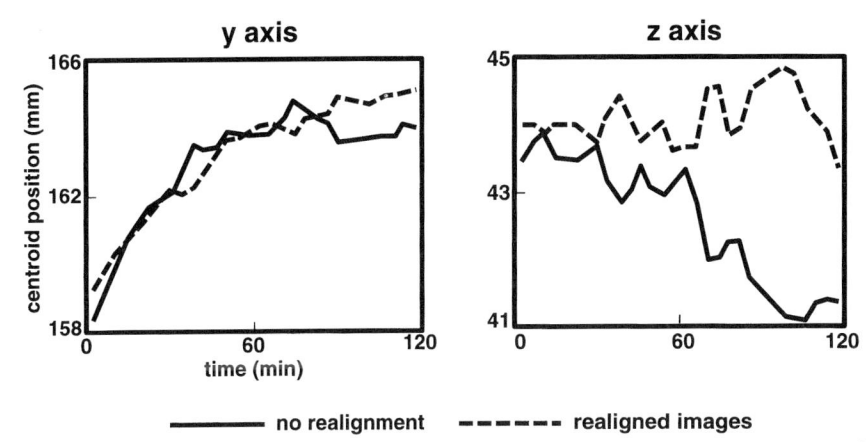

FIGURE 3 Effect of changing radioactivity distribution. These plots of the center of mass of the radioactivity distribution show that there is a change in both the y and z axis with time. This is due to the gradual loss of radioactivity in cerebellum and occipital cortex, areas poor in opiate receptors. The correction algorithm *corrects* the change in the z axis, which has the effect of pulling the images down and causing areas of false displacement (see Fig. 2). The same phenomenon is also seen in the pitch of the image (data not shown).

they assume that the radioactivity distribution in the two images to be coregistered is similar. However, in the early frames of dynamic PET scans with neuroreceptor ligands the distribution of radioactivity reflects tracer delivery, whereas in the later frames radioactivity uptake depends on the distribution of receptors for the ligand. This change over time in the distribution of radioactivity will increase the sum of squares difference between two spatially coregistered images, and the coregistration algorithm will attempt to *correct* this by moving one of the two images.

In the case of [^{11}C]diprenorphine, there is a gradual relative loss of radioactivity in the occipital lobe and cerebellum in the late time frames. This washout results in a gradual shift in the center of mass of the radioactivity distribution with time (Fig. 3). After application of the realignment algorithm, the images have been moved such that the center of mass remains constant in the z axis (Fig. 3). This implies that the brain has been pulled down by the realignment procedure. As shown in Fig. 2, this can lead to artifact (i.e., areas of *false radioligand displacement*). In this case, there is an apparent loss of radioactivity in the superior part of the thalamus, which results from the realignment algorithm *pulling* the brain down. This could be misinterpreted as reflecting opioid peptide release in the thalamus.

IV. CONCLUSION

These studies have identified two serious problems in the measurement of endogenous neurotransmitter release with PET. First, the effects of changes in CBF causing changes in ligand binding must be taken into account. Whatever method one uses to measure neurotransmitter release, simulation studies will be helpful in determining the potential confounding effect of vascular changes. Second, standard motion correction algorithms cannot be applied to dynamic ligand studies if they are based on the assumption that radioactivity distribution remains constant during the entire scan.

References

Bartenstein, P. A., Duncan, J. S., Prevett, M. C., Cunningham, V. J., Fish, D. R., Jones, A. K., Luthra, S. K., Sawle, G. V., and Brooks, D. J. (1993). Investigation of the opioid system in absence seizures with positron emission tomography. *J. Neurol., Neurosurg. Psychiatry* 56: 1295–1302.

Breier, A., Su, T. P., Saunders, R., Carson, R. E., Kolachana, B. S., de Bartolomeis, A., Weinberger, D. R., Weisenfeld, N., Malhotra, A. K., Eckelman, W. C., and Pickar, D. (1997). Schizophrenia is associated with elevated amphetamine-induced synaptic dopamine concentrations: Evidence from a novel positron emission tomography method. *Proc. Natl. Acad. Sci. U.S.A.* 94: 2569–2574.

Carson, R. E., Breier, A., de Bartolomeis, A., Saunders, R. C., Su, T. P., Schmall, B., Der, M. G., Pickar, D., and Eckelman, W. C. (1997). Quantification of amphetamine-induced changes in [^{11}C] raclopride binding with continuous infusion. *J. Cereb. Blood Flow Metab.* 17: 437–447.

Chen, J. L., Wei, L., Bereczki, D., Hans, F. J., Otsuka, T., Acuff, V., Ghersi Egea, J. F., Patlak, C., and Fenstermacher, J. D. (1995). Nicotine raises the influx of permeable solutes across the rat blood-brain barrier with little or no capillary recruitment. *J. Cereb. Blood Flow Metab.* 15: 687–698.

Dewey, S. L., Smith, G. S., Logan, J., Brodie, J. D., Yu, D. W., Ferrieri, R. A., King, P. T., MacGregor, R. R., Martin, T. P., Wolf, A. P., Volkow, N. D., Fowler, J. S., and Meller, E. (1992). GABAergic inhibition of endogenous dopamine release measured in vivo with ^{11}C-raclopride and positron emission tomography. *J. Neurosci.* 12: 3773–3780.

Dewey, S. L., Smith, G. S., Logan, J., Brodie, J. D., Fowler, J. S., and Wolf, A. P. (1993a). Striatal binding of the PET ligand ^{11}C-raclopride is altered by drugs that modify synaptic dopamine levels. *Synapse* **13**: 350–356.

Dewey, S. L., Smith, G. S., Logan, J., Brodie, J. D., Simkowitz, P., MacGregor, R. R., Fowler, J. S., Volkow, N. D., and Wolf, A. P. (1993b). Effects of central cholinergic blockade on striatal dopamine release measured with positron emission tomography in normal human subjects. *Proc. Natl. Acad. Sci. U.S.A.* **90**: 11816–11820.

Dewey, S. L., Smith, G. S., Logan, J., Alexoff, D., Ding, Y. S., King, P., Pappas, N., Brodie, J. D., and Ashby, C. R., Jr. (1995). Serotonergic modulation of striatal dopamine measured with positron emission tomography (PET) and in vivo microdialysis. *J. Neurosci.* **15**: 821–829.

Duelli, R., and Kuschinsky, W. (1993). Changes in brain capillary diameter during hypocapnia and hypercapnia. *J. Cereb. Blood Flow Metab.* **13**: 1025–1028.

Evans, A. C., Marrett, S., Neelin, P., Collins, L., Worsley, K., Dai, W., Milot, S., Meyer, E., and Bub, D. (1992). Anatomical mapping of functional activation in stereotactic coordinate space. *NeuroImage* **1**: 43–63.

Farde, L., Eriksson, L., Blomquist, G., and Halldin, C. (1989). Kinetic analysis of central [^{11}C]raclopride binding to D2-dopamine receptors studied by PET—a comparison to the equilibrium analysis. *J. Cereb. Blood Flow Metab.* **9**: 696–708.

Field, L. M., Dorrance, D. E., Krzeminska, E. K., and Barsoum, L. Z. (1993). Effect of nitrous oxide on cerebral blood flow in normal humans. *Br. J. Anaesth.* **70**: 154–159.

Finck, A. D., Samaniego, E., and Ngai, S. H. (1995). Nitrous oxide selectively releases Met5-enkephalin and Met5-enkephalin-Arg6-Phe7 into canine third ventricular cerebrospinal fluid. *Anesth. Analg.* (*Cleveland*) **80**: 664–670.

Friston, K. J., Frith, C. D., Liddle, P. F., and Frackowiak, R. S. (1991). Comparing functional (PET) images: The assessment of significant change. *J. Cereb. Blood Flow Metab.* **11**: 690–699.

Friston, K. J., Ashburner, J., Frith, C. D., Poline, J.-B., Heather, J. D., and Frackowiak, R. S. J. (1995). Spatial registration and normalization of images. *Hum. Brain Mapp.* **2**: 165–189.

Goldstein, A. (1964). "Biostatistics, An Introductory Text." Macmillan, New York.

Guo, T. Z., Poree, L., Golden, W., Stein, J., Fujinaga, M., and Maze, M. (1996). Antinociceptive response to nitrous oxide is mediated by supraspinal opiate and spinal alpha 2 adrenergic receptors in the rat. *Anesthesiology* **85**: 846–852.

Gyulai, F. E., Firestone, L. L., Mintun, M. A., and Winter, P. M. (1996). In vivo imaging of human limbic responses to nitrous oxide inhalation. *Anesth. Analg.* (*Cleveland*) **83**: 291–298.

Gyulai, F. E., Firestone, L. L., Mintun, M. A., and Winter, P. M. (1997). In vivo imaging of nitrous oxide-induced changes in cerebral activation during noxious heat stimuli. *Anesthesiology* **86**: 538–548.

Holthoff, V. J., Koeppe, R. A., Frey, K. A., Paradise, A., and Kuhl, D. E. (1991). Differentiation of radioligand delivery and binding in the brain: Validation of a two-compartment model for [^{11}C]flumazenil. *J. Cereb. Blood Flow Metab.* **11**: 745–752.

Hormann, C., Schmidauer, C., Kolbitsch, C., Kofler, A., and Benzer, A. (1997). Effects of normo- and hypocapnic nitrous-oxide-inhalation on cerebral blood flow velocity in patients with brain tumors. *J. Neurosurg. Anesthesiol.* **9**: 141–145.

Innis, R. B., Malison, R. T., al Tikriti, M., Hoffer, P. B., Sybirska, E. H., Seibyl, J. P., Zoghbi, S. S., Baldwin, R. M., Laruelle, M., Smith, E. O., Charney, D. S., Heninger, G., Elsworth, J. D., and Roth, R. H. (1992). Amphetamine-stimulated dopamine release competes in vivo for [^{123}I]IBZM binding to the D2 receptor in nonhuman primates. *Synapse* **10**: 177–184.

Jones, A. K. P., Cunningham, V. J., Ha-Kawa, S.-K., Fujiwara, T., Liyii, Q., Luthra, S. K., Ashburner, J., Osman, S., and Jones, T. (1994). Quantitation of [^{11}C]diprenorphine cerebral kinetics in man acquired by PET using presaturation, pulse-chase and tracer-only protocols. *J. Neurosci. Methods* **51**: 123–134.

Lammertsma, A. A., and Hume, S. P. (1996). Simplified reference tissue model for PET receptor studies. *NeuroImage* **4**: 153–158.

Laruelle, M., Abi Dargham, A., van Dyck, C. H., Rosenblatt, W., Zea Ponce, Y., Zoghbi, S. S., Baldwin, R. M., Charney, D. S., Hoffer, P. B., Kung, H. F., and Innis, R. B. (1995). SPECT imaging of striatal dopamine release after amphetamine challenge. *J. Nucl. Med.* **36**: 1182–1190.

Laruelle, M., Abi-Dargham, A., van Dyck, C. H., Gil, R., CD, D. S., Erdos, J., McCance, E., Rosenblatt, W., Fingado, C., Zoghbi, S. S., Baldwin, R. M., Seibyl, J. P., Krystal, J. H., Charney, D. S., and Innis, R. B. (1996). Single photon emission computerized tomography imaging of amphetamine-induced dopamine release in drug-free schizophrenic subjects. *Proc. Natl. Acad. Sci. U.S.A.* **93**: 9235–9240.

Logan, J., Fowler, J. S., Vokow, N. D., Wolf, A.P., Dewey, S. L., Schyler, D. J., and MacGregor, R. (1990). Graphical analysis of reversible radioligand binding from time-activity measurements applied to [N-^{11}C-methyl]-(-)-cocaine PET studies in human subjects. *J. Cereb. Blood Flow Metab.* **10**: 740–747.

Logan, J., Dewey, S. L., Wolf, A. P., Fowler, J. S., Brodie, J. D., Angrist, B., Volkow, N. D., and Gatley, S. J. (1991). Effects of endogenous dopamine on measures of [^{18}F]N-methylspiroperidol binding in the basal ganglia: Comparison of simulations and experimental results from PET studies in baboons. *Synapse* **9**: 195–207.

Logan, J., Volkow, N., Fowler, J. S., Wabg, G. J., Dewey, S. L., MacGregor, R., Schyler, D., and Gatley, S. J. (1994). Effects of blood flow on [^{11}C]raclopride binding in the brain: Model simulations and kinetic analysis of PET data. *J. Cereb. Blood Flow Metab.* **14**: 995–1010.

Quock, R. M., Kouchich, F. J., and Tseng, L. F. (1985). Does nitrous oxide induce release of brain opioid peptides? *Pharmacology* **30**: 95–99.

Quock, R. M., Curtis, B. A., Reynolds, B. J., and Mueller, J. L. (1993). Dose-dependent antagonism and potentiation of nitrous oxide antinociception by naloxone in mice. *J. Pharmacol. Exp. Ther.* **267**: 117–122.

Reinstrup, P., Ryding, E., Algotsson, L., Berntman, L., and Uski, T. (1994). Effects of nitrous oxide on human regional cerebral blood flow and isolated pial arteries. *Anesthesiology* **81**: 396–402.

Tedroff, J., Pedersen, M., Aquilonius, S. M., Hartvig, P., Jacobsson, G., and Langstrom, B. (1996). Levodopa-induced changes in synaptic dopamine in patients with Parkinson's disease as measured by [^{11}C]raclopride displacement and PET. *Neurology* **46**: 1430–1436.

Villringer, A., and Dirnagl, U. (1995). Coupling of brain activity and cerebral blood flow: Basis of functional neuroimaging. *Cerebrovasc. Brain Metab. Rev.* **7**: 240–276.

Woods, R. P., Cherry, S. R., and Mazziotta, J. C. (1992). Rapid automated algorithm for aligning and reslicing PET images. *J. Comput. Assist. Tomogr.* **16**: 620–633.

Young, L. T., Wong, D. F., Goldman, S., Minkin, E., Chen, C., Matsumura, K., Scheffel, U., and Wagner, H. N., Jr. (1991). Effects of endogenous dopamine on kinetics of [^{3}H]N-methylspiperone and [^{3}H]raclopride binding in the rat brain. *Synapse* **9**: 188–194.

CHAPTER 68

Imaging Receptor Occupancy by Endogenous Transmitters in Humans[1]

MARC LARUELLE, ANISSA ABI-DARGHAM, and ROBERT B. INNIS

Department of Psychiatry
Yale University School of Medicine
New Haven, Connecticut and
Veteran Administration Medical Center
West Haven, Connecticut 06516

The aim of this study was to test the feasibility of measuring neuroreceptor occupancy by endogenous transmitters in the living human brain. The dopaminergic system was chosen because of preclinical data showing significant level of D_2 receptor occupancy by dopamine in rodents. The impact of endogenous dopamine on in vivo measurement of D_2 receptors in humans was evaluated with single photon emission computerized tomography (SPECT) by comparing the binding potential (BP) of the selective D_2 radiotracer [^{123}I]IBZM before and after acute dopamine depletion. Dopamine depletion was achieved by administration of the tyrosine hydroxylase inhibitor α-methyl-p-tyrosine (AMPT), given orally at a dose of 1 g every 6 hr for 2 days. AMPT increased [^{123}I]IBZM BP by 28 ± 16% (±SD, n = 9). Experiments in rodents suggested that this effect was due to the removal of endogenous dopamine rather than D_2 receptor upregulation. Synaptic dopamine concentration was estimated as 45 ± 25 nM, in agreement with values reported in rodents. The amplitude and the variability of the AMPT effect suggested that competition by endogenous dopamine introduces a significant error in measurement of D_2 receptors in vivo with positron emission tomography or SPECT. However, these results also imply that D_2 receptor imaging coupled with acute dopamine depletion might provide estimates of synaptic dopamine concentration in the living human brain.

I. INTRODUCTION

Since the early 1990s, several groups have documented that *in vivo* measurement of D_2 receptors is affected by changes in the concentration of endogenous dopamine. For example, increasing synaptic dopamine concentration with amphetamine or methylphenidate acutely reduced the striatal specific binding of several positron emission tomography (PET) and single photon emission computerized tomography (SPECT) radiotracers in baboons (Logan *et al.*, 1991; Innis *et al.*, 1992; Dewey *et al.*, 1993; Kessler *et al.*, 1993; Laruelle *et al.*, 1997b) and in humans (Farce *et al.*, 1992; Volkow *et al.*, 1994; Laruelle *et al.*, 1995). These observations opened the possibility to use PET or SPECT neuroreceptor imaging to gain information about changes in neurotransmitter concentration in the "synaptic space" or, more exactly, in the vicinity of the receptors, as not all D_2 receptors are located at synaptic junctions (Yung *et al.*, 1995).

Specifically, the amphetamine or methylphenidate challenges coupled with D_2 receptor imaging provide assessments of relative increase in dopamine release. Using this paradigm, an increase in the displacement of the selective dopamine D_2 receptor radiotracer [^{123}I]IBZM following amphetamine injection (0.3 mg/kg) in drug-free patients with schizophrenia compared to controls was reported (Laruelle *et al.*, 1996). This observation suggested that amphetamine-induced dopamine release is increased in patients with schizophrenia. This

---

[1] Transcripts of the BRAINPET97 discussion of this chapter can be found in Section VIII.

result has been replicated by another group using [¹¹C]raclopride and a lower dose of amphetamine (0.2 mg/kg). Volkow *et al.* (1997) showed a blunting of methylphenidate-induced displacement of striatal [¹¹C]raclopride binding in recently detoxified cocaine users. This observation suggested that cocaine abuse leads to decreased responsiveness of dopamine neurons following methylphenidate exposure. These studies demonstrate the usefulness of this method to evaluate functional responsivity of DA neurons associated with pathological states.

However, the interpretation of these challenges is limited as they do not allow measurement of dopamine concentration at baseline (i.e., in the absence of pharmacological intervention). In contrast, a pharmacological intervention that would induce a rapid and complete depletion of synaptic dopamine could theoretically provide a measure of D_2 receptor occupancy by dopamine at baseline and, providing knowledge of the *in vivo* affinity of dopamine for D_2 receptors, an absolute quantification of synaptic dopamine concentration. Such a measurement would allow the correlation of DA concentration with the clinical state of the subjects in the absence of any pharmacological intervention.

Preclinical studies indicated that such a measurement might be feasible for the dopamine system, as *in vivo* striatal D_2 receptor occupancy by dopamine is far from negligible. Acute dopamine depletion induced a 10–60% increase in the *in vivo* binding of D_2 receptor radioligands in rodents (Van der Werf *et al.*, 1986; Ross and Jackson, 1989a, b; Seeman *et al.*, 1990; Ross, 1991; Young *et al.*, 1991). In baboons, a 30% increase in the specific binding of the D_2 receptor radiotracer [¹²³I](S)-(-)-3-iodo-2-hydroxy-6-methoxy-N-[(1-ethyl-2-pyrrolidinyl)methyl]benzamide ([¹²³I]IBZM) was observed following acute exposure to the tyrosine hydroxylase inhibitor α-methyl-p-tyrosine (AMPT) (Laruelle *et al.*, 1997b).

The purpose of the present investigation was to estimate D_2 receptor occupancy by dopamine in the resting state in young healthy volunteers. For each subject, [¹²³I]IBZM binding potential (BP) at baseline was compared to that during dopamine depletion as achieved by oral administration of AMPT (1 g every 6 hr for 2 days). AMPT is a competitive and reversible inhibitor of tyrosine hydroxylase, the rate-limiting enzyme for dopamine and norepinephrine synthesis (Spector *et al.*, 1965; Udenfriend *et al.*, 1965). AMPT was selected as the depleting agent because it is approved for human use and because, in contrast to reserpine, AMPT effects are rapidly reversible. The dose and frequency of AMPT administration were selected to provide and maintain significant inhibition of tyrosine hydroxylase activity (Engelman *et al.*, 1968). AMPT was given for 2 days, based on the expectation that this duration of treatment would be adequate to produce marked dopamine depletion but too short to induce significant D_2 receptor upregulation.

II. MATERIALS AND METHODS

A. Human Subjects

The study was approved by the Yale institutional review board for human investigation. Eleven healthy volunteers participated in this study (males, age 25 ± 4 years, with these and subsequent values given as mean ± SD).

B. Depletion Regimen and Clinical Monitoring

The study lasted 4 days. Each subject was scanned twice, at 48-hr intervals, in the baseline state (scan 1, day 1) and after dopamine depletion (scan 2, day 3). Dopamine depletion was induced by oral administration of AMPT, 1 g every 6 hr, for 48 hr. The clinical status of the subjects was evaluated daily at 10 AM from day 1 to day 4. Five subjective feelings, known to be affected by monoamine depletion, were rated by the subjects on a visual analog scale ranging from 1 ("not at all") to 10 ("most ever"): sleepiness, happiness, anxiety, energy, and restlessness. Subjects were also evaluated for the presence of extrapyramidal side effects with the Simpson Angus scale (Simpson and Angus, 1970).

C. Scan Protocol

[¹²³I]IBZM was prepared as described previously (Kung and Kung, 1990). The previously described radiotracer constant infusion technique was used to perform the experiments under sustained equilibrium-binding conditions (Laruelle *et al.*, 1994, 1995). An intravenous catheter was inserted in each arm of the subjects for drug administration and blood sampling, respectively. A total [¹²³I]IBZM dose of 5.79 ± 0.76 mCi was given as a bolus (2.57 ± 0.33 mCi) followed by a continuous infusion at a rate of 0.66 ± 0.08 mCi/hr (bolus to hourly infusion rate of 3.90 ± 0.02 hr) for the duration of the experiment (294 ± 13 min, with these and all subsequent times given with reference to the beginning of the radiotracer administration). This bolus-to-infusion ratio was originally calculated from kinetic experiments performed in baboons. This protocol of administration was shown, in previous experiments, to induce a state of sustained [¹²³I]IBZM binding equilibrium: both striatal and nonspecific activities remained at a constant level from 150 min to the end of the experiment (Laruelle *et al.*, 1995). Each subject received the same dose of [^{123}I]IBZM for scans 1 and 2. During the first 210 min of the infusion, subjects were allowed to relax in a comfortable setting, in a room adjacent to the camera room. SPECT data were acquired with the three-headed PRISM 3000 camera (Picker, Cleveland, OH) equipped with high resolution fan beam collimators (resolution at full width half-maximum, 11 mm; ^{123}I point source sensitivity, 16.5 counts/sec/μCi). Continuous scanning was performed during 80 min from 214 ± 13 to 294 ± 13 min. Ten acquisitions of 8 min each were obtained. Venous plasma samples were obtained every 20 min during the imaging session ($n = 4$) to measure plasma [^{123}I]IBZM, homovanillic acid (HVA), and methoxy-4-hydroxyphenelethyleneglycol (MHPG). In addition, AMPT plasma levels were measured on day 3.

D. Image Analysis

SPECT images were reoriented to the canthomeatal line as visualized by four external fiducial markers. The four slices with highest striatal uptake were summed and attenuation was corrected, assuming uniform attenuation within an ellipse drawn around the subject's head. For each subject, the same attenuation ellipse was used for scans 1 and 2. Striatal and occipital regions of interest were positioned on the summed images. Standard regions of interest of constant size (striatum 556 mm^2, occipital 2204 mm^2) were used to analyze all studies. Right and left striatal regions were averaged. Specific binding was calculated as striatal minus occipital activity. The occipital region was selected as the background region because (1) the density of dopamine D$_2$ receptors is negligible in this region compared to the striatum (Lidow *et al.*, 1989); (2) this region can be identified with greater reliability than the cerebellum; and (3) in humans, [^{123}I]IBZM activity in the occipital region is equal to the nonspecific activity in the striatum (Seibyl *et al.*, 1992). The quality of the equilibrium state was assessed by measuring the change over time of regional activity, expressed relative to the average regional activity.

At equilibrium, the [^{123}I]IBZM specifically bound concentration (B) is related to the [^{123}I]IBZ-free concentration (F), the maximal number of D$_2$ receptors (B_{max}), and the equilibrium dissociation rate constant of [^{123}I]IBZM for D$_2$ receptors (K_D) by the Michaelis–Menten equation

$$B = \frac{B_{max} F}{K_D + F}. \quad (1)$$

Because [^{123}I]IBZM was given at tracer doses, F was negligible relative to K_D, and this equation simplified to

$$B = \frac{B_{max} F}{K_D}. \quad (2)$$

The [^{123}I]IBZM-binding potential (BP, ml g^{-1}), corresponding to the product of the receptor density (B_{max}, nM or pmol per g of brain tissue) and affinity ($1/K_D$, nM^{-1}, or ml of plasma per pmol), was thus equal to the B over F ratio. B (μCi per g of brain tissue) was measured as the difference between striatal total activity (STR) and occipital activity (OCC).

$$BP = \frac{B_{max}}{K_D} = \frac{B}{F} = \frac{(STR - OCC)}{F}. \quad (3)$$

Because [^{123}I]IBZM crosses the blood–brain barrier by passive diffusion, the free [^{123}I]IBZM concentration equilibrates on both sides of the blood–brain barrier under sustained equilibrium conditions (Kawai *et al.*, 1991; Laruelle *et al.*, 1994). Therefore, the intracerebral free [^{123}I]IBZM concentration was estimated by the steady-state free unmetabolized plasma [^{123}I]IBZM concentration ($f_1 C_{SS}$, μCi per ml of plasma).

In the presence of a competitive inhibitor such as endogenous dopamine, Eq. (1) becomes

$$B = \frac{B_{max} F}{K_D(1 + F_{DA}/K_I) + F}, \quad (4)$$

where F_{DA} (nM) is the temporal average concentration of free dopamine in the vicinity of receptors and K_I is the inhibition constant of dopamine for [^{123}I]IBZM binding to D$_2$ receptors. Assuming complete dopamine depletion during scan 2, BP$_1$ ([^{123}I]IBZM BP measured during scan 1, i.e., in the presence of endogenous dopamine) and BP$_2$ ([^{123}I]IBZM BP measured during scan 2, i.e., in the absence of dopamine) are given by

$$BP_1 = \frac{B_{max}}{K_D(1 + F_{DA}/K_I)} \quad \text{and} \quad BP_2 = \frac{B_{max}}{K_D}.$$

(5 and 6)

The percentage occupancy of D$_2$ receptors by dopamine during scan 1 was calculated as 100*[(BP$_2$ − BP$_1$)/BP$_2$]. The free synaptic dopamine concentration (F_{DA}) was obtained by combining and rearranging Eqs. (6) and (7) and defining "a" as the BP$_2$/BP$_1$ ratio,

$$F_{DA} = K_I(a - 1). \quad (7)$$

Thus F_{DA} can be expressed as a fraction of K_I. To calculate the absolute value of F_{DA}, the value of K_I must be known. For this calculation, we used the K_I

value of dopamine for [^{125}I]IBZM binding measured *in vitro* at room temperature in the presence of sodium ions and antioxidants was used (160 n*M*; Brücke *et al.*, 1988). This model assumes competitive inhibition at one receptor site. *In vitro*, D$_2$ receptors present high and low affinity states for dopamine (Sibley *et al.*, 1982; Hamblin *et al.*, 1984), and because D$_2$ antagonists have similar affinity for both sites, the interaction between dopamine and D$_2$ antagonists appears noncompetitive under certain conditions (Sibley and Creese, 1980). However, the proportion of high versus low affinity agonist-binding states of D$_2$ receptors might be much higher *in vivo* (90% versus 10%, for the high and low affinity states, respectively) than *in vitro* (50% for each subtype) (Richfield *et al.*, 1986). The uncertainty about the exact proportion of high versus low affinity states of D$_2$ receptors *in vivo* precluded incorporating these factors in the model.

III. RESULTS

A. Clinical Effects of AMPT

Nine out of 11 subjects completed the protocol. AMPT effects on clinical ratings were analyzed with repeated measures ANOVA in the 9 subjects who completed the protocol. Sleepiness, happiness, restlessness, and Parkinsonism ratings were significantly affected by AMPT treatment (Fig. 1). No significant effects of AMPT were noted on anxiety or energy ratings. Sleepiness increased steadily from baseline, peaked on day 3, and resolved on day 4. Happiness significantly decreased from day 1 to day 3 and regained the baseline level on day 4. Restlessness increased, which persisted 16 hr after the last AMPT dose (day 4). Clinical examination revealed mild signs of Parkinsonism in all subjects. These signs were still noted on day 4. Face akinesia, hypo/bradykinesia, and resting tremors were the extrapyramidal signs most often noted. Two subjects experienced an acute dystonic reaction (spasms of muscles of the neck and jaw) during the evening of day 3. In one case, this reaction necessitated injection of 50 mg diphenhydramine im, with complete and immediate response.

B. Plasma HVA, MHPG, and AMPT

AMPT significantly decreased plasma HVA by 70 ± 12% (repeated measures ANOVA, $p < 0.0001$, range from 54 to 84%) and plasma MHPG by 66 ± 6% ($p < 0.0001$, range from 57 to 73%). Plasma HVA and MHPG were correlated at baseline ($r^2 = 0.65$, $p = 0.008$), but not after AMPT ($r^2 < 0.01$, $p = 0.98$). AMPT levels were obtained in seven out of nine subjects. The average AMPT plasma concentration was 21 ± 7 μg/ml (range 13 to 31 μg/ml).

C. SPECT Results

A significant increase in [^{123}I]IBZM BP following AMPT administration was observed (repeated measures ANOVA, $p = 0.0003$, Fig. 2). The average increase was 28 ± 16% (range from 4 to 54%). Comparing the 10 pre-AMPT with the 10 post-AMPT acquisitions, a significant increase was observed in [^{123}I]IBZM BP in seven out of nine subjects. Assuming

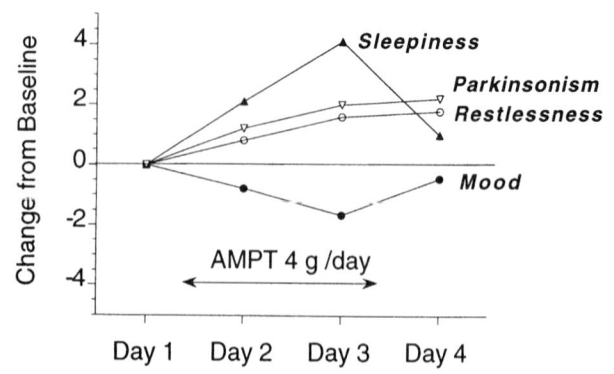

FIGURE 1 Effect of AMPT on clinical ratings obtained from day 1 (prior to AMPT administration) to day 4 (16 hr after last AMPT dose). Data are presented in change from baseline (day 1) ratings. Sleepiness, restlessness, and mood were self-rated on a 10-point scale from 0 ("not at all") to 10 ("most ever"). Parkinsonism was rated using the Simpson and Angus scale. Error bars were omitted for clarity. Symptoms presented on the graph showed a significant time effect (repeated measure ANOVA, $p < 0.05$).

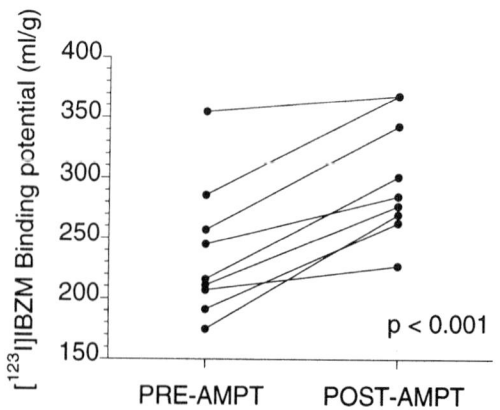

FIGURE 2 [^{123}I]IBZM BP measured at baseline (pre-AMPT) and at the end of AMPT administration (1 g four times a day for 2 days). AMPT-induced depletion in endogenous dopamine resulted in a significant increase in [^{123}I]IBZM BP (repeated measures ANOVA, $p = 0.0003$). The average increase was 28 ± 16%.

a complete depletion of dopamine following AMPT, these results indicate a baseline D_2 receptor occupancy by dopamine of $21 \pm 10\%$, which would correspond to a synaptic dopamine concentration equal to 0.28 ± 0.16 K_I (or 45 ± 25 nM, using a K_I value of 160 nM). AMPT plasma levels on day 3 were not correlated with an AMPT-induced change in [^{123}I]IBZM BP ($r^2 = 0.18$, $p = 0.33$).

IV. DISCUSSION

This study showed that AMPT administration (1 g every 6 hr for 2 days) induced a significant increase ($28 \pm 16\%$) in measured [^{123}I]IBZM BP in nine young healthy volunteers. The magnitude of this effect far exceeds the test–retest variability for repeated [^{123}I]IBZM BP measurements in humans. In a previous study (Laruelle et al., 1995), seven subjects were scanned twice at 2-week intervals under identical conditions ([^{123}I]IBZM constant infusion). The average difference in [^{123}I]IBZM BP between test and retest was $-0.1 \pm 6.4\%$ (range from -6.6 to $\pm 9.4\%$). The average absolute difference was $5.6 \pm 2\%$. This reproducibility corresponded to an intraclass coefficient of correlation close to unity (0.96).

The AMPT effect on [^{123}I]IBZM BP suggested that endogenous dopamine occupied at least $20 \pm 10\%$ of D_2 receptors in these subjects. Before adopting this conclusion, potential sources of artifactual results must be considered. (1) The increase in [^{123}I]IBZM BP measured after AMPT might have reflected receptor upregulation rather than removal of endogenous dopamine, yet in vitro measurement of D_2 receptor density in rodents following 2 days of AMPI administration failed to reveal significant receptor upregulation (Laruelle et al., 1997a). This experiment confirmed an earlier study that could not demonstrate significant D_2 receptor upregulation after 1 week of dopamine depletion induced by 6-hydroxydopamine (Narang and Wamsley, 1995). Similarly, acute pretreatment of mice with reserpine (5 mg/kg) resulted in a significant increase in the in vivo specific binding of [^{3}H]raclopride whereas no changes in D_2 receptor density could be detected in vitro (Ross and Jackson, 1989a). Thus, to the extent that these results can be extrapolated to humans, receptor upregulation does not seem to play a major role in the observed effect. (2) AMPT administration might have increased cerebral blood flow and accelerated the radiotracer delivery to the brain, yet [^{123}I]IBZM BP measurement were obtained under sustained equilibrium conditions, i.e., when no net radiotracer transfer occurs across the blood–brain barrier. Therefore,

[^{123}I]IBZM BP measurements could not have been affected by potential effects of AMPT on cerebral blood flow. (3) AMPT and fluids administration, by increasing [^{123}I]IBZM peripheral clearance, might have affected the quality of equilibrium achieved during scan 2, and such a factor might bias the results of the equilibrium analysis (Carson et al., 1993). However, changes over time in striatal and occipital [^{123}I]IBZM activities were minimal (around 1% hr) during both scans 1 and 2 and were not affected by the AMPT administration. This observation indicated that an appropriate state of equilibrium was reached and maintained during the scanning session for both pre- and post-AMPT treatment. Together, these considerations clearly implicate the reduction in endogenous dopamine as the main, if not the only, mechanism responsible for the increase in [^{123}I]IBZM BP.

Plasma AMPT levels measured in this study were in good agreement with pharmacokinetic data published previously. Engelman et al. (1968) showed that peak AMPT plasma concentrations of 12–14 μg/ml were achieved about 2 hr after oral ingestion of a single dose of 1 g AMPT and that plasma AMPT declined with a half-life of about 4 hr. Therefore, the administration of AMPT every 6 hr was expected to yield plasma concentrations of about 19–22 μg/ml at steady state. Values observed in this study (21 ± 7 μg/ml) were in accordance with this expectation. In this AMPT concentration range, the large variation observed between subjects (10–30 μg/ml) would not be expected to have significant consequences for the pharmacodynamic effect of AMPT. Given an average plasma tyrosine level of 10 μg/ml, a tyrosine affinity constant for tyrosine hydroxylase of 62.5 μM, and an AMPT affinity constant for tyrosine hydroxylase of 17 μM (Udenfriend et al., 1965), AMPI concentrations of 13 μg/ml (smallest value observed in this study), 21 μg/ml (mean value) and 31 μg/ml (highest value) are expected to produce about the same degree of enzymatic inhibition (68, 78, and 83%, respectively). The lack of correlation between AMPT plasma levels and the AMPT-induced increase in [^{123}I]IBZM BP is in agreement with this prediction.

To calculate D_2 receptor occupancy by dopamine ($21 \pm 10\%$), it is assumed that this AMPT regimen induced a complete depletion of synaptic dopamine. This assumption might not be valid, as AMPT does not completely block tyrosine hydroxylase. Because dopamine synthesis was not completely blocked, the magnitude of striatal synaptic dopamine depletion achieved by this regimen is uncertain and the D_2 receptor occupancy by dopamine might be more than $21 \pm 10\%$.

The derivation of absolute synaptic dopamine concentration used the published K_I value of dopamine for [^{123}I]IBZM binding (160 nM), measured at room temperature (60 min incubation), in the presence of sodium ions and antioxidants (Brücke et al., 1988). This seemed a logical choice, given the large range of values of dopamine affinity at D$_2$ receptors that are reported in the literature (Seeman, 1993). The in vitro measurement of dopamine K_D for D$_2$ receptors is very sensitive to the conditions of the assay and is also affected by the choice of the radiotracer. For example, dopamine has lower affinity for D$_2$ receptors in the presence than in the absence of sodium ions (Hamblin et al., 1984; Watanabe et al., 1985) and is more potent at competing with [^{3}H]raclopride than [^{3}H]spiperone binding to the D$_2$ receptors (Hall et al., 1990). This situation is further complicated by the existence of high and low affinity states of D$_2$ receptors for dopamine and dopamine agonists in vitro (Sibley et al., 1982) and by the absence of information on the proportion of high versus low affinity rates in vivo (Richfield et al., 1986).

Despite these limitations, the estimates of synaptic dopamine concentration (from 45 ± 25 nM, assuming complete depletion) found here were very close to values measured previously in rodents using a radiolabeled agonist. Comparing the in vivo binding of the D$_2$ receptor agonist [^{3}H]N-n-propylnorapomorphine in mouse striatum at baseline and following dopamine depletion induced by γ-butyrolactone or reserpine, the synaptic dopamine concentration in mice striatum was estimated to be in the 40–60 nM range (Ross and Jackson, 1989b; Ross, 1991), in good agreement with values derived from this study (45–72 nM).

In conclusion, this study demonstrates that competition by endogenous dopamine results in a 20–30% underestimation of D$_2$ receptor density in vivo and suggests the possibility to take advantage of this phenomenon to measure striatal synaptic dopamine concentration in humans. With the recently developed ability to measure extrastriatal D$_2$ receptors (Halldin et al., 1995), this technique might also allow measurement of dopamine activity in extrastriatal regions, such as the cerebral cortex. Such a method would be very valuable to better understand regional alterations of dopamine activity in schizophrenia and other neuropsychiatric illnesses and to relate these alterations to the clinical symptomatology.

Acknowledgments

Supported by the Scottish Rite Schizophrenia Research Program, the Department of Veterans Affairs (Schizophrenia Research Center), and the Public Health Service (Yale Mental Health Clinical Research Center).

References

Brücke, T., Tsai, Y., McLellan, C., Singhanyom, W., Kung, H., Cohen, R., and Chiueh, C. (1988). in vitro binding properties and autoradiographic imaging of 3-iodobenzamide ([^{125}I] IBZM): A potential imaging ligand for D-2 dopamine receptors in SPECT. Life Sci. 42: 2097–2104.

Carson, R. E., Channing, M. A., Blasberg, R. G., Dunn, B. B., Cohen, R. M., Rice, K. C., and Herscovitch, P. (1993). Comparison of bolus and infusion methods for receptor quantification: Application to [^{18}F]cyclofoxy and positron emission tomography. J. Cereb. Blood Flow Metab. 13: 24–42.

Dewey, S. L., Smith, G. S., Logan, J., Brodie, J. D., Fowler, J. S., and Wolf, A. P. (1993). Striatal binding of the PET ligand ^{11}C-raclopride is altered by drugs that modify synaptic dopamine levels. Synapse 13: 350–356.

Engelman, K., Jequier, E., Udenfriend, S., and Sjoerdsma, A. (1968). Metabolism of alpha-methyltyrosine in man: Relationship to its potency as an inhibitor of catecholamine biosynthesis. J. Clin. Invest. 47: 568–576.

Farde, L., Nordström, A. L., Wiesel, F. A., Pauli, S., Halldin, C., and Sedvall, G. (1992). Positron emission tomography analysis of central D$_1$ and D$_2$ dopamine receptor occupancy in patients treated with classical neuroleptics and clozapine. Arch. Gen. Psychiatry 49: 538–544.

Hall, H., Wedel, I., Halldin, C. J. K., and Farde, L. (1990). Comparison of the in vitro binding properties of N-[^{3}H]methylspiperone and [^{3}H]raclopride to rat and human brain membranes. J. Neurochem. 55: 2048–2057.

Halldin, C., Farde, L., Hogberg, T., Mohell, N., Hall, H., Suhara, T., Karlsson, P., Nakashima, Y., and Swahn, C. G. (1995). Carbon-11-FLB 457: A radioligand for extrastriatal D2 dopamine receptors. J. Nucl. Med. 36: 1275–1281.

Hamblin, M. W., Lef, S. F., and Creese, I. (1984). Interactions of agonists with D-2 dopamine receptors: Evidence for a single receptor population existing in multiple agonist affinity-states in rat striatal membranes. Biochem. Pharmacol. 33: 877–887.

Innis, R. B., Malison, R. T., Al-Tikriti, M., Hoffer, P. B., Sybirska, E. H., Seibyl, J. P., Zoghbi, S. S., Baldwin, R. M., Laruelle, M. A., Smith, E., Charney, D. S., Heninger, G., Elsworth, J. D., and Roth, R. H. (1992). Amphetamine-stimulated dopamine release competes in vivo for [^{123}I]IBZM binding to the D$_2$ receptor in non-human primates. Synapse 10: 177–184.

Kawai, R., Carson, R. E., Dunn, B., Newman, A. H., Rice, K. C., and Blasberg, R. G. (1991). Regional brain measurement of B$_{max}$ and K$_D$ with the opiate antagonist cyclofoxy: Equilibrium studies in the conscious rat. J. Cereb. Blood Flow Metab. 11: 529–544.

Kessler, R. M., Votaw, J. R., de Paulis, T., Bingham, D. R., Ansari, S. M., Mason, N. S., Holburn, G., Schmidt, D. E., Votaw, D. B., Manning, R. G., and Ebert, M. H. (1993). Evaluation of 5-[^{18}F]fluoropropylepidepride as a potential PET radioligand for imaging dopamine D$_2$ receptors. Synapse 15: 169–176.

Kung, M.-P., and Kung, H. (1990). Peracetic acid as a superior oxidant for preparation of [^{123}I]IBZM, a potential dopamine D-2 receptor imaging agent. J. Labelled Compel. Radiopharm. 27: 691–700.

Laruelle, M., Abi-Dargham, A., Al-Tikriti, M. S., Baldwin, R. M., Zea-Ponce, Y., Zoghbi, S. S., Charney, D. S., Hoffer, P. B., and Innis, R. B. (1994). SPECT quantification of [^{123}I]iomazenil binding to benzodiazepine receptors in nonhuman primates. II. Equilibrium analysis of constant infusion experiments and correlation with in vitro parameters. J. Cereb. Blood Flow Metab. 14: 453–465.

Laruelle, M., Abi-Dargham, A., van Dyck, C. H., Rosenblatt, W., Zea-Ponce, Y., Zoghbi, S. S., Baldwin, R. M., Charney, D. S.,

Hoffer, P. B., Kung, H. F., and Innis, R. B. (1995). SPECT imaging of striatal dopamine release after amphetamine challenge. *J. Nucl. Med.* **36**: 1182–1190

Laruelle, M., Abi-Dargham, A., van Dyck, C. H., Gil, R., De Souza, C. D., Erdos, J., Mc Cance, E., Rosenblatt, W., Fingado, C., Zoghbi, S. S., Baldwin, R. M., Seibyl, J. P., Krystal, J. H., Charney, D. S., and Innis, R. B. (1996). Single photon emission computerized tomography imaging of amphetamine-induced dopamine release in drug free schizophrenic subjects. *Proc. Natl. Acad. Sci. U.S.A.* **93**: 9235–9240.

Laruelle, M., D'Souza, C. D., Baldwin, R. M., Abi-Dargham, A., Kanes, S. J., Fingado, C. L., Seibyl, J. P., Zoghbi, S. S., Bowers, M. B., Jatlow, P., Charney, D. S., and Innis, R. B. (1997a) Imaging D2 receptor occupancy by endogenous dopamine in humans. *Neuropsychopharmacology* **17**: 162–174.

Laruelle, M., Iyer, R. N., Al-Tikriti, M. S., Zea-Ponce, Y., Malison, R., Zoghbi, S. S., Baldwin, R. M., Kung, H. F., Charney, D. S., Hoffer, P. B., Innis, R. B., and Bradberry, C. W. (1997b). Microdialysis and SPECT measurements of amphetamine-induced dopamine release in nonhuman primates. *Synapse* **25**: 1–14.

Lidow, M. S., Goldman-Rakic, P. S., Rakic, P., and Innis, R. B. (1989). Dopamine D_2 receptors in the cerebral cortex: Distribution and pharmacological characterization with [^{3}H]raclopride. *Proc. Natl. Acad. Sci. U.S.A.* **86**: 6412–6416.

Logan, J., Dewey, S. L., Wolf, A. P., Fowler, J. S., Brodie, J. D., Angrist, B., Volkow, N. D., and Gatley, S. J. (1991). Effects of endogenous dopamine on measures of [^{18}F]N-methylspiroperidol binding in the basal ganglia: Comparison of simulations and experimental results from PET studies in baboons. *Synapse* **9**: 195–207.

Narang, N., and Wamsley, J. K. (1995). Time dependent changes in DA uptake sites, D_1 and D_2 receptor binding and mRNA after 6-OHDA lesions of the medial forebrain bundle in the rat brain. *J. Chem. Neuroanat.* **9**: 41–53.

Richfield, E. K., Young, A. B., and Penney, J. B. (1986). Properties of D2 dopamine receptor autoradiography: High percentage of high-affinity agonist sites and increased nucleotide sensitivity in tissue sections. *Brain Res.* **383**: 121–128.

Ross, S. B. (1991). Synaptic concentration of dopamine in the mouse striatum in relationship to the kinetic properties of the dopamine receptors and uptake mechanism. *J. Neurochem.* **56**: 22–29.

Ross, S. B., and Jackson, D. M. (1989a). Kinetic properties of the *in vivo* accumulation of the dopamine D-2 receptor antagonist raclopride in the mouse brain *in vivo*. *Naunyn Schmiedeberg's Arch. Pharmacol.* **340**: 6–12.

Ross, S. B., and Jackson, D. M. (1989b). Kinetic properties of the *in vivo* accumulation of [^{3}H](-)-N-n-propylnorapomorphine in the mouse brain. *Naunyn Schmiedeberg's Arch. Pharmacol.* **340**: 13–20.

Seeman, P. (1993). "Receptor Tables. Vol. 2: Drug Dissociation Constants for Neuroreceptors and Transporters." Toronto, Canada.

Seeman, P., Niznik, H. B., and Guan, H.-C. (1990). Elevation of dopamine D2 receptors in schizophrenia is underestimated by radioactive raclopride. *Arch. Gen. Psychiatry* **47**: 1170–1172.

Seibyl, J., Woods, S., Zoghbi, S., Baldwin, R., Dey, H., Goddard, A., Zea-Ponce, Y., Zubal, G., Germinne, M., Smith, E., Heninger, G. R., Charney, D. S., Kung, H., Alavi, A., Hoffer, P., and Innis, R. (1992). Dynamic SPECT imaging of D_2 receptors in human subjects with iodine-123-IBZM. *J. Nucl. Med.* **33**: 1964–1971.

Sibley, D. R., and Creese, I. (1980). Pseudo non-competitive interactions with dopamine receptors. *Eur. J. Pharmacol.* **65**: 131–133.

Sibley, D. R., De Lean, A., and Creese, I. (1982). Anterior pituitary receptors: Demonstration of interconvertible high and low affinity states of the D_2 dopamine receptor. *J. Biol. Chem.* **257**: 6351–6361.

Simpson, G. M., and Angus, J. W. S. (1970). A rating scale for extrapyramidal sides effects. *Acta Psychiatr. Scand.* **212**: 1–58.

Spector, S., Sjoerdsma, A., and Udenfriend, S. (1965). Blockade of endogenous norepinephrine synthesis by alpha-methyl-tyrosine, an inhibitor of tyrosine hydroxylase. *J. Pharmacol. Exp. Ther.* **147**: 86–95.

Udenfriend, S., Nagatsu, T., and Zaltzman-Nirenberg, P. (1965). Inhibitors of purified beef adrenal tyrosine hydroxylase. *Biochem. Pharmacol.* **14**: 837–847.

Van der Werf, J. F., Sebens, J. B., Vaalburg, W., and Korf, J. (1986). *In vivo* binding of N-n-propylnorapomorphine in the rat brain: Regional localization, quantification in striatum and lack of correlation with dopamine metabolism. *Eur. J. Pharmacol.* **87**: 259–270.

Volkow, N. D., Wang, G.-J., Fowler, J. S., Logan, J., Schlyer, D., Hitzemann, R., Lieberman, J., Angrist, B., Pappas, N., MacGregor, R., Burr, G., Cooper, T., and Wolf, A. P. (1994). Imaging endogenous dopamine competition with [^{11}C]raclopride in the human brain. *Synapse* **16**: 255–262.

Volkow, N., Wang, G., Fowler, J., Logan, J., Gatley, S., Hitzemann, R., Chen, A., Dewey, S., and Pappas, N. (1997). Decreased striatal dopaminergic responsiveness in detoxified cocaine-dependent subjects. *Nature (London)* **386**: 830–833.

Watanabe, M., George, S. R., and Seeman, P. (1985). Dependence of dopamine receptor conversion from agonist high to low-affinity state on temperature and sodium ions. *Biochem. Pharmacol.* **34**: 2459–2463.

Young, L. T., Wong, D. F., Goldman, S., Minkin, E., Chen, C., Matsumura, K., Scheffel, U., and Wagner, H. N. (1991). Effects of endogenous dopamine on kinetics of [^{3}H]methylspiperone and [^{3}H]raclopride binding in the rat brain. *Synapse* **9**: 188–194.

Yung, K. K., Bolam, J. P., Smith, A. D., Hersch, S. M., Ciliax, B. J., and Levey, A. I. (1995). Immunocytochemical localization of D1 and D2 dopamine receptors in the basal ganglia of the rat: Light and electron microscopy. *Neuroscience* **65**: 709–730.

CHAPTER 69

Quantification of Extracellular Dopamine Release in Schizophrenia and Cocaine Use by Means of TREMBLE[1]

DEAN F. WONG,* THOMAS SØLLING,† FUJI YOKOI,* and ALBERT GJEDDE†

*Department of Radiology, Division of Nuclear Medicine
Johns Hopkins University
Baltimore, Maryland 21287
† PET Center and Department of Neurology
Aarhus University Hospitals
8000-C Aarhus, Denmark

The first purpose of this study was to quantify human intrasynaptic/extracellular dopamine release following cocaine and amphetamine. The results in 11 cocaine users, 5 controls, and 4 schizophrenic patients are reported. The second purpose of this study is to illustrate a new technique to quantify dopamine receptor density and neurotransmitter release. The method allows the calculation of an absolute receptor density (B_{max}) and an estimation of the dopamine bound to the receptor. The method was applied to normal volunteers and patients with schizophrenia in a four positron emission tomography scan paradigm. The steps for the quantification are outlined in this presentation and are generalizable to other radioligands and neurotransmitter systems.

I. INTRODUCTION

Since the first quantification of images of neuroreceptors *in vivo* (Wagner *et al.*, 1983), it has been hypothesized that extracellular and intrasynaptic neurotransmitters may affect the measurements of these receptors. Seeman *et al.* (1989) proposed that endogenous dopamine must affect the measurement of receptor density (B_{max}) depending on the radioligand employed. He found a significant effect on *in vitro* B_{max} measurements with exogenously added dopamine, as well as an increase of the measured B_{max} at low dopamine concentrations. This observation provided a hypothetical explanation of the discrepancies often noted in studies of psychosis *in vivo* using different radioligands such as [11C]N-methyl-spiperone (NMSP) and [11C]raclopride (Wong *et al.*, 1997). The hypothesis was not rejected by the studies of Dewey *et al.* (1993) in rodents and baboons. Subsequently, it has been shown that dopamine can be released by psychostimulants such as methylphenidate in humans (Volkow *et al.*, 1994).

Two psychostimulant drugs were studied amphetamine and cocaine, both believed to elicit dopamine release *in vivo*. Not only are these drugs of interest in that they perturb the system to release dopamine intrasynaptically, but they or their derivatives have become drugs of abuse; hence their mechanism of action is a public health concern as well.

Using single photon emission computerized tomography (SPECT) or positron emission tomography (PET) radioligands as markers, amphetamine was shown to increase dopamine to a greater extent on average in schizophrenia than in normals (Laruelle *et al.*, 1996; Breier *et al.*, 1997). With [11C]raclopride, the increase was shown to be correlated with increased dopamine concentration in the tissue measured by *in vivo* microdialysis (Breier *et al.*, 1997; Endres *et al.*, 1997). Cocaine is also thought to increase dopamine release; however, the release may have a small effect (Schlaepfer *et al.*, 1997), although the reinforcing properties of cocaine are believed to be mediated in part via the dopamine system. Hence quantification of ampheta-

---

[1] Transcripts of the BRAINPET97 discussion of this chapter can be found in Section VIII.

mine and cocaine effects on dopamine are of considerable medical interest as tools for determining changes of variables such as receptor density B_{max}.

II. MATERIALS

A. Theory

The first step in the method is the correct estimation of the specifically bound (M_b) and nonspecifically bound or free (M_f) quantities of the radiotracer in the tissue. The estimation is made with best fits of the operational equation to the total tissue tracer curve (M), such that the rate of tracer accumulation, dM/dt, subsequently can be determined accurately, from which the remaining computations follow. The second step analyzes the competition between the exogenous tracer and putative endogenous ligands, e.g., dopamine, as described in the section on the calculation of dopamine binding.

1. Determination of Equilibrium-Binding Potential

In vivo binding equilibrium occurs when the first derivative of the quantity of tracer specifically bound in the tissue equals zero, i.e., when $dM_b/dt = 0$. The times at which binding equilibrium is present can be identified after bolus administration of a radioligand (transient method) by solving the following equation for M_f,

$$\frac{dM(t)}{dt} - V_0\frac{dC_a(t)}{dt} = K_1 C_a(t) - k_2 M_f(t), \quad (1)$$

where K_1 is the "systemic" unidirectional clearance of the tracer, k_2 is the rate of fractional clearance of tracer from the tissue, and V_0 is the initial volume of distribution. Both M_f and M_b can be calculated from rearranging Eq. (1),

$$M_f(t) = V_e\left[C_a(t) - \frac{1}{K_1}\left(\frac{dM(t)}{dt} - V_0\frac{dC_a(t)}{dt}\right)\right] \quad (2)$$

$$M_b(t) = M(t) - M_f(t) - V_0 C_a(t), \quad (3)$$

where V_e is the partition volume equal to the ratio K_1/k_2. The binding potential pB is defined as the ratio

$$pB = \frac{M_b(t)}{M_f(t)} \quad (4)$$

for the time(s) of equilibrium when $dM_b(t)/dt = 0$. This calculation of the binding potential is referred to as the TREMBLE method (TRue EquilibriuM BoLus Estimation) as presented by Sølling *et al.* (1997).

2. Calculation of B_{max} and $K'_d V_d$

The change of tracer specifically bound in the tissue can be expressed as

$$\frac{dM_b(t)}{dt} = \frac{k_{on}}{V_d}M_f\left(B'_{max} - \frac{M_b(t)}{A_s}\right) - k_{off} M_b(t), \quad (5)$$

where A_s is the specific activity, k_{on} is the rate constant for transfer into the specifically bound compartment, V_d is the physical volume of solution, k_{off} is the rate of clearance of specifically bound tracer, and B'_{max} is the density of receptor sites available for radioligand binding, after accounting for the binding of endogenous or exogenous inhibitor (B_I). Hence $B'_{max} = B_{max} - B_I$. At binding equilibrium (at time T when $dM_b(T)/dt = 0$), Eq. (5) can be rewritten as the equation for the Eadie–Hofstee–Scatchard plot

$$\frac{M_b(T)}{A_s} = K'_d V_d \frac{M_b(T)}{M_f(T)} + B_{max}, \quad (6)$$

where $K'_d V_d$ is the affinity constant, equal to the term $V_d \cdot (k_{off}/k_{on}) \cdot (1 + C_I/K_I)$, where C_I and K_I are the steady-state concentration and inhibition constants of competitor, respectively. Given two sets of scans (e.g., one at low and one at high specific activity), two sets of data for Eq. (6) with two unknowns allow B_{max} and $K'_d V_d$ to be determined.

3. Calculation of Bound Endogenous Ligand (e.g., Dopamine)

Assuming the dopamine saturation (σ_{DA}) of its receptors to be 25% under normal conditions (Laruelle *et al.*, 1997), it is possible to estimate the binding potential in the absence of dopamine. The relative concentration of any putative endogenous competitor, in this case dopamine (DA), in the absence of the competitors is

$$\chi_{DA} = \frac{\sigma_{DA}}{1 - \sigma_{DA}}. \quad (7)$$

Here χ_{DA} is the endogenous competitor concentration of DA relative to its own half-saturation constant. The binding potential of the tracer in the absence of endogenous dopamine, pB, can be estimated (Gjedde and Wong, 1997) as

$$pB = pB_I(1 + \chi_{DA}), \quad (8)$$

where pB_I is the binding potential in the presence of endogenous DA. The inherent affinity of the receptors toward the radioligand itself, $K_d V_d$, can be determined:

$$K_d V_d = \frac{B_{max}}{pB}. \quad (9)$$

Using Eq. (7), χ_{DA} was calculated to be 0.33 for normal volunteers and 2.75 for schizophrenic patients (Gjedde and Wong, 1997).

Finally, rearranging Eq. (5) and at equilibrium ($dM_b/dt = 0$), the dopamine binding ($B_{DA} = B_I$) is

$$B_{DA} = B_{max} - (K_d V_d)\left(\frac{M_b}{M_f}\right) - \frac{M_b}{A_s}. \quad (10)$$

B. Analytic Approach for Calculation of M_f and M_b

In order to estimate the bound and free activity to determine B_I, it is necessary to perform two fits. First the two parameters K_1 and k_2 of a simple-compartment model is fitted to the activity measured in the reference region, e.g., cerebellum. This provides an estimate of V_e, the partition volume of the tracer in cerebellum. It is assumed that V_e for the reference region (cerebellum) and the region of interest (putamen) are equal. Second, a function with a known time derivative is fitted to the measured activity in putamen, using a five-parameter equation.

1. Determining V_e

Estimates of K_1, k_2, and V_e are made by fitting a two-compartment model to the measured data in cerebellum (M_c),

$$M_c(T) = K_1 \int_0^T C_a(t)\, dt - k_2 \int_0^T M_c(t)\, dt. \quad (11)$$

2. Analytical Expression of the Derivative of the Activity, dM/dt, and the Free Activity, $M_f(t)$

Using a nonlinear least-square fitting of a five-parameter conventional three-compartment model to data measured in putamen, an analytic expression for the total tissue activity, M, is obtained. The fitted five parameters do not all represent biological rate constants and are therefore denoted (K_1, l_2, l_3, l_4, V_0) and the curve expressed with these parameters is denoted $\hat{M}$. However, the estimated parameters K_1 and V_0 from this five parameter fit represent the unidirectional blood–brain clearance (K_1) and the initial vascular distribution volume (V_0). It follows that dM/dt equals $d\hat{M}/dt$, where $d\hat{M}/dt$ is calculated analytically by differentiating $\hat{M}$. Plugging $d\hat{M}/dt$ into Eq. (2) gives the free activity, $M_f(t)$, where V_e is the quantity known from the cerebellum, and K_1 and V_0 are known from the five-parameter equations. Subsequently, the bound activity $M_b(t)$ is obtained from Eq. (3).

III. METHODS

A. Cocaine Abuse

Eleven cocaine users received saline or 48 mg cocaine iv between 0 and 10 min after the administration of a high specific activity bolus [^{11}C]raclopride (> 2 μCi/pmol). The details of the patient selection and clinical details are contained in Schlaepfer et al. (1997). To optimize the timing, the cocaine dose was given at varying times during the first 10 min. The greatest change of the time–activity curves in the caudate or putamen occurred when cocaine was given 5 min following the tracer injection.

B. Schizophrenia

Three normal volunteers and six patients with DSM IV schizophrenia received 0.3 mg/kg amphetamine iv. To test the effect of intrasynaptic dopamine release on measures of B_{max}, subjects underwent two sets of PET measurements, separated by a week. Each set consisted of one high specific activity and one low specific activity (< 0.01 μCi/pmol) iv [^{11}C]raclopride bolus administration in a single day. Each injection was preceded by iv saline in the first set or by iv amphetamine (0.3 mg/kg) before the second set of scans, 5 min prior to the injection of radiotracer. As in the case of cocaine, the timing was evaluated, and 5 min before the [^{11}C]raclopride injection was found to be the time for the greatest difference between amphetamine and saline curves. The B_{max} value estimated from the saline injection was compared to that estimated 1 week later with amphetamine.

Results of the high specific activity [^{11}C]raclopride studies alone were analyzed by comparing the saline and amphetamine cases (scan 1 versus scan 3) for healthy volunteers and schizophrenic patients.

The iv bolus of [^{11}C]raclopride was delivered over 45 sec. Subjects were imaged on the GE4096 + PET camera. Data were acquired for 90 min with 50 frames of 15-sec to 6-min durations. Radial arterial blood samples were corrected for plasma metabolites by HPLC to provide the [^{11}C]raclopride input function.

IV. RESULTS

A. Cocaine Studies

The binding potential calculated using the estimate of k_3/k_4 in caudate or putamen with K_1/k_2 constrained to be equal to that in the cerebellum showed no significant trend between the saline and iv cocaine PET scans. As predicted, these changes were small, although time–activity curves show a small trend (Schlaepfer *et al.*, 1997).

B. Studies in Normals and Schizophrenic Patients

The following results describe the application of the TREMBLE method to obtain the kinetically derived B_{max} and dopamine binding. Figure 1 illustrates the TREMBLE method for estimating total bound and free concentrations of the radiotracer and the B_{max} calculations. Shown is an example of measured data, $M(t)$, $M_b(t)$, and $M_f(t)$ in putamen for all four cases of high and low specific activity and saline versus amphetamine. Table 1 lists comparisons of binding potential in putamen calculated using Eq. (4) and nonequilibrium solutions provided by Gjedde (1982), Logan *et al.* (1990), and Cunningham and Jones (1993). Mean B_{max} for the three subjects was found to be 26 ± 4 pmol/cm^3 (SEM) when calculated from the binding potential. Table 2 illustrates the bound over free estimation of the binding potential (M_b/M_f) at high and low specific activity for the saline and amphetamine cases using the TREMBLE method. Table 3 summarizes the results of the B_{max} calculation using the

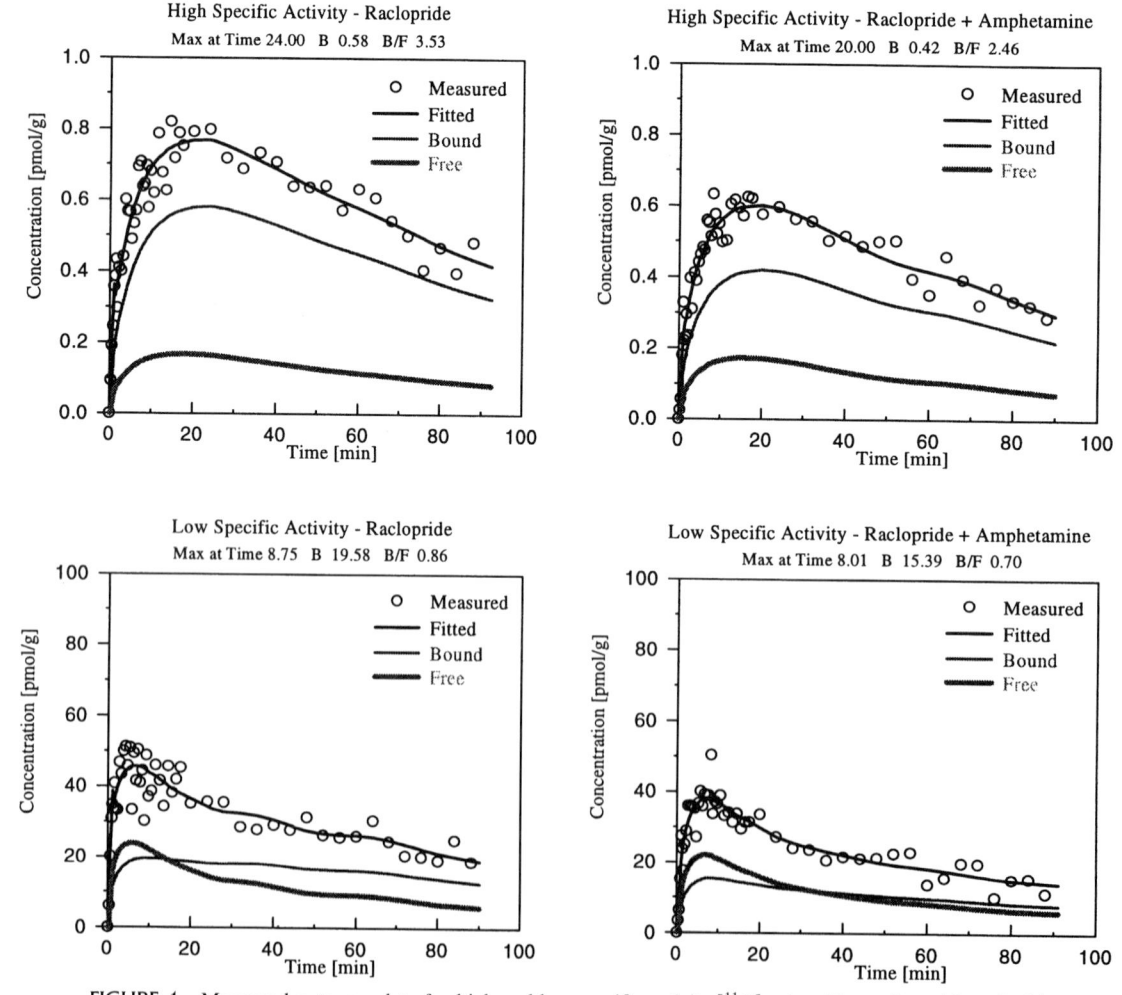

FIGURE 1 Measured putamen data for high and low specific activity [^{11}C]raclopride studies with and without amphetamine.

TABLE 1 Comparison of Binding Potential Methods

Specific activity	TREMBLE method (M_b/M_f)	Cunningham et al. (1993) method	Logan et al. (1990) method	Gjedde (1982) method
High	3.1 ± 0.2	3.4 ± 0.3	3.3 ± 0.2	3.3 ± 0.2
Low	0.8 ± 0.1	1.0 ± 0.3	1.0 ± 0.3	1.1 ± 0.1

Binding potential [ratio] (mean ± SEM) ($n = 3$)

TREMBLE method for normal volunteers and schizophrenic patients.

V. DISCUSSION

The findings of an inconsistent change with iv cocaine studies using the constrained k_3/k_4 method is not unexpected. Further work needs to be done to estimate the relatively low intrasynaptic dopamine release that is obtained. Although this is expected to be dose related, the dose of iv cocaine that was given to the subjects was fixed at 48 mg.

The TREMBLE method, which was applied only to the schizophrenic patients and the corresponding normal controls, represents a new alternative to computation of B_{max} and potentially intrasynaptic dopamine measures. As shown in Table 1, the method for calculation of binding potentials agrees with a number of other methods. However, this specific procedure attempts to incorporate a kinetic method for calculating B_{max} even in the presence of drug perturbations such as with iv amphetamine. The underlying implicit assumption is that the B_{max} is calculated at the peak of M from which the rate of change of the bound radiotracer provides the best estimate of the baseline B_{max}. In the presence of amphetamine, the observed B_{max} will change because of the presence of increased release of intrasynaptic dopamine. Again, the implicit assumption is that the apparent B_{max} that can be calculated in this fashion is also faithfully determined using the TREMBLE method.

The next unique component of this effort is the estimate of the bound dopamine in both normals and patients in the presence and absence of amphetamine or low specific activity [^{11}C]raclopride. It is anticipated that not only will amphetamine cause release of dopamine, but that low specific activity raclopride itself will cause an increased release of dopamine as described in animal studies, e.g., Walters et al. (1990). The four PET scan paradigms allow examination of this.

An important assumption that is made in the calculations of bound dopamine in Table 4 is the value of χ_{DA}. This value has been derived previously from other studies in normals and schizophrenic patients undergoing studies with [^{11}C]NMSP using the method of Wong et al. (1997). In Gjedde and Wong (1997) and Wong et al. (1991), it was argued that the apparent affinity of the D_2 receptor for dopamine is substantially reduced. This study provided the estimated dopamine binding in the schizophrenic patients in two scenarios in Table 4 where it was assumed that the χ_{DA} is the same as that in normals. In this case it can be observed that there are some modest differences in the average dopamine binding in the amphetamine and low specific activity cases. However, should the findings described in Gjedde and Wong (1997) be subsequently confirmed, the more appropriate χ_{DA} of 2.75 results in dramatic differences in the dopamine binding B_{DA} to the D_2 receptors for all four conditions of high and low specific activity and saline versus amphetamine cases (Table 4). Further validation of the χ_{DA} is required, but the results are provocative.

TABLE 2 Binding Potential Using TREMBLE in Normals and Patients

Specific activity	Normals ($n = 3$) Control	Normals ($n = 3$) Amphetamine	Patients ($n = 6$) Control	Patients ($n = 6$) Amphetamine
High	2.7 ± 0.3	2.3 ± 0.1	3.1 ± 0.1	3.1 ± 0.3
Low	1.1 ± 0.2	0.8 ± 0.2	0.9 ± 0.1	1.0 ± 0.1

Binding potential [ratio] (mean ± SEM)

TABLE 3 B_{max} Comparison in Normals and Patients

Condition	Normals ($n = 3$) B_{max} [pmol/g] (mean ± SEM)	Schizophrenia ($n = 6$) B_{max} [pmol/g] (mean ± SEM)
Control	26 ± 4	25 ± 3
Amphetamine	29 ± 11	30 ± 2

TABLE 4 Dopamine Binding (B_{DA}) in Normals and Schizophrenia

Specific activity	B_{DA} [pmol/g] (mean ± SEM)					
	Normals $\chi_{DA} = 0.33$ Control	Normals $\chi_{DA} = 0.33$ Amphetamine	Patients $\chi_{DA} = 0.33^a$ Control	Patients $\chi_{DA} = 0.33^a$ Amphetamine	Patients $\chi_{DA} = 2.75^b$ Control	Patients $\chi_{DA} = 2.75^b$ Amphetamine
High	6 ± 1	10 ± 3	6 ± 1	6 ± 2	18 ± 2	18 ± 3
Low	2 ± 1	1 ± 3	1 ± 0.4	3 ± 2	5 ± 1	7 ± 3

[a] Assuming the χ_{DA} relative concentration of dopamine is the same in patients as normals.
[b] Assuming the χ_{DA} relative concentration of dopamine is elevated in schizophrenia (Gjedde and Wong, 1997).

In summary, these procedures represent a novel approach to the calculation of B_{max} that can be applied to situations involving drug perturbations, such as with iv amphetamine. Such methods can then lead to estimates of the intrasynaptic dopamine binding and free dopamine as illustrated here.

Acknowledgments

This work was supported in part by USPHS Grants MH42821, DA09482, RR00052-37, NARSAD and Danish SSVF Grants 9305246, and 9305247. Special thanks go to Dr. David Wilson, who made key suggestion in the fitting of equations. Acknowledgments are to S. Kim, S. M. Schlaepfer, and G. Pearlson, who worked on the clinical and other related aspects of the cocaine studies, and to G. Grunder, C. Hong, S. Szymanski, G. Nestadt, and K. Neufeld, who all participated in the studies involving the schizophrenic patients.

References

Breier, A., Su, T. P., Saunders, R., Carson, R. E., Kolachana, B. S., de Bartolomeis, A., Weinberger, D. R., Weisenfeld, N., Malhotra, A. K., Eckelman, W. C., and Pickar, D. (1997). Schizophrenia is associated with elevated amphetamine-induced synaptic dopamine concentrans: Evidence from a novel positron emission tomography method. *Proc. Natl. Acad. Sci. U.S.A.* **94**: 2569–2574.

Cunningham, V. J., and Jones, T. J. (1993). Spectral analysis of dynamic PET studies. *J. Cereb. Blood Flow Metab.* **13**: 15–23.

Dewey, S. L., Smith, G. S., Logan, J., Brodie, J. D., Fowler, J. S., and Wolf, A. P. (1993). Striatal binding of the PET ligand 11C-raclopride is altered by drugs that modify synaptic dopamine levels. *Synapse* **13**(4): 350–356.

Endres, C. J., Kolachana, B. S., Saunders, R. C., Su, T., Weinberger, D., Breier, A., Eckelman, W. C., and Carson, R. E. (1997). Kinetic modeling of [11C]Raclopride: Combined PET-microdialysis studies. *J. Cereb. Blood Flow Metab.* **17**: 932–942.

Gjedde, A. (1982). Calculation of cerebral glucose phosphorylation from brain uptake of glucose analogs in vivo: A re-examination. *Brain Res.* **257**: 237–274.

Gjedde, A., and Wong, D. F. (1997). Psychotic propensity associated with fourfold elevated dopamine binding to D2-like receptors in caudate nucleus. *NeuroImage* **5**: A44.

Laruelle, M., Abi-Dargham, A., van Dyck, C. H., Gil, R., D'Souza, C. D., Erdos, J., McCance, E., Rosenblatt, W., Fingado, C., Zoghbi, S. S., Baldwin, R. M., Seibyl, J. P., Krstal, J. H., Charney, D. S., and Innis, R. B. (1996). Single photon emission computerized tomography imaging of amphetamine-induced dopamine release in drug-free schizophrenic subjects. *Proc. Natl. Acad. Sci. U.S.A.* **93**: 9235–2940.

Laruelle, M., D'Sonza, C. D., Baldwin, R. M., Abi-Dargham, A., Kanes, S. J., Fingado, C. L., Seibyl, J. P., Zoghbi, S. S., Bowers, M. B., Jatlow, P., Charney, D. S., and Innis, R. B. (1997). Imaging D_2 receptor occupancy by endogenous dopamine in humans. *Neuropsychopharmacology* **17**: 162–174.

Logan, J., Fowler, J. S., Volkow, N. B., Wolf, A. P., Dewey, S. L., Schlyer, D. J., MacGregor, R. R., Hitzemann, R., Bendriem, B., Gatley, S. J., and Christman, D. R. (1990). Graphical analysis of reversible radioligand binding from time-activity measurements applied to [N-11C-methyl]-(-)-cocaine PET studies in human studies. *J. Cereb. Blood Flow Metab.* **10**: 740–747.

Schlaepfer, T. E., Pearlson, G. D., Wong, D. F., Marenco, S., and Dannals, R. F. (1997). PET study of competition between intravenous cocaine and [11C]raclopride at dopamine receptors: in human subjects. *Am. J. Psychiatry* **154**: 1209–1213.

Seeman, P., Guan, H. C., and Niznik, H. B. (1989). Endogenous dopamine lowers the dopamine D2 receptor density as measured by [3H]raclopride: Implications for positron emission tomography of the human brain. *Synapse* **3**: 96–97.

Sølling, T., Brust, P., Cunningham, V., Wong, D. F., and Gjedde, A. (1997). True Equilibrium Bolus Estimation (TREMBLE) confirms rapid transient equilibrium. *NeuroImage* **5**: A29.

Volkow, N. D., Wang, G. J., Fowler, J. S., Logan, J., Schlyer, D., Hitzemann, R., Lieberman, J., Angrist, B., Pappas, N., MacGregor, R., Burr, G., Cooper, T., and Wolf, A. P. (1994). Imaging endogenous dopamine competition with [11C]raclopride in the human brain. *Synapse* **16**: 255–262.

Wagner, H. N., Jr., Burns, H. D., Dannals, R. F., Wong, D. F., Langstrom, B., Duelfer, T., Frost, J. J., Ravert, H. T., Links, J. M., Rosenbloom, S. B., Lukas, S. E., Kramer, A. V., and Kuhar, M. J. (1983). Imaging dopamine receptors in the human brain by positron tomography. *Science* **221**: 1264–1266.

Walters, D., Chapman, C., and Howard, S. (1990). Development of haloperidol-induced dopamine release in the rat striatum using intracerebral dialysis. *J. Neurochem.* **54**: 181–186.

Wong, D. F., Gjedde, A., Young, D., Tune, L., Pearlson, G., Shaya, E., Young, T., Chan, B., Burckhardt, D., Wilson, P. D., Dannals, R. F., Wilson, A. A., Ravert, H. T., Natarajan, T. K., and Wagner, H. N., Jr. (1991). PET demonstrates reduced dopamine receptor affinity in psychosis in vivo. *J. Cereb. Blood Flow Metab.* **11**(Suppl. 2): S650.

Wong, D. F., Yokoi, F., Grunder, G., Hong, C., Szymanski, S., Dogan, A., Nestadt, G., Neuteld, K., and Gjedde, A. (1997). The effect of intrasynaptic dopamine release on measuring B_{max} and B_{max}/K_D in schizophrenia by PET. *J. Nucl. Med.* **38**: 11P.

SECTION VIII

BRAINPET97 DISCUSSION

SECTION VIII
BRAINPET97 Discussion

Approximately half of the chapters in this book correspond to oral presentations that were given at the BRAINPET97 conference (June 20–22, 1997). The following is an edited transcript of the discussion that followed these oral presentations.

CHAPTER 1

Brain Imaging in Small Animals Using MicroPET

T. Jones: The challenge of imaging small animals is very different from imaging selected parts of the human body. When imaging mice, for example, maybe we should be thinking of surrounding the mouse completely with detectors in order to get both high resolution and high sensitivity. In the next 5 years, do you see this technology compatible with 4-pi geometry to stop all the gamma rays?

S. Cherry: It's what your budget is. There's absolutely no reason why you can't do that. We have to move to very high sensitivity and very high resolution if you want to do those kinds of studies. Between our two institutions, we've gone from clinical PET scanners: with your rat PET system, you've pushed the sensitivity pretty high, and with our system we've pushed the resolution. Now we need to bring those two things together to create a very high sensitivity, very high resolution system.

There's no reason why you can't surround the animal with detectors. It's just cost, and unfortunately the detectors are relatively expensive because of the multi-channel photomultiplier tubes that we are using. But they have already come down by a factor of three in cost in the last 3 years since we started this project, so we can now build a three-ring system for the same cost we built that one-ring system. So I think it's only a matter of time and money before we can do what you suggest.

A. Lammertsma: I am concerned about whether these types of systems will have the sensitivity needed for dynamic studies with, for example, a 200-μCi injection in a rat. How will you solve that?

S. Cherry: That's a good point. The sensitivity of this device is not as high as we would like, and the reason is that we only have one ring of detectors. It would be very easy to add more rings of detectors to increase the sensitivity.

The other thing to realize on those rat images is that we were doing multibed positions so the imaging time at each position was only a few minutes. If we were to image dynamically over only one part of the body, we'd actually be doing pretty well.

The sensitivity is roughly the same as some of the older clinical systems, like the ECAT 931 in 2D mode. So relative to our new 3D clinical systems, the sensitivity is low, but it's a matter of expense again. There's no reason why you can't build multiple rings. This is a first step. We only have money to build one ring just to prove the principle. Obviously now we're facing options as to whether we add more rings onto that system or build a completely new system. We've decided to go ahead and build a new system with even higher resolution, but that system will have multiple rings, which will give us better axial coverage and much better sensitivity. We predict that we should be able to go up almost an order of magnitude in sensitivity for the next system.

M. Graham: I would think sensitivity would be actually one of the parameters you could trade off to try to gain resolution because of the abilities to inject larger amounts of radioactivity. My bigger concerns would be whether the tracers will have high enough specific activity to avoid mass effects and whether the count rate capability will be there to collect all the counts.

S. Cherry: First of all, in answer to your comment about the specific activity, I agree with you. This needs to be a team effort. Part of the team are the instrumentation people designing the scanners, and part of the team are the chemists. The challenge that we're going to lay before them is to increase the specific activity of the tracers as much as possible because that's obviously going to help us. That's another way to gain net sensitivity, so it is an important issue.

In terms of performance of the current system, our peak noise equivalent counts are somewhere around the range of 20,000 or 25,000 counts per second. It depends on the object, though, obviously. Our peak coincidence count rates can be 50,000–60,000 counts per second. So they are reasonable count rates. We have not yet run into any count rate limitations on the system.

R. Carson: Ideally these devices will allow us to screen new receptor-binding tracers in the rat brain. To do this we will need to collect time–activity curves in small brain structures with reasonable statistics. Will we be able to do this?

S. Cherry: What's reasonable statistics? Obviously, another option is one that we're exploring, and I know you're also exploring, which is to use iterative reconstruction on these data sets to try to get a higher signal to noise.

R. Carson: Iterative reconstruction will move us along the noise-resolution curve, but we need the technology to improve both.

S. Cherry: We're going to have to see. It's the early days. We've not gotten to the point yet where we've tried to do quantitative dynamic studies. We're obviously headed down that road, and once we get there we'll have a better idea of how far away we are in terms of sensitivity with the current device. But if it turns out that the current device does not have enough sensitivity for certain studies, that's not going to worry me because we can dramatically improve the sensitivity by putting more detectors into the system. So how much sensitivity do you need? I think ultimately Terry Jones is right, you need to collect every pair of photons that come out, and for small animals, that's feasible.

M. Senda: I found your images acquired with [^{68}Ga]EDTA much worse in quality than those acquired with ^{18}F tracers. Is it due to the positron range effect or is it due to low count statistics?

S. Cherry: It's actually a combination of both. Gallium has sufficient positron range that we probably do see a slight degradation in the images. But also, rabbits are surprisingly big, and very little of the activity got into the tumor in those images. So they were quite statistics-limited as well.

D. Bailey: Would you like to comment on corrections for attenuation, scatter, etc.?

S. Cherry: Obviously, you need to correct for a lot of things, and attenuation is one of the major corrections you have to do. There are several alternatives. The one that we're most likely going to pursue is to use an annular source to get a transmission scan, reconstruct that, segment it, and then forward project. Attenuation correction is a little less critical in the small animals because the attenuation factor is much smaller than it is in humans, but it still needs to be done and it still needs to be done well.

We've also implemented scatter correction and we're working on deadtime. Normalization doesn't seem to be a problem with these detectors. That's actually an interesting issue because the efficiency of the detectors varies in a much more random pattern than it does with block detectors. So we don't get a lot of those patterns that you get using conventional PET systems that cause problems in the images. So normalization is actually quite easy on our system.

P. Esser: How well do these detectors operate at higher count rates?

S. Cherry: The count rate of these devices is not as good as with single photomultiplier tubes. There is no doubt about that, but we're not in the range where we see problems. Typically, each tube can go up to somewhere in the range of 50,000 to 100,000 counts per second in singles before we start to see a drift in the gain, which starts to become apparent at those kind of rates. But since we have 30 detectors in the system, that's actually a very high singles rate. The whole system can accommodate it, so it's not been a limiting factor at this point. It was something we looked at carefully when we chose those tubes because it is a known problem with those devices.

CHAPTER 2

Design of a High-Resolution, High-Sensitivity PET Camera for Human Brains and Small Animals

T. Jones: With this high-speed detector, you begin to lose some of the advantages if you bring it in close to the brain because you're having to deal with more flux from outside the field of view and you're busy dealing with that rather than just the solid angle.

Please qualify how long this detector is going to be in terms of surveying the brain. Will it look at the whole brain in terms of the solid angle? Because it's very important, I think, in modern tomograph design for brain that we see the whole brain with uniform efficiency.

I'd also like you to discuss what Simon Cherry showed. What crisper images are we going to get from LSO, because it looks as if the positioning in the detector is going to be far crisper than what we're seeing with BGO. So I'd like you to share with us your insight into how sharp these images are going to be with the LSO detector itself.

W. Moses: The first question is how long is the axial field of view. We are proposing a 15-cm axial field of view, which is very similar to the ECAT EXACT and the GE Advance. Given that field of view, we had the amount of out-of-field activity that I showed here, which I believe is very similar to the ECAT EXACT or to conventional 3D systems. So I don't think that is going to be too much of a problem.

However, if we do expand the axial field of view more, if we try to go to 25 cm, it is going to be much more of a problem. The out-of-field activity is going to be much more of a problem for a small diameter ring than it is for a large diameter ring. I think it's a question of solid angles. It's going to be more of a problem, but I can't quantify it for you.

The other thing that you asked about is the crispness of the images with LSO. The detector that Simon Cherry has put together and the detector that I'm proposing use an individual coupling scheme in order to look at the crystals, as opposed to a block detector, which more or less uses Anger logic. A block detector can give you a lot of mispositioning, and that is the lack of crispness that you're talking about.

I have a hard time quantifying how much better it is going to be. Simon, if you're shown a point source at one element of yours, I presume that you'd identify it correctly maybe 80% of the time. That's what I have seen in our detector modules, whereas with Anger logic, the block detector tends to get it right about 50% of the time.

S. Strother: I'm curious to know where these LSO detectors will go. Simon Cherry's were, I think, 4 square mm, and yours are 3 square mm. I wonder if you or Simon can look into the crystal ball and tell us how many square millimeters we're going to be dealing with in terms of the individual detector element at the next BRAINPET meeting. What's the fundamental problem we're going to run into? You've solved depth of penetration, but you just talked about the problem of decoding unless you go to the direct coupling that Simon is using. So are the problems just going to mount up so that there's some foreseeable limit that you guys can see in your next grant proposals?

S. Cherry: As you cut the crystals smaller and smaller, they get harder to handle, and the cost starts to go up. And, of course, the number of elements you need to cover a certain solid angle also rises very quickly. But I think there's still some way to go. We've done some simulations and you can probably take crystal sizes down to about 1×1 mm and still fairly accurately identify the crystals with the kind of one-to-one coupling that we're talking about, and also have most of the interactions occur in the right crystal. After that, we'll have to see. But I think 1×1 mm may be feasible, and that's what we're looking at for our next system.

W. Moses: I basically agree with Simon. I think one other issue is that you have some sort of a reflector or coating on the outside of the crystal, and when you start getting close to 1-mm crystals, this volume starts to become relatively significant. So that can get in the way a bit. Also, the cost per channel for the electronics becomes an issue.

Finally, you said that you'd done simulations down to 1×1 mm. I've done measurements down to 1.3 mm square and basically agree that you can get a significant amount of light. So I think that 1 or maybe 1.5 mm is doable.

I. Kanno: So far, we use photomultiplier tubes. Is there any other solid-state photodetector?

W. Moses: I'm using photodiodes and regular pin photodiodes. Avalanche photodiodes are also potentially there. However, I'd say avalanche photodiodes unfortunately have been the detector of tomorrow ever since they were invented in 1968, and they have seemed to remain the detector of tomorrow.

S. Cherry: I wanted to ask you a question about the packing fraction of your detectors. How much dead space do you have around that photodiode array, if any? Presumably you have to bring wires out between the detector modules. So can you give us some sense of how much space there's going to be between the detector modules and whether you feel you're going to lose sensitivity because of that?

W. Moses: That is an issue, and it depends on how you do your packaging. In most detector modules you're inside of a can, and so we're thinking of putting more than one detector module inside of a can because most of the dead space is coming from the can. There are some wires going to the outside, but we're using a Kapton circuit board in order to do that. That doesn't take up very much room, but it does take up something; when you're talking about a 1-mm crystal, that's a quarter of a crystal.

CHAPTER 3

The Road to Simultaneous PET / MR Images of the Brain

T. Videen: It's very impressive technology. However, with the human studies, you are clearly sacrificing sensitivity with the optical coupling in the long fiber optic cables. On the animal studies, it's possible to rigidly bolt the holder to the animal's head and use fiducials, which are the current gold standard for alignment. For what kind of studies you see this design as providing an advantage over the standard methodology.

Y. Shao: Well, obviously, sensitivity is a problem for the current system because we use optical fiber coupling, so it is difficult to reach multiple rings. However, for animal studies, I think it still could be useful, even with this kind of single-ring, single-slice system. Since it's an MR-compatible system, you could use MR, look at where you're interested in the image plane, then you line up the detector ring and just look at this interesting image plane. In this case, you can still get your region of interest inside the image plane.

For the patient studies, I think we have to do more work. We have to build a multiple-ring system. But I also think if we carefully select where the image plane will be, this still can be useful. Basically, you get very good anatomic structure information from the MR.

S. Cherry: As to Tom Videen's question, we don't have all the answers yet as to what you might be able to do with this device. I think the most interesting applications are probably in the temporal correlation of PET with MR spectroscopy. For the registration issue, I think the most interesting application is probably outside the brain. If you were doing rat imaging, cardiac imaging, or tumor imaging where the rat might be in very different orientations in the PET and the MR, and then you'd have to use warping registration algorithms and it gets quite complicated.

Those might be some of the applications, but we're going to have to see. It certainly is a big question as to whether this is just a fun toy for physicists to produce or whether it will actually provide some useful biological information. We hope it's the latter, and we think we have some good applications, but time will tell.

M. Graham: One of the interesting effects of PET imaging in a high magnetic field is that you get a circular movement of the positron as it comes out of the nucleus, and thus you limit the range of the positron, potentially. With high energy emitters like rubidium, it might actually have a real effect, but I don't know if at 1.5 Tesla that's significant or not. Can you tell me if that's going to actually improve resolution because you're doing it in a high magnetic field?

Y. Shao: Certainly it's going to help the PET image. Reduction of the positron range is quite small for 1.5 Tesla. You have to go higher, let's say 5 or 8 Tesla. Then you can see a significant change in the positron range. That's why this is not the main motivation for us, but I think this is certainly helpful. We look at the image carefully, but it seems like this kind of change is quite small.

However, very interestingly, we find the count rate seems to change when we put it inside the MR magnet. This is kind of surprising. But when we look at the literature, it seems that other groups also find that light output has been improved when you put it inside the MR magnet, and with certain threshold, certainly you get more current. So this is the kind of thing that's positive for these kinds of applications. We haven't seen any negatives yet.

A. Lammertsma: The first question after this talk has me worried again because sensitivity in human studies was set off against something else in animal studies. I think this is a misunderstanding that sensitivity is not a problem in animal studies because of the specific activity that Mike Graham mentioned. It is a major problem and I cannot see what you're going to do with this system if you can only take a long image over 30 or 60 min to get a high field of view image. Why not just chop the head off and take an autoradiogram?

Y. Shao: That is a problem for doing this kind of thing. That's also the kind of information we try to collect from the audience to see whether this system can be useful. Obviously the sensitivity is very low. For doing patient studies it's probably not that practical, but maybe there is some specific information for a single-ring system.

Also, we have to think that if down the road we try to build a patient system, then we really have to build multiple rings. Right now we have to think about if there are other ways to couple the optical fibers or other intelligent ways to get the multiple rings. But certainly this should be compatible with the MR system, and to have to do that technically, it is a challenge.

K. Herholz: You have quite nicely shown that there are no obvious artifacts in the MR images. My understanding is that magnetic resonance spectroscopy, which is probably the more reasonable thing to combine it with, is much more sensitive to small field inhomogeneities than the imaging. Did you look at the problem of whether inserting the PET device distorts magnetic resonance spectroscopy experiments?

Y. Shao: We did some experiments with this prototype system and NMR, so we do get simultaneous MR images and spectroscopy. They're very nicely matched to each other, and we have some results. We haven't really looked at the artifacts based on that system yet and we are preparing to do that. If we go to, say a 10-Tesla magnet, it might have some distortion.

But I think these crystals and optical fibers are of quite low magnetic susceptible material. So this kind of inhomogeneity distortion is quite small. In principle it can be corrected by shimming the magnetic field. This effect probably is quite small, although we haven't done experiments yet. One would intuitively think that even if it has artifacts, probably it's at quite a low level. From here we see it's only 5%. But, if we go to a high magnetic field environment, in a large system with many crystals and optical fibers, how large will this distortion be? We don't have the answer yet. We have to do more experiments.

I. Kanno: This is quite an important approach because the MRI people want to validate their values with PET. I hope you will get much higher sensitivity and better quantification in the future.

CHAPTER 4

Brain PET Studies with a High Sensitivity Fully Three-Dimensional Tomograph

S. Strother: If you just took this machine and replaced all the crystal blocks with LSO, I know we're going to solve randoms and dead time, but can you make some prediction about the dose levels that we might then be injecting for similar signal to noise?

D. Bailey: Yes. In a sort of half joking sense, we think this is perhaps one of the best brain tomographs that we've been able to produce. But for the whole body, we've clearly got problems of extra activity coming from outside the field of view. We can't use the shields. So if we replace these BGO crystals with LSO, we think we'd get a very, very good whole body machine as well. With BGO, that's still to be proven. I suspect that if we replace the detectors with LSO, looking at the figures that people like Bill Moses are producing—a four times increase in noise equivalent counts with LSO—we'd probably be able to administer the doses that we currently give on our previous generation 3D machine, the 953B, i.e., 10 or 12 mCi for an [^{15}O]water bolus.

S. Strother: Or you'd go the other way presumably and keep the signal to noise you have now and halve the dose yet again.

D. Bailey: In effect, that's right. Now we can play that tradeoff as well. We can give something like 50 injections of water with this machine and stay within our national guidelines.

T. Videen: I wonder if you could comment on the quantitative uniformity of the scanner. We in St. Louis have an EXACT HR, and we have been disappointed with the factory software when we look at an off-center uniform phantom when applying all of the standard corrections. We find uniformity variations in high count phantoms of 10% or more, and I'm wondering if this scanner and this software are working better than that.

D. Bailey: The scanner software is identical to what you would be using, I imagine, the second release of the latest software from CTI, and I suspect we suffer from the same problems. But one thing I should say is that we can't use the standard normalization procedure for this machine because we don't have rotating rod sources anymore. So we're accepting what it is at the moment and saying we're only going to improve on it.

I maybe could just add that the U.K. PET group, which is all the centers in the U.K. getting together, have done a survey of what their reconstructions look like, and we have produced results across different centers that look very similar to what you've shown in the past. So I think it's a consistent finding, and I suspect we'd see the same thing on our machine.

A. Lammertsma: Basically, you've got a whole body machine which you've trimmed down for brain studies. If you had to design a brain scanner, would you stick with the same axial and transaxial resolution or would you make it smaller?

D. Bailey: No. I think along the lines of what Bill Moses is doing, if it's just a brain machine, you would surely think about bringing in the detectors, but you've got to be very cognizant of what happens when you decrease the ring diameter. You must limit the way you open up that penumbra to single photons. Bill's shielding seems to be protecting against that, and we've discussed that at length. I think the issue, of course, is cost. This machine with LSO would be a brilliant brain machine, and we understand some of the out of field-of-view problems now. I guess we would keep it the same if we could, because then it becomes a general purpose machine, but I'm not sure that our budget permits that at the moment.

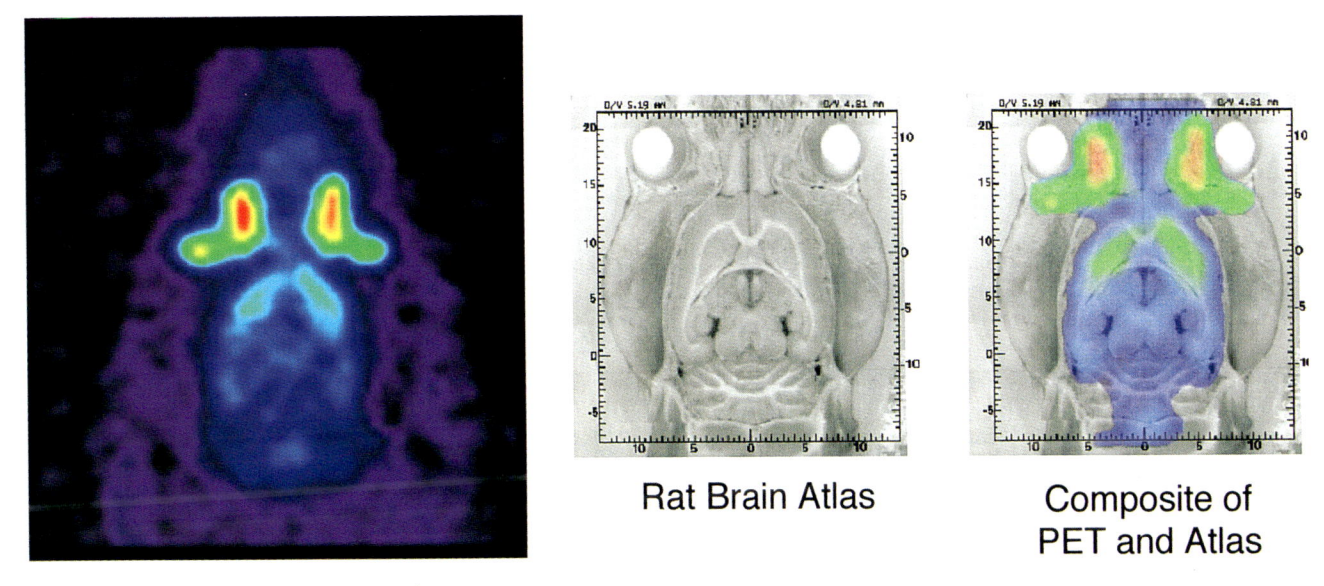

CHAPTER 1, FIGURE 7   FDG distribution in the rat head. The brain is clearly visualized, although uptake is highest in the glands that reside behind the eyes. Manual registration with a rat brain atlas shows excellent correspondence of the PET signal with the anatomy. The injected dose was 2.5 mCi, and the images were acquired over 8 min starting 60 min postinjection.

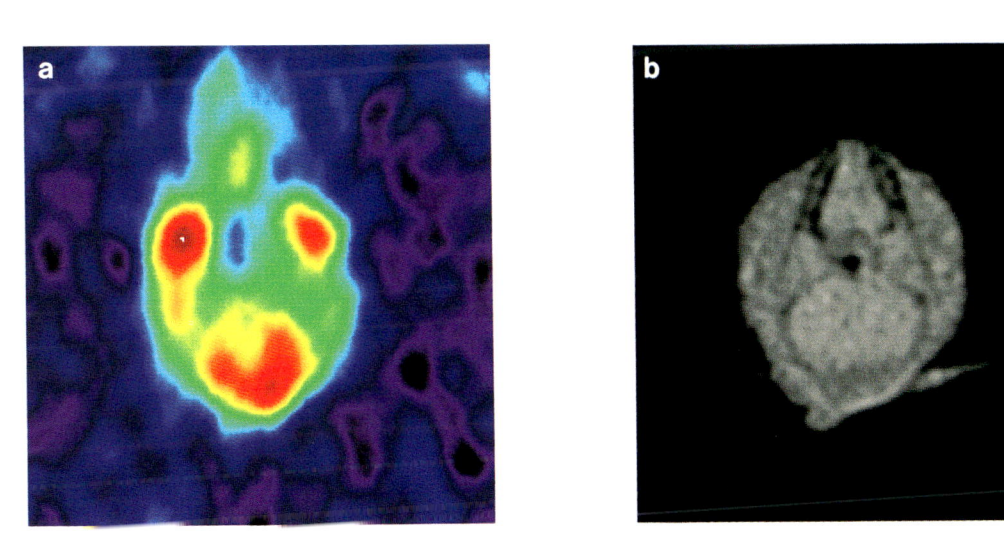

CHAPTER 3, FIGURE 3   Simultaneously acquired (a) FDG PET image and (b) MR image of a rat brain.

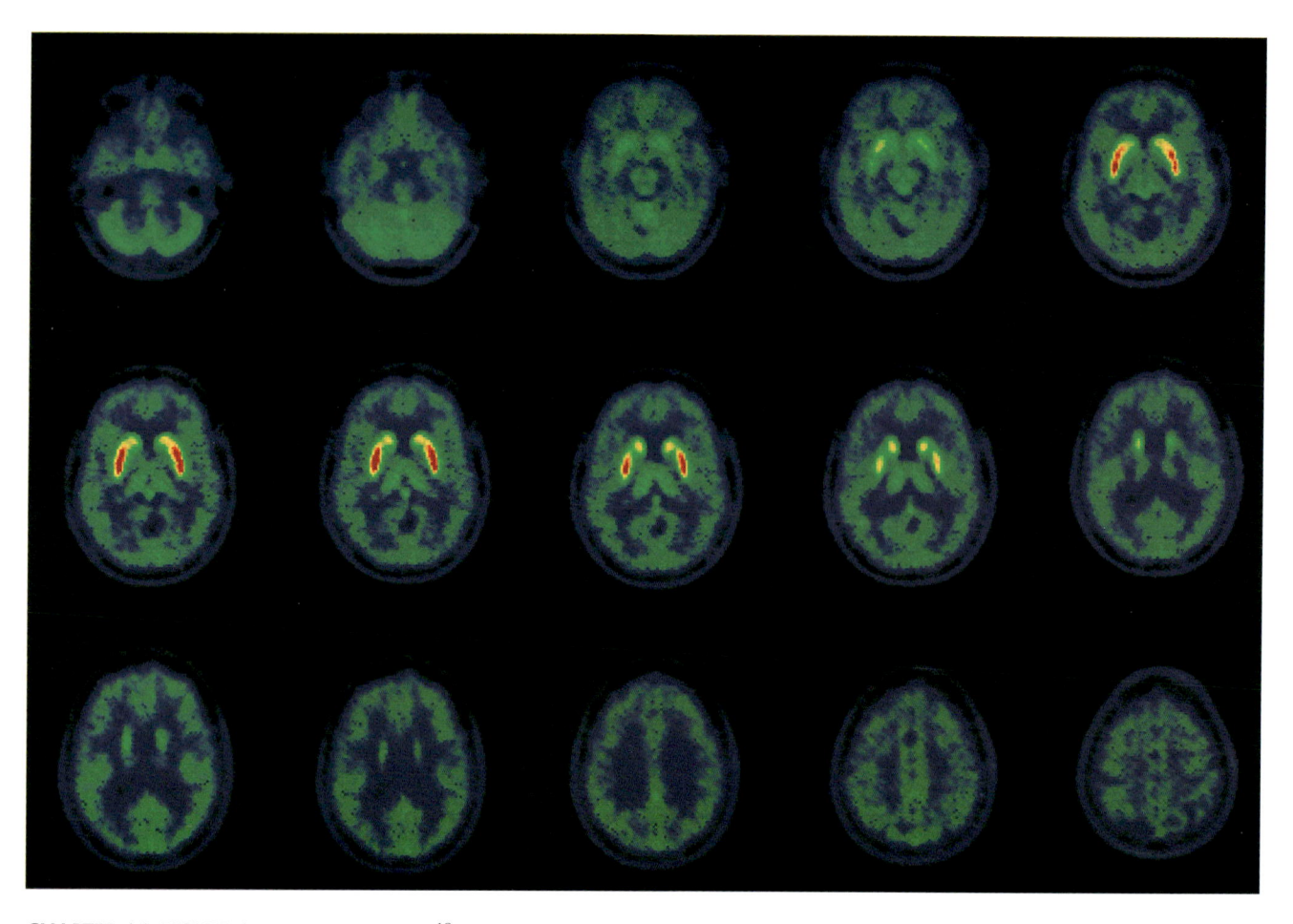

CHAPTER 16, FIGURE 1 Three-dimensional [^{18}F]fluorodopa "add image" (0 to 90 min) from a normal subject. PET scan performed on the advanced EXACT 3D PET scanner. Selected transverse planes.

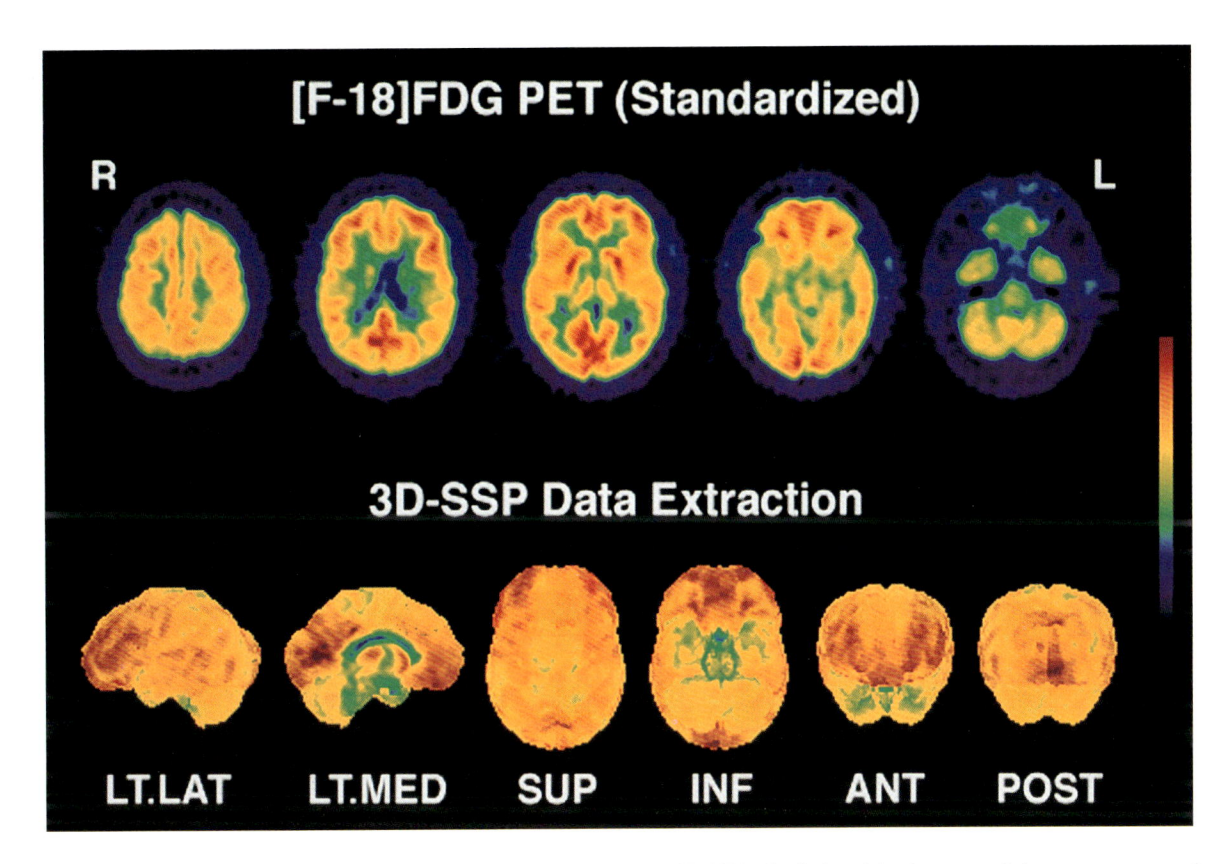

CHAPTER 18, FIGURE 1 Three-dimensional stereotactic surface projections (3D-SSP). Cortical activity is extracted from a reconstructed and anatomically standardized PET image set (upper row) into the 3D-SSP format. Three-dimensional SSP-extracted data (bottom row) can be viewed from any direction, including lateral (LAT), medial (MED), superior (SUP), inferior (INF), anterior (ANT), and posterior (POST) aspects of the brain.

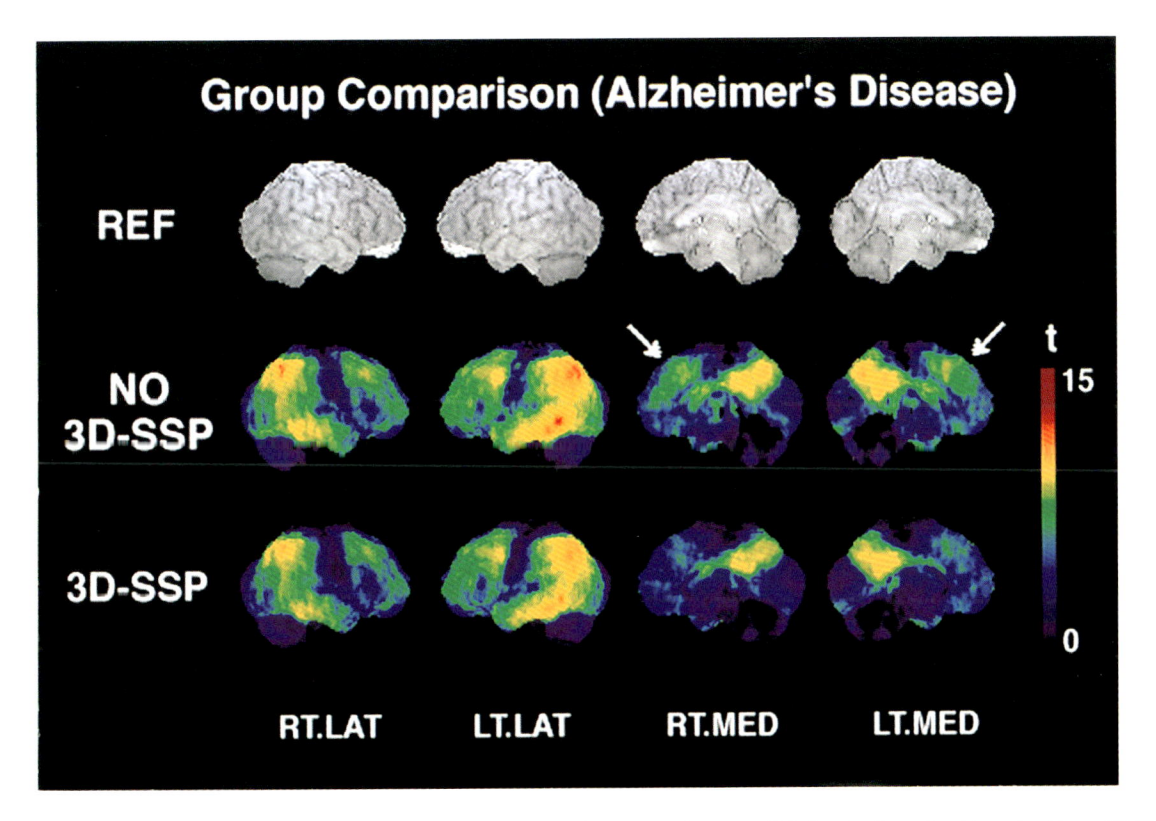

CHAPTER 18, FIGURE 5 Pixel-by-pixel two-sample *t*-statistic comparison (normal versus Alzheimer's disease) with and without 3D-SSP data extraction. Arrows indicate subtraction artifacts resulting from slightly widened interhemispheric space due to cortical atrophy.

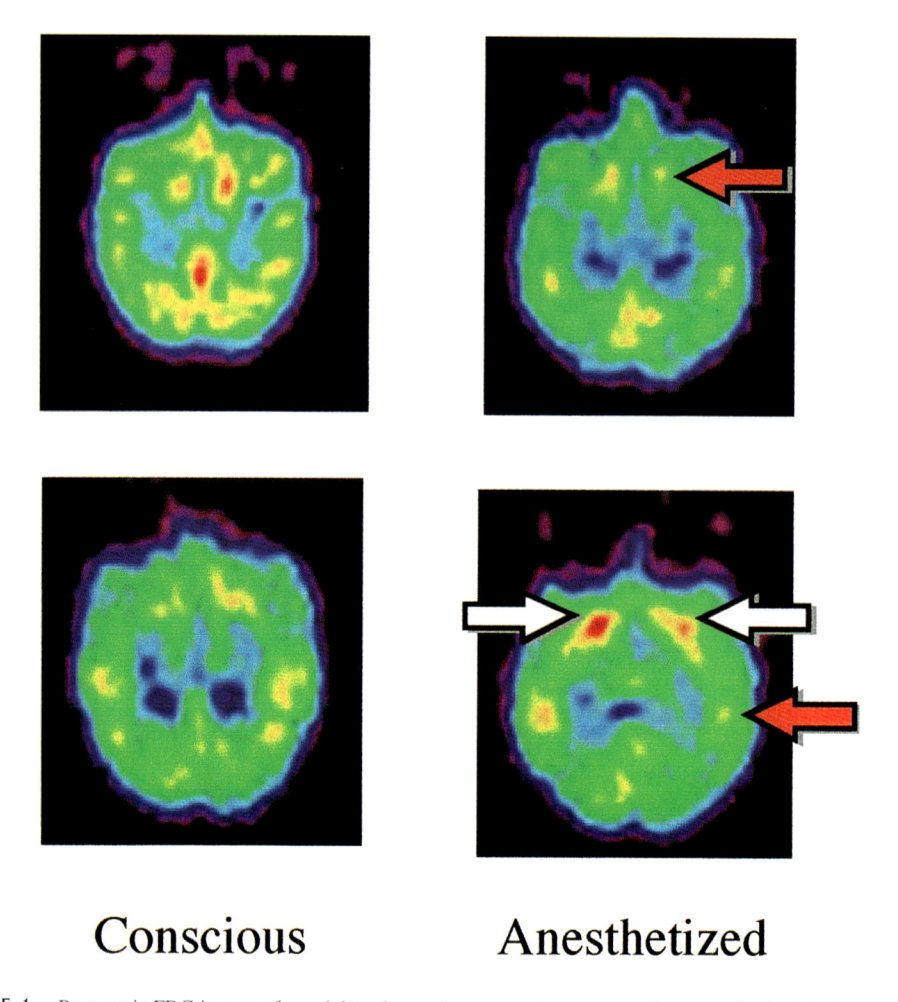

CHAPTER 26, FIGURE 1 Parametric FDG images of an adult male monkey scanned on separate days: anesthetized during tracer uptake and conscious during tracer uptake. (Top) Level of striatum, (bottom) level of temporal cortex/midbrain. Arrows indicate changes ≥ 20% from conscious values of cerebral metabolic rate of glucose.

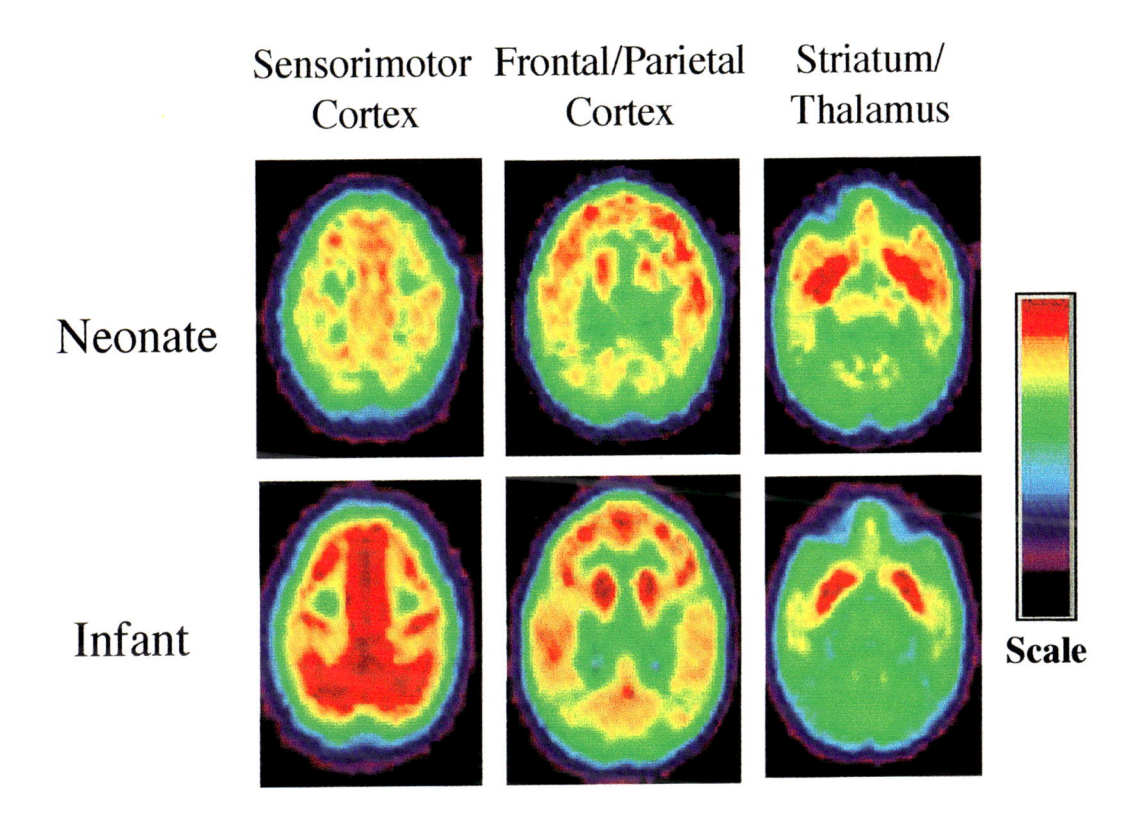

CHAPTER 26, FIGURE 5 Representative FDG–PET images of conscious monkeys. The contrast in pattern and intensity of cerebral glucose uptake between neonatal (≤ 4 months) and infant (6–8 months) monkeys is illustrated.

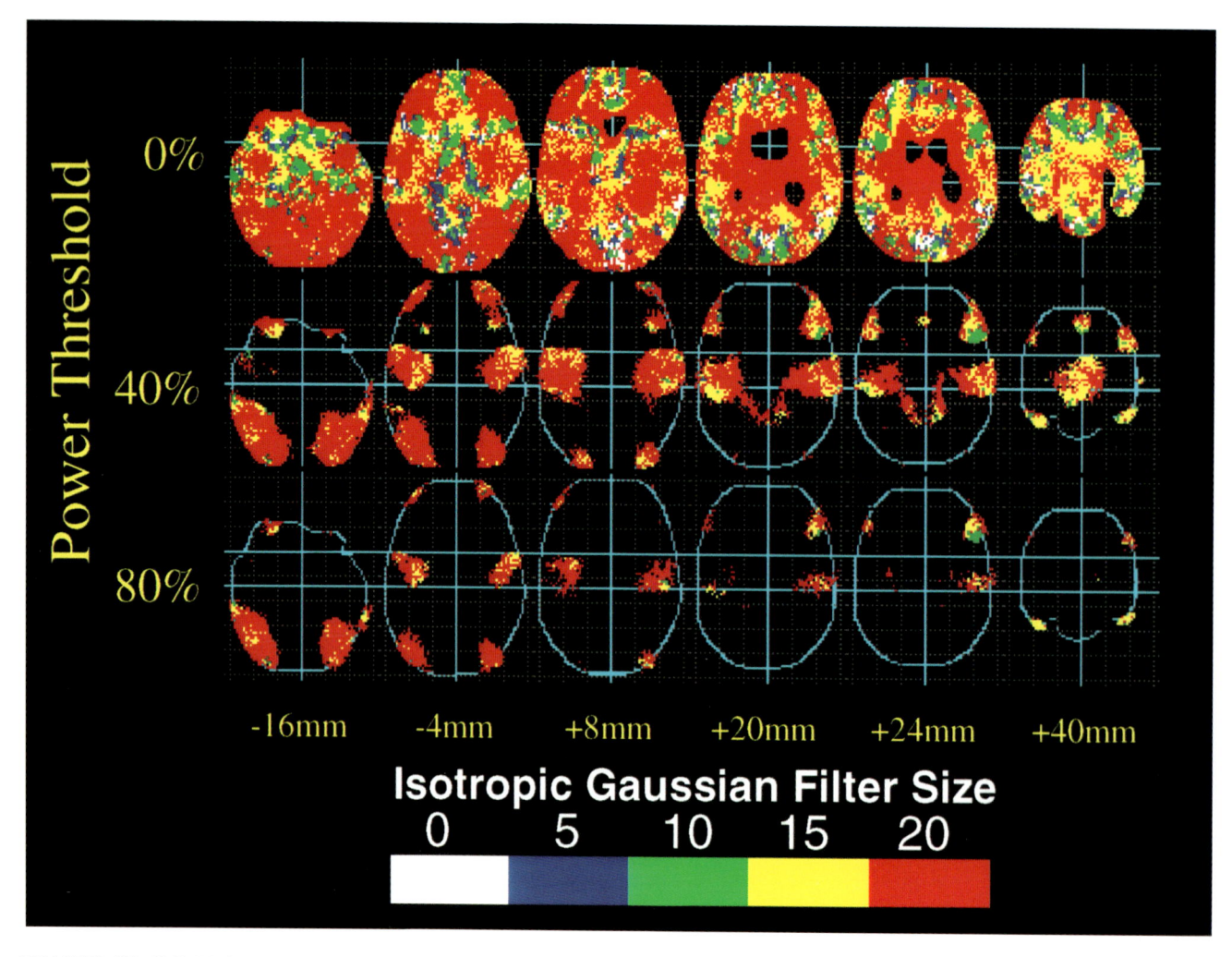

CHAPTER 35, FIGURE 3 The spatial variation in the smoothing filter size required to obtain maximal effect size and power for the $N = 40$ sample. Larger filter size tended to be associated with brain areas of large structural extent, whereas smaller filter sizes tended to be associated with smaller brain structures, a result consistent with "the matched filter theorem." The blue reference outline defines the edge of the 0% power threshold filter size map. Brain areas remaining after thresholding at 80% power indicate those brain areas typically observed as being activated during the WCS paradigm but also that filter sizes in the range of 15–20 mm best resolve significant changes in those areas.

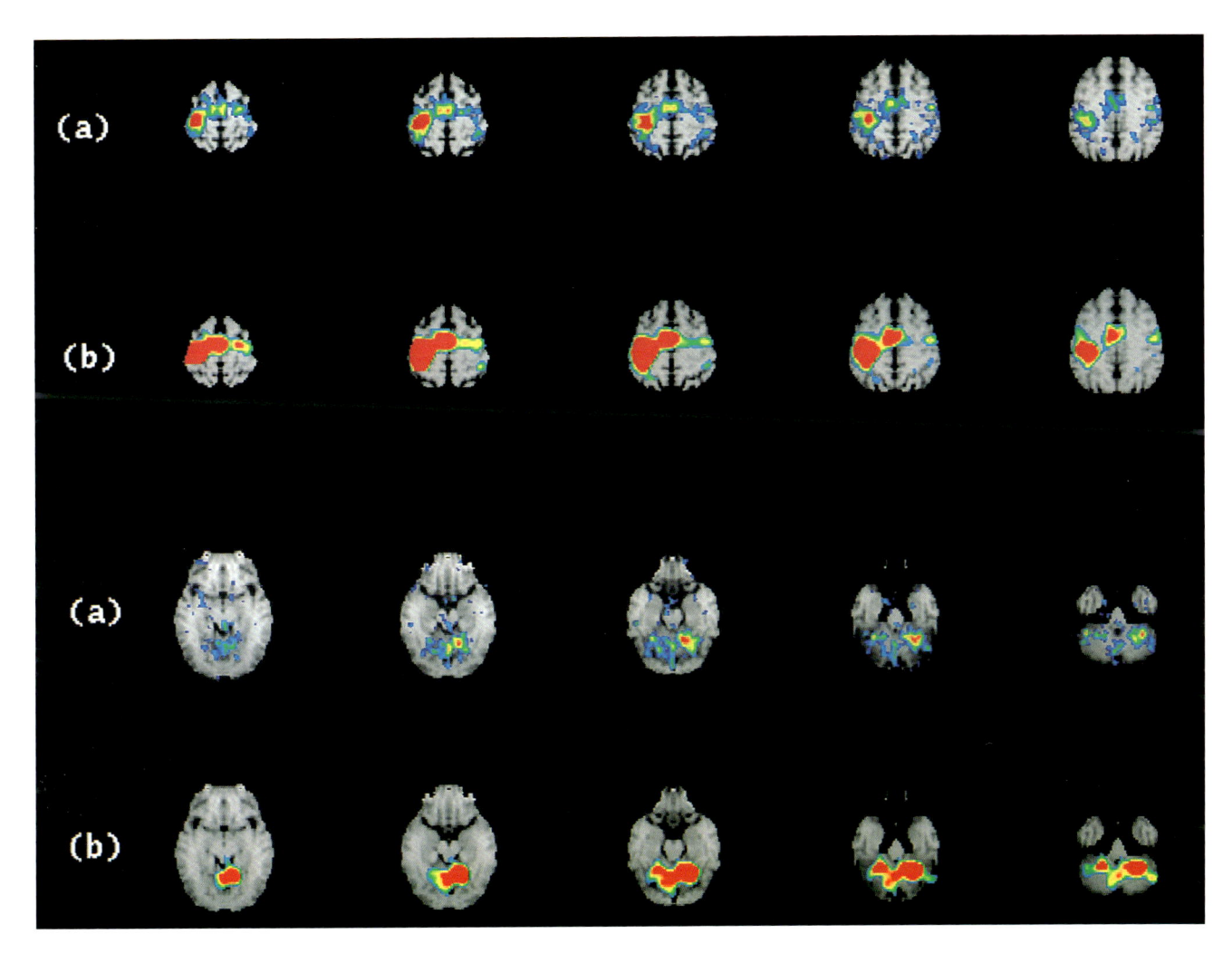

CHAPTER 38, FIGURE 4 Activated regions ($p = 0.01$) obtained using (a) PCA, FLDA, and nonparametric testing and (b) pooled variance t statistics and parametric testing based on a Gaussian random field model.

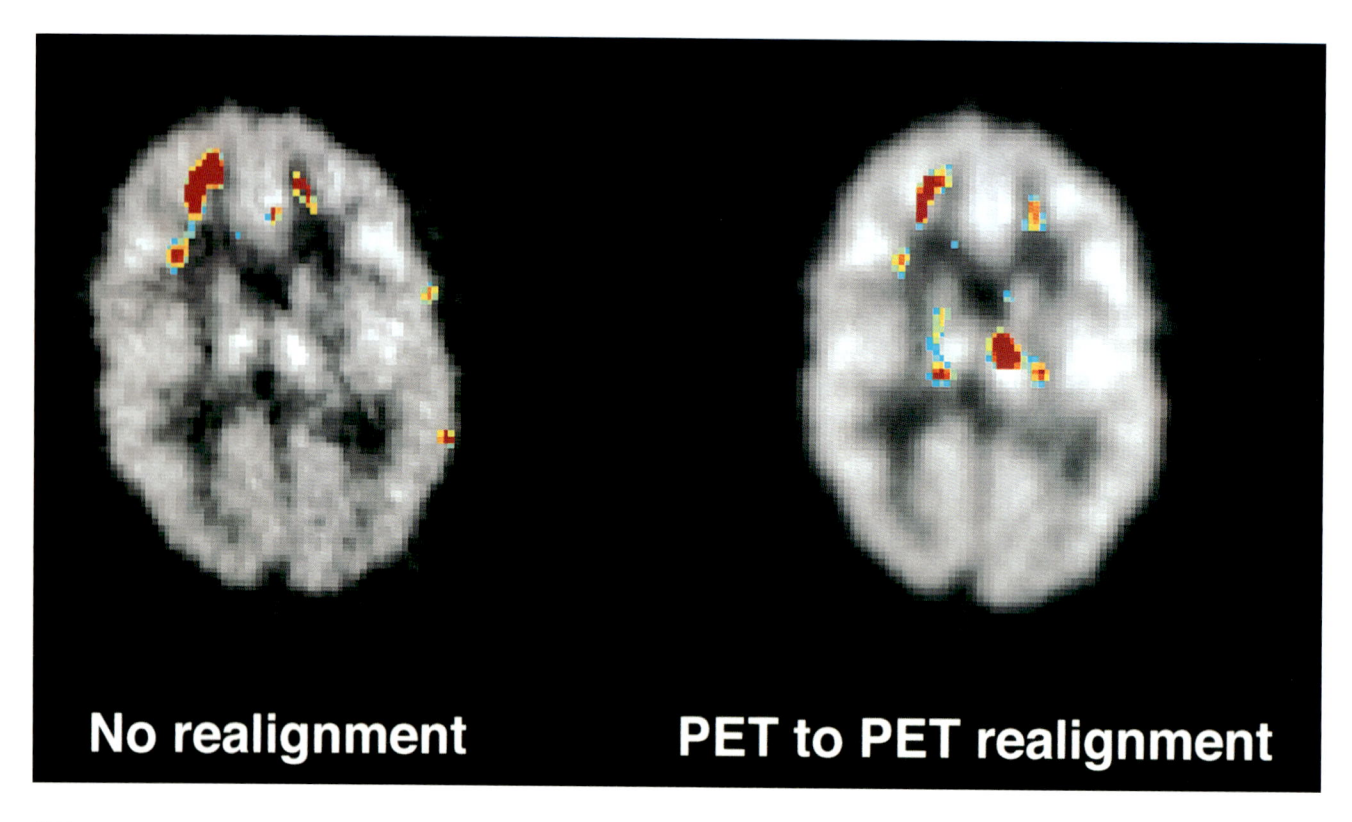

CHAPTER 67, FIGURE 2 Pain study with [^{11}C]diprenorphine and capsaicin cream. The colored areas represent the t statistical map, thresholded for statistical significance ($t > 3.5$), which is overlaid on the radioactivity distribution image (in black and white). On the left, data uncorrected for patient movement show obvious motion artifacts. On the right, after application of the coregistration algorithm to the dynamic images, there is a reduction in motion artifact, but also a new area of tracer displacement in the thalamus. This does not represent opioid release, but rather an artifact of the coregistration technique, which pulls the later time frames of the PET scan inferiorly, leading to a false-positive result (see Fig. 3).

A. Lammertsma: If you do a dynamic study, at some stage you have high activity and you have a very sensitive machine to follow detail, but you might have problems at the peak. Are we full circle again? Are we going back to steady-state studies?

D. Bailey: Yes, possibly, but one of the other ideas that we've entertained is that one might give a priming dose of perhaps 5% of what you intend to give as a full dose, just to get the initial rise of the curve, and then maybe 5 min into the study deliver the other 95% of the dose, throw away the first 5 min, and then join the two studies together. I think we have to work more closely with the modeling people to work ways to use this machine in its dynamic range, which is fairly limited, to get the most out of it. Steady state is obviously one option.

P. Maguire: I have a question about the design. You describe one that is cylindrical geometry, so what would have been the advantages of perhaps a more spherical geometry in the axial direction in terms of axial resolution and sensitivity?

D. Bailey: You want to wrap it around a little bit more and increase the solid angle. Is that right? I haven't looked at it. I guess this is Bill Moses' and Steve Derenzo's area maybe. We did want to use standard detectors and a very standard way of constructing the machine because we realized the problems of a fully 3D machine were going to keep us busy enough.

W. Moses: It doesn't do you very much good at all to try to wrap it around a little bit. The extra coincidences that you would be getting are sort of from nonexistent detectors. You'd be trying to get coincidences from detectors that are inside the body.

V. Sossi: What is the actual reconstructed resolution for patients? Do you have to do some filtering?

D. Bailey: The resolution in the patients is identical to what's measured in any acceptance test because we only use a ramp filter in the reconstruction. We cut it off at the Nyquist frequency, which is about 2.5 cycles per centimeter. So that's why I say with confidence it's 4.5 mm in those images.

V. Sossi: Since the axial extent is so long, do you have any practical problems in pushing the patient in enough so as to be able to get in middle of the scanner?

D. Bailey: The practicalities of putting somebody in the machine with this axial extent is actually not that great. It's a whole body machine, remember, so it's quite wide open. When we've added the lead shields and brought them in to 35 cm, we have had to drop the position of a screen if we're doing a visual stimulation paradigm. It's less than optimal in that people are looking down their nose a little bit at the stimulus rather than being able to just relax and look up, as they were before we had the shields. Despite that, we're still using it very effectively for that purpose.

T. Jones: It's not a steady-state machine. It's a dynamic machine. If we take 10 mCi of raclopride and inject it into somebody and look at the noise equivalent counts over time and compare what you would have had if the machine were perfect in terms of dead time and randoms, we lose about 20% of NEC over 1.5 hr. So this machine could be improved by 20% by a perfect detector. So it is a dynamic machine

dealing with standard injection doses for D_1-D_2 receptor studies.

CHAPTER 7

Parametric Image Reconstruction Using Spectral Analysis of (Rebinned) Three-Dimensional Projection Data

S. Strother: You're using iterative reconstruction on your parametric images, and I guess you didn't mention any form of regularization. So you have a spatially varying, sort of nonlinear signal problem now using ordered subsets. So you've thrown away one of the advantages of going through your original set of linear processes. Is that a concern? Now your resolution is locally dependent on the structure and the number of iterations you've done. So while the signal to noise is certainly improving, it's doing it in a sort of funny, nonlinear way.

S. Meikle: Are you saying that we can't add in those regularizing constraints?

S. Strother: No, you can definitely add them in, but Jeff Fessler has shown us how to keep the resolution uniform, but the images you showed us didn't have Jeff's uniform resolution regularizer, I take it.

S. Meikle: No, you're right. There's no reason why we couldn't add in those extra regularizing parameters. In fact, the other important point is that we don't need to use the EM algorithm at all. I've talked to Jeff Fessler about using alternative algorithms. I wasn't so concerned with using resolution models. I was more concerned about whether the EM algorithm was even appropriate given that our noise structure, once you've done kinetic modeling, is nothing like what you'd expect from Poisson data. Jeff's advice there was just to try using unweighted least-squares reconstruction.

So there are a few things that I want to explore a bit further, but certainly I do want to try various different approaches to reconstruction.

R. Carson: I was wondering how important is the single slice rebinning. Would you have a problem if you bothered to apply the spectral analysis for all the lines of response in 3D, for example, due to noise enhancement?

S. Meikle: You're saying why don't we apply spectral analysis directly on 3D data? I think it's just time. Spectral analysis at the moment is taking around 10 or 15 min to process on the rebinned projection data. To apply that on the 953B scanner, you've got 256 sinograms and 20 or 30 dynamic frames. It's just a huge volume of data to process, but in principle there's no reason that you couldn't do that. I suppose you would then get the benefit of reducing that data from a dynamic series down to the actual parameters you're interested in, so I guess you could argue about exactly where you fit these different processes in. But I think the way we've done it is probably more efficient than that approach.

S. Cherry: I find it very sad that we take these 3D data sets and then we reduce them to 2D data sets. Shouldn't we do the right thing with the data? I mean, we're talking mainly about research studies at this conference and trying to get regular quantification. So shouldn't we put the events in the right place?

S. Meikle: I take your point that, particularly in research studies, it may not matter how much time you spend processing these data. The only answer I'd give to that is that it's still enough to stop people from doing it. I don't necessarily understand that myself, but I think the two issues that stop people from adopting 3D for dynamics are the issues of the quantitation and the issues of the volume of data that needs to be processed. I think that the quantitative issues have largely been solved. So why aren't people doing it? The only reason I can offer is that they're still somewhat put off by the size of these data sets. So time is not important, but it still seems to be a factor. I'm not sure why.

S. Cherry: If you do the single slice rebinning, then in the iterative reconstruction you have an opportunity to at least model that rebinning process in the reconstruction process and maybe get something back.

S. Meikle: If you're going to make the various approximations that I've talked about, there's certainly no reason why you couldn't model those approximations and potentially correct for some of the inaccuracies that may propagate through that process.

A. Dagher: One area that a lot of people are interested in is correcting for patient movement during the scans. Suppose I had a method which could tell you exactly where the head was at all times. How might you plug that information into your algorithm to correct for patient movement without going to the impossible method of just collecting tons of data and changing the frame every time the patient moved? Could you imagine any hardware or software to do that?

S. Meikle: Actually, I don't think it would be a problem at all. The experience we've had with doing motion correction in Sydney was mainly with SPECT data, where it's essentially a 2D reconstruction problem. All that happens when you've got the ability to measure head movement is that you turn the 2D problem into a 3D reconstruction problem. So, in principle, I think with 3D PET data, that's still the case. All you're doing is using the measurements you've made of the head movement to reassign the positioning of the lines of response. It's still a 3D reconstruction problem; you've just moved the lines of response around.

So, in principle, I don't see any problem with incorporating head movement information into 3D reconstruction. I believe there's some moves both at CTI and at Hammersmith to do something along those lines.

P. Maguire: I was very interested to see your choice of spectral analysis. We've done something similar with the Patlak analysis, and I wondered how sensitive your technique is to the assumed β-values in the operational equation.

S. Meikle: There's been quite a bit of work done on that at the Hammersmith. The traditional approach is to assign the βs logarithmically across the dynamic range that you're interested in. Personally, I haven't tried various different approaches, but I know that Julian Matthews has explored different schemes for mapping those β values, and I think

I'm right in saying that he didn't see a huge benefit from doing it. But there certainly are different approaches to mapping those predefined beta values.

P. Maguire: When we presented our work on applying Patlak analysis to projection data at the last conference, the question of variance images cropped up. Have you tried to calculate images of the variance of the parametric images?

S. Meikle: No, I haven't done that. That seems like a good suggestion to do that. I guess the best way to do that would be through Monte Carlo. So far, my analysis has always involved comparing this approach versus the traditional approach, but I think that's probably a good suggestion.

V. Cunningham: With respect to Alain Dagher's question, in a perfect world, we would be collecting list mode data, monitoring the movement at the same time, and then just resorting the data off line or afterwards.

With respect to the point about the range of β values, as long as you've got enough of them, it doesn't really matter, and there are quite a lot of them.

Participant: You wouldn't care to define "enough" I suppose.

(Laughter.)

V. Cunningham: Anywhere between 64 and 100. One hundred is a good round number, isn't it?

S. Meikle: Applying the spectral analysis to projection data, I found that you can even reduce the number of basis functions quite substantially, down to around 30. You get a time saving, and I don't see any bias in the values of the parametric images that I get.

The point about list mode data is a good one, too. The method we were using in Sydney did not use list mode data, but list mode data are really ideal. You can exactly incorporate the timing of the head movement. So it's an ideal way to incorporate head motion information.

CHAPTER 9

Absolute PET Quantification with Correction for Partial Volume Effects within Cerebral Structures

T. Videen: Presumably your algorithm is sensitive to the assumptions in the segmentation process. How would you validate that you have a good probabilistic segmentation?

M. Graham: You were sorting out maps of gray matter and white matter, and yet you say there's no *a priori* estimate of tracer concentration. It seems to me that the whole point of sorting out gray matter and white matter is that there has to be an assumption that there is a certain concentration in gray matter, and a concentration in white matter. How can you reconcile those two concepts, that you're sorting these out and yet you say there's no *a priori* assumption of concentration?

V. Cunningham: When Claire Labbé said there was no *a priori* assumption of concentration, what she meant is that the calculation does not require you to put in a seed value for, say, the white matter. You are essentially solving an equation that expresses the final PET image as a combination of concentrations times the various segments. So if you've segmented gray, white, or various VOIs, you're multiplying each of those by an assumed homogeneous radioactivity concentration to give you your PET image. That was what was meant by no *a priori* assumptions.

M. Graham: I can see how that will work out nicely in the normal brain, but if you've got asymmetric uptake in the caudate, say, and you're assuming caudate is gray matter, then how does this work?

V. Cunningham: It's assumed that each of the segments, be they gray or white VOIs, are homogeneous. There's homogeneous radioactivity within each segment, a segment being either gray or white, or a VOI a bit of gray or a bit of white.

M. Graham: So they're really a whole lot of blobs of homogeneous activity.

V. Cunningham: Yes. It's spatially compartmentalized in the sense that you assume that each of these many segments are homogeneous. So you are correcting for the edge effects on the partial volume.

N. Alpert: The question of the probabilistic map seems to be the key element, and perhaps somebody can clarify this. It seems to me in some sense that using a probabilistic map that's drawn from a very large population would protect you and it should do well. But I would like somebody to explain this a bit more, in the sense that everything will tend to the proper mean when you have a large group.

S. Strother: Are you segmenting each subject's MR into a probabilistic map, or are you using the average Montreal probabilistic map built from many MR scans of normal subjects?

C. Labbé: We have a lot of possibilities. We could choose either the individual MRI or a large number of subchoices from Montreal, or it could be our own template to segment into probabilistic maps. But here, we have used data from Montreal.

O. Rousset: From a mathematical viewpoint, could you tell me what is the difference between solving a system of linear equations by linear least square and solving this set of linear equations by matrix inversion?

V. Cunningham: The matrix inversion is part of the least-squares solution. When you've got $Ax = b$, the solution is $x = A^{-1}b$, and when you do that, you're effectively doing the least-squares solution. So the matrix inversion, or actually the pseudo-inverse, is part of the least-squares fit.

O. Rousset: Mathematically they are equivalent, right?

V. Cunningham: Yes.

S. Minoshima: I have one philosophical question about MRI-based correction because in the introduction you said that we're interested in cellular function or physiological functioning. But the MR-segmented image only shows the thickness of the gray matter, but most or all of the tracer has some layer-specific distribution in the cerebral cortex and you're measuring mostly the thickness of the cortex. So in some diseases, maybe the tracer distributes in layers 3 and 5, but maybe the thickness is affected at layer 1, and you're correcting this thickness using this kind of technique. So when applying this MR-based correction, we have to know what kind of tracer we're using and how this tracer distributes in the cortex and how diseases affect specific aspects of the cells in the cerebral cortex.

S. Houle: When you mentioned that you're using a homogeneous distribution in each of the regions, how do you deal, for example, with Parkinson's disease, in which there appears to be a very gradual change in the putamen, from anterior to posterior putamen, where obviously the activity is not going to be uniform? Do you subdivide into a small region, or how do you deal with that issue?

V. Cunningham: I guess you'd have to subdivide into smaller regions because it's part of the way the calculation is done that you need homogeneous regions.

M. Senda: I'm interested in the number and size of volumes of interest. Your method assumes homogeneous radioactivity concentration within the single volume of interest, but that, of course, is not always true. But if you can subdivide the volumes of interest smaller and smaller until the size reaches the pixel, you can get rid of your assumption.

C. Labbé: We assumed that each VOI contained either gray or white matter, or CSF, or a combination of these three tissue classes. We assumed that each VOI was homogeneous. Of course, if we subdivide a lot of VOIs into subvolumes, we will reduce the error because we are taking a small region. If we take a large region, like the hippocampus, I have subdivided into three subvolumes. But when we look at the total hippocampus mean value after partial volume effect, it was in fact a mean value of the three subvolumes.

CHAPTER 10

Pixel- versus Region-Based Partial Volume Correction in PET

S. Cherry: I was wondering if anybody has looked at the sensitivity of the partial volume correction to errors in the registration between the PET and the MR. I think for very small structures that could be a significant problem.

O. Rousset: We have demonstrated earlier, in '93, that the actual misregistration error is on the order of 5% per millimeter of misregistration. This is in terms of absolute activity values. We have also shown that this error, as propagated through the mathematical model, was on the order of 10 to 15% in K_1 and k_3 values.

T. Videen: What would you consider an adequate validation of a gray–white segmentation?

O. Rousset: Since we have the ability to create simulated images, I would correct first for partial volume, get the true activity levels, go through the PET simulator, and then generate a simulated PET image that should look like the real PET image. Then you do a subtraction, for example, between the real PET and the simulated PET, and probably your areas of missegmentation are going to show up in the difference image.

C. Meltzer: Have you looked at the sensitivity of your method to errors in segmentation?

O. Rousset: We haven't looked at the error of missegmentation and its consequence on quantitation, and I think that's a very important issue. I think the accuracy of such techniques really depends on the accuracy of the segmentation.

C. Meltzer: What type of MR pulse sequences do you use for segmentation?

O. Rousset: We use a SPGR sequence to have the best distinction between the gray and white and CSF.

V. Cunningham: Does it make much difference as to whether you apply the point spread correction to the dynamic data and then do fits, or would it be possible just to do the fits in a regular way and then apply the correction to the functional images at the end? Does the order in which you do it make any difference?

O. Rousset: Like a parametric image, for example. That could be a possibility. I think we should look into this.

K. Herholz: To that question, it's related to the issue of heterogeneity within tissue used for evaluating dynamic curves. That might have quite a large effect if the model is highly nonlinear. So I would add caution with nonlinear models.

J. C. Baron: It is very important to do corrections for partial volume effects and atrophy. But, as Dr. Minoshima said, from the pathophysiological point of view, one has to be very cautious when using these tools. For example, take a structure that has lost part of its volume but the receptor number, for example, has stayed the same. Then you will find, by correction, an increased density, whereas the number has not changed. So this means that when you do these kinds of corrections, you have to take into account the volume of the structure that you have corrected. This also applies to Alzheimer's disease and all the diseases where atrophy is part of the disease. So if you correct for atrophy, you may miss the effects of pathology you're actually interested in.

With the new generation machines which have such a high resolution, do we still need correction for these effects?

O. Rousset: I think so.

CHAPTER 13

Registration of Multitracer PET Data

S. Strother: What you're doing looks a little bit reminiscent of mutual information, except you're actually choosing some discrete classes. I'm wondering if you've had any experience using mutual information as a relative comparison to what you're doing.

J. Andersson: I've had no experience using it and I couldn't really tell you what are the similarities and what are the differences because I don't understand the mutual information theory enough.

S. Strother: You've got lots of tuning parameters in the early phases, particularly in your clustering segmentation process. Can you tell us a little bit about the sensitivity of the algorithm, or how many actual parameters are there that you're having to tune and how sensitive are your results to what you choose for those parameters?

J. Andersson: There are two tuning parameters. One is how many principal components you want to retain, and the other is how many classes you want to use. For the first one, the number of principal components, you use more than necessary because the principal component algorithm will scale the variance such that, in the subsequent classification step, the majority of the classifications or the splits will be in the first principal component, there will be one or two in the second, and very rarely one in the third. So that's not really a tuning parameter.

The other one is how many classes you want to use, and I've been using eight for quite some time now and it's working all the time.

S. Strother: Have you looked at the sensitivity to using 5 versus 16?

J. Andersson: No, I haven't, actually. I tried more in the beginning. There's one rather important thing: you shouldn't use too many classes. You should have relatively few classes. If you have a class that is a patch, sort of a little blot, and you move that around in the other image, you'll see nothing. But if you have a class that's like a torus that goes around the cortex and you move that around, only one class, you'll be able to use that, even one class, for registration. So it's important you don't have too many classes because then it would be too patchy. You want these rather large clusters, and you want them to go around the brain pretty much.

A. Dagher: One problem that a lot of people are interested in is coregistering different frames of a single dynamic PET study to correct for patient movement. The problem with those scans is that the radioactivity distribution changes with time. I wonder if you think PCA could be used to differentiate normal change in radioactivity from patient movement and correct for that.

J. Andersson: Actually, we're doing this dynamic reorientation quite routinely. But we take the second frame and reorient that to the first frame, then the third frame is reoriented to the second frame, and multiply the transformation matrix with the first one you obtained, etc. We're doing that quite routinely, and we have no problem with that. So I have actually not even tried this method for that because just a simple registration algorithm is enough.

A. Dagher: I've used that method, and it doesn't work for tracers where there's a loss of radioactivity with time. For example, with dopamine tracers like raclopride, you see a loss of activity in the occipital cortex and in the cerebellum as the scan goes on. So the programs tend to interpret that as movement and tend to move the brain down. The radioactivity can change for two reasons. It changes normally, and then it can change for patient movement, and you have to separate those two factors out.

J. Andersson: Actually, we haven't encountered that problem. I've seen other people discuss this as a problem and I've never understood why because we haven't seen it.

CHAPTER 16

Methodology for Statistical Parametric Mapping of [¹⁸F]Fluorodopa Uptake Rate Using Three-Dimensional PET

M. Senda: Do you have morphological anatomical evidence of validation for the spatial normalization within the putamen and caudate nucleus?

J. Rakshi: We haven't directly validated it. I'm not sure quite how we would go about that. The fact that using the conventional region of interest approach produces similar results is reassuring, but we haven't actually been able to directly validate this approach of applying spatial normalization of images to the standard template.

J. Matthews: In your statistical analysis, are you making an assumption of normality? And if so, do you think that's reasonable for a K_i map?

J. Rakshi: Essentially, we're attempting to do a simple unpaired *t*-test on a voxel-by-voxel basis. Again, I'm not sure of the statistics behind it, but the fact that we are getting consistent results with a conventional approach would suggest that it probably is a statistically sound approach.

J. Matthews: It's more the case of how significant it is. The values of 5 and 6 you're getting, is that the true value? Is that an artifact of the assumptions? I've no doubt that it's significant, but how significant it is?

J. Rakshi: From what I understand, a normal degree of significance would be above Z scores of 2.3, and these are considered to be very high Z scores and to be highly statistically significant and don't necessarily need to be hypothesis driven.

J. Matthews: But if you have a distribution that's not normal, that has a very long tail, the change of getting a very extreme outlier is a lot greater, which would tend to bias your statistics to maybe identifying a significant region when there isn't one.

J. Rakshi: While I didn't show the individual K_i values after SPM analysis, there is actually clear discrimination between all values in the normal group and in the PD groups when looking at the putamen or the anterior cingulate gyri, and there are no outliers. So all the individual values are within a relatively tight range.

J. Matthews: Hence, normality may be reasonable.

M. Laruelle: When you construct your map of K_i, I guess you have a lot of pixels that have zero value or very low value, especially when you use the occipital input function, and I assume that they were the thresholded out of the analysis. What criteria do you use to decide which pixel you threshold in versus the one you threshold out? How do you define those criteria?

D. Bailey: The values themselves are influx rate constants. They're K_i's; they're not K_1's. So it's the slope of the linear regression from a Patlak analysis on a pixel-by-pixel basis. As I understand it, there's no thresholding on data that's gone into the SPM apart from the so-called gray matter threshold, which is used just to segment out the brain from the rest of the data.

J. C. Baron: One easy way to check the stereotaxic registration of your method would be to do some water studies just before the injection. Have you done that?

J. Rakshi: We haven't done that, but that is one suggestion that we could certainly look at.

J. C. Baron: I also had a question regarding the occipital reference. How do you do that? Do you use regions of interest?

J. Rakshi: Yes. Essentially what we do is we place two large regions over the occipital tissue at the same level, the same transverse planes as the striatal regions, and they are 32 mm in diameter.

J. Frost: We had similar results using the dopamine transporter ligand WIN35428 with SPM analysis, some of which are presented at the last SNM meeting, similar results to yours, although in the midbrain in Parkinson's patients, contralateral to the symptomatic side, we see a reduction in transporter binding. It's interesting that you see an increase in fluorodopa synthesis in that region but not in the basal ganglia. Do you have any ideas on why the midbrain binding might be able to compensate for metabolism when the projection areas in the basal ganglia do not, in the face of neuronal loss in both areas?

J. Rakshi: The actual increase is in quite a specific region. It's not a general increase in fluorodopa uptake in the midbrain. It's actually at the level of the superior colliculus.

J. Frost: Do you see changes in the nigra?

J. Rakshi: We don't in early Parkinson's disease. I know that in your paper you found a significant reduction. I think it was around about 80%.

J. Frost: Yes, that was in the 1993 paper with Stage II PD. But in Stage I, we're seeing a unilateral reduction, which we think is linked to neuronal density.

J. Rakshi: One possible explanation that has been proposed is that it may represent downregulation of these reuptake sites.

J. Frost: Yes, or they're on the dendrites of those nigral cells, so it may also represent actual loss of cells, and maybe fluorodopa metabolism is able to compensate, again showing the differences between what fluorodopa is measuring and what the DAT ligands are measuring.

R. Carson: I'm a little confused about the spatial normalization against the CBF template. Why does it work? What is it about the large contrast around the basal ganglia that it can happily ignore?

J. Rakshi: If you look at the CBF template in transverse plane, the striata are well defined because they have extensive blood flow. With the increased sensitivity in resolution of 3D fluorodopa PET imaging, you also have cortical detail. In fact, there is quite a close similarity if you look at all the transverse planes in both the CBF template and in the fluorodopa PET image.

R. Carson: But is it in the nature of the spatial normalization function that it doesn't matter that there's a great change in local contrast? There is a huge difference in contrast between your images and what's in the CBF template.

J. Rakshi: I agree, and that would appear to be the case, because if it did matter, we wouldn't be able to do it.

D. Bailey: The volume of the brain occupied by the basal ganglia is tiny compared to the whole brain that we're matching here, and I think that's the reason. There's a huge contrast between striatum and the rest of the brain compared to the CBF template, but that's a very tiny proportion of what we're actually matching.

R. Carson: So if there was local distortion around the basal ganglia, your match would be wrong, it wouldn't adapt to that.

D. Bailey: I think it has a 5% effect.

R. Carson: But a lot of people with fluorodopa are interested in quantitating the basal ganglia. So are you're saying the match will do everything right except the basal ganglia?

D. Bailey: It will match the whole brain the best it can.

P. Maguire: You said it wasn't quantifiable, your technique, but you could apply the transformation matrix to all the dynamic frames in your study and then generate automatic regions of interest within the areas of significant change to give you automatic time–activity curves, which you can then use to fit for any parameters you're interested in.

K. Herholz: It's quite usual for degenerative diseases to have an enlargement of the ventricles, and I'm not certain how this method would account for that.

J. Rakshi: In Parkinson's disease, it is well recognized that there's no gross morphological abnormalities in both the striatum and cortical regions. In general idiopathic Parkinson's disease, the brain is normal, although I accept that comment.

N. Alpert: Could I ask you to clarify the nature of your reference region technique? Are you using the method of Hoshi?

J. Rakshi: All I know is that it's a multiple-time graphical analysis approach.

A. Gjedde: The method is a variation of the Patlak and Blasberg paper from 1985 in which you use a tissue input function to determine the slope, which essentially is a somewhat imperfect estimate of k_3 and has nothing to do, per se, with K_i, which is an estimate of the K-complex, which is used when you use a plasma input function.

J. Andersson: When you're registering data to an atlas like this, where you know where the structures are, you might in principle use a masking which you apply over the basal ganglia, thereby removing them from this fitting. Do you think you would benefit from doing that?

J. Rakshi: That's an interesting option. We could try to see if it makes any difference to spatial normalization, and then compare the two before and after to see what difference it makes. The other thing that has been suggested is that you could in some way reduce the intensity of the basal ganglia to make it fit more closely or match more closely to the CBF template.

J. Andersson: Regarding the normality assumption, have you tried the permutation test by Andrew Holmes to see if you get similar results?

J. Rakshi: No.

CHAPTER 18

Data Extraction from Brain PET Images Using Three-Dimensional Stereotactic Surface Projections

S. Strother: I'm wondering how sure you can be that you're actually moving in normal to the cortical surface when you extract your maximum in your SSP process. If, for example, you go through your warping process on an MRI scan and then you have a template surface that you move in, are most of those lines actually normals to the cortical surface? Because you worry about crossing gyri halfway down the sulcus. In which direction are you going? Are you going into the sulcus, or are you actually coming across the gyrus toward one of the other sulci?

S. Minoshima: We examined how the algorithm is searching the cortical activity using MRI. Regionally, for example, near the lateral sulcus, this search is not truly searching peak activity perpendicular to the surface, and there may be some bias. But for most of the lateral neocortex or free surface of the neocortex, this algorithm detects fairly accurately the peak activity in the gray matter.

The depth of the sulcus is the other issue. This presentation gives you the impression that we are only searching the superficial part of the sulcus. But if the algorithm searches into the brain 14 or 15 mm from the nearest free surface of the cortex, this actually covers most of the depth of the sulcus. Although there are some sulci that appear deep in the brain on transaxial images, this search scheme is picking up the peak activity from most of those areas, as well. The limitation of this method is obviously deep gray matter without adjacent free surface such as the insular cortex or putamen.

T. Videen: You showed a very convincing comparison of the advantage of this method to traditional images. I wonder how you think it would compare with fully partial volume corrected PET images.

S. Minoshima: Actually, we are currently testing that. We acquired MRI scans and segmented into gray and white matter, as was discussed this morning. Assuming that FDG has a relatively uniform distribution within the cortical layers, we can actually apply this correction factor, if necessary, prior to the data extraction or after the data extraction and make the correction directly for the 3D-SSP extracted activity.

CHAPTER 20

Sequential Experimental PET: Voxel-Based Analysis Reveals Spatiotemporal Dynamics of Perfusion in Transient Focal Cerebral Ischemia

J. Andersson: Scatter plots seem like a useful tool. It seems to me, however, that if you took all the time points, created a covariance matrix of them, and performed a principal component or a factor analysis, you'd be able to pick out all the various components, and in a more objective manner at the same time. Could you comment on that?

R. Graf: Obviously, quantification is a problem in these many slices within these changeable patterns. So the threshold approach is the one that we favor now because we can set physiological thresholds. However, that is quite difficult to perform consistently. So our idea is to go into cluster analysis. But, as I learn from colleagues who know more about that, normal hierarchical cluster analysis would not help a lot, so one would have to find some other ways, and we are on the way to discuss that in our lab.

What I wanted to show is that regional patterns change over time. You may find, for example, changes of patterns where the basal ganglia are hyperperfused only at a certain time point. In the clinic, you often get a single time point and you are left with an image that does not tell you what the sequence of events was in that particular case. The images in the clinic are very heterogeneous.

J. Andersson: You mean you don't have all these time points in the clinical situation, and therefore a principal components analysis would not be useful?

R. Graf: What we are trying here is to find some kind of signature to have an idea of what the images at certain time points in the clinic might mean.

CHAPTER 25

Frequency-Dependent Changes in Cerebral Metabolic Rate of Oxygen during Activation of Human Visual Cortex Studied by PET

L. Junck: The region that you're examining is very close to the superior sagittal sinus, and my question is whether your model is adequately accounting for residual ^{15}O activity within that large venous structure?

M. Vafaee: Yes. We have tested our model with other kinds of stimulus. For example, we have done vibrotactile stimulation using our model, and there is no oxygen change in the vibrotactile stimulation. So, yes, we are sure it is not a vascular artifact. The model has been tested.

A. Gjedde: The interesting finding here, among others, is that for much of the blood flow increase, there was no oxygen increase. So the residual oxygen trace could not explain the fact that for some of the frequencies there was no increase in oxygen. It was only 4 Hz, which I find could not be explained by some deficiency of the model.

L. Sokoloff: We have looked at the changes in glucose utilization in response to stimulation of afferent nerves. So, for example, we stimulated the cervical sympathetic trunk at different frequencies—5, 10, 15, and 20 Hz—and we also stimulated the sciatic nerve at 10, 20, and 30 Hz and looked at glucose utilization in the projection regions of the pathways. In the case of the cervical sympathetic stimulation, we looked at cervical ganglion, and in the case of sciatic nerve stimulation, we looked in the dorsal root ganglion and in the terminal zones in the dorsal horn of the lumbar.

There was, in both cases, linear increases in glucose utilization with spike frequency. In the case of sciatic nerve stimulation, this went through the entire range, up to 30 Hz. In the cervical sympathetic system, we found a linear increase up to 15 Hz, and then it flattened out. Fortunately, we also recorded the electrical activity at the projection site, the compound action potentials, and that didn't increase either.

So there was a flattening out of the neural response that couldn't follow the higher frequency above 15 Hz, and I was wondering, since your simulation frequency applies to the retina but the striate cortex is looking at what's coming from the lateral geniculate, is it possible that there wasn't a linear following of the spike frequency at the site where you were measuring the oxygen consumption?

M. Vafaee: That's possible.

M. Senda: When you showed the blue and yellow pattern, the yellow color looked much brighter than the blue color. Was that the case for the actual stimuli?

M. Vafaee: No, we tried to keep the luminescence or brightness constant through all of the study.

M. Senda: Is it because you wanted to look at the response to a change in color instead of the change in brightness? If the yellow color and blue color are of the same brightness, the stimuli to the subject is to change color, not to change the brightness. It is very different from the black and white checkerboard stimuli.

M. Vafaee: Yes. If you use black and white, then you change the luminescence. You have to be careful. The reason we chose the checkerboard is because no matter what frequency you use in a checkerboard, the luminescence or the integral over the period stays the same, the period of time. The number of photons you get by presentation, let's say at 3 min, no matter what frequency you use, stays the same. So that's why we chose a checkerboard.

B. Horwitz: Although you thought you were just changing color, in fact you were also changing form, because the brain has a number of mechanisms for extracting form from color. So the neuron populations in V1 that you're affecting by your stimulus are not just the ones that use cytochrome oxidase.

M. Vafaee: Yes, I agree with that. As a matter of fact, we have done the study using the stimulus, and we've been able to activate V4 and V5 using this stimulus also.

P. Herscovitch: You pointed out that this stimulus increased both CBF and $CMRO_2$, whereas simple flashes only increased CBF. Did you have the opportunity to compare those two types of visual stimuli with fMRI to see if there is a difference in the bold deoxyhemoglobin signal?

M. Vafaee: This is an ongoing project in our lab in collaboration with our MRI people.

P. Herscovitch: That was the same answer we got in 1993.

I. Kanno: If blood flow increases, for example, due to changes in pCO_2, is it possible that your method underestimates oxygen extraction during functional hyperemia?

A. Gjedde: Again, the point is that oxygen consumption did not increase for many of the frequencies that caused blood flow to increase. So this failure cannot be explained by a deficiency of the model, which then also explains the increase for oxygen at 4 Hz.

CHAPTER 29

Use of Two- and Three-Dimensional PET and [^{11}C](R)-PK11195 to Image Focal and Regional Brain Pathology

L. Junck: Some of the areas of high uptake that you have shown are in areas of concentration of venous structures of the brain, in particular in the area of the thalamus and the visual cortex. Can you exclude the possibility that the remaining intravascular activity is contributing to the signal? In fact, can you convince us that this is different from findings in normals?

R. Myers: I didn't show any normal studies. But clearly, if you look at thalamus, for example, we don't see anything like the magnitude of signals seen in patients. We do see increased binding in the thalamus relative to other structures. So I think there is a component of what you're saying there. But in the patients, the elevated signals are much, much higher. Also, when you see things like the unilateral thalamic signals, we know this isn't just a generalized effect. This is which is syndrome related. I should also add that these were actually late images, so I think the vascular effect should be relatively small.

L. Junck: Can you tell us what pathophysiology you believe corresponds with the tracking that you describe?

R. Myers: Well, in cases like the medial longitudinal fasciculus, I think the pathophysiology there is obvious. In other cases, clearly we don't know, and there may be no overt pathology associated with the actual tracking itself. Certainly the neurologists at Hammersmith now are particularly interested in using this almost as if it were an activation study to actually track the progress of damage through the system. The actual pathophysiologic implications of the tracking itself isn't clear, but the actual end point may be.

L. Junck: By tracking, you're not referring to the movement of the ligand but rather to axonal damage?

R. Myers: These are neurons transmitting a pathophysiological stimulus down the line, which we know they do from experimental studies, and the microglia respond to that. There is no actual movement itself.

R. Woods: One of the interesting aspects of quantification, especially in diseases like multiple sclerosis, is the notion of following something over time. I wonder if you could address the dosimetry issues that are related to that and any technical issues of trying to adjust some sort of ratio to normalize the images.

R. Myers: Second point first. The normalization of the images to each other is a real issue, and I wish I had an answer to that. We have tried many types of normalization and at the moment we're working on getting a decent input function. So that really rests in this quantification challenge that I mentioned, and I have no clear answer to that.

As far as the dosimetry, at the moment we are allowed to scan patients three times, and one could argue that that's really not enough if you look at some of the MR serial studies that have been done where patients can be scanned every couple of weeks. We would like to be able to scan them more frequently. But clearly serial studies is where the real power of this is going to be, in looking at the relationship between microglial activity and the development and etiology of the pathophysiology.

A. Dagher: The conditions you have shown are likely to have alterations in blood flow, like stroke, or blood:brain barrier function, like MS active plaques. How are you planning to differentiate delivery of tracer from binding in the brain? I think that's crucial if you're going to make sense of these images.

R. Myers: The actual delivery of the ligand itself is not blood:brain barrier limited, and we know this from experimental studies. If you look at a brain early after injection, it's full of PK11195. So I think that we don't have so much of an effect from permeability of the blood:brain barrier. It may effect the K_1, the flow times extraction component, and I wish we were far enough down the quantification road to be able to know what kind of effect that was, but at the moment we simply don't.

A. Dagher: Have you tried various graphical methods to make images?

R. Myers: No, we haven't.

S. Raja: What is the time course of inflammation you can follow through the chronology of acute and chronic?

R. Myers: From the time course point of view, microglial activation occurs very, very rapidly. We can see it usually within 24 hr of any kind of stimulus in experimental studies. How long it continues depends on the type of stimulus involved. We haven't done extensive time course studies on this, but certainly the activation itself is very rapid.

S. Raja: How does it compare with [^{11}C]diprenorphine, which also images damaged neurons?

R. Myers: We haven't done that sort of study, but clearly there's going to be a correlation between the inflammatory cell response and any neuronal death that you get.

S. Raja: Do you have any experience with epilepsy?

R. Myers: We have done some epilepsy studies. It depends on what type of epilepsy you're looking at. If you take hippocampal sclerosis, for example, interestingly we see no PK11195 signal there, and that actually correlates very well with the literature reports that there is no microglial activation. It's purely astrocytic. In patients with Rasmussen's encephalitis, however, which is characterized by a unilateral inflammatory cell reaction occupying a large area in one hemisphere, we see a pattern of PK11195 binding that precisely matches this pattern of pathology.

CHAPTER 30

Noninvasive Imaging of Serotonin Synthesis Rate Using PET and α-Methyltryptophan in Autistic Children

A. Thiel: I have a question about the quantification of these images, especially the input function. The input function you show and derive from the ventricle had no measurement points on the ascending part. The first point was the top. Does this influence the quantitation?

O. Muzik: Usually we started the acquisition before so you could see the zero and then the peak and the fall.

J. C. Baron: How do you prevent head motion in your subjects?

O. Muzik: All the children were sedated, which changes the absolute values. It doesn't change the pattern. We did validation study in adults, because we couldn't do it in children, where we looked at the differences between sedated and nonsedated adults in test–retest studies, and the absolute values are about 10% down. But there is no difference in the pattern of serotonin synthesis rate.

J. C. Baron: The MRs were normal?

O. Muzik: The MRs were normal. So this is a pattern you can only see with α-methyltryptophan (AMT).

A. Gjedde: In the *K*-complex images that you showed, you didn't indicate how you solved the problem of the protein binding of AMT. As the curve declines, the protein binding is a very complex process about which you need to make a number of assumptions to actually compute the free AMT that enters into the generation of the uptake, and therefore into the generation of the so-called *K*-complex.

O. Muzik: Yes, you're absolutely right. This topic is actually very controversial as to what fraction of tryptophan is available to go into tissue. To my knowledge, the AMT doesn't bind to proteins.

M. Laruelle: First I'd like to compliment you for all these efforts to do these PET scan in children. I think it's very important. I have a question which might be a question of semantics, but when you stated that you measure a decrease in the serotonin synthesis rate, I don't think you measured that because you didn't measure tryptophan. So you cannot calculate the serotonin synthesis rate. What you might have shown is that there is a lower capacity to synthesize serotonin for a given tryptophan level, but it might well be that autistic children have a higher tryptophan plasma level and therefore would have a higher serotonin synthesis rate.

O. Muzik: We measured tryptophan and it was very stable. That's the reason why we fast these children 6 hr prior and do the scans at the same time of day, to be independent of circadian rhythm.

This might be somewhat controversial, but we really don't believe that the formula, where you multiply the *K*-complex by the cold tracer and divide by the lumped constant, is applicable to serotonin. I spoke with Dr. S. Shoaf and she found that the tryptophan level might change quite considerably from day to day even if you don't do anything. But does that mean that your serotonin synthesis rate changes in the brain? That's not very likely.

So this simple model might not be that applicable, but that's a field of controversy. Our intention was to get standardized conditions where we fast the subjects, scan at the same time of day, and say this is the information we get from dynamic PET. If you derive all your information from a systemic value, then you don't need PET.

D. Stout: Since tryptophan goes across the blood:brain barrier via the large neutral amino acid transporter, it seems that you would want to measure your large neutral amino acid pool from one of your blood samples, because I think you would find that the normal is quite variable among different subjects.

O. Muzik: You mean in the fasting state? Well, Dr. M. Diksic, who developed the method tries to go in this direction. But he's worked at least a year on this, on measuring large neutral amino acids and trying to correlate it with the findings of *K*-complex, and somehow it doesn't fit. So it's more complicated. That's why, although it might be simplistic, we tried to standardize, to apply the method to children, to get going in a clinical environment.

S. Shoaf: One of the major hypotheses of the serotonin synthesis rate method is that during the course of your scan, there is a steady state. So you fast the people so that the large neutral amino acid ratio between all of the amino acids is constant throughout your PET scan and during the course of your studies. That's the reason for trying to fast the people and looking at the tryptophan.

T. Jones: Why did you collect the cardiac data in 2D? Why not take it in 3D and get proper sensitivity.

O. Muzik: Well, we validated this method in 2D and we really didn't want to go into 3D. Plus, I think that the noise characteristic in 3D is more prohibitive than in 2D. So we tried to stick with what we know is working and not go into a higher level of sophistication.

I. Kanno: In taking the input function from left ventricle, your recovery coefficient is 0.9. Is it true?

O. Muzik: The correlation between the arterial input function and the left ventricle is very high. The limit of this method is the size of the ventricle. So if you go to smaller and smaller hearts, you have partial volume effects.

I. Kanno: We also have a lots of experience in getting input functions from the left ventricle. In adults, the recovery coefficient is 0.8–0.9. It's very much dependent on the subject.

O. Muzik: It really depends on where you put your region of interest. It's a small region of interest, and for most tracers that do not accumulate, it's really difficult. In cardiac studies, this method is well accepted. All of our flow studies with ammonia are based on this. So there is no reason why we shouldn't use it in brain tracers. The limitation is that you can hardly see the heart and there is no anatomical landmark.

CHAPTER 32

PET Analysis Using a Variance Stabilizing Transform

G. Pawlik: This quite nicely demonstrates that liberal decisions are bound to result when you make false assumptions. But what you may not have noticed is that with your kind of power transformation, what you apply is actually what we have been suggesting for a long time, which is essentially the Box-Cox transformation. There's only one major difference. When you make your nonlinear data conform to some formal assumptions required by parametric statistics, you need not do this across a large number of pixels.

Actually, you could do that pixel by pixel, or resel by resel. That way you might obtain transformation that's different from resel to resel, making each resel conform more strictly to the desired assumption in the sense that you perform a perfect linearization, because what you showed was a mixture of across-patients pooled with across-brains or across-resels regression, and there was some significant deviation from homogeneity or from the horizontal line that you attempted to reach.

U. Ruttimann: I'm aware that I could have used the Box-Cox transform right away, but I was not sure if that would be the right transform. So I tried to fit several functions, and it just happened that the Box-Cox transform was the most reasonable approach to pick. So I suggest in the future, you would simply apply the Box-Cox transform procedure to do that.

Second, there is no regional analysis. I did not define regions of interest. The scatter plots were pixel by pixel. Yes, you get a large scatter, but you cannot say you can get better homogeneity. If you apply a standard test for homogeneity, even with uncorrected data, the test accepts homogeneity. So you have a lot of scatter in data.

Third, I don't want to fit more parameters than I actually have to. So I don't want to fit an additional parameter for each pixel. I want to keep data as close as possible to original data for the analysis because you don't want to spend these degrees of freedom to estimate additional parameters that are not really of interest.

G. Pawlik: But then you need not be surprised when your total data just do not conform to what you, in the first place, tried to achieve.

U. Ruttimann: The data do produce what I tried to achieve.

G. Pawlik: Well, they didn't look horizontal over the full range. That's what you could achieve by performing your transformation resel by resel.

U. Ruttimann: You could argue if it was flat or not. You have so much scatter that any test for the slope will tell you that the slope was horizontal. What I'm really interested in are the differences in the PET images and to find a method that might be more acceptable. Some people simply do not accept the assumption of a homogeneous variance, so here's a way to get around it. You might find a better method, to extract more out of it, but I don't think it's worth the effort.

S. Strother: I wondered if you had any evidence that the Box-Cox or your simple power function also fits other data sets, or just the group of 13 subjects?

U. Ruttimann: I had other subjects, but that's all FDG data. With the [^{15}O]water, I'm not so sure about it.

S. Strother: I think we need to be really careful. There is a tendency to do a test on a limited group of subjects and then somehow it becomes institutionalized within the community, and we should really be retesting on new data sets all the time.

U. Ruttimann: That was the reason why I split the group into two halves: cross-validation. So if I would have another hundred patients, I would certainly try that. The point is that with FDG, this approach seems to be reasonable. I don't know about other tracers.

CHAPTER 33

Error and *t* Images Depend on ANOVA Design and Anatomical Standardization in PET Activation Analysis

R. Woods: I'm a little puzzled by your three-way analyis being worse in terms of your residual error than your two-way analysis, because my understanding is that the three-way model includes all of the terms that are included in the two-way models. So how is it that including more terms ends up with bigger errors? Or is there some normalization that I'm not aware of?

M. Senda: When you include more terms, the residual variation decreases. But the mean square error is the ratio of the residual error to the degree of freedom. The degree of freedom is small in the three-way design.

J. C. Baron: It's very reassuring to see that the main findings with the *t* maps do not depend much on the way you analyze your data. But then you said that the actual coordinates of the peaks vary and that this may not be the best way to describe the results. So could you please elaborate on this and tell us what you think could be the best way to describe the areas of the zones of significant activation if it's not by the peaks?

M. Senda: I have no good solution to that question.

J. C. Baron: Is it just by illustration, for example?

M. Senda: Yes. Presenting the image itself, the distribution itself, is more straightforward.

J. Votaw: It seems, when I speak to statisticians, that they much prefer that you decide on your analysis protocol before you do the study rather than after you do the study. So it seems wrong to me to base your analysis method on whichever technique gave you the answer closest to what you want.

M. Senda: I just wanted to show you how the result is affected by the statistical analysis method, and this is not the study to detect which area is activated by this language processing task. This is just to show the dependency of the result on the method.

G. Pawlik: What is your interpretation of the heterogeneity of the residual error? Isn't it the case that these images lend indirect support to what Dr. Ruttimann just told us, that it is a matter of heterogeneous or unstable variance, i.e. the higher the mean, the higher the variance?

M. Senda: Yes, there was a tendency, but some specific structures had even higher variance than expected from the mean CBF.

J. Votaw: I missed the point, then. Were you making a recommendation of which ANOVA technique should be used?

M. Senda: No.

B. Horwitz: I would like to make a comment about what Dr. Baron said, following up on the point about if the local maxima aren't very stable. I suggest that people publish lots of images. Since you can't often publish these in a journal, that's what Web pages are now for, because you could put many of your results on a Web page and then refer to that in an article, and then people can see the full extent of data that you have and the analyses that you've done, and you wouldn't have to restrict your attention to having just a few local maxima trying to epitomize your findings.

K. Herholz: I really would like to underscore that comment. It would be really helpful to see the error images more than we are used to in publications.

J. Andersson: I have a question regarding the use of interactions, namely the task-by-subject interaction. We want to see task effects with the largest possible sensitivity. By including an interaction, you would increase your sensitivity only if there is a significant task-subject interaction. My question is, if you have a significant task-subject interaction, does the main effect of task mean anything?

M. Senda: Statistically, if you have a large subject-task interaction, the task main effect means nothing.

J. Andersson: So essentially, if you would find higher sensitivity with the interactions included, that wouldn't really mean anything, would it? You wouldn't gain anything.

M. Senda: You are right in the statistics, but a lot of people are using this method.

J. Andersson: I know they're using it, but we should pose this question because I'm not sure we're helped by these interactions.

CHAPTER 36

Measuring Activation Pattern Reproducibility Using Resampling Techniques

J. C. Baron: I think there is one very important variable that you didn't include in your variance. It's the performance of each subject in each task, as well as what we call the cognitive style, and a lot of your variance is probably explained not only by methodological factors, but also by this meaningful variance due to each subject's brain activity during the task.

S. Strother: Well, the performance is included to the extent that it's included in any particular model, so typically you have some indicator variables, which indicate what you thought the subject was doing or what you measured that they were doing. So there is no reason why you can't use a parametric regression model to extract an activation image, which is predictive of the rate at which they were tapping their finger, for example. The point here is a process that alllows you to compare activation images that were generated by some modeling process, and that modeling process can easily include performance measures if you design it that way.

R. Woods: Your metric is reproducibility in a different group of subjects and you have not used a random effects model for your ANOVA-based analyses. Is there a component of intersubject variability that gets factored into the pixel-to-pixel variance estimate when you use the pooled variance, and might that account for the difference in the performance between those two?

S. Strother: I guess there is in the pooled estimate, but there isn't in the multivariate approach. We are actually looking for a linear discriminant vector in the space of eigenvectors, and you can look and see where that projects out. It doesn't project onto the components that contain intersubject effects, so I think the answer is no to your question.

One of the things that was a little surprising to me was that the pooled variance appears to perform so well, although we all believe it's wrong. I'm wondering whether that's a sample size issue, even though I'm using relatively large samples, given what you find in the literature—12 subjects with 8 repeat scans in each subject.

CHAPTER 39

One for All or One for Each? Matching Radiotracers and Regional Brain Pharmacokinetics

C. Mathis: When you plotted k_3, as derived *in vivo* versus the rate *in vitro*, really what you're plotting are two very different things. The k_3 has in it the concentration of the enzyme times the catalytic rate, whereas the rate *in vitro* is the second-order rate multiplied by that same $K_{catalysis}$. Are the two the same?

M. Kilbourn: The *in vivo* study is clearly the enzyme rate times the number of enzymes. The *in vitro* study was done the same way. It gives you the combined forward rate constant, the k_3 times E, so they are directly comparable. They're essentially the same value.

C. Mathis: But the two $K_{catalysis}$, the true rate of the reaction, are they the same *in vitro* and *in vivo*? How do those compare?

M. Kilbourn: Nobody knows. You can do the *in vitro* study, but I don't know how you'd do that experiment *in vivo* because you'd have to do it where the concentration of the substrate is not limiting and you can't do that *in vivo*. These radiotracers are totally insensitive to specific activity.

C. Mathis: The way to get it from *in vivo* data is to kill the animal, whether it's a rat or a monkey, and determine postmortem the true enzymatic concentration in the different regions, and compare those derived catalysis rates or rate constants.

M. Kilbourn: We could do that. I tend not to sacrifice my monkeys.

H. Namba: PMP and AMP—we call them MP4A and MP4P—are suitable tracers for cortical measurements. AMP, which has the highest reactivity rate, is the most suitable tracer for measuring acetylcholinesterase (AChE) activity of the cortex, whereas PMP is more suitable for a region with a little bit higher AChE activity, such as thalamus. So we selected AMP for human studies because we wanted to study Alzheimer's disease, in which cortical AChE activity is further reduced.

R. Koeppe: We agree that simulations show that AMP does work well in cortex. However, PMP actually has better kinetic properties and smaller coefficients of variation in simulation studies in cortex than in any other region, even in the thalamus. With intermediate hydrolysis rate tracers, you're already running into partial delivery limitation problems, and you're still best suited with PMP in the cortex. You can do estimates in thalamus. By the time you get to cerebellum or basal ganglia, you really have very poor estimates. With AMP, you run into those problems more quickly, even in the thalamus. BMP, of course, is good in basal ganglia, but it has too low a hydrolysis rate, and for the reason that Mike Kilbourn mentioned, that you have butylcholinesterase problems as well.

C. Mathis: Do you think that there is a tracer somewhere in between the two? Is there any tracer, a single tracer, that can ever work for all regions with some rate that is a compromise between all those that you've measured?

R. Koeppe: If you had to pick one that was optimal for everything, it has to be between PMP and BMP. I'm not sure where because I don't know the level of the butylcholinesterase. We were hoping, when we did the BMP study in monkeys, that we would get a larger drop in the cortical value, down closer to zero, but that didn't happen, I think, because of the butylcholinesterase. I would guess that if you had a tracer with about half the hydrolysis rate, you could get a reasonable estimate in the basal ganglia, but you would definitely be entering into the area of problems in the cortex.

M. Kilbourn: The important thing is that this isn't just for this enzyme tracer. This applies to every single radiotracer we use where there is a dynamic range of a 100-fold between two regions.

R. Koeppe: For reversible as well as irreversible tracers. This is a totally irreversible tracer, but rapidly reversible tracers might also have that high a difference.

R. Carson: If I'm interested in both the striatum and the cortex, is it better to do two studies in the same subject if I can inject twice as much and get twice the statistics. Is it worth the logistical issues of two different syntheses and double the scanner time?

M. Kilbourn: Well, if you want to measure AChE in the striatum, you're better off using the more sensitive radiotracer if you want to see a small change, if that's what your disease process was. It's just two syntheses.

B. Tavitian: What are the relative affinities of these tracers? Are the three identical?

M. Kilbourn: I'm not sure either group has done that yet, or if you go in and read the literature on AChE, it's God's gift to enzymology. It has a turnover rate that's ungodly. It's 25,000 molecules per minute.

B. Tavitian: The use of this type of tracer is based on the assumption that the end product will remain in the brain. AChE is an extracellular enzyme mostly in the brain, so I wonder what evidence you have that the end product will stay on the spot.

M. Kilbourn: If you look at all of those tissue time–activity curves, after a certain point in any animal, all the way from mouse to man, they just go flat. Just perfectly flat. That's because nothing is moving anymore. Everything is stuck in those brain regions, and there is no longer any blood input function because all the tracer in the blood gets hydrolyzed, and that goes to zero. So after about 30 min, everything is at a perfect standstill.

B. Tavitian: Well, it looks a little surprising to me because physostigmine has been labeled, as we did on the carboxyl group, but also on the same methyl link to the nitrogen like you did, and the end product would be another alcohol, and it didn't yield very good images because of clearance.

M. Kilbourn: We found a better one.

CHAPTER 41

Statistical Power Analysis of *in Vivo* Studies in Rat Brain using PET Radiotracers

M. Senda: You did the injection of saline as a sham operation for a target experiment with actual treatments, right?

S. Houle: We added some pharmacological chronic injections of other compounds that we had used, and we added injection of saline as a control to see if the injection of saline would produce any effects. So we decided to use that as one of the control groups, as well as another group of rats that did not have any intervention at all, because they were available in our data bank.

M. Senda: Did you find that the error of the pharmacologically treated animals was the same as that of the control group?

S. Houle: They were not exactly the same. With other compounds, we have noticed that there were some differences in the pharmacological response, and those were not in the baseline study, but when we looked at results of the pharmacological intervention, there were differences between the untreated and the saline control rats.

I'm sure it's a question of stressing every day with an injection of saline. In fact it was somewhat distressing because very often people use a saline as a control, but in fact just using the saline by itself was introducing changes in receptor binding. So now, we have been doing two groups of controls, one that is an untouched control, and the other group is one with saline, because we felt it was safer.

Again, there has been some variability in the literature that may be due just to the use of treated versus nonsaline-treated rats.

M. Senda: I'm asking about the difference of variance between the treated group and the control group.

S. Houle: In the baseline one, before treatment, the variance was very similar. We saw a difference after treatment. If we used, for example, amphetamine, then the variance or the change was different between the two groups, but not in the baseline.

M. Senda: So you may not be able to assume homogeneous variance between the two groups?

S. Houle: Right.

M. Kilbourn: I think you're presenting this to the wrong audience. You should have shown up at the radiochemistry meeting in Uppsala and told all the radiochemists about this.

S. Houle: Well, this is a PET methodology meeting, and we felt that it would be better received here than at a chemistry meeting, but we did have some arguments within our own group about this issue, and that's the reason why we went to the trouble of doing the whole exercise. Not everybody liked the results.

M. Kilbourn: Yes, but everybody in the chemistry field needs to hear this.

M. Graham: I was intrigued by your last comment that we could do a whole lot better statistically using PET imaging of animals. I wonder if you'd speculate on the power of using animal studies and if we could get away with an *n* of 1 in that sort of situation.

S. Houle: A lot of people in this meeting have been promoting the use of animal scanners to avoid doing a rat animal dissection. Do we expect to see better results by use of imaging than we have from dissection?

S. Cherry: I think that's exactly the question we have to ask. I don't think the data are out there yet. I don't know if the Hammersmith group has looked at that. We are certainly not at the point to do that yet, but it is one of the first things we need to do.

The key question, whatever intervention you're looking at, is how much variability it introduces in the animal model and how much gain you get by being able to look at the same animal twice. I don't think anybody knows the answer yet. Actually, the experiment to do that has to be carefully designed, so we will need quite a bit of help so that we can really convince people, but I think the whole future animal PET depends on the answer to that question.

R. Carson: I was intrigued that the variability of the striatum-to-cerebellum ratio was the same as that of the percentage injected dose per gram, so perhaps percentage injected dose per gram isn't so bad after all.

S. Houle: When I saw the result for that, I was intrigued too, but there didn't seem to be any errors. I went over and checked that.

R. Carson: Often the issue is not whether there is a difference, but, if I'm assessing a range of tracers, which one is better? Can you extend your power analysis to be able to say that, yes, I've picked the best ligand.

S. Houle: As Mike Kilbourn pointed out before, power analysis has not really been used for animal studies. Our case was a simple question where we probably would have expected between a 10 and 20% change from the pharmacological intervention. The question was, can we actually design the experiment so that we're going to show that effect or not? It was very pragmatic. Obviously, I think it would be nice to see this type of analysis expanded to other situations, like the one that you mentioned.

T. Jones: Do you think you're making the most of your *in vitro* data? You've got a lot of animals and you're using blocking doses. I wonder if you are making the most of these data in setting up a model from which you could derive binding potentials, etc., from limited data sets?

S. Houle: These are very conventional studies that have been published by many people. We wanted to look at a slightly more sophisticated way of analyzing data. We had to have a good idea of the variance in our baseline data. If we want to identify parameters, then obviously if will require a slightly more complex analysis than what I've shown.

T. Jones: I think it may be wrong to leave this very flat feeling that animals are very hard work and you need millions to get the right answer. There may be other tricks so as to use data more constructively than we have done so far.

S. Houle: I hope this type of negative result prompts others to come up with good suggestions about how we can get away from the bind that we are in if we continue to do what we have been doing with animal studies. Maybe Terry Jones would like to pick up the challenge and come up with a clever way of using data.

CHAPTER 42

A Human Liver Model of Metabolism as a Tool in the Identification of Potential PET Radiotracers

C. Mathis: You have two preparations, cytosolic and microsomal, and you see different rates for those two preparations. Which one is most faithful to the *in vivo* situation? Where would you expect most of the metabolism to occur *in vivo*, because you showed differences *in vitro*. Do you prefer one preparation over the other as being more indicative of the *in vivo* situation?

S. Houle: Well, Dr. Inaba suggests that we should probably do both preparations, because they are different sets of enzymes. In the case of CPC222, the microsome is probably closer to what we observed when we injected it in humans, because the metabolism was much more rapid, and in the cytosol we had very little metabolism. So if the liver behaves as the cytosol preparation, then we would hardly have any metabolite, but *in vivo* the metabolism was quite rapid.

Dr. Inaba's feeling, having dealt with a number of nonradioactive drugs, is that unless one knows ahead of time what to expect about metabolism, you should probably be conservative and run both assays, as they are not very difficult to carry out.

D. Kiesewetter: You have said that there's a metabolite of the CPC222. I was wondering if you had any idea what it is.

S. Houle: No, we're working on that. We're trying to produce enough compound to identify it. One hypothesis is that there may be some hydroxy groups that are being tagged on the molecules, but we have not had a chance yet to identify what these might be.

R. Blasberg: It is very nice to look at the metabolism in the test tube before administering the tracer to patients or primates. One point that I was interested in was the log *P* values that you had, which were in the range of 4 or higher. I was wondering whether you're assessing plasma protein binding and albumin binding of these compounds.

S. Houle: The protein binding for CPC222 is about 95%.

R. Blasberg: Is is rapidly exchangeable or have you looked at that issue?

S. Houle: No, we have not looked at the rate of exchange of protein binding. When we found that we had that metabolite problem, we stopped doing the human studies because of that, and began looking at other alternatives.

C. Mathis: I talked to Dr. Alan Wilson about this work. The log *P* values are calculated values, not experimentally derived log *P*'s. They're calculated by the C log *P* program. I've used that program myself and I'm not very happy with the absolute values that are derived from that program. I think it gives one a relative feeling of rank order of lipophilicity, but I would not put a lot of stock in the values of 4 or greater. I think it overestimates the apparent log *P* at pH 7.4. With other similar compounds I've seen that the absolute value given by the program is not very accurate.

S. Houle: I think the value may be slightly off, but I think that the feeling, looking at the way it gets into the brain and the way that it was sticking to some tubing, is that it's more lipophilic than WAY 100635.

R. Blasberg: It's easily measurable with the radiolabeled compounds that you are making. It is important in the analysis of data or in the kinetic modeling to understand before you get into the animal what are the potential limitations in data acquisition. A protein binding of 95% raises some concern.

S. Moerlein: In some of our work with isolated hepatocytes, we found that there is a dramatic decrease in viability of hepatocytes when they were frozen compared to freshly harvested hepatocytes. I wonder if you had any results with that, with the idea in mind that maybe if you use freshly harvested cells you would have a more sensitive assay.

S. Houle: This is probably true. Having access to live tissue is probably much more complex. Since the question that we wanted to address is do we make radiolabeled WAY 100634, I think that we can run the reaction long enough to convert all the parent compound to the metabolites. We just want to identify where metabolites are, as opposed to trying to calculate some rate of conversion where living or fresh tissue may be more important. In this case we can run the reaction to completion by incubating a little longer or by increasing the concentration. So from that point of view, it's not a problem.

S. Moerlein: Do you have any information about the subtype specificity for CPC versus the WAY compound?

S. Houle: We carried out a number of assays, and so far, it's very similar to the WAY compound. We're still trying to do a more extensive *in vitro* binding, but from what we know from rat distribution data, it behaves very much like the 635.

K. P. Wong: Yes, we have. The results are very close, except when we performed the clinical study, we found the standard deviations of the region-of-interest values were very large as compared to the computer stimulation.

M. Graham: How critical are the blood samples? If you leave those two blood samples out entirely, does the method work at all? Because that is the ideal that we are moving toward, getting rid of blood sampling entirely.

K. P. Wong: I am not sure whether we can remove the blood samples, but if we move the two blood sampling times a little bit earlier, the method can still apply.

J. Matthews: At Hammersmith, we were looking at the shapes of FDG input functions over a population of starved patients. We found that the shapes of the FDG input functions were remarkably similar within the population, and all that was required to define an FDG curve for an individual was a single arterial measurement after about 10 min. Any time from about 10 min to the end of the study was sufficient.

You are using your two samples and you are predicting that you would be able to successfully calculate the glucose rate constants. The work that I have been involved in suggests that as long as you have at least one sample, one warm blood sample, you can quite accurately predict the input function.

S. Meikle: We found something very similar in Sydney as well, that the FDG input function is very reproducible. This approach is an alternative to getting the input function non-invasively, and our intention is to compare the two techniques, but the advantage of this technique is that it could be extended to other tracers where the input function may not be so reproducible.

J. Andersson: It is a very good idea. My question is would it be sensitive to the ROIs you use? If the three regions of interest you use had too similar dynamics, would the method still work? How different do they have to be?

K. P. Wong: I think it might be difficult, because this method requires kinetics in the regions of interest to be very different.

J. Andersson: Yes, but how many different regions with different kinetics do you need? They have to be really different, these regions. So can you give advice of where you place them? Should you always take one gray matter region and one white matter region?

K. P. Wong: I think if you want to have very different kinetics, we may apply principal components to extract the regions of interest.

CHAPTER 48

Simultaneous Extraction of Physiological and Input Function Parameters from PET Measurements

A. Lammertsma: It looks interesting, but so far you have shown only simulations. Have you actually tried the method with *in vivo* data? Do you get the same results?

CHAPTER 49

Suppression of Noise Artifacts in Spectral Analysis of Dynamic PET Data

J. Frost: Once you know that you are dealing with two or three components, then what you are doing is very similar to what we see in tracer kinetic models. Is there an advantage in reducing the variance of the estimated parameters using this analysis over that which can be achieved with compartmental models?

V. Cunningham: Yes. The main problem with fitting *a priori* fixed compartmental models at the pixel level is that you are never sure as to whether you can numerically identify all the components you want. With ROIs, you can work out what is really going on, but for many models, you don't know whether it's going to work at the pixel level. Sometimes it will, sometimes it won't.

So the advantage of going back to a more flexible basis function approach is that it effectively adapts itself to the noise in data. If there is something recoverable, it will add in more components in an attempt to do it. If data are too noisy, it won't fit them. I think at the pixel level, in practice, we've found basis function approaches to be much easier and quicker to use than *a priori* fixed models.

J. Yap: One of the advantages of this method over some of the other more statistical approaches, more data-driven approaches, is that it is fundamentally based on the compartmental structure. The general question I have is how far can you take this to get information about the compartmental structure. More specifically, in the case of noise and with resolution effects, can you identify the correct number of components, and how are the accuracy and precision of the αs and the βs affected by noise?

V. Cunningham: To be realistic, spectral analysis is very good at picking out or describing what the impulse response function of the tissue is. It's not so good at consistently bringing out the right number of components. It is just a very fast way of describing data and giving you a good, smooth deconvolution.

How you interpret that impulse response function is your own business. It depends on what your other models might be, but there are certainly some properties of the impulse response function that are robust in themselves and quite useful. You have to use caution when reinterpreting those results in terms of a compartmental model. But I must emphasize again, there is flexibility. It doesn't have to be compartmentally derived. You can more or less generate response functions that you anticipate, such as a displacement.

J. Yap: If you are using it simply as a deconvolution technique, to get rid of the input function, then consider the issue that came up yesterday about the number of basis functions. You may be able to come up with a reduced set of basis functions that span the space defined by all of your exponential functions, if you're not particularly interested in getting the α and β terms themselves, but just getting the impulse response function.

V. Cunningham: Yes, but the point is the method is so fast. Steve Meikle said he uses 30 basic functions. We use 100 because it's a round number, and the pixel-by-pixel version uses 64 because that is another round number, but it's so quick, you just have to use enough. I know that that's not a very satisfactory answer, but it's quicker.

S. Houle: Would it be possible to transform your spacing to a standard sensitivity range by not having equiangular spacing?

V. Cunningham: No. The range of the sensitivity is a function of the protocol and input function. If you have a given input function and a given protocol, this generates basis functions and makes sure they are evenly angled. In other words, the dot product between successive ones is the same. This A matrix is unit length. It's been normalized. It's been weighted, so all you then have to do with that particular matrix is make sure the dot product between adjacent columns is the same. You then get this plot. Retrospectively, the plot tells you whether or not your data collection protocol is suitable for the dynamics you are interested in. It's quite a useful plot.

S. Meikle: One of the practical things that you used to suggest before implementing this noise suppression technique for generating parametric images was to come in from (i.e., avoid) the extremes of the impulse response. You would go for a 1-min image rather than the intercept. With the new noise suppression technique, would you still do that or do you think it is worthwhile to chase absolute values?

V. Cunningham: Again, it depends on the quality. What we used to do before, since extrapolation wasn't too good, was to take either a 1-min impulse response function image to give you a K_1-type picture or a 60-min image in the case of diprenorphine, because this was dominated by binding, but they're not entirely satisfactory. You get very, very good pictures, because they're well within the range that is defined by data, but it's simpler to interpret extrapolated data. If you extrapolate the V_d it is easier to interpret, but it depends on the particular data set.

S. Meikle: But with the new technique, you think that it is better suited to extrapolation?

V. Cunningham: Yes. I'd give it a go and see whether you like the picture or not.

R. Gunn: Traditionally, in spectral analysis, you tend to get up to five components. With the new noise suppression approach, it will typically bring in one or two components corresponding to the number of compartments in that region. Therefore, it allows you to use this, first of all, as a model order estimation tool for different regions of the brain, and then, because you're restricting to these numbers, it allows you to move back into the compartmental framework and map into parameters from compartmental models.

CHAPTER 54

Temporally Overlapping Dual-Tracer PET Studies

D. Bailey: This sort of approach raises a number of issues. Can you see any gains by maybe using ^{18}F and ^{11}C to try and separate these out?

R. Koeppe: Certainly that would be advantageous for the input function determination. You can wait and count the same samples at different time points to allow for the differential decay in order to separate ^{18}F and [11]C. You still have to do the metabolite separations, but the isotope differentiation problem would be alleviated.

I think that is possible. I haven't looked at any of these combinations in simulations, but one would inject the ^{11}C first, particularly for a tracer that has rapid kinetics—not just reversible. For instance, PMP is irreversible, but the kinetics are rather rapid, so it works well when administered first, particularly in the cortex. I think with ^{18}F, if you don't have very rapid kinetics, you are going to have to scan for a long time afterwards, but dual-tracer studies are certainly possible.

D. Bailey: Being a modeler, you naturally tend to put things in boxes and draw arrows between them, but if you are looking at shapes of uptake and incorporation into the brain, this seems to appeal to using factor analysis, when you have multiple compartments in the same structures. Have you thought about using that and can you support this sort of work at the pixel level?

R. Koeppe: You can do work at a pixel level under certain circumstances. You should be able to do pixel-by-pixel analysis with some smoothing. According to the simulations, even using about 8-mCi injections in 3D for the FMZ/DTBZ pair, pixel-by-pixel analysis should work quite well in some regions, such as basal ganglia. However, if you do injections in the other order [DTBZ/FMZ], then you get nice looking images in other brain regions, such as cortex. I've not thought much about principal component or factor analysis, but that certainly seems to be a possibility.

J. Frost: You propose injecting the tracer with rapid kinetic first, and that makes perfect sense. And you can get a K_1 and DV estimate, so how much of additional data from tracer 1 after you inject tracer 2 is really contributing to the estimate of the tracer 1 parameters?

R. Koeppe: The estimate involves all data at once, but I agree that there is less chance for later data to influence the tracer 1 estimates. There is an interaction, though, when you're using the simplified models, particularly if there is some bias in the fits of a single injection study. There were some interesting cases where not just DV, but K_1, became biased. The K_1 of the second tracer picked up some of the bias from the lack of fit from a two-parameter estimate from the first tracer.

J. Frost: It appears that if you really can estimate K_1 and DV from the first tracer prior to the time that you inject the second, then what you're really doing is just two separate studies very close together, and the kinetics from tracer 1 are being subtracted out from tracer 2, clearly by the determination of K_1 and DV.

R. Koeppe: I think in general that's right. However, I would guess that if you use a combined estimate you will do a little better, but if there are problems when the first tracer doesn't equilibrate rapidly, that will cause problems in the separation of the kinetics of the two tracers.

J. Frost: You still have an advantage because you're doing studies closer together.

R. Koeppe: Right; the main disadvantage is with those tracers where the metabolite separation and the separation of the two authentic components are difficult.

M. Graham: We've been doing dual-tracer studies with [^{11}C]glucose followed by FDG. That addresses Dale Bailey's question, can you do ^{18}F and ^{11}C, and it works very well. We have looked at the timing and we can come up to about 20 to 25 min between injections from our simulation studies.

A. Lammertsma: We know that the metabolite correction of the input function is quite difficult to do, so if you have two confounding tracers together, that must be quite a difficult task.

R. Koeppe: It will depend on the specific pairs of tracers used. For instance, with PMP, it may not be so bad because the metabolites occur so rapidly that the PMP input function is very nearly zero by 30 min. Thus, you will be able to measure the tail end of whatever you inject second quite well and them extrapolate backwards. We've looked at this and, in most cases, we will probably need to separate the two authentic ligands in no more than one to three samples.

CHAPTER 57

Neutral Amino Acids Influence [^{18}F]Fluorodopa Quantification

H. Kuwabara: K_1 can be expressed in terms of the Michaelis–Menten parameters of the transporter and show curvature with amino acid concentration. You can fit with this curve very well. We did something similar to this and found that that curve explains data very well.

S. C. Huang: With the reference region, you can take away the correlation with the plasma amino acid levels, but as you postulated, the amino acid level affects transport across the blood:brain barrier. In the cerebellum, for example, it would affect its crossing forward into the tissue, and presumably it would equally affect the amount that's coming out, even with no specific uptake. How would that account for the change in K_1, while in the striatum you saw that the uptake actually decreases with a higher amino acid level?

P. Vontobel: The difference is in the variation seen in K_{1d} and the plasma slope-derived fluorodopa uptake rate constants. They are clearly due to this effect of saturation of the transporter.

S. C. Huang: But you should not affect the equilibrium value in the reference region, the cerebellum.

A. Gjedde: That's not correct, because all that is happening with K_1 is that the amino acid carrier is normally saturated with these 10 large neutral amino acids, so that when you increase the amino acids, K_1 is declining. It's actually inversely proportional to the total amino acid level. What is happening in the reference region is simply that you are reducing the total amount of tracer that comes into the reference region, which means that your input function is lower. All that is happening is that with the tissue reference K_i, which is a somewhat poor estimate of the enzyme activity, the effect of the increased amino acids on the input function is removed.

S. C. Huang: Why would the time–activity curve of the cerebellum be affected by the transporter? It is related to the distribution volume of the compound.

A. Gjedde: It's the same thing as with FDG. If you increase glucose, you get less FDG into the region because the activity of the carrier is reduced. It is less capable of transporting tracer into the tissue.

S. C. Huang: That is for a specific uptake or metabolism, but for fluorodopa in the cerebellum, it is not supposed to have any specific uptake or trapping, and the level in the cerebellum should be in equilibrium with the plasma.

A. Gjedde: No. It is the total amount, because the amount that can be moved in and out is less, because of the competition with the cold amino acids. The cold amino acids will not allow that much hot fluorodopa to be transported, or as much hot 3-*O*-methyldopa to be transported into the tissue. The total amount of the tracer is reduced.

CHAPTER 58

Imaging of the Dopamine Presynaptic System by PET: 6-[^{18}F]Fluoro-L-DOPA versus 6-[^{18}F]Fluoro-L-*m*-tyrosine

A. Gjedde: I'm surprised that the 3-*O*-methyldopa accumulates so freely in muscle tissue, but the FMT, which is also an amino acid and which is present in equal amounts in plasma, does not accumulate in muscle tissue. What is it about 3-*O*-methyldopa that makes it so prone to be found in muscle tissue, unlike the other amino acids that we are also talking about?

D. Doudet: I think the picture that I showed was a little misleading for that. It does accumulate a lot, but because of the way it was displayed, it seems to accumulate more than it does. I don't know why FMT does not accumulate as much in the muscle tissue.

A. Gjedde: I was surprised about your conclusion that carbidopa had little effect on the fluorodopa content in your monkeys, because we have found that fluorodopa availability is exquisitely sensitive to the amount of carbidopa and to the dopa decarboxylase activity in the periphery. Do you have an explanation for that?

D. Doudet: The sensitivity of the two tracers is very different. We have done some studies with and without carbidopa, and we find that indeed fluorodopa is also sensitive to carbidopa, but not to the extent of FMT.

M. Senda: Could you explain how you computed the loss rate constant?

D. Doudet: First, we do a Patlak plot to get the uptake rate constant. Then if you extrapolate the line, you see that data break away from the extrapolated line, which suggested to us that activity is lost from the striatum. So we fit the data, and take the late points and bring them back to the extrapolated line. It's a very simple way of doing it and we have a paper describing the technique in press in the *Journal of Nuclear Medicine*. This is the work of Dr. Jim Holden from Wisconsin.

D. Stout: We have done a lot of work on FMT as well, and we had very similar findings, with FMT being a bit noisier than fluorodopa. We also found that the carbidopa issue is definitely there. FMT does definitely seem to be more sensitive to the carbidopa level. Our squirrel monkeys were a little more sensitive to the carbidopa than the vervets were, but we found the same thing.

R. Carson: Why is the FMT noisier? In the normal controls, it seemed there was more variability in FMT data than in fluorodopa data, and it seemed that you had everything else in your favor. You have a better input function and less *O*-methylfluorodopa. Why is it more variable? Is it measuring something different?

D. Doudet: I am not sure it was more variable, but because you don't have that background. When you calculate the Patlak plot, I think that background measures bring data in fluorodopa back down. The standard deviations were fairly similar, except that the values were much bigger in FMT data.

J. Rakshi: Have you done FMT studies in human subjects?
D. Doudet: No.

J. Rakshi: With fluorodopa, now that we're able to image the brain stem, we are finding uptake in the dorsal midbrain region, and it may be that fluorodopa isn't taken up just by dopaminergic neurons, but is also taken up by other monoaminergic neurons such as serotonergic and adrenergic.

D. Doudet: In the monkey, I can also see the brain stem, but it is of course very difficult to determine where it is, especially if you don't have MRI with FDOPA.

J. Rakshi: Have you compared using fluorodopa with entecapone or a COMT inhibitor against FMT to look at what the differences are in the influx rate constants, and the difference in discriminating between the Parkinson's disease and normals or MPTP monkeys.

D. Doudet: We have performed the studies comparing COMT inhibition with FMT, although the data are not analyzed yet. My guess from a brief look at the data, is that both methods will have comparable sensitivity.

If I had to choose between using FMT or using FDOPA with a COMT inhibitor, I would use FMT because of the simpler analysis of the peripheral metabolites, and also because some patients may have different amounts of inhibition of the enzyme, and then different amounts of methyldopa in the plasma. Also, because FMT will allow you to image the brain stem, I think FMT is probably a very good tracer for routine studies.

J. Rakshi: But if we're going to use it for clinical studies, then we'll have to make sure that we give the carbidopa at the same time before each PET scan.

D. Doudet: Yes, but then you will have a very similar problem if you do clinical studies with different patient groups. If you are using tolcapone, entecapone, or netecapone, you may have different amount of inhibition, so that is one thing that you will have to look at, as well as with FMT.

C. Mathis: What is the nature of the loss of 6-fluorodopa from the striatum. Is it 6-fluorodopa or is it a metabolite that differs from the metabolites of tyrosine. Could you explain any differences in loss between groups not from dopamine turnover, but from loss of a metabolite that is different between the two compounds?

D. Doudet: I should probably not have said fluorodopa. I should have probably said fluorodopamine, because obviously we are not losing fluorodopa, but rather fluorodopamine and fluorodopamine metabolites. We have not done studies to determine what is exiting the striatum, but it is very likely that what you are losing from a fluorodopa scan is fluoro-HVA or fluoroDOPAC. With FMT, we haven't done any of the studies. We have done HPLC on the plasma, and we find 6-FMA, but I am not sure of the metabolite clearance from striatum.

R. Blasberg: In part addressing the issue of the difference between the tracers, the graphical plots explain or at least project the differences between the two. The abscissa values are a log order difference between fluorodopa and FMT, due to the rapid clearance of fluorodopa. The conclusion that was drawn with respect to measurements of k_{loss} for FMT is explained by the graphical plots. I appreciate you presenting the graphical plots because they illustrate the point very nicely.

CHAPTER 59

Quantitative Measurement of Acetylcholinesterase Activity in Living Human Brain Using a Radioactive Acetylcholine Analog and Dynamic PET

R. Koeppe: Your plasma metabolites show something that we see with the PMP tracer, that the metabolite fraction of the first points is actually higher than a few points after that. We're postulating that it's going into the lungs and then coming back out, and it's protected from AChE, but we don't have a good explanation. Have you thought about that?

S. Nagatsuka: Our tracer is rather quickly metabolized, and it gets to the detection limit within 20 min, so we could not see such an increase.

B. Tavitian: In relation to what we have heard earlier concerning dopamine, what would be the influence of the occupancy of AChE by ACh in your k_3 measures? Do you think it is possible that you are also measuring the occupancy by the endogenous neurotransmitter?

S. Nagatsuka: We are concerned about that. This tracer should be competitively metabolized by AChE, so we're going to use the quickest tracer, and by using infusion, we could see the competitive inhibition of AChE, but that is future work.

M. Kilbourn: If you read about AChE, ACh doesn't sit at that active site at all. It is such an efficient enzyme, it chews it up and spits it out. ACh doesn't bind to that site. If you look at models of that enzyme, it's just in and out.

B. Tavitian: It's a very active enzyme, but it's also an enzyme that shows substrate inhibition, which means that there is a measurable residency time on the active site or near the active site by the substrate.

M. Kilbourn: Well, the substrates that we're using can't be inhibited. We've gone up to 10,000 times the administered dose of a tracer dose, and it has absolutely no effect on the kinetics. That's probably far beyond any endogenous levels of ACh that you might find in brain tissue.

S. Pappata: Did you look into the relationship between the changes in AChE activity with respect to aging in your normal subjects?

S. Nagatsuka: In normal subjects, we could not see any age differences. We performed studies in subjects aged 24 to 89.

S. Pappata: This tracer is very specific to detect AChE in the cortex. You have not shown activity in the hippocampus, but that would be very interesting, especially to study Alzheimer's disease.

S. Nagatsuka: Actually, we were very interested in the hippocampus, but we could get only a horizontal image, and it's very difficult to place an ROI in the horizontal image. We are now checking the results in hippocampus.

S. Pappata: Your approach is very nice to study AChE in cerebral cortex. However, other ligands, such as physostigmine, when used in normal human brain, show a good correlation between the regional concentration of tracer and the activity of the enzyme as measured postmortem. So this ligand is not sensitive to detect changes in the cerebral cortex, but allows us to look at the concentration of AChE in the whole brain.

S. Nagatsuka: As was presented earlier, we use a more slowly metabolized tracer to obtain reliable measurements in high AChE activity regions.

CHAPTER 65

Characteristics of Neurotransmitter Competition Studies Using Constant Infusion of Tracer

A. Dagher: Did you compare for a given tracer, for example, raclopride, whether the bolus or the infusion method was better or equal?

C. Endres: It is difficult for us to evaluate that because we haven't considered the effect of noise. It's easy from our noise-free simulations to determine which method will give you a larger measure, but that's not really valid without considering noise. These simulations are geared toward understanding the signal that you obtain. When you try to consider the different noise issues of the two methods, it's not appropriate to use these simulations as a basis for judging which method is superior.

A. Gjedde: You showed that the change in tissue concentration correlates with the integral, rather than with the level of the neurotransmitter itself. Doesn't that suggest that the effect on the occupancy is not directly related to the neurotransmitter level, but to some indirect or other effect of the neurotransmitter on other aspects of the system?

C. Endres: Well, the model just looks at the effect of neurotransmitter as a competitive inhibitor of the tracer, so I'm not sure from where you're drawing effects.

A. Gjedde: If it were the neurotransmitter itself, it would be expected to correlate with the neurotransmitter level, not with its integral.

C. Endres: But when you induce neurotransmitter release, we're looking at a transient change, so what neurotransmitter level would you be talking about?

A. Gjedde: That would be any one at a particular time. The fact that it correlates with the integral suggests that there are other indirect effects which are possible, such as other membrane effects, etc. Not blood flow, but certain other issues that have to do with the microenvironment in which the receptor is operating.

R. Carson: The kinetics are slow enough that the tracer is unable to rapidly follow the kinetics of the neurotransmitter. In the ideal case, we have high B_{max}, high k_{off}, sufficient binding potential, and we get a perfect negative representation of the neurotransmitter pulse, but in every real tracer we have, we lose that. But Chris Endres showed that we are still sensitive to the integral pulse. We just can't follow its kinetics. Maybe if our chemistry colleagues come up with the ideal compounds, we will be able to follow the time course better.

A. Gjedde: That's one possible interpretation. I'm just asking isn't it possible that an alternative interpretation is that it is sensitive to some indirect effect of the neurotransmitter surge?

R. Carson: That's possible, but this is simulation data and there isn't anything else here. That's what the kinetics are telling us. There may indeed be much more going on in the real system.

C. Endres: Actually, it's not just proportional to the integral of the neurotransmitter pulse, but it's proportional to the integral of the neurotransmitter pulse divided by the affinity, so if there were changes in affinity, for example, that would also affect the change in concentration that you measure.

M. Laruelle: The problem is that there are many tradeoffs. For example, a tracer with a fast k_2 would tend to be more lipophilic, and therefore to have larger nonspecific binding, and so on. So instead of analyzing one property at a time, it would be ideal if you could come up with a general model that would really predict what the ideal tracer is, because we have conflicting requirements.

CHAPTER 66

PET Measurement of Endogenous Neurotransmitter Activity Using High and Low Affinity Radiotracers

C. Endres: I think that this illustrates a theoretical result that we obtained in describing neurotransmitter competition with the bolus method. When you give amphetamine, the dopamine concentration varies with time, and as a result, the distribution volume varies with time.

As Albert Gjedde said yesterday, the beautiful thing about the Logan plot is that no matter what you do, you get a straight line, so you can interpret it as corresponding to a certain distribution volume. In your displacement study, when the distribution volume varies with time, what is that distribution volume that you get with Logan analysis? The answer is that the distribution volume is the time-varying distribution volume weighted by the free tracer concentration.

That means that it's better to have a tracer that clears quickly, because then you will be preferentially weighting the early part of the distribution volume curve, which is what you want, because that's when you give your displacement. So with raclopride, which clears quickly, you would expect to get a larger percent change in these studies, because it's weighting the early part of that curve, as opposed to fallypride, which is slower.

J. Price: I thought that was consistent with what others have found. I was concerned about the differences in protein binding, effects of anesthesia, and other factors. There are many other technical factors that are playing a role.

M. Laruelle: When you have a tracer that goes on for 3 hr, a change in flow becomes a huge problem because amphetamine would probably decrease the flow in the striatum rather early, and then the flow would come back up, so K_1 would decrease, but k_2 would be unchanged. I wonder if you tried to incorporate that in your three and a half compartment model.

J. Price: I didn't incorporate a flow component. One confusing aspect is the difference in regional kinetics, say for the cerebellum relative to the striatum. If you have early flow effects, it is possible that your distribution volumes could be affected differently in these two areas, but I'm still using the distribution volume from the cerebellum in the calculation of the binding potential. So that is something to be considered.

A. Gjedde: Fallypride is an interesting compound. I think you said it had an affinity of 0.04 nM. Couldn't your high or very varying binding potentials be because at such high affinity, the k_3 may be somewhat indeterminate, because it's not really reversible?

J. Price: I tried to shy away from indicating estimates of k_3 or k_4, partially because of the slow reversibility, but yet significant k_2. That is why there was some " battle" between a two- and three-compartment model. What I did notice, even though it was only in two baboons, was that the distribution volume-derived binding potentials normalized out to about the same values. However, I think that the instability in k_3 is a problem, and I haven't identified the appropriate constraints to pull that parameter out.

CHAPTER 67

Measuring Neurotransmitter Release with PET: Methodological Issues

S. C. Huang: In your graph showing the significant effect of flow on K_1, what is the normal K_1 value for this compound?

A. Dagher: I used the published values for both raclopride and diprenorphine, which are K_1 values in the order of 0.2 to 0.3.

S. C. Huang: So these are pretty well flow-limited tracers and that would not be a problem for compounds that have much lower K_1, for example.

A. Dagher: But all the existing tracers—raclopride, Schering, diprenorphine, flumazenil—have K_1 values in that range, so unless someone comes up with a different tracer, this is going to be a problem.

R. Carson: Any promising ideas on motion correction within the study?

A. Dagher: Yes. The best method, I think, is to detect patient motion during the scan. Chris Thompson in Montreal has developed a video-recording technique that is accurate to within 1 mm. Now, the next problem is what do you do with that information? Of course, the best thing is if you're collecting data in list mode, then every time the patient moves you can put in a little event, and then *post hoc*, when you do your reconstruction, take the movement into account. The other way would be to do simultaneous transmission/emission scans during the study and coregister the transmission scan to the MRI.

J. Andersson: One thing that we do differently is that before we apply a registration algorithm, we have a mask. Essentially, we differentiate the volume in all directions and then we only use the high derivative regions, meaning that we are only using the regions around the borders of the brain or the borders of the skull, whichever has the highest uptake.

This makes the method very insensitive to the gross changes in distribution, as you're describing. So it's relatively insensitive, and if you do that, I think you'll find that a registration algorithm will work.

A. Dagher: The problem with that, though, is that with dopamine receptor ligands, by the end of the scan, your image is essentially an image of the striatum, and it's hard to understand how one of these algorithms could realign just that sort of hot spot.

J. Andersson: Yes. You're scanning out to 2 hr. We have a 2D scanner. We cannot do that, so I don't run into that problem.

M. Laruelle: For that issue for the dopamine tracer, we had fiducial markers that are quite hot, usually hotter than the brain, so we can even threshold out the brain and just perform the Woods algorithm on the markers. That works quite well.

A. Dagher: And what are they?

M. Laruelle: Technetium.

CHAPTER 68

Imaging Receptor Occupancy by Endogenous Transmitters in Humans

A. Gjedde: This is such exciting work, that it stimulates us to delve into the details such as measurement of the binding potential. At the Society of Nuclear Medicine meeting, you alluded to the question of the very high binding potentials that you have. You even showed that an important variable in the calculation of the binding potential is to know what the partition coefficient of the tracer is. In fact, you went to some lengths to show that even though the other binding potential, the k_3/k_4 ratio, is elevated in schizophrenics, this is due to a change of the partition coefficient of your compound in some other part of the brain. I think you used cerebellum and you showed that the partition coefficient for the benzamide in the cerebellum of schizophrenics is significantly lower than it is in controls. Therefore the binding potentials that you calculate by means of the k_3/k_4 ratio happen to come out to be the same.

M. Laruelle: This question refers to my talk at the Society of Nuclear Medicine, and my only comment is that I'm really sorry and concerned to see the field of PET neuroreceptor imaging accepting as an outcome measure k_3/k_4, or what we used to call V_3'', or what is more exactly the ratio between the binding potential and the nonspecific volume of distribution, as in the reference method. I think this is very dangerous, and we showed, in those data from schizophrenics, that if you have change in the volume of distribution of your cerebellum or of your nonspecific volume of distribution, you might have artifactual results if you use that method.

A. Gjedde: Your methodology involves taking the measure of binding potential that you don't like and dividing it by a figure that you don't know. The figure that you don't know is the free fraction of your tracer in the tissue, which you can calculate by certain assumptions. Namely, if you have the partition coefficient and the free fraction in the plasma, you can combine these into a free fraction in the tissue, but what you're doing is that you're taking a fairly manageable number and dividing it with an extremely small number to get a very large number. Thus, the binding potential, which may have errors in the partition coefficient, becomes an experimentally more uncertain number, in that it is based on assumptions and measurements that are not directly made.

M. Laruelle: I think that we showed at the Society of Nuclear Medicine meeting that the measure which I call the true binding potential, which is basically the bound over free, is more reliable than the measurement of the distribution volume ratio. This ratio of bound over the free plus nonspecific binding tends to be very reproducible, but has a poor reliability in the identification of between-subject differences.

A. Gjedde: If the binding potential on AMPT is in fact very much increased in the schizophrenics, would that not imply that either the receptor number or the affinity is quite different in the schizophrenics?

M. Laruelle: I agree. We haven't done the study in schizophrenics yet, but if the finding was that there is a much larger increase in the IBZM binding potential following AMPT in schizophrenics than in controls, and if the baseline binding potential is the same, then one would have to conclude that schizophrenics have more D_2 receptors and that they might have more dopamine at baseline, but that's a study we're currently doing, so we don't have the answer yet.

J. Frost: I have a comment that I think applies to all of these talks. In 1984, we investigated the rate of displacement of opiate ligands from the opiate receptor and showed that this was dependent not only on k_{on} and k_{off}, but also on the ability to diffuse away from the synapse, which was a function of the molecular weight and partition coefficient. Therefore, in terms of Albert Gjedde's earlier question, what else in the microenvironment could cause an effect, anything that alters the diffusion rate away from the receptor could also impact the binding.

Further, we showed that it was only the affinity in the presence of sodium and GTP, but not without, that correlated with the off rates, so anything that changes affinity will alter this.

A third possible factor is the phenomenon of agonist-mediated internalization, which occurs in a number of systems. This seems to be particularly relevant to the opiate system, where some agonists and partial agonists induce internalization, but not others, for example, morphine doesn't induce internalization. Sometimes internalization occurs with the ligand and sometimes the receptor is internalized without the ligand, and also there is a whole series of intracellular receptor trafficking activities that are involved.

So I think all of these factors are important and it's not just simply endogeneous competition between ligand and receptors.

M. Laruelle: I certainly agree with these comments, especially in the case of the pulse of amphetamine, as I think Rich Carson and Chris Endres have shown. There are still a lot of unanswered questions, definitely under those conditions.

The beauty of the AMPT paradigm is that everything is at an equilibrium. We can postulate that the baseline dopamine is constant during the time of a scan and the tracer is constant, so for the two binding assays that we do at 2-day intervals, everything that binds to the receptor is at equilibrium, so we get something easier to interpret.

Also, the effect size in our hands is much bigger. We see with AMPT a 20–30% increase in the binding potential, while with the amphetamine, at best we can displace 10–15% in healthy controls, so the signal-to-noise ratio is better.

Overall, I think this is a paradigm that deals elegantly with a lot of the difficulties that you raised and that I think are very important.

CHAPTER 69

Quantification of Extracellular Dopamine Release in Schizophrenia and Cocaine Use by Means of TREMBLE

R. Carson: I don't think there is time for methodological questions, so let me ask a biological one. When we did two amphetamine injections in rhesus monkeys with microdialysis, we noticed that the baseline dopamine increased following the first pulse, and stayed higher up to 3 hr later when we gave our second pulse. How would that affect the calculations you made, which would have assumed that dopamine returned to the original baseline?

D. Wong: There's obviously a possibility that it could affect it. The timing between the first and second amphetamine injections was typically 6 hr because we used the first patient slot in the morning and the last slot at the end of the day, because of medical safety issues. So it is something that we have had to take into consideration, and we should probably estimate it. However, since we did our studies at the beginning and the end of day, this will allow the dopamine levels to come down further.

A. Lammertsma: I have a more general point which may relate to your cocaine studies. If you give a challenge, any change that you measure is being taken as a neurotransmitter release. In studies from Hammersmith, dopa was given acutely to rats, and an increase in intracellular dopamine was measured with *in vivo* microdialysis. However, the measured binding potential increased despite the increase in intracellular dopamine level. Further studies revealed that there was a change in K_D.

D. Wong: So you're suggesting a change in affinity state during the dopamine injection. This is a possibility and this particular analysis allows us to look for steady-state changes in affinity. I think that's probably the next challenge, to detect a change in affinity state between the psychostimulant injection.

A. Lammertsma: Taking it one step further, suppose you change affinity and you happen to have two different populations with two different affinities, then you have to be careful using Scatchard analysis with two points.

D. Wong: That's right. One way that has been suggested to examine this is to estimate the population of affinity states with new ligands. I've had this discussion with Phil Seeman a number of times, and I think some dopamine agonists might be available looking at these populations, but I'm not sure how you would sort this problem out.

M. Laruelle: If we accept the data that the affinity of D_2 receptors for haloperidol is decreased in schizophrenia, and that it's reflecting a higher baseline dopamine concentration, then the affinity of the available D_2 receptor for dopamine would also be decreased. Therefore, seeing a larger displacement after an amphetamine injection in schizophrenics would indicate a very large dopamine release because the displacement would be larger, despite the fact that they had a lower affinity for dopamine. Thus we might underestimate the dopamine release when we do the amphetamine challenge in schizophrenics compared to controls.

D. Wong: But wouldn't it really depend on your AMPT experimental results in the schizophrenic patients, because we can argue the other way as well. We really need to see the results of your experiment to address that.

A. Gjedde: If the affinity for haloperidol is lower in schizophrenia because there is more baseline dopamine, this calculates to about a threefold higher dopamine level at the baseline. This higher dopamine, nonetheless, gives the binding potential in the normal state, which is the same. This means that the basic affinity for raclopride that is being used to measure the binding potential must be much higher. In fact, it must mean that the two populations have different affinities in the basic nondopamine occupied state.

The calculations that we did were with dopamine levels relative to the unknown affinity, which nobody knows. This creates a problem because, as was pointed out by Adriaan Lammertsma, dopamine itself may induce an increase in its own affinity. This means that if you tried to calculate true levels of dopamine not only relative to its unknown affinity, you could have all kinds of changes. So what was shown here was the level of dopamine relative to its unknown affinity, which may show a considerable change as dopamine concentration changes.

Index

A

AADC, *see* Aromatic L-amino acid decarboxylase
Acetylcholinesterase, *see also* N-Methylpiperidinylpropionate; N-Methylpiperidyl-4-acetate
  activity variation within brain, 261–265, 396, 398–399
  cholinergic degeneration in disease, 393–394
  piperidinyl esters as substrates in PET studies
    data acquisition, 262
    delivery rates, 262–263
    dual tracer studies, 264
    reactivity as substrates, 262–264, 486
    synthesis, 262
    tacrine inhibition studies, 264–265
  turnover rate, 486
Activation studies, *see* Cerebral blood flow
Altanserin, fluorine-18 compound
  brain metabolism, 294–298, 428, 432
  brain uptake, 294–297
  high-performance liquid chromatography of metabolites, 294–295
  5-HT$_{2A}$ binding, 293–294, 427
  nonspecific binding analysis
    data acquisition, 428
    data analysis, 428–430, 432
    plasma analysis, 429
  synthesis, 294, 428
Alzheimer's disease
  PK11195 binding studies, 199
  stereotactic normalization of PET data, 125–130
AMT, *see* α-Methyltryptophan
Analysis of variance (ANOVA), design in PET activation studies
  comparison of designs, 225, 227–228
  data acquisition, 224
  overview, 223–224
  registration of PET and magnetic resonance imaging data, 224
  statistical analysis, 224–225, 484–485
Anesthesia, *see* Ketamine; Pentobarbitol
ANOVA, *see* Analysis of variance
Aphasia, magnetic resonance imaging-guided language activation PET
  application with abnormal brain anatomy, 159
  data analysis, 160
  foci of activation, 160–163
  tasks, 160
Approximate kinetic imaging, theory, 208, 210
Aromatic L-amino acid decarboxylase (AADC), PET assay, 285–286, 290–291, 387–391
Artificial neural network
  advantages in parameter estimation, 347, 350
  architecture for parametric image generation, 347–348
  image generation, 349
  nodes, 348–349
  noise tolerance, 349–350
  performance, 349–350, 352
  scaling, 348
  training procedure, 349–350
Attenuation correction, 15
Autism, α-methyltryptophan PET in children
  arterial input function extraction, 203, 205
  data acquisition, 202, 483
  dosimetry, 202, 205
  nonstationary spatial filter, 203, 205
Axial smoothing, *see* Feature-matching axial smoothing

B

BGO, *see* Bismuth germanate
B/I method, *see* Bolus/Infusion method

Binding potential (BP)
  caveats in determining receptor number or occupancy, 407
  correction for ligand mass effect
    cases, 408–409
    comparison of graphical and kinetic methods, 410–413
    discrepancies between compartment models, 411
    inclusion of nonspecific binding, 412
    simulations, 409–410
  definition, 408
  dependence on ligand mass, 407–409
  setoperone binding potential determination
    data acquisition, 422
    kinetic modeling, 423
    nonlinear least-squares fitting, 423–425
    reference region graphical analysis, 423–424
Bismuth germanate (BGO)
  crystal size in detectors, 472–473
  performance compared to lutetium oxyorthosilicate, 11–12, 472, 474
Bolus/Infusion (B/I) method
  d-amphetamine pretreatment in raclopride imaging, 445, 447
  compartmental modeling, 435–436
  maximal percentage change in concentration, interpretation, 437
  noise effects, 438
  simulation studies, 436–437, 492–493
  tracer kinetic effects on sensitivity to displacement, 437–438
BP, *see* Binding potential
Brain tumor, PET activation studies in patients
  average activation volumes, 156–157
  data acquisition and analysis, 155–156
  filtering of images, 156, 158
  surgical planning application, 155

C

CBF, see Cerebral blood flow
Cerebral blood flow (CBF), $H_2^{15}O$ measurement with PET
  activation pattern detection
    data acquisition, 254
    Fisher's linear discriminate analysis, 253, 255–257
    nonparametric testing, 255–256
    overview, 253
    preprocessing of images, 254
    principal component analysis, 254–256
  analysis of variance design in activation studies
    comparison of designs, 225, 227–228
    data acquisition, 224
    overview, 223–224
    registration of PET and magnetic resonance imaging data, 224
    statistical analysis, 224–225
  brain networks in motor behavior
    covariance pattern interpretation, 168, 172
    data acquisition, 166–167
    statistical parametric mapping analysis, 167–168
    subprofile scaling model statistical analysis, 165–168
    tasks, 165–166
  brain tumor patients
    average activation volumes, 156–157
    data acquisition and analysis, 155–156
    filtering of images, 156, 158
    surgical planning application, 155
  coupling to neuronal activity, 173
  effects in neurotransmitter release displacement studies, 451, 453
  ischemic flow estimation without arterial blood sampling
    applications, 151, 153
    data acquisition and analysis, 151–152
    overview of method, 151, 153
    regions of interest, 152
  multifiltering signal detection in activation studies
    data acquisition, 238
    registration, 238
    smoothing, 238, 240
    statistical power, 237–240
  parameters in single-compartment model
    covariance matrix calculation, 309–310
    Kety–Schmidt blood flow modeling, 310, 367, 369
    Monte Carlo simulations, 310–311
    normalization, 308–309
    PET protocol assessment, 311, 313
    sensitivity, 308
  probability of site activation
    correlated Gaussian images, 230–231
    correlated t statistic images, 231–234
    covariance in data, determination, 232, 234
    implementation of techniques, precautions, 232–233
    multidimensional t statistic probability function, derivation, 234–235
    overview of techniques for calculation, 229–230
  reproducibility of activation patterns, measurement using resampling techniques
    overview, 241–242
    performance variability of individuals, 485
    principal component analysis, 242–245
    subprofile scaling model, 242, 244
  reproducibility of regional covariance patterns in Parkinson's disease
    principal component analysis, 247
    prospective discrimination, 248, 250–251
    scaled subprofile model, 247–250
    topographic profile rating, 248, 250–251
  signal-to-noise ratio enhancement in activation studies
    cold bolus utilization, 371–372, 375–376
    data acquisition, 375
    image analysis, 375
    implementation of cold bolus/switched protocol, 377
    simulation studies, 372–375
    task switching, 371–372, 374–376
  simultaneous estimation of perfusion and initial tracer distribution volume
    nonlinear least-squares fitting, 368–370
    two-compartment model overview, 367–368
  statistical parametric mapping in analysis, 117, 119
  transient focal cerebral ischemia
    animal models, 145–146
    voxel-based analysis of reperfusion, 146–148, 480–481
Cerebral ischemia, see Ischemia
Cerebral metabolic rate of glucose (CMRGlc), see also Fluorodeoxyglucose
  conscious vervet monkey studies
    adults, 180
    calculation of metabolic rate of glucose, 179
    data acquisition, 178–179
    data analysis, 179–180
    ketamine anesthesia, 177–178
    neonates and infants, 181
    plasma glucose measurement, 179
    region of interest selection, 179
    restraint of animals, 177–178
  determination without arterial blood sampling
    cascaded model approach, 322–323
    noise effects
      input function parameters, 324
      physiological parameters, 324, 326
    patient study, 488
    plasma time–activity curve model, 321–322
    simulation study, 323–324
    tracer kinetic model, 322
Cerebral metabolic rate of oxygen (CMR_{O_2})
  coupling with cerebral blood flow, 173–175
  visual stimulus presentation frequency effects
    data acquisition and analysis, 174, 481
    mechanisms, 175
    peak rate, 175
Cluster analysis, parametric imaging of ligand–receptor interactions with reference tissue model
  automatic extraction of reference tissue, 403–406
  basis function method, 402–403
  data acquisition, 403, 405
  simulations, 403, 405
  theory, 401–402
CMRGlc, see Cerebral metabolic rate of glucose
CMR_{O_2}, see Cerebral metabolic rate of oxygen
Cocaine, dopamine release
  activity calculations, 465
  bound dopamine calculation, 464–465
  equilibrium-binding potential determination, 464
  partition volume determination, 465
  raclopride studies, 456, 465
  receptor occupancy calculation, 464
Computational chemistry, see Quantitative structure–activity relationship
Count rate, see Noise equivalent count rate

D

Depth of interaction (DOI), camera design, 11, 13–16
Dihydrotetrabenazine (DTBZ)
  dual tracer studies
    compartmental modeling and parameter estimation, 361–362
    Hoffman brain phantom simulations, 363, 365–366
    rationale and applications, 359–360
    single region time–activity curve simulations, 362–365
  monoamine transporter assay by PET, 265
Diprenorphine, neurotransmitter release studies
  cerebral blood flow effects, 451, 453
  compartmental modeling, 450
  diprenorphine displacement, 451–452
  patient motion correction, 452–453
  scanning protocol, 451
  simulations, 452
DOI, see Depth of interaction
Dopamine receptor, see Cocaine; Fluorodopa; Fluoro-L-m-tyrosine; IBZM; Lesch-Nyhan syndrome; Parkinson's disease; Raclopride; Schizophrenia; Spiperone

Doxepin
    histamine H$_1$ receptor binding, 207
    PET aging study of histamine H$_1$ receptor decline
        approximate kinetic imaging, 208, 210, 212
        binding potential, 210, 212–213
        data acquisition and analysis, 208–209
        overview, 207
        statistical parametric mapping, 209–210
        tracer delivery, 212–213
DTBZ, see Dihydrotetrabenazine
Dual tracer studies, see Dihydrotetrabenazine

E

Emisssion–transmission realignment, see Postinjection simultaneous emisssion–transmission scan
Error, type I versus type II, 273
EXACT 3D
    clinical studies, 28
    data acquisition, 26
    data reconstruction, 26
    detector, 25–26
    impact of out of field view radioactivity, 26–27, 31
    point source, 26
    scatter correction, 27
    sensitivity, 25–26, 28, 474–475
    spatial resolution, 25
    transmission scanning, 26–27

F

Fallypride, fluorine-18 compound
    affinity for D$_2$ receptor, 493
    d-amphetamine pretreatment in D$_2$ imaging
        binding potential determination, 442–444, 447
        data acquisition, 442
        data analysis, 442–443
        plasma analysis, 443
        rationale, 441
Feature-matching axial smoothing
    comparison to conventional axial smoothing, 88, 90
    computer simulations, 85–88
    Hoffman brain phantom studies, 86, 88
Filter
    brain tumor images in activation studies, 156, 158
    Gaussian filters, 237
    multifiltering signal detection in activation studies
        data acquisition, 238
        registration, 238
        smoothing, 238, 240
        statistical power, 237–240
    nonstationary spatial filter, 203, 205
Fisher's information matrix, 307–308
Fisher's linear discriminant analysis (FLDA), activation pattern detection, 253, 255–257
Flumazenil
    benzodiazepine receptor antagonist, 360
    dual tracer studies
        compartmental modeling and parameter estimation, 361–362
        Hoffman brain phantom simulations, 363, 365–366
        rationale and applications, 359–360
        single region time–activity curve simulations, 362–365
    partial volume effect correction, 62–66
Fluorodeoxyglucose, see also Cerebral metabolic rate of glucose
    cerebral metabolism studies in developing vervet monkey
        adults, 180
        calculation of metabolic rate of glucose, 179
        data acquisition, 178–179
        data analysis, 179–180
        ketamine anesthesia, 177–178
        neonates and infants, 181
        plasma glucose measurement, 179
        region of interest selection, 179
        restraint of animals, 177–178
    MicroPET imaging
        rat, 6–7
        vervet monkey, 6
    optimal image sampling schedule
        data acquisition, 316
        generalized linear least-squares algorithm
            simulation study, 343–345
            theory, 339–341, 343
        performance of technique, 320
        resampling temporal frame data, 316, 318, 320
        theory, 315–316
    postinjection simultaneous emisssion–transmission scan, 35, 37, 40
    reproducibility of regional covariance patterns in Parkinson's disease
        principal component analysis, 247
        prospective discrimination, 248, 250–251
        scaled subprofile model, 247–250
        topographic profile rating, 248, 250–251
    variance stabilizing transform, application to PET data
        comparison of thresholding methods, 220
        data acquisition, 218
        estimation and application of transform, 218–220
        registration of PET and magnetic resonance imaging data, 218
        statistical analysis, 219
Fluorodopa
    blood–brain barrier transporter
        large neutral amino acid competition with fluorodopa, 381–382, 384, 490–491
        rate constants, 380–381
        saturation, 379, 490
    comparison to fluoro-L-m-tyrosine in dopamine system imaging, 387–391, 491–492
    dopamine metabolism study overview, 187, 285
    high-performance liquid chromatography, 188
    ketamine/midazolam effects on PET kinetics in monkeys
        data collection, 183–184
        recovery time, 184–185
    o-methylation, 387
    partial volume effect correction, 72–75
    pig brain metabolism
        adults
            animal preparation, 188
            compartment model, 191
            data acquisition and analysis, 188–189
            metabolites, 189, 191
            net influx rate constant, 187–189, 191
            partial volume effect correction, 192
            plasma clearance rate, 187–188, 191
        neonatal pigs
            animal preparation, 286
            aromatic L-amino acid decarboxylase assay, 285–286, 290–291
            blood–brain barrier permeability, 291
            high-performance liquid chromatography, 286–289
            metabolism by brain region, 289–291
        postinjection simultaneous emisssion–transmission scan, 35, 39
        synthesis of radiolabeled compound, 286
    uptake rate in Parkinson's disease, statistical parametric mapping, 118
        advanced bilateral cases, 121
        early left hemi cases, 120–121
Fluoroethylspiperone
    MicroPET imaging in vervet monkey, 6
    nonspecific binding, 419
Fluoro-L-m-tyrosine
    aromatic amino acid decarboxylase substrate, 387, 389
    comparison to fluorodopa in dopamine system imaging, 387–391, 491–492
    monkey PET studies, 388
    synthesis, 388
    uptake rate in brain, 389, 491

G

Generalized linear least-squares algorithm, see Optimal image sampling schedule
Gaussian filter, see Filter

H

Histamine H$_1$ receptor, *see* Doxepin

I

IBZM
  D$_2$ imaging
    binding potential determination, 457, 459
    dopamine depletion, 456, 458–459, 494–495
    image analysis, 457–458
    plasma analysis, 458
    SPECT data acquisition, 456–457
  dopamine competition, 460
Image processing, *see* Feature-matching axial smoothing; Registration; Statistical parametric mapping; Stereotactic normalization; Time–activity curve; Volume of interest
Impulse response function (IRF), parametric image reconstruction, 47
Information technology
  data bases, 268, 270
  tracer development, 267–268
Ischemia
  ischemic flow estimation without arterial blood sampling
    applications, 151, 153
    data acquisition and analysis, 151–152
    overview of method, 151, 153
    regions of interest, 152
  transient focal cerebral ischemia
    animal models, 145–146
    voxel-based analysis of reperfusion, 146–148, 480–481

J

Java
  application programming interfaces, 354
  byte code, 354–355
  computer platform versatility, 353, 355
  PET data analysis, 354–355
  program types, 354
  simplicity, 354
  speed of program execution, 355

K

Ketamine
  fluorodeoxyglucose studies in developing vervet monkey, 177–178
  ketamine/midazolam effects on fluorodopa PET kinetics in monkeys
    data collection, 183–184
    recovery time, 184–185
Kety blood flow modeling, 310, 367, 369

L

L-703,717
  animals and tracer administration, 300
  blood and tissue sampling, 300–303
  brain uptake index, 300–301
  N-methyl-D-aspartate receptor binding, 299–300
  plasma protein binding, 301
  synthesis of radiolabeled compound, 300
Lesch-Nyhan syndrome, partial volume effect correction impact on dopamine ligand time–activity curves, 77–80, 82
Liver metabolism, testing of PET tracers
  BPA, 280–281
  CPA, 280–281
  CPC, 280–282
  high-performance liquid chromatography, 280–282
  microsome incubations, 280–281, 487
  WAY 100635, 279–281, 488
Local cerebral metabolic rate of glucose, *see* Cerebral metabolic rate of glucose
LSO, *see* Lutetium oxyorthosilicate
Lutetium oxyorthosilicate (LSO)
  crystal size in detectors, 472–473
  MicroPET detector, 4
  performance compared to bismuth germanate, 11–12, 16, 472, 474
  utilization for simultaneous magnetic resonance imaging, 19–20

M

Magnetic resonance imaging (MRI)
  functional magnetic resonance imaging, limitations, 162
  guided language activation PET
    application with abnormal brain anatomy, 159
    data analysis, 160
    foci of activation, 160–163
    tasks, 160
  registration with PET data, 51–53, 57–58, 64, 91, 96, 99–105, 140, 160, 196, 218, 224
  segmentation of data, 51–53, 58, 60, 64, 165
  simultaneous imaging with PET
    artifact and distortion studies, 20–21, 474
    experimental setup, 20
    fluorodeoxyglucose imaging in rat, 20–22
    optimization, 22
    rationale, 19
    scanner design, 19–20
    sensitivity, 473
    single-slice system development, 22–23
  surface rendering of data in functional delineation
    activation studies, 141–142
    alignment of data sets, 140
    data acquisition, 139–140
    segmentation, 140
    visualization, 140–141
    volume of interest definition, 141
Mass effect, *see* Binding potential
MDL 105,519
  animals and tracer administration, 300
  blood and tissue sampling, 300–303
  brain uptake index, 300–301
  N-methyl-D-apartate receptor binding, 299–300
  plasma protein binding, 301
  synthesis of radiolabeled compound, 300
N-Methyl-D-aspartate receptor, *see* L-703,717
N-Methylpiperidinylpropionate (PMP)
  acetylcholinesterase substrate, 361, 485–486
  dual tracer studies
    compartmental modeling and parameter estimation, 361–362
    Hoffman brain phantom simulations, 363, 365–366
    isotope separations, 489–490
    rationale and applications, 359–360
    single region time–activity curve simulations, 362–365
N-Methylpiperidyl-4-acetate (MP4A)
  acetylcholinesterase assay with PET
    aging study, 492
    blood sampling, 394
    data acquisition, 394
    hydrolysis rates in various regions, 396, 398–399
    input function, 395
    parameter estimation, 395–396
    scanning duration and time–activity curve, 396
  brain uptake, 394
  synthesis, 394
N-Methylspiperone, *d*-amphetamine pretreatment in D$_2$ imaging, 441
α-Methyltryptophan (AMT)
  autistic children PET studies
    arterial input function extraction, 203, 205
    data acquisition, 202, 483
    dosimetry, 202, 205
    nonstationary spatial filter, 203, 205
  K complex, 201–202
  serotonin metabolism study overview, 201
MicroPET
  count rate performance, 5, 7
  data correction, 8
  detector, 4, 472
  EDTA imaging of rabbit brain tumor, 6
  fluorodeoxyglucose imaging
    rat, 6–7
    vervet monkey, 6
  fluoroethylspiperone imaging in vervet monkey, 6
  scanner features, 3–5
  sensitivity enhancement, 7, 471
  spatial resolution, 3, 5, 8, 471–472
Midazolam, *see* Ketamine

Molecular modeling
  algorithms, 269
  comparative molecular fields analysis, 269–270
  structure-based drug design, 269
  tracer development, 269–270
MP4A, see N-Methylpiperidyl-4-acetate
MRI, see Magnetic resonance imaging
MS, see Multiple sclerosis
Multiple sclerosis, PK11195 PET study
  data acquisition, 196
  magnetic resonance imaging coregistration, 196
  three-dimensional data, 198–199
  two-dimensional data, 196, 198

N

Neurotransmitter release
  cerebral blood flow effects, 451, 453
  compartmental modeling, 450
  diprenorphine displacement, 451–452
  displacement versus competition studies, 450
  importance of study, 449
  patient motion correction, 452–453, 494
  scanning protocol, 451
  simulations, 436–437, 452
Noise
  artifact suppression with spectral analysis
    advantages, 489
    basis functions, range and spacing, 330–332, 489
    data acquisition and analysis, 331
    impulse response function, 330, 333
    nonnegative least-squares algorithm, 329–330, 333
    overview, 330–331
    performance, 332–333, 489
    simulations, 331
  effect on parameter uncertainty, 307
Noise equivalent count rate
  ECAT EXACT HR-47 PET camera, 42, 44
  MicroPET, 5, 7
  optimization, 11, 14
Noise equivalent sensitivity
  MicroPET, 5, 7
  optimization of camera design, 12–15
Nonparametric testing, activation pattern detection, 255–256

O

Optimal image sampling schedule (OISS)
  data acquisition, 316
  generalized linear least-squares algorithm
    simulation study, 343–345
    theory, 339–341, 343

performance of technique, 320
resampling temporal frame data, 316, 318, 320
theory, 315–316, 340

P

Parametric images
  impulse response function, 47
  ligand–receptor interactions with cluster analysis and reference tissue model
    automatic extraction of reference tissue, 403–406
    basis function method, 402–403
    data acquisition, 403, 405
    simulations, 403, 405
    theory, 401–402
  overview, 45
  signal-to-noise ratio of images, 45, 49
  spectral analysis, 46–47
  temozolomide study, 47–49
  three-dimensional rebinning, 46, 475–476
Parkinson's disease
  dual tracer studies, 360
  fluorodopa uptake rate, statistical parametric mapping, 118
    advanced bilateral cases, 121
    early left hemi cases, 120–121
  reproducibility of regional glucose metabolism covariance patterns
    principal component analysis, 247
    prospective discrimination, 248, 250–251
    scaled subprofile model, 247–250
    topographic profile rating, 248, 250–251
Partial volume effect correction
  components of partial volume effects, 77
  computer simulation, 61–62
  flumazenil scanning in humans, 62–66
  Hoffman brain phantom imaging, 61–63
  image acquisition, 60
  impact on time–activity curves
    brain phantom creation, 78
    data acquisition, 78–79
    image data analysis, 79
    kinetic modeling, 79–80, 82
    Lesch-Nyhan syndrome subjects, 78
    simulated PET data generation, 79
  importance, 59
  large data set handling, 61
  linear least-squares approach, 59–66
  magnetic resonance imaging
    registration with PET, 64, 477
    segmentation, 59–60, 64, 476–478
  multidata analysis processing, 61
  pixel-based correction
    algorithms, 67–68
    fluorodopa study in humans, 72–75
    implementation, 68–70
    regional spread function image generation, 68
    theory, 68
    tracer uptake simulation, 68, 71–72
  point spread function inhomogeneity, 64–65

region of interest-based correction
  algorithms, 67–68
  fluoro-L-dopa study in humans, 72–75
  implementation, 70–71
  regional spread function image generation, 68
  theory, 68
  tracer uptake simulation, 68, 71–72
PCA, see Principal component analysis
Pentobarbital
  effects on fluorodopa PET kinetics, 184–185
  recovery time, 183
PET chemistry data base, features and goals, 270–271
Photogrammetry
  derived surfaces in multimodal registration
    advantages, 103–104
    calibration, 100–101
    comparison to voxel correlation methods, 99–100, 103, 105
    laser scanning, 101
    limitations, 104
    mirror placement and attenuation, 104–105
    surface fitting, 101
    validation, 101–102
  overview, 100
PITsim, see Postinjection simultaneous emisssion–transmission scan
PK11195
  multiple sclerosis PET study
    data acquisition, 196
    magnetic resonance imaging coregistration, 196, 482
    three-dimensional data, 198–199
    two-dimensional data, 196, 198
  peripheral benzodiazepine-binding site
    ligand binding and applications, 195–196, 199
    microglia expression, 196
    synthesis of carbon-11 compound, 196
PMP, see N-Methylpiperidinylpropionate
Postinjection simultaneous emisssion–transmission scan (PITsim)
  human fluorodeoxyglucose scan
    data acquisition, 35
    data analysis, 35, 37, 40
  human raclopride scan
    data acquisition, 35
    data analysis, 35, 39
  monkey scans
    data acquisition, 35
    data analysis, 35
    fluorodopa, 39
    raclopride, 39–40
  patient motion effects, 33
Postinjection simultaneous emisssion–transmission scan (PITsim) (Continued)
  phantom study
    data acquisition, 34
    data analysis, 34–35, 37

postinjection transmission method, 34
rationale, 33
realignment method, 34
validation, 33
Principal component analysis (PCA)
  activation pattern detection
    data acquisition, 254
    Fisher's linear discriminant analysis, 253, 255–257
    nonparametric testing, 255–256
    overview, 253
    preprocessing of images, 254
    principal component analysis, 254–256
  brain networks in motor behavior
    covariance pattern interpretation, 168, 172
    data acquisition, 166–167
    statistical parametric mapping analysis, 167–168
    subprofile scaling model, 165–168
    tasks, 165–166
  pixel classification using *a priori* kinetic factors, 107–110, 114
  reproducibility of activation patterns, measurement using resampling techniques
    overview, 241–242
    principal component analysis, 242–245
    subprofile scaling model, 242, 244
  reproducibility of regional covariance patterns in Parkinson's disease
    principal component analysis, 247
    prospective discrimination, 248, 250–251
    scaled subprofile model, 247–250
    topographic profile rating, 248, 250–251

Q

Quantitative structure–activity relationship (QSAR), tracer development, 268–271

R

Raclopride
  d-amphetamine pretreatment in D_2 imaging
    binding potential determination, 442–444, 447
    data acquisition, 442
    data analysis, 442–443
    plasma analysis, 443
    rationale, 441
  binding potential, correction for ligand mass effect
    cases, 408–409
    comparison of graphical and kinetic methods, 410–413
    discrepancies between compartment models, 411
    inclusion of nonspecific binding, 412
    simulations, 409–410
  bolus/infusion method
    d-amphetamine pretreatment, 445, 447
    compartmental modeling, 435–436
    maximal percentage change in concentration, interpretation, 437
    noise effects, 438
    simulation studies, 436–437
    tracer kinetic effects on sensitivity to displacement, 437–438
  dopamine competition, 460
  dopamine release studies
    cocaine study, 456, 465
    schizophrenia, 465–467
  ligands, 77
  parametric imaging with cluster analysis and reference tissue model
    automatic extraction of reference tissue, 403–406
    basis function method, 402–403
    data acquisition, 403, 405
    simulations, 403, 405
    theory, 401–402
  partial volume effect correction, impact on time–activity curves, 77–80, 82
  postinjection simultaneous emisssion–transmission scan, 39–40
  synthesis, 78
Radial elongation, minimization, 13–14
Regional cerebral blood flow, *see* Cerebral blood flow
Regional cerebral metabolic rate of oxygen, *see* Cerebral metabolic rate of oxygen
Registration
  magnetic resonance imaging and PET data, 51–53, 57–58, 64, 91, 96, 99–105, 160, 196, 218, 224, 476–478
  multitracer PET data
    applications, 91–92
    cost function evaluation
      extraction of tracer dynamics, 92–93
      object delineation, 93
      outside the object class, 93
      reference volume, 93
      Wilks' Λ evaluation, 93–94
    nomifensine and DOPA registration, 94–95
    overview of algorithm, 92, 478
    search algorithm, 94, 96
    validation, 94
  photogrammetrically-derived surfaces in multimodal registration
    advantages, 103–104
    calibration, 100–101
    comparison to voxel correlation methods, 99–100, 103, 105
    laser scanning, 101
    limitations, 104
    mirror placement and attenuation, 104–105
    photogrammetry overview, 100
    surface fitting, 101
    validation, 101–102
Resolution, *see* Spatial resolution

S

SCH 23390
  brain distribution in rats, 274

parametric imaging with cluster analysis and reference tissue model
  automatic extraction of reference tissue, 403–406
  basis function method, 402–403
  data acquisition, 403, 405
  simulations, 403, 405
  theory, 401–402
  statistical power analysis in competition experiments, 274, 276–277, 486–487
Schizophrenia, dopamine release
  activity calculations, 465
  amphetamine effects, 463
  bound dopamine calculation, 464–465, 467–468
  equilibrium-binding potential determination, 464
  partition volume determination, 465
  raclopride studies, 465–467
  receptor occupancy calculation, 464, 466–467
  TREMBLE analysis, 466–467, 495
Sensitivity, *see* Noise equivalent sensitivity
Serotonin receptor, *see also* Altanserin; Setoperone; WAY 100635
  disorders in disease, 421
  ligands for imaging, 422
Setoperone
  binding potential determination
    data acquisition, 422
    kinetic modeling, 423
    nonlinear least-squares fitting, 423–425
    reference region graphical analysis, 423–424
  5-HT$_{2A}$ receptor binding, 415, 421–422
  multicompartmental modeling, 415–416
  nonspecific binding analysis
    arterial blood sampling, 416
    baboon studies, 416–419
    control versus challenge studies, 417–418
    data acquisition, 416
    data analysis, 417
    human studies, 416–419
  synthesis, 422
Single photon emission computerized tomography (SPECT), D_2 imaging with IBZM
  binding potential determination, 457, 459
  dopamine depletion, 456, 458–459
  image analysis, 457–458
  plasma analysis, 458
  SPECT data acquisition, 456–457
SKF 82957
  brain distribution in rats, 274
  statistical power analysis in competition experiments, 274, 276–277, 486–487
Spatial resolution
  estimation, 13
  MicroPET, 3, 5, 8
  radial elongation artifact, 13–14
SPECT, *see* Single photon emission computerized tomography

Spectral analysis
    fixed partition volume inclusion, 335–337
    modifications accommodating model constraints, 335–337
    noise artifact suppression
        basis functions, range and spacing, 330–332
        data acquisition and analysis, 331
        impulse response function, 330, 333
        nonnegative least-squares algorithm, 329–330, 333
        overview, 330–331
        performance, 332–333
        simulations, 331
        theory, 335–336
Spiperone
    dopamine competition, 460
    fluoroethylspiperone, see Fluoroethylspiperone
    N-methylspiperone, see N-Methylspiperone
SPM, see Statistical parametric mapping
SSM, see Subprofile scaling model
Statistical parametric mapping (SPM)
    doxepin aging study of histamine H_1 receptor decline, 209–210
    fluorodopa uptake rate mapping
        comparison to region of interest approach, 121–122
        data acquisition, 118
        data analysis, 118–120
        image reconstruction, 118
        Parkinson's disease patients, 118, 479–480
            advanced bilateral cases, 121
            early left hemi cases, 120–121
            scanning protocol, 118
            spatial normalization of add images, 122, 479–480
            unpaired t test validity, 122, 479
    ligand-binding analysis, 117–118
    regional cerebral blood flow analysis, 117, 119, 167–168
Stereotactic normalization
    data acquisition, 126
    stereotactic atlas template, 125–126
    Volterra nonlinear analysis
        Alzheimer's disease patients, 125–130
        kernel analysis, 127–128
        theory, 126–127
Stereotatic surface projections
    three-dimensional stereotactic surface projections, 133–137
    volume of interest analysis, 133–137
Subprofile scaling model (SSM), see Principal component analysis

T

Three-dimensional stereotactic surface projection (3D-SSP), data extraction
    brain phantom studies, 134
    cortical atrophy effects on data analysis, 135–136
    image noise effects on data analysis, 136
    overview, 133–134, 480
    partial volume effects on data analysis, 134–136
    pixel-by-pixel group comparison, 133, 137
    region size effects on analysis, 134
    tracer kinetic analysis, 136
Time–activity curve
    partial volume effect correction, impact on time–activity curves, 77–80, 82
    pixel classification using a priori kinetic factors
        cell proliferation studies with thymidine tracer, 110, 113
        fallypride binding analysis, 110
        fluorodeoxyglucose studies, 110
        Hoffman brain phantom simulations, 110
        masking and residual factors, 109–110
        model-based approach for model factor definition, 108
        population-based approach for model factor definition, 108, 114–115
        principal component analysis, 107–110, 114
        R-PK11195 binding analysis, 110
        single subject approach for model factor definition, 108
        tissue classification, 108–109, 113
TREMBLE, dopamine release analysis in schizophrenia, 466–467, 495

V

Variance stabilizing transform
    Box-Cox transform, 483–484
    fluorodeoxyglucose study
        comparison of thresholding methods, 220
        data acquisition, 218
        estimation and application of transform, 218–220
        registration of PET and magnetic resonance imaging data, 218
        statistical analysis, 219
        rationale, 217–218
VOI, see Volume of interest
Volterra nonlinear analysis, see Stereotactic normalization
Volume of interest (VOI)
    construction of three-dimensional data, 53
    data extraction
        brain phantom studies, 134
        cortical atrophy effects on data analysis, 135–136
        image noise effects on data analysis, 136
        partial volume effects on data analysis, 134–136
        region size effects on analysis, 134
        tracer kinetic analysis, 136
    magnetic resonance imaging registration with PET, 51–53, 57–58
    partial volume effect correction, see Partial volume effect correction
    PET data acquisition, 52
    quantitation of PET activity, 53–57
    surface rendering of magnetic resonance imaging data in functional delineation
        activation studies, 141–142
        alignment of data sets, 140
        data acquisition, 139–140
        segmentation, 140
        visualization, 140–141
        volume of interest definition, 141

W

WAY 100635
    analogs, liver metabolism testing
        BPA, 280–281
        CPA, 280–281
        CPC, 280–282
        high-performance liquid chromatography, 280–282
        microsome incubations, 280–281
    5-HT1A antagonism, 279
    liver metabolism to WAY 100634, 279–281, 488
    parametric imaging with cluster analysis and reference tissue model
        automatic extraction of reference tissue, 403–406
        basis function method, 402–403
        data acquisition, 403, 405
        simulations, 403, 405
        theory, 401–402
Water
    dose optimization, 42–44
    duration of acquisition in activation studies, 41–44
    noise equivalent count rate calculation, 42, 44
    optimization of $H_2^{15}O$ dose and data acquisition
    signal-to-noise ratio enhancement in activation studies
        cold bolus utilization, 371–372, 375–376
        data acquisition, 375
        image analysis, 375
        implementation of cold bolus/switched protocol, 377
        simulation studies, 372–375
        task switching, 371–372, 374–376
    task for analysis, 42
WIN 35,428
    carbon-11 compound synthesis, 78
    partial volume effect correction, impact on time–activity curves, 77–80, 82